The World Formula

The World Formula

A Late Recognition of David Hilbert's Stroke of Genius

Norbert Schwarzer

JENNY STANFORD
PUBLISHING

Published by

Jenny Stanford Publishing Pte. Ltd.
Level 34, Centennial Tower
3 Temasek Avenue
Singapore 039190

Email: editorial@jennystanford.com
Web: www.jennystanford.com

British Library Cataloguing-in-Publication Data
A catalogue record for this book is available from the British Library.

The World Formula: A Late Recognition of David Hilbert's Stroke of Genius

ISBN 978-981-4877-20-6 (Hardcover)
ISBN 978-1-003-14644-5 (eBook)

Dedication

To the victims of ignorant politicians.
We will not forget. We will not forgive.

Contents

REPETITION

USE

Acknowledgment

Thank you, reader, for your interest in my work!

About the Book

No, David Hilbert's work "on the fundaments of physics" [$\alpha \leftrightarrow \omega$] is not unknown. This is—by no means—what we meant to say when stating in the title of this book that here we intend to give a fairly "late recognition" to his work. In fact, there was a lot recognition over the past decades already. But the true meaning of Hilbert's work, and thus his true stroke of genius, obviously was not discovered yet. It seems that Hilbert had already written down the world formula over 100 years ago.

Even though this author still considers the book a draft, we think that it is time to bring it out, simply because we want to have some basis for discussion.

After a brief motivation, thereby reprinting one of the stories which actually brought this author to start working on this book in the first place, we will derive a, or rather write down, the world formula. If truth be told, this apparently huge task isn't much more than representing the Einstein–Hilbert action [$\alpha \leftrightarrow \omega$], which already contained it all. We only needed to dig a little bit deeper than Einstein and Hilbert had done.

Then, directly from the Einstein–Hilbert action we will extract the theory of relativity, quantum theory, thermodynamics (here meaning the second law of thermodynamics), the principle forces of evolution, interaction, and more.

Surprisingly, in connection with evolution, it is thereby found that the second law of thermodynamics fundamentally hides the basic driving forces of evolution, which means evolution comes with the second law of thermodynamics and the second law comes with evolution. That is not an option for the two, but a must.

Or still shorter: "Life and death belong together and are coded in only one metric term."

Taking the old wisdom of many ancient natural religions, this actually is not very new, though, but still it appears to be a nice finding if one sees it in an equation coming out from something as fundamental as the Einstein–Hilbert action.

Finally, we will consider a variety of potential applications, show how to derive the classical quantum equations from Hilbert's formula, and present a list of project ideas using a world formula approach.

Reference

[$\alpha \leftrightarrow \omega$] D. Hilbert, Die Grundlagen der Physik, Teil 1, *Göttinger Nachrichten*, 1915, 395–407.

Personal Motivation

Why the Classical Explanations Do Not Suffice

How to explain the world to my dying child?

The most natural motivation to try for a better understanding of
EVERYTHING can be personal **loss** [1, 2]!

Perhaps we want to be ready when our child needs us and we want to
explain to him/her the world before he/she has to leave it.

Along the way, then, we may find some unexpected results:

 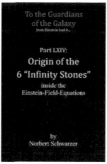  

And then we learned:

**Instead of making up your mind,
you could calculate an opinion, but you also
have to accept that there is a principle
limit to all models.**

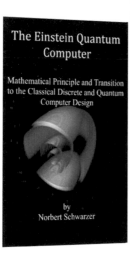 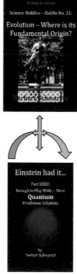 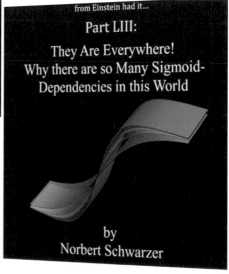

Some Fundamental Motivation

Why—apart from the philosophical question—would we need a quantum gravity or, as one also calls it, a Theory of Everything approach for everything?

Well, because problems in many fields and applications are often extremely complex and stretch over such wide ranges of properties, physical fields, processes and mechanisms, relaxation, retardation, degradation, all sorts of evolution varieties and parameter dependencies that it simply appears much more convenient to have a general approach and potentially reduce it towards the application in question, rather than making a simple approach ever more complicated. The latter approach, which is also known as bottom-up technique, always bears the risk of missing something which only reveals itself as important after an unfortunate disaster has occurred or—what often is the much greater disadvantage—the model creator deliberately or unintentionally (this does not matter) somehow incorporates his very own personal feelings and preferences into the model, thereby rendering the whole model in principle biased and, thus, useless. All current modern climate social science, political, and many economic and socioeconomic models, at least if financed with public funds, are flawed in this way.

But would a fundamental Theory of Everything help humankind to avoid such stupid mistakes?

Well, here the author is very skeptical and holds it with Einstein who once said:

"Two things are infinite: the universe and human stupidity; and I'm not sure about the universe."

And together with stupidity always also comes the human greed. Thereby the latter isn't actually a typical human thing, but the simple—often short-sighted—opportunism of making use of parasitic niches. However, a comprehensive coverage of this topic would lead too far here as it requires a Theory of Everything approach in the field of theoretical biology like evolution, evolutionary stable strategies, co-evolution, and behavior biology. The author simply needs to point out that gender gaga, equality "science," most social and political science, the climate lies, and much similar nonsense is no science at all, but rubbish which has only entered our Western universities and institutions of education in order to live on the tax payers' money with absolutely no use to the latter whatsoever. On the contrary, these parasites produce nothing but more costs, damage, failure, defects, and severe deformation and degradation of the very society that is stupid enough to let them prosper and scrounge its resources. With such forces feasting like maggots within the brain of the society, usually helped by corrupt and completely useless politicians, media and parasitic industries, no holistic and fundamental theoretical approach can prevent a domesticated, sedated, or stupefied society to fall for such liars.

Let us take the example of the famous hockey stick curve, which was created by some Mr. M. Mann in order to show us that man-made CO_2 is responsible for a global temperature increase since man started to burn fossil fuels. After Mr. Mann lost his case against T. Ball, who had accused him of being a fraud (famous are his words that Michael Mann "belongs in the state pen, not Penn State," a comical reference to the fraudulent "hockey stick" graph that knowledgeable scientists knew to be fakery [A1]), many people slowly—very slowly indeed—started to realize how much this obviously wrong and made-up "man-made CO_2 = global warming hypothesis" has cost humanity. Billions and billions of dollars, euros, yen, and whatever spent for nothing but a crude idea, a "scientist" having no scruples to fake data and a sufficiently big number of opportunistic followers with no backbone but big lust for good positions, totally pointless and idle "science projects," and no conscience, inner compass, or at least logical orientation whatsoever. Thereby it was rather obvious that the climate community and their doing had very little to do with science—any science at all, if truth be told [A2]. This, combined with the always omnipresent corrupt politicians jumping the bandwagon and some clever entrepreneurs smelling a nice chance for easy profit and personal gain on the cost of the society, gave the right mix for the financial, political, and educational disaster of the century. But the worst of it was, and still is, that all these resources spent on the nonsensical fight against a harmless gas are lost for some reasonable measures against the process of man definitively destroying the planet. One simple glance on the trash, most especially the dangerous plastics, swimming around in the world's seas and another one on the forest destruction, wetland "cultivation" and disruption of the hydrological cycle, annihilation of natural water buffer systems, the killing of species, etc., should suffice to get the gist of the dimension of the true and important problems.

Definitively, a holistic model, showing clearly and early enough that the Mann-made-up hockey stick was anything but a lie, would have been a nice weapon against the parasites. But would it also have helped enough to prevent this disaster?

The answer is a clear NO, because the best scientific tools are of no use against cheats, ideologists, and parasites if the latter are too dominant inside the society, if they have conquered the scientific institutions and infiltrated the education system, and if they dominate the mainstream media and spread their rubbish among the stupid, story-hungry journalists and even dimmer politicians, who are always willing to take on even the worst ideology or religion as long as it helps them to gain power over the people they are keen to govern. And let's face it, all this gender rubbish, climate hype, and social "science" claptrap is nothing else but a pseudoreligion. It has not a jot of scientific substance whatsoever. It comes with its very own, very cruel, and most inhuman Sharia, and it is meanwhile dominating the Western education system at large.

However, having a holistic fundamental theoretical approach at hand, which also properly covers socioeconomic space-times, which, in other words, can make predictions about whole societies and their interactions [A10], and seeing it in connection with a healthy education system and truly scientific institutions, combined with a science community consisting of scientists instead of frauds and parasites, would have made the difference. Perhaps it might even have helped a lot in preventing this disaster. At least it would not have allowed it to ever grow to the scale we see today, a state where whole countries like Merkel's Germany go down the drain not least because of Mann's rubbish hockey-stick figure [A2] and dim politicians either stupidly believing or intentionally abusing such fraudulent pseudoscience to trigger hysteria and fanaticism to help their own cause.

Already talking about Germany and the climate hysteria, new speech, thought control and gender fanaticism haunting the country, the author sees some great parallels to the racial fanaticism of the Hitler regime or the Kulak prosecution under Stalin. Most definitively, the presence of a fundamental, universal, and holistic science model plus the existence of a working, politically independent and truthful scientific community and its dominance in the awareness of society would have rendered it completely impossible for Hitler's racial "scientists" to even climb the first steps in getting any social attention, not to talk about getting recognition. The same holds with respect to the gender-Gaga-climate-Merkelism, which—this author is convinced—will probably cost more victims than the Nazi fascism and the Stalin-Mao-Pol Pot-NK socialism experiments put together. Merkelism is nothing else but the worst genocide program against the autochthone Europeans there ever was. The slaughtering has already started, the killing fields are full with blood, the victims are numerous already, all hushed up or just named "Einzelfälle" (isolated cases*), and, as happened during the Holocaust, the world does not want to know, preferring to look the other way. The problem in ignoring the Merkel regime's totalitarianism, however, will be worse than any result Pol Pot, Mao, Stalin, or Hitler could have ever produced. The German political "elite" is currently creating its very own dystopian *1984*, and if it is not stopped now and confined to the not even remotely sexy, but evil-smelling city of Berlin, it is guaranteed to proceed on, gradually infecting the whole Europe. But this, the Merkel-1984ism, is going to be more destructive to civilization than any global killing asteroid could ever be. Some might argue that the Merkel regime—even though rotten to its core—can by far not be as disastrous as the above-mentioned dictatorships have been. Here we need to remind ourselves that we are living in a much denser world now. Everything is connected, and thus there will be a huge chain reaction.

Another example comes from material science, where almost around the time Mann's lie took off, another fraudulent scientist made himself quite a name as the man who had produced materials much harder than diamond. This nonsense became known as "ultra-hardness" [A3]. When it was finally shown that the whole story was made up and the data were cleverly manipulated, when the "ultra-hard man" (the celebrated hero of material science) later explained that the materials could not be tested

*I. Laufer: "Refugees are up to 20 times more criminal than Germans. The extreme criminality of refugees in numbers, facts and proofs." In German: "Flüchtlinge sind bis zu 20 Mal krimineller als Deutsche. Die extreme Flüchtlingskriminalität in Zahlen, Fakten und Beweisen." https://www.fischundfleisch.com/ineslaufer/fluechtlinge-sind-bis-zu-20-mal-krimineller-als-deutsche-die-extreme-fluechtlingskriminalitaet-in-31542

by independent parties, because they had all—miraculously—degraded, completely lost their "ultra-hardness," not to say their hardness, entirely and could not be reproduced by anybody in the whole wide world, when it even had to be admitted that the original data were not to be found, again lots of public and private money had been wasted on a "scientific finding" that was nothing but a hoax. In the end, an open dispute cleared up all the dust and it could not only be demonstrated how severe the fraud had been [A3], but also that the whole "ultra-hard story" had violated the law of conserved energy [A4]. Having had this fundamental insight at hand quite a few years earlier would have probably saved not only some gullible investors into spending "dumb money" but also spared the material science community a very embarrassing situation. But here, too, the fundamental model revealing the nakedness of the "emperor of ultra-hardness" would have also needed an independent, constantly vigilant scientific community and here was the main and very fundamental problem. Already severely weakened by all the Genders, Thought-Controllers, Social and Political "Scientists", the frauds in the natural science fields had by far too easy play.

But there is more, of course.

We all know about the deadly Lysenkoism which not only completely ruined all reasonable genetics and agriculture science in the Soviet Union, USSR, during the Stalin and Khrushchev era, but also sent many thousands of talented scientists into the communist death camps. This author is convinced that nowadays gender and climate freaks who call themselves scientists, plus their cronies from corresponding parasitic industries and politics, would very much like to do the same and kill us all in the well-remembered Lysenko style and manner.

Well, we can only try and make it as difficult as possible for them. One way is, naturally, to kick them out of the universities, the whole education sector, and all publically financed institutions. The other way is to always try and give the better example, avoid any form of parasitism, and give our best in providing better models and research results. Results—and one cannot stress that enough—have to be evaluated by their use to the society that is financing them.

But how to do this in a fair and neutral manner?

Here, one might see the closing of the circle. Good politics and society planning requires optimum decision making [A6, A7]. This, on the other hand, requires knowledge, a fundamental understanding of the reality, and critical observation of the uncertainties residing in each and every model.

Only then the accumulation of stupidities like gender gaga, social justice worrier-shipping, and climate nonsense can be prevented. Only then the degeneration and denaturalization of bigger amounts of tissues within the society due to infectious diseases and parasitic or oncogene growth will not be able to spread, because they are detected by a healthy organism and forced into apoptosis by its inbuilt immune system.

Starting with our own research, which was *not financed* by the society, by the way, we have already tried to find applications in healthcare [A7], socioeconomics [A6], theoretical physics (see main references), quantum computing [A8], and material science [A9].

Especially in electronic, computer, and power supply applications, the complexity of important influencing parameters is often so overwhelming that classical models even fail with the simplest design and optimization or failure-analyzing tasks. One simply takes the example of the development of properly working fusion reactors, to say nothing of the process itself or the technical problems in developing quantum computers (adiabatic or entangled). A comprehensive fundamental approach seems to be by far the better approach.*

It is for this reason that we will use a most fundamental material science approach starting on the level of quantum gravity for our motivation section here.

But before we are going to do that, we will try to explain the origin of our own and very personal motivation.

References

A1. www.quora.com / What-does-Michael-Mann-s-court-battle-loss-mean-to-the-notion-of-climate-change.

A2. P. Frank, Propagation of error and reliability of global air temperature projection, *Front. Earth Sci.*, 2019, **7**, 223, https://doi.org/10.3389/feart.2019.00223 or www.frontiersin.org/articles/10.3389/fcart.2019.00223/full.

A3. A. C. Fischer-Cripps, St. J. Bull, N. Schwarzer, Critical review of claims for ultra-hardness in nanocomposite coatings, *Philos. Mag.*, 2012, **92**(13), 1601–1630.

A4. N. Schwarzer, Short note on the effect of pressure induced increase of Young's modulus, *Philos. Mag.*, 2012, **92**(13), 1631–1648.

---

*The interested reader may find an illustrative example here: https://youtu.be/eRWfikGxhuQ

A5. https://rationalwiki.org/wiki/Lysenkoism.

A6. N. Schwarzer, *Einstein had it, But He Did not See it, Part LXVIII: Most Fundamental Tools for Optimum Decision-Making Based on Quantum Gravity*, www.amazon.com, ASIN: B07KDFDZVZ.

A7. N. Schwarzer, *Einstein had it, But He Did not See it, Part LXIX: The Hippocratic Oath in Mathematical Form and Why – So Often – It Will Be of No Use*, www.amazon.com, ASIN: B07KDSMNSK.

A8. N. Schwarzer, *Is There an Ultimate, Truly Fundamental and Universal Computer Machine?* www.amazon.com, ASIN: B07V52RB2F.

A9. N. Schwarzer, *From Quantum Gravity to a Quantum Relative Material Science – A Very First Principle Material Science Concept*, www.amazon.com, ASIN: B07X8H9Y7Z.

A10. P. Heuer-Schwarzer, N. Schwarzer, An unorthodox, almost shocking way of demonstrating the proximity of material interaction and human societies using the example of the battle of Midway, https://youtu.be/SngNIa6lt3g.

The Eighth Day
(From T. Bodan, with Thanks)

Preface to *The Eighth Day*

Yes, my child has died. I mean, his body has died. Nevertheless, I cannot get rid of the feeling that still I have to keep my promise. Whatever and wherever my child is now, I feel that it measures me on this. No matter the dead body. I have given that promise and so I have a responsibility. After all, I cannot know whether my child still listens to my every word and perhaps longingly waits for me to fulfil that task. It might wait for me to finish the work we started together a bit more than seven days ago, trying to explain the world. For all I know, information cannot die or disappear in this universe, and so, I think, the set of information which once defined my child still exists. It is there, somewhere. It has dignity and deserves attention. Thus, I will finish the job now. Interestingly, it isn't much I have to do myself here. The task was solved by a boy called Samuel and his father about 70 years ago. Both were killed by the Nazis in one of the gas chambers in Auschwitz. They died together in that chamber on Christmas Eve in 1944 and they were the greatest scientists of all time.

It was very difficult for me to collect all their astounding work, because often it wasn't more than some scribbled notes on the rim of an old newspaper. Something written with shaking hands in the middle of publications of other scientists of their time. More was scratched on the walls of the miserable places where they were forced to live. The most important pieces of their work, however, were photographed from the interior of that *Reichsbahn* wagon which transported them to the KZ, and finally there was this unobtrusive sketch in that dismal room where they both died together, a gas chamber in Auschwitz.

One word about the translation from the German original into English: The translation was done by a colleague and good friend of the author. Unfortunately, this colleague isn't a professional translator from German into English. He isn't even good at writing. True, he has written quite a few articles and has successfully submitted them to scientific journals, but this probably doesn't count when it comes to literature, does it? The author thinks that it is the knowledge that counts more here rather than smooth formulations and high-flying text passages. The author thinks his book, after all, is more a scientific work, rather than a story, even though it tells one. In short, the author was happy with the translation. He believes it serves its purpose, namely, to describe the inner structure and workings of the world.

Preface to *The First Seven Days*

("Seven Days, or How to Explain the World to My Dying Child?" Available in German: "Sieben Tage oder wie erkläre ich meinem sterbenden Kind die Welt")

This book is difficult to read and probably also difficult to understand. This shouldn't come as a surprise, because—after all—the world is not explained easily. Learning and true understanding are never easy. Learning and understanding are tasks that are as complex, multiple, and demanding as life itself. This is because life is learning. Dying belongs to that. This too one has to learn and it does belong to life.

This is the story of a child, an astounding and courageous child, who had mastered all these tasks. And thus, his life, no matter how brief, never was in vain, never without meaning. And thus, his life was important and of great value.

With every piece of knowledge I'm allowed to give forward to others I do something to not forget this little hero and all the other courageous children who were gone much too early, because we had been too stupid to help them. But none of these children was unimportant. This book elaborates why.

There is no logic in our existence if there wouldn't be a task for us. Probably we haven't been very good in seeing that very task, not to talk

about performing it. Surely, we are going to leave this world without any great and real contribution—each one of us. A contribution to both the world and the task, which, in essence, is the same.

Shouldn't we try and help our children to do it better than we have done?

Why

A good friend of mine owns a small summer cottage at the seaside. One day, she had the idea to make this holiday home available free of charge to families with children suffering from cancer.

Little children, who are not lucky enough to live for long.

Eighth Day

My dear child, this will now be the last and probably also the most difficult part of our "course" in trying to understand the world. This time we will not leave the math aside. On the contrary, this time, we are going to use it. Like the true big scientists, we are going to write down the equations and let them do their magic. We will try our best to let them evolve, one might say. Because, after all, we want to know how the world is and not how we would like it to be, right?

Thereby, I do not want to adorn myself with borrowed plumes. The keys to understand the world were given to me many years ago. First there was this box. It was full with extremely old reprints of publications, scientific publications, some of which looked so yellowed that it was almost impossible to read the original texts. But I didn't need to anyway. I knew most of these papers already. I had them in my collection. The interesting part about these papers was the handwriting along the margins and between the lines. This was still pretty readable. The topmost paper was one of those predecessor papers of Einstein's general theory of relativity. An old woman, not herself knowing how she had come by that box, had given it to me. She had found the carton in her attic. Opening the box and seeing the first paper, she suspected the whole content to be of a scientific character. And so she considered it a nice gift for me, as I was just about to become a physicist those days—a very lazy physicist by the way and , above all, extremely slow on the uptake.

As this whole Einstein theory was by far too complicated for me I only took the box out of sheer politeness and intended to get rid of it as soon as possible. A nice fire seemed to be just the right thing. I thought Einstein's theory of relativity would still be far beyond my grasp if Samuel and his father hadn't helped. The two had left so many hints and additional elaborations on these papers that at some point it was almost easy to get the gist. This, however, I was not to learn for a very long time. For some funny reason, something I'm absolutely unable to explain, I did not burn the box. I kept it with me wherever I moved. It learned to know almost all of my various girlfriends and I might even add that it learned to know some of them better than I knew them myself. It crisscrossed all over Germany, stood in many cellars and lofts, and spent a lot of time in various car trunks. Why, in the end, this unobtrusive box did manage to survive in my possession, I cannot tell. I'm neither a believer not a fatalist, and so, I think, it was just my laziness or the same kind of accident which one day brought me to open the very box a second time, but this time to look a bit more closely, to show a little bit more respect and to open my heart, or whatever was necessary, a tiny bit more.

There was this one day where an official guy from the local tax office had announced himself to check on our "working rooms." I had no idea what he wanted to "check on." But I thought that all rooms we had declared as working rooms should also better look like some. I saw absolutely no problems anywhere except for our so-called server-cum-archive room. Not that I suspected anything wrong there, but, well, as the name says, it is a room where we keep the servers nobody looks at, except there's something wrong. And there is the stuff being stored there for safe keeping, which usually just means for good. With all things you rarely see, you don't exactly know how they look like, and so I wasn't sure what impression this room would make to an overcritical taxman, especially one who was keen on justifying his job by "finding something." Thus, I decided to check on the room myself before the taxman and see whether it needed a bit of "structural optimization." Most of my anxiety, by the way, comes from the fact that you and your siblings have played there, even when it was forbidden. In fact, I found rather impressively huge and fairly intact ecosystems of dinosaurs, knights, and Native Americans there. Wonderful adventures must have taken place there, in that—at first sight—unimpressive room. And it is a pity that you can't tell us these stories anymore. Forcing all those knights, Natives, and dinosaurs into a bag, I suddenly saw the box. You had used it as a small platform on which the Native Americans had built their home. I immediately recognized it, and it almost made me feel guilty somewhere back in my brain. And so, this time I did not just put it to another place, but opened it.

I didn't expect anything. After all, science meanwhile was years ahead and what on earth was there to learn reading old papers that others had smeared on. If I had known that there also were old newspaper rims scribbled full with funny equations in a style so very strange to me, I would probably have chucked the whole package straight into the fire. Instead I stared on the first sheet of paper in amazement and wasn't sure whether to trust my eyes. There, in a funny, old-fashioned German handwriting, were the two words *dimension* and *Hilbert* followed by a question mark. Each of these two words on its own would have meant nothing to me, but together on this Einstein paper they not only made sense but . . .

Something wasn't right here. This was an old box, nobody had opened it for decades, and still there was a hint about fractal multidimensional spaces directly in that Einstein paper. This couldn't be just an accident. Now my attention was caught.

As you well know, my little child, I have the technical possibilities to check on the age of things, especially when they are made from or contain organic substances. So, extremely careful in order not to put in impurities, I took some samples. As any material from outside the box was of young origin, such an impurity would have led to a younger dating, not to an overestimation of the age.

By the way, I had meanwhile completely forgotten about the taxman and so was completely perplexed when suddenly the bell rang and a young, well-dressed man stood in the door, showing me his ID. When slowly my brain locked into gear again, I made the guy a latte macchiato and rather boldly told him to "feel like home," because "I had no time" and was "working on something important," while I hurried back to that box checking its content and sorting it. I didn't even realize the guy following me. I remember him asking me something about "permission to look over my shoulder," but I can't even tell whether or not I ever gave that permission, or in fact even answered at all. In fact, he was a bit like a second and very quiet shadow, probably even quieter than my first one, because I did not become aware of him before he announced himself satisfied and thanked me "for such an excellent demonstration of my typical work." Meanwhile I was so deeply immersed in the content of that box that I probably didn't even say goodbye, which, apparently, the taxman did not mind at all. A few days later we received an official-looking letter from the tax office telling us—surprisingly—that there are absolutely no complaints and everything is in best order. Your mother gave me a huge hug that day for the "good job I had done about this tax thing" and I never revealed to her that—if truth be told—I had not done anything to deserve that hug. So, one should note: a room declared as a working room for the tax office is most convincing if you can absentmindedly demonstrate how to work in it.

One day before the official letter, the results from the age determination measurement arrived:

"Older than 80 years—both, paper and handwriting!"

I had expected anything, but definitively not such a number. It meant that the scribbling was made at a time when nobody even thought about things like strings and fractal spaces, to say nothing about discussing it or making notes and evaluations about such ideas. Now I was definitively intrigued. I wanted to know more about those people who had left these unbelievable messages.

Reichskristallnacht

It happened in the night from November 9 to 10, 1938, the night hundreds of Jews were killed, tens of thousands maltreated, disseized of their property, and thrown into concentration camps. All this happened under the watchful eyes and protection of the German police and a usually passive, often jubilant mass of ordinary German citizens, some of whom even actively supported the pogroms. That night, Schmuel went to a meeting with old colleagues. They wanted to discuss the Einstein–Podolsky–Rosen paradox in quantum mechanics and as an old friend of the family he was invited to bring his son Samuel and his daughter Judith. Early in 1934 Schmuel and his son had discussed this problem with Einstein directly. As belonging to the "unwavering defenders of their home," as Einstein had called them, he had even sent them a rather personal draft with comments of the later, very famous paper treating the "EPR paradox," as it would later go into history. The three authors of that paper (A. Einstein, B. Podolsky, and N. Rosen) had claimed that quantum mechanics can't be correct or has to be incomplete, or both, because it predicted a "spooky remote action effect" between well-separated spatial regions faster than light. Nowadays we know that this effect is real. Many experiments have shown what we now call entanglement, but in the days of Einstein, Podolsky, and Rosen it was only a thought experiment and the three considered its result complete nonsense and couldn't believe that there was a theory allowing such a thing.

While Judith, still almost a baby, played in a neighboring room with the two children of Schmuel's old colleague, Schmuel and Samuel followed the discussion of the group with keen interest. However, even though very knowledgeable in the topic themselves, they didn't actually contribute. This had nothing to do with them feeling anyhow superior regarding the topic, but it was simply that they were very conservative when it came to suggesting new and rather revolutionary approaches, and in their case the approach they had in mind was nothing short of revolutionary. Some would have even declared it absolutely insane those days. Nevertheless, they felt a bit bad about not telling anyone in that room what they had found out, or at least suspected, especially as there was a small pinch of guilt towards their host. After all, it was not without risk in Hitler's Germany, the Reich, to still stick to one's former friends and colleagues if they were Jews. But when looking at each other, both father and son silently agreed that the time wasn't ripe to reveal that, taking their approach, the EPR paradox would not come as a question or a paradox but as a structural necessity of space and time.

When the riots began, their host suggested that Schmuel, Samuel, and Judith simply stay in his house until everything was over. One might even fetch Schmuel's wife when the night was over. Schmuel felt uneasy about this, but his host didn't want to let him go and consoled him:

"It is extremely un-German to attack a lonely wife, and even those bastards from the SA won't dare to commit such an ignominy."

How very wrong he was he would not learn before the next day and he would carry on this failed judgment for the rest of his life.

But for the moment, everyone busied her- or himself in the attempt to make the home as cozy and comfortable as possible for the three unexpected guests. The host's wife conjured a wonderful late supper, and at the time everyone went to bed, the atmosphere was surprisingly at ease, almost ignorantly relaxed.

In the middle of the night, Samuel, who was sleeping alone on the sofa in the living room, awoke because of a timid knock on the door. After a while their host opened and a very excited voice could be heard whispering from the entrance. What Samuel now had to listen to made his insides freeze. The host and his informant quickly agreed that one must not inform their guests right now as it would probably make things worse. Then the door was closed and heavy, muffled steps sounded through the house when the host went back to bed.

Samuel's heart was racing when he silently dressed and tiptoed out of the house. He had taken the house key from a hook next to the door in order not to disturb anybody when coming back.

It was only a few hours later when he used this very key, but in these few hours he had changed completely. It was a brachial methamorphosis from a young innocent scientist into a tough and deadly warrior.

A Courageous Jewess

They only knocked once. Sarah, even though a mother of two and in her mid-forties, was still a very pretty woman, and rather athletic, too. She barely had time to grab her dressing gown when the door burst open under the heavy strokes of a sledgehammer and an axe. Men in civil clothes stormed into the house.

They were from the SA, the last, the most stupid, and the most disgusting kind of human beings the German community had to offer—the meanest of beasts one would find when lifting the dirtiest of stones of an already rotten society. They immediately sprang on Sarah while in the background more SA henchmen pushed in, smashing what they saw and what would not fit into those bags they had brought for valuables. When the fourth of these monsters threw himself on Sarah, she got hold of one of those wooden pieces of furniture the SA dunderheads had smashed before. With all the force she could master she hammered it into the head of her rapist, who collapsed straight away.

"Hey, this Jewish bitch is still fighting!" exclaimed one of those other SA men. And then they came and like crazy they beat her with all they had. But in all their hatred they carefully avoided to beat her on the head or into any vital organs. Instead, almost systematically, they smashed her legs, arms, shoulders, and pelvis. They were nothing but cruel German monsters.

Then they put a cord around the middle of her mutilated body and hung her outside on an iron latch of the gate. Thereby they jeered and chanted their dim Nazi songs.

"Anyone who tries to help this rag bag will be treated the same way as her, understood?" they bawled. And indeed nobody even tried to step in. Not even the policemen who stood nearby and like so many "ordinary German citizens" only greedily stared at the naked mutilated body. Horrible moaning and heavy rasping breathing coming from the poor woman moved nobody there.

Suddenly, a loud shot rang through the night. Sarah's head was pushed to one side. She was dead in an instant. The sound of this shot was still ringing between the houses when there was the next, and the next, and the next. SA man after SA man was hit, and they fell like rotten apples from the tree. It was so fast, so precise, that even after five perfect hits there was almost no reaction among the bystanders, the police, or the other SA men who had come to watch the hanging woman. This was a professional doing his job. Finally, there fell the two policemen and three of those disgusting gapers before the panic started and all hell broke loose. Simultaneously with this public tumult the shooting ended and everything was nothing more than panic running, screaming, and stumbling away.

Later the German police would find the alleged murder weapon, a standard SA carbine, near the corpse of a high-ranking SA officer. Just as all other of his fellows that night, he too was in civil clothes. His head was smashed, but the most peculiar thing about the situation was that usually SA officers did not carry carbines. They had pistols and sabers. The whole case was so mysterious that in the end the Gestapo took over and covered everything up. The dead "SA Volksgenossen" were declared victims of a freak accident, and the dead woman was registered as a victim, too. In fact it was the only Jewish victim properly recorded by both police and Gestapo, while the latter even named her as a "potentially internal NSDAP victim" of some high-up order.

The Two Brothers and the Einstein–Hilbert Action

It was extremely difficult for Schmuel to get over the death of his wife. And as there was no other distraction, he threw himself into the only task left: the explanation of the world. Samuel fully understood his father's sorrow. He also understood and accepted the way his father went in order to overcome the grief. He himself would have liked to spend a bit more of his time with this task, but now it was on him to care for the family, to try and protect them, to feed them in these difficult times. This was an almost impossible task. Even though it was only him, his beloved sister Judith, who he always called his "Julchen" and his father, there was barely enough food for one of them alone. The Nazi regime had decided that Jews must not get more than 200 calories per day, which was nothing else but a cruel sentence to slow and steady starvation. For comparison: even a sick person who is put on a strict diet would still get about 1000 calories.

But somehow Samuel managed to keep his small family alive. With amazing resourcefulness and irrepressible will he was able to perform the impossible. Every day. Anew. At least his sister he was able to feed so that she was having enough almost every day. He and his father on the other hand had to tighten their belts and despite all the risks and the efforts it was very difficult for Samuel to get all of them enough to eat. During his forays Samuel saw almost unbelievable cruelty and horror and several times he thought that he could no longer stand it, that he would rather die than endure more of this horrible world they had to live in.

There was this one day, for instance, when he came back from his black market business outside the Ghetto, when a small boy asked him for a piece of bread. The little boy sat right next to this pile of rubble behind which the whole outside world was hidden. Obviously the kid knew that here the "jumpers" came back in. There were others like Samuel who did little trading business in order to smuggle bits of food into the Ghetto to avoid starvation under this horrible dictatorship of the SS.

Next to the boy sat another one, apparently his brother. He was even smaller, even thinner than the first one, almost a skeleton. They both squatted in a pile of tatters and both seemed too exhausted, too lethargic to move. The moment the bigger of the two became aware of Samuel's presence, a small thin hand came out of the pile of rags. It looked as if it was done with the last strength he could muster. The thin skeletal hand moved towards Samuel and he simply stood there, shocked. The boy didn't even move his head. It was as if he already knew that they wouldn't get anything. It seemed like hopelessness and despair would have been born into this world as children and now they crouched here on this wall, just about to die. Samuel just stood there and stared. He simply couldn't move, couldn't tear himself away.

Suddenly, however, the smaller of the two children, even though weaker in appearance, nothing but skin and bone and huge, huge eyes, looked at him. He looked at Samuel with eyes that stared into this world from a place far, far away, and they looked at it as something of unbelievable beauty and nameless cruelty going hand in hand. Something this little boy never had the chance to grasp, to explore. It even seemed like this boy internally already knew he would never ever get this chance, but still he was not ready to accept this.

For a very brief moment Samuel felt a shiver running through his body, moving him. He felt the loaf of bread in his jacket and suddenly thought of his mother. His mother would rather have died than leave these two brothers— than pass them without breaking her bread and hand a piece of it over to them. But then, in an instant, Samuel's innermost hardened again and he thought: *Yes, and because of that she is dead now!*

Then he forced the picture of his little sister, his little Julchen, into his head and it became almost easy to turn away from the sight of the two starving brothers in their dirty pile of rags. A scrawny hand slipped back under the tatters, but a pair of huge eyes was still not ready to accept the death sentence. While Samuel was turning away, a mouth opened, a mouth with lips thin like paper, and a funny, slightly familiar noise reached Samuel's ears. Samuel accelerated but the boy called again. Again and again the boy called and Samuel was almost about to run, when he heard the commanding outcry of a SS man:

"Shut your damn ugly mouth, you Jewish brat!" it rang through the street and Samuel's heart beat like crazy.

Shut up, little boy, he thought and knew the same moment it was no good.

Slowly Samuel walked on and behind him the little starving boy called again and again. There was something peculiar about his voice. At the beginning the noise it had made was hoarse, almost like rasping, but with every call it became clearer and brighter and now it was so pure that it reminded Samuel of a very fine, high-pitched and perfectly tuned bell.

"I said, keep your bloody mouth shut! Didn't you get it?" bellowed the SS man and Samuel couldn't help it. He had to turn and watch.

There were two SS men, huge and terrifying with their boots, their leather coats and their machine guns.

"Hey man, look what an awful carcass this is!" Samuel heard the voice of the second SS man. "Disgusting, this vermin."

"Why do you think you can shout into my ears?" The bigger of the two had grabbed the boy and pulled him easily out of the pile of rags. After all it was not much weight the SS man had to lift.

"Is this your damn Jewish language?" The voice was trembling in the most devilish threat Samuel had ever heard. He knew he had to go. now! But he could not take his gaze off that bizarre scene in front of him.

As if there were no morning and no danger in the world, the little boy made this peculiar noise again and again.

"Are you mocking me?" the big SS man bellowed. "Just you wait a second!" Then he grabbed those skinny ankles of the boy and tossed him through the air like a wooden stick. With a dull thud the boy's skull hit the wall and burst.

"Bäh, what a mess!" said the other SS man rather quietly. "Bloody hell, throw this away, man!"

The addressee was grinning. He threw the headless corpse into the lap of the other boy, the brother, and jeered:

"Well, Jew-boy, have a look! Now you have something to eat!" And laughing, the two went away.

Samuel didn't hear any crying or sobbing, nothing. There was nothing except endless resignation and silence. Very slowly he turned and went home.

"I have to be stronger than my mother was!" he said to himself over and over again while he was walking.

It was not before he was at home when he allowed himself to think about the noise this little boy had made. Without a word Samuel put the loaf of bread on the table and then seated himself near the stove. Now the words hidden in this bell-like noise of the little boy were clearly in his head and his heart tightened when he finally realized their meaning. It was simply:

"Hevenu shalom aleichem!"

The boy had only wished him peace. He had wished him, Samuel Stamler, who did not want to share his food with the two starving boys, peace. This little boy had done nothing but shouted the Jewish rite of peace—and had been killed for it.

Samuel's father immediately saw that something was wrong. He looked up from a stack of crumpled papers, put the stub of pencil aside, stood up and went to his son without a word. He took him in his still powerful arms and hugged him.

Now, it broke out of Samuel. Sobbing, he reported what he had seen and his father cradled him like a small child. The little sister came into the room, but the father only nodded to her and so she went back to the sleeping room. After all, they had plenty of space now since they became less and less in the Ghetto.

When Samuel had calmed down a bit, his father stroked his head and said:

"Come son, let us see whether we still can help!" With this he stood up, pulled on his old coat, with the Yellow Star plainly visible on his chest, and went to the door. Samuel did not understand this, but he followed all the same. They went down the stairs and out on the street.

It already was curfew and if seen they would be shot straightaway, but Samuel's father suddenly radiated such power, such reliance, that Samuel simply couldn't help it, but calmed down completely.

When they reached the spot, the headless corpse of the little boy was still lying over the lap of his bigger brother. Careful, almost as if they did not want to make contact at all, Schmuel's strong hands touched the bigger boy's shoulders and shook him almost imperceptibly. Suddenly, the boy's torso collapsed and fell forward covering the corpse of his little brother. They both were dead now. Samuel began to sob again. No matter how much he had hardened himself after he had to shoot his own mother in order to save her from further pain, this was too much for his still very young soul. Schmuel slowly, almost graciously erected the body again and leaned him backwards on the wall. The head fell onto the right shoulder and moonlight shone on the staved features of the boy. It was a picture of utmost resignation and sorrow, infinite sadness and pain. Schmuel let go of the little body, stood up and laid his arms around his son. He felt him shivering and said softly:

"We will give them a proper burial and bemoan them like they were our own sons and brothers."

With this he bent down and carefully wrapped some of the rags in which the little boys had sat around the two corpses. Then he lifted the bigger package on his strong shoulder. Without hesitation Samuel took the smaller one and very fondly, almost like a holy object, carried it over his arms in front of him. It was extremely light. They walked slowly, like being in a procession.

"Halt!" a voice bellowed suddenly in front of them.

In the moonlight one could just make out the silhouette of an SS uniform.

"It is curfew, you damn Jews!" a second voice yelled. Samuel immediately recognized that voice. Even though the man was still in the shadows and rather far away, Samuel recognized it easily. He thought he would always recognize it. Almost no matter the volume, no matter whether his senses might not be able to detect the sound, it wouldn't help his soul from still sensing this man. He was sure that they would die now. Die like the two poor boys, whose dead bodies they carried through the streets. They would be shot or battered to death by those two monsters who directly came from hell and who still spoke the same language as he and his father, the same wonderful language they called their mother tongue. Samuel reflected that, he didn't know why, he somehow liked the idea of dying just now. Right in that moment he thought about all those horrible pictures in his head, the unbearable screaming, begging and those many inhuman scenes he had to witness, and death appeared to offer such a wonderful salvation, such an easy way out of all this misery. He wanted to have this weight put off his heart and die. He only felt sorry for his little sister, his Julchen, who would also have to die without him or father. Perhaps she would even die like the two little boys had died or she would just starve in their little flat in the middle of the Ghetto—another nameless Jewish child. Also the wonderful interesting work he and his father had begun. They would not be able to finish it and as they never would be given a chance to publish their results, their amazing insights, in proper form, it would simply be lost. Ready to die, Samuel looked at his father's broad back and wondered. He wondered how this strong man could still walk on and calmly, as if it were the most ordinary thing in the world he was doing, lifted up one hand.

"Hauptscharführer!" he shouted with the steadiest voice, knowing full well that such a rank was at least five grades above any of those two SS men.

"Please do not shoot us before we have been able to finish the order issued by your SS administration!"

Even if the sentence would have given rise to suspicion, as Schmuel said it, it did not sound ironic—not in the slightest.

"What order?" asked the voice of the SS man deep in the shadows and one could hear that there was a bit of unease ringing in it. These creatures didn't know anything but cruelty and obedience and one single, no matter how weak, unusual aspect brought them out of balance and took away from them the bit of confidence they had. Thus, Schmuel helped them and cited by heart from one of those recent official announcements.

"In the interest of the health and purity of the Volksgemeinschaft [*which means the people*], all potential sources of impurity and disease have to be removed immediately, Herr Hauptscharführer. Thus, we only dispose of these corpses, because at such temperatures—well, Herr Hauptscharführer, you probably know."

Of course, nothing did these creatures know except that corpses did start to rot and stink if left outside after the killing was done. But they have heard the word "Volksgemeinschaft" in connection with health and purity often enough in order to meanwhile string them together.

"Now see, a Jew who actually can work," said the SS man in the light and actually laughed. Then he added:

"Well, then, in the interest of the Volksgesundheit [*the people's health*], go ahead!"

Thereby he bowled the last two words into the night, saluted and clicked his heels so that it sounded like a sharp shot.

Then the other SS man also came to snapping attention and shouted:

"Jawoll, everything for the Volksgesundheit!"

He had just come out of the shadows and Samuel could see his face while they passed. It was by far the most stupid face Samuel had ever seen. He could not help it, but suddenly he had the overwhelming feeling that he should be happy about his fate. Happy about being a poor Jew in such a miserable situation rather than being punished for being such a dimwitted monster. For nothing in the whole world he would have swapped with this creature.

Along the way to the house was a small patch of green. Once it had been a playground for the children of the neighborhood. To this place they brought the two dead bodies. Samuel quickly fetched two spades out of the cellar of their house and then they stated to dig.

When they had put both children next to each other in the pit, they started singing the Lamentations of Jeremiah. They sang quietly, very quietly; it was more a humming. Because they knew what would happen if the SS caught them here, singing Jewish songs or alone doing this small ceremony for the two brothers. After that they closed the grave and went home.

They found Judith on the bench next to the small oven. She was fast asleep and the father said to his son:

"You did right, my son, when you preserved the loaf of bread for your little sister, but we also did right in honoring these two little boys and in lamenting their death.

"In this lies great wisdom you must know. It is the wisdom of our belief and the reason why we have been chosen above all other people. In this also lies the reason why we have been asked so much, why we have been tested so many times and so hard. But also why we have been crowned above all others. Never forget this, my son!"

"Yes, Father!" Samuel answered solemnly and felt his confidence coming back to him.

"Let us break the bread in the morning when your sister awakes. Then we will mourn the two boys again, but now I would like to show you something."

With this he pulled his son over to the small table in the middle of the room. He gestured him to sit down and took a seat himself. In the dim light of a candle he showed him what he had found.

"I think the good old Professor Hilbert has made one or two simplifications too many, do you see?!" he started, and in a jiffy father and son were lost in a world where no SS men and no Hitler regime with all its might could ever follow them. No matter how big the cruel efficiency, this Reich of Samuel and his father was not and would never be theirs.

Both sat in front of a publication of a famous mathematician from the University of Göttingen, David Hilbert. He had shown in 1915 that one could extract the Einstein field equations of gravitation, known as the general theory of relativity, out of a simple variation principle. When applying the principle, one mathematically "shakes" a term until it orders or structures itself such that it obtains a minimum of a unit called action. It is a bit like a sphere rolling around on a topographic surface with hills and valleys, losing more and more energy while moving on the surface and trying to find its deepest position within the topographic field. The sphere might even find the most perfect place, meaning the deepest position, if only shaken long and carefully enough.

Now, he, which is to say Hilbert, had never given the Einstein field equations himself. After all, he was a mathematician, and as such, he stopped his calculation the moment it was "obvious" where the whole thing would head. It was a bit like professional chess players who would not actually finish a game the moment its outcome is logically fixed.

Schmuel now showed his son the part in Hilbert's derivation where he had dismissed a term as meaningless—that is to say, zero—because it was a surface term one could omit according to the rules set by his variation principle.

"I think," he said to his son pointing at this term, "in reality we have here the expression which brings in matter into the general theory of relativity.

I mean this is what Einstein and Hilbert all the same brought in artificially later on and called it the energy–momentum tensor."

"What?" said Samuel, who couldn't believe it. "You reckon they had thrown away a term and later realized that something was missing? I mean something that has been originally there all along?"

Schmuel smiled.

"No," he said in a drawling voice, "it wasn't that easy. Unfortunately! Because taking the rules of their variation principle, it was absolutely fine to erase this term, but, as it seems, nature, or the universe, if you prefer, does not follow such a simple rule—well, not always anyway. It kind of finds ways around it and then, surprise, the term does not disappear and is—as I suspect—the reason for what we see, detect, or take as *mass* and all other kinds of matter in the universe."

"Thus, you have invented a new kind of math with a new variation principle, right?" Samuel asked in a doubtful tone.

"No," said Schmuel again, "I have not invented it! It is just the universe acting this way. The universe is simply using more degrees of freedom than Einstein and Hilbert had thought. Well. This just makes the variation more comprehensive, more holistic, more *catholic*."

And he briefly laughed about the last word.

"But how? What is the additional degree of freedom? I mean the one Hilbert did not use?"

"It is the dimension of time and space itself."

Samuel needed a while to comprehend his father's answer, to digest it. Then he asked in a tone which simultaneously revealed disbelief and the beginning of understanding:

"So, the universe variates more than Einstein and Hilbert assumed? Thus, it finds another, more global minimum and in the end reaches a different state than we have supposed, right?"

"Yes, my son," Schmuel answered. "One could put it that way. But with the later introduction of the energy–momentum tensor, both had made up for this little 'blunder' well enough, but this way matter comes in artificially while with the correct variation it would already be there."

"And when the surface term is not thrown away and the variation is performed differently, then matter does come in automatically?"

"I'm convinced of it, my son, but on this aspect we still have to do some work," Schmuel answered. Then he showed his son what he had derived so far and with verve they started to work, forgetting hunger, exhaustion, and misery. The goal was set. They had to find those missing pieces in the universal puzzle and then they had to put it all together.

Judith, the Little Sister

Father and son still sat over those papers and excitedly discussed the new world expanding itself before their eyes, when suddenly a small hand was laid on Schmuel's arm and a timid little voice asked:

"Papa, I'm a bit hungry. Only a little bit. Is there something today?"

Completely surprised, Schmuel and Samuel turned their heads, eyed the small girl right next to them, and asked themselves how long Judith might already have been standing there. Both father and son looked with fondness at the girl. It was as if the sun would rise again this day. Despite all the privation and suffering she was still a very beautiful little lady. She had the bright hair of her mother and together with her blue eyes she might have made the perfect Aryan, if only there weren't those Nürnberg racial laws, which made her a Jewess.

"It is really only a little hunger," Julchen repeated almost ruefully.

Immediately there came life into the two men. The father shoved the loaf of bread in front of the eyes of the little girl and answered simply:

"Here, my beautiful lady!"

Judith's eyes widened in happiness. But then her gaze came to rest on her brother's hand, which had just slipped into the pocket of his jacket. Seeing what he had brought for them, she almost yelled and clapped her hands.

"Oh good Lord!" the father exclaimed. "Ham and cheese! How on earth did you do that again, my son?"

"Well," the boy answered in a mock voice, "the ordinary Jew, as they say, is after all sly and pretty good at doing business."

Father and daughter laughed, but they were completely stunned when Samuel conjured a small glass of honey out of the other pocket of his jacket.

When they had put all the goodies on the table and Schmuel had made a tea out of nettle leaves, he became silent and contemplative.

"I would like us," he began, "to savor our food in a properly kosher manner, even though the ham most likely is anything but kosher."

With this he winked at Samuel, who smiled and made a skeptical face, as if to say, 'Who knows?!'

"Let us think about all the other Jews while we eat! Let us think about those who do not have such a luxurious meal as we do. Above all, let us think about those children the Nazis have driven into horrible starvation right in front of our eyes and about those we could not help, *no matter how much we might have wanted to.*"

These last words he spoke loud and clear. His daughter looked surprised, almost shocked, but Samuel understood that the words were for him and sensed how much better they made him feel. Then his father spoke the thanksgiving prayer and he blessed his children and the meal—a meal, he explicitly mentioned in his somewhat unorthodox prayer, Samuel "had so cleverly gained" for the family.

Brief Hours of Scholarship

No matter how unbelievable it sounds, but Samuel succeeded for a surprisingly long time in keeping the family alive. During the war his trading business got ever better. As astonishing as it seems, especially as shortage outside the Ghetto grew, the business opportunities for clever guys like Samuel became better because the black market became more and more important for ordinary German people, too. He sometimes even made a bit of profit, which would have allowed him to flee the country. Some of his partners had offered him that. Getting out of Germany, to safety. But he would never have the heart to leave his father and his sister. And so he stayed, did his business, and took good care of his little family. It had already happened several times that his partners had warned him when there was another raid or "evacuation." This had always allowed him to hide until the storm was over and then proceed with the daily struggles of life. Usually, it wasn't a big deal for him to get new papers telling any ordinary SS man that they were still allowed to be here, but this also was the limit of helpfulness he could expect from his partners. After all, they were ordinary Germans and he was "only" a Jew. But when it came to such limits he still was a pure businessman, never asking more than what was "reasonable."

Soon, thanks to the war and the deficits it had brought, his little business was so successful that he again found time to help his father with the other important and much more interesting, much more satisfying task, the solution of the world's principal riddles. Especially when it came to complex and tedious mathematical derivations, his father usually trusted him to cope with them. And truly, he almost never made a mistake, no matter how long and cumbersome those evaluations were. He simply had a knack for these things and his unbelievable brain allowed him to visualize stretches of equations others would have needed to write over many, many sheets of papers, while he didn't even need to write them down but just noted the final results for his father.

His father, meanwhile, often sat in a corner with Julchen and taught her something. She too was an extremely bright child. When Samuel was stuck in a calculation he always loved to watch the two "doing school" and he marveled at those astute blue eyes of his little sister, which were glued to his father's lips. She listened carefully to every word those lips spoke and took them in, in her most individual manner. Watching them very often, Samuel had an idea about how to proceed with his own derivations, because suddenly a new way had appeared to him, one he had not seen before. Sometimes, all this made him see his sister as an angel—a small, skinny, at times noisy but always lovely angel. He would have given everything for her, but he already knew he would not be able to save her. His only wish was that she should not die like her mother or the two boys had died, that it would be quick and painless, more or less. He knew he would not be able to save a single one of them, except for himself, but to him, this was completely out of the question. He had decided months ago that he would never leave them here alone. And thus, Samuel once again listened to his father while he explained to the little sister why extremely small things behaved so much differently compared to our scales. It was the basics of quantum mechanics that his father was just laying down in that small room and there was an almost palpable concentration radiating from the little angel to whom his father spoke.

". . . now my pretty lady, there simply is no reason why there shouldn't be cosmic structures, universes also on completely different scales, you know. Your clever brother has just proved that a Friedmann cosmos could easily be constructed in only two spatial dimensions. This cosmos, as is tradition within the fast community of the Friedmann cosmos, would also have constant curvature when it is assumed to be a spherical surface."

"Don't forget, it could also be a hyperboloid or a plane!" Samuel added without being asked.

But father only smiled and drew those various geometric patterns with a spirited movement of his huge hands into the air.

"Your brother is right, of course," he admitted, "but I prefer the sphere, because with this I can easily construct the whole of our universe, which is to say, the universe we do observe."

"How can you do this, Papa?" asked Judith, equally surprised and excited.

"Well, simply imagine the space full of such little Friedmann cosmos. They are so small that we usually don't recognize them."

"How small is that, Papa?"

"It is about 10 minus 35 meters, which is the Planck length."

"Oh," Judith said, "this Planck I have heard about. Isn't he the one who actually invented quantum mechanics?"

Samuel laughed from his watching position and mockingly confirmed:

"Oh yes, Planck invented quantum mechanics, and since he had had this wonderful idea, those stupid atoms with the electrons in their shells do know that they are allowed to exist. Before, they simply didn't know that, you must understand. And most important of all, since then we are allowed to exist, too."

"Oh come on, Samuel," his sister protested. "You know perfectly well what I meant by *invented*." But then she also had to laugh.

"And who actually was this Friedmann, by the way?" she finally asked, not without hoping to gloss over her little mishap.

"He was a Russian scientist. He was one of the first, if not to say *the* first, to actually apply the Einstein field equations to the world. Thereby he also allowed the world to behave dynamically, which was quite revolutionary during his time."

"What do you mean by dynamic, Papa?"

"Oh, it just means that the world does change. It moves and evolves and so on. Einstein wasn't very fond of that idea. He had preferred a more static world. Well, at least until Hubble discovered the expansion of our universe, this was."

"And you two," Judith pointed at her brother and nudged her father, "you two think that the Friedmann cosmos could also be extremely small and that there are so many of them that they make the crumbs of our universe, right?"

"In a sense, one could put it that way, yes," her father admitted. "Well, and to be honest, I prefer it that way, because, and that is the important point here, these little Friedmann cosmos automatically bring in properties into the universe we actually do observe and for which there are no proper explanations otherwise."

"What properties are these?" Judith asked in a slightly perky tone.

"The whole quantum mechanics, you little silly," Samuel said, chuckling.

"Now you wait!" Judith said threateningly, and jumping to her feet, she threw a small, very skinny fist against his ribs.

"And," she threatened to strike again, "how shall this explain quantum mechanics?"

Samuel pretended to be hurt very badly after his sister's heavy blow and tilted sidewards off his chair. But during the fall he grabbed his sister and pulled her with him to the floor. Down there, he started to tickle her.

"What a stupid question this was, little sister," he said, "although it is so simple. It is so that all those little Friedmen *and* Friedwomen (who knows) are extremely ticklish. Thus, they are fidgeting all the time and because of the permanent and very nasty fidgeting—Will you stop to dither now, little sister!—there are simply no fixed coordinates. Everything is constantly on the move and changes its properties. It's as if one has to move through a room full of little dithering sisters. It is absolutely impossible to walk or even crawl over a straight line in such an awful room. It's like being drunk all the time. You can't help it, you totter seesaw-like through space and time."

Giggling, the little girl saw a fly in the ointment. She intended to spill it out, but Samuel had chosen that moment to tickle her again. So, the only noise coming out was of cheering and more giggling. But finally Samuel held her fast to the ground and allowed her to speak:

"But then the whole world would permanently be drunk. However, this can't be true, because one does not see the planets and stars totter trough space, right? These objects move pretty normal, ehm sober?"

"Oh my wonderful daughter, what an amazing input!" her father said. "Now you only have to imagine that the fat old Schmuel is moving through the very room with dithering Judiths. He is so fat and heavy that those little Judiths do not bother him. He doesn't even feel them and their fidgeting and so he, I mean this clumsy oaf, lumbers though the room as if it were empty."

Meanwhile Samuel had let go of his sister and both helped each other get back on the feet.

"And this is it?" asked Judith, a bit disappointed. "This is the way you bring the Einstein theory and quantum mechanics together?"

Schmuel and his father looked at each other. Finally the father said:

"Well, one small problem we still have. We still do not know how to describe the Friedmann cosmos in such a way that it reveals its fidgety properties in the right manner."

"OK, if your Friedmen and Friedwomen do not want to dither," Judith suggested, still giggling, "then you simply have to give them a bit of space, degree of freedom, as you always say, and then you have to tickle them a little bit!"

As if petrified, father and son looked down to the little girl. Both stood there, stunned and open-mouthed, and almost simultaneously they both smote their foreheads.

The Evacuation

Then, one morning, their time was up. From all directions military trucks moved into the Ghetto and they were trapped. SS men jumped from the trucks and stormed the houses. Yelling, lashing, and shouting, they beat out the occupants. This time Samuel had not received a warning and so just had time to hide Judith inside the small hole behind the stove and cover it before the door flew open.

"Get out, you filthy Jewish vermin!" one man yelled and "Hands behind the head!" bawled another.

Samuel immediately obeyed. With his hands above his head, Schmuel also got up from his chair, but this was not quick enough for the SS bastards. So, one rammed the butt of his carbine into Schmuel's underbelly. He sank to the floor. The man who had struck him now pulled back his rifle and was just aiming when a small, almost mingy SS man appeared in the doorway and cried:

"Stop it!"

He had a horsewhip in his right hand, and even though his statue was anything but dangerous, this man was radiating pure threat.

"Not those who can still work, you idiot!" the mingy man said rather calmly and the huge SS man with the gun immediately snapped to attention.

The other SS man retreated. Both were about two heads bigger in size, but there was no doubt in the world that they feared the little man with his ridiculous whip, his by far too big pistol on the belt, and his leather clothing. In an almost polite tone the man now spoke to Samuel:

"Are there by any chance any more members of the Jewish race in that house, sir?"

"Not that I would know about, Herr Sturmbannführer!" Samuel answered quickly and without a trace of betrayal. But the mingy one had not looked at him at all. Instead he had suspiciously eyed his father and watched his reactions. Now the little man's eyes were scanning the room. And it wasn't only his eyes which seemed to work. It was like a small, but extremely tough and well-trained dog catching a scent.

"Well, well, well, the Jew does know the ranks. Excellent! And well done indeed!"

He turned to his fellow SS men and said:

"You see, SS Sturmmänner, this is the reason why you have to watch the Jews. You turn your head only once in the wrong moment and they will have grown above your head by several inches with their enormous but all the same inhuman intellect. And if you don't be careful, dunderheads like you would just have to go back to those ugly holes we rescued you from–was it a dull backyard or a stinking pigsty."

Truly featherbrained, the two only answered in unison: "Jawohl, Herr Sturmbannführer!"

But then something caught the mingy man's interest and he poked in the pile of sheets of papers spread on the table:

"Oh, what do we have here?" he said. "It looks like the Jew can even write . . . And look what he can write!"

Visibly surprised, he studied the papers for a while and then said:

"It is very interesting that you Jews, even though just Jews, have put Hilbert above Einstein. It is almost like you ordered the Jewish madcap Einstein underneath the genius German mathematician Hilbert . . . very interesting, this is very interesting indeed . . ." His voice trailed away and became an unintelligible mumbling. But suddenly it became clear again and he demanded:

"You," and he pointed to one of the SS men, "pack all this, search carefully whether you find more, and bring all of it into my car! Understood?"

"Jawohl, Herr Sturmbannführer!"

Now he turned to the other SS man:

"Escort the two Jews down and treat them well!"

When the SS man had confirmed the order and demandingly nodded to Schmuel and his son to go ahead, the mingy man called after them:

"Hey Jews, you know what? As surprising you might find it, but with me Einstein would still be ordered above Hilbert!"

With this, Samuel and Schmuel were shoved to the staircase downwards and out of the house. As it was ordered, the SS man did not push them, he did not beat them. He strictly obeyed the command of his superior. The moment they reached the street, however, a high-pitched shout came from above their heads.

"Halt, you Jews!"

Schmuel and Samuel stopped and looked up to the balcony from which the voice seemed to come.

"Didn't you forget something, gentlemen?"

The mingy SS officer held a deadly pale Judith by her wonderful blond hair and a demon-like grin spread on his face. Slowly, almost infinitely slowly, the SS man lifted Judith up, grabbed her by the ankles, and let her swing over the banister. All the time, he watched, almost studied, Schmuel and his son. Nobody moved, nobody made a sound. The scene was so unreal. Even Judith did not say anything, or scream. On the contrary, she appeared like a small heroine. Hanging there high above the stone-paved street, she knew perfectly well that it was over for her. She also knew that it would only also cost her dear brother's and father's life if she now started to cry for help. But then she would not achieve anything but kill those she loved. Almost relaxed, her eyes had found those of her father when the little man finally let go of her. Falling, she closed her eyes and made no sound. Samuel and Schmuel knew that they could not help her beloved Julchen, so they kept silent and didn't move. And silently they watched her fall and hit the ground with a dull thud. They even endured to watch silently when the little body twitched spastically and an unbearable gurgling sound came out of the little mouth. The noise and the twitching did not end before another SS man pulled out his gun and finally released the girl.

Before the End Lies a Way through Fractal Dimensions

"What a shame," Schmuel sighed many hours later, when he and his son were cowering next to each other, penned with many others in an animal wagon. "I'd have loved to let her know how much her suggestion with the 'tickled space' had brought us forward."

"Yes," Samuel answered, happy about the fact that finally his father had abandoned his crushing, pondering silence. "But in a way I'm also happy that it is over for her and that it was so quick. What a brave little heroine she was. I couldn't be more proud."

And with this, eventually, he started to cry.

His father, who was already beyond tears for a long time, pulled his son into his strong arms and said:

"You have spoken the wise words of a father, my dear reasonable son. Words I should have said instead, but I didn't need to, because I have you. I'm the most blessed father in the world and I'm so proud of both of you. You and your sister, you have made my life worth living even through all this misery. There can be no higher blessing.

"But now, let us honor your sister in the best way we can, and that is not by mourning her or praying for her. No, we can do something much more exceptional: with the help of her suggestion, we can solve the world's biggest riddle before those dimwitted Nazis kill us, too."

After his father had spoken these words, Samuel stood up, wiped his tears, and retrieved a small piece of chalk out of his pocket. His father cleaned a reasonable patch of the floor free of the dirty straw lying there. Now they had everything they needed to get started. When the first patch of floor was filled with equations and sketches, they simply moved forward to a neighboring place. The people around them never asked; they simply backed away as well as they could. Most of them even listened with interest, and although they seldom understood what was being said, they were still happy about this unusual distraction.

"This is it!" Schmuel said loudly after a somewhat elongated passage of joint evaluation on the rear wall of the Reichsbahn wagon in which they were transported towards Auschwitz. Then he almost jeered and raised his arms in jubilation. Samuel, again with tears in his eyes, but tears of joy this time, slapped his father's broad back and said:

"You are right. That is the way, and Judith's hint has made all the difference."

The father turned to his son, closed his arms around him, and whispered:

"To our little Judith, the most wonderful girl in the world!" And then, finally, he started to weep openly.

After a while, one of the older inmates cleared his throat and asked in an extremely respectful, rather old-fashioned tone:

"Please, highly honored gentlemen, may one ask what you have found out? We know perfectly well, of course, that it is not appropriate to even ask highly skilled scholars, perhaps scientists like you, but maybe ... and under the circumstances?"

Everybody had started to listen, and all heads had turned to the old man.

The old man who had asked in that peculiar way was wearing the clothes and the dignity of a Charam, of a rabbi of the Sephardic Jews. And this made his almost subservient tone and expression even more special. The deep respect he obviously showed those funny two men with their curious scribbling all around the wagon told all others that there was indeed something very special going on. That the two were conveying in the language of the slaughterer made the whole situation almost mysterious.

Samuel and his father looked around. It was as if they only just now, in this very moment, realized that they were in an animal wagon together with many other Jews—people who were freezing, starving, and were closer to death than to life. Above all, they knew they were heading towards their extermination. Still with his son in his arms, Schmuel finally answered:

"Please excuse our impertinence, our boldness to use the surfaces of this somewhat inconvenient means of transportation without asking for your permission. Thank you very much for still letting us do what we have done! Thank you for your help! Now we are ready to explain to you nothing less than the inner structure of the world, or," he tilted his head a little, "let us say, with great care, put together those bricks necessary to build up the understanding about the world's structure."

The Charam was smiling. "It looks like you are a very good Ecclesiastes, my son. So please, preach to us now! Preach of the world, its inner wisdom, and please make sure that we all can take as much as possible of this to our Lord."

And so Schmuel started his most remarkable and similarly his most important lecture of his life, a lecture being considered a sermon in the eyes and ears of his audience, the last and most important sermon of their lives. As he had taught Judith, Schmuel began with the smallest. The smallest, he elaborated, we would be able to recognize but which wasn't the smallest in the world, because we'd only be able to "see" those things on our scale. He explained that in fact there are smaller things "going much deeper inside" and bigger things "going much further out." Even when using all the resources our portion, our scale of the universe has to offer, we would not be able to leave that very scale, neither to see the SMALLER nor to reach the BIGGER, but still we do feel its influence. He stood in the middle of the wagon, sometimes holding on to one of the wooden pillars, while Samuel walked around, and pointing at various sketches and drawings the two had made earlier in order to illustratively support what his father was just saying. When they moved through the topics and Samuel had to get from one drawing to another, the people in the wagon backed away again when necessary so that everybody was able to see what Samuel was showing. Here and there, he even added something while his father was explaining certain features of their findings. It was an arrow here or a line there. The people were extremely careful not to obliterate any of those figures, equations, and sketches when moving.

"But," Schmuel added, smiling at his wonderful audience, "we can still learn a lot about these other regions, these other scales. Even though we cannot reach them ourselves, we still get influenced by the stuff which is outside our possibilities."

"But how?" asked a deep voice.

"These boundaries . . ." Schmuel made a brief break and pointed in a certain direction at the wall. Immediately this area was cleared free of the people in front. "Thank you! I appreciate that!

"These boundaries are not fixed. As you can see, my son has drawn them in a pretty scrawly way, and you have to believe me, this definitively was on purpose."

The people laughed.

"They are permanently on the move, these boundaries, they are fidgeting all the time, and this gives us what science meanwhile has named quantum mechanics. Thus, behind this funny word isn't much more than the fact that, if looking more and more closely, space is granular. And the grains are moving all the time, which is bringing about all those peculiar things like radioactivity and which also allows certain things to exist, like stable atoms, for instance, which is to say, in the end, the granular structure seems to be necessary to allow something as funny as us."

"And what about those boundaries above, I mean the bigger?" asked the deep male voice again. It belonged to a bearded giant directly next to the door.

"Is there God?" asked a child right next to him.

Schmuel smiled, and with a nod of his head he sent his son to another corner of the wagon.

"Well, little boy . . . Ah what is you name by the way, I mean if it is allowed to ask?"

"Benjamin . . . Benjamin Baum, sir," answered the little one, politely.

"Ah thank you, but I meant the little boy next to you!" and Schmuel moved his gaze from the boy to the bearded giant.

Everyone in the wagon laughed and Schmuel proceeded.

"To be honest, personally I do not think that God is lurking or even hiding behind such boundaries. I rather think he is simply everywhere and these boundaries do not exist for him, because he is defined on them or let us put it this way: *He is all this*!"

He let these words ring a short while before he went on:

"Please note, this is only what I believe, not what I know. But I know that the upper boundaries also fidget. But they do it with a tardiness and strength we cannot measure or recognize directly. We think that the fidgeting upper boundaries brought about the necessary disturbances that, in the end, formed galaxies, stars, planets, and us. Without those perturbations, the universe would still be nothing but a uniform, homogeneous, and boring gravy . . . a very boring gravy."

Again everybody in the wagon laughed.

"But as we all, I mean we all in that room, and I explicitly exclude those dimwitted Nazis outside, as we all can see . . ."

More laughter.

"Luckily, we don't have such a boring universe filled with a probably rather tasteless gravy, right Benjamin? Or do you think that we all are nothing but boring gravy?"

"No, sir!"

"Or perhaps a boring porridge . . . maybe with a little bit of sugar?"

"No, sir!" the little boy said with a surprisingly strong voice and the bearded giant tenderly stroked the boy's almost bald head.

The laughter was getting louder again and Schmuel gave them time. There was a little bit of whispering and separate discussion going on here and there. Finally, Schmuel raised his hands above his head and in an instant it became quiet again. Only the rattling of the train and the whistling of the wind from outside could be heard.

"Well, now you might like to ask how we, I mean how my son and I have derived all this, right?"

They nodded, almost in unison.

"I could make it easy for me and say: Well, you can see it scribbled all over the walls and on the floor here. You don't even need to turn pages, lift stone plates, or roll out endless sheets of paper. The only thing you need to do is turn the head here and there."

They had got wind about The Ten Commandments and the Torah and laughed again. The Charam even clapped his hands and cried out:

"What a wonderful sermon this is, my son!"

He looked so happy and his eyes were so bright that Schmuel couldn't help but think about the holy ghost of the Christians who must have gotten into the old religious Jew, and with this thought in his mind he *had* to laugh. Then he looked the old man straight in eye and simply said: "Thank you!"

Only when the word was out did he realize how good those words of the old rabbi felt.

"But of course, we are not going to make it so easy for us," Schmuel finally proceeded.

Without the need of his father's help, Samuel went to a stretch of floor in the middle of the wagon. Whereas his father was explaining, he pointed to a drawing he had made there earlier.

"To be honest, we needed quite a while to make those little Friedmann cosmos do what we liked them to do, I mean them to do what we see, to make them bring about all the quantum mechanics we detect. Then one day, I elaborated the whole thing to my wonderful seven-year-old daughter and she simply answered, 'Why not give them some space and tickle them a bit!' Yes, this is what she had suggested."

Again the people laughed, but they laughed very silently in order not to miss anything.

"We already knew that one needed to make the surfaces of the Friedmann spheres curl in order to get to the right dither, but we had no idea how to do this in a mathematically feasible manner. Which is to say, we didn't know this until our Judith, our little heroine, showed us a way. After her hint, it was almost easy. Instead of describing complicated geometrical surface variations for our Friedmen and Friedwomen"—laughter—"we simply allowed those surfaces to be anything else but surfaces. Yes, in fact to be something in between a line and a surface or a surface and space—which is to say a thing with a fractal dimension."

A murmur went through the room. The people looked at him as if he had just said the craziest thing in the world, and Schmuel, who had anticipated this reaction, simply stood there and smiled.

"This is much less crazy than you might think," he explained calmly. "Just have a closer look at those lines in our drawings here. From far away they are lines, which is one dimension, as you probably already know, and if you did not know it, you do know it now. Getting closer, you will see that these lines also have a certain width, right? Thus they are surface and no lines. And now, when you are getting ever closer and also wear the right spectacles"—general laughter—"then you will recognize that the apparent single surface is nothing but a collection of several surfaces, between which you find patches free of any color. Still even closer, and perhaps magnified under a microscope, you find ever smaller surface dots and so on until you reach the scale of molecules and atoms. And as we all know, atoms are also not just spheres, right?"

"No, sir! These are not just spheres!" little Benjamin answered and everybody laughed.

"So, instead of constructing complex and complicated curls on those surfaces of the Friedmen and Friedwomen we simply use the dimension as additional degree of freedom, allowing us to consider rather generally perforated, fractured, or roughed surfaces, if you like.

An interesting thing happened the moment we incorporated this fractality into the Einstein field equations."

Here, Schmuel made a brief pause. His gaze moved over the crowd that sat there as if petrified, completely still, but at second glance, fully alert. "Suddenly, time appeared as nothing else but a variation of dimension, you see!"

Schmuel now did not see these people as an audience anymore. They had become his students, and they were the best he had ever had.

"The fact the little Friedmann cosmos, the Friedmen, Friedwoman, or—who knows—Friedmademoiselles, as Judith liked to call them"—giggling—"the fact that these things are a bit undecided about how to actually be dimensioned on the surfaces, that is to say they permanently jiggle around the property SURFACE, which is to say around the two-dimensional being, does bring the appearance of time for us."

There was such tension now that one would have heard a hair fall to the ground if it weren't for the noises from outside and the rattling of the wagon.

"And something else is happening now: the properties that the Friedmen and Friedmademoiselles are now taking on are exactly those we observe in our quantum mechanical world."

"And above?" asked again the deep voice of the bearded giant, who meanwhile held little Benjamin in his strong arms. His face showed such tension that it gave the impression of a source of some strange light or warmth. His voice was suddenly of a supernatural kind, as if coming from everywhere. Schmuel smiled. He was so infinitely happy about this excellent audience, these wonderful students and scholars. He knew these were his last and best.

"At the boundaries of our own universe, which is our own Friedmann cosmos as a matter of fact, it is the same thing. There, too, we find fractal surfaces, a fractal hypersurface to be more specific. And this will determine the quantum mechanical properties or appearance in the world or scale above, while the signals and influence from there have brought about the structures we see here. I mean the galaxies, stars, planets, and us."

"Then we and the whole cosmos are only like the layer of an onion, sir?" little Benjamin asked, disappointment in his voice.

Samuel couldn't help it. He had to laugh out loud. But then he came to assist his father and said: "But yes, you are perfectly right little man. But why do you think this is something to be disappointed about? If truth be told, there are even more layers, we think. Infinitely many to the smaller scales and infinitely many to the bigger, and to make things even more complicated, there are probably even more of these layered onions, infinitely many probably. And all of which are likely to be full of Friedmen and Friedmademoiselles."

"And still, little man," Schmuel now added, "this doesn't make you one single bit less important. It does not make you insignificant. Although you are only a being on one scale in one of those many universes or parts of the universes, in one of those Friedmen and Friedmadmoiselles and although there are infinitely many scales up and down and so on, you are special. Because only this one being, this boy named Benjamin, has just asked this wonderful question. Only you have just created the perfect comparison with the onion. We all, my son Samuel, me, all the people in that wagon and the whole universe have to thank you for that!"

With these words, Schmuel and Samuel bowed in unison before the little boy, whose eyes now radiated like little stars and whose face brightened as if just being given the gift of internal life.

"It is those little things which make us special and important, and here and now it is you, you Benjamin Baum, who are special and important."

In this moment, Samuel remembered the words his father had spoken after they had buried the two little boys on that old playground and he added:

"In this lies great wisdom you must know. It is the wisdom of our belief and the reason why we have been chosen above all other people. In this also lies the reason why we have been asked so much, why we have been tested so many times and so hard. But also why we have been crowned above all others. Never forget this, my brethren!"

Schmuel had now finished his lecture. He was so proud of his son to have chosen this moment to repeat those words he once had said to console him. Now his son had used the same words to comfort many people about to die and he had performed it like a grown-up man, a very wise man. Schmuel knew that there was still one piece of the puzzle yet to be found, but he didn't want to spread such technical difficulties here. In this he could be rather conservative. Like an old-fashioned teacher, he took for himself the right to send his scholars away with a few questions, a few degrees of freedom or voids to be filled with individual interpretation.

This time it took a while until the audience started to move. But then, the old rabbi suddenly simply said, "Amen!"

He started to sing the *Avinu Malkeinu* and everybody in the wagon joined in immediately (www.youtube.com/watch?v=0YONAP39jVE). Yes, they even heard them join from the other wagons, and so it was like a wondrous train of salvation moving irresistibly towards death.

It was the most wonderful singing Samuel and his father had ever heard, and they started to weep again. And again it was out of pure joy and gratitude.

*

When the train reached the ramp of Auschwitz, almost all surfaces in the wagon were full with writing and sketches. And although the people were exhausted, they all tried as hard as possible not to step on those equations and symbols in order not to destroy them. They did not know at this moment that the wagon and the rest of that train were about to go back to the Reich in order to fetch more Jews. More Jews should be transported to Auschwitz to be annihilated there. But this transportation would never happen.

*

A short while after the train reached its destiny in the Reich, US forces conquered the city and the very train station in which the train had been brought. A young officer couldn't believe his eyes when he saw the inner walls of the wagon after an excited private called him to have a look. This young officer was a mathematician, and a very good one, too. He was given the chance to work on a top secret project, which later became famous as the Manhattan project, but his whole family had a military tradition. Almost every male member of that family had served in the army, and so he did not want to break with that tradition and joined the "real forces." And after having come so far, he did not regret his decision. I might sound strange to some people, but this young mathematician preferred to be on the front, in the real war rather than in a safe office at X-site, as they called the Oak Ridge Laboratory in Tennessee, where they built a super bomb.

He too was stunned when he saw the drawings and the writing all over the walls and on the floor. This was far beyond his understanding. But he recognized certain parts and sensed that there seemed to be a connection between quantum mechanics and the general theory of relativity. Thus, he called for his camera, took dozens of photographs, and noted down the original positions of the drawings within the wagon in a small, leather-bound notebook.

The Last Piece

Samuel and his father were deaf and blind to all those cruelties, the suffering and the death around them. Yes, they even almost completely ignored all the atrocities they themselves were subjected to. They were too busy to note and to care about them. Being pushed in a certain direction on the selection ramp, they had discussed a certain problem and they had deepened that discussion when being forced to a wing where they had to undress. They ignored the sanctimonious elaborations about now being ridden of lice and disinfected in the KZ shower. They knew that they had very little time left and the stupid trial to cover the cruel killing by gassing somebody with Zyklon B and call it "disinfection" only insulted their paramount intellect. They simply read the signs, combined the information at hand, knew what was coming, and still had much better things to worry about than their impending death. They did not worry more about this than they needed in order to estimate the time they might have left to finish their task. If only one of those self-proclaimed "Herrenmenschen," the dimwitted watchmen, had known what a divine stroke of genius was being performed right in front of the eyes of those blind SS men, he would have immediately shot his comrades and then asked to be gassed himself simply out of shame. But none of these monsters had even a clue. Nobody sensed anything, not even the Kapos, who were inmates, forced to help the SS doing the killing. Nobody wanted to or could see, hear, or feel anything.

And the victims?

I have no idea, my child, because I was not there. But there was a lot of space just around Samuel and his father when they later opened the gas chamber. In all this inhuman crush, the suffering and dying, the people had made room for the two. And so, they had been able to leave a final message, scratched into the inner wall of the gas chamber in which the Nazis killed them.

How do I know that?

There is a peculiar entry in the "daily Vernichtungsprotokoll", the minutes the Nazis kept with great bureaucratic accuracy. It reports about "two wondrous male cadavers." And there were also the other inmates, many of whom were in danger of being sent into the gas chambers anyday themselves. Many of them even hoped to be salvaged from the horror this way. But that very day, some of them felt a kind of deliverance stronger than any death, and when the KZ was finally liberated on January 27, 1945, they told their liberators from that day. They told them a funny story about God having shown himself to them in the unlikeliest place there was in the world, the extermination wing in Auschwitz.

This scribbling in the hard concrete wall, using a broken fragment of a glazed tile, was the coronation of their work. Proud, indescribably proud they were in the view of what they had achieved. They looked at each other, father and son, knowing that they would die now, and still they were happy and completely at ease. It was Christmas Eve in the year 1944, and there they stood, the greatest scientists of all times, father and son, looking at each other in a gas chamber of Auschwitz, and there was nothing but happiness. They had found their equation. For them it did not matter that they were not given the chance to announce their result to humankind, because they knew that information could not get lost in this universe. A time would come where somebody would recognize and combine their conclusions and complete them, making things possible which today would appear as magic. If somebody would have asked them just now whether they would have preferred a different life, they might have just answered with a plain NO. Of course, both of them might have preferred a little less misery and suffering for all those close to them, not such a sudden and cruel end for their little Judith, not this painful, horrible death of their mother and wife, the two little boys in the street, and their fellow prisoners in that wagon and here in this room. They watched little Benjamin Baum being hugged for a last time by this bearded giant, saw a very tired-looking Charam, but for them, this was all just sad. Perhaps it was this sadness and all the suffering, the cruelty of a Nazi Germany and the atrocities this system did to the Jews, that was the trigger for them to do what they had done, to achieve what they had achieved. So how could they be sorry for themselves?

Hitler and his dunderheads, they might have killed millions of them, but by doing so, the Nazis created the platform for two geniuses and made them able to perform what even the "Tausendjähriges Reich" (thousand-year-long German kingdom) and the immortal Hitler could never ever have done and what this horrible regime also could not destroy. It was the greatest contribution to mankind's task in this world, the reason why mankind was allowed to exist in the first place. This way, and pretty much unwillingly, Hitler and his Nazi henchmen helped the Jews to fulfill what was said in the

Bible, in the Old Testament—helped to truly become the chosen people, not by choice of God, however, but by their own deeds.

In this very moment, Hitler had already lost. His defeat was so complete and so deep that the outcome of the whole war would not have mattered anymore. He was nothing, a complete nobody, a necessary boundary condition, that was all. And after he and his Nazis had served their purpose, there was nothing but a stain of dirt in the universal book of history. While the Jews shone bright and clear, the Nazi regime lay dying on the floor, a set of information the universe had no use for ever again.

Almost as a byproduct, Samuel and his son had also achieved a victory over the Hitler regime no military forces could have achieved. They had destroyed its soul, its own inner reasoning. No matter when and whether mankind would see and understand this, the universe knew, and that is was what mattered.

The father took his son into his arms and kissed him on the forehead.

"Thank you, my son!" he whispered. "You have made me the greatest present in the world. Now I know I can go without fear. I know that I will see you again and we both know where this path is leading us."

Again he hugged his son and sobbed. But this was out of pure happiness.

"Now I have to make you a present in return," the father whispered. "For those Christians killing us today, this is a special day where they make presents to each other. My present will come from the depth of my soul and it is the best I can give you under these circumstances."

Samuel nodded. His head resting on his father's chest, he didn't say a word. In the background, there was the sound of a hatch opening in the ceiling. Children, mothers, fathers began to cry, here and there was the hushed whisper of prayer, people lost control of their bodies and some simply collapsed out of mortal fear. They all knew now what would come. From a small metal cylinder, the Zyklon B crystals were dumped through the hatch into a robust steel grating. Then the hatch was closed, and above all the human noise, the cries and yells, one could make out that devilish fizzling coming from the crystals exhausting their deadly load into the air.

"Thank you!" Samuel whispered and he made his weak body as relaxed as possible. He willed himself to make it easy for his father, and with a quick, powerful twist of his arm, Schmuel broke his son's neck so that he would not have to suffer. Only now the father handed himself over to his sorrow over the son he had to kill. He held the dead shell of the wonderful young man in his arms, and within all his mourning he didn't feel his body passing away. He only thought about the wonderful time they had had together and he recapitulated their great work. This made him feel light and airy and full of pure joy.

In the protocols of that day, there was a very peculiar input from the camp doctor. It was about two male corpses, probably father and son, both with such a bright and honest smile on their faces that the liquidators did not dare touch them. Then, when the Kapos came to beat them for their disobedience, they also stopped and just stood there, completely shocked. Nobody had ever seen anything like that after a gasification with Zyklon B. What was more was the fact that even though the chamber was full to bursting point, the dying people, despite their own agony, had managed to keep a lot of free space around father and son. When, finally, an SS man came to check what was all the hold-up about, he almost froze seeing them. After a while of expressionless gazing, he almost panicked and called for the camp doctor.

In that moment, a group of six detached themselves from the other awed liquidators and moved towards the two dead men. For a moment, completely ignoring the SS and the Kapos, they lifted father and son with great dignity and in an almost perfect procession they marched them to the lift. Nobody moved. At the lift the six arranged the dead bodies such that father and son could face and smile at each other. Everybody took off his hat and stood still. Even the SS man who had called for the camp doctor earlier stood in attention. It was said that he had shot himself the same evening.

TRIALS

Trials are an essential part of new developments. We therefore present here a series of trials that helped to gradually understand more and more essentials of the Theory of Everything. For those who are not interested in this phase of evolution, we suggest to directly move to the book parts "Use" and "Teaching."

Chapter 1

The History of Our Developments

Partially motivated by some personal experiences and more or less popular publications [1–7], the author, in a book and some serious papers, showed that quantum equations emerge from the Einstein–Hilbert action in various forms and at various positions within the Einstein–Hilbert action [8–136]. However, while these equations appeared more or less as perturbations in cases where we either varied classical solutions to the Einstein field equations on the level of the line element [8–29], or used the degrees of freedom offered by the surface term within the Einstein–Hilbert action [30–65], we later also saw them as exact outcomes in connection with a further variated metric tensor [66–136]. Thereby we here want to especially point out the extension of the variation of the Einstein–Hilbert action [137, 138] towards the base vectors and their transformations [114, 117, 136]. The reason for the importance of the latter approach lies in the fact that here also classical quantum equations can be derived directly from the classical Einstein–Hilbert action and that their solution automatically fulfills the whole variation condition. In other words, with these quantum solutions being active, the Einstein field equations do not need to be solved anymore as the whole Einstein–Hilbert action is already satisfied. The fact that there are still realizations of solutions to the "classical" Einstein field equations results, as it will be shown later in this book, from a structure of various scales that the universe apparently has taken on. Thereby each scale can realize different forms of satisfying the action, either as Einstein field equations or

The World Formula: A Late Recognition of David Hilbert's Stroke of Genius
Norbert Schwarzer
Copyright © 2022 Jenny Stanford Publishing Pte. Ltd.
ISBN 978-981-4877-20-6 (Hardcover), 978-1-003-14644-5 (eBook)
www.jennystanford.com

quantum equations (erupting from the metric variation) or even both. The section about the reason for the weakness of gravity in comparison with the other fundamental interactions, which are electroweak and electrostrong (see section 12) will give an illustrative example.

But, with this result already coming out from the extension of the variation of the Einstein–Hilbert action towards the metric, why bother with the other options, like the surface term or the perturbation?

Well, while the perturbation methods truly were not more than a tentative working round to a metric understanding of quantum theory, the consideration of the surface term [30–65] can simply be seen as an alternative approach to the classical Hilbert one. Hilbert dismissed the surface term of the Einstein–Hilbert action (87) with the argument that one could always use the intrinsic degree of freedom of the variation to demand the variation to vanish at a certain closed surface. As this assumption can be made, the resulting solution is simply forced into a certain direction, which is to say the solution is still correct but restricted in accordance with the Hilbert condition being demanded. By not using this boundary condition, we therefore do nothing wrong but simply extend the playground. The resulting solutions might become more complicated or more difficult to find. But also the opposite would be possible, and that is why we had to have a look and still—even after finding a more suitable approach—are not willing to dismiss this path. Thus, we are going to present it here, too.

With the quantum equations extracted from the Einstein–Hilbert action [137] just together with the Einstein field equations [137, 138], we might come to the conclusion that we have obtained a so-called Theory of Everything. With such a theory at hand, however, we should in principle be able to also tackle other problems in completely different fields. Here, just as an example, we intend to work out a general way to consider a variety of problems in connection with material science. Thereby, we want to motivate our general approach by the observation [113] that electromagnetic interaction, which also clearly dominates material behavior, can directly be derived from a suitable metric quantum approach [139]. This derivation shall be given within the next subsection, and it shall give us some confidence that the generality of our path is not only leading into the right direction, but also that it might help us to handle and understand up-to-date unsolved or insufficiently solved problems in material science, and subsequently—hopefully—elsewhere, too.

Chapter 2

An Unusual Introduction

In the following, we are going to need the Einstein field equations, the Einstein–Hilbert action, tensor transformations, and many other things. As all this is standard textbook material, the attentive and well-informed reader will not need it. Still we will give a more comprehensive introduction to these things later in this book. Here we want to keep such explanations brief and prefer to refer to the literature instead.

Our starting point shall be the classical Einstein–Hilbert action [137, 138]:

$$\delta_g W = 0 = \delta_g \int_V d^n x \left(\sqrt{-g} \left[R - 2\kappa L_M - 2\Lambda \right] \right). \tag{1}$$

Here, g denotes the determinant of the metric tensor, W and V the action and the volume of the n-dimensional space, respectively, and R the scalar curvature or Ricci scalar. $f(R)$ shall be an arbitrary function of R in case we intend to consider the following generalization of Eq. (1):

$$\delta_g W = 0 = \delta_g \int_V d^n x \left(\sqrt{-g} \left[f(R) - 2\kappa L_M - 2\Lambda \right] \right). \tag{2}$$

The last term, $2\kappa L_M$, describes the Lagrange density of matter.

Over one hundred years ago, Einstein [138] and Hilbert [137] showed that, as a result of the variation (1), a space of a given metric $g^{\alpha\beta}$ must curve and that the curvature described by the Ricci tensor $R^{\alpha\beta}$ must satisfy certain

The World Formula: A Late Recognition of David Hilbert's Stroke of Genius
Norbert Schwarzer
Copyright © 2022 Jenny Stanford Publishing Pte. Ltd.
ISBN 978-981-4877-20-6 (Hardcover), 978-1-003-14644-5 (eBook)
www.jennystanford.com

conditions. The *classical* result for the metric tensor $g^{\alpha\beta}$ of the curved space can be given as follows:

$$R^{\alpha\beta} - \frac{1}{2}Rg^{\alpha\beta} + \Lambda g^{\alpha\beta} = -\kappa T^{\alpha\beta}. \tag{3}$$

Here we have $R^{\alpha\beta}$, $T^{\alpha\beta}$, the Ricci and the energy momentum tensor, respectively, while the parameters Λ and κ are constants (usually called cosmological and coupling constant, respectively). These are the well-known Einstein field equations in n dimensions with the indices α and β running from 1 to n. The theory behind is called the General Theory of Relativity.

But even though most people connect the General Theory of Relativity only with big scales like cosmology, galaxies, planetary movements, and black holes, as illustrated in the following figure, we will demonstrate that apparently all other aspects and laws of natural science also actually reside within the action given above, be it in the classical linear form (1) or the generalization (2).

Figure 1　What we usually connect with the General Theory of Relativity.

2.1 Derivation of Electromagnetic Interaction (and Matter) via a Set of Creative Transformations

In order to allow for an easy entrance into the theoretical apparatus, we shall start with the Minkowski flat space metric tensor in Cartesian coordinates t, x, y, and z as follows:

$$g_{\alpha\beta}^{\text{flat}} = \begin{pmatrix} -c^2 & 0 & 0 & 0 \\ 0 & 1 & 0 & 0 \\ 0 & 0 & 1 & 0 \\ 0 & 0 & 0 & 1 \end{pmatrix}. \tag{4}$$

This metric solves the Einstein field equations (3) [138] and corresponds to a Ricci scalar of $R = 0$. Now we introduce the following special transformation:

$$G_{\alpha\beta} = g_{\alpha\beta}^{\text{flat}} \cdot F\left[f\left[t, x, y, z\right]\right] = \begin{pmatrix} -c^2 & 0 & 0 & 0 \\ 0 & 1 & 0 & 0 \\ 0 & 0 & 1 & 0 \\ 0 & 0 & 0 & 1 \end{pmatrix} \cdot F\left[f\left[t, x, y, z\right]\right] \tag{5}$$

and evaluate the resulting new Ricci scalar R^*:

$$R^* = \frac{R}{F[f]} + \begin{pmatrix} -\Gamma_{\sigma\alpha}^{\mu}\Gamma_{\beta\mu}^{**\sigma} + \Gamma_{\alpha\beta}^{\sigma}\Gamma_{\sigma\mu}^{**\mu} - \Gamma_{\sigma\alpha}^{**\mu}\Gamma_{\beta\mu}^{\sigma} + \Gamma_{\alpha\beta}^{**\sigma}\Gamma_{\sigma\mu}^{\mu} \\ +\Gamma_{\alpha\beta,\sigma}^{**\sigma} - \Gamma_{\beta\sigma,\alpha}^{**\sigma} - \Gamma_{\sigma\alpha}^{**\mu}\Gamma_{\beta\mu}^{**\sigma} + \Gamma_{\alpha\beta}^{**\sigma}\Gamma_{\sigma\mu}^{**\mu} \end{pmatrix} \frac{g^{\alpha\beta}}{F[f]}$$

$$\xrightarrow{R=0; g_{\alpha\beta}^{\text{flat}}} = \frac{1}{F[f]^3} \cdot \left(\left(C_{N1} \cdot \left(\frac{\partial F[f]}{\partial f}\right)^2 - C_{N2} \cdot F[f] \cdot \frac{\partial^2 F[f]}{\partial f^2} \right) \cdot \left(\tilde{\nabla}_g f\right)^2 \right. \\ \left. - C_{N2} \cdot F[f] \cdot \frac{\partial F[f]}{\partial f} \cdot \Delta_g f \right)$$

$$= \frac{1}{F[f]^3} \cdot \left(\left(\frac{3}{2} \cdot \left(\frac{\partial F[f]}{\partial f}\right)^2 - 3 \cdot F[f] \cdot \frac{\partial^2 F[f]}{\partial f^2} \right) \cdot \left(\tilde{\nabla}_g f\right)^2 - 3 \cdot F[f] \cdot \frac{\partial F[f]}{\partial f} \cdot \Delta_g f \right)$$

$$= \frac{3}{2 \cdot F[f]^3} \cdot \left(\left(\left(\frac{\partial F[f]}{\partial f}\right)^2 - 2 \cdot F[f] \cdot \frac{\partial^2 F[f]}{\partial f^2} \right) \cdot \left[\begin{array}{c} -\frac{(\partial_t f)^2}{c^2} + (\partial_x f)^2 \\ + (\partial_y f)^2 + (\partial_z f)^2 \end{array} \right] \\ -2 \cdot F[f] \cdot \frac{\partial F[f]}{\partial f} \cdot \left[-\frac{\partial_t^2}{c^2} + \partial_x^2 + \partial_y^2 + \partial_z^2 \right] f \right).$$

$$\tag{6}$$

Thereby we have used

$$\Gamma_{\alpha\beta}^{*\gamma} = \frac{g^{\gamma\sigma}}{2 \cdot F[f]} \left(\left[F[f] \cdot g_{\sigma\alpha}\right]_{,\beta} + \left[F[f] \cdot g_{\sigma\beta}\right]_{,\alpha} - \left[F[f] \cdot g_{\alpha\beta}\right]_{,\sigma} \right)$$

$$= \frac{g^{\gamma\sigma}}{2 \cdot F\,[f]} \left([F\,[f] \cdot g_{\sigma\alpha}]_{,\beta} + [F\,[f] \cdot g_{\sigma\beta}]_{,\alpha} - [F\,[f] \cdot g_{\alpha\beta}]_{,\sigma} \right)$$

$$= \frac{g^{\gamma\sigma}}{2} \left(g_{\sigma\alpha,\beta} + g_{\sigma\beta,\alpha} - g_{\alpha\beta,\sigma} \right)$$

$$+ \frac{g^{\gamma\sigma}}{2 \cdot F\,[f]} \left(F\,[f]_{,\beta} \cdot g_{\sigma\alpha} + F\,[f]_{,\alpha} \cdot g_{\sigma\beta} - F\,[f]_{,\sigma} \cdot g_{\alpha\beta} \right)$$

$$= \Gamma^{\gamma}_{\alpha\beta} + \frac{g^{\gamma\sigma}}{2 \cdot F\,[f]} \left(F\,[f]_{,\beta} \cdot g_{\sigma\alpha} + F\,[f]_{,\alpha} \cdot g_{\sigma\beta} - F\,[f]_{,\sigma} \cdot g_{\alpha\beta} \right)$$

$$\equiv \Gamma^{\gamma}_{\alpha\beta} + \Gamma^{**\gamma}_{\alpha\beta}. \tag{7}$$

The reader will find a more comprehensive elaboration about the motivation of our special transformation in the section 3.3, "The Ricci Scalar Quantization." However, it is quite entertaining to follow this simple trial and observe its amazing evolution into something rather unexpected (at least if taking its origin as a metric solution to the Einstein field equations) out of our "wrapper transformation" (5). As here the result is the essential aspect and as the various derivations are mainly repeated further below in this book, we will not go into too much detail, because this only spoils the reading fun and the punch line.

The symbol $\tilde{\nabla}_g$ in Eq. (6) denotes a first-order differential operator similar to the Nabla operator in the metric $g_{\alpha\beta}$. The symbol C_{Ni} stands for constants, which only depend on the number of dimensions. Please note that the transformation (5) is just the simplest form of a general approach like

$$G_{\alpha\beta} = F\,[f\,[t, x, y, z]]^{ij}_{\alpha\beta}\, g_{ij} \rightarrow G_{\alpha\beta} = F\,[f\,[t, x, y, z]] \cdot \delta^i_\alpha \delta^j_\beta g_{ij}. \tag{8}$$

Without loss of generality we can now demand that the first term in parenthesis in Eq. (6) would be zero, which is to say, we choose our arbitrary function F such that we have

$$C_{N1} \cdot \left(\frac{\partial F\,[f]}{\partial f} \right)^2 - C_{N2} \cdot F\,[f] \cdot \frac{\partial^2 F\,[f]}{\partial f^2} = 0. \tag{9}$$

For instance, in four dimensions this would always be the case for $F\,[f] = f^2$, giving us the Klein–Gordon-like equation

$$\left[\Delta_g + \frac{R^* \cdot f^4}{f \cdot C_{N2} \cdot 2 \cdot f} \right] f = \left[\Delta_g + \frac{R^* \cdot f^2}{2} \right] f = 0; \Delta_{g-\text{coordinates}} = \Delta_g. \tag{10}$$

Setting the solution $F\,[f] = f^2$ into our transformation starting point (5), however, would directly result in an all-scale quantum-dominated metric solution for $G_{\alpha\beta}$. This is in total contrast to the daily experiences that quantum effects are only important in smaller scales. In order to overcome this problem, we evaluate the general solution of condition (9) in four dimensions and find

$$F[f] = C_1 \cdot f + \frac{C_1^2 \cdot f^2}{4 \cdot C_2} + C_2. \tag{11}$$

Thereby the constants C_i are arbitrary.

For entertainment and further motivation, it should be pointed out here that the functional wrapper $F[f]$ from (11) could also be adapted as follows:

$$F[f] = C_1 \cdot f^2 + \frac{C_1^2 \cdot f^4}{4 \cdot C_2} + C_2. \tag{12}$$

With arbitrary constants C_i, this assures the appearance of linear Laplace operator terms according to the condition (9) for the resulting Ricci scalar of the transformed metric in Eq. (8) as follows:

$$R^* = \frac{1}{F[f]^3} \cdot \left(-C_{N2} \cdot F[f] \cdot \frac{\partial F[f]}{\partial f} \cdot \Delta_g f^2 \right) \xrightarrow{4D} R^* \cdot \frac{\left(2 \cdot C_2 + C_1 \cdot f^2\right)^3}{8 \cdot C_2 \cdot C_1}$$
$$= -3 \cdot \Delta_g f^2. \tag{13}$$

Now we assume f to be a constant, which automatically leads to the simple equation

$$R^* \cdot \frac{\left(2 \cdot C_2 + C_1 \cdot f^2\right)^3}{8 \cdot C_2 \cdot C_1} = 0. \tag{14}$$

Also assuming that the Ricci scalar curvature R^* is proportional to f^n, with an arbitrary exponent $n \geq 0$, we obtain the familiar trivial solution of $f_0 = 0$. However, from Eq. (14) we also obtain the non-trivial ground states

$$f^2 = -\frac{2 \cdot C_2}{C_1}. \tag{15}$$

As the constants C_i are arbitrary, we could simply set them as follows:

$$2 \cdot C_2 = -\mu^2; C_1 = 2 \cdot \lambda. \tag{16}$$

This gives us the additional ground state solutions directly in the classical Higgs field style [140], namely

$$(f_{1,2})^2 = \frac{\mu^2}{2 \cdot \lambda} \Rightarrow f_{1,2} = \pm\frac{\mu}{\sqrt{2 \cdot \lambda}}. \tag{17}$$

The measured value for the $f_{1,2}$ is known to be [141]

$$\left| f_{1,2} \right| = \frac{\mu}{\sqrt{2 \cdot \lambda}} = \frac{246 \text{ GeV}}{\sqrt{2} \cdot c^2}. \tag{18}$$

Rewriting Eq. (13) with the use of Eq. (16) as

$$R^* \cdot \frac{\left(-\mu^2 + 2 \cdot \lambda \cdot f^2 \right)^3}{8 \cdot \mu^2 \cdot 2 \cdot \lambda} = f^n \cdot \frac{\left(2 \cdot \lambda \cdot f^2 - \mu^2 \right)^3}{16 \cdot \mu^2 \cdot \lambda} = 3 \cdot \Delta_g f^2 \tag{19}$$

gives us the total Higgs–Ricci-curvature connection. It also, automatically, gives us a curvature value R^* at the ground state, which is

$$R^* \sim \left| f_{1,2} \right|^n = \left(\frac{\mu}{\sqrt{2 \cdot \lambda}} \right)^n = \left(\frac{246 \text{ GeV}}{\sqrt{2} \cdot c^2} \right)^n. \tag{20}$$

Now, with the Higgs mass m_H known to be $125 \text{ GeV}/c^2$ and the fact that we have [141]

$$m_H = \sqrt{2 \cdot \mu^2} = \sqrt{4 \cdot \lambda \cdot \left(f_{1,2} \right)^2}, \tag{21}$$

we can obtain

$$\lambda \simeq 0.13; \mu \simeq 88.8 \text{ GeV}/c^2. \tag{22}$$

Please note that we obtain the simple case $F[f] = f^2$, which, e.g., was applied in [91], with the condition

$$c_1^2 = 2 \cdot C_2 \tag{23}$$

and the subsequent limit procedure $C_2 \to 0$ as follows:

$$\Rightarrow F[f] = \sqrt{2 \cdot C_2} \cdot f + f^2 + C_2 \xrightarrow[C_2 \to 0]{\lim} F[f] = f^2. \tag{24}$$

Using Eq. (11) instead of $F = f^2$ in Eq. (5) gives us

$$R^* = \frac{1}{F[f]^3} \cdot \left(\begin{array}{l} \left(C_{N1} \cdot \left(\frac{\partial F[f]}{\partial f} \right)^2 - C_{N2} \cdot F[f] \cdot \frac{\partial^2 F[f]}{\partial f^2} \right) \cdot \left(\tilde{\nabla}_g f \right)^2 \\ -C_{N2} \cdot F[f] \cdot \frac{\partial F[f]}{\partial f} \cdot \Delta_g f \end{array} \right)$$

$$= -C_{N2} \cdot \frac{\frac{\partial F[f]}{\partial f} \cdot \Delta_g f}{F[f]^2} \xrightarrow{4D} = -3 \cdot \frac{\left(C_1 + \frac{c_1^2 \cdot f}{2 \cdot C_2} \right)}{\left(C_1 \cdot f + \frac{c_1^2 \cdot f^2}{4 \cdot C_2} + C_2 \right)^2} \cdot \Delta_g f. \tag{25}$$

Immediately we see that in the many cases with $R^* = 0$, we would obtain classical quantum Klein–Gordon equations. Taylor expansion for small f (which is to say at $f = 0$) for the general case in the second line in Eq. (25)

also gives us equations leading to these classical equations. For instance, if only considering up to linear f as follows:

$$\frac{R^*}{3} \cdot \frac{\left(C_1 \cdot f + \frac{C_1^2 \cdot f^2}{4 \cdot C_2} + C_2\right)^2}{C_1 + \frac{C_1^2 \cdot f}{2 \cdot C_2}} + \Delta_g f = 0 \xrightarrow{\text{small} f} \frac{R^*}{3} \cdot \left(\frac{C_2^2}{C_1} + \frac{3}{2} \cdot C_2 \cdot f + \ldots\right)$$

$$+ \Delta_g f = 0 \qquad (26)$$

and assuming constant R^*, we already have the typical structure of the Klein–Gordon equation plus a constant term $\frac{R^*}{3} \cdot \frac{C_2^2}{C_1}$. Such a constant would still not influence the principle structural character of our resulting quantum equations because a simple transformation of

$$f^* = \frac{C_2^2}{C_1} + \frac{3}{2} \cdot C_2 \cdot f; \Delta_g f^* = \Delta_g f \qquad (27)$$

does give us back the ordinary Klein–Gordon equation with

$$\frac{R^*}{3} \cdot f^* + \Delta_g f^* = 0. \qquad (28)$$

The resulting transformation in four dimensions would give the following quantum metric:

$$G_{\alpha\beta} = g_{\alpha\beta} \cdot F[f] = g_{\alpha\beta} \cdot \left(C_1 \cdot f + \frac{C_1^2 \cdot f^2}{4 \cdot C_2} + C_2\right). \qquad (29)$$

This clearly allows for the classical physics or Einstein situation with quantum effects becoming dominant only at small scales and, thus, a dominating part C_2 at bigger scales. As a result, one might separate as follows in four dimensions:

$$G_{\alpha\beta} = \overbrace{g_{\alpha\beta} \cdot C_2}^{\text{class. GTR}} + g_{\alpha\beta} \cdot \left(C_1 \cdot f \cdot \left(1 + \frac{C_1 \cdot f}{4 \cdot C_2}\right)\right). \qquad (30)$$

$$\overbrace{\phantom{g_{\alpha\beta} \cdot C_2 + g_{\alpha\beta} \cdot (C_1 \cdot f)}}^{\text{quantum gravity in 4D}}$$

$$\overbrace{\phantom{g_{\alpha\beta} \cdot \left(C_1 \cdot f \cdot \left(1 + \frac{C_1 \cdot f}{4 \cdot C_2}\right)\right)}}^{\text{quantum metric}}$$

Thereby it does not come as a surprise that we have already found so many cases where f has to be determined by classical quantum equations [82–112], that is to say where our equations give similar or even equal solutions. Normalization of the quantum gravity metric $G_{\alpha\beta}$ requires division by C_2 and gives

$$G_{\alpha\beta}^{\text{norm}} = \frac{G_{\alpha\beta}^{\text{norm}}}{C_2} = \overbrace{g_{\alpha\beta}}^{\text{class. GTR}} + g_{\alpha\beta} \cdot \left(\frac{C_1}{C_2} f \cdot \left(1 + \frac{C_1 \cdot f}{4 \cdot C_2}\right)\right). \qquad (31)$$

Now we apply the recipe outlined above to our simple flat space example in order to see how the transformation can bring about matter.

We saw that by taking the metric (4), using rule (9) with the subsequent solution (11), we obtain the new Ricci scalar curvature:

$$R^* = -3 \cdot \frac{\left(C_1 + \frac{C_1^2 \cdot f}{2 \cdot C_2}\right)}{\left(C_1 \cdot f + \frac{C_1^2 \cdot f^2}{4 \cdot C_2} + C_2\right)^2} \cdot \Delta_g f. \tag{32}$$

Now we demand that the Ricci scalar should be a conserved quantity and as we have $R = 0$ with the flat space metric (4), we shall also demand $R^* = 0$. This gives us the simplest equation:

$$0 = \Delta_g f = \left[\partial_x^2 + \partial_y^2 + \partial_z^2 - \frac{\partial_t^2}{c^2}\right] f. \tag{33}$$

Just for an increase of generality we extend Eq. (4) as follows (A, B, D are constants):

$$g_{\alpha\beta}^{\text{flat}} = \begin{pmatrix} -c^2 & 0 & 0 & 0 \\ 0 & A^2 & 0 & 0 \\ 0 & 0 & B^2 & 0 \\ 0 & 0 & 0 & D^2 \end{pmatrix}, \tag{34}$$

which changes Eq. (33) to

$$0 = \Delta_g f = \left[\frac{\partial_x^2}{A^2} + \frac{\partial_y^2}{B^2} + \frac{\partial_z^2}{D^2} - \frac{\partial_t^2}{c^2}\right] f. \tag{35}$$

Applying the separation approach $f[t, x, y, z] = T[t] * X[x] * Y[y] * Z[z]$ gives us the solutions:

$$T[t] = C_{t1} \cdot \cos[c \cdot C_t \cdot t] + C_{t2} \cdot \sin[c \cdot C_t \cdot t], \tag{36}$$

$$X[x] = C_{x1} \cdot \cos[A \cdot C_x \cdot x] + C_{x2} \cdot \sin[A \cdot C_x \cdot x], \tag{37}$$

$$Y[y] = C_{y1} \cdot \cos[B \cdot C_y \cdot y] + C_{y2} \cdot \sin[B \cdot C_y \cdot y], \tag{38}$$

$$Z[z] = C_{z1} \cdot \cos[D \cdot C_z \cdot z] + C_{z2} \cdot \sin[D \cdot C_z \cdot z], \tag{39}$$

and the following characteristic equation:

$$0 = C_x^2 + C_y^2 + C_z^2 \quad C_t^2. \tag{40}$$

The matter coded with this solution for R^* can now directly be obtained via the Einstein field equations with matter, which reads

$$R^{*\alpha\beta} - \frac{1}{2}R^* \cdot g^{\alpha\beta} + \Lambda \cdot g^{\alpha\beta} = -\kappa \cdot T^{\alpha\beta}. \tag{41}$$

Here we have $R^{\alpha\beta}$, $T^{\alpha\beta}$, the Ricci and the energy momentum tensor, respectively, while the parameters Λ and κ are constants (usually called cosmological and coupling constant, respectively). As we had per demand $R^* = R = 0$ and as we also assume the cosmological constant to be equal to zero, the following identity results:

$$R^{*\alpha\beta} = -\kappa \cdot T^{\alpha\beta}. \tag{42}$$

Thus, in order to find out what kind of matter our transformation (5) has created, we simply need to evaluate the corresponding Ricci tensor of the transformed metric $G_{\alpha\beta}$. In order to keep the presentation general* and as the evaluation of the derivatives is simple, we give the Ricci tensor with the function f. But for simplicity we simplified $F[f]$ to $F[f] = (f + C_1)^2$

$$R^{*00} = \frac{(A \cdot B \cdot c)^2 \, (f_{,z})^2 + D^2 \left(A^2 c^2 \, (f_{,y})^2 + B^2 \left(\begin{array}{c} c^2 \, (f_{,x})^2 + 3A^2 \, (f_{,t})^2 \\ -2A^2 \, (C_1 + f) \, f_{,t,t} \end{array} \right) \right)}{(A \cdot B \cdot D)^2 \, c^4 \, (C_1 + f)^6}. \tag{43}$$

$$R^{*11} = \frac{-(A \cdot B \cdot c)^2 (f_{,z})^2 + D^2 \left(-A^2 c^2 (f_{,y})^2 + B^2 \left(\begin{array}{c} 3c^2 (f_{,x})^2 \\ -2c^2 (C_1 + f) f_{,x,x} \end{array} \right) + A^2 (f_{,t})^2 \right)}{(c \cdot B \cdot D)^2 \, A^4 \, (C_1 + f)^6}. \tag{44}$$

$$R^{*22} = \frac{-(A \cdot B \cdot c)^2 \, (f_{,z})^2 + D^2 \left(\begin{array}{c} 3A^2 c^2 \, (f_{,y})^2 - 2A^2 c^2 \, (C_1 + f) \, f_{,y,y} \\ -B^2 c^2 \, (f_{,x})^2 + B^2 A^2 \, (f_{,t})^2 \end{array} \right)}{(c \cdot A \cdot D)^2 \, B^4 \, (C_1 + f)^6}. \tag{45}$$

$$R^{*33} = \frac{3(A \cdot B \cdot c)^2 \, (f_{,z})^2 + D^2 \left(\begin{array}{c} -A^2 c^2 \, (f_{,y})^2 - B^2 c^2 \, (f_{,x})^2 + B^2 A^2 (f_{,t})^2 \\ +2(C_1 + f) \, (A^2 c^2 \, f_{,y,y} + B^2 \, (c^2 \, f_{,x,x} - A^2 f_{,t,t})) \end{array} \right)}{(c \cdot A \cdot D)^2 \, B^4 \, (C_1 + f)^6}. \tag{46}$$

---

*This comes in handy the moment we want to exploit the additive character of our metric solutions. Thereby the additivity is assured by the means of condition (9). See further below in the main text.

$$R^{*01} = \frac{2\left(-2 f_{,x} f_{,t} + (C_1 + f)\, f_{,t,x}\right)}{A^2 c^2\, (C_1 + f)^6}; \quad R^{*02} = \frac{2\left(-2 f_{,y} f_{,t} + (C_1 + f)\, f_{,t,y}\right)}{B^2 c^2\, (C_1 + f)^6}$$

$$R^{*03} = \frac{2\left(-2 f_{,z} f_{,t} + (C_1 + f)\, f_{,t,z}\right)}{D^2 c^2\, (C_1 + f)^6}; \quad R^{*12} = \frac{2\left(2 f_{,y} f_{,x} - (C_1 + f)\, f_{,x,y}\right)}{A^2 B^2\, (C_1 + f)^6}$$

$$R^{*13} = \frac{2\left(2 f_{,z} f_{,x} - (C_1 + f)\, f_{,x,z}\right)}{A^2 D^2\, (C_1 + f)^6}; \quad R^{*23} = \frac{2\left(2 f_{,y} f_{,z} - (C_1 + f)\, f_{,z,y}\right)}{B^2 D^2\, (C_1 + f)^6}$$

$$\tag{47}$$

Previously we discussed corresponding solutions in spherical [109] and cylindrical [110] geometries, and it is clear that we have obtained photons or photonic forms. What is more, with condition (9) we have automatically achieved the additive character for all our solutions (36) to (40), allowing us to apply simple integral transform methods in order to construct almost arbitrary photonic matter forms.

Without applying the technique of superposition of our solutions (36) to (40), however, the subsequent photonic solution would be a plane structure with no changes along the x axis. It was shown in [20] how such a structure can be localized also in lateral directions via standard integral transform methods and that the subsequent solutions do fulfill the Maxwell equations. The connection with the energy momentum tensor for the electromagnetic field is therefore achieved via (42). The electrostatic and magnetic vector fields $\mathbf{E}$ and $\mathbf{B}$ with components E_i and B_i form the energy momentum tensor in contravariant form as follows:

$$\left(T^{\alpha\beta}\right) = \begin{pmatrix} \frac{1}{2}\cdot(\mathbf{E}\cdot\mathbf{E} + \mathbf{B}\cdot\mathbf{B}) & (\mathbf{E}\times\mathbf{B})^T \\[2mm] \mathbf{E}\times\mathbf{B} & \frac{1}{2}\cdot(\mathbf{E}\cdot\mathbf{E}+\mathbf{B}\cdot\mathbf{B})\cdot\delta_{ik} - E_i E_k - B_i B_k \end{pmatrix}$$

$$\delta_{ik} = \begin{pmatrix} 1 & 0 & 0 \\ 0 & 1 & 0 \\ 0 & 0 & 1 \end{pmatrix}. \tag{48}$$

Please note that in SI units, this tensor reads

$$\left(T^{\alpha\beta}\right) = \begin{pmatrix} \dfrac{\varepsilon_0\cdot\mathbf{E}\cdot\mathbf{E} + \dfrac{\mathbf{B}\cdot\mathbf{B}}{\mu_0}}{2} & c\cdot\varepsilon_0\cdot(\mathbf{E}\times\mathbf{B})^T \\[4mm] c\cdot\varepsilon_0\cdot\mathbf{E}\times\mathbf{B} & \dfrac{\varepsilon_0\cdot\mathbf{E}\cdot\mathbf{E}+\dfrac{\mathbf{B}\cdot\mathbf{B}}{\mu_0}}{2}\cdot\delta_{ik} - \varepsilon_0\cdot E_i E_k - \dfrac{B_i B_k}{\mu_0} \end{pmatrix}.$$

$$\tag{49}$$

The constants ε_0 and μ_0 are the electric and the magnetic field constant, respectively.

For brevity, we shall leave it to the mathematically skilled and interested reader to work out the detailed connection of the two classical fields with our quantum metric distortion. For illustration, however, we just consider the energy density of the electromagnetic field given via the component T^{00} and thus obtain

$$
\begin{aligned}
R^{*00} &= \frac{(A \cdot B \cdot c)^2 (f_{,z})^2 + D^2 \left(A^2 c^2 (f_{,y})^2 + B^2 \left(\begin{array}{c} c^2 (f_{,x})^2 + 3A^2 (f_{,t})^2 \\ -2A^2 (C_1 + f) f_{,t,t} \end{array} \right) \right)}{(A \cdot B \cdot D)^2 c^4 (C_1 + f)^6} \\
&= -\kappa \cdot T^{00} = -\kappa \cdot \frac{\varepsilon_0 \cdot \mathbf{E} \cdot \mathbf{E} + \frac{\mathbf{B} \cdot \mathbf{B}}{\mu_0}}{2} \\
&= -\frac{\kappa}{2} \cdot \left(\varepsilon_0 \cdot (E_1^2 + E_2^2 + E_3^2) + \frac{B_1^2 + B_2^2 + B_3^2}{\mu_0} \right).
\end{aligned}
\tag{50}
$$

In fact, we find quadratic terms on both sides of the equation and it appears kind of attractive to seek the connections between the classical $\mathbf{E}$ and $\mathbf{B}$ fields and the quantum metric distortion in the following way:

$$
\begin{aligned}
\frac{(f_{,x})^2 + C_{tx} \left(3 (f_{,t})^2 - 2 (C_1 + f) f_{,t,t} \right)}{(C_1 + f)^6} &= -\frac{\kappa_x}{2} \cdot \left(\varepsilon_0 \cdot E_1^2 + \frac{B_1^2}{\mu_0} \right) \\
\frac{(f_{,y})^2 + C_{ty} \left(3 (f_{,t})^2 - 2 (C_1 + f) f_{,t,t} \right)}{(C_1 + f)^6} &= -\frac{\kappa_y}{2} \cdot \left(\varepsilon_0 \cdot E_2^2 + \frac{B_2^2}{\mu_0} \right). \\
\frac{(f_{,z})^2 + C_{tz} \left(3 (f_{,t})^2 - 2 (C_1 + f) f_{,t,t} \right)}{(C_1 + f)^6} &= -\frac{\kappa_z}{2} \cdot \left(\varepsilon_0 \cdot E_3^2 + \frac{B_3^2}{\mu_0} \right)
\end{aligned}
\tag{51}
$$

Thus, obviously we have created photonic matter from a flat space vacuum solution to the Einstein field equations by the means of a simple metric transformation. The scalar Ricci curvature was thereby kept unchanged, which is to say $R = R^*$. The equations necessary to solve this were the classical quantum Klein–Gordon equations, which we directly obtain from the $R = R^*$ condition.

2.1.1 Intelligent Zero Approaches: Just One Example

It was already shown in our previous papers (especially see [8] and references given in there) that quantum equations can be derived from intelligent zeros formed from the metric solutions to the Einstein field equations. Here, we want to derive such an equation directly out of the variation of the metric to any given Riemann space in a most spectacular manner.

This time, our starting point shall be the usual tensor transformation rule for the covariant metric tensor:

$$g_{\delta\gamma} = \mathbf{g}_\delta \cdot \mathbf{g}_\gamma = \frac{\partial G^\alpha [x_k]}{\partial x^\delta} \frac{\partial G^\beta [x_k]}{\partial x^\gamma} g_{\alpha\beta}. \tag{52}$$

The base vectors $\mathbf{g}_\delta$ to a certain metric are given as

$$\mathbf{g}_\delta = \frac{\partial G^\alpha [x_k]}{\partial x^\delta} \mathbf{e}_\alpha, \tag{53}$$

where the functions $G^\alpha [\ldots]$ denote arbitrary functions of the coordinates x_k. Here the vectors e_α shall denote the base vectors of a fundamental coordinate system of the right (in principle arbitrary) number of dimensions. Thus, we have the variation for $\delta g_{\delta\gamma}$ in Eqs. (52) and (53) as follows:

$$\delta g_{\delta\gamma} = \delta \left(\mathbf{g}_\delta \cdot \mathbf{g}_\gamma \right) = \mathbf{g}_\delta \cdot \delta \mathbf{g}_\gamma + \delta \mathbf{g}_\delta \cdot \mathbf{g}_\gamma$$
$$= \frac{\partial G^\alpha [x_k]}{\partial x^\delta} \mathbf{e}_\alpha \cdot \delta \left(\frac{\partial G^\beta [x_k]}{\partial x^\gamma} \mathbf{e}_\beta \right) + \delta \left(\frac{\partial G^\alpha [x_k]}{\partial x^\delta} \mathbf{e}_\alpha \right) \cdot \frac{\partial G^\beta [x_k]}{\partial x^\gamma} \mathbf{e}_\beta. \tag{54}$$

Reconsidering the Einstein–Hilbert action in its classical form (1) and performing the variation with respect to the metric tensor $g_{\alpha\beta}$ results in the well-known Einstein field equations (3) [137] under the integral times the variated metric. Thus, we should write Eq. (1) in its full form as

$$\delta_g W = 0 = \int_V d^n x \left(R^{\delta\gamma} - \frac{1}{2} R g^{\delta\gamma} + \Lambda g^{\delta\gamma} + \kappa T^{\delta\gamma} \right) \delta g_{\delta\gamma}. \tag{55}$$

Now we want to perform the remaining variation of the metric in covariant form. The informed reader knows of course that such covariant variations of the type

$$\left(R^{\delta\gamma} - \frac{1}{2} R g^{\delta\gamma} + \Lambda g^{\delta\gamma} + \kappa T^{\delta\gamma} \right) \delta g_{\delta\gamma} = \left(R^{\delta\gamma} - \frac{1}{2} R g^{\delta\gamma} + \Lambda g^{\delta\gamma} + \kappa T^{\delta\gamma} \right)$$
$$\times g_{\delta\gamma;\beta} \cdot \delta \left(x^\beta \right) \tag{56}$$

directly give zero, because it always holds that

$$g_{\delta\gamma;\beta} = g^{\delta\gamma}{}_{;\beta} = 0. \tag{57}$$

But what about the covariant variation of the base vectors, thereby looking for intrinsic solutions, just as we saw them with the variation of the line elements to solutions of the Einstein field equations as shown in [8] and some other papers given as references there? We know that we can write Eq. (57) in its base vector decomposed form, reading

$$g_{\alpha\beta;\gamma} \cdot \delta\left(x^{\gamma}\right) = \left[\mathbf{g}_{\alpha} \cdot \left(\mathbf{g}_{\beta,\gamma} - \mathbf{g}_{\sigma}\Gamma^{\sigma}_{\gamma\beta}\right) + \mathbf{g}_{\beta} \cdot \left(\mathbf{g}_{\alpha,\gamma} - \mathbf{g}_{\sigma}\Gamma^{\sigma}_{\gamma\alpha}\right)\right] \cdot \delta\left(x^{\gamma}\right). \quad (58)$$

Now we expand just one of the addends and find

$$\mathbf{g}_{\alpha} \cdot \left(\mathbf{g}_{\beta,\gamma} - \mathbf{g}_{\sigma}\Gamma^{\sigma}_{\gamma\beta}\right) \cdot \delta\left(x^{\beta}\right) = -\frac{1}{2}\left(\mathbf{g}_{\alpha} \cdot \left(\mathbf{g}_{\gamma,\beta} - \mathbf{g}_{\beta,\gamma}\right) + \mathbf{g}_{\beta} \cdot \left(\mathbf{g}_{\alpha,\gamma} - \mathbf{g}_{\gamma,\alpha}\right)\right)$$
$$+ \mathbf{g}_{\gamma} \cdot \left(\mathbf{g}_{\alpha,\beta} - \mathbf{g}_{\beta,\alpha}\right)\right) \cdot \delta\left(x^{\gamma}\right). \quad (59)$$

We reshape the expression on the right-hand side and obtain

$$\mathbf{g}_{\alpha} \cdot \left(\mathbf{g}_{\beta,\gamma} - \mathbf{g}_{\sigma}\Gamma^{\sigma}_{\gamma\beta}\right) \cdot \delta\left(x^{\beta}\right) = -\frac{\mathbf{g}_{\gamma}}{2} \cdot \left(\delta^{\gamma}_{\alpha}\left(\mathbf{g}_{\gamma,\beta} - \mathbf{g}_{\beta,\gamma}\right) + \delta^{\gamma}_{\beta}\left(\mathbf{g}_{\alpha,\gamma} - \mathbf{g}_{\gamma,\alpha}\right)\right.$$
$$+ \left.\left(\mathbf{g}_{\alpha,\beta} - \mathbf{g}_{\beta,\alpha}\right)\right) \cdot \delta\left(x^{\gamma}\right)$$
$$= -\frac{\mathbf{g}_{\gamma}}{2} \cdot \left(\left(\mathbf{g}_{\alpha,\beta} - \mathbf{g}_{\beta,\alpha}\right) + \left(\mathbf{g}_{\alpha,\beta} - \mathbf{g}_{\beta,\alpha}\right)\right.$$
$$+ \left.\left(\mathbf{g}_{\alpha,\beta} - \mathbf{g}_{\beta,\alpha}\right)\right) \cdot \delta\left(x^{\gamma}\right)$$
$$= -3 \cdot \frac{\mathbf{g}_{\gamma}}{2} \cdot \left(\mathbf{g}_{\alpha,\beta} - \mathbf{g}_{\beta,\alpha}\right) \cdot \delta\left(x^{\gamma}\right). \quad (60)$$

We demand it to give an intrinsic zero if combined with its partnering factor from Eq. (56). In order to keep things simple, we only pick the terms with the contravariant metric tensor $\left(-\frac{1}{2}Rg^{\delta\gamma} + \Lambda g^{\delta\gamma}\right)$ from Eq. (56), which leads to

$$0 = \frac{1}{2}\left(\Lambda - \frac{1}{2}R\right)g^{\alpha\beta}\mathbf{g}_{\alpha} \cdot \left(\mathbf{g}_{\beta,\gamma} - \mathbf{g}_{\sigma}\Gamma^{\sigma}_{\gamma\beta}\right) \cdot \delta\left(x^{\beta}\right)$$
$$= 3 \cdot \left(\Lambda - \frac{1}{2}R\right)\mathbf{g}^{\alpha} \cdot \mathbf{g}^{\beta} \cdot \frac{\mathbf{g}_{\gamma}}{2} \cdot \left(\mathbf{g}_{\beta,\alpha} - \mathbf{g}_{\alpha,\beta}\right) \cdot \delta\left(x^{\gamma}\right)$$
$$= 3 \cdot \left(\Lambda - \frac{1}{2}R\right)\mathbf{g}^{\alpha} \cdot \frac{\delta^{\beta}_{\gamma}}{2} \cdot \left(\mathbf{g}_{\beta,\alpha} - \mathbf{g}_{\alpha,\beta}\right) \cdot \delta\left(x^{\gamma}\right)$$
$$= 3 \cdot \left(\Lambda - \frac{1}{2}R\right) \cdot \frac{\mathbf{g}^{\alpha}}{2} \cdot \left(\mathbf{g}_{\gamma,\alpha} - \mathbf{g}_{\alpha,\gamma}\right) \cdot \delta\left(x^{\gamma}\right). \quad (61)$$

We can also directly apply Eq. (59), which gives us the same but avoids the somewhat awkward index-switches we applied in Eq. (61):

$$-\frac{1}{2}\left(\Lambda - \frac{1}{2}R\right)g^{\alpha\beta}(\mathbf{g}_\alpha \cdot (\mathbf{g}_{\gamma,\beta} - \mathbf{g}_{\beta,\gamma}) + \mathbf{g}_\beta \cdot (\mathbf{g}_{\alpha,\gamma} - \mathbf{g}_{\gamma,\alpha}) + \mathbf{g}_\gamma \cdot (\mathbf{g}_{\alpha,\beta} - \mathbf{g}_{\beta,\alpha}))$$

$$= 0$$

$$\Rightarrow 0 = -\frac{1}{2}\mathbf{g}^\alpha \cdot \mathbf{g}^\beta \cdot (\mathbf{g}_\alpha \cdot (\mathbf{g}_{\gamma,\beta} - \mathbf{g}_{\beta,\gamma}) + \mathbf{g}_\beta \cdot (\mathbf{g}_{\alpha,\gamma} - \mathbf{g}_{\gamma,\alpha}) + \mathbf{g}_\gamma \cdot (\mathbf{g}_{\alpha,\beta} - \mathbf{g}_{\beta,\alpha}))$$

$$= -\frac{1}{2}\left(\mathbf{g}^\alpha \cdot \delta^\beta_\alpha \left(\mathbf{g}_{\gamma,\beta} - \mathbf{g}_{\beta,\gamma}\right) + \mathbf{g}^\beta \cdot \delta^\alpha_\beta \left(\mathbf{g}_{\alpha,\gamma} - \mathbf{g}_{\gamma,\alpha}\right) + \mathbf{g}^\alpha \cdot \mathbf{g}^\beta \cdot \mathbf{g}_\gamma \cdot \left(\mathbf{g}_{\alpha,\beta} - \mathbf{g}_{\beta,\alpha}\right)\right)$$

$$= -\frac{1}{2}\left(\mathbf{g}^\alpha \cdot (\mathbf{g}_{\alpha,\gamma} - \mathbf{g}_{\gamma,\alpha}) + \mathbf{g}^\alpha \cdot (\mathbf{g}_{\alpha,\gamma} - \mathbf{g}_{\gamma,\alpha}) + \mathbf{g}^\alpha \cdot \delta^\beta_\gamma \cdot (\mathbf{g}_{\alpha,\beta} - \mathbf{g}_{\beta,\alpha})\right)$$

$$= -\frac{1}{2}\left(2 \cdot \mathbf{g}^\alpha \cdot (\mathbf{g}_{\alpha,\gamma} - \mathbf{g}_{\gamma,\alpha}) + \mathbf{g}^\alpha \cdot (\mathbf{g}_{\alpha,\gamma} - \mathbf{g}_{\gamma,\alpha})\right)$$

$$= -\frac{3}{2} \cdot \mathbf{g}^\alpha \cdot (\mathbf{g}_{\alpha,\gamma} - \mathbf{g}_{\gamma,\alpha}) \Rightarrow 0 = \mathbf{g}^\alpha \cdot (\mathbf{g}_{\gamma,\alpha} - \mathbf{g}_{\alpha,\gamma}) . \tag{62}$$

Incorporation of the transformation rule for the base vectors (53) gives us

$$G^j [x_k] \equiv G^j;$$

$$0 = \mathbf{g}^\alpha \cdot \left(\left(\frac{\partial G^j}{\partial x^\gamma}\mathbf{e}_j\right)_{,\alpha} - \left(\frac{\partial G^j}{\partial x^\alpha}\mathbf{e}_j\right)_{,\gamma}\right)$$

$$= \left(\frac{\partial x^\alpha}{\partial G^i}\mathbf{e}^i\right) \cdot \left(\left(\frac{\partial G^j}{\partial x^\gamma}\mathbf{e}_j\right)_{,\alpha} - \left(\frac{\partial G^j}{\partial x^\alpha}\mathbf{e}_j\right)_{,\gamma}\right)$$

$$\simeq \delta^\alpha_i \mathbf{e}^i \cdot \left(\left(\frac{\partial G^j}{\partial x^\alpha}\mathbf{e}_j\right)_{,\gamma} - \left(\frac{\partial G^j}{\partial x^\gamma}\mathbf{e}_j\right)_{,\alpha}\right)$$

$$= \mathbf{e}^\alpha \cdot \left(\left(\frac{\partial G^j}{\partial x^\alpha}\mathbf{e}_j\right)_{,\gamma} - \left(\frac{\partial G^j}{\partial x^\gamma}\mathbf{e}_j\right)_{,\alpha}\right)$$

$$\simeq \left(\left(\frac{\partial G^j}{\partial x^\alpha}\delta^\alpha_j\right)_{,\gamma} - \left(\frac{\partial G^j}{\partial x^\gamma}\delta^\alpha_j\right)_{,\alpha}\right)$$

$$= \left(\left(\frac{\partial G^\alpha}{\partial x^\alpha}\right)_{,\gamma} - \left(\frac{\partial G^\alpha}{\partial x^\gamma}\right)_{,\alpha}\right), \tag{63}$$

where in the second and last line we have used the fact that in a nearly flat space we should be able to Taylor-expand $\frac{\partial x^\alpha}{\partial G^i}$ as follows:

$$\frac{\partial x^\alpha}{\partial G^i} = \frac{\partial x^\alpha}{\partial x^i + \partial f^i} \xrightarrow{\partial x^i \gg \partial f^i} \simeq \delta^\alpha_i , \tag{64}$$

and that we approximately can assume a small effect from the derivatives of the base vectors $\mathbf{e}^j$, meaning that we can set

$$\mathbf{e}^\alpha \cdot \left(\mathbf{e}_j\right)_{,\gamma} \simeq \left(\mathbf{e}^\alpha \cdot \mathbf{e}_j\right)_{,\gamma}. \tag{65}$$

Thereby the justification of our approximation shall for the moment only be the assumption of a nearly flat space or space-time and a small variation from the Cartesian-like x^k to the $G^j[x^k]$. We will take care of such details later in the book. Such an assumption should also allow us to fully justify the interchangeability of the derivatives in $\left(\frac{\partial G^\alpha}{\partial x^\alpha}\right)_{,\gamma}$, which automatically fulfills Eq. (63). This would not hinder us, however, to demand an intrinsic solution of the kind

$$\left(\frac{\partial G^\alpha}{\partial x^\gamma}\right)_{,\alpha} = \frac{\partial G^\alpha}{\partial x^\alpha \partial x^\gamma} = 0; \left(\frac{\partial G^\alpha}{\partial x^\alpha}\right)_{,\gamma} = \frac{\partial G^\alpha}{\partial x^\gamma \partial x^\alpha} = 0. \tag{66}$$

The experienced reader probably recognizes the differential equation for an incompressible linear elastic space. It was shown in previous papers (e.g. [139] and see references given there) that Eq. (66) can be solved with the so-called *n*-function-ansatz for $G^j[\ldots]$ consisting of harmonic functions $G[\ldots]$. For simplicity we only give the ansatz in Cartesian coordinates:

$$G^j[x_k] = g^{jl}\partial_l G[x_k]; \quad \Delta G[x_k] = 0$$
$$G^j[x_k] = x_\xi \cdot g^{jl}\partial_l G[x_k] + \alpha \cdot G[x_k]; \quad \xi \text{ any of } 0, 1, \ldots n-1, \tag{67}$$

$$G^j[x_k] = \left\{ \ldots, \overbrace{\frac{\partial G[x_k]}{\partial x_\zeta}}^{\text{pos } \xi}, \ldots, \overbrace{-\frac{\partial G[x_k]}{\partial x_\xi}}^{\text{pos } \zeta}, \ldots \right\}; \quad \circ \forall (\ldots, \ldots) = 0$$

$$G^j[x_k] = \left\{ \ldots, \overbrace{-\frac{\partial G[x_k]}{\partial x_\zeta}}^{\text{pos } \xi}, \ldots, \overbrace{\frac{\partial G[x_k]}{\partial x_\xi}}^{\text{pos } \zeta}, \ldots \right\}; \quad \circ \forall (\ldots, \ldots) = 0. \tag{68}$$

Any combination of the basic solutions (67) and (68) is also a solution. We see that we can construct non-trivial solutions to Eq. (66), which are automatically also solutions to Eq. (61), and thus Eq. (56), out of combinations of harmonic functions and their derivatives. In order to satisfy Eq. (66) in its currently given form, we have to set $\alpha = -1$.

We may consider the second type of solutions (68) spin-like. More such forms can be found easily via the following recipe:

$$G^j\,[x_k] = \left\{ \ldots, A_1 \overbrace{\frac{\partial^2 G}{\partial \xi_2 \partial \xi_3}}^{\text{pos } \xi_1}, \ldots, A_2 \overbrace{\frac{\partial^2 G}{\partial \xi_1 \partial \xi_3}}^{\text{pos } \xi_2}, \ldots, A_3 \overbrace{\frac{\partial^2 G}{\partial \xi_1 \partial \xi_2}}^{\text{pos } \xi_3}, \ldots \right\}$$

$$G^j\,[x_k] = \left\{ \ldots, A_1 \overbrace{\frac{\partial^3 G}{\partial \xi_2 \partial \xi_3 \partial \xi_4}}^{\text{pos } \xi_1}, \ldots, A_2 \overbrace{\frac{\partial^3 G}{\partial \xi_1 \partial \xi_3 \partial \xi_4}}^{\text{pos } \xi_2}, \ldots, A_3 \overbrace{\frac{\partial^3 G}{\partial \xi_1 \partial \xi_2 \partial \xi_4}}^{\text{pos } \xi_3}, \right.$$

$$\left. \ldots, A_4 \overbrace{\frac{\partial^3 G}{\partial \xi_1 \partial \xi_2 \partial \xi_3}}^{\text{pos } \xi_4} \ldots \right\}$$

$$\forall\,(\ldots, \ldots) = 0; \quad \sum_{\forall k} A_k = 0. \tag{69}$$

In a Minkowski space-time with $c = 1$ and the coordinates t, x, y, and z, we could have the following of such solutions:

$$G^j\,[x_k] = \pm \left\{ \frac{\partial G}{\partial z}, 0, 0, -\frac{\partial G}{\partial t} \right\}; G^j\,[x_k] = \pm \left\{ \frac{\partial G}{\partial y}, 0, -\frac{\partial G}{\partial t}, 0 \right\};$$

$$G^j\,[x_k] = \pm \left\{ \frac{\partial G}{\partial x}, -\frac{\partial G}{\partial t}, 0, 0 \right\} G^j\,[x_k] = \pm \left\{ 0, \frac{\partial G}{\partial z}, 0, -\frac{\partial G}{\partial x} \right\};$$

$$G^j\,[x_k] = \pm \left\{ 0, \frac{\partial G}{\partial y}, -\frac{\partial G}{\partial x}, 0 \right\}; G^j\,[x_k] = \pm \left\{ 0, 0, \frac{\partial G}{\partial z}, -\frac{\partial G}{\partial y} \right\}, \tag{70}$$

$$G^j\,[x_k] = \left\{ A_1 \frac{\partial^2 G}{\partial x \partial y}, A_2 \frac{\partial^2 G}{\partial t \partial y}, A_3 \frac{\partial^2 G}{\partial x \partial t}, 0 \right\};$$

$$G^j\,[x_k] = \left\{ A_1 \frac{\partial^2 G}{\partial x \partial z}, A_2 \frac{\partial^2 G}{\partial t \partial z}, 0, A_3 \frac{\partial^2 G}{\partial x \partial t} \right\}$$

$$G^j\,[x_k] = \left\{ A_1 \frac{\partial^2 G}{\partial z \partial y}, 0, A_2 \frac{\partial^2 G}{\partial t \partial z}, A_3 \frac{\partial^2 G}{\partial z \partial t} \right\};$$

$$G^j\,[x_k] = \left\{ 0, A_1 \frac{\partial^2 G}{\partial z \partial y}, A_2 \frac{\partial^2 G}{\partial x \partial z}, A_3 \frac{\partial^2 G}{\partial x \partial y} \right\}. \tag{71}$$

$$G^j\,[x_k] = \left\{ A_1 \frac{\partial^3 G}{\partial x \partial y \partial z}, A_1 \frac{\partial^3 G}{\partial t \partial y \partial z}, A_1 \frac{\partial^3 G}{\partial x \partial t \partial z}, A_4 \frac{\partial^3 G}{\partial x \partial y \partial t} \right\}; \sum_{\forall k} A_k = 0. \tag{72}$$

Thereby we find that the last solution in Eq. (71) with a setting of the kind $A_1 = 1/3$, $A_2 = 1/3$, and $A_3 = -2/3$ shows a peculiar closeness to the charges of quarks, the building blocks of hadrons and baryons.

Please note that, even though we can easily see that Eq. (58) is identical zero, because we have

$$
\begin{aligned}
\mathbf{g}_{\alpha\beta;\gamma} &= \mathbf{g}_\alpha \cdot \left(\mathbf{g}_{\beta,\gamma} - \mathbf{g}_\sigma \Gamma^\sigma_{\gamma\beta}\right) + \mathbf{g}_\beta \cdot \left(\mathbf{g}_{\alpha,\gamma} - \mathbf{g}_\sigma \Gamma^\sigma_{\gamma\alpha}\right) \\
&= -\frac{1}{2} \left(\mathbf{g}_\alpha \cdot \left(\mathbf{g}_{\gamma,\beta} - \mathbf{g}_{\beta,\gamma}\right) + \mathbf{g}_\beta \cdot \left(\mathbf{g}_{\alpha,\gamma} - \mathbf{g}_{\gamma,\alpha}\right) + \mathbf{g}_\gamma \cdot \left(\mathbf{g}_{\alpha,\beta} - \mathbf{g}_{\beta,\alpha}\right)\right) \\
&\quad -\frac{1}{2} \left(\mathbf{g}_\beta \cdot \left(\mathbf{g}_{\gamma,\alpha} - \mathbf{g}_{\alpha,\gamma}\right) + \mathbf{g}_\alpha \cdot \left(\mathbf{g}_{\beta,\gamma} - \mathbf{g}_{\gamma,\beta}\right) + \mathbf{g}_\gamma \cdot \left(\mathbf{g}_{\beta,\alpha} - \mathbf{g}_{\alpha,\beta}\right)\right) \\
&= -\frac{1}{2} \begin{pmatrix} \mathbf{g}_\alpha \cdot \left(\mathbf{g}_{\gamma,\beta} - \mathbf{g}_{\beta,\gamma}\right) + \mathbf{g}_\alpha \cdot \left(\mathbf{g}_{\beta,\gamma} - \mathbf{g}_{\gamma,\beta}\right) \\ +\mathbf{g}_\beta \cdot \left(\mathbf{g}_{\alpha,\gamma} - \mathbf{g}_{\gamma,\alpha}\right) + \mathbf{g}_\beta \cdot \left(\mathbf{g}_{\gamma,\alpha} - \mathbf{g}_{\alpha,\gamma}\right) \\ +\mathbf{g}_\gamma \cdot \left(\mathbf{g}_{\alpha,\beta} - \mathbf{g}_{\beta,\alpha}\right) + \mathbf{g}_\gamma \cdot \left(\mathbf{g}_{\beta,\alpha} - \mathbf{g}_{\alpha,\beta}\right) \end{pmatrix} \\
&= -\frac{1}{2} \begin{pmatrix} \mathbf{g}_\alpha \cdot \left(\mathbf{g}_{\gamma,\beta} - \mathbf{g}_{\beta,\gamma}\right) - \mathbf{g}_\alpha \cdot \left(\mathbf{g}_{\gamma,\beta} - \mathbf{g}_{\beta,\gamma}\right) \\ +\mathbf{g}_\beta \cdot \left(\mathbf{g}_{\alpha,\gamma} - \mathbf{g}_{\gamma,\alpha}\right) - \mathbf{g}_\beta \cdot \left(\mathbf{g}_{\alpha,\gamma} - \mathbf{g}_{\gamma,\alpha}\right) \\ +\mathbf{g}_\gamma \cdot \left(\mathbf{g}_{\alpha,\beta} - \mathbf{g}_{\beta,\alpha}\right) - \mathbf{g}_\gamma \cdot \left(\mathbf{g}_{\alpha,\beta} - \mathbf{g}_{\beta,\alpha}\right) \end{pmatrix} = 0, \quad (73)
\end{aligned}
$$

it does not mean that the solution to Eq. (66) could not be realized.

After all, it is not up to us to decide what type of effective zero the universe realized in order to satisfy the extremal condition (55) in the end.

Thus, all options for possible solutions of the total variation of the kind

$$
\begin{aligned}
\delta_g W = 0 &= \int_V d^n x \begin{pmatrix} R^{\alpha\beta} - \frac{1}{2} R g^{\alpha\beta} \\ +\Lambda g^{\alpha\beta} + \kappa T^{\alpha\beta} \end{pmatrix} \\
&\quad \times \left(\mathbf{g}_\alpha \cdot \left(\mathbf{g}_{\beta,\gamma} - \mathbf{g}_\sigma \Gamma^\sigma_{\gamma\beta}\right) + \mathbf{g}_\beta \cdot \left(\mathbf{g}_{\alpha,\gamma} - \mathbf{g}_\sigma \Gamma^\sigma_{\gamma\alpha}\right)\right) \\
&= -\int_V d^n x \frac{\left(R^{\alpha\beta} + \kappa T^{\alpha\beta}\right)}{2} \\
&\quad \times \left(\begin{array}{l} \left(\mathbf{g}_\alpha \cdot \left(\mathbf{g}_{\gamma,\beta} - \mathbf{g}_{\beta,\gamma}\right) + \mathbf{g}_\beta \cdot \left(\mathbf{g}_{\alpha,\gamma} - \mathbf{g}_{\gamma,\alpha}\right) + \mathbf{g}_\gamma \cdot \left(\mathbf{g}_{\alpha,\beta} - \mathbf{g}_{\beta,\alpha}\right)\right) \\ + \left(\mathbf{g}_\beta \cdot \left(\mathbf{g}_{\gamma,\alpha} - \mathbf{g}_{\alpha,\gamma}\right) + \mathbf{g}_\alpha \cdot \left(\mathbf{g}_{\beta,\gamma} - \mathbf{g}_{\gamma,\beta}\right) + \mathbf{g}_\gamma \cdot \left(\mathbf{g}_{\beta,\alpha} - \mathbf{g}_{\alpha,\beta}\right)\right) \end{array}\right) \\
&\quad -\frac{3}{2} \cdot \int_V d^n x \left(\Lambda - \frac{1}{2} R\right) \cdot \left(\mathbf{g}^\alpha \cdot \left(\mathbf{g}_{\alpha,\gamma} - \mathbf{g}_{\gamma,\alpha}\right) + \mathbf{g}^\beta \cdot \left(\mathbf{g}_{\beta,\gamma} - \mathbf{g}_{\gamma,\beta}\right)\right)
\end{aligned}
$$

$$(74)$$

have to be taken into account. Using the result from Eq. (63) and taking into account that the bases $\mathbf{e}_j$ are to be considered bases of almost flat or almost affine spaces, we can simplify the last line of Eq. (74):

$$
0 = - \int_V d^n x \frac{\left(R^{\alpha\beta} + \kappa\, T^{\alpha\beta} \right)}{2}
$$

$$
\cdot \left(\begin{array}{c} \left(\mathbf{g}_\alpha \cdot (\mathbf{g}_{\gamma,\beta} - \mathbf{g}_{\beta,\gamma}) + \mathbf{g}_\beta \cdot (\mathbf{g}_{\alpha,\gamma} - \mathbf{g}_{\gamma,\alpha}) + \mathbf{g}_\gamma \cdot (\mathbf{g}_{\alpha,\beta} - \mathbf{g}_{\beta,\alpha}) \right) \\ + \left(\mathbf{g}_\beta \cdot (\mathbf{g}_{\gamma,\alpha} - \mathbf{g}_{\alpha,\gamma}) + \mathbf{g}_\alpha \cdot (\mathbf{g}_{\beta,\gamma} - \mathbf{g}_{\gamma,\beta}) + \mathbf{g}_\gamma \cdot (\mathbf{g}_{\beta,\alpha} - \mathbf{g}_{\alpha,\beta}) \right) \end{array} \right)
$$

$$
- \frac{3}{2} \cdot \int_V d^n x \left(\Lambda - \frac{1}{2} R \right) \cdot \left(\begin{array}{c} \left(\frac{\partial x^\alpha}{\partial G^i} \mathbf{e}^i \right) \cdot \mathbf{e}_j \left(\left(\frac{\partial G^j [x_k]}{\partial x^\alpha} \right)_{,\gamma} - \left(\frac{\partial G^j [x_k]}{\partial x^\gamma} \right)_{,\alpha} \right) \\ + \left(\frac{\partial x^\beta}{\partial G^i} \mathbf{e}^i \right) \cdot \mathbf{e}_j \left(\left(\frac{\partial G^j [x_k]}{\partial x^\beta} \right)_{,\gamma} - \left(\frac{\partial G^j [x_k]}{\partial x^\beta} \right)_{,\alpha} \right) \end{array} \right).
$$

$$(75)$$

Exchanging one of the dummy indexes results in

$$
0 = - \int_V d^n x \frac{\left(R^{\alpha\beta} + \kappa\, T^{\alpha\beta} \right)}{2}
$$

$$
\cdot \left(\begin{array}{c} \left(\mathbf{g}_\alpha \cdot (\mathbf{g}_{\gamma,\beta} - \mathbf{g}_{\beta,\gamma}) + \mathbf{g}_\beta \cdot (\mathbf{g}_{\alpha,\gamma} - \mathbf{g}_{\gamma,\alpha}) + \mathbf{g}_\gamma \cdot (\mathbf{g}_{\alpha,\beta} - \mathbf{g}_{\beta,\alpha}) \right) \\ + \left(\mathbf{g}_\beta \cdot (\mathbf{g}_{\gamma,\alpha} - \mathbf{g}_{\alpha,\gamma}) + \mathbf{g}_\alpha \cdot (\mathbf{g}_{\beta,\gamma} - \mathbf{g}_{\gamma,\beta}) + \mathbf{g}_\gamma \cdot (\mathbf{g}_{\beta,\alpha} - \mathbf{g}_{\alpha,\beta}) \right) \end{array} \right)
$$

$$
- 3 \cdot \int_V d^n x \left(\Lambda - \frac{1}{2} R \right)
$$

$$
\cdot \left(\left(\frac{\partial x^\alpha}{\partial G^i} \mathbf{e}^i \right) \cdot \mathbf{e}_j \left(\left(\frac{\partial G^j [x_k]}{\partial x^\alpha} \right)_{,\gamma} - \left(\frac{\partial G^j [x_k]}{\partial x^\gamma} \right)_{,\alpha} \right) \right). \tag{76}
$$

We see that there are many possibilities to form the resulting zero and assuming a universe applying all these options, it is of little wonder why in quantum electrodynamics (or all similar classical field theories) we always have to sum over so many (infinitely many) paths in order to get the total result.

In a similar manner, we can also seek for the intrinsic solutions to Eq. (58) in a more direct manner without the excursion into the base vectors. We know that the expansion of Eq. (58) would read

$$
g_{\alpha\beta;\gamma} \cdot \delta\left(x^\gamma\right) = \left[g_{\alpha\beta,\gamma} - g_{\alpha\sigma} \Gamma^\sigma_{\gamma\beta} - g_{\beta\sigma} \Gamma^\sigma_{\gamma\alpha} \right] \cdot \delta\left(x^\gamma\right)
$$

$$
= \left[g_{\alpha\beta,\gamma} - \frac{1}{2} \left(g_{\gamma\alpha,\beta} + g_{\alpha\beta,\gamma} - g_{\gamma\beta,\alpha} \right) \right.
$$

$$
\left. - \frac{1}{2} \left(g_{\gamma\beta,\alpha} + g_{\alpha\beta,\gamma} - g_{\gamma\alpha,\beta} \right) \right] \cdot \delta\left(x^\gamma\right). \tag{77}
$$

The reader may easily prove that this gives zero in total, but here we are looking for intrinsic zeros in connection with the partnering terms from the complete variation (55), that is (56). For the reason of simplicity and brevity, we consider only the terms with the contravariant metric tensor and subsequently have to evaluate

$$\left(-\frac{1}{2}Rg^{\alpha\beta} + \Lambda g^{\alpha\beta}\right) g_{\alpha\beta;\gamma} \cdot \delta\left(x^{\gamma}\right)$$

$$= \left(\Lambda - \frac{R}{2}\right) g^{\alpha\beta} \left[g_{\alpha\beta,\gamma} - \frac{1}{2}\left(g_{\gamma\alpha,\beta} + g_{\alpha\beta,\gamma} - g_{\gamma\beta,\alpha}\right)\right.$$

$$\left. -\frac{1}{2}\left(g_{\gamma\beta,\alpha} + g_{\alpha\beta,\gamma} - g_{\gamma\alpha,\beta}\right)\right] \cdot \delta\left(x^{\gamma}\right)$$

$$= \left(\Lambda - \frac{R}{2}\right) g^{\alpha\beta} \frac{1}{2} \left[\left(g_{\gamma\beta,\alpha} - g_{\gamma\alpha,\beta}\right) + \left(g_{\gamma\alpha,\beta} - g_{\gamma\beta,\alpha}\right)\right] \cdot \delta\left(x^{\gamma}\right)$$

$$\Rightarrow g^{\alpha\beta}g_{\gamma\beta,\alpha} = 0 \cup g^{\alpha\beta}\left(g_{\gamma\beta,\alpha} + g_{\gamma\alpha,\beta}\right) = 0 = g^{\alpha\beta}\left(g_{\gamma\alpha,\beta} + g_{\gamma\beta,\alpha}\right). \quad (78)$$

Once more applying the assumption of the nearly flat space and using Eq. (64), we can write

$$\gamma^{\alpha\beta}g_{\gamma\beta,\alpha} = 0 \cup \gamma^{\alpha\beta}\left(g_{\gamma\beta,\alpha} + g_{\gamma\alpha,\beta}\right) = 0 = \gamma^{\alpha\beta}\left(g_{\gamma\alpha,\beta} + g_{\gamma\beta,\alpha}\right), \quad (79)$$

with $\gamma^{\alpha\beta}$ giving the flat space metric. The corresponding evaluation was performed as follows (we only pick the simple term $g^{\alpha\beta}g_{\gamma\beta,\alpha}$ for demonstration):

$$g^{\alpha\beta}g_{\gamma\beta,\alpha} = \frac{\partial x^{\alpha}}{\partial G^{i}}\frac{\partial x^{\beta}}{\partial G^{j}}\gamma^{ij}\left(G^{k}_{,\gamma}G^{l}_{,\beta}\gamma_{kl}\right)_{,\alpha}$$

$$= \frac{\partial x^{\alpha}}{\partial\left(x^{i}+f^{i}\right)}\frac{\partial x^{\beta}}{\partial\left(x^{j}+f^{j}\right)}\gamma^{ij}\left(G^{k}_{,\gamma}G^{l}_{,\beta}\gamma_{kl}\right)_{,\alpha}$$

$$\simeq \delta^{\alpha}_{i}\delta^{\beta}_{j}\gamma^{ij}\left(G^{k}_{,\gamma}G^{l}_{,\beta}\gamma_{kl}\right)_{,\alpha}$$

$$= \gamma^{\alpha\beta}\left(G^{k}_{,\gamma}G^{l}_{,\beta}\gamma_{kl}\right)_{,\alpha}. \quad (80)$$

Assuming Cartesian coordinates, we can further evaluate

$$0 = g^{\alpha\beta}g_{\gamma\beta,\alpha} \simeq \gamma^{\alpha\beta}\left(G^{k}_{,\gamma}G^{l}_{,\beta}\gamma_{kl}\right)_{,\alpha}$$

$$= \gamma^{\alpha\beta}\left(G^{k}_{,\gamma\alpha}G^{l}_{,\beta}\gamma_{kl} + G^{k}_{,\gamma}G^{l}_{,\beta\alpha}\gamma_{kl} + G^{k}_{,\gamma}G^{l}_{,\beta}\gamma_{kl,\alpha}\right)$$

$$= \gamma^{\alpha\beta}\left(G^{k}_{,\gamma\alpha}G^{l}_{,\beta}\gamma_{kl} + G^{k}_{,\gamma}G^{l}_{,\beta\alpha}\gamma_{kl} + G^{k}_{,\gamma}G^{l}_{,\beta}\gamma_{kl,\alpha}\right)$$

$$= G^{k}_{,\gamma}{}^{\beta}G^{l}_{,\beta}\gamma_{kl} + G^{k}_{,\gamma}\gamma^{\alpha\beta}G^{l}_{,\beta\alpha}\gamma_{kl} + G^{k}_{,\gamma}G^{l,\alpha}\gamma_{kl,\alpha}$$

$$\overbrace{}^{\approx 0}$$

$$\simeq G^{k}_{,\gamma}{}^{\beta}G_{k,\beta} + G^{k}_{,\gamma}G^{l}_{,\beta}{}^{\beta}\gamma_{kl} + G^{k}_{,\gamma}G^{l,\alpha}\gamma_{kl,\alpha}$$

$$\simeq B \cdot G_{\beta,\gamma}{}^{\beta} + G_{l,\gamma}G^{l}_{,\beta}{}^{\beta} \simeq B \cdot G_{\beta,\gamma}{}^{\beta} + b \cdot G_{\gamma,\beta}{}^{\beta} = 0 = G_{\beta,\gamma}{}^{\beta} + a \cdot G_{\gamma,\beta}{}^{\beta}, \quad (81)$$

where we recognize the fundamental equation of linear elasticity for homogenous isotropic bodies (space) with solutions (67) and (68) and $\alpha = -1 - 2^*a$. The finding of the structural elements of the basic equation of elasticity (e.g. [152], p. 166), which can be given for an isotropic material with the Poisson's ratio ν as

$$\left(\cdot \Delta G^\delta \, [x_k] + \left(\frac{\partial^2 G^\delta \, [x_k]}{\partial x^\gamma \partial x^\delta} \right) \right) = 0, \tag{82}$$

within our derivation (81) leaves us with an interesting analogy.

So we ask: Even though we here only considered two very simple forms of metric variation, we ended up in the typical equations for spatial deformation, with fundamental types of solutions containing typical harmonic quantum solutions. Thus, could it be that quantum theory is just some deformation or deformational fluctuation of an already gravitationally pre-stressed space or space-time?

This wouldn't be too surprising, as we already saw quantum equations emerge from simple variations of intelligent zeros formed from the line elements [8].

Before we jump to conclusions, however, we intend to find out whether there are also other possibilities to extract quantum equations from extended forms of the Einstein–Hilbert action.

2.1.2 Introduction Summed Up

Even though we only halfheartedly did some trial and error with the variation of metric solutions to the Einstein field equations and extended (a bit) the Einstein–Hilbert action with respect to potential intrinsic solutions of the latter, we already found equations so close to the classical quantum equations (or even directly found those quantum equations) that it is quite possible that we found a way to unify the General Theory of Relativity and quantum theory to a Theory of Everything.

Chapter 3

How Many Theories of Everything Are There?

3.1 About the Theory of Everything

There are reasons why we gave this part of the book the title "Trials." There are many possibilities to come to a quantum gravity or Theory of Everything, and as we will see, it is not straightforward to make a pick and select one as the best way or strategy. Consequently, we have to cover a great number of options and even though some may look strange or even obsolete, we better keep them for completeness and later discussion.

The construction of a Theory of Everything requires the combination of Einstein's General Theory of Relativity (GTR) and quantum theory (QT). There have been quite some attempts to achieve this either by quantizing the GTR or by constructing a metrically based QT. The fact that there was no success for neither of the two ways led to the suspicion that a unification of the two great theories is not possible simply because it is not needed. Assuming, namely, that one of the two theories already resided inside the other, only that the math wasn't found so far to show the joint, provides a very direct explanation of why a combination is not achievable.

In order to find the answer to the question—if our assumption is considered to be correct—which of the two theories is more fundamental and therefore could contain the other, we simply have a look at the degrees of freedom that the solutions to each one of the two theories provide. We

The World Formula: A Late Recognition of David Hilbert's Stroke of Genius
Norbert Schwarzer
Copyright © 2022 Jenny Stanford Publishing Pte. Ltd.
ISBN 978-981-4877-20-6 (Hardcover), 978-1-003-14644-5 (eBook)
www.jennystanford.com

see that there seems to be a greater flexibility within the GTR solutions than we find in the QT-counterparts. As we know that all solutions of the Einstein field equations (EFE) can be subjected to coordinate transforms without changing the principle results, we think that we have found the lever which is necessary to bring quantum character into GTR solutions. On the other hand, we were already able to show that the EFE resulted from the most fundamental starting point we were able to think of. This starting point is much more fundamental than the classical or usual concept of space and time, but still it results in Einstein's equations,* that is, its fundament, the Einstein–Hilbert action. These two facts together, namely,

(a) the metric GTR solutions and their fundament, the Einstein–Hilbert action, provide the flexibility which is necessary for a quantization process of these solutions, that is, their origin (the action),
(b) the principle equations governing the GTR, the so-called Einstein field equations, erupt from a most fundamental principle, that is, the most basic origin one could think of,

give us a great deal of confidence that we indeed have a nice starting point for the construction of the Theory of Everything.

3.2 A Most Fundamental Starting Point and How to Proceed from There

As the introduction to the method of quantizing metric solutions of the Einstein field equations is quite lengthy and was already presented in a variety of papers [82–113], we refrain from presenting it here again. Instead, we will only give a brief verbal description:

(1) Let there be a set of properties, perhaps even just one property.
(2) Let us further assume that, if accepting the possible existence of one property, we might just consider the smallest variations or fluctuations (like shades of grey) within this very property and allow for the split-up of some of the fluctuations as perpendicular (sufficiently different) properties. In this sense, we might even assume that the one

---

*Thereby we did not just consider the classical Einstein–Hilbert action as *the* minimum principle but also more general forms. The principle result was the same. As considering this in general form would cost too much space in this book, we refer to our previous work [82–113].

primary property sprouted fluctuations, which might not necessarily appear perpendicular to the primary property, because they are just its own derivatives or offspring, but which are perpendicular among themselves. Then, we have an ensemble of such linear, independent secondary properties.

(3) There is also always the possibility of an intelligent zero being split up into two deconstructive wave functions (property functions). Together they—that is, the sums of these functions—give back the intelligent zero, but considered separately they are just what they are, namely coherent entities with the ability to evolve. An important example was given in connection with one-dimensional spaces subjected to the Einstein field equations (see Eq. (83) further below). Classically—so we learned—the Einstein field equations are not meaningful in one-dimensional spaces as they are always automatically fulfilled. Well, this may be true, but it does not mean that there are no (potentially intrinsic) solutions still leading to very interesting constellations. In fact, this author found 6 such fundamental solutions [93]. It appears to be a funny coincidence that this is also the number of the "infinity stones" [96] that the Marvel© heroes chase higgledy-piggledy through the whole universe in order to hinder a certain bad guy from abusing these mighty building blocks of the whole as some kind of universal decimation or cleaning-up instrument. Seeing the theoretical origin of these "infinity stones" as just the "genetic code of everything," one might indeed find that such things should not fall into the hands of bad people (Marvel maniacs, ideological or religious fanatics, and today's politicians alike).

(4) Now, we consider space as an ensemble of properties.

(5) These properties could just be seen as degrees of freedom and, thus, dimensions, which are subjected to a Hamilton extremal principle.

(6) This leads to the Einstein–Hilbert action [137] and subsequently to the Einstein field equations [137, 138].

(7) Until now, which is to say in our earlier papers (e.g., [8–81]), we assumed time to be somehow different from the other dimensions: "Time seems to take a special place among the properties as it is not such a property itself, but consists of all other properties' internal changes and variations. Applying the Einstein field equations on the internal degrees of freedom of each single property as a one-dimensional space (cf. [93]), gives exactly 6 solutions [96] among which

we always also find oscillations. These internal periodic processes (changes) inside each and every property or dimension are realized as time from an external observer ... time, which itself forces other properties to change. Thus, starting as an internal property (solution) within each and every dimension, time not only is change, but also brings change about. Apparently, time is the most fractal and self-similar thing there is in this universe."

Then, in a pair of previous papers of the series "Science Riddle" (No. 4 and 6 [121, 123]), we saw that there could also be another explanation for the effect of time. Thereby time arises as a pure quantum effect, which distinguishes time from other degrees of freedom or dimensions only by its appearance among the quantum gravity solutions. Thus, it depends on the problem in question, its solution and the viewpoint of the observer which dimension is time-like and whether there is a time at all. It was shown that in principle any dimension could take on the "position" or functionality of time in a certain system. Apparently, it is a simple transformation of the metric solutions to the Einstein field equations which not only brings about matter and quantum effects, but coming along with the latter also makes certain dimensions (not necessarily only one) to stand out as time or time-like. This finding was further investigated within the papers of our "Science Riddles" series [123, 124]. Thereby it was found that time obviously is an effect quantum systems show, respectively have, as long as they are "in limbo" or—more physically—in coherence between certain "macroscopic" states. Fixing systems to a certain state robes them of their ability to evolve and to have the property of time. In [131] we could show that solutions to the Einstein field equations always develop time-like coordinates when at least two of them are entangled. Please note that for this just the classical Einstein field equations were of need and no additional quantization was necessary (for more, please see [131]).

(8) The quantum effects and the entanglement solutions leading to such time dimensions result from the interesting situation that solutions to the Einstein field equations are not necessarily unique. Derived from the Einstein–Hilbert action as an extremal principle, the starting quantity is the so-called Ricci scalar R, being the essential kernel of this action. This results in a metric solution to the Einstein field equations, being subsequently derived from the Einstein–Hilbert action. Most

interestingly, there are also infinitely many solutions to just one given Ricci kernel. Just as an example, one might take the flat Minkowski space and the Schwarzschild vacuum metric [142]. Both have a vanishing Ricci scalar $R = 0$, but while the first describes empty space, the second, even though being a vacuum solution, contains a gravitational "object" of spherical symmetry. And yes, this all comes out from a variational kernel of $R = 0$. Thus, so our conclusion, there seem to be quite some degrees of freedom regarding the choice of metric solutions to just one Ricci scalar curvature.

(9) Tickling metric solutions with respect to this degree of freedom, which is to say to perform a variation (or transformation) of metric solutions, thereby treating the Ricci scalar as a conserved quantity, gives us classical quantum equations and thus quantum theory.

(10) Very interesting results are also obtained for an extension of the variation with respect to the number of dimensions [145].

(11) Another possible way to result in classical quantum equation emerges from an extension of the classical Einstein–Hilbert action, usually variated with respect to the metric, now being variated with respect to the base vectors or even the coordinates. The resulting quantum equations are factors within the variational integrand. Solving them, subsequently solves the whole variation problem and therefore quantum equations and Einstein field equations alike [114, 117, 136].

(12) Last but not the least, we have also considered a more general Einstein–Hilbert action with non-linear Ricci scalar dependencies as Lagrange functions [114]. In fact there, too, one could end up with classical Einstein field equations plus quantum equations if assuming large-scale differences between certain Ricci scalar curvatures [114].

In the following sections we shall consider all options to obtain quantum equations from our general "property basis assumption" and the subsequent Einstein–Hilbert action.

3.3 The Ricci Scalar Quantization

Our first path shall be the application of rather general transformations to metric solutions of the Einstein field equations. Essentially one finds quantum gravity as transformations to metrics solving the Einstein field equations:

$$R^{\alpha\beta} - \frac{1}{2}Rg^{\alpha\beta} + \Lambda g^{\alpha\beta} = -\kappa T^{\alpha\beta}. \tag{83}$$

Here we have: $R^{\alpha\beta}$, $T^{\alpha\beta}$ the Ricci- and the energy–momentum tensor, respectively, while the parameters Λ and κ are constants (usually called cosmological and coupling constant, respectively). These are the well-known Einstein field equations in n dimensions with the indices α and β running from 0 to $n - 1$. The theory behind is called the General Theory of Relativity. In the classical theory it was assumed that no-matter or vacuum solutions with $T^{\alpha\beta} = 0$ would only suffice to describe true vacuum cases, but we have been able to show that this assumption was wrong (e.g., [108–112]). While Einstein had no fundamental explanation for the appearance of the energy–momentum tensor and had to postulate it, we clearly found matter equations when subjecting vacuum metrics to certain—and quite simple— transformations.

Thereby we can use external or internal degrees of freedom. While in most of our previous papers (e.g., [112] as this is most compact) we more or less concentrated on external or wrapper-like transformations of the kind

$$G_{\alpha\beta} = F\,[f\,[t,\,x,\,y,\,z]]_{\alpha\beta}^{ij}\,g_{ij} \rightarrow G_{\alpha\beta} = F\,[f\,[t,\,x,\,y,\,z]]\cdot\delta_\alpha^i\delta_\beta^j g_{ij}, \tag{84}$$

we also introduced inner or Killing-like approaches in [113]. Most interestingly, these inner quantum gravity transformations led to the Dirac equation (see also [133, 134] with the classical Dirac approach given in [143]).

In conclusion we might state that quantum theory is just the inner degree or fluctuation of metric solutions to the Einstein field equations.

The interested reader will find a fairly compact mathematical presentation of the above recipe in [112].

The question which is still unanswered, however, is why the degree of freedom needs to "be activated" and what forces it to obey the known quantum theoretical rules and equations?

3.3.1 About the Origin of Matter

Within the short series "Science Riddles" [120–132], we essentially concentrated on the following question:

How can we understand transformations of metric tensors like Eq. (84), which do not sum up to an identical transformation?

In order to answer this question we want to start with a very simple change of the metric tensor. In [100] and some previous papers of the newer series "Science Riddles" it was shown that a metric $g_{\alpha\beta}$, which fulfills the vacuum Einstein field equations (EFEs), can be generalized to a quantized state via an approach of the form

$$G_{\alpha\beta} = g_{\alpha\beta} \cdot F[f]. \tag{85}$$

Please do not mix the f of $F[f]$ with the $f[R]$ in our previous paper and the generalized Einstein–Hilbert action [144]:

$$\delta_g W = 0 = \delta_g \int_V d^n x \left(\sqrt{-g} \, [f[R] - 2 \cdot \kappa \cdot L_M - 2 \cdot \Lambda] \right). \tag{86}$$

For completeness we also give the classical form [137]

$$\delta_g W = 0 = \delta_g \int_V d^n x \left(\sqrt{-g} \, [R - 2\kappa L_M - 2\Lambda] \right). \tag{87}$$

However, as demonstrated in [87], it can be quite complicated to find holistic, EFE-compatible solutions even for the simplest cases. But what would be the meaning of transformations leading to metrics not fulfilling the vacuum Einstein field equations anymore?

For generality, we here want to repeat the derivation from [87], where it was shown that for arbitrary numbers n of dimensions, our approach Eq. (85) yields a certain Ricci scalar R^*.* Now, while $g_{\alpha\beta}$ shall satisfy the vacuum Einstein field equations with the classical linear dependency on R, this does not hold for R^*. This gives us two Einstein–Hilbert actions with respect to the two metrics $g_{\alpha\beta}$ and $G_{\alpha\beta}$ plus a third one giving the variation of R^* with respect to $g_{\alpha\beta}$:[†]

$$\delta_g W = 0 = \delta_g \int_V d^n x \left(\sqrt{-g} \cdot \Phi_R[R] \right) = \delta_g \int_V d^n x \left(\sqrt{-g} \cdot R \right)$$

$$\delta_G W = \, ?? = \delta_G \int_V d^n x \left(\sqrt{-G} \cdot R^* \right)$$

$$\delta_g W = \, ? = \delta_g \int_V d^n x \left(\sqrt{-g} \cdot \Phi_{R^*}[R^*] \right), \tag{88}$$

Please note that this R should be seen as an auxiliary Ricci scalar as it is based on the auxiliary or "functionally wrapped" metric Eq. (85) and not the original metric $g_{\alpha\beta}$ which was assumed to fulfill the vacuum Einstein field equations. How a complete solution with $G_{\alpha\beta}$, still fulfilling the Einstein field equations, can be constructed was demonstrated in [87].

[†]In order to avoid mixing up the f function from the wrapper forms F[f], we rewrite Eq. (86) now with a function $\Phi(R)$ instead of $f(R)$.

where we do not consider any cosmological constant and also have erased the classical (artificially introduced) matter term. We also name the Ricci scalar of the auxiliary (quantized) metric $G_{\alpha\beta}$ from Eq. (85) R^*, while we leave the symbol R for the metric $g_{\alpha\beta}$.

Thereby, just as said, we assume R to be the Ricci scalar to the unperturbed (non-transformed) metric $g_{\alpha\beta}$. On the other hand, we know the Ricci scalar R to be defined by the metric tensor via:

$$R = R_{\alpha\beta}g^{\alpha\beta} = \left(\Gamma^{\sigma}_{\alpha\beta,\sigma} - \Gamma^{\sigma}_{\beta\sigma,\alpha} - \Gamma^{\mu}_{\sigma\alpha}\Gamma^{\sigma}_{\beta\mu} + \Gamma^{\sigma}_{\alpha\beta}\Gamma^{\mu}_{\sigma\mu}\right)g^{\alpha\beta}, \tag{89}$$

with

$$\Gamma^{\gamma}_{\alpha\beta} = \frac{g^{\gamma\sigma}}{2}\left(g_{\sigma\alpha,\beta} + g_{\sigma\beta,\alpha} - g_{\alpha\beta,\sigma}\right). \tag{90}$$

Similarly, for R^* we have

$$R^* = R^*_{\alpha\beta}G^{\alpha\beta} = \left(\Gamma^{*\sigma}_{\alpha\beta,\sigma} - \Gamma^{*\sigma}_{\beta\sigma,\alpha} - \Gamma^{*\mu}_{\sigma\alpha}\Gamma^{*\sigma}_{\beta\mu} + \Gamma^{*\sigma}_{\alpha\beta}\Gamma^{*\mu}_{\sigma\mu}\right)G^{\alpha\beta}$$

$$= \left(\Gamma^{*\sigma}_{\alpha\beta,\sigma} - \Gamma^{*\sigma}_{\beta\sigma,\alpha} - \Gamma^{*\mu}_{\sigma\alpha}\Gamma^{*\sigma}_{\beta\mu} + \Gamma^{*\sigma}_{\alpha\beta}\Gamma^{*\mu}_{\sigma\mu}\right)\frac{g^{\alpha\beta}}{F[f]}, \tag{91}$$

with

$$\Gamma^{*\gamma}_{\alpha\beta} = \frac{g^{\gamma\sigma}}{2\cdot F[f]}\left([F[f]\cdot g_{\sigma\alpha}]_{,\beta} + [F[f]\cdot g_{\sigma\beta}]_{,\alpha} - [F[f]\cdot g_{\alpha\beta}]_{,\sigma}\right)$$

$$= \frac{g^{\gamma\sigma}}{2\cdot F[f]}\left([F[f]\cdot g_{\sigma\alpha}]_{,\beta} + [F[f]\cdot g_{\sigma\beta}]_{,\alpha} - [F[f]\cdot g_{\alpha\beta}]_{,\sigma}\right)$$

$$= \frac{g^{\gamma\sigma}}{2}\left(g_{\sigma\alpha,\beta} + g_{\sigma\beta,\alpha} - g_{\alpha\beta,\sigma}\right)$$

$$+ \frac{g^{\gamma\sigma}}{2\cdot F[f]}\left(F[f]_{,\beta}\cdot g_{\sigma\alpha} + F[f]_{,\alpha}\cdot g_{\sigma\beta} - F[f]_{,\sigma}\cdot g_{\alpha\beta}\right)$$

$$= \Gamma^{\gamma}_{\alpha\beta} + \frac{g^{\gamma\sigma}}{2\cdot F[f]}\left(F[f]_{,\beta}\cdot g_{\sigma\alpha} + F[f]_{,\alpha}\cdot g_{\sigma\beta} - F[f]_{,\sigma}\cdot g_{\alpha\beta}\right)$$

$$\equiv \Gamma^{\gamma}_{\alpha\beta} + \Gamma^{**\gamma}_{\alpha\beta}. \tag{92}$$

Setting this into the second line in Eq. (91) yields

$$R^* = R^*_{\alpha\beta}G^{\alpha\beta} = \left(\Gamma^{*\sigma}_{\alpha\beta,\sigma} - \Gamma^{*\sigma}_{\beta\sigma,\alpha} - \Gamma^{*\mu}_{\sigma\alpha}\Gamma^{*\sigma}_{\beta\mu} + \Gamma^{*\sigma}_{\alpha\beta}\Gamma^{*\mu}_{\sigma\mu}\right)G^{\alpha\beta}$$

$$= \left(\begin{array}{c}\Gamma^{\sigma}_{\alpha\beta,\sigma} - \Gamma^{\sigma}_{\beta\sigma,\alpha} - \Gamma^{\mu}_{\sigma\alpha}\Gamma^{\sigma}_{\beta\mu} + \Gamma^{\sigma}_{\alpha\beta}\Gamma^{\mu}_{\sigma\mu} \\ -\Gamma^{\mu}_{\sigma\alpha}\Gamma^{**\sigma}_{\beta\mu} + \Gamma^{\sigma}_{\alpha\beta}\Gamma^{**\mu}_{\sigma\mu} - \Gamma^{**\mu}_{\sigma\alpha}\Gamma^{\sigma}_{\beta\mu} + \Gamma^{**\sigma}_{\alpha\beta}\Gamma^{\mu}_{\sigma\mu} \\ +\Gamma^{**\sigma}_{\alpha\beta,\sigma} - \Gamma^{**\sigma}_{\beta\sigma,\alpha} - \Gamma^{**\mu}_{\sigma\alpha}\Gamma^{**\sigma}_{\beta\mu} + \Gamma^{**\sigma}_{\alpha\beta}\Gamma^{**\mu}_{\sigma\mu}\end{array}\right)\frac{g^{\alpha\beta}}{F[f]}$$

$$= \frac{R}{F[f]} + \left(\begin{array}{c}-\Gamma^{\mu}_{\sigma\alpha}\Gamma^{**\sigma}_{\beta\mu} + \Gamma^{\sigma}_{\alpha\beta}\Gamma^{**\mu}_{\sigma\mu} - \Gamma^{**\mu}_{\sigma\alpha}\Gamma^{\sigma}_{\beta\mu} + \Gamma^{**\sigma}_{\alpha\beta}\Gamma^{\mu}_{\sigma\mu} \\ +\Gamma^{**\sigma}_{\alpha\beta,\sigma} - \Gamma^{**\sigma}_{\beta\sigma,\alpha} - \Gamma^{**\mu}_{\sigma\alpha}\Gamma^{**\sigma}_{\beta\mu} + \Gamma^{**\sigma}_{\alpha\beta}\Gamma^{**\mu}_{\sigma\mu}\end{array}\right)\frac{g^{\alpha\beta}}{F[f]}$$

$$\equiv \frac{R + R^{**}}{F[f]} = \Phi[R]. \tag{93}$$

Further evaluation of Eq. (93) is possible by rewriting Eq. (92) as follows:

$$
\begin{aligned}
\Gamma^{*\gamma}_{\alpha\beta} &= \Gamma^{\gamma}_{\alpha\beta} + \frac{g^{\gamma\sigma}}{2 \cdot F\,[f]} \left(F\,[f]_{,\beta} \cdot g_{\sigma\alpha} + F\,[f]_{,\alpha} \cdot g_{\sigma\beta} - F\,[f]_{,\sigma} \cdot g_{\alpha\beta} \right) \\
&= \Gamma^{\gamma}_{\alpha\beta} + \frac{F\,[f]_{,\beta} \cdot \delta^{\gamma}_{\alpha} + F\,[f]_{,\alpha} \cdot \delta^{\gamma}_{\beta} - g^{\gamma\sigma} \cdot F\,[f]_{,\sigma} \cdot g_{\alpha\beta}}{2 \cdot F\,[f]} \\
&= \Gamma^{\gamma}_{\alpha\beta} + F' \cdot \frac{f_{,\beta} \cdot \delta^{\gamma}_{\alpha} + f_{,\alpha} \cdot \delta^{\gamma}_{\beta} - g^{\gamma\sigma} \cdot f_{,\sigma} \cdot g_{\alpha\beta}}{2 \cdot F\,[f]} \\
&\equiv \Gamma^{\gamma}_{\alpha\beta} + \frac{F'}{F\,[f]} \cdot \Gamma^{***\gamma}_{\alpha\beta}; \qquad F' = \frac{\partial F\,[f]}{\partial f}.
\end{aligned}
\tag{94}
$$

Now we can give Eq. (93) as follows:

$$
\begin{aligned}
R^* &= \frac{R}{F\,[f]} + \begin{pmatrix} -\Gamma^{\mu}_{\sigma\alpha}\Gamma^{**\sigma}_{\beta\mu} + \Gamma^{\sigma}_{\alpha\beta}\Gamma^{**\mu}_{\sigma\mu} - \Gamma^{**\mu}_{\sigma\alpha}\Gamma^{\sigma}_{\beta\mu} + \Gamma^{**\sigma}_{\alpha\beta}\Gamma^{\mu}_{\sigma\mu} \\ +\Gamma^{**\sigma}_{\alpha\beta}{}_{,\sigma} - \Gamma^{**\sigma}_{\beta\sigma}{}_{,\alpha} - \Gamma^{**\mu}_{\sigma\alpha}\Gamma^{**\sigma}_{\beta\mu} + \Gamma^{**\sigma}_{\alpha\beta}\Gamma^{**\mu}_{\sigma\mu} \end{pmatrix} \frac{g^{\alpha\beta}}{F\,[f]} \\
&= \frac{R}{F\,[f]} + \begin{pmatrix} \frac{F'}{F[f]} \cdot \left(-\Gamma^{\mu}_{\sigma\alpha}\Gamma^{***\mu}_{\beta\mu} + \Gamma^{\sigma}_{\alpha\beta}\Gamma^{***\mu}_{\sigma\mu} - \Gamma^{***\mu}_{\sigma\alpha}\Gamma^{\sigma}_{\beta\mu} + \Gamma^{***\sigma}_{\alpha\beta}\Gamma^{\mu}_{\sigma\mu} \right) \\ +\frac{F'}{F[f]} \cdot \left(\Gamma^{***\sigma}_{\alpha\beta}{}_{,\sigma} - \Gamma^{***\sigma}_{\beta\sigma}{}_{,\alpha} \right) + \left(\frac{F'}{F[f]} \right)_{,\sigma} \cdot \Gamma^{***\sigma}_{\alpha\beta} - \left(\frac{F'}{F[f]} \right)_{,\alpha} \cdot \Gamma^{***\sigma}_{\beta\sigma} \\ +\left(\frac{F'}{F[f]} \right)^2 \cdot \left(\Gamma^{***\sigma}_{\alpha\beta}\Gamma^{***\mu}_{\sigma\mu} - \Gamma^{***\mu}_{\sigma\alpha}\Gamma^{***\sigma}_{\beta\mu} \right) \end{pmatrix} \\
&\quad \times \frac{g^{\alpha\beta}}{F\,[f]}.
\end{aligned}
\tag{95}
$$

How could this be understood in the metric picture?

To answer this question we reshape Eq. (95) and write it as follows:

$$
\begin{aligned}
R &= F\,[f] \cdot R^* - \begin{pmatrix} -\Gamma^{\mu}_{\sigma\alpha}\Gamma^{**\sigma}_{\beta\mu} + \Gamma^{\sigma}_{\alpha\beta}\Gamma^{**\mu}_{\sigma\mu} - \Gamma^{**\mu}_{\sigma\alpha}\Gamma^{\sigma}_{\beta\mu} + \Gamma^{**\sigma}_{\alpha\beta}\Gamma^{\mu}_{\sigma\mu} \\ +\Gamma^{**\sigma}_{\alpha\beta}{}_{,\sigma} - \Gamma^{**\sigma}_{\beta\sigma}{}_{,\alpha} - \Gamma^{**\mu}_{\sigma\alpha}\Gamma^{**\sigma}_{\beta\mu} + \Gamma^{**\sigma}_{\alpha\beta}\Gamma^{**\mu}_{\sigma\mu} \end{pmatrix} g^{\alpha\beta} \\
&= F\,[f] \cdot R^* \\
&\quad - \begin{pmatrix} \frac{F'}{F[f]} \cdot \left(-\Gamma^{\mu}_{\sigma\alpha}\Gamma^{***\mu}_{\beta\mu} + \Gamma^{\sigma}_{\alpha\beta}\Gamma^{***\mu}_{\sigma\mu} - \Gamma^{***\mu}_{\sigma\alpha}\Gamma^{\sigma}_{\beta\mu} + \Gamma^{***\sigma}_{\alpha\beta}\Gamma^{\mu}_{\sigma\mu} \right) \\ +\frac{F'}{F[f]} \cdot \left(\Gamma^{***\sigma}_{\alpha\beta}{}_{,\sigma} - \Gamma^{***\sigma}_{\beta\sigma}{}_{,\alpha} \right) + \left(\frac{F'}{F[f]} \right)_{,\sigma} \cdot \Gamma^{***\sigma}_{\alpha\beta} - \left(\frac{F'}{F[f]} \right)_{,\alpha} \cdot \Gamma^{***\sigma}_{\beta\sigma} \\ +\left(\frac{F'}{F[f]} \right)^2 \cdot \left(\Gamma^{***\sigma}_{\alpha\beta}\Gamma^{***\mu}_{\sigma\mu} - \Gamma^{***\mu}_{\sigma\alpha}\Gamma^{***\sigma}_{\beta\mu} \right) \end{pmatrix} g^{\alpha\beta}.
\end{aligned}
\tag{96}
$$

Now, as we know that the Ricci scalar R, which is based on the metric $g_{\alpha\beta}$, satisfies the vacuum Einstein field equations, we have to conclude that the same holds for the right hand side of Eq. (96). Without any form of simplification or approximation, we can now define the two addends on the right hand side as follows:

$$R = \overbrace{F\,[f] \cdot R^*}^{R_{\text{matter}}} - \overbrace{\begin{pmatrix} -\Gamma^{\mu}_{\sigma\alpha}\Gamma^{**\sigma}_{\beta\mu} + \Gamma^{\sigma}_{\alpha\beta}\Gamma^{**\mu}_{\sigma\mu} - \Gamma^{**\mu}_{\sigma\alpha}\Gamma^{\sigma}_{\beta\mu} + \Gamma^{**\sigma}_{\alpha\beta}\Gamma^{\mu}_{\sigma\mu} \\ +\Gamma^{**\sigma}_{\alpha\beta\,,\sigma} - \Gamma^{**\sigma}_{\beta\sigma\,,\alpha} - \Gamma^{**\mu}_{\sigma\alpha}\Gamma^{**\sigma}_{\beta\mu} + \Gamma^{**\sigma}_{\alpha\beta}\Gamma^{**\mu}_{\sigma\mu} \end{pmatrix}}^{2\cdot\kappa\cdot L_{\text{M}}+2\cdot\Lambda} g^{\alpha\beta}$$

$$\equiv R_{\text{matter}} - 2\cdot\kappa\cdot L_{\text{M}} - 2\cdot\Lambda. \tag{97}$$

Comparing this with the classical Einstein–Hilbert action Eq. (87) (which is just Eq. (86) with $f[R] = R$), thereby knowing that R from Eq. (97) fulfills the corresponding Eq. (83) in the vacuum case (with both, L_{M} and Λ being zero), we come to the conclusion that our wrapper transformation Eq. (85) kind of separates a matter-free metric $g_{\alpha\beta}$, which is represented by the matter-free Ricci scalar R, into a matter-containing Ricci scalar $R_{\text{matter}} = F\,[f] \cdot R^*$ plus the corresponding Lagrange matter plus a Λ term. Setting $\Lambda = 0$ gives us just the matter Lagrange density as follows:

$$L_{\text{M}} = \frac{g^{\alpha\beta}}{2\cdot\kappa}\begin{pmatrix} -\Gamma^{\mu}_{\sigma\alpha}\Gamma^{**\sigma}_{\beta\mu} + \Gamma^{\sigma}_{\alpha\beta}\Gamma^{**\mu}_{\sigma\mu} - \Gamma^{**\mu}_{\sigma\alpha}\Gamma^{\sigma}_{\beta\mu} + \Gamma^{**\sigma}_{\alpha\beta}\Gamma^{\mu}_{\sigma\mu} \\ +\Gamma^{**\sigma}_{\alpha\beta\,,\sigma} - \Gamma^{**\sigma}_{\beta\sigma\,,\alpha} - \Gamma^{**\mu}_{\sigma\alpha}\Gamma^{**\sigma}_{\beta\mu} + \Gamma^{**\sigma}_{\alpha\beta}\Gamma^{**\mu}_{\sigma\mu} \end{pmatrix}. \tag{98}$$

We further conclude:

(A) That all non-identical metric transformations like Eq. (84)—with Eq. (85) being its simplest form—are just acting in the same way and might therefore be considered as matter-creation operations.

(B) There should also be a reverse operation (transformation) $F[f]^{-1}$, which annihilates matter states into vacuum states.

(C) Every matter ensemble L_{M} in fact has a $R_{\text{matter}} = F\,[f] \cdot R^*$-counterpart or partner and the two always neutralize to an R providing a vacuum-solution to the Einstein field equations.

(D) Stable matter forms are more or less permanent split-ups of L_{M} and $R_{\text{matter}} = F\,[f] \cdot R^*$.

But how does this lead us to a connection with the classical quantum theory?

Even though many equations we will derive here have already been shown in the introduction part, we will still represent them here for the reason of better overview and convenience.

One can show (cf. [100] and references given there) that with Eq. (93) and "wrapped metrics" of the kind Eq. (85) in all diagonal and especially flat space metrics—where usually classical quantum theory is defined on—we result in Ricci scalars of the form

$$
R^* = \frac{1}{F\,[f]^3} \cdot \left(\begin{array}{l} \left(C_{N1} \cdot \left(\frac{\partial F[f]}{\partial f} \right)^2 - C_{N2} \cdot F\,[f] \cdot \frac{\partial^2 F[f]}{\partial f^2} \right) \cdot \left(\tilde{\nabla}_{g-\text{coordinates}}\, f \right)^2 \\ -C_{N2} \cdot F\,[f] \cdot \frac{\partial F[f]}{\partial f} \cdot \Delta_{g-\text{coordinates}}\, f \end{array} \right).
$$

(99)

This holds in an arbitrary number of dimensions. Thereby the symbol $\tilde{\nabla}_{g-\text{coordinates}}$ denotes a first-order differential operator similar to the Nabla operator in the metric $g_{\alpha\beta}$. The symbols C_{Ni} stand for constants which only depend on the number of dimensions. Without loss of generality, we can now demand that the first term in parenthesis in Eq. (99) would be zero, which is to say, we choose our arbitrary function F such that we have

$$
C_{N1} \cdot \left(\frac{\partial F\,[f]}{\partial f} \right)^2 - C_{N2} \cdot F\,[f] \cdot \frac{\partial^2 F\,[f]}{\partial f^2} = 0.
$$

(100)

For instance, in four dimensions, this would always be the case for $F[f] = f^2$, giving us the Klein–Gordon-like equation

$$
\left[\Delta_g + \frac{R^* \cdot f^4}{f \cdot C_{N2} \cdot 2 \cdot f} \right] f = \left[\Delta_g + \frac{R^* \cdot f^2}{2} \right] f = 0;\ \Delta_{g-\text{coordinates}} = \Delta_g.
$$

(101)

Many examples and subsequent F functions for other numbers of dimensions are given in [97, 110]. In the general case (arbitrary number of dimensions), with the condition Eq. (100), we obtain

$$
R^* + C_{N2} \cdot \frac{1}{F\,[f]^2} \cdot \frac{\partial F\,[f]}{\partial f} \cdot \Delta_g f = 0 \Rightarrow \left(\frac{R^* \cdot F\,[f]^2}{C_{N2} \cdot f \cdot \frac{\partial F[f]}{\partial f}} + \Delta_g \right) f = 0.
$$

(102)

We recognize the typical structure of the Klein–Gordon equation, whose original form can be given in arbitrary coordinates as

$$
\left[\Delta_g + \frac{M^2 c^2}{\hbar^2} + V \right] f = 0,
$$

(103)

with the first term in Eq. (102) (second equation) being seen as the corresponding metric equivalent for the mass and potential. This leads us to the identity

$$
\frac{M^2 c^2}{\hbar^2} + V = \frac{R^* \cdot F\,[f]^2}{f \cdot C_{N2} \cdot \frac{\partial F[f]}{\partial f}}.
$$

(104)

Thus, without any approximation, we have expressed the effective mass+potential-term of the Klein–Gordon equation as something solely dependent on the auxiliary or wrapped Ricci scalar R^* times a function of f.

This gives us now a rather basic connection between quantum theory and General Theory of Relativity in any number of dimensions. We see that no governing theory seems to be of need. Instead, the General Theory of Relativity GTR already contained the quantum theory. The Klein–Gordon equation in GTR should read

$$\left[\Delta_g + \frac{R^* \cdot F\,[f]^2}{f \cdot C_{N2} \cdot \frac{\partial F[f]}{\partial f}}\right] f = 0. \tag{105}$$

It needs to be pointed out explicitly that our choice for $F[f]$ is no restriction to the whole, but only a technical trick. Any information residing in F would be taken on by f the moment we fix F for convenience as we did with condition Eq. (100).

3.3.2 Postulation of a Constant Ricci Scalar

In the attempt to motivate the construction of the classical quantum equations out of the Ricci scalar and thus, out of solutions to the Einstein field equations, we postulated a certain behavior for the Ricci scalar in [127]. So, by postulation of a constant Ricci scalar, which is to say an $R \rightarrow F[f] * R^*$ that does not change under the transformation Eq. (84), we would result in the following general condition:

$$F\,[f] \cdot R^* - R = 0$$

$$= \left(\begin{array}{c} -\Gamma^\mu_{\sigma\alpha}\Gamma^{**\sigma}_{\beta\mu} + \Gamma^\sigma_{\alpha\beta}\Gamma^{**\mu}_{\sigma\mu} - \Gamma^{**\mu}_{\sigma\alpha}\Gamma^\sigma_{\beta\mu} + \Gamma^{**\sigma}_{\alpha\beta}\Gamma^\mu_{\sigma\mu} \\ +\Gamma^{**\sigma}_{\alpha\beta,\sigma} - \Gamma^{**\sigma}_{\beta\sigma,\alpha} - \Gamma^{**\mu}_{\sigma\alpha}\Gamma^{**\sigma}_{\beta\mu} + \Gamma^{**\sigma}_{\alpha\beta}\Gamma^{**\mu}_{\sigma\mu} \end{array}\right) g^{\alpha\beta}$$

$$\Rightarrow 0 = \left(\begin{array}{c} \frac{F'}{F[f]} \cdot \left(-\Gamma^\mu_{\sigma\alpha}\Gamma^{***\sigma}_{\beta\mu} + \Gamma^\sigma_{\alpha\beta}\Gamma^{***\mu}_{\sigma\mu} - \Gamma^{***\mu}_{\sigma\alpha}\Gamma^\sigma_{\beta\mu} + \Gamma^{***\sigma}_{\alpha\beta}\Gamma^\mu_{\sigma\mu}\right) \\ +\frac{F'}{F[f]} \cdot \left(\Gamma^{***\sigma}_{\alpha\beta,\sigma} - \Gamma^{***\sigma}_{\beta\sigma,\alpha}\right) + \left(\frac{F'}{F[f]}\right)_{,\sigma} \cdot \Gamma^{***\sigma}_{\alpha\beta} - \left(\frac{F'}{F[f]}\right)_{,\alpha} \cdot \Gamma^{***\sigma}_{\beta\sigma} \\ +\left(\frac{F'}{F[f]}\right)^2 \cdot \left(\Gamma^{***\sigma}_{\alpha\beta}\Gamma^{***\mu}_{\sigma\mu} - \Gamma^{***\mu}_{\sigma\alpha}\Gamma^{***\sigma}_{\beta\mu}\right) \end{array}\right) g^{\alpha\beta}$$

$$\Gamma^{***\gamma}_{\alpha\beta} \equiv \frac{f_{,\beta} \cdot \delta^\gamma_\alpha + f_{,\alpha} \cdot \delta^\gamma_\beta - g^{\gamma\sigma} \cdot f_{,\sigma} \cdot g_{\alpha\beta}}{2}; \, F' = \frac{\partial F\,[f]}{\partial f}, \tag{106}$$

where we have used the result Eq. (96) and the abbreviation introduced in Eq. (94). The second line in Eq. (106) might be seen as the general quantum gravity equation based on the simplest possible transformation Eq. (85) with f being a function of all coordinates. Thus, we might consider $f = f[\ldots]$ as the quantum gravity wave function. However, having just seen that the term

$$\begin{aligned}&\left(-\Gamma^{\mu}_{\sigma\alpha}\Gamma^{**\sigma}_{\beta\mu} + \Gamma^{\sigma}_{\alpha\beta}\Gamma^{**\mu}_{\sigma\mu} - \Gamma^{**\mu}_{\sigma\alpha}\Gamma^{\sigma}_{\beta\mu} + \Gamma^{**\sigma}_{\alpha\beta}\Gamma^{\mu}_{\sigma\mu} + \Gamma^{**\sigma}_{\alpha\beta,\sigma} - \Gamma^{**\sigma}_{\beta\sigma,\alpha}\right.\\ &\left.-\Gamma^{**\mu}_{\sigma\alpha}\Gamma^{**\sigma}_{\beta\mu} + \Gamma^{**\sigma}_{\alpha\beta}\Gamma^{**\mu}_{\sigma\mu}\right)\end{aligned} \tag{107}$$

is proportional to the matter Lagrange density (cf. Eq. (98)), condition Eq. (106) would always demand this term to be zero. As this seems to be a too rigid condition, we try for a different one which leaves us a bit more freedom.

Please note that our current postulation could also be one where the original R just shall be a constant. In this case we would have

$$R = \text{const}$$

$$= F\,[f]\cdot R^* - \left(\begin{array}{c}-\Gamma^{\mu}_{\sigma\alpha}\Gamma^{**\sigma}_{\beta\mu} + \Gamma^{\sigma}_{\alpha\beta}\Gamma^{**\mu}_{\sigma\mu} - \Gamma^{**\mu}_{\sigma\alpha}\Gamma^{\sigma}_{\beta\mu} + \Gamma^{**\sigma}_{\alpha\beta}\Gamma^{\mu}_{\sigma\mu}\\ +\Gamma^{**\sigma}_{\alpha\beta,\sigma} - \Gamma^{**\sigma}_{\beta\sigma,\alpha} - \Gamma^{**\mu}_{\sigma\alpha}\Gamma^{**\sigma}_{\beta\mu} + \Gamma^{**\sigma}_{\alpha\beta}\Gamma^{**\mu}_{\sigma\mu}\end{array}\right)g^{\alpha\beta}$$

$$0 = F\,[f]\cdot R^* - R$$
$$- \left(\begin{array}{c}\frac{F'}{F[f]}\cdot\left(-\Gamma^{\mu}_{\sigma\alpha}\Gamma^{***\sigma}_{\beta\mu} + \Gamma^{\sigma}_{\alpha\beta}\Gamma^{***\mu}_{\sigma\mu} - \Gamma^{***\mu}_{\sigma\alpha}\Gamma^{\sigma}_{\beta\mu} + \Gamma^{***\sigma}_{\alpha\beta}\Gamma^{\mu}_{\sigma\mu}\right)\\ +\frac{F'}{F[f]}\cdot\left(\Gamma^{***\sigma}_{\alpha\beta,\sigma} - \Gamma^{***\sigma}_{\beta\sigma,\alpha}\right) + \left(\frac{F'}{F[f]}\right)_{,\sigma}\cdot\Gamma^{***\sigma}_{\alpha\beta} - \left(\frac{F'}{F[f]}\right)_{,\alpha}\cdot\Gamma^{***\sigma}_{\beta\sigma}\\ +\left(\frac{F'}{F[f]}\right)^2\cdot\left(\Gamma^{***\sigma}_{\alpha\beta}\Gamma^{***\mu}_{\sigma\mu} - \Gamma^{***\mu}_{\sigma\alpha}\Gamma^{***\sigma}_{\beta\mu}\right)\end{array}\right)g^{\alpha\beta}, \tag{108}$$

which would require us to have an idea about the transformed Ricci scalar R^*. Now we see that matter is created in cases where the resulting Ricci difference

$$\Delta R \equiv F\,[f]\cdot R^* - R \tag{109}$$

does not vanish. In generalization of this postulation we might simply demand that the sum of Ricci scalar, Lagrange density of matter and cosmological constant as given in the kernel of the classical Einstein–Hilbert action Eq. (87) stays constant, respectively that its variation would give zero. Then condition Eq. (108) might be seen as follows:

$$R - 2\kappa L_{\mathrm{M}} - 2\Lambda = 0$$

$$= R - F\,[f]\cdot R^*$$
$$+ \left(\begin{array}{c}\frac{F'}{F[f]}\cdot\left(-\Gamma^{\mu}_{\sigma\alpha}\Gamma^{***\sigma}_{\beta\mu} + \Gamma^{\sigma}_{\alpha\beta}\Gamma^{***\mu}_{\sigma\mu} - \Gamma^{***\mu}_{\sigma\alpha}\Gamma^{\sigma}_{\beta\mu} + \Gamma^{***\sigma}_{\alpha\beta}\Gamma^{\mu}_{\sigma\mu}\right)\\ +\frac{F'}{F[f]}\cdot\left(\Gamma^{***\sigma}_{\alpha\beta,\sigma} - \Gamma^{***\sigma}_{\beta\sigma,\alpha}\right) + \left(\frac{F'}{F[f]}\right)_{,\sigma}\cdot\Gamma^{***\sigma}_{\alpha\beta} - \left(\frac{F'}{F[f]}\right)_{,\alpha}\cdot\Gamma^{***\sigma}_{\beta\sigma}\\ +\left(\frac{F'}{F[f]}\right)^2\cdot\left(\Gamma^{***\sigma}_{\alpha\beta}\Gamma^{***\mu}_{\sigma\mu} - \Gamma^{***\mu}_{\sigma\alpha}\Gamma^{***\sigma}_{\beta\mu}\right)\end{array}\right)g^{\alpha\beta}. \tag{110}$$

This interpretation shall be investigated in the next section. But before going into this we want to present a brief summary of our findings so far.

3.3.3 Intermediate Sum-Up

In principle, we have only found a triviality, namely:

(A) The transformation $G_{\alpha\beta} = F[f] \cdot \delta^i_\alpha \delta^j_\beta g_{ij}$ (seen as the simplest form of $G_{\alpha\beta} = F^{ij}_{\alpha\beta} g_{ij}$) of a metric solution g_{ij} of the vacuum Einstein field equations, in general, leads to a new metric which only solves the Einstein field equations containing matter.

(B) Thereby the matter being created is well defined via the Einstein–Hilbert–Lagrange density given in Eq. (98), respectively (without the abbreviations hiding all the F dependencies):

$$
L_\text{M} = \frac{g^{\alpha\beta}}{2 \cdot \kappa}
\begin{pmatrix}
\frac{F'}{F[f]} \cdot \left(-\Gamma^\mu_{\sigma\alpha} \Gamma^{***\sigma}_{\beta\mu} + \Gamma^\sigma_{\alpha\beta} \Gamma^{***\mu}_{\sigma\mu} - \Gamma^{***\mu}_{\sigma\alpha} \Gamma^\sigma_{\beta\mu} + \Gamma^{***\sigma}_{\alpha\beta} \Gamma^\mu_{\sigma\mu} \right) \\
+ \frac{F'}{F[f]} \cdot \left(\Gamma^{***\sigma}_{\alpha\beta}{}_{,\sigma} - \Gamma^{***\sigma}_{\beta\sigma}{}_{,\alpha} \right) + \left(\frac{F'}{F[f]} \right)_{,\sigma} \cdot \Gamma^{***\sigma}_{\alpha\beta} - \left(\frac{F'}{F[f]} \right)_{,\alpha} \cdot \Gamma^{***\sigma}_{\beta\sigma} \\
+ \left(\frac{F'}{F[f]} \right)^2 \cdot \left(\Gamma^{***\sigma}_{\alpha\beta} \Gamma^{***\mu}_{\sigma\mu} - \Gamma^{***\mu}_{\sigma\alpha} \Gamma^{***\sigma}_{\beta\mu} \right)
\end{pmatrix}.
$$

(111)

(C) The corresponding new, which is to say transformed, Ricci scalar R^* now satisfies the Einstein field equations in connection with the previous, untransformed metric $g_{\alpha\beta}$ and the matter Lagrange density L_M. The corresponding Einstein–Hilbert action Eq. (88) (last line) becomes

$$
\delta_g W = ? = 0 = \delta_g \int_V d^n x \left(\sqrt{-g} \cdot \Phi_{R^*}[R^*] \right)
$$

$$
= \delta_g \int_V d^n x \left(\sqrt{-g} \cdot \left(\overbrace{F[f] \cdot R^*}^{R_\text{matter}} - 2 \cdot \kappa \cdot L_\text{M} \right) \right).
$$

(112)

3.3.4 The Situation in n Dimensions

Now we need to generalize our results from the subsections above to the situation in n dimensions. Diagonal structured metrics will always give a Ricci scalar R^* as presented in Eq. (13) or—in general—Eq. (99). Applying condition Eq. (100) and thereby fixing $F[f]$ then always leads to structures as follows:

$$R^* + C_{N2} \cdot \frac{\frac{\partial F[f]}{\partial f} \cdot \Delta_g f}{F[f]^2} = 0. \tag{113}$$

Here we give the subsequent equations for the dimensions $n = 2$ to $n = 10$.

3.3.4.1 The 2-dimensional space

Equation (100) gives the following solution for the function $F[f]$:

$$F[f] = C_1 \cdot \exp[f \cdot C_2] \tag{114}$$

and results in the GTR-Klein–Gordon equation as follows:

$$\left[\Delta_g + \frac{R^* \cdot (C_1 \cdot \exp[f \cdot C_2])^2}{f \cdot C_{N2} \cdot C_1 \cdot C_2 \cdot \exp[f \cdot C_2]} \right] f = \left[\Delta_g + \frac{R^* \cdot C_1 \cdot \exp[f \cdot C_2]}{f \cdot C_2} \right] f$$
$$= 0. \tag{115}$$

3.3.4.2 The 3-dimensional space

Equation (100) gives the following solution for the function $F[f]$:

$$F[f] = C_2 \cdot (f - C_1)^4 \tag{116}$$

and results in the GTR-Klein–Gordon equation as follows:

$$C_2 \cdot \frac{(f - C_1)^5}{8} \cdot R^* + \Delta_g f = \left[C_2 \cdot \frac{(f - C_1)^5}{f \cdot 8} \cdot R^* + \Delta_g \right] f$$
$$= \left[C_2 \cdot \frac{(f - C_1)^4}{8} \cdot R^* + \Delta_g \right] (f - C_1) = 0. \tag{117}$$

3.3.4.3 The 4-dimensional space

(see subsections above)

3.3.4.4 The 5-dimensional space

Equation (100) gives the following solution for the function $F[f]$:

$$F[f] = C_2 \cdot (3 \cdot f - C_1)^{4/3} \tag{118}$$

and results in the GTR-Klein–Gordon equation as follows:

$$\left[\Delta_g + \frac{3 \cdot R^* \cdot C_2 \cdot (3 \cdot f - C_1)^{7/3}}{f \cdot 16}\right] f$$

$$= \left[\Delta_g + \frac{3 \cdot R^* \cdot C_2 \cdot (3 \cdot f - C_1)^{4/3}}{16}\right] (3 \cdot f - C_1) = 0. \quad (119)$$

3.3.4.5 The 6-dimensional space

Equation (100) gives the following solution for the function $F[f]$:

$$F[f] = f + C_1 \quad (120)$$

and results in the GTR-Klein–Gordon equation as follows:

$$R^* + \frac{5}{F[f]^2} \cdot \frac{\partial F[f]}{\partial f} \cdot \Delta_g f = R^* + \frac{5}{(f + C_1)^2} \cdot \Delta_g f$$

$$\Rightarrow \frac{(f + C_1)^2}{5} \cdot R^* + \Delta_g f = \left(\frac{(f + C_1)^2}{f \cdot 5} \cdot R^* + \Delta_g\right) f$$

$$= \left(\frac{f + C_1}{5} \cdot R^* + \Delta_g\right) (f + C_1) = 0. \quad (121)$$

3.3.4.6 The 7-dimensional space

Equation (100) gives the following solution for the function $F[f]$:

$$F[f] = C_2 \cdot (7 \cdot f - C_1)^{4/7} \quad (122)$$

and results in the GTR-Klein–Gordon equation as follows:

$$R^* + \frac{2}{F[f]^2} \cdot \frac{\partial F[f]}{\partial f} \cdot \Delta_g f = 0$$

$$\Rightarrow \frac{C_2 \cdot 7 \cdot (7 \cdot f - C_1)^{11/7}}{8} \cdot R^* + \Delta_g f$$

$$= \left(\frac{C_2 \cdot 7 \cdot (7 \cdot f - C_1)^{11/7}}{f \cdot 8} \cdot R^* + \Delta_g\right) f$$

$$= \left(\frac{C_2 \cdot 7 \cdot (7 \cdot f - C_1)^{4/7}}{8} \cdot R^* + \Delta_g\right) (7 \cdot f - C_1) = 0. \quad (123)$$

3.3.4.7 The 8-dimensional space

Equation (100) gives the following solution for the function $F[f]$:

$$F[f] = C_2 \cdot (3 \cdot f - C_1)^{2/3} \tag{124}$$

and results in the GTR-Klein–Gordon equation as follows:

$$R^* + \frac{1}{F[f]^2} \cdot \frac{\partial F[f]}{\partial f} \cdot \Delta_g f = 0$$

$$\Rightarrow \frac{C_2 \cdot 7 \cdot (3 \cdot f - C_1)^{5/3}}{2} \cdot R^* + \Delta_g f = \left(\frac{C_2 \cdot 7 \cdot (3 \cdot f - C_1)^{5/3}}{f \cdot 2} \cdot R^* + \Delta_g \right) f$$

$$= \left(\frac{C_2 \cdot 7 \cdot (3 \cdot f - C_1)^{2/3}}{2} \cdot R^* + \Delta_g \right) (9 \cdot f - C_1) = 0. \tag{125}$$

3.3.4.8 The 9-dimensional space

Equation (100) gives the following solution for the function $F[f]$:

$$F[f] = C_2 \cdot (7 \cdot f - C_1)^{3/7} \tag{126}$$

and results in the GTR-Klein–Gordon equation as follows:

$$R^* + \frac{8}{F[f]^2} \cdot \frac{\partial F[f]}{\partial f} \cdot \Delta_g f = 0$$

$$\Rightarrow \frac{C_2 \cdot 7 \cdot (7 \cdot f - C_1)^{10/7}}{24} \cdot R^* + \Delta_g f = \left(\frac{C_2 \cdot 7 \cdot (7 \cdot f - C_1)^{10/7}}{f \cdot 24} \cdot R^* + \Delta_g \right) f$$

$$= \left(\frac{C_2 \cdot 7 \cdot (7 \cdot f - C_1)^{3/7}}{24} \cdot R^* + \Delta_g \right) (7 \cdot f - C_1) = 0. \tag{127}$$

3.3.4.9 The 10-dimensional space

Equation (100) gives the following solution for the function $F[f]$:

$$F[f] = C_2 \cdot (2 \cdot f - C_1)^{1/2} \tag{128}$$

and results in the GTR-Klein–Gordon equation as follows:

$$R^* + \frac{9}{F[f]^2} \cdot \frac{\partial F[f]}{\partial f} \cdot \Delta_g f = 0$$

$$\Rightarrow \frac{C_2 \cdot 2 \cdot (2 \cdot f - C_1)^{3/2}}{9} \cdot R^* + \Delta_g f = \left(\frac{C_2 \cdot 2 \cdot (2 \cdot f - C_1)^{3/2}}{f \cdot 9} \cdot R^* + \Delta_g \right) f$$

$$= \left(\frac{C_2 \cdot 2 \cdot (2 \cdot f - C_1)^{1/2}}{9} \cdot R^* + \Delta_g \right) (2 \cdot f - C_1) = 0. \tag{129}$$

3.3.5 Periodic Space-Time Solutions

Now we are interested in learning how simplest oscillating solutions could be obtained for a variety of settings to R^*. Only assuming Minkowski-like flat spaces the Ricci scalar R of the original metric $g_{\alpha\beta}$ before the transformation would be zero, of course. Picking the most simple example from above, which was found to be the 6-dimensional space-time, we assume a flat space of 6 dimensions with the coordinates $t, x, y, z, u,$ and w. With the following ansatz for the Ricci scalar and the function f in 6 dimensions:

$$R^* = \frac{5}{(f + C_1)} \cdot \sum_{k=1}^{6} R_k^*; \quad f + C_1 = \prod_{k=1}^{6} (f_k [\xi_k] + C_{1k});$$
$$\xi_k = \{t, x, y, z, u, w\} \tag{130}$$

and the trace of the metric $g_{\alpha\beta}$ being $\{1,1,1,1,1,1\}$ (please note that, for simplicity, we ignore the special character of the time coordinate by setting $c = 1$ and not applying any sign convention as it does not matter here), we obtain the following sets of ordinary differential equations from our quantum gravity Klein–Gordon equation Eq. (121) (last line):

$$\left(R_k^* + \frac{\partial^2}{\partial \xi_k^2} \right) (f_k [\xi_k] + C_{1k}) = 0. \tag{131}$$

The subsequent solutions for all these equations in Eq. (131) read

$$f_k [\xi_k] + C_{1k} = C_{k1} \cdot \sin \left[\xi_k \cdot \sqrt{R_k^*} \right] + C_{k2} \cdot \cos \left[\xi_k \cdot \sqrt{R_k^*} \right]. \tag{132}$$

Applying the rule of generalization Eq. (390), which is valid for any arbitrary solution of the Einstein field equations, we can generalize our solutions Eq. (132) to

$$f_k [\xi_k] + C_{1k} = C_{k1} \cdot \sin \left[g_k [\xi_k] \cdot \sqrt{R_k^*} \right] + C_{k2} \cdot \cos \left[g_k [\xi_k] \cdot \sqrt{R_k^*} \right]. \tag{133}$$

Please note that by using this transformation rule as follows:

$$g_{ij} [x_0, x_1, \ldots, x_n] \Rightarrow G_{ij}$$
$$= g'_{x_i} [x_i] \cdot g'_{x_j} [x_j] \cdot g_{ij} [g_{x_0} [x_0], g_{x_1} [x_1], \ldots, g_{x_n} [x_n]]$$

with:

$$g'_{x_i} [x_i] \equiv \frac{\partial g_{x_i} [x_i]}{\partial x_i}, \tag{134}$$

this will only change the Ricci scalar R^* as follows:

$$R^* \lfloor f_k [\xi_k] + C_{1k}] \Rightarrow R^* [f_k [g_k [\xi_k]] + C_{1k}], \tag{135}$$

which means that there is no principle structural change, as the g_k stay intrinsic and the special choice of R^* Eq. (130) in our 6-dimensional space becomes

$$R^* = \frac{5}{\prod_{k=1}^{6} (f_k [g_k [\xi_k]] + C_{1k})} \cdot \sum_{k=1}^{6} R_k^*. \qquad (136)$$

Thus, in order to obtain R^* free of singularities, we chose our constants C_{1k} such that we always have

$$f_k [\xi_k] + C_{1k} \neq 0. \qquad (137)$$

Obviously, there are infinitely many options fulfilling the conditions above. Applying the transformation rule Eq. (85) and using the periodic solutions we have obtained, yields

$$G_{\alpha\beta} = g_{\alpha\beta} \cdot F [f] = g_{\alpha\beta} \cdot (f + C_1)$$

$$= g_{\alpha\beta} \cdot \prod_{k=1}^{6} (f_k [g_k [\xi_k]] + C_{1k})$$

$$= g_{\alpha\beta} \cdot \prod_{k=1}^{6} C_{1k} + g_{\alpha\beta} \cdot \left[\prod_{k=1}^{6} (f_k [g_k [\xi_k]] + C_{1k}) - \prod_{k=1}^{6} C_{1k} \right].$$
$$(138)$$

Normalizing the result by dividing it by $\prod_{k=1}^{6} C_{1k}$ results in the following split up of classical General Theory of Relativity metric $g_{\alpha\beta}$ and its quantized part:

$$\frac{G_{\alpha\beta}}{\prod\limits_{k=1}^{6} C_{1k}} = \overbrace{g_{\alpha\beta}}^{\text{classical GTR}} + g_{\alpha\beta} \cdot \overbrace{\left[\prod_{k=1}^{6} (f_k [g_k [\xi_k]] + C_{1k}) - \prod_{k=1}^{6} C_{1k} \right]}^{\text{Quantum Gravity}}. \qquad (139)$$

Thus, we ended up with a quantized, but still completely Einstein field equation-compatible metric solution in 6 dimensions. In [97] it was demonstrated how similar (periodic) solutions in other numbers of dimensions can be obtained. Here now we intend to investigate a variety of symmetries in connection with our quantum gravity approach. In order not to overload the book chapter, however, we thereby concentrate only on cases in 4 dimensions.

So it can be easily seen that our transformed, that is, quantized, metric $G_{\alpha\beta}$ describes states with matter, which is to say, the corresponding Einstein–Hilbert action—effectively—contains a non-zero Lagrange density term L_M for matter. As our starting point always was a flat space without anything in it, we have to conclude that the quantum transformation $F[f]$ can be used to create matter from Einstein–Hilbert vacua, which is to say, from nothing.

3.3.6 Spherical Coordinates

The case with spatial spherical coordinates was partially investigated in connection with the Schwarzschild metric in [108] and with the Robertson-Walker metric in [109]. Here we shall directly start with the vacuum case:

$$g_{\alpha\beta}^{flat} = \begin{pmatrix} -c^2 & 0 & 0 & 0 \\ 0 & 1 & 0 & 0 \\ 0 & 0 & r^2 & 0 \\ 0 & 0 & 0 & r^2 \cdot \sin^2 \vartheta \end{pmatrix}. \tag{140}$$

For convenience we here want to repeat the case $R^* = 0$, which was investigated in [108]. From Eq. (13) we obtain the simple 4D Laplacian:

$$R^* = -3 \cdot \frac{\left(C_1 + \frac{C_1^2 \cdot f}{2 \cdot C_2}\right)}{\left(C_1 \cdot f + \frac{C_1^2 \cdot f^2}{4 \cdot C_2} + C_2\right)^2} \cdot \Delta_g f = 0$$

$$= \Delta_g f = \Delta_{\text{3D-sphere}} f - \frac{f^{(2,0,0,0)}}{c^2}. \tag{141}$$

Applying the separation approach $f[t, r, \vartheta, \varphi] = g[t] * h[r] * Y[\vartheta, \varphi]$ gives us the solution Eq. (141) for $g[t]$:

$$g[t] = C_1 \cdot \cos[c \cdot C_t \cdot t] + C_2 \cdot \sin[c \cdot C_t \cdot t], \tag{142}$$

and the spherical Bessel functions $j_n[z]$ and $y_n[z]$ for $h[r]$ as already derived in [108], only that this time the solution reads as follows:

$$h[r] = C_j \cdot j_k[C_t \cdot r] + C_y \cdot y_k[C_t \cdot r]; k = \frac{\sqrt{1 + 4\omega^2} - 1}{2}. \tag{143}$$

Solutions, not producing singularities or growing infinitely with respect to r, are only found for the spherical Bessel function of the first kind $j_k[z]$ with the usual $\omega^2 = l * (l + 1)$ and $l = 1, 2, 3, \ldots$. Also, as discussed in [108] (there see subsection "The $r_s = 0$-Case Now Without Feeding from Classical quantum theory") and very different to the Schrödinger hydrogen solution, we have no main quantum number.

This has quite some effect on the possible selection for the quantum numbers l and m for the spherical harmonics $Y_l^m[\vartheta, \varphi]$ (be it for the usual spherical metric or our hyperbolical spheres, which can be given as follows (simply set $r_s = \Lambda = 0$ in order to have the Einstein–Hilbert vacuum situation there, too):

$$
g_{\alpha\beta}^{\text{hyp}} =
\begin{pmatrix}
h \cdot g_t'\,[t]^2 \cdot \alpha\,[r] & 0 & 0 & 0 \\
0 & a \cdot G_r'\,[r]^2\,\alpha\,[r]^{-1} & 0 & 0 \\
0 & 0 & g_{22} & 0 \\
0 & 0 & 0 & g_{33}
\end{pmatrix}
$$

$$
g_{22} = b \cdot g_\vartheta'\,[c_\vartheta\vartheta]^2\,G_r\,[r]^2 \cdot
\begin{Bmatrix} \sinh[g_\vartheta\,[c_\vartheta\vartheta]]^{-2} \\ \cosh[g_\vartheta\,[c_\vartheta\vartheta]]^{-2} \end{Bmatrix}
$$

$$
g_{33} = d \cdot g_\varphi'\,[c_\varphi\varphi]^2 \cdot G_r\,[r]^2 \cdot
\begin{Bmatrix} \sinh[g_\vartheta\,[c_\vartheta\vartheta]]^{-2} \\ \cosh[g_\vartheta\,[c_\vartheta\vartheta]]^{-2} \end{Bmatrix}
$$

$$
\alpha\,[r] = \left(C_1 + \frac{r_s}{G_r\,[r]} + \frac{\Lambda}{3 \cdot c^2} \cdot G_r\,[r]^2 \right) .) \tag{144}
$$

For one thing we have the condition that $-|l-1| \le m \le |l-1|$ and for a second we find that $l = 0$ and $l = 1$ are providing the same trivial result, namely $Y_{l=0,1}^0\,[\vartheta, \varphi] = 1$. Thereby we note that only for $l = 0$ also $m = l$ gives a non-vanishing solution. We also note that, in contrast to the Schwarzschild metric we have considered in [108], it does not matter which spherical harmonics we choose. Both forms:

$$
Y_l^m\,[\vartheta, \varphi] = e^{i \cdot m \cdot \varphi} \cdot \left(C_{j_P} \cdot P_{l_j}^{m_j}\,[\cos\vartheta] + C_{j_Q} \cdot Q_{l_j}^{m_j}\,[\cos\vartheta] \right), \tag{145}
$$

$$
Y_l^m\,[\vartheta, \varphi] = e^{i \cdot m \cdot \varphi} \cdot \left(C_{j_P} \cdot P_{l_j}^{m_j}\,[\tanh\vartheta] + C_{j_Q} \cdot Q_{l_j}^{m_j}\,[\tanh\vartheta] \right), \tag{146}
$$

being the spherical and the spherical-hyperbolical one, respectively, give reasonable distributions with the conditions for the quantum numbers m and l as demanded in the paragraph above. However, if we insist on vanishing values on the poles, we should also exclude the quantum number $m = 0$ (except in the case $l = 0$, where the result is a constant anyway).

With respect to spherical harmonics, it should be noted that classically one learns that l have to be integers, but it will be shown in section "3.4 The Other Hydrogen" that we can also have $l = k/2$ solutions ($k = 1, 3, 5, \ldots$).

It should be noted that the quantum solutions of our transformation are principally additive in the case of $R^* = 0$. It was demonstrated in [109] that, while the Einstein field equations are not additive with respect to the metric, we find that our differential equations for gravity quantum fields Eq. (141) are perfectly linear operations on f. This gives us the option of superposition. Thus, this time applying the separation approach in form of a sum of functions $f = \sum_{j=1}^{\Omega} f_j$, with all $f_j [t, r, \vartheta, \varphi] = g_j[t] * h_j[r] * Y_j[\vartheta, \varphi]$ and solving the angular part in the usual way (cf. [170]) makes Eq. (141) to

$$\Delta_g f = \Delta_f \sum_{j=1}^{\Omega} f_j = 0. \tag{147}$$

Now we simply demand that each of the summands gives zero and we have

$$\Delta_g f = \Delta_g \sum_{j=1}^{\Omega} f_j = \sum_{j=1}^{\Omega} \Delta_g f_j = 0 \Rightarrow \Delta_g f_j = 0. \tag{148}$$

The separation calculation with our form $f = \sum_{j=1}^{\Omega} f_j$ from above has now simply to be performed for each function f_j separately. Subsequently we result in a set of solutions $Y_{jl}{}^m [\vartheta, \varphi]$, $h_j[r]$ and $g_j[t]$ with parameters C_{jt} and ω_j respectively, l_j and m_j as follows:

$$Y_{jl}{}^m [\vartheta, \varphi] = e^{i \cdot m_j \cdot \varphi} \cdot \left(C_{jp} \cdot P_{l_j}^{m_j} [\cos \vartheta] + C_{jQ} \cdot Q_{l_j}^{m_j} [\cos \vartheta] \right), \tag{149}$$

$$Y_{jl}{}^m [\vartheta, \varphi] = e^{i \cdot m_j \cdot \varphi} \cdot \left(C_{jp} \cdot P_{l_j}^{m_j} [\tanh \vartheta] + C_{jQ} \cdot Q_{l_j}^{m_j} [\tanh \vartheta] \right), \tag{150}$$

for the usual spherical metric and the spherical hyperbolical metric (144), respectively, and

$$g_j [t] = C_{j1} \cdot \cos \left[c \cdot C_{jt} \cdot t \right] + C_{j2} \cdot \sin \left[c \cdot C_{jt} \cdot t \right] \tag{151}$$

for the time functions and

$$h_{jl} [r] = C_{jj} \cdot j_{jk} \left[C_{jt} \cdot r \right] + C_{jy} \cdot y_{jk} \left[C_{jt} \cdot r \right] ; k_j = \frac{\sqrt{1 + 4\omega_j^2} - 1}{2} \tag{152}$$

for the radius functions.

An important simplification is obtained in the case $l = 0$, where, as an interesting side-effect of our superposition approach, we find that the radial distribution in the case of $l = 0$ and the limit of the flat space can be made a Dirac delta distribution. From Eq. (152), setting $C_{jj} = 1$ and $C_{jy} = 0$, we derive the limit $l = 0$ as follows:

$$h_{j0} [r] = \frac{\sin \left[C_{jt} \cdot r \right]}{C_{jt} \cdot r}. \tag{153}$$

The following integrals could be seen as superpositions of solutions $h_{j0}[r]$ for certain distributions of C_{jt} according to Eq. (152):

$$h_{\Sigma \infty 0} [r] = \int_{-\infty}^{\infty} \frac{\sin [C_t \cdot r]}{C_t \cdot r} \cdot dC_t = \frac{\pi}{r}$$

$$\tag{154}$$

$$h_{\Sigma 0} [r] = \int_{-A}^{A} \frac{\sin [C_t \cdot r]}{C_t \cdot r} \cdot dC_t - 2 \cdot \frac{\mathrm{Si} [A \cdot r]}{r}; \quad \mathrm{Si} [z] = \int_{0}^{z} \frac{\sin [u]}{u} \cdot du$$

Thereby it needs to be pointed out that the integration should be performed over the whole function $f[t, r, \vartheta, \varphi]$ and with $g[t]$ also depending on C_t this needs to be taken into account. Here, for simplicity, however, we pick a $t = t_0$ for which we would have $g[t_0] = \text{const}$ and concentrate solely on the radial dependency. It requires a skilled mathematician to work out the situation in the general cases, a task we therefore leave to the interested reader. The limits for $r = 0$ at $t = t_0$ are

$$\lim_{r \to 0} h_{\Sigma \infty 0} [r] = \lim_{r \to 0} \int_{-\infty}^{\infty} \frac{\sin [C_t \cdot r]}{C_t \cdot r} \cdot dC_t = \lim_{r \to 0} \frac{\pi}{r} = \infty$$

$$\lim_{r \to 0} h_{\Sigma 0} [r] = \lim_{r \to 0} \int_{-A}^{A} \frac{\sin [C_t \cdot r]}{C_t \cdot r} \cdot dC_t = \lim_{r \to 0} 2 \cdot \frac{\text{Si} [A \cdot r]}{r} = 2 \cdot A \tag{155}$$

The interesting fact here is that the second superposition, one might see as finite, gives finite values at $r = 0$, while the first—infinite—superposition results in a singularity at $r = 0$. This does not only leave some interesting options for the construction of space out of miniscule FLRW-universes with $k = 0$ as suggested in [109]*, but also opens up a completely new possibility to explain gravity in the limit to the Newton cases without resorting to the Schwarzschild singularity. We discussed this in [111] and will come to it again in the chapter about "10 The Origin of Time."

For the reason of generality, we should point out that the superposition above can be extended as follows:

$$h_{\Sigma \infty 0} [r] = \int_{-\infty}^{\infty} C_j [C_t] \cdot \frac{\sin [C_t \cdot r]}{C_t \cdot r} \cdot dC_t;$$

$$h_{\Sigma 0} [r] = \int_{-A}^{A} C_j [C_t] \cdot \frac{\sin [C_t \cdot r]}{C_t \cdot r} \cdot dC_t. \tag{156}$$

Further discussion for our special vacuum quantum gravity solutions to the spatial spherical symmetry is given in [108].

Things are getting more difficult (but also more interesting) in the case of $R^* \neq 0$. From Eq. (141) we then have

$$\frac{R^*}{3} \cdot \frac{\left(C_1 \cdot f + \frac{C_1^2 \cdot f^2}{4 \cdot C_2} + C_2 \right)^2}{C_1 + \frac{C_1^2 \cdot f}{2 \cdot C_2}} + \Delta_g f = 0 \tag{157}$$

---

*…and thereby itself resulting from an idea in a fictive story [1, 2, 3].

and Taylor expending with respect to f (at $f = 0$) we get

$$\xrightarrow{\text{Taylor at } f=0} \frac{R^*}{3} \cdot \left(\frac{C_2^2}{C_1} + \frac{3}{2} \cdot C_2 \cdot f + \dots \right) + \Delta_g f = 0. \tag{158}$$

Now we introduce a simple transformation of the following form:

$$f^* = \frac{C_2^2}{C_1} + \frac{3}{2} \cdot C_2 \cdot f; \quad \Delta_g f^* = \Delta_g f \tag{159}$$

and obtain the ordinary Klein–Gordon equation with

$$\frac{R^*}{3} \cdot f^* + \Delta_g f^* = 0. \tag{160}$$

In [108] we investigated a variety of assumptions regarding the curvature term R^* of which we here shall only represent the basic results.

With the separation approach $f^*[t, r, \vartheta, \varphi] = g[t] * h[r] * Y[\vartheta, \varphi]$ we are able to solve this equation for quite some assumptions regarding R^*. Solving the angular part in the usual way (cf. [170]) gives us

$$0 = R^* + \frac{3}{f^*} \cdot \left(\frac{2 \cdot f^{*(0,1,0,0)} + r \cdot f^{*(0,2,0,0)}}{r} - \frac{\omega^2}{r^2} \cdot f^* - \frac{f^{*(2,0,0,0)}}{c^2} \right);$$

$$R_6 \equiv \frac{R^*}{3}$$

$$0 = \left(R_6 - \frac{\omega^2}{r^2} \right) \cdot f^*[t, r, \vartheta, \varphi]$$

$$+ \left(g[t] \cdot \left(\frac{2}{r} \cdot \frac{\partial h[r]}{\partial r} + \frac{\partial^2 h[r]}{\partial r^2} \right) - h[r] \cdot \frac{\partial^2 g[t]}{c^2 \cdot \partial t^2} \right) \cdot Y[\vartheta, \varphi]. \tag{161}$$

Thereby $Y[\vartheta, \varphi]$ denotes the spherical harmonics, which are usually been given with the quantum numbers l and m and written as $Y_l^m[\vartheta, \varphi]$. The quantum numbers l, m are integers with the conditions $l = 1, 2, 3, \dots$ and $-l \leq m \leq +l$. Division by $f^*[\dots]$ on both sides of Eq. (161) yields

$$0 = \left(R_6 - \frac{\omega^2}{r^2} \right) + \left(\frac{1}{h[r]} \cdot \left(\frac{2}{r} \cdot \frac{\partial h[r]}{\partial r} + \frac{\partial^2 h[r]}{\partial r^2} \right) - \frac{1}{g[t]} \cdot \frac{\partial^2 g[t]}{c^2 \cdot \partial t^2} \right)$$

$$\frac{1}{g[t]} \cdot \frac{\partial^2 g[t]}{c^2 \cdot \partial t^2} = \left(R_6 - \frac{\omega^2}{r^2} \right) + \frac{1}{h[r]} \cdot \left(\frac{2}{r} \cdot \frac{\partial h[r]}{\partial r} + \frac{\partial^2 h[r]}{\partial r^2} \right). \tag{162}$$

Assuming R_6 not be a function of t or the angles, meaning that we just have $R_6 = R_6[r]$, demands the terms on both sides to be a constant. Thus, we can write

$$\frac{1}{g[t]} \cdot \frac{\partial^2 g[t]}{c^2 \cdot \partial t^2} \equiv C_t^2 \equiv \left(R_6 - \frac{\omega^2}{r^2} \right) + \frac{1}{h[r]} \cdot \left(\frac{2}{r} \cdot \frac{\partial h[r]}{\partial r} + \frac{\partial^2 h[r]}{\partial r^2} \right). \tag{163}$$

We shall start with the assumption of R_6 just being a constant. While the solution for $g[t]$ is a simple periodic function

$$g[t] = C_1 \cdot \cos[c \cdot C_t \cdot t] + C_2 \cdot \sin[c \cdot C_t \cdot t], \tag{164}$$

we obtain the spherical Bessel functions $j_n[z]$ and $y_n[z]$ for $h[r]$ as follows:

$$h[r] = C_j \cdot j_k\left[-i \cdot r \cdot \sqrt{-C_t^2 - R_6}\right] + C_y \cdot y_k\left[-i \cdot r \cdot \sqrt{-C_t^2 - R_6}\right];$$

$$k = \frac{\sqrt{1 + 4\omega^2} - 1}{2}. \tag{165}$$

As before, solutions, which are not producing singularities or growing infinitely with respect to r, are only found for the spherical Bessel function of the first kind $j_k[z]$. Please note that in contrast to the Schrödinger hydrogen solution, we have no main quantum number.

We get much closer to the classical Schrödinger hydrogen by choosing more suitable Ricci curvature terms, which automatically bring matter into our quantum gravity picture (cf. section "3.3.3 Intermediate Sum-Up").

However, in order to simplify the comparison with the Schrödinger case, we divide Eqs. (161), (162) by $\frac{12 \cdot M}{\hbar^2}$ and obtain

$$0 = \frac{\hbar^2}{2 \cdot M} \cdot R_6 + \frac{\hbar^2}{2 \cdot f^* \cdot M} \cdot \left(\Delta_{3D-sphere} f^* - \frac{f^{*(2,0,0,0)}}{c^2}\right)$$

$$= R_6^* + \frac{\hbar^2}{2 \cdot f^* \cdot M} \cdot \left(\Delta_{3D-sphere} f^* - \frac{f^{*(2,0,0,0)}}{c^2}\right). \tag{166}$$

A very interesting setting for R_6^* would be as follows:

$$R_6^* = R_6^*[r] = R_{60} + \frac{R_{61}}{r} + \frac{R_{62}}{r^2}. \tag{167}$$

This namely reproduces the Schrödinger hydrogen problem with Eq. (162) being changed to

$$0 = \left(R_{60} + \frac{R_{61}}{r} + \frac{R_{62}}{r^2} - \frac{\omega^2}{r^2}\right) + \frac{\hbar^2}{2 \cdot M}$$

$$\times \left(\frac{1}{h[r]} \cdot \left(\frac{2}{r} \cdot \frac{\partial h[r]}{\partial r} + \frac{\partial^2 h[r]}{\partial r^2}\right) - \frac{1}{g[t]} \cdot \frac{\partial^2 g[t]}{c^2 \cdot \partial t^2}\right)$$

$$\frac{\hbar^2}{2 \cdot M} \cdot \frac{1}{g[t]} \cdot \frac{\partial^2 g[t]}{c^2 \cdot \partial t^2} = \left(R_{60} + \frac{R_{61}}{r} + \frac{R_{62}}{r^2} - \frac{\omega^2}{r^2}\right) + \frac{\hbar^2}{2 \cdot M} \cdot \frac{1}{h[r]}$$

$$\times \left(\frac{2}{r} \cdot \frac{\partial h[r]}{\partial r} + \frac{\partial^2 h[r]}{\partial r^2}\right). \tag{168}$$

Comparison with the Schrödinger derivation (e.g., [170], pp. 155) and assuming that R_{60} is a constant gives us

$$\overbrace{\frac{\hbar^2}{2 \cdot M} \cdot \frac{1}{g\,[t]} \cdot \frac{\partial^2 g\,[t]}{c^2 \cdot \partial t^2}}^{-E_{n-\text{Schrödinger}}} = -R_{60} - C_t^2$$

$$= \left(\overbrace{\frac{R_{61}}{r}}^{-V/r} + \overbrace{\frac{R_{62}}{r^2} - \frac{\omega^2}{r^2}}^{-\omega^2_{\text{Schrödinger}}/r^2} \right) + \frac{\hbar^2}{2 \cdot M} \cdot \frac{1}{h\,[r]} \cdot \left(\frac{2}{r} \cdot \frac{\partial h\,[r]}{\partial r} + \frac{\partial^2 h\,[r]}{\partial r^2} \right). \quad (169)$$

Using the result for $g[t]$ from above, which is to say Eq. (163) and Eq. (164) with the following adaptation:

$$g\,[t] = C_1 \cdot \cos\left[c \cdot C_{tt} \cdot t\right] + C_2 \cdot \sin\left[c \cdot C_{tt} \cdot t\right]; C_{tt} = \frac{\sqrt{2 \cdot M \cdot \left(R_{60} + C_t^2\right)}}{\hbar}, \quad (170)$$

we can now rewrite Eq. (169) as follows:

$$E_{n-\text{Schrödinger}} = R_{60} + C_t^2 = -\left(\overbrace{\frac{R_{61}}{r}}^{-V/r} + \overbrace{\frac{R_{62}}{r^2} - \frac{\omega^2}{r^2}}^{-\omega^2_{\text{Schrödinger}}/r^2} \right) - \frac{\hbar^2}{2 \cdot M} \cdot \frac{1}{h\,[r]}$$

$$\times \left(\frac{2}{r} \cdot \frac{\partial h\,[r]}{\partial r} + \frac{\partial^2 h\,[r]}{\partial r^2} \right) \quad (171)$$

and thus, get the classical radial Schrödinger solution for the radial part of Eq. (166).

By the way, the classical wave function solution to the hydrogen atom can be given as follows [170]:

$$\Psi_{n,l,m}\,[r, \vartheta, \varphi] = e^{i\,m\,\varphi} \cdot P_l^m\,[\cos \vartheta] \cdot R_{n,l}\,[r]$$

$$= \sqrt{\left(\frac{2}{n \cdot a_0}\right)^3 \frac{(n-l-1)!}{2 \cdot n \cdot (n+l)!}} \cdot e^{-\rho/2} \cdot \rho^l \cdot L_{n-l-1}^{2l+1}\,[\rho] \cdot Y_l^m\,[\vartheta, \varphi]; \quad \rho = \frac{2 \cdot r}{n \cdot a_0}. \quad (172)$$

The constant a_0 is denoting the Bohr radius with (m_e = electron rest mass, ε_0 = permittivity of free space, Q_e = elementary charge):

$$a_0 = \frac{4 \cdot \pi \cdot \varepsilon_0 \cdot \hbar^2}{m_e \cdot Q_e^2} = 5.292 \cdot 10^{-11} \text{ meter.} \quad (173)$$

The functions P, L, and Y denote the associated Legendre function, the Laguerre polynomials and the spherical harmonics, respectively. The wave function Eq. (406) directly solves our flat space limit Eq. (166) for static cases where f does not depend on t. The extension to time-dependent f^* according to Eq. (166) thereby is simple. By using the results from above, we obtain

$$f^*_{n,l,m} [t, r, \vartheta, \varphi] = g[t] \cdot \Psi_{n,l,m} [r, \vartheta, \varphi] = g[t] \cdot e^{i \cdot m \cdot \varphi} \cdot P_l^m [\cos \vartheta] \cdot R_{n,l} [r]$$

$$= (C_1 \cdot \cos[c \cdot C_{tt} \cdot t] + C_2 \cdot \sin[c \cdot C_{tt} \cdot t]) \cdot N \cdot e^{-\rho/2} \cdot \rho^l$$

$$\times L_{n-l-1}^{2l+1} [\rho] \cdot Y_l^m [\vartheta, \varphi]$$

$$\rho = \frac{2 \cdot r}{n \cdot a_{00}}; \quad N = \sqrt{\left(\frac{2}{n \cdot a_0}\right)^3 \frac{(n-l-1)!}{2 \cdot n \cdot (n+l)!}}; \quad C_{tt} = \frac{\sqrt{2 \cdot M \cdot (R_{60} + C_t^2)}}{\hbar}$$

$$C_1 = \pm C_2 = 1; \quad a_{00} = \frac{\hbar^2}{M} \cdot \frac{1}{R_{61}}; \quad E_n = -[R_{60} + C_t^2]_n = -\frac{R_{61}^2}{n^2} \cdot \frac{M}{2 \cdot \hbar^2};$$

$$\omega^2 - R_{62} = l^2 + l$$

$$\Rightarrow f_{n,l,m} [t, r, \vartheta, \varphi] = e^{\pm i \cdot c \cdot C_{tt} \cdot t} \cdot N \cdot e^{-\rho/2} \cdot \rho^l \cdot L_{n-l-1}^{2l+1} [\rho] \cdot Y_l^m [\vartheta, \varphi]. \quad (174)$$

Regarding the conditions for the quantum numbers n, l, and m, naturally, we have the usual

$$\{n, l, m\} \in \mathbb{Z}; \quad n \geqslant 0; \quad l < n; \quad -l \leqslant m \leqslant +l. \quad (175)$$

However, in section "3.4 The Other Hydrogen" we will learn that there are also half-l solutions.

Assuming an omnipresent constant curvature R_{60} does not seem to make much sense in our current system and thus, we should set $R_{60} = 0$. Doing the same with the curvature parameter R_{62} just gives us the usual condition for the spherical harmonics, namely $\omega^2 = l \cdot (l + 1) = l^2 + l$. Thus, we have metrically derived some kind of a "hydrogen atom". In addition to the Schrödinger structure, our form also sports a time-dependent factor clearly showing the options for matter and antimatter via the $\pm$-sign in the $g[t]$ function.

Please note that according to [67] we should not expect to have modelled the true hydrogen atom here. What we have considered was a four-dimensional space with a three-dimensional central symmetry and not a true two-body problem. According to [67], a fully described problem consisting of two three-dimensional objects would require 6 spatial dimensions plus time. The reason for this lies in the fact that in the metric theory not the apparent spatial dimensions are the dimensions for the metric, but that in the latter we have to consider all degrees of freedom of the system we intent to investigate. Thus, the true metric hydrogen atom would require 7 dimensions instead of 4. Things are getting simpler, of course, if one could assume the objects as point-like and without inner properties. It also helps a lot that the electron is so much lighter than the proton.

We will see later in this book in connection with an extended variation of the Einstein–Hilbert action that the complete quantum gravity solutions for the "Hydrogen Problem" are always containing (or rather creating) a potential with a repulsive part very near the center. This will be shown in section "8 Anti-gravity."

We therefore conclude that the metric transformation Eq. (85) applied on a spherically symmetric flat space given via Eq. (140) produces energy-states which are similar to the ones of the Schrödinger hydrogen problem if we assume certain (rather simple) r-dependencies Eq. (167) for the Ricci-like curvature R^*. Thereby the connection to the Einstein field equations with matter Eq. (83) and the corresponding Einstein–Hilbert action Eq. (112) has been given as follows:

$$\frac{\hbar^2}{12 \cdot M} \cdot F\,[f] \cdot R^* = \frac{\hbar^2}{2 \cdot M} \cdot \frac{R^{**}}{6} = \frac{\hbar^2}{2 \cdot M} \cdot R_6 = R_6^* = R_6^*\,[r] = R_{60} + \frac{R_{61}}{r} + \frac{R_{62}}{r^2}. \tag{176}$$

Now we can also give the metric of the hydrogen atom (matter and antimatter case).

Applying Eq. (97) and combining it with Eq. (159) yields the normalized quantum gravity metric including our—metric—wave function f^*:

$$G_{\alpha\beta}^{\mathrm{norm}} = \frac{G_{\alpha\beta}}{C_2} = \overbrace{g_{\alpha\beta}}^{\text{class. GTR}} + g_{\alpha\beta} \cdot \overbrace{\left(\frac{C_1}{C_2} \cdot f \cdot \left(1 + \frac{C_1 \cdot f}{4 \cdot C_2} \right) \right)}^{\substack{\text{quantum gravity in 4D} \\ \text{quantum metric}}}$$

$$= \overbrace{g_{\alpha\beta}}^{\text{class. GTR}} + \frac{g_{\alpha\beta}}{9} \cdot \overbrace{\left(\frac{C_1 \cdot f^* \cdot \left(4 \cdot C_2^2 + C_1 \cdot f^* \right)}{C_2^4} - 5 \right)}^{\substack{\text{quantum gravity in 4D} \\ \text{quantum metric}}}. \tag{177}$$

Please note that no quantum equation was postulated or constructed. Instead, we directly derived our "quantum hydrogen" solution from the scalar Ricci curvature of a transformed metric.

3.3.7 Cartesian Coordinates

Similar to the 6-dimensional case which we considered in the section "3.3.5 Periodic Space-Time Solutions," we can also construct a periodic solution in 4 dimensions.

3.3.7.1 A somewhat more general case

Thereby we assume a flat space of 4 dimensions with the coordinates t, x, y, z. The equation we have to solve would be Eq. (157). As in the section before, we assume f to be small against C_2 and Taylor expansion gives us the simple Eq. (160). With the following ansatz for the Ricci scalar and the function f in 4 dimensions:

$$\frac{R^*}{3} = \sum\nolimits_{k=1}^{4} R_k^*; \quad f^* = \prod\nolimits_{k=1}^{4} f_k\,[\xi_k]; \quad \xi_k = \{t, x, y, z\}, \tag{178}$$

and the trace of the metric $g_{\alpha\beta}$ being $\{-c^2, 1, 1, 1\}$, we obtain the following sets of ordinary differential equations from our quantum gravity Klein–Gordon equation (121) (last line):

$$\left(R_1^* - \frac{\partial^2}{c^2 \cdot \partial\xi_1^2}\right) f_1\,[\xi_1] = 0; \quad \left(R_k^* + \frac{\partial^2}{\partial\xi_k^2}\right) f_k\,[\xi_k] = 0 \,\forall\, k > 1. \tag{179}$$

The subsequent solutions for all these equations in (131) read

$$k = 1: \quad f_k\,[\xi_k] = C_{k1} \cdot \sin\left[\xi_k \cdot c \cdot \sqrt{-R_k^*}\right] + C_{k2} \cdot \cos\left[\xi_k \cdot c \cdot \sqrt{-R_k^*}\right]$$

$$k > 1: \quad f_k\,[\xi_k] = C_{k1} \cdot \sin\left[\xi_k \cdot \sqrt{R_k^*}\right] + C_{k2} \cdot \cos\left[\xi_k \cdot \sqrt{R_k^*}\right]. \tag{180}$$

Applying the rule of generalization Eq. (390), which is valid for any arbitrary solution of the Einstein field equations, we can generalize our solutions Eq. (132) to

$$k = 1: \quad f_k\,[\xi_k] = C_{k1} \cdot \sin\left[g_k\,[\xi_k] \cdot c \cdot \sqrt{-R_k^*}\right] + C_{k2} \cdot \cos\left[g_k\,[\xi_k] \cdot c \cdot \sqrt{-R_k^*}\right]$$

$$k > 1: \quad f_k\,[\xi_k] = C_{k1} \cdot \sin\left[g_k\,[\xi_k] \cdot \sqrt{R_k^*}\right] + C_{k2} \cdot \cos\left[g_k\,[\xi_k] \cdot \sqrt{R_k^*}\right]. \tag{181}$$

Please note that by using this transformation rule as follows:

$$g_{ij}\,[x_0, x_1, \ldots, x_n] \Rightarrow G_{ij} = g'_{x_i}\,[x_i] \cdot g'_{x_j}\,[x_j] \cdot g_{ij}\,[g_{x_0}\,[x_0], g_{x_1}\,[x_1], \ldots, g_{x_n}\,[x_n]]$$

$$\text{with:} \quad g'_{x_i}\,[x_i] \equiv \frac{\partial g_{x_i}\,[x_i]}{\partial x_i}, \tag{182}$$

will only change the Ricci scalar R^* as follows:

$$R^*\,[f_k\,[\xi_k]] \Rightarrow R^*\,[f_k\,[g_k\,[\xi_k]]], \tag{183}$$

which means that there is no principle structural change, as the g_k stay intrinsic and the special choice of R^* Eq. (130) is maintained.

Obviously, there are infinitely many options fulfilling the conditions above. One might consider these solutions as quantum gravity waves. Applying the transformation rule Eq. (85) and using the period solutions we have obtained, yields

$$
G_{\alpha\beta}^{\text{norm}} = \frac{G_{\alpha\beta}}{C_2} = \overbrace{g_{\alpha\beta}}^{\text{class. GTR}} + \overbrace{g_{\alpha\beta} \cdot \left(\frac{C_1}{C_2} \cdot f \cdot \left(1 + \frac{C_1 \cdot f}{4 \cdot C_2} \right) \right)}^{\overbrace{\text{quantum metric}}^{\text{quantum gravity in 4D}}}
$$

$$
= \overbrace{g_{\alpha\beta}}^{\text{class. GTR}} + g_{\alpha\beta} \cdot \overbrace{\left(\frac{C_1 \cdot f^* \cdot (4 \cdot C_2^2 + C_1 \cdot f^*)}{C_2^4} - 5 \right)}^{\overbrace{\text{quantum metric}}^{\text{quantum gravity in 4D}}}
$$

$$
= \overbrace{g_{\alpha\beta}}^{\text{class. GTR}} + \frac{g_{\alpha\beta}}{9} \cdot \overbrace{\left(\frac{C_1 \cdot \prod_{k=1}^{4} f_k \, [\xi_k] \cdot \left(4 \cdot C_2^2 + C_1 \cdot \prod_{k=1}^{4} f_k \, [\xi_k] \right)}{C_2^4} - 5 \right)}^{\overbrace{\text{quantum metric}}^{\text{quantum gravity in 4D}}}.
$$

$$(184)$$

3.3.7.2 The 1D harmonic oscillator

More or less for completeness we pick our general approach form the last subsection again, but this time demand that the various R_k^* are as follows:

$$R_1^* = \text{const} = 0; \; R_2^* = -C_x \cdot x^2; \; R_3^* = R_4^* = 0. \qquad (185)$$

The subsequent differential equation would read

$$
-\frac{\partial^2 f_1 \, [\xi_1]}{c^2 \cdot f_1 \cdot \partial \xi_1^2} + \left(\frac{\partial^2 f_2 \, [\xi_2]}{f_2 \cdot \partial \xi_2^2} - C_x \cdot x^2 \right) + \left(\frac{\partial^2 f_3 \, [\xi_3]}{f_3 \cdot \partial \xi_3^2} + \frac{\partial^2 f_4 \, [\xi_4]}{f_4 \cdot \partial \xi_4^2} \right) = 0
$$

$$
\Rightarrow \frac{\partial^2 f_1 \, [\xi_1]}{c^2 \cdot f_1 \cdot \partial \xi_1^2} = -C_t^2 = \left(\frac{\partial^2 f_2 \, [\xi_2]}{f_2 \cdot \partial \xi_2^2} - C_x \cdot x^2 \right) + \left(\frac{\partial^2 f_3 \, [\xi_3]}{f_3 \cdot \partial \xi_3^2} + \frac{\partial^2 f_4 \, [\xi_4]}{f_4 \cdot \partial \xi_4^2} \right)
$$

$$
= 0
$$

$$\Rightarrow \frac{\partial^2 f_1 [\xi_1]}{c^2 \cdot f_1 \cdot \partial \xi_1^2} = -C_t^2 = \frac{\partial^2 f_2 [\xi_2]}{f_2 \cdot \partial \xi_2^2} - C_x \cdot x^2 + C_{yz}^2 = 0$$

with:
$$\left(\frac{\partial^2 f_3 [\xi_3]}{f_3 \cdot \partial \xi_3^2} + \frac{\partial^2 f_4 [\xi_4]}{f_4 \cdot \partial \xi_4^2} \right) - C_{yz}^2 = 0. \qquad (186)$$

The solution for f_1 was already obtained above and reads

$$f_1 [t] = C_{11} \cdot \cos [c \cdot C_t \cdot t] + C_{12} \cdot \sin [c \cdot C_t \cdot t]. \qquad (187)$$

The third line in Eq. (186) gives us the differential equations for $x = \xi_2$ with

$$\Rightarrow \frac{\partial^2 f_2 [\xi_2]}{f_2 \cdot \partial \xi_2^2} - C_x \cdot x^2 + C_{yz}^2 + C_t^2 = 0, \qquad (188)$$

where we immediately recognize the equation for the one-dimensional harmonic quantum oscillator. The difference to the classical quantum theoretical considerations (e.g., [174]) is to be found in the fact, that this time no quantum equations had to be postulated and everything resulted from the Einstein field equations plus a simple transformation Eq. (85) and subsequent calculation of the corresponding Ricci scalar. It needs to be emphasized that, just as with the classical harmonic oscillator, the x^2 term can only be the approximation of a local potential for a minimum. Outside of this minimum, the infinite growth of the x^2 term has no place anymore. Then one might just see that the minimum in fact was the one of a periodic function (for instance) and the x^2 only approximated its center part. In the metric picture, x^2 would then approximate the minimum of a local—quite symmetric—arch along x.

Only for convenience and a bit of later discussion in connection with the purely metric origin of our oscillator, we here repeat the classical solution.

With the classical solutions of the harmonic oscillator known to be [174]*

$$\psi_n (x) := \left(\frac{m \cdot \omega}{\pi \cdot \hbar} \right)^{\frac{1}{4}} \frac{1}{\sqrt{2^n n!}} H_n [w] * e^{\left[\frac{-w^2}{2} \right]}; w = x \cdot \sqrt{\frac{m \cdot \omega}{\hbar}} \qquad (189)$$

and the corresponding—classical—Schrödinger equation:

$$\Rightarrow \left(\frac{m \cdot \omega^2}{2} \cdot x^2 - \frac{\hbar^2 \cdot \partial^2 \psi [x]}{2 \cdot m \cdot \psi [x] \cdot \partial x^2} \right) = E_n = \omega \cdot \hbar \cdot \left(n + \frac{1}{2} \right) = 0;$$
$$n = 1, 2, 3, \ldots, \qquad (190)$$

we are able to construct the connection to the metric picture.

---

*with $H_n(z)$ denoting the Hermite polynomial of order n and argument z.

Comparison of Eq. (190) with Eq. (188) yields

$$\Rightarrow C_t^2 + C_{yz}^2 = \frac{2 \cdot m \cdot E_n}{\hbar^2} = \frac{2 \cdot m \cdot \omega}{\hbar} \cdot \left(n + \frac{1}{2}\right); \quad C_x = \frac{m^2 \cdot \omega^2}{\hbar^2}. \quad (191)$$

Not surprisingly, we find the energy states being connected with the separation constants, especially the one from the separation of the time-dependency. It is also understandable that the curvature parameter C_x is connected with "spring-constant" of the classical oscillator (cf. [174]). Thus, the x-part solution for our metric 1D quantum gravity operator reads

$$f_{2n}[x] = \left(\frac{1}{\pi} \cdot \frac{C_t^2 + C_{yz}^2}{2 \cdot n + 1}\right)^{\frac{1}{4}} \frac{1}{\sqrt{2^n n!}} H_n[w] \cdot e^{\left[\frac{-w^2}{2}\right]}; \quad w = x \cdot \sqrt{\frac{C_t^2 + C_{yz}^2}{2 \cdot n + 1}}. \quad (192)$$

Thereby we gave C_t and C_{yz} the role of absolute constants. Alternatively and probably also more reasonably one better assumes them to be quantized parameters, equivalent to the classical energy E_n in Eq. (190). Then we should write Eq. (191) as follows:

$$\Rightarrow C_{tn}^2 + C_{yz}^2 = \frac{2 \cdot m \cdot E_n}{\hbar^2} = \frac{2 \cdot m \cdot \omega}{\hbar} \cdot \left(n + \frac{1}{2}\right) \equiv C_{tyz0}^2 \cdot (2 \cdot n + 1), \quad (193)$$

where we might like to assume only the C_t to be quantized, while C_{yz} shall truly be an absolute constant. The resulting solution then reads

$$f_{2n}[x] = \left(\frac{C_{tyz0}^2}{\pi}\right)^{\frac{1}{4}} \frac{1}{\sqrt{2^n n!}} H_n[w] \cdot e^{\left[\frac{-w^2}{2}\right]}; \quad w = x \cdot \sqrt{C^2_{tyz0}}. \quad (194)$$

A bit discussion is needed in connection with the parameter C_{yz} as this one does not appear in the classical theory. In the metric picture, however, the parameter can be split up into C_y and C_z, solving the last line in Eq. (186) as follows:

$$\left(\frac{\partial^2 f_3[\xi_3]}{f_3 \cdot \partial \xi_3^2} + \frac{\partial^2 f_4[\xi_4]}{f_4 \cdot \partial \xi_4^2}\right) - C_{yz}^2 = \overbrace{\frac{\partial^2 f_3[\xi_3]}{f_3 \cdot \partial \xi_3^2} - C_y^2}^{=0} + \overbrace{\frac{\partial^2 f_4[\xi_4]}{f_4 \cdot \partial \xi_4^2} - C_z^2}^{=0} = 0$$

$$\Rightarrow f_3[t] = C_{31} \cdot \exp\left[-C_y \cdot y\right] + C_{32} \cdot \exp\left[C_y \cdot y\right]$$

$$\Rightarrow f_4[t] = C_{41} \cdot \exp\left[-C_z \cdot z\right] + C_{42} \cdot \exp\left[C_z \cdot z\right] \quad (195)$$

and thus, allowing us to always assume an exponentially vanishing of the harmonic excitations with the distance y-z from the oscillator.

3.3.7.3 The total vacuum case

Now we assume $R^* = 0$ again and solve the subsequent equation resulting from Eq. (157) as a simple:

$$\frac{R^*}{3} \cdot \frac{\left(C_1 \cdot f + \frac{C_1^2 \cdot f^2}{4 \cdot C_2} + C_2\right)^2}{C_1 + \frac{C_1^2 \cdot f}{2 \cdot C_2}} + \Delta_g f = 0 \xrightarrow{R^*=0} \Delta_g f = 0. \tag{196}$$

Applying the separation approach as before:

$$f = \prod_{k=1}^{4} f_k [\xi_k]; \quad \xi_k = \{t, x, y, z\} \tag{197}$$

this time results in

$$\frac{1}{f} \left(\sum_{k=2}^{4} \frac{\partial^2}{\partial \xi_k^2} - \frac{\partial^2}{c^2 \cdot \partial \xi_1^2}\right) f = \sum_{k=2}^{4} \frac{\partial^2 f_k}{f_k \cdot \partial \xi_k^2} - \frac{\partial^2 f_1}{f_1 \cdot c^2 \cdot \partial \xi_1^2} = 0$$

$$\Rightarrow \sum_{k=2}^{4} \frac{\partial^2 f_k}{f_k \cdot \partial \xi_k^2} = -C_t^2 = \frac{\partial^2 f_1}{f_1 \cdot c^2 \cdot \partial \xi_1^2}. \tag{198}$$

As before the time function $f_1 = f_1[\xi_1] = f_1[t]$ gives

$$f_1[t] = C_{11} \cdot \cos[c \cdot C_t \cdot t] + C_{12} \cdot \sin[c \cdot C_t \cdot t], \tag{199}$$

while for convenience, we split up the constant C_t with respect to the spatial coordinates, into

$$-C_t^2 = -C_x^2 - C_y^2 - C_z^2 = -C_2^2 - C_3^2 - C_4^2 \tag{200}$$

and allocate as follows:

$$\frac{\partial^2 f_k}{f_k \cdot \partial \xi_k^2} = -C_k^2. \tag{201}$$

This gives the solutions

$$f_k[\xi_k] = C_{k1} \cdot \cos[C_k \cdot \xi_k] + C_{k2} \cdot \sin[C_k \cdot \xi_k]. \tag{202}$$

3.3.8 Cylindrical Coordinates

In cylindrical coordinates with the metric:

$$g_{\alpha\beta}^{\text{flat}} = \begin{pmatrix} -c^2 & 0 & 0 & 0 \\ 0 & 1 & 0 & 0 \\ 0 & 0 & 1 & 0 \\ 0 & 0 & 0 & r^2 \end{pmatrix}; \quad \xi_k = \{t, x, r, \varphi\}, \tag{203}$$

we will only concentrate on the case $R^* = 0$. As this leads to the simple 4D Laplacian:

$$\frac{R^*}{3} \cdot \frac{\left(C_1 \cdot f + \frac{C_1^2 \cdot f^2}{4 \cdot C_2} + C_2\right)^2}{C_1 + \frac{C_1^2 \cdot f}{2 \cdot C_2}} + \Delta_g f = 0 \xrightarrow{R^*=0} \Delta_g f = 0 \tag{204}$$

$$\Rightarrow \Delta_{3D-\text{Cylindrical}} f - \frac{\partial^2 f}{c^2 \cdot \partial t^2} = 0$$

again and gives us, apart from the time-derivative, just the well-known cylindrical Laplacian in 3D, we will not show the evaluation in detail but only present the result. A solution can be found via a separation approach $f[t, x, r, \varphi] = g[t] * f_x[x] * h[r] * f_\varphi[\varphi]$ with:

$$g[t] = C_1 \cdot \cos[c \cdot C_t \cdot t] + C_2 \cdot \sin[c \cdot C_t \cdot t], \tag{205}$$

$$f_x[x] = C_{x1} \cdot e^{x \cdot i \cdot \sqrt{C_t^2 + \omega_{r\varphi}^2}} + C_{x2} \cdot e^{-x \cdot i \cdot \sqrt{C_t^2 + \omega_{r\varphi}^2}}, \tag{206}$$

$$f_\varphi[\varphi] = C_{\varphi 1} \cdot \cos[\omega_\varphi \cdot \varphi] + C_{\varphi 2} \cdot \sin[\omega_\varphi \cdot \varphi], \tag{207}$$

$$h[r] = C_J \cdot J_k[\omega_{r\varphi} \cdot r] + C_Y \cdot Y_k[\omega_{r\varphi} \cdot r] \; ; k = \omega_\varphi. \tag{208}$$

Thereby we have the Bessel functions $J_n[z]$ and $Y_n[z]$ as well as the separation parameters C_t, ω_φ, $\omega_{r\varphi}$. We immediately realize that we have derived a rather fundamental photonic solution. As due to the linearity of Eq. (204) (last line) our solution Eq. (205) to Eq. (208) is additive (cf. text between Eq. (146) and Eq. (155)), many photon-forms could be constructed via simple superposition.

Again, we see that, out of pure Einstein–Hilbert vacuum states, we have been able to obtain states of photonic energy and thus, matter.

3.3.9 Schwarzschild Metric in Its Quantum Transformed Form

Now we want to apply our findings on the—apparently—simple example of the Schwarzschild metric [142].

For convenience and as this is such an important solution to the Einstein field equations, we will here repeat the most important equations we are going to use in this section:

$$R^* + C_{N2} \cdot \frac{1}{F[f]^2} \cdot \frac{\partial F[f]}{\partial f} \cdot \Delta_g f = 0 \Rightarrow \left(\frac{R^* \cdot F[f]^2}{C_{N2} \cdot f \cdot \frac{\partial F[f]}{\partial f}} + \Delta_g\right) f = 0. \tag{209}$$

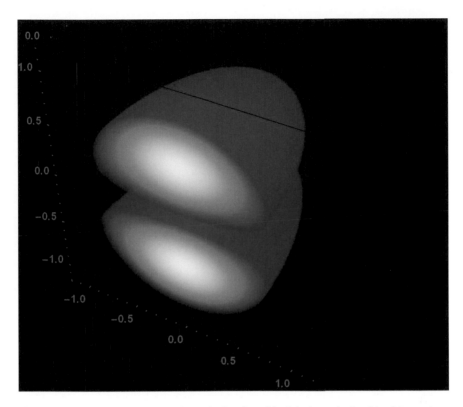

Figure 3.1 Inner quantum gravity solution for a black hole as derived in this section.

$$C_{N1} \cdot \left(\frac{\partial F \, [f]}{\partial f} \right)^2 - C_{N2} \cdot F \, [f] \cdot \frac{\partial^2 F \, [f]}{\partial f^2} = 0. \tag{210}$$

$$R = \overbrace{F \, [f] \cdot R^*}^{R_{\text{matter}}} - \overbrace{\left(\begin{array}{c} -\Gamma^{\mu}_{\sigma\alpha} \Gamma^{**\sigma}_{\beta\mu} + \Gamma^{\sigma}_{\alpha\beta} \Gamma^{**\mu}_{\sigma\mu} - \Gamma^{**\mu}_{\sigma\alpha} \Gamma^{\sigma}_{\beta\mu} + \Gamma^{**\sigma}_{\alpha\beta} \Gamma^{\mu}_{\sigma\mu} \\ + \Gamma^{**\sigma}_{\alpha\beta\,,\sigma} - \Gamma^{**\sigma}_{\beta\sigma\,,\sigma} - \Gamma^{**\mu}_{\sigma\alpha} \Gamma^{**\sigma}_{\beta\mu} + \Gamma^{**\sigma}_{\alpha\beta} \Gamma^{**\mu}_{\sigma\mu} \end{array} \right)}^{2 \cdot \kappa \cdot L_M + 2 \cdot \Lambda} g^{\alpha\beta}$$

$$\equiv R_{\text{matter}} - 2 \cdot \kappa \cdot L_M - 2 \cdot \Lambda. \tag{211}$$

The simplicity regarding the Schwarzschild metric holds because we have $R = 0$ (as in the flat space, too). Now using Eq. (93) plus moving to four dimensions we can write Eq. (209) also as:

$$\frac{R^{**}}{F \, [f]} = R^* \Rightarrow R^{**} = -C_{N2} \cdot \frac{1}{F \, [f]} \cdot \frac{\partial F \, [f]}{\partial f} \cdot \Delta_g f \xrightarrow{4D} \left[\frac{R^{**}}{6} + \Delta_g \right] f = 0, \tag{212}$$

where we immediately see the proximity to the classical quantum Klein–Gordon equation again. Of course this is just a trivial identity* but assuming that one would have some physical knowledge about the value of R^{**} it would become non-trivial and could be applied for the determination of the function f. Here we intend to consider the matter-free case with the condition $\Delta R = 0$, leading to equation Eq. (106).

Knowing that the Ricci scalar of the Schwarzschild solution is $R = 0$ and applying the simplification rule Eq. (210), we have from Eq. (209) (cf. Eq. (212)):

$$F[f] \cdot R^* = R^{**} = -\frac{6}{f} \cdot \Delta_g f = -\frac{6}{f} \cdot \Delta_{SS} f$$

$$= -\frac{6}{f} \cdot \left(\Delta_{\text{sphere}} f - r_s \cdot \left(\frac{f^{(0,1,0,0)} + r \cdot f^{(0,2,0,0)}}{r^2} + \frac{f^{(2,0,0,0)}}{c^2 \, (r - r_s)} \right) \right)$$

$$= -\frac{6}{f} \cdot \left(\Delta_{\text{3D-sphere}} f - \frac{f^{(2,0,0,0)}}{c^2} \right.$$

$$\left. - r_s \cdot \left(\frac{f^{(0,1,0,0)} + r \cdot f^{(0,2,0,0)}}{r^2} + \frac{f^{(2,0,0,0)}}{c^2 \, (r - r_s)} \right) \right)$$

$$f^{(0,1,0,0)} = \frac{\partial f \, [t, r, \vartheta, \varphi]}{\partial r}; \quad f^{(0,2,0,0)} = \frac{\partial^2 f \, [t, r, \vartheta, \varphi]}{\partial r^2};$$

$$f^{(2,0,0,0)} = \frac{\partial^2 f \, [t, r, \vartheta, \varphi]}{\partial t^2}. \tag{213}$$

Thereby we denoted the Laplace operator in spherical polar coordinates with $\Delta_{\text{3D-sphere}}$. It is connected with the Laplace operator in spherical Minkowski coordinates as follows:

$$\Delta_{\text{sphere}} f = \Delta_{\text{3D-sphere}} f - \frac{f^{(2,0,0,0)}}{c^2}. \tag{214}$$

Please note that the functional wrapper $F[f]$ reading:

$$G_{\alpha\beta} = g_{\alpha\beta} \cdot F[f] = g_{\alpha\beta} \cdot (C_1 + f)^2 = g_{\alpha\beta} \cdot (C_1 + f \, [t, r, \vartheta, \varphi])^2 \tag{215}$$

fulfills the necessary condition Eq. (210) in 4 dimensions and was applied in Eq. (213).

---

*Please prove by setting in the corresponding expressions from Eq. (93) and evaluating for arbitrary flat metrics!

In order to construct the complete Schwarzschild Laplacian, we took the Schwarzschild metric from [142] as follows:

$$g_{\alpha\beta}^{\text{Schwarzschild}} = \begin{pmatrix} -c^2\left(1 - \frac{2\cdot G\cdot M}{c^2\cdot r}\right) & 0 & 0 & 0 \\ 0 & \left(1 - \frac{2\cdot G\cdot M}{c^2\cdot r}\right)^{-1} & 0 & 0 \\ 0 & 0 & r^2 & 0 \\ 0 & 0 & 0 & r^2\cdot\sin^2\vartheta \end{pmatrix}. \quad (216)$$

Here r gives the spherical radius, M stands for the gravitational mass of the Schwarzschild object and the constants c and G are giving the speed of light in vacuum and the gravitational or Newton constant, respectively. One easily sees that the solution above has two singularities, namely one at $r = r_s = \frac{2\cdot M\cdot G}{c^2}$, which is a result of our choice of coordinates and could be transformed away, and at $r = 0$. The parameter r_s is called the Schwarzschild radius.

Taking the result from Eq. (211) and inserting it in Eq. (213), we would obtain the following Lagrange density L_M:

$$R = 0 = \overbrace{F\,[f]\cdot R^*}^{R_{\text{matter}}} - \overbrace{\begin{pmatrix} -\Gamma_{\sigma\alpha}^{\mu}\Gamma_{\beta\mu}^{**\sigma} + \Gamma_{\alpha\beta}^{\sigma}\Gamma_{\sigma\mu}^{**\mu} - \Gamma_{\sigma\alpha}^{**\mu}\Gamma_{\beta\mu}^{\sigma} + \Gamma_{\alpha\beta}^{**\sigma}\Gamma_{\sigma\mu}^{\mu} \\ +\Gamma_{\alpha\beta,\sigma}^{**\sigma} - \Gamma_{\beta\sigma,\alpha}^{**\sigma} - \Gamma_{\sigma\alpha}^{**\mu}\Gamma_{\beta\mu}^{**\sigma} + \Gamma_{\alpha\beta}^{**\sigma}\Gamma_{\sigma\mu}^{**\mu} \end{pmatrix}}^{2\cdot\kappa\cdot L_M + 2\cdot\Lambda\,(here:\,\Lambda=0)}\,g^{\alpha\beta}$$

$$\equiv R_{\text{matter}} - 2\cdot\kappa\cdot L_M - 2\cdot\Lambda$$

$$\Rightarrow = R^{**} + \overbrace{\frac{6}{f}\cdot\left(\triangle_{\text{sphere}}f - r_s\cdot\left(\frac{f^{(0,1,0,0)} + r\cdot f^{(0,2,0,0)}}{r^2} + \frac{f^{(2,0,0,0)}}{c^2\,(r - r_s)}\right)\right)}^{2\cdot\kappa\cdot L_M}$$

$$\Rightarrow L_M = \frac{3}{\kappa\cdot f}\cdot\left(\triangle_{\text{sphere}}f - r_s\cdot\left(\frac{f^{(0,1,0,0)} + r\cdot f^{(0,2,0,0)}}{r^2} + \frac{f^{(2,0,0,0)}}{c^2\,(r - r_s)}\right)\right). \quad (217)$$

The metric for the Laplace operator $\triangle_{\text{sphere}}$ as used above can be obtained from Eq. (366) simply by setting $M = 0$. Setting $r_s = 0$ in Eq. (213) gives us the Klein–Gordon equation for the central symmetric problem:

$$0 = \left(\frac{R^{**}}{6} + \triangle_{\text{sphere}}\right) f. \quad (218)$$

This becomes the classical quantum mechanical central field problem for time-independent f and R^{**} chosen as follows (E_n denotes the energy eigenvalues)*:

$$\frac{R^{**}}{6} = \frac{M^2 c^2}{\hbar^2} + V - E_n. \quad (219)$$

---

*Here we have the reduced Planck constant h, the rest mass M and the speed of light in vacuum c. V denotes the classical potential (cf. classical text books like [170]).

Inserting Eq. (219) into Eq. (218), thereby assuming $f = f[r, \vartheta, \varphi]$ (but not t), directly gives us the Schrödinger hydrogen equation [170] (or appendix of this book):

$$0 = \left(\frac{R^{**}}{6} + \Delta_{\text{sphere}} \right) f \xrightarrow{f_t = 0} 0 = \left(\frac{M^2 c^2}{\hbar^2} + \frac{V_r}{r} - E_n + \Delta_{\text{3D-sphere}} \right) f. \tag{220}$$

As a kind of add-on bonbon we can also evaluate the corresponding Einstein–Hilbert Lagrange density for the "Schrödinger hydrogen", being:

$$\Rightarrow L_{\text{M}} = \frac{3}{\kappa \cdot f} \cdot \Delta_{\text{sphere}} f, \tag{221}$$

which is obtained from Eq. (217) by setting $r_s = 0$. In the case of time-independent f, we exactly reproduce the Schrödinger H-atom problem and could write:

$$\Rightarrow L_{\text{M}} = \frac{3}{\kappa \cdot f} \cdot \Delta_{\text{3D-sphere}} f. \tag{222}$$

Setting Eq. (220) into Eq. (222) finally connects the Einstein–Hilbert Lagrange matter density with the classical quantum theoretical mass, potential and energy terms as follows:

$$\Rightarrow L_{\text{M}} = \frac{3}{\kappa \cdot f} \cdot \Delta_{\text{3D-sphere}} f = \frac{3}{\kappa \cdot f} \cdot \left(E_n - \frac{M^2 c^2}{\hbar^2} - \frac{V_r}{r} \right) f$$
$$= \frac{3}{\kappa} \cdot \left(E_n - \frac{M^2 c^2}{\hbar^2} - \frac{V_r}{r} \right). \tag{223}$$

Not exactly unexpected, the result still mirrors the classical matter-energy structure in a surprisingly close manner. Naturally, this has to be the result for the simplification of a vanishing Schwarzschild radius r_s and is explicitly incorporating the classical relation.

Now we intent to solve the Laplace equation for Eq. (213) under the condition of vanishing Ricci scalars R and R^* ($\rightarrow R^{**} = 0$).

Applying the separation approach $f[t, r, \vartheta, \varphi] = g[t] * h[r] * Y[\vartheta] * Z[\varphi]$ with the separation parameters C_t and m (compare for the subsection "Flat Space Situation" in [112]) gives us the following immediate solution for $g[t]$ and $Z[\varphi]$:

$$g[t] = C_{t1} \cdot \cos[c \cdot C_t \cdot t] + C_{t2} \cdot \sin[c \cdot C_t \cdot t], \tag{224}$$

$$Z_m[\varphi] = C_{\varphi 1} \cdot \cos[m \cdot \varphi] + C_{\varphi 2} \cdot \sin[m \cdot \varphi]. \tag{225}$$

The two remaining differential equations can be constructed as follows:

$$-\frac{m^2}{\sin[\vartheta]^2} + \frac{\cot[\vartheta]\,Y'[\vartheta] + Y''[\vartheta]}{B^2 Y[\vartheta]} = l\,(1+l), \qquad (226)$$

$$\frac{l\,(1+l)}{r^2}\,(r_s - r) - \frac{((r_s - 2\cdot r)\cdot h'[r] + (r_s - r)\cdot r\cdot h''[r])}{r^2\cdot h[r]} = \frac{r\cdot C_t^2}{c^2} - \frac{R^*}{6}. \qquad (227)$$

While the first equation of the two gives the usual solution with the associate Legendre Polynomials P_l^m and Q_l^m, as also known from the Schrödinger hydrogen solution [115] (but please see also the next section 3.4 about half-l solutions):

$$Y_l^m[\vartheta] = C_{P\vartheta}\cdot P_l^m[\cos[\vartheta]] + C_{Q\vartheta}\cdot Q_l^m[\cos[\vartheta]], \qquad (228)$$

we have a rather non-Schrödinger-like solution with respect to the radius coordinate in the case of $R^* = C_t = 0$, namely:

$$h[r] = C_{\mathrm{Pr}}\cdot P_l\left[2\cdot\frac{r}{r_s} - 1\right] + C_{Qr}\cdot Q_l\left[2\cdot\frac{r}{r_s} - 1\right]. \qquad (229)$$

This time we have the simple Legendre Polynomials P_l and Q_l.

Regarding the angular functions within $f[\dots]$ we might assume that, just as known from the Schrödinger hydrogen problem, we have the conditions $l = 0, 1, 2, 3, \dots$ and $-l \le m \le l$. However, it was shown in [112] that we also have options for singularity free solutions with $l = \{1/2, 3/2, 5/2, \dots\}$. We realize that our solution—so far—has no main quantum number as the classical Schrödinger solution (see further below) does. However, it was shown that such a number appears with the introduction of a potential $V[r] \sim 1/r$ (the Schrödinger way), the breaking of symmetry (appendix of subsection "3.4 The Other Hydrogen") or the assumption of a boundary in r (e.g., [112]). It was also shown that the number of dimensions can be effectively seen as the presence of a potential [121].

In the case of $l = 0$ the solutions for $Y[\vartheta]$ and $h[r]$ read:

$$Y[\vartheta] = C_{\vartheta 1}\cdot \cosh\left[B\cdot m\left(\ln\left[\cot\left[\frac{\vartheta}{2}\right]\right]\right)\right]$$

$$-i\cdot C_{\vartheta 2}\cdot \sinh\left[B\cdot m\left(\ln\left[\cot\left[\frac{\vartheta}{2}\right]\right]\right)\right], \qquad (230)$$

$$h[r] = C_{r1} + \frac{C_{r2}}{r_s}\cdot\ln\left[\frac{r - r_s}{r}\right]. \qquad (231)$$

3.3.9.1 Discussion

We immediately realize that with $R^* = C_t = C_{r2} = l = 0$ the quantum gravity wave function just reproduces the original Schwarzschild metric because we have $f = $ constant.

The more general $l = 0$ solution Eq. (231) with $C_{r2} \neq 0$, on the other hand, only gives reasonable pure-real values for $r > r_s$ and we might immediately ask what the rest of the space with $r \leq r_s$ would be doing. Most interestingly, if simplifying Eq. (231) as follows:

$$h[r] = 1 + C_2 \cdot \ln \left[\frac{r - r_s}{r} \right] \tag{232}$$

and evaluating the resulting metric components of the quantum transformed metric $G_{\alpha\beta}$:

$$G_{\alpha\beta}^{Q-SS} = \left(1 + C_2 \cdot \ln \left[\frac{r - r_s}{r} \right] \right)^2 \cdot \left(Z_m \cdot Y_0^m \right)^2$$

$$\times \begin{pmatrix} -c^2 \left(\frac{r - r_s}{r} \right) & 0 & 0 & 0 \\ 0 & \left(\frac{r - r_s}{r} \right)^{-1} & 0 & 0 \\ 0 & 0 & r^2 & 0 \\ 0 & 0 & 0 & r^2 \cdot \sin^2 \vartheta \end{pmatrix} \tag{233}$$

gives us—almost—back the classical solution (Figs. 3.3.1 and 3.3.2, where for simplicity we have set $c = 1$). Thereby we have set $C_{r1} = 1$ and $C_1 = 0$ (cf. Eq. (215)). At least with respect to the well-known experimental tests of the Schwarzschild solution and the General Theory of Relativity, we would obtain significant differences only in cases of big $C_2 \gg 1$. Thus, our quantum transformation does not change the outside solution in such a way that it would differ from the classical one in an experimentally detectable manner.

Now we stick to our assumption that only purely real solutions should be allowed and add the condition that no structural singularities shall be contained in the total quantum gravity solution. As we know that the singularity at $r = r_s$ is only a result of our choice of coordinates, we can generously ignore the infinite growth of G_{11} in Fig. 3.3.2 for $r \rightarrow r_s$. With the means currently at hand, the only solution of non-singular character at the principle Schwarzschild singularity at $r = 0$ is obtained with Eq. (229) for $C_{Ql} = 0$, $C_1 = 0$ and certain $l \neq 0$. A more complete solution will give us better options. Figures 3.3.3 to 3.3.6 show a variety of such completely singularity-free constellations.

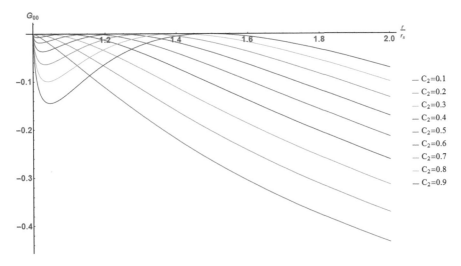

Figure 3.3.1 Metric time component of the quantized black hole solution for the outside region according to Eq. (233).

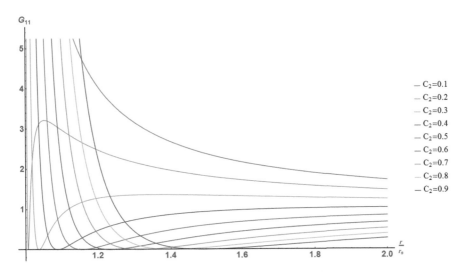

Figure 3.3.2 Metric radius component of the quantized black hole solution for the outside region according to Eq. (233).

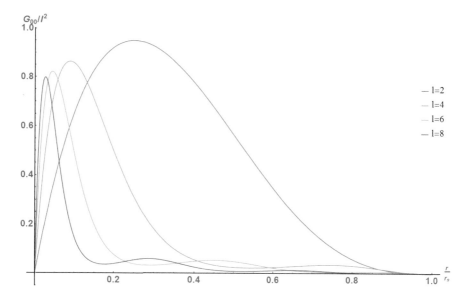

Figure 3.3.3 Metric time component of the quantized black hole solution for the inside region according to Eq. (234) for some even angular quantum numbers l.

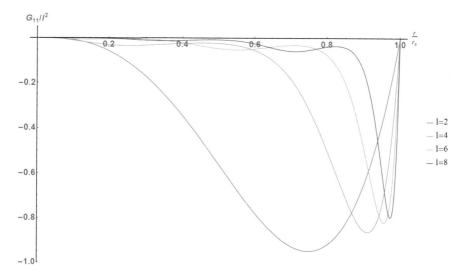

Figure 3.3.4 Metric radius component of the quantized black hole solution for the inside region according to Eq. (234) for some even angular quantum numbers l.

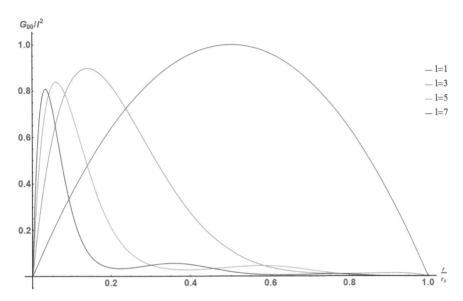

Figure 3.3.5 Metric time component of the quantized black hole solution for the inside region according to Eq. (234) for some odd angular quantum numbers l.

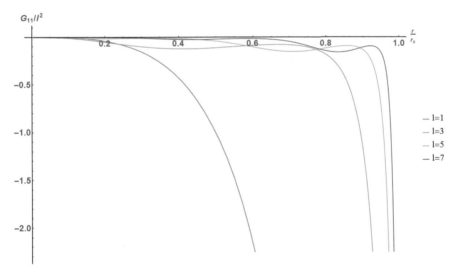

Figure 3.3.6 Metric radius component of the quantized black hole solution for the inside region according to Eq. (234) for some even angular quantum numbers l.

These results give reason to the suspicion that for black holes at $r = r_s$ space kind of shears itself off, thereby forming the core or inside of the black hole. The inner metric solution is thereby given due to:

$$G_{\alpha\beta}^{Q-SS} = \left(P_l \left[2 \cdot \frac{r}{r_s} - 1 \right] - (-1)^l \right)^2 \cdot \left(Z_m \cdot Y_l^m \right)^2$$

$$\times \begin{pmatrix} -c^2 \left(\frac{r-r_s}{r} \right) & 0 & 0 & 0 \\ 0 & \left(\frac{r-r_s}{r} \right)^{-1} & 0 & 0 \\ 0 & 0 & r^2 & 0 \\ 0 & 0 & 0 & r^2 \cdot \sin^2 \vartheta \end{pmatrix}. \tag{234}$$

The non-zero (even or odd) integer quantum numbers for the angular momentum $l = 1, 2, 3, 4, \ldots$ assure the non-singularity of the solution for the inside metric.

3.3.9.2 The other quantum number

Just for completeness it should be pointed out that by setting the curvature $R*$ in Eq. (227) as follows:

$$\frac{l(1+l)}{r^2} (r_s - r) - \frac{((r_s - 2 \cdot r) \cdot h'[r] + (r_s - r) \cdot r \cdot h''[r])}{r^2 \cdot h[r]} = \frac{r \cdot C_t^2}{c^2} + \frac{r_s^2 \cdot R_0^2}{4 \cdot r} \tag{235}$$

and thereby making it similar to the central field (potential $V[r] \sim 1/r$) problem of the Schrödinger hydrogen, we obtain the following solution for $h[r]$ in the case of $C_t = 0$:

$$h[r] = C_{\mathrm{Pr}} \cdot P_l^{R_0} \left[2 \cdot \frac{r}{r_s} - 1 \right] + C_{Qr} \cdot Q_l^{R_0} \left[2 \cdot \frac{r}{r_s} - 1 \right]. \tag{236}$$

The corresponding non-singular inner solutions for $R0 \neq 0$ read:

$$G_{\alpha\beta}^{Q-SS} = \left(P_l^{R_0} \left[2 \cdot \frac{r}{r_s} - 1 \right] \cdot Z_m \cdot Y_l^m \right)^2 \cdot \begin{pmatrix} -c^2 \left(\frac{r-r_s}{r} \right) & 0 & 0 & 0 \\ 0 & \left(\frac{r-r_s}{r} \right)^{-1} & 0 & 0 \\ 0 & 0 & r^2 & 0 \\ 0 & 0 & 0 & r^2 \cdot \sin^2 \vartheta \end{pmatrix}. \tag{237}$$

Just as with the angular coordinate ϑ (see Eq. (228)) we now have the associate Legendre Polynomials again. Similar to the situation there we have the conditions $l = 0, 1, 2, 3, \ldots$ and $-l \leq R_0 \leq l$, we see that for each non-zero quantum momentum we have $2 * l + 1$ options for the curvature term R_0. Setting the curvature proportional to r_s, like $R_0 = c_0 * r_s$, which appears quite reasonable as the mass surely can be connected with the curvature of space-time, immediately gives us an opportunity to quantize mass. We also might have obtained an explanation for certain restrictions of the masses of elementary particles due to the limiting rule $-l \leq R_0 \leq l \longrightarrow -l \leq c_0 * r_s \leq l$.

For completeness we should also give the solution in the case of $l = 0$, which reads:

$$h[r] = C_{cr} \cdot \cosh\left[\frac{R_0}{2} \cdot \ln\left[\frac{r}{r-r_s}\right]\right] + C_{sr} \cdot \sinh\left[\frac{R_0}{2} \cdot \ln\left[\frac{r}{r-r_s}\right]\right]. \quad (238)$$

Here we have the especially interesting cases where either the time or the radial metric component of the classical Schwarzschild metric are completely neutralized by the quantum factor (see Figs. 3.3.7 to 3.3.12).

Of special interest thereby would be the constellation with $R_0 = 1$ as it gives the inverse of the square root of metric time component and would therefore neutralize the principle Schwarzschild singularity at $r = 0$, because we have:

$$h[r]^2 = C_{cr}^2 \cdot \left(\cosh\left[\frac{R_0}{2} \cdot \ln\left[\frac{r}{r-r_s}\right]\right]\right.$$
$$\left. + \sinh\left[\frac{R_0}{2} \cdot \ln\left[\frac{r}{r-r_s}\right]\right]\right)^2 \xrightarrow{R_0=1} \frac{r}{r-r_s}. \quad (239)$$

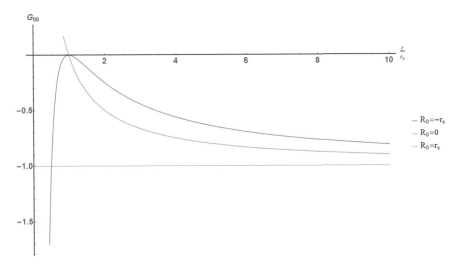

Figure 3.3.7 Metric time component of the quantized black hole solution for a "Ricci potential" $\sim R_0/r$ resulting in $h[r]$ according to Eq. (238) with $C_{cr} = C_{sr}$.

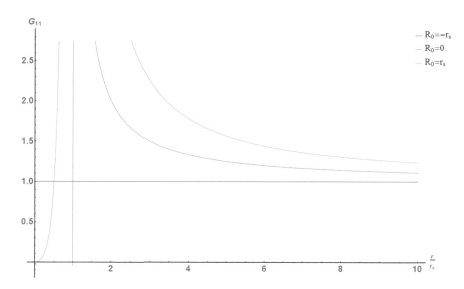

Figure 3.3.8 Metric radius component of the quantized black hole solution for a "Ricci potential" $\sim R_0/r$ resulting in $h[r]$ according to Eq. (238) with $C_{cr} = C_{sr}$.

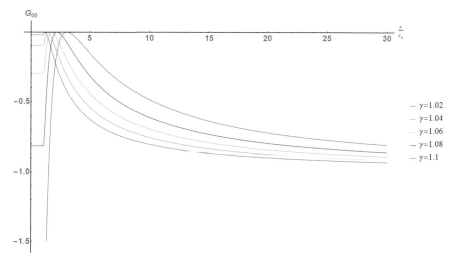

Figure 3.3.9 Metric time component of the quantized black hole solution with an inner-outer approach for $h[r]$ according to Eq. (240) with various values for the boundary parameter γ (see text).

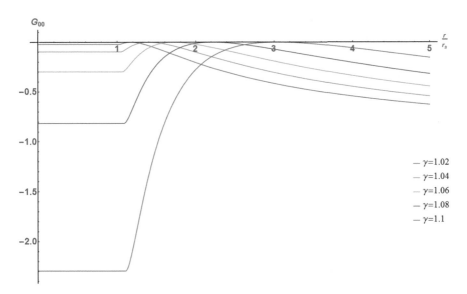

Figure 3.3.10   Metric time component of the quantized black hole solution with an inner-outer approach for $h[r]$ according to Eq. (240) with various values for the boundary parameter γ (see text).

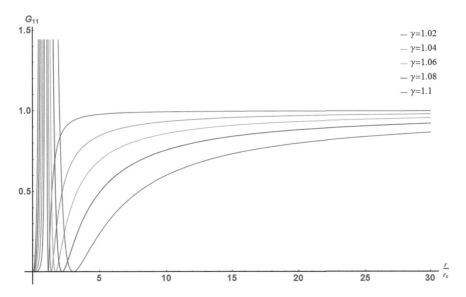

Figure 3.3.11   Metric radius component of the quantized black hole solution with an inner-outer approach for $h[r]$ according to Eq. (240) with various values for the boundary parameter γ (see text).

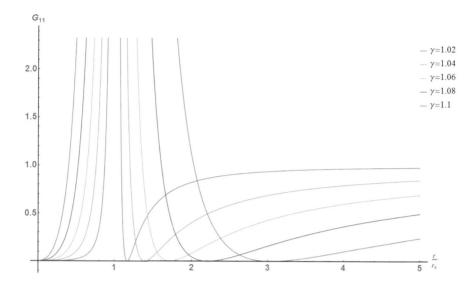

Figure 3.3.12 Metric radius component of the quantized black hole solution with an inner-outer approach for $h[r]$ according to Eq. (240) with various values for the boundary parameter γ (see text).

Combining this with solution Eq. (231) as outside solution at a certain $r = \gamma * r_s$ and demanding smoothness by the means of suitable constants C_{r1}, C_{r2} and C_{cr} gives the following total solution:

$$
h\,[r]^2 = \frac{1}{C_N} \cdot
\begin{cases}
\frac{r}{r-r_s} & \text{for } r \leq \gamma \cdot r_s \\
\left(\frac{2+\ln\left[\frac{\gamma-1}{\gamma}\right]}{2\sqrt{\frac{\gamma-1}{\gamma}}} + \frac{\sqrt{\gamma}}{2\sqrt{\gamma-1}} \cdot \ln\left[\frac{r-r_s}{r}\right] \right)^2 & \text{for } r \leq \gamma \cdot r_s
\end{cases}
$$

$$
C_N = \frac{\gamma \cdot \left(2 + \ln\left[\frac{\gamma-1}{\gamma}\right] \right)^2}{4\,(\gamma - 1)}. \tag{240}
$$

The normalization constant C_N has been chosen such that for $r \to \infty$ $h[r]^2$ goes to 1. Figures 3.3.9 and 3.3.10 show the convergence behavior against the flat space value, the smooth transition at the boundary $r = \gamma * r_s$ and the now non-singular situation at $r = 0$.

Please note that the condition $R_0 = 1$ from above (necessary to reach the singularity neutralization in Eq. (240)) automatically fixes the Ricci-potential in Eq. (235) to $\frac{r_s^2 \cdot R_0^2}{4 \cdot r} \xrightarrow{R_0=1} \frac{r_s^2}{4 \cdot r}$. Thus, a matter creation also assuring vanishing of the fundamental Schwarzschild singularity at $r = 0$ fixes the

Schwarzschild radius. Here we only kept it flexible by introducing the free boundary parameter γ.

3.4 The Other Hydrogen

Because of its importance for spherically symmetric problems, we will here reconsider the Schrödinger hydrogen problem and discuss some extentions and additional solutions. For convenience we will repeat all essential equations, but are going to refer to their earlier appearances within this book, wherever possible in order to point out possible connections with the applications of the extended hydrogen problem.

From the principles elaborated above it is clear that treating the hydrogen problem would require more than only 4 degrees of freedom. Even if restricting the electron and the proton to point-like objects, their combined individuality sums up to 6 spatial degrees of freedom plus time. There might be clever ways of reducing this problem due to the fact that electron and proton—if truly point-like—might move in plane manifolds around each other, but this shall not be our intention here. We want to revisit the classical Schrödinger problem and mirror it as close as possible with our metric approach. We therefore start with the following metric:

$$g_{\alpha\beta} = \begin{pmatrix} -H^2 \cdot c^2 & 0 & 0 & 0 \\ 0 & A^2 & 0 & 0 \\ 0 & 0 & B^2 \cdot r^2 & 0 \\ 0 & 0 & 0 & D^2 \cdot \sin[\vartheta]^2 \cdot r^2 \end{pmatrix}. \tag{241}$$

The only condition to solve the Einstein field equations is:

$$\Lambda = 0. \tag{242}$$

The resulting Ricci scalar R^* of the transformed metric using recipe Eq. (84) and the corresponding 4D wrapper-function $F[f = f[t, r, \vartheta, \varphi]] = (C_1 + f)^2$ reads:

$$R^* = \frac{1}{F[f]^3} \cdot \begin{pmatrix} \left(C_{N1} \cdot \left(\frac{\partial F[f]}{\partial f} \right)^2 - C_{N2} \cdot F[f] \cdot \frac{\partial^2 F[f]}{\partial f^2} \right) \cdot (\tilde{\nabla}_g f)^2 \\ -C_{N2} \cdot F[f] \cdot \frac{\partial F[f]}{\partial f} \cdot \Delta_g f \end{pmatrix}$$

$$= -C_{N2} \cdot \frac{\frac{\partial F[f]}{\partial f} \cdot \Delta_g f}{F[f]^2} \xrightarrow{\text{4D}} = -3 \cdot \frac{(C_1 + 2 \cdot f)}{\frac{1}{4} \cdot (C_1 + 2 \cdot f)^2} \cdot \Delta_g f = -\frac{12 \cdot \Delta_g f}{C_1 + 2 \cdot f}. \tag{243}$$

$$\Rightarrow -\frac{C_1 + 2 \cdot f}{12} \cdot R^* = \Delta_g f = \Delta_{3D-g} f - \frac{f^{(2,0,0,0)}}{H^2 \cdot c^2}. \tag{244}$$

Many examples and subsequent F-functions for other numbers of dimensions are given in [97, 110]. In the general case (arbitrary number of dimensions), applying the condition:

$$C_{N1} \cdot \left(\frac{\partial F[f]}{\partial f} \right)^2 - C_{N2} \cdot F[f] \cdot \frac{\partial^2 F[f]}{\partial f^2} = 0, \tag{245}$$

we obtain:

$$R^* + C_{N2} \cdot \frac{1}{F[f]^2} \cdot \frac{\partial F[f]}{\partial f} \cdot \Delta_g f = 0 \Rightarrow \left(\frac{R^* \cdot F[f]^2}{C_{N2} \cdot f \cdot \frac{\partial F[f]}{\partial f}} + \Delta_g \right) f = 0. \tag{246}$$

Now we substitute as follows in Eq. (244):

$$-\frac{C_1 + 2 \cdot \tilde{f}}{12} \cdot R^* = \Delta_g \tilde{f} = \Delta_{3D-g} \tilde{f} - \frac{\tilde{f}^{(2,0,0,0)}}{H^2 \cdot c^2} \quad f \equiv \frac{C_1 + 2 \cdot \tilde{f}}{12}$$

$$\Rightarrow -f \cdot R^* = \Delta_g \left(6 \cdot f - \frac{C_1}{2} \right) = 6 \cdot \Delta_g f$$

$$= 6 \cdot \left(\Delta_{3D-g} f - \frac{f^{(2,0,0,0)}}{H^2 \cdot c^2} \right). \tag{247}$$

Applying the separation approach $f[t, r, \vartheta, \varphi] = g[t] * h[r] * Y[\vartheta] * Z[\varphi]$ with the previously used separation parameters C_t and m (compare for the subsection "Flat Space Situation" in [112]) in Eq. (244) gives us the following immediate solution for $g[t]$ and $Z[\varphi]$:

$$g[t] = C_{t1} \cdot \cos[H \cdot c \cdot C_t \cdot t] + C_{t2} \cdot \sin[H \cdot c \cdot C_t \cdot t], \tag{248}$$

$$Z_m[\varphi] = C_{\varphi 1} \cdot \cos[D \cdot m \cdot \varphi] + C_{\varphi 2} \cdot \sin[D \cdot m \cdot \varphi]. \tag{249}$$

The two remaining differential equations can be constructed as follows:

$$-\frac{m^2}{\sin[\vartheta]^2} + \frac{\cot[\vartheta] Y'[\vartheta] + Y''[\vartheta]}{B^2 Y[\vartheta]} = l(1+l), \tag{250}$$

$$\frac{l(1+l)}{r^2} - \frac{(2 \cdot h'[r] + r \cdot h''[r])}{B^2 \cdot r \cdot h[r]} = \frac{B^2 \cdot C_t^2}{A^2} - \frac{R^*}{6}. \tag{251}$$

While the first equation of the two gives the usual solution with the Legendre Polynomials, as also known from the Schrödinger hydrogen solution [170], only in slightly generalized manner (watch the asymmetry-parameter B):

$$Y_k^m[\vartheta] = C_{Pv} \cdot P_k^{B \cdot m}[\cos[\vartheta]] + C_{Qv} \cdot Q_k^{B \cdot m}[\cos[\vartheta]];$$

$$k = \frac{1}{2}\left(\sqrt{1 + 4B^2 l(1+l)} - 1 \right)$$

$$B^2 l(1+l) \equiv L(1+L), \tag{252}$$

we have a rather non-Schrödinger-like solution with respect to the radius coordinate in the case of $R^* = 0$:

$$h[r] = C_j \cdot j_k \left[\frac{B^2}{A} \cdot C_t \cdot r \right] + C_y \cdot y_k \left[\frac{B^2}{A} \cdot C_t \cdot r \right] ;$$

$$k = \frac{1}{2} \left(\sqrt{1 + 4B^2 l (1 + l)} - 1 \right) . \tag{253}$$

Regarding the angular functions within $f[\ldots]$ we might assume that, just as known from the Schrödinger hydrogen problem, we have the conditions $L = 0, 1, 2, 3, \ldots$ and $-L \le B * m \le L$. However, it was shown in [112] that we also have options for singularity free solutions with $L = \{1/2, 3/2, 5/2, \ldots\}$. For convenience we repeat the evaluation in the appendix of this section. We realize that our solution has no main quantum number as the classical Schrödinger solution (see further below) does. However, it was shown that such a number appears with the introduction of a potential $V[r] \sim 1/r$ (the Schrödinger way), the breaking of symmetry (appendix) or the assumption of a boundary in r (e.g., [112]). The latter case will be considered also with respect to half spin states in one of the following issues of this series.

In the case of $l = L = 0$ the solutions for $Y[\vartheta]$ and $h[r]$ read:

$$Y[\vartheta] = C_{\vartheta 1} \cdot \cosh \left[B \cdot m \left(\ln \left[\cot \left[\frac{\vartheta}{2} \right] \right] \right) \right]$$

$$- i \cdot C_{\vartheta 2} \cdot \sinh \left[B \cdot m \left(\ln \left[\cot \left[\frac{\vartheta}{2} \right] \right] \right) \right] , \tag{254}$$

$$h[r] = \frac{C_{r1}}{r} \cdot e^{-i \cdot r \cdot \sqrt{\frac{B^4 \cdot C_t^2}{A^2}}} + \frac{C_{r2}}{r} \cdot e^{i \cdot r \cdot \sqrt{\frac{B^4 \cdot C_t^2}{A^2}}} . \tag{255}$$

In the case of $C_t = 0, l \ne 0$ the solution for $h[r]$ would read:

$$h[r] = C_{r1} \cdot r^{\frac{1}{2} \left(\sqrt{1 + 4 \cdot B^2 l (1 + l)} - 1 \right)} + C_{r2} \cdot r^{\frac{1}{2} \left(-\sqrt{1 + 4 \cdot B^2 l (1 + l)} - 1 \right)} . \tag{256}$$

Further simplification of $h[r]$ is obtained in the case $C_t = l = 0$:

$$h[r] = C_{r1} - \frac{C_{r2}}{r} . \tag{257}$$

3.4.1 Mainly Geometric Interpretation

Only allowing for solutions free of singularities, we systematically investigate the solutions above. Starting with solution Eq. (257), we have no time dependency with $f[\ldots]$ and it was shown in [112] section "Inner and Outer

Approaches" that the apparent $1/r$-singularity cannot only be transformed away, but that the resulting total metric also mirrors the Schwarzschild metric close enough to give the same experimental results and practical (gravitational) observations. Thus, solution Eq. (257) might just be the "true" gravitating object with no explicit time-dependency. In contrast to the Schwarzschild solution [142], however, it is a pure quantum gravity solution, originated from the flat space by a quantum gravity transformation of the kind Eq. (84). With this solution resulting from the condition $C_t = l = 0$, we do not only have a time-independency (cf. Eq. (224)), but also a peculiar dependency with respect to the angle ϑ. This dependency, originally, which is to say taking the Schrödinger solution, interpreted as constant behavior ($l = 0$ in Eq. (228)), has in fact a few more degrees of freedom, because it results in solution Eq. (230). There, in fact we have a constant solution for the case $m = 0$, but it should explicitly be pointed out that non-singular real solutions are also possible with imaginary terms $B * m$. Admittedly, towards the poles ($\vartheta = 0$ and $\vartheta = \pi$) this leads to infinite oscillations but the present author cannot see why this should be a problem. As we do not want to compromise the $Z[\varphi]$-function, we assume the separation parameter m to be real and demand B to become imaginary. This will not compromise our metric solution Eq. (241) with respect to the Einstein field equations (the interested reader may prove that also imaginary B would satisfy these equations), but results in a peculiar line element with one negative spatial component for the angular coordinate ϑ. One might say that the ϑ-coordinate becomes time-like. In Fig. 3.4.1 we have tried to illustrate the geometry.

Regarding solution Eq. (256) we have from the corresponding solution Eq. (228) the condition $L = 0, 1, 2, 3, \ldots$ and $-L \leq B * m \leq L$. This solution was already discussed in [111]. Thereby it was of special interest that we have the option of superposition and thus, could describe quite a variety of different states of time-independent solutions for $f[\ldots]$ with the quantum number L:

$$h_L[r] = C_{Lr1} \cdot r^{-\frac{1}{2}+\frac{1}{2}\sqrt{1+4\cdot L\cdot(L+1)}} + C_{Lr2} \cdot r^{-\frac{1}{2}-\frac{1}{2}\sqrt{1+4\cdot L\cdot(L+1)}}$$
$$\xrightarrow{L\in N} h_L[r] = C_{Lr1} \cdot r^L + C_{Lr2} \cdot r^{-L-1}. \tag{258}$$

A most interesting superposition of solutions would be one with $L = 0$ and $L = 1$, giving:

$$h_0[r] = C_{0r1} + C_{0r2} \cdot r^{-1} \quad \& \quad h_1[r] = C_{1r1} \cdot r + C_{1r2} \cdot r^{-2}. \tag{259}$$

Setting $C_{1r1} = 0$ we would have the radius behavior of the time component of the Reissner-Nordström metric [164, 165]. The classical Reissner-

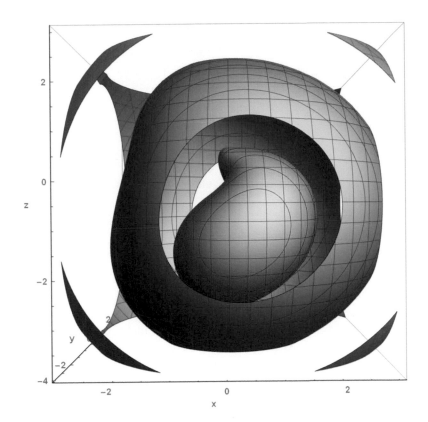

Figure 3.4.1 Peculiar space geometry in the case of quantum gravity solution Eq. (230).

Nordström metric can be given as follows:

$$g_{\alpha\beta} = \begin{pmatrix} \left(\frac{r_s}{r} - \left(\frac{r_q}{r}\right)^2 - 1\right) \cdot c^2 & 0 & 0 & 0 \\ 0 & \left(1 - \frac{r_s}{r} + \left(\frac{r_q}{r}\right)^2\right)^{-1} & 0 & 0 \\ 0 & 0 & r^2 & 0 \\ 0 & 0 & 0 & r^2 \cdot \sin^2 \vartheta \end{pmatrix};$$

$$r_q = \frac{K \cdot G \cdot Q^2}{c^4}, \tag{260}$$

where K, G and Q denote the Coulomb constant, the Newton constant and the electric charge of the object, respectively.

Unfortunately, the complete solution $f[\dots]$ would also have to be multiplied with the corresponding $Y[\dots]$ and $Z[\dots]$ dependencies. Thus, the total superposed solution with an r^{-2}-term applying Eq. (258) and the corresponding superposed L-forms would read:

with: $f_{L\mu}[r, \vartheta, \varphi] = h_L[r] \cdot Y_k^m[\vartheta] \cdot Z_m[\varphi]$

$\Rightarrow f_{0\mu}[r, \vartheta, \varphi] + f_{1\mu}[r, \vartheta, \varphi] = C_{0r1} + C_{0r2} \cdot r^{-1} + C_{1r2} \cdot r^{-2} \cdot Y_1^m[\vartheta] \cdot Z_m[\varphi]$

$L(1 + L) \equiv B^2 l (1 + l); \quad \mu \equiv B \cdot m$

$Y_L^\mu[\vartheta] = C_{Pv} \cdot P_k^\mu [\cos[\vartheta]] + C_{Qv} \cdot Q_k^\mu [\cos[\vartheta]]$

$Z_m[\varphi] = C_{\varphi 1} \cdot \cos[D \cdot m \cdot \varphi] + C_{\varphi 2} \cdot \sin[D \cdot m \cdot \varphi].$ \hfill (261)

Avoiding singularities with the angular functions, we can simplify our superposed solution to:

$\Rightarrow f_{0\mu}[r, \vartheta, \varphi] + f_{1\mu}[r, \vartheta, \varphi] = C_{0r1} + C_{0r2} \cdot r^{-1}$

$+ C_{1r2} \cdot r^{-2} \cdot P_1^{B \cdot m} [\cos[\vartheta]] \cdot (C_{\varphi 1} \cdot \cos[D \cdot m \cdot \varphi] + C_{\varphi 2} \cdot \sin[D \cdot m \cdot \varphi])$

with: $m = -1, 0, 1.$ \hfill (262)

In was demonstrated in [111] how Eq. (262) can be brought into an asymptotical Reissner-Nordström form by a simple coordinate transformation with respect to the radius coordinate, which is been transformed for fixed angles. The transformation leads to a new radius coordinate $g_r[r]$, which has to be evaluated from the following differential equation:

$$g_r[r]^2 = r^2 \cdot \left(1 - \frac{r_s}{r} + \left(\frac{r_q}{r}\right)^2\right); \quad \frac{\partial g_r[r]}{\partial r} \cdot \left(1 - \frac{r_s}{g_r[r]} + \frac{r_q^2}{g_r[r]^2}\right) - 1 = 0$$

$$\Rightarrow \frac{\partial g_r[r]}{\partial r} \cdot \left(1 - \frac{r_s}{r \cdot \sqrt{1 - \frac{r_s}{r} + \left(\frac{r_q}{r}\right)^2}} + \frac{r_q}{r \cdot \left(1 - \frac{r_s}{r} + \left(\frac{r_q}{r}\right)^2\right)}\right) - 1 = 0.$$

\hfill (263)

The solution is extremely lengthy and therefore it will not be presented here. Solving Eq. (263) leads to an integrational constant we shall name C_r.

It was also shown in [112] that such a procedure converges to the classical solutions for bigger radii. As an example we only present the asymptotic agreement for the metric time and radius components in Fig. 3.4.2.

Therefore, it could be concluded that charged particles might also just be seen as flat space with a certain (rather special) quantum gravity signature or fluctuation, which, for a distant observer, appears as charge.

As it was shown in [112] that also half L (respectively l in case of $B = 1$) lead to reasonable singularity-free solutions, we should also consider

radius behavior of g_{00} & g_{11}

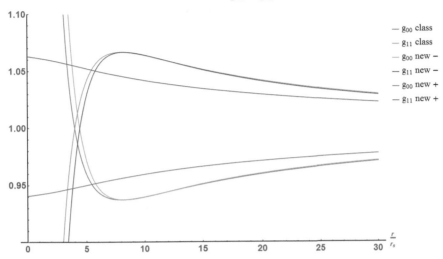

Figure 3.4.2 Metric components for time and radius coordinate of our approximated quantum gravity solution with a radius function obtained from (263) in comparison with the classical form Eq. (260). Presented was only the real part of the new solution. The parameter choice was $r_s = 1$, $r_q = 2$, $C_r = -5$ ("new –") and $C_r = 5$ ("new +").

these cases. We find that the exponents in $h[r]$ would lead to the following additional radius functions:

$$h_L[r] = C_{Lr1} \cdot r^{-\frac{1}{2}+\frac{1}{2}\sqrt{1+4\cdot L\cdot(L+1)}} + C_{Lr2} \cdot r^{-\frac{1}{2}-\frac{1}{2}\sqrt{1+4\cdot L\cdot(L+1)}}$$

$$\xrightarrow{L=\frac{1}{2},\frac{3}{2},\frac{5}{2},\frac{7}{2},\dots} h_L[r] = C_{Lr1} \cdot r^L + C_{Lr2} \cdot r^{-L-1}. \tag{264}$$

Obviously, we have options for inner ($C_{Lr2} = 0$) and outer ($C_{Lr1} = 0$) solutions. With the option of superposition "activated" in our quantum gravity package via the transformation Eq. (84) and condition Eq. (210) we have a huge portfolio of options for half spin combinations. Unfortunately, even though now having a spin $L = 1/2$-solution at hand, the spherical symmetric solution gives no geometric explanation to the Pauli principle [175], where the combination of parallel and antiparallel spin direction of $m = \pm 1/2$ clearly favors the antiparallel pairing (cf. [112]). The situation changes, however, the moment we distort the symmetry and move from spherical to elliptical coordinates. The derivation is shown in the appendix. While the time and angular solutions will not be influenced by the symmetry breaking, we obtain a completely different solution for the new radius coordinate, which we here denote with u instead of r in order to properly distinguish the spherical and the elliptical case (cf. appendix of this subsection). For $C_t = 0$ the solution for the elliptical radius function $h[u]$ can be given in closed form and reads:

$$h_L[r] \xrightarrow{\text{elliptic}} h_L^m[u] = \left(C_{Pu} \cdot P_k^q[\tanh[u]] + C_{Qu} \cdot Q_k^q[\tanh[u]]\right)$$
$$\times \left(\cosh[u]^{-2}\right)^{1/4}$$
$$k = \frac{1}{2}(2 \cdot m - 1); q = \frac{1}{2}(1 + 2 \cdot L), \tag{265}$$

where we kept the L instead of l in order to still have the option for some B-parametrization later on.

Now, just as evaluated in [112], we see that the combination of parallel $\frac{1}{2}$-spins requires more distorted space than the pairing of $m = +1/2$ and $m = -1/2$, which is just the metrical explanation for the Pauli principle [175]. A few example calculations are presented in the appendix.

A similarly intensive discussion has been performed with respect to solution Eq. (231) in [112]. To this case, however, it should be added that just as we had it with solution Eq. (257) we have a possible ϑ-dependency of the kind Eq. (230) with potentially imaginary B (see above and Fig. 3.4.1).

Last but not least we have the complete solution Eq. (229), which is, as already said, very much different from the classical Schrödinger radius function. The latter would be:

$$h_{nL}[r] = e^{-r} \cdot r^l \cdot L_{n-l-1}^{2l+1}[r]. \tag{266}$$

Thereby the complete classical wave function solution to the hydrogen atom can be given as follows [170]:

$$\Psi_{n,l,m}[r, \vartheta, \varphi] = e^{i \cdot m \cdot \varphi} \cdot P_l^m[\cos \vartheta] \cdot R_{n,l}[r]$$
$$= \sqrt{\left(\frac{2}{n \cdot a_0}\right)^3 \frac{(n-l-1)!}{2 \cdot n \cdot (n+l)!}} \cdot e^{-\rho/2} \cdot \rho^l \cdot L_{n-l-1}^{2l+1}[\rho] \cdot Y_l^m[\vartheta, \varphi];$$
$$\rho = \frac{2 \cdot r}{n \cdot a_0}. \tag{267}$$

The constant a_0 is denoting the Bohr radius with (m_e = electron rest mass, ε_0 = permittivity of free space, Q_e = elementary charge):

$$a_0 = \frac{4 \cdot \pi \cdot \varepsilon_0 \cdot \hbar^2}{m_e \cdot Q_e^2} = 5.292 \cdot 10^{-11} \text{ meter.} \tag{268}$$

The functions P, L and Y denote the associated Legendre function, the Laguerre polynomials and the spherical harmonics, respectively. This solution appears the moment we establish certain conditions to the curvature R^*, which was demonstrated in [108]. For convenience, we here repeat the essentials and start with an approach for R^* as follows:

$$\frac{R^*}{6} = R_6^* = R_6^*[r] = R_{60} + \frac{R_{61}}{r} + \frac{R_{62}}{r^2}. \tag{269}$$

Setting this into the last line of Eq. (247) gives:

$$-f \cdot \frac{R^*}{6} = -f \cdot \left(-R_{60} + \frac{R_{61}}{r} + \frac{R_{62}}{r^2}\right) = \Delta_{3D-g} f - \frac{f^{(2,0,0,0)}}{H^2 \cdot c^2}. \tag{270}$$

Now the wave function Eq. (406) directly solves our quantum gravity equation (270) for static cases where f does not depend on t. The extension to time-dependent f according to Eq. (270) is simple. However, in order to avoid too many parameters and work out the connection to the classical Schrödinger evaluation, we set $H = A = B = D = 1$. By using the results from above we can reproduce the Schrödinger hydrogen problem from Eq. (270) being changed to:

$$\frac{\hbar^2}{2 \cdot M} \cdot \frac{1}{g[t]} \cdot \frac{\partial^2 g[t]}{c^2 \cdot \partial t^2} = \left(\frac{R_{61}}{r} + \frac{R_{62}}{r^2} - R_{60} - \frac{L(L+1)}{r^2}\right)$$
$$+ \frac{\hbar^2}{2 \cdot M} \cdot \frac{1}{h[r]} \cdot \left(\frac{2}{r} \cdot \frac{\partial h[r]}{\partial r} + \frac{\partial^2 h[r]}{\partial r^2}\right). \tag{271}$$

Thereby we have introduced the term $\frac{\hbar^2}{2 \cdot M}$ (reduced Planck constant squared $\hbar^2$ divided by mass M) in order to truly mirror the classical Schrödinger equation. Multiplying with M would give us:

$$\frac{\hbar^2}{2} \cdot \frac{1}{g[t]} \cdot \frac{\partial^2 g[t]}{c^2 \cdot \partial t^2} = \left(\frac{R_{61}}{r} + \frac{R_{62}}{r^2} - R_{60} - \frac{L(L+1)}{r^2}\right) \cdot M$$
$$+ \frac{\hbar^2}{2} \cdot \frac{1}{h[r]} \cdot \left(\frac{2}{r} \cdot \frac{\partial h[r]}{\partial r} + \frac{\partial^2 h[r]}{\partial r^2}\right) \tag{272}$$

and connects the mass M with the curvature terms R_6 and the momentum. In fact, we might just interpret this as "mass is curvature" and consequently reformulate Eq. (272) as follows:

$$\frac{R_{ML62}}{r^2} = \left(\frac{R_{62}}{r^2} - \frac{L(L+1)}{r^2}\right) \cdot \frac{2 \cdot M}{\hbar^2}; \frac{R_{M61}}{r} = \frac{2 \cdot R_{61} \cdot M}{\hbar^2 r};$$

$$R_{M60} = 2 \cdot \frac{R_{60} \cdot M}{\hbar^2}$$

$$\frac{1}{g[t]} \cdot \frac{\partial^2 g[t]}{c^2 \cdot \partial t^2} = \left(\frac{R_{M61}}{r} + \frac{R_{ML62}}{r^2} - R_{M60}\right) + \frac{1}{h[r]} \cdot \left(\frac{2}{r} \cdot \frac{\partial h[r]}{\partial r} + \frac{\partial^2 h[r]}{\partial r^2}\right). \tag{273}$$

Thus, if we want to, we have the connection of the classical Schrödinger form with the curvature terms of the metric hydrogen form.

Comparison of Eq. (271) with the Schrödinger derivation (e.g., [170], pp. 155) and assuming that R_{60} is a constant gives us:

$$
\overbrace{\frac{\hbar^2}{2 \cdot M} \cdot \frac{1}{g[t]} \cdot \frac{\partial^2 g[t]}{c^2 \cdot \partial t^2}}^{-E_{n-\text{Schrödinger}}} + R_{60}
$$

$$
= \left(\overbrace{\frac{R_{61}}{r}}^{-V/r} + \overbrace{\frac{R_{62}}{r^2} - \frac{L(L+1)}{r^2}}^{-\omega^2_{\text{Schrödinger}}/r^2} \right) + \frac{\hbar^2}{2 \cdot M} \cdot \frac{1}{h[r]} \cdot \left(\frac{2}{r} \cdot \frac{\partial h[r]}{\partial r} + \frac{\partial^2 h[r]}{\partial r^2} \right).
$$

$$(274)$$

Using the result for $g[t]$ from above, which is to say:

$$
\overbrace{R_{60} - \frac{\hbar^2 \cdot C_t^2}{2 \cdot M}}^{-E_{n-\text{Schrödinger}}} = \left(\overbrace{\frac{R_{61}}{r}}^{-V/r} + \overbrace{\frac{R_{62}}{r^2} - \frac{L(L+1)}{r^2}}^{-\omega^2_{\text{Schrödinger}}/r^2} \equiv \omega^2 \right)
$$

$$
+ \frac{\hbar^2}{2 \cdot M} \cdot \frac{1}{h[r]} \cdot \left(\frac{2}{r} \cdot \frac{\partial h[r]}{\partial r} + \frac{\partial^2 h[r]}{\partial r^2} \right), \qquad (275)
$$

we can now rewrite Eq. (169) as follows:

$$
E_{n-\text{Schrödinger}} = - \left(\overbrace{\frac{R_{61}}{r}}^{-V/r} + \overbrace{\frac{R_{62}}{r^2} - \frac{\omega^2}{r^2}}^{-\omega^2_{\text{Schrödinger}}/r^2} \right) - \frac{\hbar^2}{2 \cdot M} \frac{1}{h[r]} \cdot \left(\frac{2}{r} \cdot \frac{\partial h[r]}{\partial r} + \frac{\partial^2 h[r]}{\partial r^2} \right)
$$

$$(276)$$

and thus, get the classical radial Schrödinger solution for the radial part of $f[\ldots]$:

$$
f_{n,l,m}[t, r, \vartheta, \varphi] = g[t] \cdot \Psi_{n,l,m}[r, \vartheta, \varphi] = g[t] \cdot e^{i \cdot m \cdot \varphi} \cdot P_l^m[\cos \vartheta] \cdot R_{n,l}[r]
$$

$$
= (C_1 \cdot \cos[c \cdot C_{tt} \cdot t] + C_2 \cdot \sin[c \cdot C_{tt} \cdot t])
$$

$$
\times N \cdot e^{-\rho/2} \cdot \rho^l \cdot L_{n-l-1}^{2l+1}[\rho] \cdot Y_l^m[\vartheta, \varphi]
$$

$$\rho = \frac{2 \cdot r}{n \cdot a_{00}}; \quad N = \sqrt{\left(\frac{2}{n \cdot a_0}\right)^3 \frac{(n-l-1)!}{2 \cdot n \cdot (n+l)!}};$$

$$C_1 = \pm C_2 = 1;$$

$$a_{00} = \frac{\hbar^2}{M} \cdot \frac{1}{R_{61}}; \quad E_n = \left[\frac{\hbar^2 \cdot C_t^2}{2 \cdot M} - R_{60}\right]_n = \frac{R_{61}^2}{n^2} \cdot \frac{\hbar^4}{16 \cdot M^2};$$

$$\omega^2 - R_{62} = l^2 + l$$

$$\Rightarrow f_{n,l,m}[t, r, \vartheta, \varphi] = e^{\pm i \cdot c \cdot C_t \cdot t} \cdot N \cdot e^{-\rho/2} \cdot \rho^l \cdot L_{n-l-1}^{2l+1}[\rho] \cdot Y_l^m[\vartheta, \varphi]. \quad (277)$$

Regarding the conditions for the quantum numbers n, l and m, we not only have the usual:

$$\{n, l, m\} \in \mathbb{Z}; \quad n \geqslant 0; \quad l < n; \quad -l \leqslant m \leqslant +l, \quad (278)$$

but also found the suitable solutions for the half spin forms as discussed above and derived in the appendix. The corresponding main quantum numbers for half spin l-numbers with $l = 1/2, 3/2, \ldots$ are simply (just as before with the integers) $n = l + 1 = 3/2, 5/2, 7/2, \ldots$.

It should explicitly be noted, however, that the usual spherical harmonics are inapplicable in cases of half spin. For $\{n, l, m\} = \{1/2, 3/2, 5/2, 7/2, \ldots\}$ the wave function Eq. (277) has to be adapted as follows:

$$f_{n,l,m}[t, r, \vartheta, \varphi] = e^{\pm i \cdot c \cdot C_t \cdot t} \cdot N \cdot e^{-\rho/2} \cdot \rho^l \cdot L_{n-l-1}^{2l+1}[\rho] \cdot Z_m[\varphi]$$

$$\times \left\{ \begin{array}{l} P_l^{m<0}[\cos \vartheta] \\ Q_l^{m>0}[\cos \vartheta] \end{array} \right\}$$

$$= e^{\pm i \cdot c \cdot C_t \cdot t} \cdot N \cdot e^{-\rho/2} \cdot \rho^l \cdot L_{n-l-1}^{2l+1}[\rho] \cdot \left\{ \begin{array}{l} \cos[m \cdot \varphi] \\ \sin[m \cdot \varphi] \end{array} \right\}$$

$$\times \left\{ \begin{array}{l} P_l^{m<0}[\cos \vartheta] \\ Q_l^{m>0}[\cos \vartheta] \end{array} \right\}. \quad (279)$$

Thereby we will elaborate below that in fact the sin- and the cos-functions seem to make the Pauli exclusion (see [175]) and not the "+" and "−" of the m. However, in order to have the usual Fermionic statistic we can simply define as follows:

$$f_{n,l,m}[t, r, \vartheta, \varphi] = e^{\pm i \cdot c \cdot C_t \cdot t} \cdot N \cdot e^{-\rho/2} \cdot \rho^l \cdot L_{n-l-1}^{2l+1}[\rho] \cdot \left\{ \begin{array}{l} \sin[m \cdot \varphi]_{m<0} \\ \cos[m \cdot \varphi]_{m>0} \end{array} \right\}$$

$$\times \left\{ \begin{array}{l} P_l^{m<0}[\cos \vartheta] \\ Q_l^{m>0}[\cos \vartheta] \end{array} \right\}. \quad (280)$$

As discussed above and derived in [112] plus the appendix, the resolution of the degeneration with respect to half spin requires a break of the symmetry, which we here achieved by introducing elliptical geometry instead of the spherical one.

Assuming an omnipresent constant curvature R_{60} does not seem to make much sense in our current system and thus, we should set $R_{60} = 0$. Doing the same with the curvature parameter R_{62} just gives us the usual condition for the spherical harmonics, namely $\omega^2 = L(1 + L) = l^2 + l$.

$$\rho = \frac{2 \cdot r}{n \cdot a_{00}}; \quad N = \sqrt{\left(\frac{2}{n \cdot a_0}\right)^3 \frac{(n - l - 1)!}{2 \cdot n \cdot (n + l)!}};$$

$$a_{00} = \frac{\hbar^2}{M} \cdot \frac{1}{R_{61}}; \quad n^2 = \frac{\hbar^2}{8} \cdot \left[\frac{R_{61}^2}{M \cdot C_t^2}\right]_n; \quad \omega^2 = l^2 + l$$

$$\Rightarrow f_{n,l,m}[t, r, \vartheta, \varphi] = e^{\pm i \cdot c \cdot C_t \cdot t} \cdot N \cdot e^{-\rho/2} \cdot \rho^l \cdot L_{n-l-1}^{2l+1}[\rho] \cdot Y_l^m[\vartheta, \varphi]. \quad (281)$$

Thus, we have metrically derived a "hydrogen atom." In addition to the Schrödinger structure, our form also sports a time-dependent factor clearly showing the options for matter and antimatter via the $\pm$-sign in the $g[t]$-function. We also found the half spin states and were able to resolve the spin-degeneration via a simple symmetry break by switching from spherical to elliptical coordinates.

With this, we might consider the metric hydrogen problem completed, but there are still a variety of questions. Especially with respect to the character and origin of the curvature, necessary to mirror the classical Schrödinger solution, its connection with the parameter mass and the interpretation of the additional solutions regarding $C_t = 0$ and $l = 0$ options we have obtained along the way, we will need further discussion. What is more: As in connection with these latter solutions no external potential or excitation was of need, we suspect particle solutions to have emerged here.

3.4.2 Particles

Instead of interpreting the solutions obtained above as electron orbitals of a simple hydrogen atom, we might also see options to use them as descriptions for a variety of particles. However, as such a consideration, if performed holistically, would by far exceed the intentions of this chapter, which after all was thought to just treat the hydrogen atom, we will only take one example in some extended form.

As one can deduce from the solution Eq. (265) to the flat space Laplace equation (286) in elliptical coordinates Eq. (285), the situation $R^* = 0$ and $C_t = 0$ requires an integer outcome for $(1 + 2 * L)/2$, because otherwise the result would lead to singularities. In order to simultaneously avoid singularities or trivial solutions also for the function $Y[v]$ (cf. Eq. (291)), where we have $-L \leq m \leq +L$, we find that L, m being integer is not possible with both conditions $R^* = 0$ and $C_t = 0$. Either curvature or mass/energy (connected with C_t) have to be non-zero.

But what about half spin values for L?

We find that with $1/2, 3/2, 5/2, \ldots$ and so on we indeed obtain $(1 + 2 * L)/2 =$ integer. On the other hand we have to demand half integer $|m| = 1/2, 3/2, 5/2, \ldots$ with $|m| > L$ for the function $h[u]$ as given in Eq. (265) in order to avoid trivial solutions. This, however, directly clashes with the non-triviality condition for $Y[v]$ as given in Eq. (291).

We conclude that, in elliptical coordinates, non-trivial solutions with simultaneous conditions $R^* = 0$ and $C_t = 0$ are not possible. Thus, keeping condition $C_t = 0$, we set:

$$R^* \sim \frac{R_{60}}{a^2 \cdot \left(\sinh [u]^2 + \cos [v]^2 \right)} = \Delta_g f. \tag{282}$$

Thereby we choose the peculiar u-v-dependency only in order to obtain a solvable differential equation. In connection with the metric Eq. (285) this gives us the following solution for $h[u]$ (cf. appendix of this subsection about the evaluation):

$$h_L^m [u] = \left(C_{Pu} \cdot P_k^q [\tanh [u]] + C_{Qu} \cdot Q_k^q [\tanh [u]] \right) \cdot \left(\cosh [u]^{-2} \right)^{1/4}$$

$$k = \frac{1}{2} (2 \cdot m - 1); \quad q = \frac{1}{2} \sqrt{(1 + 2 \cdot L)^2 - 2 \cdot R_{60}}. \tag{283}$$

Now it is the quantum number q which needs to be an integer in order to have non-trivial solutions without singularities. This can always be achieved by a suitable choice for the curvature parameter R_{60}.

Unfortunately, as we are unable to solve the resulting Laplace equation for reasonable u-v-dependencies with respect to the curvature, we resort to our half spin solution in spherical coordinates Eq. (279) in order to illustrate a variety of formations. Unfortunately the resolved degeneration for $m = -1/2$ and $m = 1/2$ states cannot be shown in these coordinates (cf. appendix). Only in order to keep things familiar and potentially compare with the classical integer hydrogen states Eq. (281), we keep the normalization and the scale of the Bohr radius a_0. One might perhaps call the resulting objects

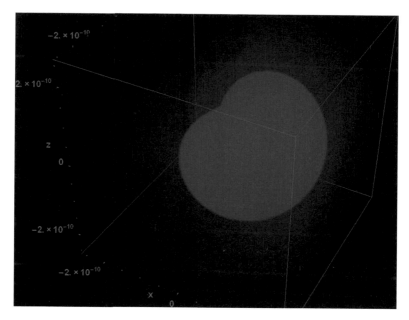

Figure 3.4.3   Absolute space geometry of distortion (or wave) function in the case of quantum gravity solution Eq. (279) for the state $n = 3/2, l = 1/2, m = 1/2$. Please note that in spherical coordinates the sign of $m = \pm 1/2$ does not matter with respect to the resulting spatial deformation. Resolving this degeneration requires a change of the φ-function with the sign of the m-value (see Eq. (280)) or a symmetry break (cf. [112] and appendix of this paper).

"half spin hydrogen atoms." With only one exception (Fig. 3.4.4a) we start with the general setting of $m > 0 (\rightarrow Z[\varphi] = \cos[m * \varphi])$. It can easily be seen with the simplest half spin states ($n = 3/3, l = 1/2$ and $|m| = 1/2$) as presented in Figs. 3.4.3 and 3.4.4 that the gross of deformed space-time is to be found on the right hand side of the $x = 0$-plane. Now choosing $n = 3/3$, $l = 1/2$ and $m = -1/2$ and applying the statistic rule defined in (280), leading to $Z[\varphi] = \sin[m * \varphi]$, gives us a concentration of deformed space-time on the other side of the $x = 0$-plane (Fig. 3.4.4a). It appears intuitive to assume that the combination of objects with deformation maxima on the left and right (anti-parallel spin) is easier than the combination of objects having the maximum deformations on the same sides of the $x = 0$-plane (parallel spin combination). We might see this as the geometric manifestation of the Pauli exclusion principle [175].

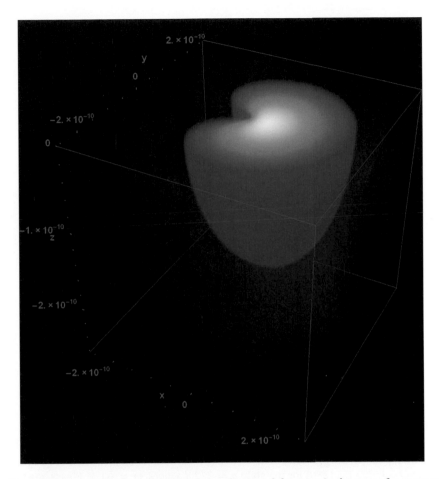

Figure 3.4.4 Space geometry of distortion (or wave) function in the case of quantum gravity solution Eq. (279) for the state $n = 3/2, l = 1/2, m = 1/2$. Please note that in spherical coordinates the sign of $m = \pm 1/2$ does not matter with respect to the resulting spatial deformation. Resolving this degeneration requires a change of the φ-function with the sign of the m-value (see Eq. (280)) or a symmetry break (cf. [112] and appendix of this paper). This time the inside of the distribution is shown.

As negative m would only bring in a mirror effect at the $x = 0$-plane to the deformation fields in our graphics, we will present further examples of half spin states only with positive m-quantum numbers according to our statistic rule as defined in Eq. (280).

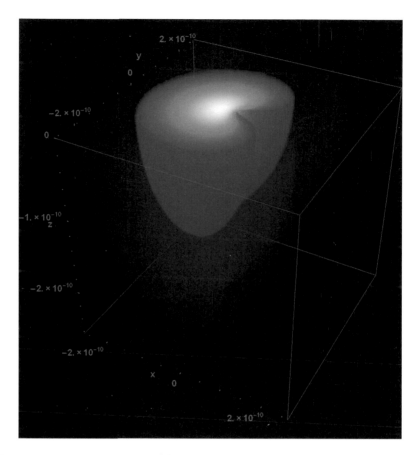

Figure 3.4.4a   Space geometry of distortion (or wave) function in the case of quantum gravity solution Eq. (280) for the state $n = 3/2, l = 1/2, m = -1/2$. Please see text and figure captions of Fig. 3.4.3 and 3.4.4.

3.4.3 Quarks?

Observing Fig. 3.4.7 and rescaling it as follows (Fig. 3.4.17), raises the suspicion that this in fact is the quark distribution. Not only the finding that we have three cores of activities within our metric distribution (Fig. 3.4.17) does justify our assumption, but also the simple fact that we have just one object supports our idea. Being one object, namely—of course—makes the cores inseparable. No matter how well-distinguished and therefore apparently separated they appear in the density plot below, the cores are

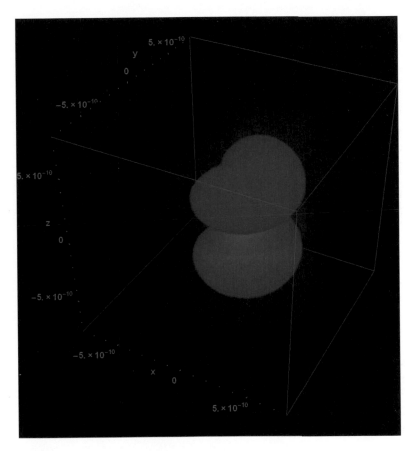

Figure 3.4.5 Space geometry of distortion (or wave) function in the case of quantum gravity solution Eq. (279) for the state $n = 5/2, l = 3/2, m = 1/2$.

still one object. Separating these cores means splitting the one particle and of course, this may cost more energy than it needs to actually just create more cores. In other words, if I was to try and split the object shown in Fig. 3.4.17, I might just end up in:

(a) increasing its energy from current $n = 5/2$ to $n = 7/2$, leaving everything else constant (Fig. 3.4.18) and then
(b) also increasing the momentum $l = 3/2 \rightarrow l = 5/2$, resulting in a doubling of equally proportioned cores, just like a doubling of the original particle (Fig. 3.4.9).

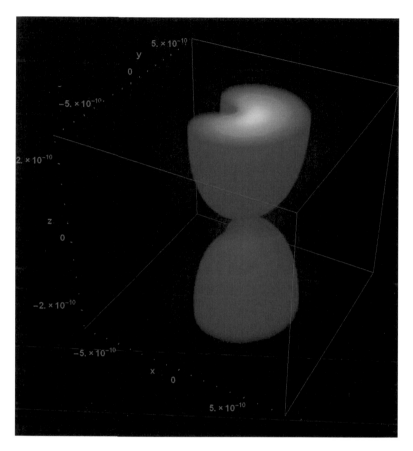

Figure 3.4.6 Space geometry of distortion (or wave) function in the case of quantum gravity solution Eq. (279) for the state $n = 5/2, l = 3/2, m = 1/2$. Upper distribution was cut open to show the metric distribution inside.

3.4.4 Time Independent Fermions?

It was worked out in the section "3.4.2 Particles" that in elliptical coordinates no suitable integer or half spin solution to the simultaneous conditions $R^* = C_t = 0$ can be found.

But what about spherical symmetry, where, unless there would be no coupling provoked by a boundary (cf. [112] and one of the following issues of this series), we have no main quantum number and therefore no combination of the momentum number l with the radius solution $h[r]$? As we have already

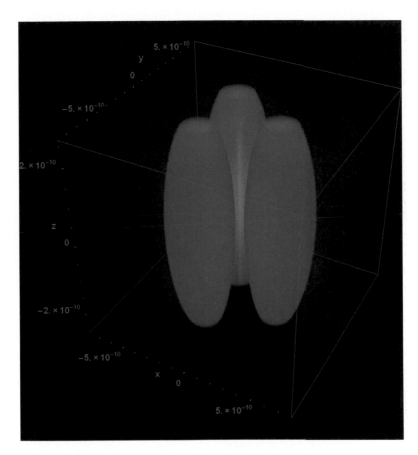

Figure 3.4.7 Space geometry of distortion (or wave) function in the case of quantum gravity solution Eq. (279) for the state $n = 5/2, l = 3/2, m = 3/2$.

considered integer spins (see sections above and [112]), we now want to observe half spin solutions. The corresponding h[r] solution to the condition $R^* = C_t = 0$ is given in Eq. (264) and thus the total solution for $f[\ldots]$ in the case of $\{l, |m|\} = 1/2, 3/2, 5/3, \ldots$ does read:

$$f_{l,m}[r, \vartheta, \varphi] = h_l[r] \cdot \begin{Bmatrix} \sin[m \cdot \varphi]_{m<0} \\ \cos[m \cdot \varphi]_{m>0} \end{Bmatrix} \cdot \begin{Bmatrix} P_l^{m<0}[\cos \vartheta] \\ Q_l^{m>0}[\cos \vartheta] \end{Bmatrix}$$

$$= (C_{lr1} \cdot r^l + C_{lr2} \cdot r^{-l-1}) \cdot \begin{Bmatrix} \sin[m \cdot \varphi]_{m<0} \\ \cos[m \cdot \varphi]_{m>0} \end{Bmatrix} \cdot \begin{Bmatrix} P_l^{m<0}[\cos \vartheta] \\ Q_l^{m>0}[\cos \vartheta] \end{Bmatrix}.$$

$$(284)$$

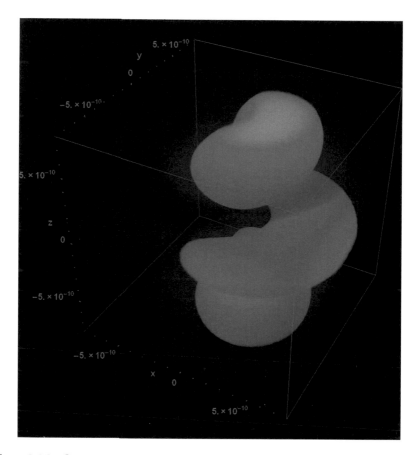

Figure 3.4.8 Space geometry of distortion (or wave) function in the case of quantum gravity solution Eq. (279) for the state $n = 7/2, l = 5/2, m = 1/2$.

By applying an inner outer solution one could always assure a solution free of singularities and infinite growth. Already the simplest of such solutions with $l = 1/2$ have quite some interesting properties and we have therefore discussed them in a bit more detail within our series "Einstein had it...."

3.4.5 The Other Hydrogen: Conclusions

It was found that the classical equations of treating the hydrogen problem in the Schrödinger way can directly be obtained from the Einstein field equations. However, in addition to the classical solutions we also found states

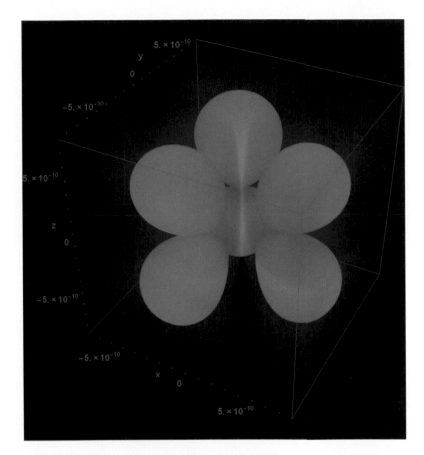

Figure 3.4.9 Space geometry of distortion (or wave) function in the case of quantum gravity solution Eq. (279) for the state $n = 7/2, l = 5/2, m = 3/2$.

with half integer energy and momentum quantum numbers. These additional solutions have the makings of particles rather than just hydrogen atom states. Along the way we also constructed an alternative (metric) hypothesis for the quark character and their confinement.

3.4.6 Appendix: About Spin 1/2, 3/2, 5/2 and so on

Instead of the perfect spherical symmetry, we now assume an elliptical one and introduce the following flat space metric with coordinates t, u, v, φ:

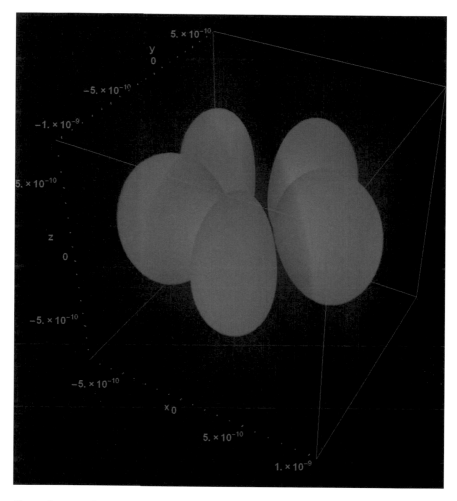

Figure 3.4.10 Space geometry of distortion (or wave) function in the case of quantum gravity solution Eq. (279) for the state $n = 7/2, l = 5/2, m = 5/2$.

$$g_{\alpha\beta}^{\text{flat}} =$$
$$\begin{pmatrix} -c^2 & 0 & 0 & 0 \\ 0 & a^2 \cdot \left(H^2 = \sinh{[u]}^2 + \cos{[v]}^2\right) & 0 & 0 \\ 0 & 0 & a^2 \cdot H^2 & 0 \\ 0 & 0 & 0 & a^2 \cdot \cosh{[u]}^2 \cdot \sin^2{[v]} \end{pmatrix}.$$

$$(285)$$

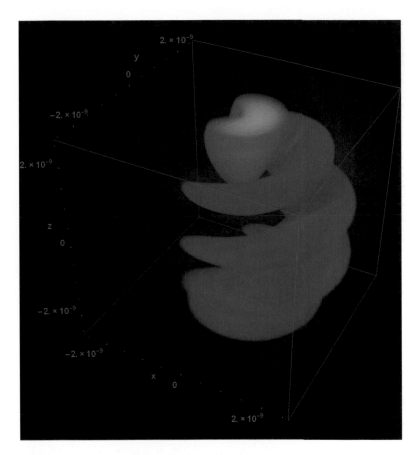

Figure 3.4.11 Space geometry of distortion (or wave) function in the case of quantum gravity solution Eq. (279) for the state $n = 13/2, l = 11/2, m = 1/2$.

Here "a" shall stand for a simple scaling factor. For convenience we apply $F[f] = (f + C_f)^2$ to the case $R^* = 0$. From (13) we then obtain the following 4D-Laplacian:

$$R^* = 0 = \Delta_g f = \Delta_{3D-g} f - \frac{f^{(2,0,0,0)}}{c^2} \cdot a^2 \cdot \left(\sinh [u]^2 + \cos [v]^2 \right). \quad (286)$$

This symmetry-breaking compared to our spherical considerations so far will not compromise our results for the angular functions, but allows additional insight with respect to the Pauli principle [175].

Applying the separation approach $f[t, u, v, \varphi] = g[t] * h[u] * Y[v] * Z[\varphi]$ with the previously used separation parameters C_t and m (compare with the

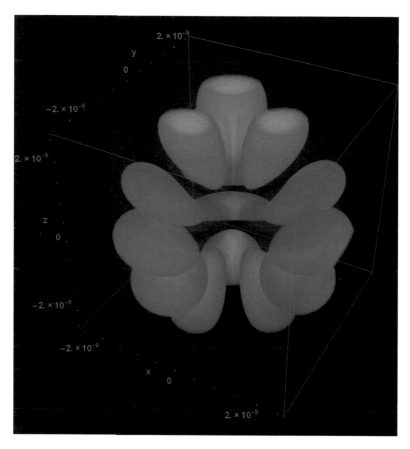

Figure 3.4.12 Space geometry of distortion (or wave) function in the case of quantum gravity solution Eq. (279) for the state $n = 13/2, l = 11/2, m = 3/2$.

main article) gives us the following immediate solution for $g[t]$ and $Z[\varphi]$:

$$g[t] = C_{t1} \cdot \cos[c \cdot C_t \cdot t] + C_{t2} \cdot \sin[c \cdot C_t \cdot t], \qquad (287)$$

$$Z[\varphi] = C_{\psi 1} \cdot \cos[m \cdot \varphi] + C_{\psi 2} \cdot \sin[m \cdot \varphi]. \qquad (288)$$

The two remaining differential equations can be constructed as follows:

$$\frac{1}{\sin[v]} \cdot \frac{\partial(\sin[v] \cdot Y'[v])}{\partial v} + \left(l \cdot (l+1) - \frac{m^2}{\sin[v]^2}\right) \cdot Y[v] = 0, \qquad (289)$$

$$\frac{1}{\cosh[u]} \cdot \frac{\partial(\cosh[u] \cdot h'[u])}{\partial u} - \left(l \cdot (l+1) - \frac{m^2}{\cosh[u]^2}\right) \cdot h[u] = -C_t^2 \cdot a^2 \cdot H^2. \qquad (290)$$

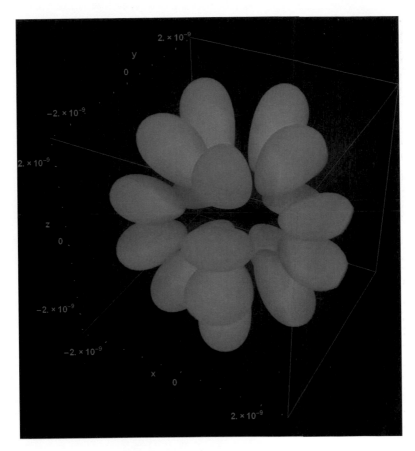

Figure 3.4.13 Space geometry of distortion (or wave) function in the case of quantum gravity solution Eq. (279) for the state $n = 13/2, l = 11/2, m = 5/2$.

While the first equation of the two gives the usual solution:
$$Y_l^m [v] = C_{Pv} \cdot P_l^m [\cos [v]] + C_{Qv} \cdot Q_l^m [\cos [v]], \qquad (291)$$
we see no way to also solve Eq. (290) in closed form. However, near the equator $v = 0$ and close to the origin $u = 0$, we have $H = 1$ and the subsequent approximated solution would read:
$$h_l^m [u] = \left(C_{Pu} \cdot P_k^q [\tanh [u]] + C_{Qu} \cdot Q_k^q [\tanh [u]] \right) \cdot \left(\cosh [u]^{-2} \right)^{1/4}$$
$$k = \frac{1}{2} (2 \cdot m - 1); \quad q = \frac{1}{2} \sqrt{(1 + 2 \cdot l)^2 - 2 \cdot a^2 \cdot C_t^2}. \qquad (292)$$
Just as known from the Schrödinger hydrogen problem, we also find here the conditions $C_{Qv} = 0, l = 0, 1, 2, 3, \ldots$ and $-l \leq m \leq l$. Only allowing

Figure 3.4.14 Space geometry of distortion (or wave) function in the case of quantum gravity solution Eq. (279) for the state $n = 13/2, l = 11/2, m = 7/2$.

for solutions free of singularities also in Eq. (292), we need to assure $q = 0$ or $q =$ integer for any pairing of m, l. This, most interestingly, automatically quantizes C_t just as we had found before with the perfect spherical symmetry constrained by a radial boundary $r = b$ in [112].

Thus, braking the symmetry (or adding boundaries) automatically quantizes the wave numbers C_t for the time-function $g[t]$. In contrast to our spherical situation as considered above, however, where we needed to add a boundary to get some C_t-quantization at all (for derivation see [112]), but having this quantization completely independent from the angular one, this time the quantization of the wave-numbers C_t is clearly coupled to the

Figure 3.4.15 Space geometry of distortion (or wave) function in the case of quantum gravity solution Eq. (279) for the state $n = 13/2, l = 11/2, m = 9/2$.

quantum number l via:

$$(1 + 2 \cdot l)^2 - 2 \cdot a^2 \cdot C_t^2 = 0 \Rightarrow C_{tl} = \pm \frac{1 + 2 \cdot l}{a \cdot \sqrt{2}}. \tag{293}$$

Obviously, there is a non-zero positive solution for C_t, which is to be found for $l = m = 0$ and gives:

$$C_{t0} = \pm \frac{1}{a \cdot \sqrt{2}}. \tag{294}$$

By comparing with the Dirac exponent of the classical solution for a particle at rest (e.g., [143]) we find that:

$$C_{tl} = \frac{M \cdot c}{\hbar}. \tag{295}$$

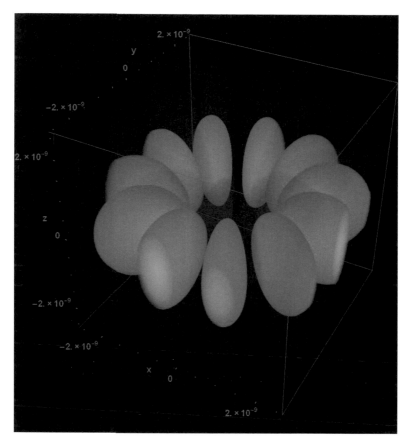

Figure 3.4.16 Space geometry of distortion (or wave) function in the case of quantum gravity solution Eq. (279) for the state $n = 13/2, l = 11/2, m = 11/2$.

In order to stay in the metric picture, we substitute the mass m by the Schwarzschild radius $r_s = \frac{2 \cdot M \cdot G}{c^2}$ and obtain:

$$C_{jl} = r_s \cdot \frac{c^3}{2 \cdot \hbar \cdot G} = \frac{r_s}{2 \cdot \ell_P^2}. \tag{296}$$

Here ℓ_P denotes the Planck length of about $1.616 * 10^{-35}$ meters. This directly connects the quantum number l, also called the "angular momentum quantum number", with the rest mass of the object or its effective Schwarzschild radius r_s as follows:

$$C_{tl} = r_s \cdot \frac{c^3}{2 \cdot \hbar \cdot G} = \frac{r_s}{2 \cdot \ell_P^2} \quad \Rightarrow \quad r_s = (\pm) \frac{\ell_P^2}{a} \cdot \sqrt{2} \cdot (1 + 2 \cdot l). \tag{297}$$

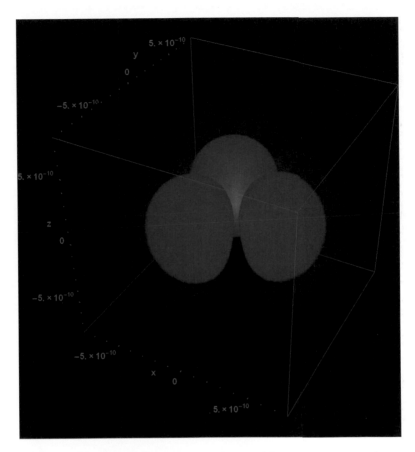

Figure 3.4.17 Space geometry of distortion (or wave) function in the case of quantum gravity solution Eq. (279) for the state $n = 5/2, l = 3/2, m = 3/2$.

Thereby we have the inverse of the scaling factor "a" times the Planck length squared as proportionality constant. Thus, the lowest possible—potentially effective—mass of an elliptic object with scaling factor a corresponds to a Schwarzschild radius of:

$$r_{s0} = (\pm)\sqrt{2} \cdot \frac{\ell_P^2}{a}. \tag{298}$$

Observing the solution $Y[v]$ more closely, we find that there exist spin $l = n/2$-solutions for $m = -n/2$ with $n = 1, 3, 5, 7\ldots$ in the case of $C_{Pv} \neq 0$, $C_{Qv} = 0$ and for $m = +n/2$ in the case of $C_{Pv} = 0, C_{Qv} \neq 0$. Figures 3.4.19 and

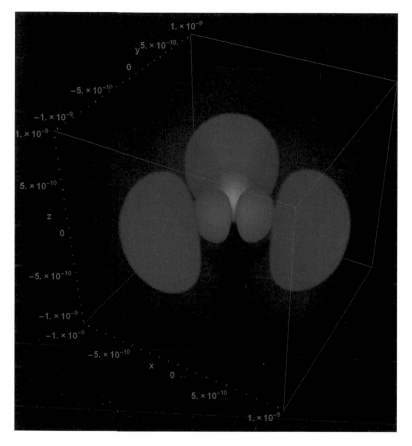

Figure 3.4.18 Space geometry of distortion (or wave) function in the case of quantum gravity solution Eq. (279) for the state $n = 7/2, l = 3/2, m = 3/2$. It is assumed that the additional smaller cores could just be the result of a particle creation reaction as answer to the attempt to separate the object shown in Fig. 3.4.17. Further adding of energy results in an increase of the momentum and six equally proportioned cores (cf. Fig. 3.4.9), which would essentially be a doubling of the original object shown in Fig. 3.4.17. This then, however, would mirror the behavior of quarks.

3.4.20 illustrate the corresponding distribution within the definition range for the angle v from 0 to π.

Now we are interested in finding the corresponding distributions for the functions $h[u]$. Interestingly, we find infinitely many solutions in the case $C_{Pu} \neq 0, C_{Qu} = 0$ and the value for C_l appears to be arbitrary (Fig. 3.4.21).

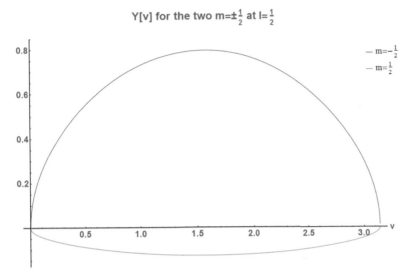

Figure 3.4.19 Spin $l = \frac{1}{2}$ situation with the two possible spin states $m = \pm 1/2$ according to our angular quantum gravity solution Eq. (228), Eq. (291). For better illustration and comparability, we divided the "Q-Legendres" by 10.

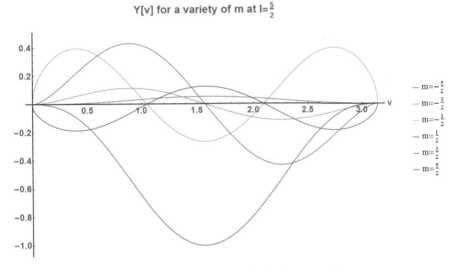

Figure 3.4.20 Spin $l = 5/2$ situation with the five possible spin states $m = \pm\{1/2, 3/2, 5/2\}$ according to our angular quantum gravity solution Eq. (228), Eq. (291). Regarding the evaluation, see text. For better illustration we divided the "Q-Legendres" by 10.

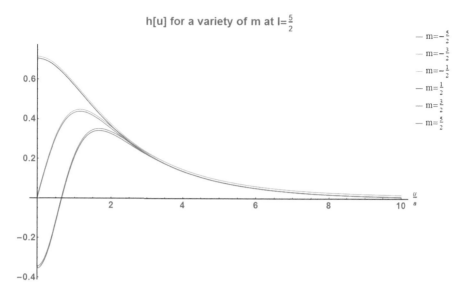

Figure 3.4.21 Distribution of function $h[u]$ from Eq. (292) for $m = -5/2$ to $m = 5/2$ with $l = 5/2$ with $q = 0$ (see condition Eq. (293)) and $C_{Pu} = 1$ and $C_{Qu} = 0$. As before with integer (Boson-like) solutions, the quantum states $m = +k$ and $m = -k$ appear degenerated. To make them still distinguishable we subtracted a small constant from the curves with $-m$.

On the other hand, however, with $C_{Qu} = 0$ we have the peculiar situation that $q = 0$ (cf. equation for the function $h[u]$ (292)) is not allowed if one intends to avoid singular behavior (Fig. 3.4.22). For simplicity, we concentrated on the real parts of the solutions.

As our solutions are just contributions to a quantum deformation of space-time, it appears logic to assume that processes, like bringing certain objects (particles) together, leading to an increase of deformation, should result in repulsion. Similarly, a decrease of deformation should bring attraction. Let us now assume that the objects with $m = -k/2$ should be brought together with their partners $m = +k/2$. As our solutions for the over-all function $f[t, u, v, \varphi]$ are additive (the corresponding differential equation is linear in f, cf. Eq. (244)), we can simply add the corresponding $Y[v]$ and $h[u]$-functions up. It is clear that for antiparallel m-pairings the angular deformation decreases because we add negative to positive curvature (cf. Figs. 3.4.19 and 3.4.20). Parallel pairing, on the other hand,

h[u] for a variety of m

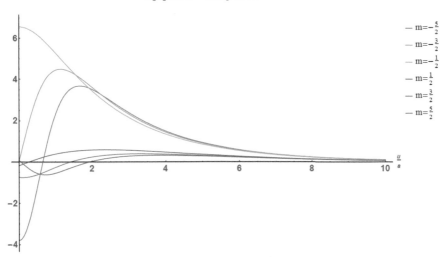

$$— \ m=-\tfrac{5}{2}$$
$$— \ m=-\tfrac{3}{2}$$
$$— \ m=-\tfrac{1}{2}$$
$$— \ m=\tfrac{1}{2}$$
$$— \ m=\tfrac{3}{2}$$
$$— \ m=\tfrac{5}{2}$$

Figure 3.4.22 Distribution of function $h[u]$ from Eq. (292) for $m = -5/2$ to $m = 5/2$ with $l = 5/2$ with the important condition $q \neq 0$ (here $q = 0.1$, see text above) and $C_{Pu} = 0$ and $C_{Qu} = 1$. In difference to the $l =$ integer (Bosonic) situation, the quantum states $m = +k$ and $m = -k$ appear not degenerated. Thus, spin obviously resolves the degeneration of the quantum number m (just as known from the spectral analysis of the hydrogen atom).

always increases the curvature. With respect to the radius-like distribution along u we have derived a few examples and presented them in Fig. 3.4.23.

Please note that the spin or spin-like solutions are also possible in perfect spherical symmetry as the $Y[v]$- and $Y[\vartheta]$-functions are identical (cf. Eq. (291) and Eq. (228)). The radius functions $h[r]$ for the spherical (see Eq. (229)) and $h[u]$ for the elliptic case Eq. (292), however, are very different. For one thing, we find no differences for the positive or negative m quantum numbers in connection with $h[r]$, while there was quite some significance with respect to $h[u]$. For completeness we shall still present the r-dependency for some fermionic l-numbers, which we do so in Fig. 3.4.24.

Thus, we can conclude that the pure spherical case does not provide a good metric explanation for the so called Pauli principle [175], while the elliptic situation obviously does. This also agrees with the fact that we already obtained hints for a metric "realization" of the Pauli principle in connection with a spherical metric with angular shear [102]. This, too, could

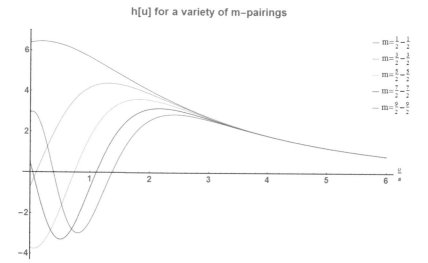

Figure 3.4.23 Distribution of function $h[u]$ from Eq. (292) for pairings of antiparallel $m = 1/2$ to $m = 9/2$ with $l = 9/2$ with the important condition $q \neq 0$ (here $q = 0.1$, see text above) and $C_{Pu} = 0$ and $C_{Qu} = 1$. In total, the deformation of space-time decreases with the anti-parallel pairing compared to the single particle state or parallel pairing situation.

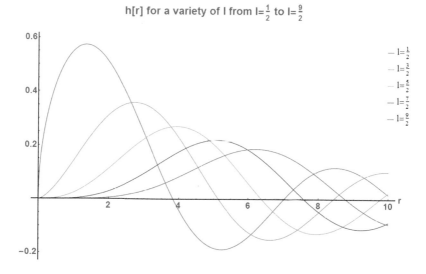

Figure 3.4.24 Distribution of function $h[r]$ from Eq. (229) for a variety of fermionic L-numbers.

just be seen as a—potentially small—break of symmetry. More discussion in connection with [175] shall be presented elsewhere (probably later in another publication).

3.5 Consideration of the Einstein–Hilbert Surface Term

It was shown in [137, 138] how to evaluate the integrand in Eq. (87). The result was given as follows (be aware of the fact that in the following the classical matter term, the Lagrange matter density, and the cosmological constant Λ were set to zero):

$$
\begin{aligned}
\delta_g W = 0 = \delta_g \int_V d^n x \sqrt{-g} R &= \delta_g \int_V d^n x \sqrt{-g} g^{\alpha\beta} R_{\alpha\beta} \\
&= \int_V d^n x \delta_g \left(\sqrt{-g} g^{\alpha\beta} R_{\alpha\beta} \right) \\
&= \int_V d^n x \left[\delta_g \left(\sqrt{-g} g^{\alpha\beta} \right) R_{\alpha\beta} + \sqrt{-g} g^{\alpha\beta} \delta_g \left(R_{\alpha\beta} \right) \right] \\
&= \int_V d^n x \sqrt{-g} \left[-G^{\kappa\lambda} \delta_g g_{\kappa\lambda} + g^{\alpha\beta} \delta_g R_{\alpha\beta} \right],
\end{aligned}
\tag{299}
$$

with $G^{\alpha\beta} = R^{\alpha\beta} - \frac{1}{2} R g^{\alpha\beta}$ denoting the Einstein tensor. It has to be noted that the result to the law of gravity does not change when we add a constant Λ, giving us $G^{\alpha\beta} + \Lambda g^{\alpha\beta} = R^{\alpha\beta} - \frac{1}{2} R g^{\alpha\beta} + \Lambda g^{\alpha\beta}$.

As mentioned above, it was assumed by Hilbert [137] that the second term in the last line in the integrand of Eq. (299), which is to say the term $g^{\alpha\beta} \delta_g R_{\alpha\beta}$, could be made a surface integral and thus, disregarded. However, as demonstrated in [53], such a surface term could still influence the outcome of a more holistic approach if, for example, our space is considered as a manifold, which is embedded in a space of higher dimension.

In order to show this, we here repeat the corresponding evaluation of the Einstein–Hilbert action in a very general manner.

After performing the variation in Eq. (299) we have:

$$
\begin{aligned}
0 = &\int_V d^n x \sqrt{-g} \left[R^{\kappa\lambda} - \frac{1}{2} R g^{\kappa\lambda} + \Lambda g^{\kappa\lambda} \right] \delta_g g_{\kappa\lambda} \\
&+ \int_V d^n x \sqrt{-g} \left[\left(\left(g^{\alpha\beta} \delta_g \Gamma^\rho_{\alpha\beta} - g^{\alpha\rho} \delta_g \Gamma^\gamma_{\alpha\gamma} \right)_{,\rho} \right) \right]
\end{aligned}
$$

$$= \int_V d^n x \sqrt{-g} \left[R^{\kappa\lambda} - \frac{1}{2} R g^{\kappa\lambda} + \Lambda g^{\kappa\lambda} \right] \delta_g g_{\kappa\lambda}$$

$$+ \int_{\partial V} d^{n-1} x \sqrt{-h} \left[\left(g^{\alpha\beta} \delta_g \Gamma^\rho_{\alpha\beta} - g^{\alpha\rho} \delta_g \Gamma^\gamma_{\alpha\gamma} \right) V_\rho \right]. \qquad (300)$$

Applying the following identity:

$$\left(g^\kappa_\alpha g^\lambda_\sigma g^\rho_\beta + g^\kappa_\sigma g^\lambda_\beta g^\rho_\alpha - g^\kappa_\alpha g^\lambda_\beta g^\rho_\sigma \right) \equiv G X^{\kappa\lambda\rho}_{\alpha\beta\sigma}, \qquad (301)$$

which allows us to give the variation of the affine connection as follows:

$$\delta_g \left(\Gamma^\gamma_{\alpha\beta} \right) = \frac{1}{2} g^{\gamma\sigma} \left(\delta g_{\sigma\alpha;\beta} + \delta g_{\sigma\beta;\alpha} - \delta g_{\alpha\beta;\sigma} \right) = \frac{g^{\gamma\sigma}}{2} G X^{\kappa\lambda\rho}_{\alpha\beta\sigma} \delta g_{\kappa\lambda;\rho}, \qquad (302)$$

we can give Eq. (300) in the following form:

$$0 = \int_V d^n x \sqrt{-g} \left[R^{\kappa\lambda} - \frac{1}{2} R g^{\kappa\lambda} + \Lambda g^{\kappa\lambda} \right] \delta g_{\kappa\lambda}$$

$$+ \int_{\partial V} d^{n-1} x \sqrt{-h} \left[\left(g^{\alpha\beta} \delta_g \Gamma^\rho_{\alpha\beta} - g^{\alpha\rho} \delta_g \Gamma^\gamma_{\alpha\gamma} \right) V_\rho \right]$$

$$= \int_V d^n x \sqrt{-g} \left[R^{\kappa\lambda} - \frac{1}{2} R g^{\kappa\lambda} + \Lambda g^{\kappa\lambda} \right] \delta g_{\kappa\lambda}$$

$$+ \int_{\partial V} d^{n-1} x \frac{\sqrt{-h}}{2} \left(\overbrace{\left(g^{\alpha\beta} g^{\rho\sigma} G X^{\kappa\lambda\omega}_{\alpha\beta\sigma} - g^{\alpha\rho} g^{\gamma\sigma} G X^{\kappa\lambda\omega}_{\alpha\gamma\sigma} \right)}^{\equiv M B^{\kappa\lambda\omega\rho}} \delta g_{\kappa\lambda;\omega} \right) V_\rho. \qquad (303)$$

At this point we have two options to avoid the surface term and evolve into the pure Einstein field equations:

Option A is to set the—until now—arbitrary Vector V_ρ identical zero, which is just the classical way of Hilbert [137] who argued that one could define the value of the integrant to be zero on the surface boundary ∂V.

Option   B   would   be   to   investigate   the   term $\overbrace{\left(g^{\alpha\beta} g^{\rho\sigma} G X^{\kappa\lambda\omega}_{\alpha\beta\sigma} - g^{\alpha\rho} g^{\gamma\sigma} G X^{\kappa\lambda\omega}_{\alpha\gamma\sigma} \right)}^{\equiv M B^{\kappa\lambda\omega\rho}} \delta g_{\kappa\lambda;\omega}$ in order to see whether this can be made zero by conditions not compromising possible solutions of the first integrant (the one of the volume integral), which is presenting the classical Einstein field equations. We considered this option in [87] in the section "Einstein Field Equation Solutions are Already Quantized."

Yet another option would be to just live with the term and to evaluate its outcome a little bit further. Integration by parts, namely, allows us to get rid of the $\delta_g g_{\kappa\lambda;\omega}$-term:

$$
0 = \int_V d^n x \sqrt{-g} \left[R^{\kappa\lambda} - \frac{1}{2} R g^{\kappa\lambda} + \Lambda g^{\kappa\lambda} \right] \delta g_{\kappa\lambda}
$$

$$
+ \int_{\partial V} d^{n-1} x \frac{\sqrt{-h}}{2} V_\rho M B^{\kappa\lambda\omega\rho} \delta g_{\kappa\lambda;\omega}
$$

$$
= \int_V d^n x \sqrt{-g} \left[R^{\kappa\lambda} - \frac{1}{2} R g^{\kappa\lambda} + \Lambda g^{\kappa\lambda} \right] \delta g_{\kappa\lambda}
$$

$$
+ \int_{\partial^2 V} d^{n-2} x \frac{\sqrt{-h_2}}{2} V_\rho M B^{\kappa\lambda\omega\rho} v_\omega \delta_g g_{\kappa\lambda}
$$

$$
- \int_{\partial V} d^{n-1} x \frac{\sqrt{-h}}{2} V_\rho \overbrace{M B^{\kappa\lambda\omega\rho}{}_{;\omega}}^{=0} \delta_g g_{\kappa\lambda}. \tag{304}
$$

We used the fact that the covariant derivative of the tensor $M B^{\kappa\lambda\omega\rho}$ must be zero and thus, $M B^{\kappa\lambda\omega\rho}{}_{;\omega} = 0$. Here V_σ denotes the normal vector on the surface ∂V, while h gives the determinant of the induced metric, which can be evaluated via:

$$
h_{ab}(\xi) = g_{\alpha\beta}(x(\xi)) \frac{\partial x^\alpha(\xi)}{\partial \xi^a} \frac{\partial x^\beta(\xi)}{\partial \xi^b}. \tag{305}
$$

On the other hand, v_σ denotes the normal vector on the sub-surface $\partial^2 V$, while h_2 gives the determinant of the induced sub-metric. This one should be given as:

$$
h_{2ab}(\xi_2) = h_{\alpha\beta}(x(\xi_2)) \frac{\partial x^\alpha(\xi_2)}{\partial \xi_2{}^a} \frac{\partial x^\beta(\xi_2)}{\partial \xi_2{}^b}. \tag{306}
$$

Now, by referring to [53] and the hypothesis that surface terms from higher dimensions might couple into our manifold, we simply take the second to last term in Eq. (304) as an additional contribution to the classical action variation. We assume such a surface term coming from a space of dimension $n + 2$ with the metric γ_{ab} to couple into the action of an embedded manifold with the induced metric $g_{\alpha\beta}$. If also assuming that the n-surface-term couples into the subspace $n - 2$, which is to say the n-space's own surface term shall contribute at two "stories below the n-th floor", the action integral for the

n-manifold now reads:

$$0 = \int_V d^n x \sqrt{-g} \left[R^{\kappa\lambda} - \frac{1}{2} R g^{\kappa\lambda} + \Lambda g^{\kappa\lambda} \right] \delta_g g_{\kappa\lambda}$$

$$+ A \int_V d^n x \sqrt{-g} \left(g^{\alpha\beta} \delta_g \Gamma^\rho_{\alpha\beta} - g^{\alpha\rho} \delta_g \Gamma^\gamma_{\alpha\gamma} \right)_{;\rho}$$

$$+ B \int_{V_{N+2}} d^{n+2} x \sqrt{-\gamma} \left[\left(\overbrace{\left(\gamma^{ab} \delta_\gamma \Gamma^c_{ab} - \gamma^{ac} \delta_\gamma \Gamma^d_{ad} \right)_{;c}}^{\text{dim}=N+2 \rightarrow \text{dim}=N} \longrightarrow M B \gamma^{\kappa\lambda\omega\rho} \right) \right]. \qquad (307)$$

Above we have assumed that by variation of the upper space, the embedded one also was variated and that therefore the metric γ_{ab} already contains all the $g_{\alpha\beta}$ variations. Thus, we only need to transform the $\delta_\gamma \gamma_{ab}$ of the upper space into $\delta_g g_{\kappa\lambda}$. Insertingly, the expansion of the variation of the affine connections and the following transformations:

$$\gamma_{ab}\left(\xi\left(x\right)\right) = g_{\alpha\beta}\left(x\left(\xi\right)\right) \frac{\partial x^\alpha\left(\xi\right)}{\partial \xi^a} \frac{\partial x^\beta\left(\xi\right)}{\partial \xi^b} = g_{\alpha\beta} \frac{\partial x^\alpha}{\partial \xi^a} \frac{\partial x^\beta}{\partial \xi^b}$$

$$\delta\gamma_{ab}\left(\xi\left(x\right)\right) = \delta g_{\alpha\beta}\left(x\left(\xi\right)\right) \frac{\partial x^\alpha\left(\xi\right)}{\partial \xi^a} \frac{\partial x^\beta\left(\xi\right)}{\partial \xi^b} = \delta g_{\alpha\beta} \frac{\partial x^\alpha}{\partial \xi^a} \frac{\partial x^\beta}{\partial \xi^b}, \qquad (308)$$

$$\gamma^{ab}\left(\xi\left(x\right)\right) = g^{\alpha\beta}\left(x\left(\xi\right)\right) \frac{\partial \xi^a\left(x\right)}{\partial x^\alpha} \frac{\partial \xi^b\left(x\right)}{\partial x^\beta} = g^{\alpha\beta} \frac{\partial \xi^a}{\partial x^\alpha} \frac{\partial \xi^b}{\partial x^\beta} \qquad (309)$$

would give us:

$$0 = \int_V d^n x \sqrt{-g} \left[R^{\kappa\lambda} - \frac{1}{2} R g^{\kappa\lambda} + \Lambda g^{\kappa\lambda} \right] \delta g_{\kappa\lambda}$$

$$+ A \int_{\partial^\lambda V} d^{n-2} x \frac{\sqrt{-h_2}}{2} U_\rho M B^{\kappa\lambda\omega\rho} u_\omega \delta_g g_{\kappa\lambda}$$

$$+ \frac{B}{2} \int_V d^n x \sqrt{-g} V_\rho M B \gamma^{\kappa\lambda\omega\rho} v_\omega \delta g_{\kappa\lambda}, \qquad (310)$$

where we have boldly assumed to be allowed to apply the transformation rules (308) and Eq. (309) onto both vectors inside the expression $[(\gamma^{ab}\delta_g\Gamma^c_{ab} - \gamma^{ac}\delta_g\Gamma^d_{ad})V_c]$. Please note that we have introduced the new vectors U_ρ and u_ρ only to distinguish them from V_ρ and v_ρ. The parameters A and B are just numbers accounting for the degree of freedom with respect to the transformation rules Eq. (305) and Eq. (306), respectively Eq. (308), Eq. (309).

This gives us the following set of extended Einstein field equations, with an assumed additional (in addition to the usual one) Hilbert surface term from space of dimension $n + 2$ (part a in Eq. (311)). We also obtain another field equation with Ricci tensor $\mathfrak{R}^{\kappa\lambda}$, Ricci scalar $\mathfrak{R}$ and metric $h^{\alpha\beta}$ for the sub-space with dimension $\dim = n - 2$ (part b in Eq. (311)):

(a) $\delta g_{\kappa\lambda}$, V :

$$0 = R^{\kappa\lambda} - \frac{1}{2}Rg^{\kappa\lambda} + \Lambda g^{\kappa\lambda} + \frac{1}{2}B \cdot V_\rho MB\gamma^{\kappa\lambda\omega\rho}v_\omega; \{\kappa, \lambda, \omega, \rho\} = 1, 2, \ldots, N$$

(b) $\delta g_{\kappa\lambda}$, $\partial^2 V$:

$$0 = \mathfrak{R}^{\kappa\lambda} - \frac{1}{2}\mathfrak{R}h^{\kappa\lambda} + \Lambda h^{\kappa\lambda} + \frac{1}{2}A \cdot U_\rho MB^{\kappa\lambda\omega\rho}u_\omega; \{\kappa, \lambda, \omega, \rho\} = 1, 2, \ldots, N - 2.$$

$$(311)$$

We see that we have obtained an additional (dimension-entanglement-) term of the form:

$$MU^{\kappa\lambda} \equiv \frac{1}{2}B \cdot V_\rho MB\gamma^{\kappa\lambda\omega\rho}v_\omega, \tag{312}$$

for the first (main) field equation providing a connection with the upper dimensions $n + 2$, plus an obviously non-zero term to the surface field equation:

$$\partial^2 MU^{\kappa\lambda} = \frac{1}{2}A \cdot U_\rho MB^{\kappa\lambda\omega\rho}u_\omega, \tag{313}$$

which is giving the connection with the lower dimensional space.

We conclude that the consideration of the surface-term, originally being ignored by both Einstein and Hilbert, only contributes in form of vectorial connections to external dimensions. Thus, only if we consider a manifold of dimension M being part of a bigger space of higher dimension n, we need to worry about such connections (dimensional entanglements). If directly grabbing the whole space with all its dimensions and not considering internal surfaces (manifolds), we are perfectly fine with the classical Einstein field equations.

It should be noted that the connection with possible higher dimensions in the case of a four-dimensional space has been investigated earlier in our series of "Einstein had it…." Especially for the problem of central symmetry [53] we have already considered various forms of entangled "spaces". Later in [87] we intended to find a systematic explanation for the connection with certain quantum theoretical solutions and $R = 0$-solutions in General Theory of Relativity.

3.6 As an Example: Consideration in 4 Dimensions

With the assumption that the Einstein field equations, respectively the Einstein–Hilbert action already contain(s) quantum theory, it should be possible to construct constellations where we obtain the classical Quantum Equations like the Klein–Gordon equation (KG-equation) at least as approximations of the Einstein field equation solutions. One way to quantize existing solutions of the Einstein field equations was shown in [8]. The technique applied there was named "intelligent zero method" and only consisted of a procedure to subsequently quantize already derived solutions to the Einstein field equations in a more or less typical quantum theoretical manner (staying linear, for instance). Here we intend to demonstrate that subsequent quantization is not of need as the Einstein field equation solutions are already quantized... respectively that they already contain everything which is of need in order to quantize them if this is required. The evaluation follows our presentation from the "unusual introduction", but as we here intent to connect it with the Heisenberg uncertainty principle and the principle limit of the speed of light, we will repeat the evaluation in a slightly different manner.

As the classical KG equation is defined on a flat space, we choose a metric approach as follows:

$$
g_{\alpha\beta}^{\text{flat}} = \begin{pmatrix} T \cdot f^2 \cdot g_t'\,[t]^2 & 0 & 0 & 0 \\ 0 & A \cdot f^2 \cdot g_x'\,[x]^2 & 0 & 0 \\ 0 & 0 & B \cdot f^2 \cdot g_y'\,[y]^2 & 0 \\ 0 & 0 & 0 & C \cdot f^2 \cdot g_z'\,[z]^2 \end{pmatrix}
$$
$$
f = f\left[g_t\,[t],\, g_x\,[x],\, g_y\,[y],\, g_z\,[z]\right]. \tag{314}
$$

In classical quantum theory, we only have one equation and one main function (the wave function) solving it. Therefore our approach Eq. (314), which contains only one function $f[\ldots]$ for all diagonal metric components.

For simplicity and clear Cartesian appearance we set all functions $g_i[x_i] = x_i$. The only scalar geometric quantity in the Einstein–Hilbert apparatus apart from the metric determinant is the Ricci scalar R. With the approach Eq. (314) and our simple setting for the $g_i[x_i]$ we obtain:

$$
R = \frac{6}{f^3} \cdot \left[\frac{\partial_t^2}{T} + \frac{\partial_x^2}{A} + \frac{\partial_y^2}{B} + \frac{\partial_z^2}{C} \right] f. \tag{315}
$$

Demanding that $R = 0$ and $A = B = C = -1$, we have:

$$\left[\frac{\partial_t^2}{T} - \Delta_{\text{Cartesian}}\right] f = 0; \quad \Delta_{\text{Cartesian}} = \partial_x^2 + \partial_y^2 + \partial_z^2. \tag{316}$$

As before with the 2-dimensional quantum well, we find some interesting similarities with the Klein–Gordon equation, which is given as:

$$\left[\frac{\partial_t^2}{c^2} - \Delta_{\text{Cartesian}} + \frac{M^2 c^2}{\hbar^2}\right] f = 0. \tag{317}$$

With the speed of light in vacuum c, the reduced Planck constant $\hbar$ and the mass M, we find that our equation $R = 0$ would mirror the Klein–Gordon equation in the case of $M = 0$ and $T = c^2$. From this, we deduce that metric approaches of the form:

$$g_{\alpha\beta} = \begin{pmatrix} c^2 \cdot F[f] & 0 & 0 & 0 \\ 0 & -F[f] & 0 & 0 \\ 0 & 0 & -F[f] & 0 \\ 0 & 0 & 0 & -F[f] \end{pmatrix}$$

$$f = f\left[g_t[t], g_x[x], g_y[y], g_z[z]\right] \tag{318}$$

with general functions g_i will give us mass-free Klein–Gordon equations in arbitrary coordinates with functions F reading:

$$F[f] = C_1 \cdot f + \frac{C_1^2 \cdot f^2}{4 \cdot C_2} + C_2. \tag{319}$$

Thereby the constants C_i are arbitrary. Please note that we obtain the simple case $F[f] = f^2$ with the conditions:

$$C_1^2 = 2 \cdot C_2 \Rightarrow F[f] = \sqrt{2 \cdot C_2} \cdot f + f^2 + C_2 \xrightarrow[C_2 \to 0]{\lim} F[f] = f^2. \tag{320}$$

The derivation of the Schrödinger equation is known to be only a technical variation (cf. [8]) of the derivation of the Klein–Gordon equation, while the Dirac equation is a special form of root extraction also from the Klein–Gordon equation (see [8] and [143]). Thus, we can conclude to have obtained typical quantum equations for the case $M = 0$ directly from the condition $R = 0$ and a homogeneous metric approach Eq. (318) with a function setting for $F[f]$ as Eq. (319).

From here, we may automatically conclude that in the case of more general $F[f]$, also equations with mass respectively $M \neq 0$ would be obtained. The general result for R does read:

$$R = \frac{3}{2 \cdot F[f]^3} \cdot \left(\begin{array}{c} \left(\left(\frac{\partial F[f]}{\partial f} \right)^2 - 2 \cdot F[f] \cdot \frac{\partial^2 F[f]}{\partial f^2} \right) \cdot \left[\frac{(\partial_t f)^2}{T} - (\partial_x f)^2 - (\partial_y f)^2 - (\partial_z f)^2 \right] \\ -2 \cdot F[f] \cdot \frac{\partial F[f]}{\partial f} \cdot \left[\frac{\partial_t^2}{T} - \partial_x^2 + \partial_y^2 + \partial_z^2 \right] f \end{array} \right). \tag{321}$$

This, however, also means that all terms in the equations $R = 0$ not belonging to what we had in Eq. (316) should somehow be connected with the mass term in the classical Klein–Gordon equations. Thus, comparing Eq. (321) with (317), we find that the mass term would be equal to:

$$\left[\frac{M^2 c^2}{\hbar^2} \right] f = -\frac{\left(\left(\frac{\partial F[f]}{\partial f} \right)^2 - 2 \cdot F[f] \cdot \frac{\partial^2 F[f]}{\partial f^2} \right)}{2 \cdot F[f] \cdot \frac{\partial F[f]}{\partial f}}$$
$$\times \left[\frac{(\partial_t f)^2}{T} - (\partial_x f)^2 - (\partial_y f)^2 - (\partial_z f)^2 \right]. \tag{322}$$

For simplicity we demand that $F''[f] = 0$, which simplifies Eq. (322) to:

$$\left[\frac{M^2 c^2}{\hbar^2} \right] f = -\frac{C_F}{2 \cdot f} \cdot \left[\frac{(\partial_t f)^2}{T} - (\partial_x f)^2 - (\partial_y f)^2 - (\partial_z f)^2 \right];$$
$$C_F = \frac{\partial F[f]}{\partial f}. \tag{323}$$

A special solution to this partial differential equation can be found via the ansatz:

$$f = e^{\tilde{M} \cdot (c_t \cdot t + c_x \cdot x + c_y \cdot y + c_z \cdot z)}; \quad \tilde{M}^2 = \frac{2 \cdot M^2 c^2}{C_F \cdot \hbar^2}, \tag{324}$$

where we easily obtain the solution with the condition:

$$1 + \frac{c_t^2}{T} - c_x^2 - c_y^2 - c_z^2 = 0. \tag{325}$$

In order to fulfill the whole equation of $R = 0$ Eq. (321), we would also have to satisfy the condition:

$$\frac{c_t^2}{T} + c_x^2 + c_y^2 + c_z^2 = 0. \tag{326}$$

Adding Eq. (325) and Eq. (326) gives:

$$1 + 2 \cdot \frac{c_t^2}{T} = 0. \tag{327}$$

This fixes c_t to $c_t^2 = -\frac{T}{2}$ and leaves us with:

$$c_x^2 + c_y^2 + c_z^2 = \frac{1}{2}. \tag{328}$$

In fact, we have obtained a typical wave function solution directly from the Ricci scalar with the condition $R = 0$. We also found an expression for the mass.

On the other hand, with the condition $\Lambda = 0$ we obtain from the extended Einstein field equations Eq. (311) part a):

$$0 = R^{\kappa\lambda} - \frac{1}{2}R \cdot g^{\kappa\lambda} + \frac{1}{2}B \cdot V_\rho MB\gamma^{\kappa\lambda\omega\rho}v_\omega; \quad \{\kappa, \lambda\} = 1, 2, \ldots, N. \tag{329}$$

As so far there was no cosmological constant experimentally detected, we consider our setting fully justified.

Now we have the interesting constellation that with $R = 0$ the kernel of the integral in the Einstein–Hilbert action is zero. However, its variation does not vanish, because the approach Eq. (318) and Eq. (324) set into Eq. (329) gives:

$$0 = R^{\kappa\lambda} + \frac{1}{2}B \cdot V_\rho MB\gamma^{\kappa\lambda\omega\rho}v_\omega \tag{330}$$

and it is impossible to satisfy $R^{\kappa\lambda} = 0$ with the approach (318) and Eq. (324) in a nontrivial way. Only in the case of a vanishing mass M we would also have the case where the wave function metric Eq. (318), Eq. (324) solves both, the vacuum Einstein field equation and the Klein–Gordon equation. Thus, condition Eq. (330) fixes the surface term in the case $M \neq 0$.

Now we intend to solve $R^{\kappa\lambda} = 0$ directly and apply a metric approach:

$$g_{\alpha\beta} = \begin{pmatrix} a & s & u & v \\ s & b & \gamma & \delta \\ u & \gamma & \alpha & \sigma \\ v & \delta & \sigma & \beta \end{pmatrix} \cdot f\,[t, x, y, z] = \begin{pmatrix} a & s & u & v \\ s & b & \gamma & \delta \\ u & \gamma & \alpha & \sigma \\ v & \delta & \sigma & \beta \end{pmatrix} \cdot f \equiv c_{\alpha\beta} \cdot f. \tag{331}$$

With the function f as introduced in Eq. (324) we are able to find a variety of solutions with the following conditions with respect to the constant components in $c_{\alpha\beta}$:

$$(-\alpha\beta + \sigma^2)c_x^2 + (-b\beta + \delta^2)c_y^2 + 2(-\gamma\delta + b\sigma)c_y c_z$$
$$+ (-b\alpha + \gamma^2)c_z^2 + 2c_x\left((\beta\gamma - \delta\sigma)c_y + (\alpha\delta - \gamma\sigma)c_z\right) = 0, \quad (332)$$

$$((-v\alpha + u\sigma)c_x + (v\gamma + u\delta - 2s\sigma)c_y)c_z = c_y((u\beta - v\sigma)c_x$$
$$+ (-s\beta + v\delta)c_y) + (-s\alpha + u\gamma)c_z^2 + c_t((-\alpha\beta + \sigma^2)c_x$$
$$+ (\beta\gamma - \delta\sigma)c_y + (\alpha\delta - \gamma\sigma)c_z), \quad (333)$$

$$c_x((u\beta - v\sigma)c_x + (-s\beta + v\delta)c_y) + ((v\gamma - 2u\delta + s\sigma)c_x$$
$$+ (-bv + s\delta)c_y)c_z + (bu - s\gamma)c_z^2 + c_t((-\beta\gamma + \delta\sigma)c_x$$
$$+ (b\beta - \delta^2)c_y + (\gamma\delta - b\sigma)c_z) = 0, \quad (334)$$

$$(v\alpha - u\sigma)c_x^2 + c_y((bv - s\delta)c_y + (-bu + s\gamma)c_z)$$
$$+ c_x((-2v\gamma + u\delta + s\sigma)c_y + (-s\alpha + u\gamma)c_z) + c_t((-\alpha\delta + \gamma\sigma)c_x$$
$$+ (\gamma\delta - b\sigma)c_y + (b\alpha - \gamma^2)c_z) = 0, \quad (335)$$

$$(\alpha\beta - \sigma^2)c_t^2 + (-v^2 + a\beta)c_y^2 + 2(uv - a\sigma)c_y c_z + (-u^2 + a\alpha)c_z^2$$
$$= 2c_t((u\beta - v\sigma)c_y + (v\alpha - u\sigma)c_z), \quad (336)$$

$$(\beta\gamma - \delta\sigma)c_t^2 + (-v^2 + a\beta)c_x c_y + ((uv - a\sigma)c_x + (sv - a\delta)c_y)c_z$$
$$+ (-su + a\gamma)c_z^2 + c_t((-u\beta + v\sigma)c_x + (-s\beta + v\delta)c_y$$
$$+ (-2v\gamma + u\delta + s\sigma)c_z) = 0, \quad (337)$$

$$(\alpha\delta - \gamma\sigma)c_t^2 + c_y((uv - a\sigma)c_x + (-sv + a\delta)c_y)$$
$$+ ((-u^2 + a\alpha)c_x + (su - a\gamma)c_y)c_z + c_t((-v\alpha + u\sigma)c_x$$
$$+ (v\gamma - 2u\delta + s\sigma)c_y + (-s\alpha + u\gamma)c_z) = 0, \quad (338)$$

$$(b\beta - \delta^2)c_t^2 + (-v^2 + a\beta)c_x^2 + 2(sv - a\delta)c_x c_z + (ab - s^2)c_z^2$$
$$= 2c_t((s\beta - v\delta)c_x + (bv - s\delta)c_z), \quad (339)$$

$$(\gamma\delta - b\sigma)c_t^2 + c_x((uv - a\sigma)c_x + (-sv + a\delta)c_y)$$
$$+ ((-su + a\gamma)c_x + (-ab + s^2)c_y)c_z$$
$$= c_t((v\gamma + u\delta - 2s\sigma)c_x + (-bv + s\delta)c_y + (-bu + s\gamma)c_z), \quad (340)$$

$$(b\alpha - \gamma^2)c_t^2 + (-u^2 + a\alpha)c_x^2 + 2(su - a\gamma)c_x c_y + (ab - s^2)c_y^2$$
$$= 2c_t((s\alpha - u\gamma)c_x + (bu - s\gamma)c_y). \quad (341)$$

As the vanishing Ricci tensor also guarantees $R = 0$, we have obtained a perfect solution of the Einstein field equations, also—most interestingly—satisfying the quantum mechanical Klein–Gordon equation, if we would add the condition:

$$\tilde{M}^2 \cdot \left(\frac{c_t^2}{c^2} - (c_x^2 + c_y^2 + c_z^2) \right) + \frac{M^2 c^2}{\hbar^2} = 0. \tag{342}$$

However, it should be emphasized that—as it will be shown later in the book—we do not need this additional condition to be satisfied, simply because it turns out that the quantum theoretical Klein–Gordon equation is just an approximated form of the Einstein field equations. Thus, the approach Eq. (331) and the conditions Eq. (332) to Eq. (341) totally suffice for the complete quantum theoretical Einstein field equation solution.

Thus, we can conclude that the solution to the quantum theoretical Klein–Gordon equation is already hidden inside a special solution to the Einstein field equations. What it actually describes, however, is a slightly different story. De facto, we have obtained a rather wiggly space-time with mass, perfectly compatible with the quantum Klein–Gordon equation AND we had NO matter term within the Einstein field equations!

With the functions $g_i[x_i]$ being still arbitrary (this follows from the well-known tensor transformation rules, where any (potentially classic) metric solution g_{ij} to the Einstein field equations EFE can be extended in the following way:

$$g_{ij}[x_0, x_1, \ldots, x_n] \Rightarrow G_{ij} = g'_{x_i}[x_i] \cdot g'_{x_i}[x_i] \cdot g_{ij}[g_{x_0}[x_0], g_{x_1}[x_1], \ldots, g_{x_n}[x_n]]$$
$$\text{with:} \quad g'_{x_i}[x_i] \equiv \frac{\partial g_{x_i}[x_i]}{\partial x_i}, \tag{343}$$

leaving us with infinitely many options for the functions $g_i[\ldots]$) and comparing the solution Eq. (331) with our results from [8] (chapter "The Photon"), we see strong evidence that we have derived a fully GTR-compatible photon. We further investigated this in a variety of issues of the series of "Einstein had it..." (e.g., [106–110]).

We have also found, however, a solution, which is fit for providing the right space-time-background field scenario, explaining the quantum uncertainties we experience on smaller scales. This aspect will be discussed in the next section.

3.7 Heisenberg Uncertainty [166] due to the Wiggly Background

Considering the space-time solution which we have found above as some kind of background solution, we now want to investigate its properties in the case of the most simple assumption for the generalized functions $g_i[x_i]$ as given in Eq. (343), leading to normal distributions. We introduce the following $g_i[x_i]$:

$$g_t[t] = t[\tau] = -\tau^2; x = -\xi^2; y = -\eta^2; z = -\zeta^2. \qquad (344)$$

The metric Eq. (331), still solving the Einstein field equations due to Eq. (343), now reads:

$$g_{\alpha\beta} = \begin{pmatrix} a \cdot \tau^2 & s \cdot \tau \cdot \xi & u \cdot \tau \cdot \eta & v \cdot \tau \cdot \zeta \\ s \cdot \tau \cdot \xi & b \cdot \xi^2 & \gamma \cdot \xi \cdot \eta & \delta \cdot \xi \cdot \zeta \\ u \cdot \tau \cdot \eta & \gamma \cdot \xi \cdot \eta & \alpha \cdot \eta^2 & \sigma \cdot \eta \cdot \zeta \\ v \cdot \tau \cdot \zeta & \delta \cdot \xi \cdot \zeta & \sigma \cdot \eta \cdot \zeta & \beta \cdot \zeta^2 \end{pmatrix} \cdot f \equiv \tilde{c}_{\alpha\beta} \cdot f. \qquad (345)$$

Studying the metric Eq. (345), we see that we have obtained one complete space-time variance. Integration of the metric over the whole space, namely, would directly give us the second moment of the metric and we easily recognize the moments of normal distributions, allowing us to evaluate the Heisenberg uncertainty principle directly from the volume integral over the metric. Interpreting τ as time and the other coordinates as spatial ones, the metric in Eq. (345) directly gives us the reason for the quantum behavior of our universe, if we assume to have such a solution as a kind of universal background wiggling of the space-time. Similar to our approach in [39], we simply assume our space to be made of many pieces of such elements or particles as given in Eq. (345) now with piecewise coordinates as follows:

$$\tau \to \tau - \tau_i; \quad \xi \to \xi - \xi_j; \quad \eta \to \eta - \eta_k; \quad \zeta \to \zeta - \zeta_n. \qquad (346)$$

Being forced seamlessly together, we might take the resulting structure not only for our space-time, but also obtain the quantum background field. It needs to be pointed out that we might not need to introduce individual, localized τ. Then it might suffice to just have:

$$\tau \to \tau; \quad \xi \to \xi - \xi_j; \quad \eta \to \eta - \eta_k; \quad \zeta \to \zeta - \zeta_n. \qquad (347)$$

This quantum background is directly given throughout the metric, which was shown to be completely quantum compatible (either it solves the

Klein–Gordon equation directly or the Klein–Gordon equation is just an approximation of the EFE). Thus, the moment we have such a $R^{\kappa\lambda} = 0$-background, it provides a variance field, which gives us the Heisenberg principle, the speed of light limit and a minimum speed limit as shown in [39] for a photonic background field and in [46] for a background of strings.

In order to prove this, we only need to evaluate the following integrals:

$$\iiint\limits_V \int g_{\alpha\beta} \cdot \mathbf{dx}^4; \quad \mathbf{dx}^4 \equiv d\tau \cdot d\xi \cdot d\eta \cdot d\zeta \tag{348}$$

and take the necessary uncertainties $\Delta\tau$, $\Delta\xi$ etc. directly from the resulting tensor of Eq. (348). Now we use the Cauchy-Schwarz or Hölder inequality, which is given as

$$|E\,[x \cdot y]|^2 \leq E\,[x^2] \cdot E\,[y^2], \tag{349}$$

and result in an extended Heisenberg uncertainty principle [36, 39, 46]. Here we repeat the central evaluation on the example of $\Delta\tau$ and $\Delta\xi$ (all other coordinate variances are similar):

$$\Delta\xi \cdot \Delta\tau = \sqrt{\iiint\limits_V \int g_{11} \cdot \mathbf{dx}^4} \cdot \sqrt{\iiint\limits_V \int g_{00} \cdot \mathbf{dx}^4} \geq \iiint\limits_V \int g_{01} \cdot \mathbf{dx}^4$$

$$\Delta\xi \cdot \Delta\tau = \sqrt{\iiint\limits_V \int b \cdot \xi^2 \cdot \mathbf{dx}^4} \cdot \sqrt{\iiint\limits_V \int a \cdot \tau^2 \cdot \mathbf{dx}^4}$$

$$\geq \iiint\limits_V \int s \cdot \tau \cdot \xi \cdot \mathbf{dx}^4$$

$$s^2 \leq a \cdot b. \tag{350}$$

From there it is only a technicality to obtain the Heisenberg uncertainty in the usual form (see [36]). Still, we want to repeat the evaluation with the new metric distribution that we here have obtained from our quantum-compatible $R^{\kappa\lambda} = 0$-background field solution.

3.7.1 Connection to the Classical Heisenberg Uncertainty Principle

We already stated in [36] that with the form Eq. (350), being defined on the metric components, we immediately realize that there will be an individual uncertainty principle for every metric. However, in contrast to [36], this time the background metric in question is not arbitrary, but resulted from an explicit $R^{\kappa\lambda} = 0$-solution. Thus, wherever in the universe such a background field is present, it directly leads to an uncertainty principle Eq. (350)... be it Heisenberg's or somebody else's.

In order to find the connection to the classical form, we start with a simple dimensional analysis. At first, similar to the procedure in [36], we write down the uncertainty for a product of two spatial components, giving us:

$$\Delta\xi \cdot \Delta\eta = \sqrt{\iiint_V \int g_{11} \cdot \mathbf{dx}^4} \cdot \sqrt{\iiint_V \int g_{22} \cdot \mathbf{dx}^4} \geq \iiint_V \int g_{12} \cdot \mathbf{dx}^4.$$

(351)

We see that the term $\iiint_V \int g_{12} \cdot \mathbf{dx}^4$ in Eq. (351) must have the dimension of length squared. In the Heisenberg picture the length limit would be the Planck length with $l_P = \sqrt{\frac{\hbar \cdot G}{c^3}} \approx 1.616229 \cdot 10^{-35}$ m. Thus, we have

$$\left[\iiint_V \int g_{12} \cdot \mathbf{dx}^4 \right]_{\text{Heisenberg}} = \left[\iiint_V \int g_{12} \cdot \mathbf{dx}^4 \right]_H = \frac{\hbar \cdot G}{c^3}. \qquad (352)$$

As we already have pointed out in [36], this does only hold for the classical Heisenberg principle, not for its general metric amendment, but it will suffice here in order to show the connection with the classical principle. We have tried to make this clear by the index "H". Now we want to derive the classical Heisenberg uncertainty principle and remember that the momentum p is just the time derivative of a spatial coordinate times mass m. Thus, we can write (treating p here as a quantity of dimension length):

$$\Delta\xi \cdot \Delta p_\xi = \Delta\xi \cdot \frac{r_s}{c} \cdot \Delta\dot{\xi} = \Delta\xi \cdot r_s \cdot \Delta\left(\frac{\partial\xi}{c \cdot \partial\tau}\right) \qquad (353)$$

and keep the geometrical form and interpretation by using the Schwarzschild radius equivalent for the mass m. Using our results from above and applying Eq. (351), we have:

$$\Delta\xi \cdot \Delta p_\xi = \Delta\xi \cdot r_s \cdot \Delta\left(\frac{\partial\xi}{c \cdot \partial\tau}\right) \geq \left[\iiint_V \int g_{12} \cdot \mathbf{dx}^4 \right]_H = \frac{\hbar \cdot G}{c^3}. \qquad (354)$$

Setting the equation for the Schwarzschild radius $r_s = \frac{2 \cdot m \cdot G}{c^2}$ into our result, finally—almost—gives us the classical Heisenberg principle:

$$\Delta\xi \cdot r_s \cdot \Delta\left(\frac{\partial\xi}{c \cdot \partial\tau}\right) = \Delta\xi \cdot \frac{2 \cdot m \cdot G}{c^2} \cdot \Delta\left(\frac{\partial\xi}{c \cdot \partial\tau}\right)$$

$$\geq \frac{\hbar \cdot G}{c^3} \Rightarrow \Delta\xi \cdot \Delta\left(\frac{\partial\xi}{\partial\tau}\right) \geq \frac{\hbar}{2 \cdot m}, \qquad (355)$$

where we clearly see that the mass matters with respect to resolution achievable regarding certain observables. Now we might say that in fact this is the Heisenberg uncertainty principle, because we can transform Eq. (355) into:

$$\Delta \xi \cdot \Delta \left(\frac{\partial \xi}{\partial \tau} \right) \geq \frac{\hbar}{2 \cdot m} \Rightarrow \Delta \xi \cdot \Delta \left(m \cdot \frac{\partial \xi}{\partial \tau} \right) \geq \frac{\hbar}{2} \Leftrightarrow \Delta \xi \cdot \Delta p_\xi \geq \frac{\hbar}{2}. \quad (356)$$

In contrast to the "flat space photonic" background field, which we had used in [36], this time we have not lost the "="-sign along the way of our derivation from the extended Einstein field equations.

Thus, the Heisenberg uncertainty has directly been obtained from a background field solution of the type $R^{\kappa \lambda} = 0$, also satisfying the Klein–Gordon equation. Thereby the Klein–Gordon equation appeared from the condition $R = 0$ (cf. Eq. (321)). Now we want to see whether this could also help us to derive other important limits to this universe.

3.7.2 Finding Other Principle Limits

Now we want to go on from there and intend to obtain other principle limits of the space-time containing such a field.

Division of both sides in Eq. (350) by $\Delta \xi^2$ leads to:

$$\frac{\Delta \tau}{\Delta \xi} = \frac{\sqrt{\iiint_V \int g_{00} \cdot \mathbf{dx}^4}}{\sqrt{\iiint_V \int g_{11} \cdot \mathbf{dx}^4}} \geq \frac{\iiint_V \int g_{01} \cdot \mathbf{dx}^4}{\iiint_V \int g_{11} \cdot \mathbf{dx}^4}. \quad (357)$$

Forming the reciprocal on both sides changes the direction of the >-sign and leaves us with:

$$\frac{\Delta \xi}{\Delta \tau} = \frac{\sqrt{\iiint_V \int g_{11} \cdot \mathbf{dx}^4}}{\sqrt{\iiint_V \int g_{00} \cdot \mathbf{dx}^4}} \leq \frac{\iiint_V \int g_{11} \cdot \mathbf{dx}^4}{\iiint_V \int g_{01} \cdot \mathbf{dx}^4} \equiv c. \quad (358)$$

This way, we would have formulated the c- or speed of light in vacuum limit for a given scale with an underlying $R^{\kappa \lambda} = 0$-background metric of the form Eq. (345).

As already shown earlier [36], there is also another way to form the quotient $\frac{\Delta \xi}{\Delta \tau}$ and this will give us a minimum speed instead of a maximum as which we consider the speed of light (within our scale, we should add). Division of Eq. (350) by $\Delta \tau^2$ leads to:

$$\frac{\Delta \xi}{\Delta \tau} = \frac{\sqrt{\iiint \int g_{11} \cdot \mathbf{dx}^4}}{\sqrt{\iiint \int g_{00} \cdot \mathbf{dx}^4}} \geq \frac{\iiint \int g_{01} \cdot \mathbf{dx}^4}{\iiint \int g_{00} \cdot \mathbf{dx}^4} \equiv v_{\text{low}}. \qquad (359)$$

We will reconsider such limits in connection with the question about "11 The Origin of Time" later in this book.

3.8 Variation with Respect to the Number of Dimensions

As stated at the end of the section "3.3 The Ricci Scalar Quantization", we still have not found a way to fundamentally derive the need for the quantum equations out of our conditions to the Ricci scalar. The simple fact that these equations fall out after certain transformations on the metric is no justification for actually doing this transformation. It could only be seen as a piece of circumstantial evidence.

We now want to achieve a further reasoning for the metric trans-formation procedure by extending the variation of the Einstein–Hilbert action Eq. (87). Thereby, instead of variating only with respect to the metric tensor $g_{\alpha\beta}$, we also variate with respect to another suitable parameter.

It appears as kind of obvious choice to start with the number of dimensions n. Performing this additional variation via:

$$\delta_{n,g} W = \left(\frac{\partial W}{\partial n} \cdot \delta n + \frac{\partial W}{\partial g_{\alpha\beta}} \cdot \delta g_{\alpha\beta} \right) = 0$$

$$\Rightarrow 0 = \delta_n \int_V d^n x \left(\sqrt{-g} \left[R - 2\kappa L_{\text{M}} - 2\Lambda \right] \right)$$

$$+ \delta_g \int_V d^n x \left(\sqrt{-g} \left[R - 2\kappa L_{\text{M}} - 2\Lambda \right] \right), \qquad (360)$$

does not only reproduce the classical Einstein field equations (83), but also delivers:

$$
\begin{aligned}
0 = &\int_S n \cdot d^{n-1}x \left(\sqrt{-g} \left[R - 2\kappa L_M - 2\Lambda \right] \right) \cdot \delta n \\
&+ \int_V d^n x \frac{\partial \left(\sqrt{-g} \left[R - 2\kappa L_M - 2\Lambda \right] \right)}{\partial n} \cdot \delta n \\
\Rightarrow 0 = R - 2\kappa L_M - 2\Lambda \Rightarrow &\frac{\partial \left(\sqrt{-g} \left[R - 2\kappa L_M - 2\Lambda \right] \right)}{\partial n} = 0, \quad (361)
\end{aligned}
$$

which contains, as we have seen above, the classical quantum equations.

Thus, with the second line in Eq. (361) and the subsequent variation in Eq. (86) or Eq. (86) leading to the classical Einstein field equations Eq. (83), the latter should then simply be seen as the variation of an intelligent zero (cf. [10]). However, as we could also apply the Hilbert argument of the surface integral being chosen such that the result would give zero on the rim S (after all, the surface on which the integration is performed would be arbitrary), we would only be left with the condition in the third line. However, as this condition could be split up into:

$$
\begin{aligned}
0 = &\frac{\partial \left(\sqrt{-g} \left[R - 2\kappa L_M - 2\Lambda \right] \right)}{\partial n} = \left[R - 2\kappa L_M - 2\Lambda \right] \cdot \frac{\partial \sqrt{-g}}{\partial n} \\
&+ \sqrt{-g} \cdot \frac{\partial \left[R - 2\kappa L_M - 2\Lambda \right]}{\partial n}, \quad (362)
\end{aligned}
$$

we would again be back with the second line in Eq. (361) and thus, our intelligent zero variation.

3.8.1 Schwarzschild Metric as an Example

The reader should not be disturbed by the fact that we are going to repeat this example in connection with "10 The Origin of Time". There, however, we are going to use a slightly varied form of the evaluations here and will also deepen the discussion in connection with the question of what actually is time.

When observing the integral in Eq. (361), we see that—in principle—we also seek for a maximum volume for a given dimension or, taking the radius of a Schwarzschild object, look for the corresponding dimension making the volume integral an extremum. As the determinant g of the Schwarzschild metric is just equal to the one of an n-sphere with the additional time-dimension to be integrated, we can easily use the volume integral of n-spheres, which reads:

$$V_N = V_{n+1} = \frac{\pi^{\frac{(n)}{2}}}{\Gamma\left[\frac{(n+2)}{2}\right]} \cdot r^n. \tag{363}$$

Please note that due to the time-coordinate t we have V_{n+1} instead of V_n. Thereby the integration via t in Eq. (363) is assumed to be performed such that it would give 1. In general we might take care about this part of the integration via a proportional constant T which we could even consider to be n-dependent $T[n]$ and thus:

$$V_N = V_{n+1} = T[n] \cdot \frac{\pi^{\frac{(n)}{2}}}{\Gamma\left[\frac{(n+2)}{2}\right]} \cdot r^n. \tag{364}$$

Now we evaluate the various dimensions for which, for a given radius r of the n-sphere, we would obtain extrema. The results are shown as dots in Fig. 3.8.1 below. There we have illustrated the resulting r_{max} as functions of the dimensions $N = n + 1$ (note: $n = n$-sphere dimension, $N = t + n$-sphere dimension).

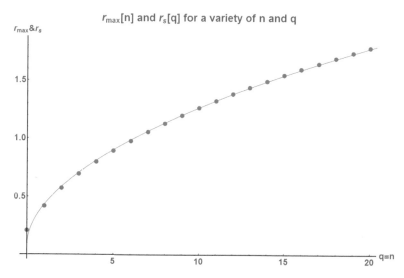

Figure 3.8.1 Radius r_{max} for which at a certain number of dimensions the n-sphere has maximum surface in dependence on n compared with the increase of the Schwarzschild radius r_s of a black hole in dependence on the number of bits q thrown into it by using the Hawking-Bekenstein formula (e.g., [131]). We find that $q = n$.

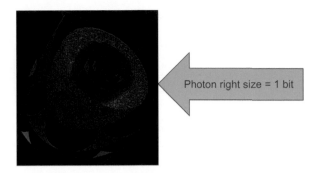

Figure 3.8.2   Illustration of the radius of the Bekenstein thought experiment.

Further discussion of this interesting finding, including a comprehensive consideration of the Bekenstein-Bit-Problem [146, 147] (see Fig. 3.8.2) is been given in [116, 145] with severe consequences regarding our notion of space, time and the question whether there might be an absolute scale in our universe [131, 132, 135].

3.8.2 Solving the Singularity Problem for Black Holes

It has to be pointed out that our discovery of the whereabouts of the Bekenstein-bits within a black hole automatically also solves the singularity problem of these objects. As shown in [131], by simply using the discovery above, namely that a black hole can be built up bit by bit, that only this way it can grow and automatically—along the way—must increase the number of its intrinsic dimensions, we will get rid of the singularity at the black hole's center.

We start with the metric of an n-dimensional spherical object (potentially a black hole):

$$g_{\alpha\beta}^n = \begin{pmatrix} -c^2 \cdot g\,[r] & 0 & 0 & 0 & \cdots & 0 \\ 0 & \frac{1}{g[r]} & 0 & 0 & \cdots & 0 \\ 0 & 0 & r^2 & 0 & \cdots & 0 \\ 0 & 0 & 0 & r^2 \cdot \sin^2 \varphi_1 & \cdots & 0 \\ \cdots & \cdots & \cdots & \cdots & \ddots & 0 \\ 0 & 0 & 0 & 0 & 0 & g_{nn} \end{pmatrix}$$

$$g_{44} = r^2 \cdot \sin^2 \varphi_1 \cdot \sin^2 \varphi_2; \, g_{nn} = r^2 \cdot \prod_{j=1}^{n-2} \sin^2 \varphi_j; \, g\,[r] = 1 - \frac{r_s^{n-3}}{r^{n-3}}. \quad (365)$$

In four dimensions the metric would be [142]:

$$
g_{\alpha\beta}^{\text{Schwarzschild}} =
\begin{pmatrix}
-c^2 \left(1 - \frac{2\cdot G\cdot M}{c^2\cdot r}\right) & 0 & 0 & 0 \\
0 & \left(1 - \frac{2\cdot G\cdot M}{c^2\cdot r}\right)^{-1} & 0 & 0 \\
0 & 0 & r^2 & 0 \\
0 & 0 & 0 & r^2\cdot \sin^2\vartheta
\end{pmatrix}.
\tag{366}
$$

Here r gives the spherical radius, M stands for the gravitational mass of the Schwarzschild object and the constants c and G are giving the speed of light in vacuum and the gravitational or Newton constant, respectively. One easily sees that the solution above has two singularities, namely one at $r = r_s = \frac{2\cdot M\cdot G}{c^2}$, which is a result of our choice of coordinates and could be transformed away, and at $r = 0$. The parameter r_s is called the Schwarzschild radius.

We see that the classical time component $g_{00} = -c^2\cdot g[r]$ vanishes at $r = r_s$ and becomes space-like for $r < r_s$, which is inside the black hole. There, however, that r-component g_{11} would become time-like and thus, we would still have a "classical" time dimension behind the event horizon.

But how could it be understood that a black hole of a certain mass or radius should contain a corresponding number of dimensions? Obviously the 4-dimensional black hole would not be a very massive and therefore not a very stable object (cf. [116] and see Schwarzschild radius for $n = q = 4$ in Fig. 3.8.1).

The way to overcome that particular problem would be to say goodbye to the idea of black holes of a certain radius existing independent on the number of dimensions they contain. Thereby especially the number of dimensions making out the Schwarzschild n-sphere is of importance. The simplest way to construct objects being compatible with the Bekenstein-observation and our finding of the $r_s[q = n]$-dependency would be a radius-dependent change of dimensions. When moving towards the black hole, the dimension near the event horizon increases until it reaches the necessary n at the corresponding $r_s[q = n]$. While it appears intuitive to assume an $n[r]$-dependency being simply inverse to the function distributed in Fig. 3.8.1* (note that we have $q = n$) for $r < r_s$, we can only guess† what might be a suitable approach for the outside region $r > r_s$ for r towards the classical event horizon. For simplicity we here assume an exponential behavior. Let the inverse function to $r_s[n]$ (Fig. 3.8.1) be denoted as $n[r_s]$, we approach the total and

---

*The explicit function is given in [131] or in section "12 The Origin of Time".
†Respectively experimentally determine.

$n[r]$-variable metric via the Schwarzschild function:

$$g[r] = 1 - \frac{r_s^{n[r]-3}}{r^{n[r]-3}}; n[r] = \begin{cases} n[r] \text{ for } r < r_s \\ C \cdot \left(n[r_s] \cdot e^{-\left(\frac{r}{r_s}\right)^p} + 4 \right) \text{ for } r \geq r_s \end{cases}$$

$$\text{with: } \left[C \cdot \left(n[r_s] \cdot e^{-\left(\frac{r}{r_s}\right)^p} + 4 \right) \right]_{r=r_s} = n[r_s]; p > 1. \qquad (367)$$

This way we would end up with the usual outside behavior for spherical massive objects, but avoid the singularity for $r = 0$. In addition we obtain a solution being able to store the Bekenstein-bits q by the means of dimensions n via the simple relation $n = q$.

As said at the beginning of section 3.8.1, because if its importance also with respect to the universal Turing machine, we are going to reconsider the Bekenstein-Bit problem in chapter "11 The Origin of Time."

3.9 Using the General Einstein–Hilbert Action with $f[R]$

Another and—what is more—apparently simple and straight forward way to achieve a general connection between GTR and quantum theory might also be seen in the application of the generalized Einstein–Hilbert action Eq. (86) and its subsequent variation as demonstrated in [114]. Only in order to give one example, let us assume to have $f[R] = R_0 + F[R_1]$. The variation Eq. (86) would then read:

$$R_{0\mu\nu} - \frac{1}{2} R_0 \cdot g_{\mu\nu} + F'[R_1] \cdot R_{1\mu\nu} - \frac{1}{2} F[R_1] \cdot g_{\mu\nu}$$
$$+ F''[R_1] \left[R_{1;\mu\nu} - \Delta_g R_1 \cdot g_{\mu\nu} \right] + F'''[R_1] \left[R_{1;\mu} R_{1;\nu} - R_1^{;\sigma} R_{1;\sigma} \cdot g_{\mu\nu} \right]$$
$$= \begin{cases} 0 \dots \text{"vacuum"} \\ -8\pi G T_{\mu\nu} \dots \text{postulated matter} \end{cases}. \qquad (368)$$

Thereby we have used the following definition for the Ricci curvatures R_0 and R_1:

$$g^{\mu\nu} R_{0\mu\nu} \equiv R_0; \quad g^{\mu\nu} R_{1\mu\nu} \equiv R_1. \qquad (369)$$

Now we assume that there is a huge scale difference between the Ricci curvatures R_0 and R_1. While R_0 can only be registered over huge distance, we find that R_1 only is visible on extremely small scales. We demand that the various scales are so distant, that we can separate them and solve our field equations for each scale differently. In addition we say that the postulated matter is just the connector between the two scales, which gives us:

$$
-\left(R_{0\mu\nu} - \frac{1}{2}R_0 \cdot g_{\mu\nu}\right) = 8\pi G T_{\mu\nu} = F'[R_1] \cdot R_{1\mu\nu} - \frac{1}{2}F[R_1] \cdot g_{\mu\nu}
$$
$$
+ F''[R_1]\left[R_{1;\mu\nu} - \Delta_g R_1 \cdot g_{\mu\nu}\right] + F'''[R_1]\left[R_{1;\mu}R_{1;\nu} - R_1^{;\sigma}R_{1;\sigma} \cdot g_{\mu\nu}\right].
$$
(370)

This gives us the usual Einstein vacuum or matter equation for the bigger scales, namely:

$$
R_{0\mu\nu} - \frac{1}{2}R_0 \cdot g_{\mu\nu} = \begin{cases} 0 \ldots \text{"vacuum"} \\ -8\pi G T_{\mu\nu} \ldots \text{matter} \end{cases}
$$
(371)

and the following additional equation with respect to the smaller scales:

$$
F'[R_1] \cdot R_{1\mu\nu} - \frac{1}{2}F[R_1] \cdot g_{\mu\nu} + F''[R_1]\left[R_{1;\mu\nu} - \Delta_g R_1 \cdot g_{\mu\nu}\right]
$$
$$
+ F'''[R_1]\left[R_{1;\mu}R_{1;\nu} - R_1^{;\sigma}R_{1;\sigma} \cdot g_{\mu\nu}\right] = \begin{cases} 0 \ldots \text{"vacuum"} \\ 8\pi G T_{\mu\nu} \ldots \text{matter} \end{cases}.
$$
(372)

Please note that the matter term is now not postulated but just the separator between the two scales. In order to easier recognize typical quantum equations within Eq. (372), we choose $F[\ldots]$ such that F'' should be— almost—a constant, thereby making $F'''[R_1] \cong 0$ and leaving us with:

$$
\frac{F'[R_1] \cdot R_{1\mu\nu} - \frac{1}{2}F[R_1] \cdot g_{\mu\nu} - 8\pi G T_{\mu\nu}}{F''[R_1]} + R_{1;\mu\nu} - \Delta_g R_1 \cdot g_{\mu\nu} = 0. \quad (373)
$$

Now we "scalarize" the equation via (n = number of dimensions):

$$
\frac{g^{\mu\nu}}{n} \cdot \left(\frac{F'[R_1] \cdot R_{1\mu\nu} - \frac{\kappa}{2}F[R_1] \cdot g_{\mu\nu} - 8\pi G T_{\mu\nu}}{F''[R_1]} + R_{1;\mu\nu} - \kappa \cdot \Delta_g R_1 \cdot g_{\mu\nu}\right)
$$
$$
= \underbrace{\frac{F'[R_1] \cdot R_{1\mu\nu}g^{\mu\nu} - \frac{\kappa \cdot n}{2}F[R_1] - 8\pi G T_{\mu\nu}g^{\mu\nu}}{\kappa \cdot n \cdot G''[R_1]} + \frac{R_{1;\mu\nu}g^{\mu\nu}}{\kappa \cdot n}}_{\text{mass \& potential}} - \Delta_g R_1 = 0
$$

(374)

and recognize a Klein–Gordon-like equation with respect to the Ricci scalar R_1. We note that this Ricci scalar R_1 here takes on the position of the wave function of the classical quantum theory, while the complex terms

$$
\frac{F'[R_1] \cdot R_{1\mu\nu}g^{\mu\nu} - \frac{\kappa \cdot n}{2}F[R_1] - 8\pi G T_{\mu\nu}g^{\mu\nu}}{\kappa \cdot n \cdot F''[R_1]} + \frac{R_{1;\mu\nu}g^{\mu\nu}}{\kappa \cdot n} = \text{mass \& potential}
$$
(375)

would stand for the classical mass, energy and other matter fields. The constant κ was introduced in order to account for the dimensional summation of the co- and contravariant metric. We see from Eq. (371) that a choice of $\kappa = 2/n$ would always lead to a vanishing scalarized classical vacuum Einstein field equation and thus, Eq. (374) would become:

$$\frac{1}{2} \cdot \left(\frac{F'[R_1] \cdot R_{1\mu\nu}g^{\mu\nu} - F[R_1] - 8\pi\,GT_{\mu\nu}g^{\mu\nu}}{F''[R_1]} + R_{1;\mu\nu}g^{\mu\nu} \right) - \Delta_g R_1 = 0.$$
(376)

In the following, however, we shall leave $\kappa = 1$.

The corresponding field equation to the Klein–Gordon equation would then be Eq. (373) and not the classical Dirac equation anymore. The simplest version can be obtained for $F[R_1] = (R_1)^2$, leading to:

$$R_1 \cdot R_{1\mu\nu} - \frac{(R_1)^2}{4} \cdot g_{\mu\nu} - 4\pi\,GT_{\mu\nu} + R_{1;\mu\nu} - \Delta_g R_1 \cdot g_{\mu\nu} = 0 \qquad (377)$$

and:

$$R_1 \cdot R_1 - \frac{(R_1)^2}{4} \cdot n - 4\pi\,GT_{\mu\nu}g^{\mu\nu} + R_{1;\mu\nu}g^{\mu\nu} - \Delta_g R_1 \cdot n$$

$$= (R_1)^2 \left(1 - \frac{n}{4} \right) - 4\pi\,GT_{\mu\nu}g^{\mu\nu} + R_{1;\mu\nu}g^{\mu\nu} - \Delta_g R_1 \cdot n = 0.$$

$$\xrightarrow{4D} \left(\frac{R_{1;\mu\nu}}{4} - \pi\,GT_{\mu\nu} \right) g^{\mu\nu} - \Delta_g R_1 = 0. \qquad (378)$$

Interestingly we obtain vanishing Ricci scalars in the case of 4 dimensions and only the derivatives of R_1 remain in our Klein–Gordon equation.

Connecting Eq. (377) and Eq. (378) with the corresponding form of Eq. (371) results in:

$$R_1 \cdot R_{1\mu\nu} - \frac{(R_1)^2}{4} \cdot g_{\mu\nu} + \frac{R_{0\mu\nu}}{2} - \frac{1}{4}R_0 \cdot g_{\mu\nu} + R_{1;\mu\nu} - \Delta_g R_1 \cdot g_{\mu\nu} = 0 \quad (379)$$

and:

$$(R_1)^2 \left(1 - \frac{n}{4} \right) + \left(\frac{R_{0\mu\nu}}{2} - \frac{1}{4}R_0 \cdot g_{\mu\nu} \right) g^{\mu\nu} + R_{1;\mu\nu}g^{\mu\nu} - \Delta_g R_1 \cdot n$$

$$= (R_1)^2 \left(1 - \frac{n}{4} \right) + \frac{R_0}{2} - \frac{1}{4}R_0 \cdot n + R_{1;\mu\nu}g^{\mu\nu} - \Delta_g R_1 \cdot n$$

$$= (R_1)^2 \left(1 - \frac{n}{4} \right) + R_0 \cdot \left(\frac{1}{2} - \frac{n}{4} \right) + R_{1;\mu\nu}g^{\mu\nu} - \Delta_g R_1 \cdot n = 0$$

$$\xrightarrow{4D} \frac{R_{1;\mu\nu}}{4}g^{\mu\nu} - \frac{R_0}{2} - \Delta_g R_1 = 0. \qquad (380)$$

We see that the simple square approach for $F[R_1]$ does not lead us to the usual quantum eigenequations. This would only be the case if the terms:

$$(R_1)^2 \left(1 - \frac{n}{4}\right) + R_0 \cdot \left(\frac{1}{2} - \frac{n}{4}\right) + R_{1;\mu\nu} g^{\mu\nu} \xrightarrow{4D} \frac{R_{1;\mu\nu}}{4} g^{\mu\nu} - \frac{R_0}{2} \quad (381)$$

could somehow be approximated by expressions like:

$$(R_1)^2 \left(1 - \frac{n}{4}\right) + R_0 \cdot \left(\frac{1}{2} - \frac{n}{4}\right) + R_{1;\mu\nu} g^{\mu\nu} \approx G_n [x_0, x_1, \ldots] \cdot R_1$$

$$\xrightarrow{4D} \frac{R_{1;\mu\nu}}{4} g^{\mu\nu} - \frac{R_0}{2} \approx G_4 [x_0, x_1, \ldots] \cdot R_1. \quad (382)$$

We therefore now leave the function $F[\ldots]$ general and demand:

$$R_{1;\mu} R_{1;\nu} - R_1^{;\sigma} R_{1;\sigma} \cdot g_{\mu\nu} \simeq 0 \quad (383)$$

or at least:

$$\left(R_{1;\mu} R_{1;\nu} - R_1^{;\sigma} R_{1;\sigma} \cdot g_{\mu\nu}\right) g^{\mu\nu} = R_{1;\mu} R_{1;\nu} g^{\mu\nu} - R_1^{;\sigma} R_{1;\sigma} \cdot n \simeq 0. \quad (384)$$

Now we set $F[\ldots]$ to be:

$$F[R_1] = C_2 + C_1 \cdot e^{R_1/C_R}. \quad (385)$$

Together with Eq. (383) this makes Eq. (373) to:

$$C_R \cdot R_{1\mu\nu} + C_R^2 \cdot \left(\frac{g_{\mu\nu}}{2} + e^{-R_1/C_R} \cdot \left(\frac{C_2 \cdot g_{\mu\nu} - 16\pi \, GT_{\mu\nu}}{2 \cdot C_1}\right)\right) + R_{1;\mu\nu} - \Delta_g R_1 \cdot g_{\mu\nu} = 0. \quad (386)$$

Contraction via the contravariant metric tensor leads to:

$$C_R \cdot R_1 + C_R^2 \cdot \left(\frac{n}{2} + e^{-R_1/C_R} \cdot \left(\frac{C_2 \cdot n}{2 \cdot C_1} - \frac{8\pi \, GT_{\mu\nu} g^{\mu\nu}}{C_1}\right)\right) + R_{1;\mu\nu} g^{\mu\nu} - \Delta_g R_1 \cdot n$$

$$= C_R \cdot R_1 + C_R^2 \cdot \left(\frac{n}{2} + e^{-R_1/C_R} \cdot \left(\frac{C_2 \cdot n}{2 \cdot C_1} + R_0 \frac{(1 - \frac{n}{2})}{C_1}\right)\right) + R_{1;\mu\nu} g^{\mu\nu} - \Delta_g R_1 \cdot n$$

$$= \frac{C_R \cdot R_1 + R_{1;\mu\nu} g^{\mu\nu}}{4} + C_R^2 \cdot \left(\frac{1}{2} + e^{-R_1/C_R} \cdot \left(\frac{C_2 \cdot 1}{2 \cdot C_1} + R_0 \frac{(2 - n)}{2 \cdot n \cdot C_1}\right)\right) - \Delta_g R_1 = 0$$

$$\xrightarrow{4D} \frac{C_R \cdot R_1}{4} + C_R^2 \cdot \left(\frac{1}{2} + e^{R_1/C_R} \cdot \left(\frac{C_2}{2 \cdot C_1} - R_0 \frac{1}{4 \cdot C_1}\right)\right) + \frac{R_{1;\mu\nu} g^{\mu\nu}}{4} - \Delta_g R_1 = 0 \quad (387)$$

and in fact, gives us an eigenequation for the "Ricci curvature wave function" R_1.

It should be pointed out that—of course—there could be not just two very different scales of curvature, but many. In consequence we'd have only one unified theory on the cost of a two respectively a multifold of scale levels of curvatures, one providing the gravity effects and the other the quantum effects. More scale levels could be responsible for an even greater variety of forces and interactions with even bigger differences to the known interactions.

3.10 The Variation of the Metric Tensor: Brief Introduction

If now seeing the properties-action-extremal-approach as "the most principle one." Quantum theory should to be found somewhere inside it. It was shown above that there are in fact possibilities to derive quantum equations from metric solutions to the Einstein field equations via special transformations, but the question remains whether there is a more general joint. This connection was first presented in [114] and it consists of a simple extension of the variation of the Einstein–Hilbert action with respect to the base vectors and coordinates instead of the classical Hilbert variation [137], which ends with the metric tensor. The variation result for the action Eq. (86) was presented in [114] in the following way:

$$
\delta_g W = 0 = \int_V d^n x \overbrace{\left(\begin{array}{c} f'\,[R]\,R^{\delta\gamma} - \dfrac{1}{2} f\,[R]\,g^{\delta\gamma} + f''\,[R]\left[R^{;\delta\gamma} - \Delta_g R \cdot g^{\delta\gamma} \right] \\[2mm] + f'''\,[R]\left[R^{;\delta}\,R^{;\gamma} - R^{;\sigma}\,R_{;\sigma} \cdot g^{\delta\gamma} \right] + \underbrace{8\pi\,G}_{\kappa}\,T^{\delta\gamma} \end{array} \right)}^{\text{Relativity}}
$$

$$
\times \underbrace{\left(\begin{array}{c} \delta g_{\delta\gamma} = \delta\left(g_\delta \cdot g_\gamma\right) = g_\delta \cdot \delta g_\gamma + \delta g_\delta \cdot g_\gamma \\[2mm] = \dfrac{\partial G^\alpha\,[x_k]}{\partial x^\delta} e_\alpha \cdot \delta\left(\dfrac{\partial G^\beta\,[x_k]}{\partial x^\gamma} e_\beta \right) + \delta\left(\dfrac{\partial G^\alpha\,[x_k]}{\partial x^\delta} e_\alpha \right) \cdot \dfrac{\partial G^\beta\,[x_k]}{\partial x^\gamma} e_\beta \end{array} \right)}_{\text{Quantum}} \cdot
$$

(388)

Thereby, in the case of the classical Einstein–Hilbert simplification [137, 138] with $f[R] = R$ as given in Eq. (87), we could write:

$$\delta_g W = 0 = \int_V d^n x \overbrace{\left(R^{\delta\gamma} - \frac{1}{2} Rg^{\delta\gamma} + \Lambda g^{\delta\gamma} + \kappa T^{\delta\gamma} \right)}^{\text{Relativity}}$$

$$\times \underbrace{\left(\begin{array}{c} \delta g_{\delta\gamma} = \delta(\mathbf{g}_\delta \cdot \mathbf{g}_\gamma) = \mathbf{g}_\delta \cdot \delta\mathbf{g}_\gamma + \delta\mathbf{g}_\delta \cdot \mathbf{g}_\gamma \\ = \frac{\partial G^\alpha [x_k]}{\partial x^\delta} \mathbf{e}_\alpha \cdot \delta\left(\frac{\partial G^\beta [x_k]}{\partial x^\gamma} \mathbf{e}_\beta \right) + \delta\left(\frac{\partial G^\alpha [x_k]}{\partial x^\delta} \mathbf{e}_\alpha \right) \cdot \frac{\partial G^\beta [x_k]}{\partial x^\gamma} \mathbf{e}_\beta \end{array} \right)}_{\text{Quantum}}. \quad (389)$$

Now we were able to show in [117, 136] how to derive classical quantum equations from the variated metric tensor $\delta g_{\delta\gamma}$.

We realize that neither gravity solutions need to be quantized, nor do we have to make quantum solutions compatible to the General Theory of Relativity. One solution to either of the two satisfies all.

In other words: We have what one might call a "world formula."

Because of the obvious importance of this approach, we want to investigate it in more detail in the following main chapter.

Before we go there, however, we want to reconsider the variation of the metric tensor in additive form as introduced and discussed in [102, 105, 107].

3.11 The Additive Variation of the Metric Tensor

3.11.1 A Slightly Philisophical Starting Point

Even though the "honest philosophical engineer in the making" [105] might not see it, we already have derived rather fundamental equations to every system of arbitrary numbers of dimensions (properties). As a problem one intends to solve is just nothing else but a system of properties and the task usually is to find the optimum answer to this very problem we have—in principle—found the equations governing everything. There might be the argument that "our" equations (in fact, these are just Einstein's equations) do not hold true in fractal spaces or non-smooth coordinates of distributive character. This, however, is not true, because, as our solutions are tensors, a simple coordinate transformation like:

$$g_{ij} \left[\forall x_i = x_0, x_1, \dots, x_n \right]$$

$$\Rightarrow G_{ij} = G'_{x_i} \left[x_{\forall i} \right] \cdot G'_{x_j} \left[x_{\forall j} \right] \cdot g_{ij} \left[G_{x_0} \left[x_{\forall i} \right], G_{x_1} \left[x_{\forall i} \right], \dots, G_{x_n} \left[x_{\forall i} \right] \right]$$

$$\text{with: } G'_{x_i} \left[x_{\forall i} \right] \equiv \frac{\partial G_{x_i} \left[x_{\forall i} \right]}{\partial x_i} \tag{390}$$

would allow us to generalize any smooth metric solution g_{ij} to a potentially distributed or nearly* fractal one G_{ij}.

We were even able to show ways for the incorporation of uncertainty. We know that there is just this one field in science, the quantum theory, namely, which only exists because of all the uncertainties we have in this very universe of ours. Thus, the task to find a truly holistic guideline for answering the philosophical question of what actually is just everything (including right and good [105]) in completely mathematical form, requires us to combine the Einstein field equations in classical (3) or extended form Eq. (368) with quantum theory.

Even though we already have quite a selection of methods to achieve this, it is both, illustrative and instructive, to also consider, respectively reconsider some earlier approaches with additive metric variation in combination with zero sum-assumptions as briefly discussed in the subsection "2.2 Intelligent Zero Approaches—Just one Example."

3.11.2 Combination with Quantum Theory via the Variation of Base Vectors—Getting Started

We have shown in our previous papers that there exists a variety of ways to quantize solutions of the Einstein field equations EFE. Usually, however, it is been considered the Holy Grail for a physical theory when it can be derived from a minimum principle. Thus, here we want to demonstrate in the simplest, which is to say first-order-derivative case (for other options see [102]), how a variation approach for the base vectors of a given metric γ_{ij} leads to a quantization of that very metric, respectively its corresponding EFE solution[†]. A reasonable and very fundamental starting point would be

---

*"Nearly" here means that—of course—the coordinates would appear whole and continuous, but on certain scales, they could be made to appear as fractal. It simply requires a suitable selection of coordinate functions $G_{x_i}[x_{\forall i}]$ to mirror all possible distributions one could think of.

[†]Please note that the metric is the solution of the Einstein field equations EFE and thus, our formulation might be seen as kind of redundant, but we simply wanted to point out that the

the application of the tensor transformation rules:

$$G^{\alpha\beta}\left[\forall G^m\left[x_{\forall m}\right]\right] = \left(\frac{\partial G^i\left[x_{\forall i}\right]}{\partial x^{\alpha}}\right)^{-1} \cdot \left(\frac{\partial G^j\left[x_{\forall i}\right]}{\partial x^{\beta}}\right)^{-1} \cdot \gamma^{ij}\left[x_{\forall i}\right]$$

$$G_{\alpha\beta}\left[\forall G^m\left[x_{\forall m}\right]\right] = \frac{\partial G^i\left[x_{\forall i}\right]}{\partial x^{\alpha}} \cdot \frac{\partial G^j\left[x_{\forall i}\right]}{\partial x^{\beta}} \cdot \gamma_{ij}\left[x_{\forall i}\right]$$

$$\{i, j, m, \alpha, \beta\} = 1\dots N; \; N = \text{Dimension of space.} \qquad (391)$$

From there we construct a Lagrange density as follows:

$$L = G^{\alpha\beta}\gamma_{\alpha\beta} - G_{\alpha\beta}\gamma^{\alpha\beta} = \frac{\partial x^{\alpha}}{\partial G^i}\frac{\partial x^{\beta}}{\partial G^j}\gamma^{ij}\gamma_{\alpha\beta} - \frac{\partial G^i}{\partial x^{\alpha}}\frac{\partial G^j}{\partial x^{\beta}}\gamma_{ij}\gamma^{\alpha\beta}. \qquad (392)$$

Even though—on first sight—this looks quite different from our approach as presented in the main sections of [102], already very few most reasonable assumptions will lead us to the same structure. For one thing, we want the Lagrangian, which we consider a pure quantum effect Lagrangian, to vanish in the case there is no quantum structure. This automatically demands that with vanishing quantum structure all derivatives should just be unit matrices. We symbolize the vanishing quantum nature of a universe by the limiting-process of (vanishing off the reduced Planck constant $\hbar$) and demand:

$$\lim_{\hbar\to 0}\frac{\partial x^{\alpha}}{\partial G^i} \to \delta_i^{\alpha}; \; \lim_{\hbar\to 0}\frac{\partial x^{\beta}}{\partial G^j} \to \delta_j^{\beta}; \; \lim_{\hbar\to 0}\frac{\partial G^i}{\partial x^{\alpha}} \to \delta_{\alpha}^i; \; \lim_{\hbar\to 0}\frac{\partial G^j}{\partial x^{\beta}} \to \delta_{\beta}^j. \qquad (393)$$

Then, just as demonstrated in [102] rather extensively, we have a choice of options to construct suitable differential dependencies with function vectors f_n and their corresponding derivatives (covariant derivatives possibly), in order to reduce the search for a dependency, fulfilling either the variation condition for $\delta L = 0$ or even the more Dirac-like condition $L = 0$ with just one such function vector. The subsequent simplest result could either be:

$$\frac{\partial G_+^i\left[x_{\forall i}\right]}{\partial x^m} = \left\{\begin{array}{l}\left(\delta_m^i \pm C^i \cdot f_m\right)\\\left(\delta_m^i \pm f^i \cdot C_m\right)\end{array}\right\}; \quad \frac{\partial x^i}{\partial G_+^m\left[x_{\forall i}\right]} = \left\{\begin{array}{l}\left(\delta_m^i \pm f_m^{;i}\right)\\\left(\delta_m^i \pm f^i{}_{;m}\right)\end{array}\right\}, \qquad (394)$$

or:

$$\frac{\partial G_+^i\left[x_{\forall i}\right]}{\partial x^m} = \left\{\begin{array}{l}\left(\delta_m^i \pm f_m^{;i}\right)\\\left(\delta_m^i \pm f^i{}_{;m}\right)\end{array}\right\}; \quad \frac{\partial x^i}{\partial G_+^m\left[x_{\forall i}\right]} = \left\{\begin{array}{l}\left(\delta_m^i \pm C^i \cdot f_m\right)\\\left(\delta_m^i \pm f^i \cdot C_m\right)\end{array}\right\}. \qquad (395)$$

---

quantization procedure also works for other metrics, which are not necessarily solving the classical EFE (being of first order in R). This might become important in cases where the Lagrange density for the Einstein–Hilbert action would be of the type $f[R]$ instead of just a linear R.

Then, as demonstrated in [102], we look for combinations of these variated base vectors in such a way that practicable equations would result with respect to our function vectors f_n. There is a variety of options but the most intuitive might be seen in the following structure (thereby applying Eq. (394)):

$$L = \left(\gamma^{mn} \left\{ \frac{\forall \left(e_m^+ \cdot e_n^- \right)}{4} \right\} - \frac{\gamma_{mn}}{4} \left\{ \forall \left(e_+^m \cdot e_-^n \right) \right\} \right)$$

$$= \left(\begin{array}{l} \frac{\gamma^{mn}}{4} \left(\begin{array}{l} e_i \left(\delta_m^i + C^i \cdot f_m \right) \cdot e_j \left(\delta_n^j - C^j \cdot f_n \right) + e_i \left(\delta_m^i + f^i \cdot C_m \right) \cdot e_j \left(\delta_n^j - f^j \cdot C_n \right) \\ + e_i \left(\delta_m^i + C^i \cdot f_m \right) \cdot e_j \left(\delta_n^j - f^j \cdot C_n \right) + e_i \left(\delta_m^i + f^i \cdot C_m \right) \cdot e_j \left(\delta_n^j - C^j \cdot f_n \right) \end{array} \right) \\ - \frac{\gamma_{mn}}{4} \left(\begin{array}{l} e^i \left(\delta_i^m + f_i^{;m} \right) \cdot e^l \left(\delta_l^n - f_l^{;n} \right) + e^i \left(\delta_i^m + f^m_{;i} \right) \cdot e^l \left(\delta_l^n - f^n_{;l} \right) \\ + e^i \left(\delta_i^m + f_i^{;m} \right) \cdot e^l \left(\delta_l^n - f^n_{;l} \right) + e^i \left(\delta_i^m + f^m_{;i} \right) \cdot e^l \left(\delta_l^n - f_i^{;n} \right) \end{array} \right) \end{array} \right).$$

$$= \left(\begin{array}{l} \frac{\gamma^{mn}}{4} \left(e_i \cdot e_j \left(\begin{array}{l} 4\delta_m^i \delta_n^j - C^i \cdot f_m C^j \cdot f_n - f^i \cdot C_m C^j \cdot f_n \\ -C^i \cdot f_m f^j \cdot C_n - f^i \cdot C_m f^j \cdot C_n \end{array} \right) \right) \\ - \frac{1}{4} \left(\gamma_{mn} e^i \cdot e^l \left(4\delta_i^m \delta_l^n - f_i^{;m} f_i^{;n} - f^m_{;i} f^n_{;l} - f_i^{;m} f^n_{;l} - f^m_{;i} f_i^{;n} \right) \\ + f_i^{;i} - f^n_{;n} + f^m_{;m} - f_i^{;l} \right) \end{array} \right)$$

(396)

Thereby the ";" denotes the covariant derivative. Simplification yields:

$$L = \frac{1}{2} \left(f^j_{;n} f^{;n}_j + f^{n;j} d_{j;n} - f^n C_j C^j f_n - C^m f_m f^j C_j \right),$$

(397)

where, with respect to the Dirac approach, we assume that we are always able to find a setting for the C_n allowing us to set $L = 0$.

With the following definition of the covariant variational derivative (D_k shall stand for a suitable form of covariant derivatives):

$$\frac{\delta L}{\delta \phi} = \frac{\partial L}{\partial \phi_i} - \partial_\mu \left(\frac{\partial L}{\partial \left(\partial_\mu \phi_i \right)} \right)$$

$$\Rightarrow \left\{ \begin{array}{l} \dfrac{\delta L}{\delta f_X} = \dfrac{\partial L}{\partial f_X} - D_k \left(\dfrac{\partial L}{\partial \left(f_X^{;k} \right)} \right) - D^k \left(\dfrac{\partial L}{\partial \left(f_{X;k} \right)} \right) \\ \dfrac{\delta L}{\delta f^X} = \dfrac{\partial L}{\partial f^X} - D^k \left(\dfrac{\partial L}{\partial \left(f^X_{;k} \right)} \right) - D_k \left(\dfrac{\partial L}{\partial \left(f^{X;k} \right)} \right) \end{array} \right. ,$$

(398)

we obtain two separate results for the co- and the contravariant form of the function vectors f_n and f^n, respectively:

$$\frac{\delta L}{\delta f_j} \Rightarrow \frac{1}{2} \left\{ -\sqrt{\gamma} f^j C_n C^n - \sqrt{\gamma} C^j f^n C_n - D_m \sqrt{\gamma} \gamma^{mn} d^j{}_{;n} - D_j \sqrt{\gamma} f^{n;j} \right.$$

$$\frac{\delta L}{\delta f^j} \Rightarrow \frac{1}{2} \left\{ -\sqrt{\gamma} C_n C^n f_j - \sqrt{\gamma} C^n f^n C_j - D_n \sqrt{\gamma} f_j{}^{;n} - D_m \sqrt{\gamma} \gamma^{mn} f_{j;n} \right..$$

$$(399)$$

Applying the rule:

$$\gamma^{ij;k} = \gamma_{mn;k} = \gamma^{ij}{}_{;k} = \gamma_{mn}{}^{;k} = 0, \tag{400}$$

we can further reshape two of the terms in (399):

$$D_j \sqrt{\gamma} f^{n;j} = D_j \sqrt{\gamma} \gamma^{jm} f^n{}_{;m}; \quad D_n \sqrt{\gamma} f_j{}^{;n} = D_n \sqrt{\gamma} \gamma^{nm} f_{j;m}, \tag{401}$$

where we immediately recognize the Laplace-Beltrami operator plus curvature terms. Please note that we would obtain a true Laplace-Beltrami operator for the operator D_n being seen as:

$$\frac{1}{\sqrt{\gamma}} D_n \sqrt{\gamma} \gamma^{nm} f_{;m} = (f^{;m})_{;n} = (\gamma^{nm} f_{;m})_{;n} = \gamma^{nm} (f_{;m})_{;n}$$

$$= \frac{1}{\sqrt{\gamma}} \left(\sqrt{\gamma} \gamma^{nm} f_{,m} \right)_{,n}$$

$$= \gamma^{nm} (f_{,m})_{,n} + (\gamma^{nm})_{,n} f_{,m} + \frac{1}{2} \gamma^{nm} \gamma^{ij} (\gamma_{ij})_{,n} f_{,m}$$

$$= \gamma^{nm} (f_{,m})_{,n} - \gamma^{nm} \Gamma^j_{nm} f_{,j}. \tag{402}$$

The meaning of D_n in the various constellations and the comparison with the classical (scalar) to the vector-character of the function f_n is been elaborated in section "3.11.3 Classical Solutions in Connection with the Flat Space Limit". There it will be shown that our current approach must be seen as inconsistent and thus, incomplete. How the whole can be brought to a consistent end will be demonstrated in section "3.11.5 Extension and Generalization of the Quantum Transformation Rules".

The equations as result of our variation read:

(I) $\quad -\sqrt{\gamma} f^j C_n C^n - \sqrt{\gamma} C^j f^n C_n - D_m \sqrt{\gamma} \gamma^{mn} f^j{}_{;n} - D_j \sqrt{\gamma} \gamma^{jm} f^n{}_{;m}$
$\quad = -\sqrt{\gamma} f^j C_n C^n - \sqrt{\gamma} C^j f^n C_n - 2 D_m \sqrt{\gamma} \gamma^{mn} f^j{}_{;n} \equiv A$

(II) $\quad -\sqrt{\gamma} C_n C^n f_j - \sqrt{\gamma} C^n f_n C_j - D_n \sqrt{\gamma} \gamma^{nm} f_{j;m} - D_m \sqrt{\gamma} \gamma^{mn} f_{j;n}$
$\quad = -\sqrt{\gamma} C_n C^n f_j - \sqrt{\gamma} C^n f_n C_j - 2 D_m \sqrt{\gamma} \gamma^{mn} f_{j;n} \equiv B$

$\quad A + B = 0. \tag{403}$

In the transition to the flat space $f_j = f^j$ and assuming only one quantum theoretical constant like $\hbar$ with subsequently $f^j C_n C^n - C^j f^n C_n = C_\Sigma f^j - C_\hbar \cdot \delta_n^j f^n$, respectively $f_j C_n C^n - C^n f_n C_j = C_\Sigma f_j - C_\hbar \cdot \delta_j^n f_n$, we result in two equal equations of the form:

$$
\begin{aligned}
\text{(I)} \quad & = -C_\Sigma f^j - C_\hbar \cdot \delta_n^j f^n - \frac{2}{\sqrt{\gamma}} D_m \sqrt{\gamma} \gamma^{mn} f^j_{\;;n} = A \\
& C_\Sigma = C_\hbar \Rightarrow -A = 2\left(C_\hbar \cdot f^j + \Delta f^j\right) \\
\text{(II)} \quad & = -C_\Sigma f_j - C_\hbar \cdot \delta_j^n f_n - \frac{2}{\sqrt{\gamma}} D_m \sqrt{\gamma} \gamma^{mn} f_{j;n} \\
& -B = 2\left(C_\hbar \cdot f_j + \Delta f_j\right),
\end{aligned}
\tag{404}
$$

which can be summed up to the classical Klein Gordon equation:

$$
C_\hbar \cdot f_j + \Delta f_j = 0 \tag{405}
$$

for four functions f_j just as it classically comes out of the Dirac approach [143]. There should be more discussion about the D_n-operator, but as this requires quite some derivation, we leave it for a later in this chapter (subsection "3.11.4 Symmetry Issues"). Even though there already was some consideration on the matter*, it will be shown that there is a much more elegant way to deal with the variation task.

We conclude that the coordinate variation of metric solutions to the Einstein field equations (also in their potentially extended form Eq. (368); or just see first factor of the integrand of Eq. (388)) as done above provides the classical quantum equations in the transition to the flat space case. Seeing this variation and the subsequent quantum equations as an additional condition to the Einstein field equations, one might consider the resulting outcome as restrictions to the coordinate frames an observer could pick to describe physical objects (or anything at all) within our universe. An extended discussion of such an interpretation and more results were presented in e.g., [102, 103–106].

3.11.3 Classical Solutions in Connection with the Flat Space Limit

From Eq. (404) and Eq. (405) we can directly conclude that for the flat space limit we should be able to apply classical quantum theoretical solutions and also solve the new equations under suitable conditions, which is to say proper settings of the free constants and parameters (especially C_i and f_i).

---

*E.g. [103–106] and especially when briefly considering the Schrödinger hydrogen problem [170] in connection with the Schwarzschild metric [142] (cf. especially [106] section "A Brief Discussion of Classical Solutions in Connection with the Flat Space Limit").

As it had quite some importance to the development of quantum theory, we considered the Schrödinger hydrogen atom (e.g., [170]) in our most recent issue of our series "Einstein had it…" [106].

The classical wave function solution to the hydrogen atom can be given as follows [8]:

$$\Psi_{n,l,m}[r, \vartheta, \varphi] = e^{i \cdot m \cdot \varphi} \cdot P_l^m[\cos \vartheta] \cdot R_{n,l}[r]$$

$$= \sqrt{\left(\frac{2}{n \cdot a_0}\right)^3 \frac{(n - l - 1)!}{2 \cdot n \cdot (n + l)!}} \cdot e^{-\rho/2} \cdot \rho^l \cdot L_{n-l-1}^{2l+1}[\rho] \cdot Y_l^m[\vartheta, \varphi] ; \rho = \frac{2 \cdot r}{n \cdot a_0}.$$

(406)

The constant a_0 is denoting the Bohr radius with (m_e = electron rest mass, ε_0 = permittivity of free space, Q_e = elementary charge):

$$a_0 = \frac{4 \cdot \pi \cdot \varepsilon_0 \cdot \hbar^2}{m_e \cdot Q_e^2} = 5.292 \cdot 10^{-11} \text{ meter.}$$

(407)

The functions P, L and Y denote the associated Legendre function, the Laguerre polynomials and the spherical harmonics, respectively. The wave function Eq. (406) directly solves our flat space limit Eq. (405) for static cases where f does not depend on t.

It needs to be emphasized that the solution is based on the spherical flat space metric. We here want to give it as limit of the spherical Schwarzschild metric, which can be written as follows [58, 90]:

$$g_{\alpha\beta}^{class} = \begin{pmatrix} h \cdot g_t'[t]^2 \cdot \alpha[r] & 0 & 0 & 0 \\ 0 & a \cdot G_r'[r]^2 \alpha[r]^{-1} & 0 & 0 \\ 0 & 0 & g_{22} & 0 \\ 0 & 0 & 0 & g_{33} \end{pmatrix}$$

$$g_{22} = b \cdot g_\vartheta'[c_\vartheta \vartheta]^2 G_r[r]^2$$

$$g_{33} = d \cdot g_\varphi'[c_\varphi \varphi]^2 \cdot G_r[r]^2 \cdot \begin{Bmatrix} \sin[g_\vartheta[c_\vartheta \vartheta]]^2 \\ \cos[g_\vartheta[c_\vartheta \vartheta]]^2 \end{Bmatrix}$$

$$\alpha[r] = \left(C_1 + \frac{r_s}{G_r[r]} + \frac{\Lambda}{3 \cdot c^2} \cdot G_r[r]^2\right),$$

(408)

where we have the following condition to solve the Einstein field equations:

$$b \cdot C_1 - a \cdot c_\vartheta^2 = 0 \quad \text{for} \quad \cos \& \sin.$$

(409)

Please note that the classical (original and pure) Schwarzschild solution is valid only in the case of $\Lambda = 0$. As this is what we intend to consider here, we have to substitute the function $\alpha[r]$ in Eq. (408) by:

$$\alpha[r] = \left(C_1 + \frac{r_s}{G_r[r]} + \frac{\Lambda}{3 \cdot c^2} \cdot G_r[r]^2\right) \Rightarrow \alpha[r] = \left(1 - \frac{r_s}{G_r[r]}\right), \quad (410)$$

where, only for the reason of comfort, we have already adjusted the constant C_1 and the sign of the G_r-function in accordance with the classical solution as given in [105] (see also Eq. (411) below). Please also note that the solution Eq. (408) is a generalized form [58] of the classical Schwarzschild metric [105]:

$$g_{\alpha\beta}^{\text{Schwarzschild}} = \begin{pmatrix} -c^2\left(1 - \frac{2\cdot G\cdot m}{c^2\cdot r}\right) & 0 & 0 & 0 \\ 0 & \left(1 - \frac{2\cdot G\cdot m}{c^2\cdot r}\right)^{-1} & 0 & 0 \\ 0 & 0 & r^2 & 0 \\ 0 & 0 & 0 & r^2\cdot\sin^2\vartheta \end{pmatrix}. \quad (411)$$

Here r gives the spherical radius, m stands for the mass of the Schwarzschild object and the constants c and G are giving the speed of light in vacuum and the gravitational or Newton constant, respectively. One easily sees that the solution above has two singularities, namely one at $r = r_s = \frac{2\cdot m\cdot G}{c^2}$, which is a result of our choice of coordinates and could be transformed away, and at $r = 0$. The parameter r_s is called the Schwarzschild radius.

As already said, it is evident that the Schrödinger solution solves our flat space Klein–Gordon limit equation Eq. (405) for all f (not $f_n(!)$) not depending on t, but what about our function vector f_n?

It was shown in [106] how we could obtain at least the radial Schrödinger wave equation for a suitable constellations of C_i and f_i. For this, we need to start with

$$f_n[r] := h[r]^*\{a, b, p^*r, d^*(r^*\sin[\vartheta])\}$$

and by consequently applying a similar approach for the constants $C_n = \{C_0, C_1, C_2^*r, C_3^*(r^*\sin[\vartheta])\}$, thereby setting $b = p = d = 0$ and—of course— vanishing Schwarzschild radius $r_s = 0$. Then we evaluate that, while the complete ($r_s \neq 0$-situation) looks as follows:

$$\frac{1}{\sqrt{\gamma}}D_n\sqrt{\gamma}\gamma^{nm}f_{i;m} = \begin{Bmatrix} \frac{a\left(\frac{r_s^2 h[r]}{r-r_s} + 2r\left((4r-3r_s)h'[r]+2r(r-r_s)h''[r]\right)\right)}{4r^3} \\ \frac{a\cdot r_s^2 h[r]}{4r\cdot(r_s-r)^3 c^2} \\ 0 \\ 0 \end{Bmatrix}, \quad (412)$$

the vanishing Schwarzschild radius gives us the classical result:

$$\frac{1}{\sqrt{\gamma}}D_n\sqrt{\gamma}\gamma^{nm}f_{i;m} \xrightarrow{\lim_{r_s\to 0}} \begin{Bmatrix} a\left(2\cdot\frac{h'[r]}{r} + h''[r]\right) \\ 0 \\ 0 \\ 0 \end{Bmatrix}. \quad (413)$$

Extending our current approach for the function vector f_n to an arbitrary spatial dependency:

$$f_n = f_n [r, \vartheta, \varphi] = \{f [r, \vartheta, \varphi], 0, 0, 0\} = \{h [r] \cdot Y [\vartheta, \varphi], 0, 0, 0\}, \quad (414)$$

gives us the operator Eq. (412) in the following form:

$$\frac{1}{a \cdot \sqrt{\gamma}} D_n \sqrt{\gamma} \gamma^{nm} f_{i;m}$$

$$= \left\{ \begin{array}{c} \dfrac{\frac{r_s^2 h[r]}{r - r_s} + 2r \left((4r - 3r_s) h'[r] + 2r(r - r_s) h''[r] \right)}{4r^3} \cdot Y [\vartheta, \varphi] + \frac{h[r]}{r^2} \cdot \Delta_{\vartheta, \varphi} Y [\vartheta, \varphi] \\ \dfrac{r_s^2 h[r] \cdot Y [\vartheta, \varphi]}{4r \cdot (r_s - r)^3 c^2} \\ 0 \\ 0 \end{array} \right\}.$$

$$(415)$$

The operator $\Delta_{\vartheta, \varphi} Y[\vartheta, \varphi]$ stands for the angular part of the spherical Laplacian and can be solved in the usual way by a separation approach as given above in Eq. (406), where we have:

$$Y_l^m [\vartheta, \varphi] = e^{i \cdot m \cdot \varphi} \cdot P_l^m [\cos \vartheta]. \quad (416)$$

It should explicitly be noted that there is also another metric of spherical symmetry, which can be given as follows:

$$g_{\alpha\beta}^{hyp} = \begin{pmatrix} h \cdot g_t' [t]^2 \cdot \alpha [r] & 0 & 0 & 0 \\ 0 & a \cdot G_r' [r]^2 \alpha [r]^{-1} & 0 & 0 \\ 0 & 0 & g_{22} & 0 \\ 0 & 0 & 0 & g_{33} \end{pmatrix}$$

$$g_{22} = b \cdot g_\vartheta' [c_\vartheta \vartheta]^2 G_r [r]^2 \cdot \left\{ \begin{array}{c} \sinh [g_\vartheta [c_\vartheta \vartheta]]^{-2} \\ \cosh [g_\vartheta [c_\vartheta \vartheta]]^{-2} \end{array} \right\}$$

$$g_{33} = d \cdot g_\varphi' [c_\varphi \varphi]^2 \cdot G_r [r]^2 \cdot \left\{ \begin{array}{c} \sinh [g_\vartheta [c_\vartheta \vartheta]]^{-2} \\ \cosh [g_\vartheta [c_\vartheta \vartheta]]^{-2} \end{array} \right\}$$

$$\alpha [r] = \left(C_1 + \frac{r_s}{G_r [r]} + \frac{\Lambda}{3 \cdot c^2} \cdot G_r [r]^2 \right), \quad (417)$$

where, as with the generalized Schwarzschild solution, we have the following condition:

$$b \cdot C_1 - a \cdot c_\vartheta^2 = 0 \quad \text{for} \quad \cosh$$
$$b \cdot C_1 + a \cdot c_\vartheta^2 = 0 \quad \text{for} \quad \sinh. \quad (418)$$

The reader may prove that the metric also solves the Einstein field equations.

Evaluating the operator Eq. (412) for this metric gives us the same result as Eq. (415) but with a slightly different angular Laplacian (cf. [90]). Its solution is similar to the classical spherical harmonics and reads:

$$Y_l^m\left[\vartheta, \varphi\right] = e^{i \cdot m \cdot \varphi} \cdot P_l^m\left[\tanh \vartheta\right].\tag{419}$$

A brief discussion can be found in [90].

It was already shown that this metric, too, describes objects of central symmetry [65, 90], but in contrast to the usual spherical coordinates in a non-periodic manner. Which means that the "angles" are running from minus to plus infinity before they fill the surface of a sphere. For convenience, we repeat this demonstration regarding the spherical character of our metric in the appendix.

3.11.4 Symmetry Issues

So far we concentrated on the investigation of the case $L = 0$, which we consider Dirac-like. There we always had a proper (symmetric) mix of the co- and the contravariant function vectors f_n and f^n. When moving on to δL, however, we had to realize that there are symmetry issues (cf. discussion in [102] and equations Eq. (398) regarding our variation approach).

As an example one might simply repeat our evaluation regarding the variated L from above (equation (412) and following) for the contravariant form of our function vector f^n. We find improper asymmetries between the operator forms for co- and contra-variance:

$$\frac{1}{\sqrt{\gamma}} D_n \sqrt{\gamma} \gamma^{nm} f_{i;m} \neq \frac{1}{\sqrt{\gamma}} D_n \sqrt{\gamma} \gamma^{nm} f^i{}_{;m}\tag{420}$$

in the variated L and conclude that our approach needs to be honed. This we will do in the next section.

3.11.5 Extension and Generalization of the Quantum Transformation Rules—Symmetric in Co- and Contra-Variance

It is evident that with an asymmetric approach like Eq. (394) or Eq. (395) we cannot expect to achieve a consistent and totally symmetric outcome. We also need to solve the mystery about the scalar character of functions in the classical Schrödinger and Klein–Gordon equation and the vector character in the Dirac equation in connection with our approach. Resolving this problem without the need to resort—as Dirac had to—to the extraction of the root of the Klein–Gordon operator via quaternions, but sticking to geometrical concepts should be our goal.

We find that a co- and contravariant symmetric generalization of Eq. (394) or Eq. (395) in the following form

$$\frac{\partial G_+^i\,[x_{\forall i}]}{\partial x^m} = \{\delta_m^i| \pm |C_m^i \cdot f| \pm |i \cdot C^i \cdot f_{,m}| \pm |C \cdot \gamma^{in} f_{,mn}| \pm |\ldots\}$$

$$\cong \{\delta_m^i| \pm |C_m^i \cdot f| \pm |i \cdot C^i \cdot f_{,m}\} = \begin{cases} \delta_m^i + C_m^i \cdot f + i \cdot C^i \cdot f_{,m} \\ \delta_m^i + C_m^i \cdot f - i \cdot C^i \cdot f_{,m} \\ \delta_m^i - C_m^i \cdot f + i \cdot C^i \cdot f_{,m} \\ \delta_m^i - C_m^i \cdot f - i \cdot C^i \cdot f_{,m} \end{cases}$$

$$\frac{\partial x^i}{\partial G_+^m[x_{\forall i}]} = \{\delta_m^i| \pm |i \cdot C_m^i \cdot f| \pm |C^i \cdot f_{,m}| \pm |i \cdot C \cdot \gamma^{in} f_{,mn}| \pm |\ldots\}$$

$$\cong \{\delta_m^i| \pm |i \cdot C_m^i \cdot f| \pm |C^i \cdot f_{,m}\} = \begin{cases} \delta_m^i + i \cdot C_m^i \cdot f + C^i \cdot f_{,m} \\ \delta_m^i + i \cdot C_m^i \cdot f - C^i \cdot f_{,m} \\ \delta_m^i - i \cdot C_m^i \cdot f + C^i \cdot f_{,m} \\ \delta_m^i - i \cdot C_m^i \cdot f - C^i \cdot f_{,m} \end{cases},$$

$$(421)$$

in the $f = $ scalar case and

$$\frac{\partial G_+^i\,[x_{\forall i}]}{\partial x^m} = \{\delta_m^i| \pm |C^i \cdot f_m| \pm |i \cdot C \cdot \gamma^{in} f_{m,n}| \pm |\ldots\}$$

$$\cong \{\delta_m^i| \pm |C^i \cdot f_m| \pm |i \cdot C \cdot \gamma^{in} f_{m,n}\} = \begin{cases} \delta_m^i + C^i \cdot f_m + i \cdot C \cdot \gamma^{in} f_{m,n} \\ \delta_m^i + C^i \cdot f_m - i \cdot C \cdot \gamma^{in} f_{m,n} \\ \delta_m^i - C^i \cdot f_m + i \cdot C \cdot \gamma^{in} f_{m,n} \\ \delta_m^i - C^i \cdot f_m - i \cdot C \cdot \gamma^{in} f_{m,n} \end{cases}$$

$$\frac{\partial x^i}{\partial G_+^m[x_{\forall i}]} = \{\delta_m^i| \pm |i \cdot C^i \cdot f_m| \pm |C \cdot \gamma^{in} f_{m,n} \ldots\}$$

$$\cong \{\delta_m^i| + |i \cdot C^i \cdot f_m| \pm |C \cdot \gamma^{in} f_{m,n}\} = \begin{cases} \delta_m^i + i \cdot C^i \cdot f_m + C \cdot \gamma^{in} f_{m,n} \\ \delta_m^i + i \cdot C^i \cdot f_m - C \cdot \gamma^{in} f_{m,n} \\ \delta_m^i - i \cdot C^i \cdot f_m + C \cdot \gamma^{in} f_{m,n} \\ \delta_m^i - i \cdot C^i \cdot f_m - C \cdot \gamma^{in} f_{m,n} \end{cases},$$

$$(422)$$

in the vector case is much more suitable than our original approach from [102]. In addition, it allows us the use of the classical variation principle with just ordinary derivatives. Please note the result of the scalar product of our vectors [...]. For instance, multiplication of two vectors of type Eq. (421)

would give us:

$$\frac{\partial G_+^i\,[x_{\nabla i}]}{\partial x^m} \cdot \frac{\partial G_+^j\,[x_{\nabla i}]}{\partial x^n}\gamma_{ij}\gamma^{mn}$$

$$= \left\{ \begin{array}{l} \delta_m^i + C_m^i \cdot f + i \cdot C^i \cdot f_{,m} \\ \delta_m^i + C_m^i \cdot f - i \cdot C^i \cdot f_{,m} \\ \delta_m^i - C_m^i \cdot f + i \cdot C^i \cdot f_{,m} \\ \delta_m^i - C_m^i \cdot f - i \cdot C^i \cdot f_{,m} \end{array} \right\} \left\{ \begin{array}{l} \delta_n^j + C_n^j \cdot f + i \cdot C^j \cdot f_{,n} \\ \delta_n^j + C_n^j \cdot f - i \cdot C^j \cdot f_{,n} \\ \delta_n^j - C_n^j \cdot f + i \cdot C^j \cdot f_{,n} \\ \delta_n^j - C_n^j \cdot f - i \cdot C^j \cdot f_{,n} \end{array} \right\} \gamma_{ij}\gamma^{mn}$$

$$= 4 \cdot \left(\delta_m^i\delta_n^j + C_m^iC_n^j \cdot f^2 + i^2 \cdot C^iC^j \cdot f_{,m}f_{,n} \right)\gamma_{ij}\gamma^{mn}$$

$$= 4 \cdot \left(\delta_m^i\delta_n^j + C_m^iC_n^j \cdot f^2 - C^iC^j \cdot f_{,m}f_{,n} \right)\gamma_{ij}\gamma^{mn}. \tag{423}$$

It should be pointed out explicitly that we are—in principle—only interested in the scalar product as given in the second last and last line of Eq. (423). As we will see later, this structure gives us a connection to the classical quantum equations. The vector form as chosen in Eq. (421) and Eq. (422)—at the moment—only provides a very convenient way to achieve the desired outcome, namely classic quantum equations in the transition to the flat space limit. Here we have a parallel to the introduction of quaternions in the Dirac theory. Seen mathematically, the vector forms above are obviously very useful, but it is not clear what they actually mean. For here and now it shall suffice to work out the connection with the classical quantum equations.

Subsequently our Lagrangian from Eq. (392) does read:

$$L = G^{\alpha\beta}\gamma_{\alpha\beta} - G_{\alpha\beta}\gamma^{\alpha\beta} = \frac{\partial x^\alpha}{\partial G^i}\frac{\partial x^\beta}{\partial G^j}\gamma^{ij}\gamma_{\alpha\beta} - \frac{\partial G^i}{\partial x^\alpha}\frac{\partial G^j}{\partial x^\beta}\gamma_{ij}\gamma^{\alpha\beta} \equiv S_{\text{I}} - S_{\text{II}}$$

$$= 8 \cdot \left(C_m^iC_n^j \cdot f^2 - C^iC^j \cdot f_{,m}f_{,n} \right)\gamma_{ij}\gamma^{mn} \tag{424}$$

in the f-scalar case and

$$L = G^{\alpha\beta}\gamma_{\alpha\beta} - G_{\alpha\beta}\gamma^{\alpha\beta} = \frac{\partial x^\alpha}{\partial G^i}\frac{\partial x^\beta}{\partial G^j}\gamma^{ij}\gamma_{\alpha\beta} - \frac{\partial G^i}{\partial x^\alpha}\frac{\partial G^j}{\partial x^\beta}\gamma_{ij}\gamma^{\alpha\beta} \equiv V_{\text{I}} - V_{\text{II}}$$

$$= 8 \cdot \left(C^iC^j \cdot f_m f_n - C^2 \cdot \gamma^{i\alpha} f_{m,\alpha}\gamma^{j\beta} f_{n,\beta} \right)\gamma_{ij}\gamma^{mn} \tag{425}$$

in the f-vector case.

Certainly, our approach can further be extended in both, the scalar and the vector form. Thus, for instance, the scalar might also—more completely—be chosen as follows:

$$\frac{\partial G_+^i [x_{\forall i}]}{\partial x^m} = \{\delta_m^i| \pm |C_m^i \cdot f| \pm |i \cdot C^i \cdot f_{,m}| \pm |C_m^{i\alpha} \cdot f_{,\alpha}| \pm |i \cdot C \cdot \gamma^{in} f_{,mn}| \pm |\ldots\}$$

$$\cong \{\delta_m^i| \pm |C_m^i \cdot f| \pm |i \cdot C^i \cdot f_{,m}| \pm |C_m^{i\alpha} \cdot f_{,\alpha}\}$$

$$= \left\{ \begin{array}{l} \delta_m^i + C_m^i \cdot f + i \cdot C^i \cdot f_{,m} + C_m^{i\alpha} \cdot f_{,\alpha}, \; \delta_m^i + C_m^i \cdot f + i \cdot C^i \cdot f_{,m} - C_m^{i\alpha} \cdot f_{,\alpha} \\ \delta_m^i + C_m^i \cdot f - i \cdot C^i \cdot f_{,m} + C_m^{i\alpha} \cdot f_{,\alpha}, \; \delta_m^i + C_m^i \cdot f + i \cdot C^i \cdot f_{,m} - C_m^{i\alpha} \cdot f_{,\alpha} \\ \delta_m^i - C_m^i \cdot f + i \cdot C^i \cdot f_{,m} + C_m^{i\alpha} \cdot f_{,\alpha}, \; \delta_m^i + C_m^i \cdot f + i \cdot C^i \cdot f_{,m} - C_m^{i\alpha} \cdot f_{,\alpha} \\ \delta_m^i - C_m^i \cdot f - i \cdot C^i \cdot f_{,m} + C_m^{i\alpha} \cdot f_{,\alpha}, \; \delta_m^i + C_m^i \cdot f + i \cdot C^i \cdot f_{,m} - C_m^{i\alpha} \cdot f_{,\alpha} \end{array} \right\}$$

$$\frac{\partial x^i}{\partial G_+^m [x_{\forall i}]} = \{\delta_m^i| \pm |i \cdot C_m^i \cdot f| \pm |C^i \cdot f_{,m}| \pm |i \cdot C_m^{i\alpha} \cdot f_{,\alpha}| \pm |C \cdot \gamma^{in} f_{,mn}| \pm |\ldots\}$$

$$\cong \{\delta_m^i| \pm |i \cdot C_m^i \cdot f| \pm |C^i \cdot f_{,m}| \pm |i \cdot C_m^{i\alpha} \cdot f_{,\alpha}\}$$

$$= \left\{ \begin{array}{l} \delta_m^i + i \cdot C_m^i \cdot f + C^i \cdot f_{,m} + i \cdot C_m^{i\alpha} \cdot f_{,\alpha}, \; \delta_m^i + i \cdot C_m^i \cdot f + C^i \cdot f_{,m} - i \cdot C_m^{i\alpha} \cdot f_{,\alpha} \\ \delta_m^i + i \cdot C_m^i \cdot f - C^i \cdot f_{,m} + i \cdot C_m^{i\alpha} \cdot f_{,\alpha}, \; \delta_m^i + i \cdot C_m^i \cdot f + C^i \cdot f_{,m} - i \cdot C_m^{i\alpha} \cdot f_{,\alpha} \\ \delta_m^i - i \cdot C_m^i \cdot f + C^i \cdot f_{,m} + i \cdot C_m^{i\alpha} \cdot f_{,\alpha}, \; \delta_m^i + i \cdot C_m^i \cdot f + C^i \cdot f_{,m} - i \cdot C_m^{i\alpha} \cdot f_{,\alpha} \\ \delta_m^i - i \cdot C_m^i \cdot f - C^i \cdot f_{,m} + i \cdot C_m^{i\alpha} \cdot f_{,\alpha}, \; \delta_m^i + i \cdot C_m^i \cdot f + C^i \cdot f_{,m} - i \cdot C_m^{i\alpha} \cdot f_{,\alpha} \end{array} \right\}.$$

$$(426)$$

Please also note that one would have options for even higher orders of functional approaches like second-order tensors with:

$$\frac{\partial G_+^i [x_{\forall i}]}{\partial x^m} = \{\delta_m^i| \pm |C \cdot f_m^i| \pm |i \cdot C^\alpha \cdot f_{m,\alpha}^i| \pm |C^{\alpha\beta} f_{m,\alpha\beta}^i| \pm |\ldots\}$$

$$\cong \{\delta_m^i| \pm |C \cdot f_m^i| \pm |i \cdot C^\alpha \cdot f_{m,\alpha}^i\} = \left\{ \begin{array}{l} \delta_m^i + C \cdot f_m^i + i \cdot C^\alpha \cdot f_{m,\alpha}^i \\ \delta_m^i + C \cdot f_m^i - i \cdot C^\alpha \cdot f_{m,\alpha}^i \\ \delta_m^i - C \cdot f_m^i + i \cdot C^\alpha \cdot f_{m,\alpha}^i \\ \delta_m^i - C \cdot f_m^i - i \cdot C^\alpha \cdot f_{m,\alpha}^i \end{array} \right\}$$

$$\frac{\partial x^i}{\partial G_+^m [x_{\forall i}]} = \{\delta_m^i| \pm |i \cdot C \cdot f_m^i| \pm |C^\alpha \cdot f_{m,\alpha}^i| \pm |i \cdot C^{\alpha\beta} f_{m,\alpha\beta}^i| \pm |\ldots\}$$

$$\cong \{\delta_m^i| \pm |i \cdot C \cdot f_m^i| \pm |C^\alpha \cdot f_{m,\alpha}^i\} = \left\{ \begin{array}{l} \delta_m^i + i \cdot C \cdot f_m^i + C^\alpha \cdot f_{m,\alpha}^i \\ \delta_m^i + i \cdot C \cdot f_m^i - C^\alpha \cdot f_{m,\alpha}^i \\ \delta_m^i - i \cdot C \cdot f_m^i + C^\alpha \cdot f_{m,\alpha}^i \\ \delta_m^i - i \cdot C \cdot f_m^i - C^\alpha \cdot f_{m,\alpha}^i \end{array} \right\}.$$

$$(427)$$

Here the corresponding scalar product similar to Eq. (424) and the subsequent L-form yields:

$$L - G^{\alpha\beta} \gamma_{\alpha\beta} - G_{\alpha\beta} \gamma^{\alpha\beta} = \frac{\partial x^\alpha}{\partial G^i} \frac{\partial x^\beta}{\partial G^j} \gamma^{ij} \gamma_{\alpha\beta} - \frac{\partial G^i}{\partial x^\alpha} \frac{\partial G^j}{\partial x^\beta} \gamma_{ij} \gamma^{\alpha\beta} \equiv T_{\mathrm{I}} - T_{\mathrm{II}}$$

$$= 8 \cdot \left(C^2 \cdot f_m^i f_n^j - C^\alpha C^\beta \cdot f_{m,\alpha}^i f_{n,\beta}^j \right) \gamma_{ij} \gamma^{mn}.$$

$$(428)$$

As with the extension to the scalar form from Eq. (426), one might also consider the following degrees of freedom with a second-order approach:

$$\frac{\partial G^i_+ [\chi_{\vee i}]}{\partial x^m} = \{\delta^i_m| \pm |C \cdot f^i_m| \pm |i \cdot C^\alpha f^i_{m,\alpha}| \pm |C^\alpha f^i_{\alpha,m}| \pm |i \cdot C_\alpha \gamma^{i\beta} F^\alpha_{m,\beta}$$

$$|\pm |C^{\alpha\beta} f^i_{m,\alpha\beta}| \pm |\ldots\}$$

$$\cong \{\delta^i_m| \pm |C \cdot f^i_m| \pm |i \cdot C^\alpha \cdot f^i_{m,\alpha}| \pm |C^\alpha \cdot f^i_{\alpha,m}| \pm |i \cdot C_\alpha \gamma^{i\beta} f^\alpha_{m,\beta}| \pm |\}$$

$$\frac{\partial x^i}{\partial G^m_+ [\chi_{\vee i}]} = \{\delta^i_m| \pm |i \cdot C \cdot f^i_m| \pm |C^\alpha f^i_{m,\alpha}| \pm |i \cdot C^\alpha f^i_{\alpha,m}| \pm |C_\alpha \gamma^{i\beta} f^\alpha_{m,\beta}$$

$$|\pm |i \cdot C^{\alpha\beta} f^i_{m,\alpha\beta}| \pm |\ldots\}$$

$$\cong \{\delta^i_m| \pm |i \cdot C \cdot f^i_m| \pm |C^\alpha \cdot f^i_{m,\alpha}| \pm |i \cdot C^\alpha f^i_{\alpha,m}| \pm |C_\alpha \gamma^{i\beta} f^\alpha_{m,\beta}\}.$$

$$(429)$$

We realize that there are many options for ever more diverse and higher-order forms. These latter extensions, however, shall not be considered here and therefore were only mentioned for completeness.

Always only considering Lagrangian forms with zero and first-order derivatives allows us to just apply the classical variational derivative. Thus, instead of our peculiar form Eq. (399), which we had to introduce in order to take care about the co- and contravariant f-vectors, we can now just distinguish between a field of the kind f, f_n or f^i_m and so on and its first derivatives with respect to the coordinates. Taking into account that this has to be done under the variational integral (cf. [102]), we obtain the following variation from the scalar form Eq. (424):

$$\frac{\delta L}{\delta \phi} = \frac{\partial L}{\partial \phi_i} - \partial_\mu \left(\frac{\partial L}{\partial (\partial_\mu \phi_i)} \right) \Rightarrow \left\{ \frac{\delta L}{\delta f} = \frac{\partial L}{\partial f} - \partial_k \left(\frac{\partial L}{\partial (f_{,\mu})} \right). \right. \quad (430)$$

Applying this rule on L as given in Eq. (424) yields:

$$\delta L \equiv \frac{\delta L}{\delta \phi} = \frac{\partial L}{\partial \phi_i} - \partial_\mu \left(\frac{\partial L}{\partial (\partial_\mu \phi_i)} \right) \Rightarrow \left\{ \frac{\delta L}{\delta f} = \frac{\partial L}{\partial f} - \partial_k \left(\frac{\partial L}{\partial (f_{,\mu})} \right) \right.$$

$$= 16 \cdot \left(\sqrt{\gamma} C^i_m C^j_n \gamma_{ij} \gamma^{mn} \cdot f + \frac{1}{2} \left[(C^i C^j \gamma_{ij} \sqrt{\gamma} \gamma^{mn} \cdot f_{,m})_{,n} \right. \right.$$

$$+ (C^i C^j \gamma_{ij} \sqrt{\gamma} \gamma^{mn} \cdot f_{,n})_{,m} \bigg] \bigg). \quad (431)$$

Please note that the square root $\sqrt{\gamma}$ with γ denoting the determinant of the metric tensor $\gamma_{\alpha\beta}$ thereby comes from the variational integral (cf. [102]).

With the most simple setting $C^i C^j \gamma_{ij} = 1$, we immediately result in the classical Klein–Gordon equation:

$$\delta L = 0 = 16 \cdot \left(\sqrt{\gamma} \, \overbrace{c_m^i C_n^j \gamma_{ij} \gamma^{mn}}^{\equiv C_\Sigma^2} \cdot f + \frac{1}{2} \left[\left(\sqrt{\gamma} \gamma^{mn} \cdot f_{,m} \right)_{,n} \left(\sqrt{\gamma} \gamma^{mn} \cdot f_{,n} \right)_{,m} \right] \right)$$

$$\Rightarrow 0 = C_\Sigma^2 \cdot f + \frac{1}{\sqrt{\gamma}} \left(\sqrt{\gamma} \gamma^{mn} \cdot f_{,n} \right)_{,m} = C_\Sigma^2 \cdot f + \Delta f. \tag{432}$$

Not using the simplification $C^i C^j \gamma_{ij} = 1$, our result changes to:

$$\delta L = 16 \cdot \left(\sqrt{\gamma} \cdot C_\Sigma^2 \cdot f + \frac{1}{2} \left[\left(C^i C^j \gamma_{ij} \sqrt{\gamma} \gamma^{mn} \cdot f_{,m} \right)_{,n} + \left(C^i C^j \gamma_{ij} \sqrt{\gamma} \gamma^{mn} \cdot f_{,n} \right)_{,m} \right] \right)$$

$$= 16 \cdot \left(\sqrt{\gamma} \cdot C_\Sigma^2 \cdot f + \frac{1}{2} \left[\begin{matrix} \left(C^i C^j \gamma_{ij} \right)_{,n} \cdot \sqrt{\gamma} \gamma^{mn} \cdot f_{,m} + C^i C^j \gamma_{ij} \cdot \left(\sqrt{\gamma} \gamma^{mn} \cdot f_{,m} \right)_{,n} \\ + \left(C^i C^j \gamma_{ij} \right)_{,m} \cdot \sqrt{\gamma} \gamma^{mn} \cdot f_{,n} + C^i C^j \gamma_{ij} \cdot \left(\sqrt{\gamma} \gamma^{mn} \cdot f_{,n} \right)_{,m} \end{matrix} \right] \right)$$

$$\Rightarrow \quad \boxed{ \begin{matrix} 0 = C_\Sigma^2 \cdot f + \left(C^i C^j \gamma_{ij} \right)_{,m} \cdot \gamma^{mn} \cdot f_{,n} + \dfrac{C^i C^j \gamma_{ij}}{\sqrt{\gamma}} \left(\sqrt{\gamma} \gamma^{mn} \cdot f_{,n} \right)_{,m} \\[2mm] = C_\Sigma^2 \cdot f + \left(C^i C^j \gamma^{ij} \right)_{,m} \cdot \gamma^{mn} \cdot f_{,n} + C^i C^j \gamma_{ij} \cdot \Delta f \end{matrix} }$$

$$\Rightarrow \quad \boxed{ 0 = \dfrac{C_\Sigma^2}{C^2 C^j \gamma_{ij}} \cdot f + \dfrac{\left(C^i C^j \gamma_{ij} \right)_{,m} \cdot \gamma^{mn} \cdot f_{,n}}{C^i C^j \gamma_{ij}} + \Delta f } \tag{433}$$

Thus, we have obtained a Klein–Gordon equation of scalar order with respect to the function f. The original Klein–Gordon equation reads:

$$\left[\frac{M^2 c^2}{\hbar^2} + V + \Delta_{\text{coordinates}} \right] f = 0. \tag{434}$$

Here we have the reduced Planck constant $\hbar$, the rest mass M and the speed of light in vacuum c. V denotes the classical potential (cf. classical text books like [170]). Now in our metric Klein–Gordon equation (433) (last line), the term $\frac{(C^i C^j \gamma_{ij})_{,m} \cdot \gamma^{mn} \cdot f_{,n}}{C^i C^j \gamma_{ij}}$ takes on the functionality of the classical potential V while $\frac{C_\Sigma^2}{C^i C^j \gamma_{ij}}$ stands for the mass, respectively:

$$\frac{M^2 c^2}{\hbar^2} = \frac{C_\Sigma^2}{C^i C^j \gamma_{ij}}; \quad V = \frac{\left(C^i C^j \gamma_{ij} \right)_{,m} \cdot \gamma^{mn} \cdot f_{,n}}{C^i C^j \gamma_{ij}}. \tag{435}$$

Due to our symmetrized scalar L-form Eq. (424) we do get the same result for the S_I and the S_{II} terms, which are standing for the transformation rules of the co- and the contravariant metric tensor. We conclude therefore that we have solved the symmetry problem regarding the scalar f.

We already saw in [102–106] that our approach Eq. (397), which we so far used as f-vector form, despite of its imperfection regarding certain symmetry aspects, gave quite some reasonable results under the simple condition $L = 0$. In fact, one could get the impression that the $L = 0$-equation for our Lagrangian Eq. (397) somehow presents the geometric form of the Dirac equation, which, after all, also sports a function vector f_n, although in a slightly different context. Taking the (simplest) L-form with our vector function Eq. (425) and variating it according to the classical recipe yields:

$$\delta L \equiv \frac{\delta L}{\delta \phi} = \frac{\partial L}{\partial \phi_i} - \partial_\mu \left(\frac{\partial L}{\partial(\partial_\mu \phi_i)} \right) \Rightarrow \left\{ \frac{\partial L}{\partial f_n} = \frac{\partial L}{\partial f_n} - \partial_k \left(\frac{\partial L}{\partial(f_{n,\mu})} \right) \right.$$

$$= 8 \cdot \left(\sqrt{\gamma} C^i C^j \gamma_{ij} \gamma^{mn} \cdot (f_m + f_n) + C^2 \cdot \left(\left(\sqrt{\gamma} \gamma^{mn} \gamma^{j\beta} f_{n,\beta} \right)_{,j} \right. \right.$$

$$\left. + \left(\sqrt{\gamma} \gamma^{mn} \gamma^{j\beta} f_{m,j} \right)_{,\beta} \right) \right). \tag{436}$$

Thereby we have used the fact that one can reshape Eq. (425) as follows:

$$L = G^{\alpha\beta} \gamma_{\alpha\beta} - G_{\alpha\beta} \gamma^{\alpha\beta} = \frac{\partial x^\alpha}{\partial G^i} \frac{\partial x^\beta}{\partial G^j} \gamma^{ij} \gamma_{\alpha\beta} - \frac{\partial G^i}{\partial x^\alpha} \frac{\partial G^j}{\partial x^\beta} \gamma_{ij} \gamma^{\alpha\beta} \equiv V_I - V_{II}$$

$$= 8 \cdot \left(C^i C^j \cdot f_m f_n - C^2 \cdot \gamma^{i\alpha} f_{m,\alpha} \gamma^{j\beta} f_{n,\beta} \right) \gamma_{ij} \gamma^{mn}$$

$$= 8 \cdot \left(C^i C^j \gamma_{ij} \cdot f_m f_n - C^2 \cdot \left\{ \begin{array}{c} \delta_j^\alpha f_{m,\alpha} \gamma^{j\beta} f_{n,\beta} \\ \text{or} \\ \gamma^{i\alpha} d_{m,\alpha} \delta_i^\beta f_{n,\beta} \end{array} \right\} \gamma^{mn} \right)$$

$$= 8 \cdot \left(C^i C^j \gamma_{ij} \cdot f_m f_n - C^2 \cdot \left\{ \begin{array}{c} f_{m,j} \gamma^{j\beta} f_{n,\beta} \\ \text{or} \\ f_{n,i} \gamma^{i,\alpha} f_{m,\alpha} \end{array} \right\} \gamma^{mn} \right). \tag{437}$$

If, as before with the scalar form, we would set $C^i C^j \gamma_{ij} = 1$, we again obtain rather classical Klein–Gordon terms:

$$\delta L = 0 = 8 \cdot \left(\sqrt{\gamma}\gamma^{mn} \cdot (f_m + f_n) + C^2 \cdot \left(\left(\sqrt{\gamma}\gamma^{mn}\gamma^{j\beta} f_{n,\beta} \right)_{,j} + \left(\sqrt{\gamma}\gamma^{mn}\gamma^{j\beta} f_{m,j} \right)_{,\beta} \right) \right)$$

$$\Rightarrow 0 = \gamma^{mn} \cdot (f_m + f_n) + \frac{C^2}{\sqrt{\gamma}} \cdot \left(\begin{array}{l} \sqrt{\gamma}\gamma^{j\beta} f_{n,\beta} \cdot \gamma^{mn}{}_{,j} + \gamma^{mn} \left(\sqrt{\gamma}\gamma^{j\beta} f_{n,\beta} \right)_{,j} \\ + \sqrt{\gamma}\gamma^{j\beta} f_{m,j} \cdot \gamma^{mn}{}_{,\beta} + \gamma^{mn} \left(\sqrt{\gamma}\gamma^{j\beta} f_{m,j} \right)_{,\beta} \end{array} \right)$$

$$= \gamma^{mn} \cdot (f_m + f_n) + C^2 \cdot \gamma^{j\beta} \left(f_{n,\beta}\gamma^{mn}{}_{,j} + f_{m,j}\gamma^{mn}{}_{,\beta} \right) + \frac{C^2}{\sqrt{\gamma}}\gamma^{mn}$$

$$\times \left(\left(\sqrt{\gamma}\,\gamma^{j\beta} f_{n,\beta} \right)_{,j} + \left(\sqrt{\gamma}\,\gamma^{j\beta} f_{m,j} \right)_{,\beta} \right)$$

$$= \gamma^{mn} \cdot (f_m + f_n) + C^2 \cdot \left(\gamma^{j\beta} \left(f_{n,\beta}\gamma^{mn}{}_{,j} + f_{m,j}\gamma^{mn}{}_{,\beta} \right) + \gamma^{mn} \cdot \Delta (f_m + f_n) \right)$$

$$\boxed{\begin{array}{l} \Rightarrow 0 = \gamma^{mn} \cdot \left[(f_m + f_n) + C^2 \cdot \Delta(f_m + f_n) \right] + C^2 \cdot \gamma^{j\beta} \left(f_{n,\beta}\gamma^{nm}{}_{,j} + f_{m,j}\gamma^{mn}{}_{,\beta} \right) \\[2mm] = \left\{ \begin{array}{l} \gamma^{mn} \cdot \left[f_m + C^2 \cdot \Delta f_m \right] + C^2 \cdot \gamma^{j\beta} f_{m,j}\gamma^{mn}{}_{,\beta} \\[1mm] \gamma^{mn} \cdot \left[f_n + C^2 \cdot \Delta f_n \right] + C^2 \cdot \gamma^{j\beta} f_{n,j}\gamma^{mn}{}_{,j} \end{array} \right\} \\[3mm] = \gamma^{mk} \cdot \left[f_m + C^2 \cdot \Delta f_m \right] + C^2 \cdot \gamma^{j\beta} f_{m,j}\gamma^{mk}{}_{,\beta} \end{array}} \quad . \text{(438)}$$

Thus, we have obtained a Klein–Gordon equation of a vector plus affine connection terms similar to our previous results in [102–106]. This time, however, our results are equal to the co- and contravariant transformation terms as the same variation result is also obtained for the terms V_{I} and V_{II} if varied separately (the reader may easily prove that). Thereby one has to note that the Kronecker symbols δ_m^i are presenting constants and therefore vanish during the variation.

It needs to be pointed out that in Cartesian coordinates or any other coordinate system with constant metric components, Eq. (438) simplifies to:

$$\delta L = 0 = \gamma^{mn} \cdot \left[(f_m + f_n) + C^2 \cdot \Delta (f_m + f_n) \right]$$

$$= \left\{ \begin{array}{l} f_m + C^2 \cdot \Delta f_m \\ f_n + C^2 \cdot \Delta f_n \end{array} \right\} = f_k + C^2 \cdot \Delta f_k. \quad (439)$$

As this is the classical Klein–Gordon equation without potential, we have to conclude that the term $C^2 \cdot \gamma^{j\beta}(f_{n,\beta}\gamma^{mn}{}_{,j} + f_{m,j}\gamma^{mn}{}_{,\beta})$ should be taken as the equivalent to the classical potential V. Thus, the presence of any potential in classical theories would be equivalent to curvature of space here. There will be more discussion about this in connection with the Schrödinger equation further below.

3.11.6 Extension and Generalization of the Quantum Transformation Rules—Asymmetry in Co- and Contra-Variance

In the section above, we have only considered quantum transformations which effectively could be considered zero-transformations, because we have from Eq. (421)*:

$$\sum_{\forall \Omega} \left\{ \frac{\partial G_+^i \, [x_{\forall i}]}{\partial x^m} \right\}_\Omega = \sum_{\forall \Omega} \begin{cases} \delta_m^i + C_m^i \cdot f + i \cdot C^i \cdot f_{,m} \\ \delta_m^i + C_m^i \cdot f - i \cdot C^i \cdot f_{,m} \\ \delta_m^i - C_m^i \cdot f + i \cdot C^i \cdot f_{,m} \\ \delta_m^i - C_m^i \cdot f - i \cdot C^i \cdot f_{,m} \end{cases}_\Omega = 4\delta_m^i$$

$$\sum_{\forall \Omega} \left\{ \frac{\partial x^i}{\partial G_+^m \, [x_{\forall i}]} \right\}_\Omega = \sum_{\forall \Omega} \begin{cases} \delta_m^i + i \cdot C_m^i \cdot f + C^i \cdot f_{,m} \\ \delta_m^i + i \cdot C_m^i \cdot f - C^i \cdot f_{,m} \\ \delta_m^i - i \cdot C_m^i \cdot f + C^i \cdot f_{,m} \\ \delta_m^i - i \cdot C_m^i \cdot f - C^i \cdot f_{,m} \end{cases}_\Omega = 4\delta_m^i.$$

$$(440)$$

In addition, we made sure that the transformations were symmetric in co- and contra-variance. Now we might also investigate the option for asymmetric and potentially effective non-zero quantum transformations.

At first, we say goodbye to our simple symmetry for co- and contra-variance. Only concentrating on mathematical simplicity for one of the two, meaning either the co- or contra-variant side, we make the following approaches for a covariant metric transformation:

$$\frac{\partial G_+^i \, [x_{\forall i}]}{\partial x^m} = \{\delta_m^i | \pm |C_m^i \cdot f| \pm |i \cdot C^i \cdot f_{,m}| \pm |C \cdot \gamma^{in} f_{,mn}| \pm | \ldots \}$$

$$\cong \{\delta_m^i | \pm |C_m^i \cdot f| \pm |i \cdot C^i \cdot f_{,m}\} = \begin{cases} \delta_m^i + C_m^i \cdot f + i \cdot C^i \cdot f_{,m} \\ \delta_m^i + C_m^i \cdot f - i \cdot C^i \cdot f_{,m} \\ \delta_m^i - C_m^i \cdot f + i \cdot C^i \cdot f_{,m} \\ \delta_m^i - C_m^i \cdot f - i \cdot C^i \cdot f_{,m} \end{cases}$$

---

*This is the scalar form. The reader may prove that this also holds for the vector and all other of our L-forms.

$$\frac{\partial x^i}{\partial G^m_+ [x_{\forall i}]} = \left\{\delta^i_m| \pm |D^i_m \cdot h| \pm |i \cdot D^i \cdot h_{,m}| \pm |D \cdot \gamma^{in} h_{,mn}| \pm |\ldots\right\}$$

$$\cong \left\{\delta^i_m| \pm |D^i_m \cdot h| \pm |i \cdot D^i \cdot h_{,m}\right\} = \begin{Bmatrix} \delta^i_m + D^i_m \cdot h + i \cdot D^i \cdot h_{,m} \\ \delta^i_m + D^i_m \cdot h - i \cdot D^i \cdot h_{,m} \\ \delta^i_m - D^i_m \cdot h + i \cdot D^i \cdot h_{,m} \\ \delta^i_m - D^i_m \cdot h - i \cdot D^i \cdot h_{,m} \end{Bmatrix},$$

$$\tag{441}$$

which shall represent the base vectors in the $f =$ scalar case and in the vector case we apply:

$$\frac{\partial G^i_+ [x_{\forall i}]}{\partial x^m} = \left\{\delta^i_m| \pm |C^i \cdot f_m| \pm |i \cdot C \cdot \gamma^{in} f_{m,n}| \pm |\ldots\right\}$$

$$\cong \left\{\delta^i_m| \pm |C^i \cdot f_m| \pm |i \cdot C \cdot \gamma^{in} f_{m,n}\right\} = \begin{Bmatrix} \delta^i_m + C^i \cdot f_m + i \cdot C \cdot \gamma^{in} f_{m,n} \\ \delta^i_m + C^i \cdot f_m - i \cdot C \cdot \gamma^{in} f_{m,n} \\ \delta^i_m - C^i \cdot f_m + i \cdot C \cdot \gamma^{in} f_{m,n} \\ \delta^i_m - C^i \cdot f_m - i \cdot C \cdot \gamma^{in} f_{m,n} \end{Bmatrix}$$

$$\frac{\partial x^i}{\partial G^m_+ [x_{\forall i}]} = \left\{\delta^i_m| \pm |D^i \cdot h_m| \pm |i \cdot D \cdot \gamma^{in} h_{m,n} \ldots\right\}$$

$$\cong \left\{\delta^i_m| \pm |D^i \cdot h_m| \pm |i \cdot D \cdot \gamma^{in} h_{m,n}\right\} = \begin{Bmatrix} \delta^i_m + D^i \cdot h_m + i \cdot D \cdot \gamma^{in} h_{m,n} \\ \delta^i_m + D^i \cdot h_m - i \cdot D \cdot \gamma^{in} h_{m,n} \\ \delta^i_m - D^i \cdot h_m + i \cdot D \cdot \gamma^{in} h_{m,n} \\ \delta^i_m - D^i \cdot h_m - i \cdot D \cdot \gamma^{in} h_{m,n} \end{Bmatrix}.$$

$$\tag{442}$$

Please note that we still have the convenient result of the scalar product of our vectors [...] as derived in Eq. (423). Subsequently our new Lagrangians from Eq. (392) should read:

$$L_{\mathrm{I}} = G^{\alpha\beta}\gamma_{\alpha\beta} - 4 \cdot \gamma_{\alpha\beta}\gamma^{\alpha\beta} = \frac{\partial x^\alpha}{\partial G^i}\frac{\partial x^\beta}{\partial G^j}\gamma^{ij}\gamma_{\alpha\beta} - 4 \cdot \gamma_{\alpha\beta}\gamma^{\alpha\beta} \equiv S_{\mathrm{I}} - 4 \cdot \gamma_{\alpha\beta}\gamma^{\alpha\beta}$$

$$= 4 \cdot \left(D^i_m D^j_n \cdot h^2 - D^i D^j \cdot h_{,m}h_{,n}\right)\gamma_{ij}\gamma^{mn}$$

$$L_{\mathrm{II}} = G_{\alpha\beta}\gamma^{\alpha\beta} - 4 \cdot \gamma_{\alpha\beta}\gamma^{\alpha\beta} = \frac{\partial G^i}{\partial x^\alpha}\frac{\partial G^j}{\partial x^\beta}\gamma_{ij}\gamma^{\alpha\beta} - 4 \cdot \gamma_{\alpha\beta}\gamma^{\alpha\beta} \equiv S_{\mathrm{II}} - 4 \cdot \gamma_{\alpha\beta}\gamma^{\alpha\beta}$$

$$= 4 \cdot \left(C^i_m C^j_n \cdot f^2 - C^i C^j \cdot f_{,m}f_{,n}\right)\gamma_{ij}\gamma^{mn} \tag{443}$$

in the f-scalar case and

$$L_{\mathrm{I}} = G^{\alpha\beta}\gamma_{\alpha\beta} - 4 \cdot \gamma_{\alpha\beta}\gamma^{\alpha\beta} = \frac{\partial x^\alpha}{\partial G^i}\frac{\partial x^\beta}{\partial G^j}\gamma^{ij}\gamma_{\alpha\beta} - 4 \cdot \gamma_{\alpha\beta}\gamma^{\alpha\beta} \equiv V_{\mathrm{I}} - 4 \cdot \gamma_{\alpha\beta}\gamma^{\alpha\beta}$$

$$= 4 \cdot \left(D^i D^j \cdot h_m h_n - D^2 \cdot \gamma^{i\alpha} h_{m,\alpha}\gamma^{j\beta} h_{n,\beta}\right)\gamma_{ij}\gamma^{mn}$$

$$L_{II} = G_{\alpha\beta}\gamma^{\alpha\beta} - 4 \cdot \gamma_{\alpha\beta}\gamma^{\alpha\beta} = \frac{\partial G^i}{\partial x^\alpha}\frac{\partial G^j}{\partial x^\beta}\gamma_{ij}\gamma^{\alpha\beta} - 4 \cdot \gamma_{\alpha\beta}\gamma^{\alpha\beta} \equiv V_{II} - 4 \cdot \gamma_{\alpha\beta}\gamma^{\alpha\beta}$$

$$= 4 \cdot \left(C^i C^j \cdot f_m f_n - C^2 \cdot \gamma^{i\alpha} f_{m,\alpha}\gamma^{j\beta} f_{n,\beta} \right) \gamma_{ij}\gamma^{mn} \tag{444}$$

in the f-vector case.

As mentioned before, our approach can further be extended in all forms, the scalar, the vector form and forms of higher orders, too.

Please also note that with respect to the evaluation of the functions f or h only one would have to be evaluated as we still have the zero-sum of the transformation vectors due to Eq. (440) and thus, the effective transformation is an identical one.

As a consequence, we obtain the same variation results just as derived before only that this time we also have equations in D and h, which we can just ignore and concentrate only on the $C - f$-forms. Thus, in essence there is nothing new with our asymmetric approach. We still result in the same equations and zero-sum quantum transformations.

Non-zero-sum quantum transformations have already been introduced earlier in our series of "Einstein had it..." as functional wrappers:

$$G_{\alpha\beta} = g_{\alpha\beta} \cdot (F[f] = F[f[\text{coordinates}]]). \tag{445}$$

The reader will find a longer discussion in [100]. Finding completely Einstein field equation-compatible solutions is still possible as it was demonstrated in [87], but the evaluations are extremely complicated.

As such approaches require a more comprehensive investigation regarding the compromising of the Einstein field equations, we will not consider them here any further but refer to issue LXXV of our "Einstein had it..." series [108].

As said above in connection with equation (423), it should be pointed out explicitly that we are—in principle—only interested in the scalar product as given in the second and the last lines of Eq. (443) and Eq. (444). We saw that variation of these structures gives us a connection to the classical quantum equations. Thus, the vector forms as chosen in Eq. (421), Eq. (422), Eq. (441) and Eq. (442) should only be seen as a convenient way or—a tool similar to the quaternions in the Dirac theory—to achieve the desired outcome, namely metrically based classic quantum equations in the transition to the flat space limit.

3.11.7 Dirac's Peculiar "Accident"

We conclude that we might see Eq. (438) as the metric vector or first-order form of the quantum theoretical Klein–Gordon equation, while the scalar partner Eq. (432) should be interpreted as Klein–Gordon equation of scalar or zero order. Consequently, we would obtain the second-order Klein–Gordon equation from an approach like Eq. (427). With the usual recipe (e.g., [8]) one can extract the Schrödinger equation from the Klein–Gordon equation and so we also automatically have various orders of Schrödinger equations with either scalars, vectors or higher-order forms of functions f.

We already hinted that the $L = 0$ equation of the form Eq. (437) should be seen as the geometric equivalent of the Dirac equation. With the functional approach in Eq. (437) being a vector, we should take it as the first-order Dirac equation. Then immediately the question pops up:

How can we show the connection of our metrical originated "Dirac" equation and the original one from Dirac himself?

In order to answer the question we should recapitulate Dirac's work. He intended to extract the "root" of the classical Klein–Gordon equation. Repeating his strategy on a simplified 2-dimensional form shall make things easier for us.

In order to demonstrate not only what Dirac had done, but also to show where he had (perhaps accidently) chosen exactly those paths leading him to this outstanding outcome, we will apply the Dirac method to a problem with only three components. At first, we need to find a suitable structure for the objects α and β allowing us to write the following:

$$A^2 + (B^2 - C^2) = (A + i \cdot \sqrt{B^2 - C^2})(\alpha \cdot A - i \cdot \sqrt{B^2 - C^2})$$
$$= (A + i \cdot (\alpha \cdot B + i \cdot \beta \cdot C)) \cdot (A - i \cdot (\alpha \cdot B + i \cdot \beta \cdot C)).$$
$$(446)$$

Dirac solved the problem as follows: He first defined a new form representing the number "one" and another one representing "zero." Thereby he set:

$$I = \begin{pmatrix} 1 & 0 \\ 0 & 1 \end{pmatrix} \text{ „ONE" and } \begin{pmatrix} 0 & 0 \\ 0 & 0 \end{pmatrix} \text{ „NULL"!.} \qquad (447)$$

After that, he introduced* the two objects:

$$\alpha = \begin{pmatrix} 0 & 1 \\ 1 & 0 \end{pmatrix} \quad \text{and} \quad \beta = \begin{pmatrix} 0 & -i \\ i & 0 \end{pmatrix} \tag{448}$$

and the reader may prove that this gives the desired result for Eq. (446), because we have the following identities:

$$\alpha^2 = \begin{pmatrix} 0 & 1 \\ 1 & 0 \end{pmatrix}\begin{pmatrix} 0 & 1 \\ 1 & 0 \end{pmatrix} = \begin{pmatrix} 0 \cdot 0 + 1 \cdot 1 & 0 \cdot 1 + 1 \cdot 0 \\ 1 \cdot 0 + 0 \cdot 1 & 1 \cdot 1 + 0 \cdot 0 \end{pmatrix} = \begin{pmatrix} 1 & 0 \\ 0 & 1 \end{pmatrix} = I$$

$$\beta^2 = \begin{pmatrix} 0 & -i \\ i & 0 \end{pmatrix}\begin{pmatrix} 0 & -i \\ i & 0 \end{pmatrix} = \begin{pmatrix} 0 \cdot 0 + i \cdot -i & 0 \cdot -i - i \cdot 0 \\ i \cdot 0 + 0 \cdot i & -i \cdot i + 0 \cdot 0 \end{pmatrix} = \begin{pmatrix} 1 & 0 \\ 0 & 1 \end{pmatrix} = I$$

$$(\alpha \cdot \beta + \beta \cdot \alpha) = \begin{pmatrix} 0 & 1 \\ 1 & 0 \end{pmatrix}\begin{pmatrix} 0 & -i \\ i & 0 \end{pmatrix} + \begin{pmatrix} 0 & -i \\ i & 0 \end{pmatrix}\begin{pmatrix} 0 & 1 \\ 1 & 0 \end{pmatrix} = \begin{pmatrix} 0 & 0 \\ 0 & 0 \end{pmatrix} = 0. \tag{449}$$

By applying the rule Eq. (446) a second time to the term A^2, thereby again applying our matrix objects α and β, we obtain the total form:

$$\sqrt{A^2 + B^2 - C^2} = \alpha_A \cdot A + (\beta_B \cdot B + \beta_C \cdot C). \tag{450}$$

Now the new matrices are of the 3×3-type of course, but otherwise the situation is just as it was above. All terms containing f or derivatives of f are linear in the new equation (450).

Dirac, however, did not use a two-fold application of his matrix-technique but used the third binomial formula as "outer multiplicator" instead. Thus, his final equation actually reads:

$$A^2 + B^2 - C^2 = (\alpha_A \cdot A + (\beta_B \cdot B + \beta_C \cdot C))(\alpha_A \cdot A - (\beta_B \cdot B + \beta_C \cdot C)). \tag{451}$$

At this point Dirac concluded (no matter that he came from a different origin than we did, he still reached the same outcome and thus, came to the same conclusion) that he could easily go back to his 2×2 matrix-structure of the $\alpha - \beta$-objects by rewriting Eq. (451) as follows:

$$A^2 + B^2 - C^2 = (A + (\alpha \cdot B + i \cdot \beta \cdot C))(A - (\alpha \cdot B + i \cdot \beta \cdot C)). \tag{452}$$

Now we want to apply the original terms from the Klein–Gordon equation just as Dirac did. However, for illustration and clarity we are going to present

---

*In fact Dirac did all this for the complete Klein-Gordon equation having 5 terms instead of three and he also did not use combined expressions as $B^2 - C^2$, but otherwise the technique in principle is the same and so we present it as if Dirac had done it on our simpler L-form like the ones we'd obtain from Eq. (424) (see [106]).

the original Dirac approach on the simplified Klein–Gordon equation for Cartesian Coordinates t, x, with just the time and the x dependency for the function f. This Klein–Gordon equation would read:

$$f[t, x] \cdot C_\hbar^2 - \frac{\partial^2 f[t, x]}{\partial x^2} + \frac{1}{c^2} \frac{\partial^2 f[t, x]}{\partial t^2} = \left(c_\hbar^2 - \frac{\partial^2}{\partial x^2} + \frac{1}{c^2} \frac{\partial^2}{\partial t^2} \right) f[t, x] = 0.$$

(453)

Here now Dirac at first applied the third binomial formula and wrote (cf. first step in Eq. (446)):

$$\left(C_\hbar + i \cdot \sqrt{\frac{1}{c^2} \frac{\partial^2}{\partial t^2} - \frac{\partial^2}{\partial x^2}} \right) \left(C_\hbar - i \cdot \sqrt{\frac{1}{c^2} \frac{\partial^2}{\partial t^2} - \frac{\partial^2}{\partial x^2}} \right) f[t, x].$$

(454)

Then the next step was the introduction of the matrices α and β which made the equation (454) to:

$$\left(C_\hbar + i \cdot \left(\alpha \cdot \frac{1}{c} \frac{\partial}{\partial t} + i \cdot \beta \cdot \frac{\partial}{\partial x} \right) \right) \left(C_\hbar - i \cdot \left(\alpha \cdot \frac{1}{c} \frac{\partial}{\partial t} + i \cdot \beta \cdot \frac{\partial}{\partial x} \right) \right) f[t, x] = 0.$$

(455)

We see that solutions to each of the two equations:*

$$\left(C_\hbar + i \cdot \left(\alpha \cdot \frac{1}{c} \frac{\partial}{\partial t} + i \cdot \beta \cdot \frac{\partial}{\partial x} \right) \right) f[t, x] = 0$$

$$\left(C_\hbar - i \cdot \left(\alpha \cdot \frac{1}{c} \frac{\partial}{\partial t} + i \cdot \beta \cdot \frac{\partial}{\partial x} \right) \right) f[t, x] = 0$$

(456)

would also solve Eq. (455). However, with α and β being matrices we have a discrepancy in rank. Either the term $C_\hbar$ is also a matrix of the same rank as α and β or f becomes a vector f_n.

It should explicitly be noted that there are options to introduce the objects α and β and depending on the way of their definition, different outcomes are possible. In connection with the Dirac approach, this led to the description of different particles and therefore it is important to point these options out (cf. text books about quantum electro dynamics, e.g., references given in [170]).

It also should be noted that Dirac—apparently on purpose—left out the option of developing his approach as a pure matrix technology. So, instead of Eq. (454) he would also have had the option of writing:

$$\left(\alpha_A \cdot C_\hbar + \beta_B \cdot \frac{1}{c} \frac{\partial}{\partial t} + \beta_C \cdot \frac{\partial}{\partial x} \right) \left(\alpha_A \cdot C_\hbar + \beta_B \cdot \frac{1}{c} \frac{\partial}{\partial t} + \beta_C \cdot \frac{\partial}{\partial x} \right) f[t, x] = 0.$$

(457)

---

*This looks a bit different from the original Dirac approach, but we should not forget that Dirac [143] applied his matrix-object root extraction onto a differential operator of second order with a full t, x, y, and z dependency in order to make it first order rather than a quadratic equation in order to make it linear.

On first sight, this would have been the more obvious choice, because it was just one technique and not a mix leading to the rank discrepancy. Meanwhile we know that Dirac had his reasons, but still we ask ourselves:

What did Dirac miss by making this special, if not to say "peculiar" choice?

We think that the answer is rather obvious and pretty simply. Dirac missed the option of f also being a an object of non-vector rank like a scalar or matrix or higher. As this merits some discussion, we will consider this aspect in one of the "Repetition" parts of this book. Here now we only will try to investigate the option coming into play with a Dirac equation of scalar character in the function f.

However, before we start this investigation we should try to find the connection between the Dirac derivation and our metric one. As we now know that Dirac had simply done some kind of root extraction from the Klein–Gordon equation, the question needs to be answered how his result would be mirrored by our approach. We start—of course—with our $L = 0$-version where we have f as vector f_n, because in the classical Dirac equation [143] f also appears as such a vector. Dirac-like reshaping of Eq. (425) gives us:

$$L = 8 \cdot \left(C^i C^j \cdot f_m f_n - C^2 \cdot \gamma^{i\alpha} f_{m,\alpha} \gamma^{j\beta} f_{n,\beta} \right) \gamma_{ij} \gamma^{mn}$$
$$= 8 \cdot \left(C^i \cdot f_m - C \cdot \gamma^{i\alpha} f_{m,\alpha} \right) \left(C^j \cdot f_n + C \cdot \gamma^{j\beta} f_{n,\beta} \right) \gamma_{ij} \gamma^{mn}. \quad (458)$$

Now we define a suitable vector D_i allowing us to write $D_i \cdot K_j^{mn} = \gamma_{ij} \gamma^{mn}$. Setting this into Eq. (458) yields:

$$L = 8 \cdot \left(C^i \cdot f_m - C \cdot \gamma^{i\alpha} f_{m,\alpha} \right) \left(C^j \cdot f_n + C \cdot \gamma^{j\beta} f_{n,\beta} \right) D_i \cdot K_j^{mn}$$
$$= 8 \cdot \left(D_i \cdot C^i \cdot f_m - C \cdot D_i \cdot \gamma^{i\alpha} f_{m,\alpha} \right) \left(C^j \cdot f_n + C \cdot \gamma^{j\beta} f_{n,\beta} \right) K_j^{mn}. \quad (459)$$

As our equation shall give $L = 0$, we demand that the first factor realizes the zero outcome via:

$$D_i \cdot C^i \cdot f_m - C \cdot D_i \cdot \gamma^{i\alpha} f_{m,\alpha} = 0 = DC \cdot f_m - C \cdot D^\alpha \cdot f_{m,\alpha}$$
$$\underset{\text{Dirac}}{}$$
$$= DC \cdot f_m - C \cdot D^\alpha \cdot \partial_\alpha f_m \overset{\text{Dirac}}{\equiv\!\Rightarrow\!\equiv} DC \cdot f_m - C \cdot \tilde{D}_m^{n\alpha} \cdot \partial_\alpha f_n; \quad DC = D_i \cdot C^i.$$
$$(460)$$

Thus, the Dirac task, respectively the task to mirror the Dirac equation, is to find a suitable set of matrices $\tilde{D}_m^n$ for each α (giving the three-form $\tilde{D}_m^{n\alpha}$), allowing us to substitute the contravariant vector D^α in Eq. (460) as follows:

$$D^\alpha \cdot \partial_\alpha f_m \overset{\text{Dirac}}{\equiv\!\Rightarrow\!\equiv} \tilde{D}_m^{n\alpha} \cdot \partial_\alpha f_n. \quad (461)$$

Thereby we need to know that Dirac (of course) had no intention in satisfying condition Eq. (461), but only wanted to find matrices allowing him the root extraction from the Klein–Gordon equation as demonstrated above. In other words, his intention was to find $\tilde{D}_m^{n\alpha}$ such that the product:

$$\left(DC - C \cdot \tilde{D}_m^{n\alpha} \cdot \partial_\alpha \right) \left(DC + C \cdot \tilde{D}_m^{n\alpha} \cdot \partial_\alpha \right) \tag{462}$$

would give the Klein–Gordon equation in Cartesian Coordinates. We leave it to the reader to show that "accidently" this yields almost the same outcome as the intention Eq. (461).

Please note that one could also use the scalar form Eq. (424) as starting point by evaluating as follows:

$$
\begin{aligned}
L &= 8 \cdot (C_m^i \cdot f + C^i \cdot f_{,m})(C_n^j \cdot f - C^j \cdot f_{,n})\gamma_{ij}\gamma^{mn} \\
&= 8 \cdot (C_m^i \cdot f + C^i \cdot f_{,m})(C_n^j \cdot f - C^j \cdot f_{,n})D_i \cdot K_j^{mn} = 0 \\
\Rightarrow 0 &= (C_n^j \cdot f - C^j \cdot f_{,n})K_j^{mn} = K \cdot f^m - C^j K_j^{mn} \cdot f_{,n} \\
&= K \cdot f^m - \tilde{K}_q^{mn} \cdot \partial_n f^q
\end{aligned}
$$
$$\text{with:} \quad K \cdot f^m \equiv C_n^j \cdot f \cdot K_j^{mn}; \quad C^j K_j^{mn} \cdot f_{,n} \equiv \tilde{K}_q^{mn} \cdot \partial_n f^q. \tag{463}$$

We see no need at this point to perfectly establish the connection of our L-forms Eq. (424) and Eq. (425) with the classical Dirac equation (and its many different appearances) and so we will not further investigate this path.

What we should be interested in, however, is the meaning of the $L = 0$ equation as the scalar form Eq. (424) for the function f: Can there be a Dirac equation of scalar character, that is, of zero order? And if so, what does it represent?

In order to answer these questions we simplify Eq. (424)

$$
\begin{aligned}
L &= 8 \cdot \left(C_m^i C_n^j \cdot f^2 - C^i C^j \cdot f_{,m} f_{,n} \right) \gamma_{ij}\gamma^{mn} \\
&= 8 \cdot \left(C_j^n C_n^j \cdot f^2 - C_j C^j \cdot f_{,m}\gamma^{mn} f_{,n} \right) \\
&= 8 \cdot \left(C_j^n C_n^j \cdot f^2 - f_{,m}\gamma^{mn} f_{,n} \right) \tag{464}
\end{aligned}
$$

(we have used the setting $C^i C^j \gamma_{ij} = 1$ in the last line again) and applied it to a variety of examples within our series of "Einstein had it …."

3.11.8 The Special Case of the Schrödinger Equation—Part I

As for many people THE quantum equation is been seen in the Schrödinger equation, our next task for this chapter should be to give a most general way to obtain the Schrödinger equation out of our metric Klein–Gordon equations

Eq. (432) and Eq. (438). This time we do not intend to obtain the classical Schrödinger equation, but its covariant analogon.

For simplicity, we assume a metric with none of the components actually being time dependent. In order to distinguish this from the general case in our considerations above, we use the usual "g" to denote the metric. The component g^{00} shall be of the form $g^{00} = C1/c^2$, with $C1$ being a constant and all other metric time components shall be $g^{0i} = 0 (i = 1, 2, 3)$. we start with the scalar Klein–Gordon form Eq. (432) (last line) and instead of the symbol f we apply the classical Greek symbol Ψ. Then we can separate the time derivative as follows:

$$0 = \left[\frac{C_\Sigma^2}{C^i C^j g_{ij}} + \frac{(C^i C^j d_{ij})_{,m} \cdot g^{mn}}{C^i C^j g_{ij}} \cdot \partial_n + \Delta \right] \Psi$$

$$= \left[\frac{C_\Sigma^2}{C^i C^j g_{ij}} + \frac{(C^i C^j \gamma_{ij})_{,m} \cdot g^{mn}}{C^i C^j g_{ij}} \cdot \partial_n + \left(C1 \cdot \underbrace{\frac{\partial_t^2}{c^2} + \frac{1}{\sqrt{g}} \partial_\alpha \sqrt{g} \cdot g^{\alpha\beta} \partial_\beta}_{\text{3D-}\Delta\text{-Operator}} \right) \right].$$

$$(465)$$

Please note that our Latin indices i, j, n, m are running from 0 to n (with n giving the dimension of space-time, which here is assumed to be 4), while the Greek indices α and β are running only from 1 to n. Now we introduce a function $\Psi = \Phi + X$ and demand the following additional condition:

$$\partial_t \Psi = c^2 \cdot C2 \cdot (\Phi - X). \tag{466}$$

Together with Eq. (585) we obtain:

$$0 = \frac{C_\Sigma^2}{C^i C^j g_{ij}} (\Phi + X) + \frac{(C^i C^j g_{ij})_{,m} \cdot g^{mn}}{C^i C^j g_{ij}} \cdot \partial_n (\Phi + X)$$

$$+ \left(C1 \cdot C2 \cdot \partial_t (\Phi - X) + \frac{1}{\sqrt{g}} \partial_\alpha \sqrt{g} \cdot g^{\alpha\beta} \partial_\beta (\Phi + X) \right). \tag{467}$$

The following two equations summed up would result in Eq. (587):

$$0 = \frac{C_\Sigma^2}{C^i C^j g_{ij}} \Phi + \frac{(C^i C^j g_{ij})_{,m} \cdot g^{mn}}{C^i C^j g_{ij}} \cdot \partial_n \Phi + \left(C1 \cdot C2 \cdot \partial_t \Phi + \frac{1}{\sqrt{g}} \partial_\alpha \sqrt{g} \cdot g^{\alpha\beta} \partial_\beta \Phi \right)$$

$$0 = \frac{C_\Sigma^2}{C^i C^j g_{ij}} X + \frac{(C^i C^j g_{ij})_{,m} \cdot g^{mn}}{C^i C^j g_{ij}} \cdot \partial_n X + \left(\frac{1}{\sqrt{g}} \partial_\alpha \sqrt{g} \cdot g^{\alpha\beta} \partial_\beta X - C1 \cdot C2 \cdot \partial_t X \right).$$

$$(468)$$

Comparing with the original Schrödinger equation as given in the form below:

$$\left[-i \cdot \hbar \cdot \partial_t - \frac{\hbar^2}{2M} \Delta_{\text{Schrödinger}} + V_{\text{Schrödinger}} \right] \Psi = 0, \qquad (469)$$

does not only give us the matter and antimatter solutions again (cf. [8, 103]), but also shows us—as seen and discussed before [21]—that mass and the potential $V_{\text{Schrödinger}}$ is now been taken on by the metric and the proportional factors C_m via the operator terms $\frac{C_\Sigma^2}{C^iC^jg_{ij}} \frac{(C^iC^jg_{ij})_{,m} \cdot f^{mn}}{C^iC^jg_{ij}} \cdot \partial_n$. Thus, a distorted metric acts like an effective potential in the Schrödinger approximation and vice versa, which is to say: what in classical physics is been described as a potential would now become a distorted metric providing the necessary interaction. Similarly, we have to formulate for the mass M, that what appears as (rest) mass to us is just permanently and locally curved space. Disregarding the antimatter solution here, setting the constants $C1, C2$ and reshaping the classical Schrödinger equation as:

$$C1 \cdot C2 = \frac{i \cdot 2 \cdot M}{\hbar}; X = 0; 0 = \left[\frac{i \cdot 2 \cdot M}{\hbar} \cdot \partial_t + \Delta_{\text{Schrödinger}} - 2 \cdot V_{\text{Schrödinger}} \right] \Psi, \tag{470}$$

gives us the proportionality:

$$\Rightarrow \left[\Delta_{3D} - 2 \cdot V_{\text{Schrödinger}} \right] \Psi \stackrel{\wedge}{=} \frac{C_\Sigma^2}{C^iC^jg_{ij}} \Phi + \frac{\left(C^iC^jg_{ij} \right)_{,m} \cdot g^{mn}}{C^iC^jg_{ij}} \cdot \partial_n \Phi$$

$$+ \frac{1}{\sqrt{g}} \partial_\alpha \sqrt{g} \cdot g^{\alpha\beta} \partial_\beta \Phi$$

$$\Rightarrow \Delta_{3D} \Psi \stackrel{\wedge}{=} \Delta_{3D} \Phi$$

$$\Rightarrow V_{\text{Schrödinger}} \stackrel{\wedge}{=} \frac{1}{2 \cdot \Psi} \cdot \left(\frac{C_\Sigma^2}{C^iC^jg_{ij}} \Phi + \frac{\left(C^iC^jg_{ij} \right)_{,m} \cdot g^{mn}}{C^iC^jg_{ij}} \cdot \partial_n \Phi \right). \tag{471}$$

This way we can now link classical Schrödinger solutions and the corresponding Schrödinger potentials with their metric analogue.

Things are getting a bit more interesting when assuming that V was chosen such that we have $\Psi = \Psi_{\text{Schrödinger}} = \Phi_{\text{new}} = \Phi$. Then we can directly write down the equation for the determination of the potential and its connection to the metric distortion:

$$V_{\text{Schrödinger}} \stackrel{\wedge}{=} -\frac{1}{2} \cdot \left(\frac{C_\Sigma^2}{C^iC^jg_{ij}} + \frac{1}{\Psi} \cdot \frac{(C^iC^jg_{ij})_{,m} \cdot g^{mn}}{C^iC^jg_{ij}} \cdot \partial_n \Psi \right). \tag{472}$$

We note: What classically is the potential V gives a metric distortion in quantum gravity.

For illustration, we apply the Schrödinger solution for the hydrogen ground state where with Eq. (406) we have:

$$\Psi_{n=1,l=0,m=0}[r, \vartheta, \varphi] = e^{i \cdot m \cdot \varphi} \cdot P_l^m[\cos \vartheta] \cdot R_{n,l}[r] = e^{-\left(\frac{r}{a_0}\right)}. \tag{473}$$

Assuming that only one of the metric components shall play a role, we set C_Σ^2 to be constant and C^i shall only have one component, namely C^1. This gives us:

$$
\begin{aligned}
V_{\text{Hydrogen}} &\stackrel{\wedge}{=} -\frac{1}{2} \cdot \left(\frac{C_\Sigma^2}{C^1 C^1 g_{11}} - \frac{1}{a_0} \cdot \frac{(C^1 C^1 g_{11})_{,1}}{(C^1)^2} \right) \\
&= \frac{1}{2} \cdot \left(\frac{1}{a_0} \cdot (C^1 C^1 g_{11})_{,1} - \frac{C_\Sigma^2}{(C^1)^2 g_{11}} \right).
\end{aligned}
\tag{474}
$$

Taking the Schwarzschild metric with $g_{11} = (1 - r_s/r)^{-1}$ and the Schwarzschild radius r_s would give us:

$$
\begin{aligned}
V_{\text{Hydrogen}} &= \frac{1}{2} \cdot \left(\frac{1}{a_0} \cdot (g_{11})_{,1} - \frac{C_\Sigma^2}{(C^1)^2 g_{11}} \right) \\
&= \frac{1}{2} \cdot \left(-\frac{1}{a_0} \cdot \frac{r_s}{(r - r_s)^2} + \frac{C_\Sigma^2}{(C^1)^2} \left(\frac{r_s}{r} - 1 \right) \right).
\end{aligned}
\tag{475}
$$

Expansion into a Taylor series for bigger r and ignoring all powers of r^{-k} for $k > 1$ yields:

$$V_{\text{Hydrogen}} \simeq \frac{1}{2} \cdot \frac{C_\Sigma^2}{(C^1)^2} \left(\frac{r_s}{r} - 1 \right). \tag{476}$$

As the constant term adds no information, we can ignore it, too and obtain:

$$V_{\text{Hydrogen}} \simeq \frac{1}{2} \cdot \frac{C_\Sigma^2}{(C^1)^2} \cdot \frac{r_s}{r}. \tag{477}$$

This is just the expected outcome with a potential of central symmetry and approximately a $1/r$-proportionality (cf. e.g., [170], pp. 165).

3.11.9 Klein–Gordon and Dirac Equations of Zero, First, Second, and n-th Order

For completeness we now want to give the L-, metric Klein–Gordon und metric Dirac equations for scalar f, vector f_n and tensor f_n^j.

3.11.9.1 L-Equations

For convenience we here repeat the L-equations from above and simplify where possible without loss of generality.

Scalar:

$$
\begin{aligned}
L = G^{\alpha\beta}\gamma_{\alpha\beta} - G_{\alpha\beta}\gamma^{\alpha\beta} &= \frac{\partial x^\alpha}{\partial G^i}\frac{\partial x^\beta}{\partial G^j}\gamma^{ij}\gamma_{\alpha\beta} - \frac{\partial G^i}{\partial x^\alpha}\frac{\partial G^j}{\partial x^\beta}\gamma_{ij}\gamma^{\alpha\beta} \equiv S_{\mathrm{I}} - S_{\mathrm{II}} \\
&= 8 \cdot \left(C_m^i C_n^j \cdot f^2 - C^i C^j \cdot f_{,m} f_{,n} \right) \gamma_{ij}\gamma^{mn} \\
&= 8 \cdot \left(C_j^n C_n^j \cdot f^2 - C^j C_j \gamma^{mn} f_{,m} f_{,n} \right)
\end{aligned}
\tag{478}
$$

Vector:

$$
\begin{aligned}
L = G^{\alpha\beta}\gamma_{\alpha\beta} - G_{\alpha\beta}\gamma^{\alpha\beta} &= \frac{\partial x^\alpha}{\partial G^i}\frac{\partial x^\beta}{\partial G^j}\gamma^{ij}\gamma_{\alpha\beta} - \frac{\partial G^i}{\partial x^\alpha}\frac{\partial G^j}{\partial x^\beta}\gamma_{ij}\gamma^{\alpha\beta} \equiv V_{\mathrm{I}} - V_{\mathrm{II}} \\
&= 8 \cdot \left(C^i C^j \cdot f_m f_n - C^2 \cdot \gamma^{i\alpha} f_{m,\alpha}\gamma^{j\beta} f_{n,\beta} \right) \gamma_{ij}\gamma^{mn} \\
&= 8 \cdot \left(C_j C^j \cdot f^n f_n C^2 \cdot \gamma^{mn} f_{m,j}\gamma^{j\beta} f_{n,\beta} \right)
\end{aligned}
\tag{479}
$$

Second-order:

$$
\begin{aligned}
L = G^{\alpha\beta}\gamma_{\alpha\beta} - G_{\alpha\beta}\gamma^{\alpha\beta} &= \frac{\partial x^\alpha}{\partial G^i}\frac{\partial x^\beta}{\partial G^j}\gamma^{ij}\gamma_{\alpha\beta} - \frac{\partial G^i}{\partial x^\alpha}\frac{\partial G^j}{\partial x^\beta}\gamma_{ij}\gamma^{\alpha\beta} \equiv T_{\mathrm{I}} - T_{\mathrm{II}} \\
&= 8 \cdot \left(C^2 \cdot f_m^i f_n^j - C^\alpha C^\beta \cdot f_{m,\alpha}^i f_{n,\beta}^j \right) \gamma_{ij}\gamma^{mn} \\
&= 8 \cdot \left(C^2 \cdot f_j^n f_n^j - C^\alpha C^\beta \cdot \gamma_{ij}\gamma^{mn} f_{m,\alpha}^i f_{n,\beta}^j \right)
\end{aligned}
\tag{480}
$$

3.11.9.2 Klein–Gordon equations

Scalar:

$$
\delta L = 0 = \frac{C_\Sigma^2}{C^i C^j \gamma_{ij}} \cdot f + \frac{\left(C^i C^j \gamma_{ij} \right)_{,m} \cdot \gamma^{mn} \cdot f_{,n}}{C^i C^j \gamma_{ij}} + \Delta f
\tag{481}
$$

Vector (this time without the simplification $C^i C^j \gamma_{ij} = 1$):

$$
\delta L = 0 = 8 \cdot \left(\sqrt{\gamma}\, C^i C^j \gamma_{ij}\gamma^{mn} \cdot (f_m + f_n) + C^2 \cdot \left(\left(\sqrt{\gamma}\,\gamma^{mn}\gamma^{j\beta} f_{n,\beta} \right)_{,j} + \left(\sqrt{\gamma}\,\gamma^{mn}\gamma^{j\beta} f_{m,j} \right)_{,\beta} \right) \right)
$$

$$
\Rightarrow 0 = C^i C^j \gamma_{ij}\gamma^{mn} \cdot (f_m + f_n) + \frac{C^2}{\sqrt{\gamma}} \cdot \left(\begin{array}{l} \sqrt{\gamma}\,\gamma^{j\beta} f_{n,\beta} \cdot \gamma^{mn}_{,j} + \gamma^{mn}\left(\sqrt{\gamma}\,\gamma^{j\beta} f_{n,\beta} \right)_{,j} \\ + \sqrt{\gamma}\,\gamma^{j\beta} f_{m,j} \cdot \gamma^{mn}_{,\beta} + \gamma^{mn}\left(\sqrt{\gamma}\,\gamma^{j\beta} f_{m,j} \right)_{,\beta} \end{array} \right)
$$

$$
- C^i C^j \gamma_{ij}\gamma^{mn} \cdot (f_m + f_n) + C^2 \cdot \gamma^{j\beta} \left(f_{n,\beta}\gamma^{mn}_{,j} + f_{m,j}\gamma^{mn}_{,\beta} \right) + \frac{C^2}{\sqrt{\gamma}}\gamma^{mn} \cdot \left(\begin{array}{l} \left(\sqrt{\gamma}\,\gamma^{j\beta} f_{n,\beta} \right)_{,j} \\ + \left(\sqrt{\gamma}\,\gamma^{j\beta} f_{m,j} \right)_{,\beta} \end{array} \right)
$$

$$
= C^i C^j \gamma_{ij}\gamma^{mn} \cdot (f_m + f_n) + C^2 \cdot \left(\gamma^{j\beta}\left(f_{n,\beta}\gamma^{mn}_{,j} + f_{m,j}\gamma^{mn}_{,\beta} \right) + \gamma^{mn} \cdot \Delta(f_m + f_n) \right)
$$

$$
\boxed{
\begin{aligned}
\Rightarrow 0 &= C^i C^j \gamma_{ij}\gamma^{mn} \cdot [(f_m + f_n) + C^2 \cdot \Delta(f_m + f_n)] + C^2 \cdot \gamma^{j\beta}(f_{n,\beta}\gamma^{mn}_{,j} + f_{m,j}\gamma^{mn}_{,\beta}) \\
&= \left\{ \begin{array}{l} \gamma^{mk} \cdot [C^i C^j \gamma_{ij} f_m + C^2 \cdot \Delta f_m] + C^2 \cdot \gamma^{j\beta} f_{m,j}\gamma^{mn}_{,\beta} \\ \gamma^{mk} \cdot [C^i C^j \gamma_{ij} f_n + C^2 \cdot \Delta f_n] + C^2 \cdot \gamma^{j\beta} f_{n,\beta}\gamma^{mn}_{,j} \end{array} \right\} \\
&= \gamma^{mk} \cdot [C^i C^j \gamma_{ij} f_m + C^2 \cdot \Delta f_m] + C^2 \cdot \gamma^{j\beta} f_{m,j}\gamma^{mk}_{,\beta}
\end{aligned}
}
\tag{482}
$$

Second-order:

$$\delta L = 0 = 16 \cdot \left(\sqrt{\gamma} C^2 \cdot f_j^n + \left(\sqrt{\gamma} C^\alpha C^\beta \cdot \gamma_{ij} \gamma^{mn} f_{m,\alpha}^i \right)_{,\beta} \right)$$
$$\Rightarrow 0 = \sqrt{\gamma} C^2 \cdot f_j^n + \left(\sqrt{\gamma} C^\alpha C^\beta \cdot \gamma_{ij} \gamma^{mn} f_{m,\alpha}^i \right)_{,\beta} \tag{483}$$

3.11.9.3 Dirac equations

Applying Dirac's trick and using the third binomial equation, one can decompose the L-equations and derive at Dirac-like linear, first-order differential equations. Here we will do this for the scalar, the vector and the 2-tensor form of L.

Scalar:

$$L = 8 \cdot \left(C_m^i C_n^j \cdot f^2 - C^i C^j \cdot f_{,m} f_{,n} \right) \gamma_{ij} \gamma^{mn}$$
$$= 8 \cdot \left(C_n^j \cdot f - C^j f_{,n} \right) \left(C_m^i \cdot f + C^i f_{,m} \right) \gamma_{ij} \gamma^{mn} \tag{484}$$

Now one of the two equations:

I) $0 = \left(C_n^j \cdot f - C^j f_{,n} \right) \gamma_{ij} \gamma^{mn} = C_i^m \cdot f - C_i \gamma^{mn} f_{,n}$

II) $0 = \left(C_m^i \cdot f + C^i f_{,m} \right) \gamma_{ij} \gamma^{mn} = C_j^n \cdot f + C_j \gamma^{mn} f_{,m}$ (485)

would assure us a $L = 0$ outcome.

Vector:

$$L = 8 \cdot \left(C^i C^j \cdot f_m f_n - C^2 \cdot \gamma^{i\alpha} f_{m,\alpha} \gamma^{j\beta} f_{n,\beta} \right) \gamma_{ij} \gamma^{mn}$$

$$= 8 \cdot \left(C^i f_m - C \cdot \gamma^{i\alpha} f_{m,\alpha} \right) \left(C^j f_n + C \cdot \gamma^{j\beta} f_{n,\beta} \right) \gamma_{ij} \gamma^{mn}$$

$$\Rightarrow \begin{cases} \text{I) } 0 = \left(C^i f_m - C \cdot \gamma^{i\alpha} f_{m,\alpha} \right) \gamma_{ij} \gamma^{mn} = f^n C_j - C \cdot \delta_j^\alpha \gamma^{mn} f_{m,\alpha} = f^n C_j - C \cdot \gamma^{mn} f_{m,j} \\ \text{II) } 0 = \left(C^j f_n + C \cdot \gamma^{j\beta} f_{n,\beta} \right) \gamma_{ij} \gamma^{mn} = f^m C_i + C \cdot \delta_i^\beta \gamma^{mn} f_{n,\beta} = f^m C_i + C \cdot \gamma^{mn} f_{n,i} \end{cases}$$

$$\tag{486}$$

Second-order:

$$L = 8 \cdot \left(C^2 \cdot f_m^i f_n^j - C^\alpha C^\beta \cdot f_{m,\alpha}^i f_{n,\beta}^j \right) \gamma_{ij} \gamma^{mn}$$

$$= 8 \cdot \left(C \cdot f_m^i - C^\alpha f_{m,\alpha}^i \right) \left(C \cdot f_n^j + C^\beta f_{n,\beta}^j \right) \gamma_{ij} \gamma^{mn}$$

$$\Rightarrow \begin{cases} \text{I) } 0 = \left(C \cdot f_m^i - C^\alpha f_{m,\alpha}^i \right) \gamma_{ij} \gamma^{mn} = C \cdot f_j^n - C^\alpha \gamma_{ij} \gamma^{mn} f_{m,\alpha}^i \\ \text{II) } 0 = \left(C \cdot f_n^j + C^\beta f_{n,\beta}^j \right) \gamma_{ij} \gamma^{mn} = C \cdot f_i^m + C^\beta \gamma_{ij} \gamma^{mn} f_{n,\beta}^j \end{cases}$$

$$\tag{487}$$

The question might be asked why also the option $L = 0$ hast to be taken into account and why not just the variated form should suffice?

At this point the author could—for the time being—simply answers that the Schwarzschild solution, too, is only a $L = 0$ solution. The corresponding variation integral, which is the Einstein–Hilbert action (1), namely has the Ricci scalar R as Lagrange function. In the Schwarzschild case, however, we have $R = 0$. However, as we will see later in the book in connection some simple and more direct forms of metric variations under the Einstein–Hilbert action integral, one might just argue that any constant $K = L$ should also give a zero outcome for the whole variation. This then renders the equation above for the scalar to:

$$L = K = 8 \cdot \left(C_n^j \cdot f - C^j f_{,n} \right) \left(C_m^i \cdot f + C^i f_{,m} \right) \gamma_{ij} \gamma^{mn} \qquad (488)$$

and leads to the two equations:

$$\text{I)} \quad K_i^m = \left(C_n^j \cdot f - C^j f_{,n} \right) \gamma_{ij} \gamma^{mn} = C_i^m \cdot f - C_i \gamma^{mn} f_{,n}$$
$$\text{II)} \quad K_j^n = \left(C_m^i \cdot f + C^i f_{,m} \right) \gamma_{ij} \gamma^{mn} = C_j^n \cdot f + C_j \gamma^{mn} f_{,m}. \qquad (489)$$

This, too, would assure us a $\delta L = 0$ outcome. The corresponding equations in the vector and the matrix read:

Vector:

$$L = K = 8 \cdot \left(C^i f_m - C \cdot \gamma^{i\alpha} f_{m,\alpha} \right) \left(C^j f_n + C \cdot \gamma^{j\beta} f_{n,\beta} \right) \gamma_{ij} \gamma^{mn}$$
$$\Rightarrow \begin{cases} \text{I)} \quad K_j^n = \left(C^i f_m - C \cdot \gamma^{i\alpha} f_{m,\alpha} \right) \gamma_{ij} \gamma^{mn} = f^n C_j - C \cdot \delta_j^\alpha \gamma^{mn} f_{m,\alpha} = f^n C_j - C \cdot \gamma^{mn} f_{m,j} \\ \text{II)} \quad K_i^m = \left(C^j f_n + C \cdot \gamma^{j\beta} f_{n,\beta} \right) \gamma_{ij} \gamma^{mn} = f^m C_i + C \cdot \delta_i^\beta \gamma^{mn} f_{n,\beta} = f^m C_i + C \cdot \gamma^{mn} f_{n,i} \end{cases}$$
$$(490)$$

Second-order:

$$L = 8 \cdot \left(C^2 \cdot f_m^i f_n^j - C^\alpha C^\beta \cdot f_{m,\alpha}^i f_{n,\beta}^j \right) \gamma_{ij} \gamma^{mn}$$
$$= 8 \cdot \left(C \cdot f_m^i - C^\alpha f_{m,\alpha}^i \right) \left(C \cdot f_n^j + C^\beta f_{n,\beta}^j \right) \gamma_{ij} \gamma^{mn}$$
$$\Rightarrow \begin{cases} \text{I)} \quad K_j^n = \left(C \cdot f_m^i - C^\alpha f_{m,\alpha}^i \right) \gamma_{ij} \gamma^{mn} = C \cdot f_j^n - C^\alpha \gamma_{ij} \gamma^{mn} f_{m,\alpha}^i \\ \text{II)} \quad K_i^m = \left(C \cdot f_n^j + C^\beta f_{n,\beta}^j \right) \gamma_{ij} \gamma^{mn} = C \cdot f_i^m + C^\beta \gamma_{ij} \gamma^{mn} f_{n,\beta}^j \end{cases}.$$
$$(491)$$

3.11.10 World Formulae (?)—Summing Up and Repetition of the Simplest (Scalar) Form

Summing up what we have derived so far, we state that arbitrary properties, forming ensembles, can be seen as spaces with dimensions equal to the number of linear independent properties (orthogonality). Their behavior is described by a minimum principle given via the Einstein–Hilbert action in its classical (1) [137] or its generalized form (2) [114]. Evaluation of this action leads to the generalized Einstein field equations (3) (or first factor of integrand of Eq. (388) with respect to the generalized form). The Lagrange function of this variation is been given by a function of the Ricci curvature $f(R)$. Thereby it is widely assumed that the function is linear in R, meaning $f(R) = R$ [137]. The resulting field equations in this simple case are then just the well-known classical Einstein field equations:

$$R^{\alpha\beta} - \frac{1}{2} R \cdot \gamma^{\alpha\beta} + \Lambda \cdot \gamma^{\alpha\beta} = -\kappa T^{\alpha\beta} = 0, \tag{492}$$

where, in contrast to Einstein [138], we have explicitly set the energy–momentum tensor zero, because the matter it presents does not occur in our model connected with the approach of the additive metric or base vector variation [102–106]. In fact, it also does not occur in Einstein's model, but was artificially postulated. We assume the matter appears due to the quantization of the solutions of the Einstein field equations. In other words: matter is just a special form of warped space or entangled dimensions as we derived in [127–132]. We, which is to say the observer in this universe, only notice these perturbations of space-time as matter, because we cannot make out their true origin and nature, which, so our finding, is just waves, oscillations and (as we will see later) entanglements of and among the properties, respectively the dimensions. What should be considered, however, is the cosmological constant. The matter-free complete action should read:

$$\delta_\gamma W = 0 = \delta_\gamma \int_V d^n x \left(\sqrt{-\gamma} \, [R - 2\Lambda] \right). \tag{493}$$

However, the story does not end here, because the observer (us) cannot resolve structures to arbitrary small scales as he is always restricted to the means available in the universe he lives in. This leads to a restriction of the coordinates the metric solutions of Eq. (492) can be represented with. Formulating this restriction via variations of the metric base vectors in additive form as done in this section directly leads us to classical quantum equations and additional formulae, which are of similar structure but sport higher-order wave functions. The scalar, the simplest form, thereby leads to transformation rules for the metric solution of the Einstein field equations:

$$G_{mn} \sim \frac{\partial G^i_+[x_{\forall i}]}{\partial x^m} \cdot \frac{\partial G^j_+[x_{\forall i}]}{\partial x^n} \gamma_{ij}$$

$$= \begin{Bmatrix} \delta^i_m + C^i_m \cdot f + i \cdot C^i \cdot f_{,m} \\ \delta^i_m + C^i_m \cdot f - i \cdot C^i \cdot f_{,m} \\ \delta^i_m - C^i_m \cdot f + i \cdot C^i \cdot f_{,m} \\ \delta^i_m - C^i_m \cdot f - i \cdot C^i \cdot f_{,m} \end{Bmatrix} \begin{Bmatrix} \delta^j_n + C^j_n \cdot f + i \cdot C^j \cdot f_{,n} \\ \delta^j_n + C^j_n \cdot f - i \cdot C^j \cdot f_{,n} \\ \delta^j_n - C^j_n \cdot f + i \cdot C^j \cdot f_{,n} \\ \delta^j_n - C^j_n \cdot f - i \cdot C^j \cdot f_{,n} \end{Bmatrix} \gamma_{ij}$$

$$= 4 \cdot \left(\delta^i_m \delta^j_n + C^i_m C^j_n \cdot f^2 + i^2 \cdot C^i C^j \cdot f_{,m} f_{,n} \right) \gamma_{ij}$$

$$= 4 \cdot \left(\delta^i_m \delta^j_n + C^i_m C^j_n \cdot f^2 - C^i C^j \cdot f_{,m} f_{,n} \right) \gamma_{ij}$$

$$= 4 \cdot \left(\gamma_{mn} + C_{mj} C^j_n \cdot f^2 - C_j C^j \cdot f_{,m} f_{,n} \right). \tag{494}$$

Thereby it should be pointed out that the meaning of the vector form (first line, right-hand side in Eq. (494)), which had been introduced in order to have a convenient way for the derivation of classical quantum equations out of our metric concept, is not clear yet.

This means, in the scalar case the quantum transformed metric G_{mn} can be evaluated out of the metric solution of the Einstein field equations γ_{ij} via the transformation above. For normalization purposes we might like to assure the outcome $G_{mn} = \gamma_{mn}$ in the trivial case of $f = 0$. Therefore, we add a suitable factor to the transformation rule above and write:

$$G_{mn} = \frac{\partial G^i_+[x_{\forall i}]}{\partial x^m} \cdot \frac{\partial G^j_+[x_{\forall i}]}{\partial x^n} \gamma_{ij}$$

$$= \begin{Bmatrix} \delta^i_m + C^i_m \cdot f + i \cdot C^i \cdot f_{,m} \\ \delta^i_m + C^i_m \cdot f - i \cdot C^i \cdot f_{,m} \\ \delta^i_m - C^i_m \cdot f + i \cdot C^i \cdot f_{,m} \\ \delta^i_m - C^i_m \cdot f - i \cdot C^i \cdot f_{,m} \end{Bmatrix} \begin{Bmatrix} \delta^j_n + C^j_n \cdot f + i \cdot C^j \cdot f_{,n} \\ \delta^j_n + C^j_n \cdot f - i \cdot C^j \cdot f_{,n} \\ \delta^j_n - C^j_n \cdot f + i \cdot C^j \cdot f_{,n} \\ \delta^j_n - C^j_n \cdot f - i \cdot C^j \cdot f_{,n} \end{Bmatrix} \frac{\gamma_{ij}}{4}$$

$$= \left(\delta^i_m \delta^j_n + C^i_m C^j_n \cdot f^2 + i^2 \cdot C^i C^j \cdot f_{,m} f_{,n} \right) \gamma_{ij}$$

$$= \left(\delta^i_m \delta^j_n + C^i_m C^j_n \cdot f^2 - C^i C^j \cdot f_{,m} f_{,n} \right) \gamma_{ij} = \left(\gamma_{mn} + C_{mj} C^j_n \cdot f^2 - C_j C^j \cdot f_{,m} f_{,n} \right). \tag{495}$$

Please note that this normalization does not change any of our results from above.

The function f has to satisfy either the condition

$$L = K = C^n_j C^j_n \cdot f^2 - C^i C_j \gamma^{mn} f_{,m} f_{,n} \tag{496}$$

or

$$\delta L = 0 = \frac{C_{\Sigma}^2}{C^i C^j \gamma_{ij}} \cdot f + \frac{\left(C^i C^j \gamma_{ij}\right)_{,m} \cdot \gamma^{mn} \cdot f_{,n}}{C^i C^j \gamma_{ij}} + \Delta f, \qquad (497)$$

where we recognize the structural features of the classical Klein–Gordon equation.

While the condition Eq. (496) provides a zero-sum or constant transformation of the metric tensor and thus, not necessarily a transformation at all*, its varied form Eq. (497) actually changes the metric and results in the production of matter. This was worked out in [108–112] and presented here in connection with "3.3. The Ricci Scalar Quantization." Thereby, we realized that we already had such a result in connection with the "wrapper-approach" Eq. (85) and our corresponding evaluations as shown in [87]. Equation (496) could be seen as an equation seeking for intrinsic solutions, respectively inner degrees of freedom. The same holds for our Dirac-like equations (485) to (487). Thereby it is not clear yet how we should interpret the vectors forms (e.g., first line, right hand side in Eq. (494)), which we had only introduced in order to have a suitable metric starting point for the derivation of classical quantum equations. Thus, even though the introduction of the vector form has to be seen as artificial, it seems to be worthwhile to think about its meaning.

3.12 Connection with the Extended Einstein–Hilbert Variation Eq. (389)

In order to find a suitable connection of our—until now—pure intelligent zero approach with the additive metric, respectively base vector variation with the extended Einstein–Hilbert variation, we rewrite Eq. (389) as follows:

---

*Please note that vectors in fact would change, but this depends on the handling of the Ω-vectors (e.g., first line, right-hand side in Eq. (494), which we had only introduced in order to have a suitable metric starting point for the derivation of classical quantum equations. Thus, the scalar product of the Ω-vectors gives the desired result for the transformation of 2-forms like the metric tensor, but so far, we have no idea what this structure does to vectors (1-forms) and how the subsequent results should be interpreted.

$$\delta_\gamma W = 0 = \int_V d^n x \overbrace{\left(R^{\delta\gamma} \left[\gamma^{\delta\gamma} \right] - \frac{1}{2} R \cdot \gamma^{\delta\gamma} + \Lambda \cdot \gamma^{\delta\gamma} + \kappa \cdot T^{\delta\gamma} \left[\gamma^{\delta\gamma} \right] \right)}^{\text{Relativity}}$$

$$\times \left(\begin{array}{c} \delta g_{\delta\gamma} = \delta \left(\dfrac{\partial G^i_+ [x_{\forall i}]}{\partial x^\delta} \cdot \dfrac{\partial G^j_+ [x_{\forall i}]}{\partial x^\gamma} \gamma_{ij} \right) = \gamma_{ij} \cdot \delta \left(\dfrac{\partial G^i_+ [x_{\forall i}]}{\partial x^\delta} \cdot \dfrac{\partial G^j_+ [x_{\forall i}]}{\partial x^\gamma} \right) \\[4mm] = \delta \left(\left\{ \begin{array}{l} \delta^i_\delta + C^i_\delta \cdot f + i \cdot C^i \cdot f_{,\delta} \\ \delta^i_\delta + C^i_\delta \cdot f - i \cdot C^i \cdot f_{,\delta} \\ \delta^i_\delta - C^i_\delta \cdot f + i \cdot C^i \cdot f_{,\delta} \\ \delta^i_\delta - C^i_\delta \cdot f - i \cdot C^i \cdot f_{,\delta} \end{array} \right\} \left\{ \begin{array}{l} \delta^j_\gamma + C^j_\gamma \cdot f + i \cdot C^j \cdot f_{,\gamma} \\ \delta^j_\gamma + C^j_\gamma \cdot f - i \cdot C^j \cdot f_{,\gamma} \\ \delta^j_\gamma - C^j_\gamma \cdot f + i \cdot C^j \cdot f_{,\gamma} \\ \delta^j_\gamma - C^j_\gamma \cdot f - i \cdot C^j \cdot f_{,\gamma} \end{array} \right\} \dfrac{\gamma_{ij}}{4} \right) \\[4mm] = \delta \left(\delta^i_\delta \delta^j_\gamma + C^i_\delta C^j_\gamma \cdot f^2 + i^2 \cdot C^i C^j \cdot f_{,\delta} f_{,\gamma} \right) \gamma_{ij} \\[2mm] \underbrace{= \delta \left(\delta^i_\delta \delta^j_\gamma + C^i_\delta C^j_\gamma \cdot f^2 - C^i C^j \cdot f_{,\delta} f_{,\gamma} \right) \gamma_{ij} = \delta \left(\gamma_{\delta\gamma} + C_{\delta j} C^j_\gamma \cdot f^2 - C_j C^j \cdot f_{,\delta} f_{,\gamma} \right)}_{\text{Quantum}} \end{array} \right)$$

$$\tag{498}$$

Thereby we explicitly used the transformed coordinates only in connection with the variated metric, because this leads to simple products of the untransformed metric $\gamma_{\delta\gamma}$. The alternative

$$\delta_g W = 0 = \int_V d^n x \overbrace{\left(R^{\delta\gamma} \left[g^{\delta\gamma} \right] - \frac{1}{2} R \cdot g^{\delta\gamma} + \Lambda \cdot g^{\delta\gamma} + \kappa \cdot T^{\delta\gamma} \left[g^{\delta\gamma} \right] \right)}^{\text{Relativity}}$$

$$\times \left(\begin{array}{c} \delta g_{\delta\gamma} = \delta \left(\dfrac{\partial G^i_+ [x_{\forall i}]}{\partial x^\delta} \cdot \dfrac{\partial G^j_+ [x_{\forall i}]}{\partial x^\gamma} \gamma_{ij} \right) = \gamma_{ij} \cdot \delta \left(\dfrac{\partial G^i_+ [x_{\forall i}]}{\partial x^\delta} \cdot \dfrac{\partial G^j_+ [x_{\forall i}]}{\partial x^\gamma} \right) \\[4mm] \underbrace{= \left(\delta^i_\delta \delta^j_\gamma + C^i_\delta C^j_\gamma \cdot f^2 - C^i C^j \cdot f_{,\delta} f_{,\gamma} \right) \gamma_{ij} = \left(\gamma_{\delta\gamma} + C_{\delta j} C^j_\gamma \cdot f^2 - C_j C^j \cdot f_{,\delta} f_{,\gamma} \right)}_{\text{Quantum}} \end{array} \right)$$

$$\tag{499}$$

results in a total form where we later on have to take into account the contravariant transformed metric tensor, which renders the final sum more complicated, because we then have to solve something like

$$\delta_g W = 0 = \int_V d^n x \left(\begin{array}{c} R^{\delta\gamma} \left[\gamma^{kl} \dfrac{\partial x^\delta}{\partial G^k_+} \cdot \dfrac{\partial x^\gamma}{\partial G^l_+} \right] - \dfrac{1}{2} R \cdot \gamma^{kl} \dfrac{\partial x^\delta}{\partial G^k_+} \cdot \dfrac{\partial x^\gamma}{\partial G^l_+} \\[3mm] + \Lambda \cdot \gamma^{kl} \dfrac{\partial x^\delta}{\partial G^k_+} \cdot \dfrac{\partial x^\gamma}{\partial G^l_+} + \kappa \cdot T^{\delta\gamma} \left[\gamma^{kl} \dfrac{\partial x^\delta}{\partial G^k_+} \cdot \dfrac{\partial x^\gamma}{\partial G^l_+} \right] \end{array} \right)$$

$$\times \left(\delta g_{\delta\gamma} = \delta \left(\dfrac{\partial G^i_+ [x_{\forall i}]}{\partial x^\delta} \cdot \dfrac{\partial G^j_+ [x_{\forall i}]}{\partial x^\gamma} \gamma_{ij} \right) \right.$$

$$\left. = \gamma_{ij} \cdot \delta \left(\dfrac{\partial G^i_+ [x_{\forall i}]}{\partial x^\delta} \cdot \dfrac{\partial G^j_+ [x_{\forall i}]}{\partial x^\gamma} \right) \right).$$

$$\tag{500}$$

Even with our simple additive settings for the $\frac{\partial G_+^j[x_{\forall i}]}{\partial x^\gamma}$ this would become a difficult task and also add the problem about defining $\frac{\partial x^\delta}{\partial G_+^i}$. Considering only the terms $-\frac{1}{2}R \cdot \gamma^{kl} \frac{\partial x^\delta}{\partial G_+^k} \cdot \frac{\partial x^\gamma}{\partial G_+^l} + \Lambda \cdot \gamma^{kl} \frac{\partial x^\delta}{\partial G_+^k} \cdot \frac{\partial x^\gamma}{\partial G_+^l}$ from the relativity factor already yields

$$\left(\Lambda - \frac{1}{2}R\right) \cdot \gamma^{kl} \frac{\partial x^\delta}{\partial G_+^k} \cdot \frac{\partial x^\gamma}{\partial G_+^l} \gamma_{ij} \cdot \delta \left(\frac{\partial G_+^i[x_{\forall i}]}{\partial x^\delta} \cdot \frac{\partial G_+^j[x_{\forall i}]}{\partial x^\gamma}\right)$$

$$= \left(\Lambda - \frac{1}{2}R\right) \cdot \gamma^{kl} \frac{\partial x^\delta}{\partial G_+^k} \cdot \frac{\partial x^\gamma}{\partial G_+^l} \gamma_{ij} \cdot \delta$$

$$\times \left(\begin{Bmatrix} \delta_\delta^i + C_\delta^i \cdot f + i \cdot C^i \cdot f_{,\delta} \\ \delta_\delta^i + C_\delta^i \cdot f - i \cdot C^i \cdot f_{,\delta} \\ \delta_\delta^i - C_\delta^i \cdot f + i \cdot C^i \cdot f_{,\delta} \\ \delta_\delta^i - C_\delta^i \cdot f - i \cdot C^i \cdot f_{,\delta} \end{Bmatrix} \begin{Bmatrix} \delta_\gamma^j + C_\gamma^j \cdot f + i \cdot C^j \cdot f_{,\gamma} \\ \delta_\gamma^j + C_\gamma^j \cdot f - i \cdot C^j \cdot f_{,\gamma} \\ \delta_\gamma^j - C_\gamma^j \cdot f + i \cdot C^j \cdot f_{,\gamma} \\ \delta_\gamma^j - C_\gamma^j \cdot f - i \cdot C^j \cdot f_{,\gamma} \end{Bmatrix} \right). \quad (501)$$

We find that simple solutions are only possible in the case where we can approximate the contravariant tensor transformation matrices as follows:

$$\frac{\partial x^\delta}{\partial G_+^k} \cdot \frac{\partial x^\gamma}{\partial G_+^l} = \frac{\partial x^\delta}{\partial \left(x^k + F^k\left[x^k\right]\right)} \cdot \frac{\partial x^\gamma}{\partial \left(x^l + F^l\left[x^k\right]\right)} \xrightarrow{x^{k:l} \gg F^{k:l}\left[x^k\right]} \simeq \delta_k^\delta \delta_l^\gamma. \quad (502)$$

We will discuss this in more detail in the next main section.

Yes, another option appears when we apply the directly variated base vectors like

$$\delta_\gamma W = 0 = \int_V d^n x \left(R^{\delta\gamma}\left[\gamma^{\delta\gamma}\right] - \frac{1}{2}R \cdot \gamma^{\delta\gamma} + \Lambda \cdot \gamma^{\delta\gamma} + \kappa \cdot T^{\delta\gamma}\left[\gamma^{\delta\gamma}\right]\right)$$

$$\times \begin{pmatrix} \delta\gamma_{\delta\gamma} = \gamma_{ij}\delta\left(\frac{\partial G_+^i[x_{\forall i}]}{\partial x^\delta} \cdot \frac{\partial G_+^j[x_{\forall i}]}{\partial x^\gamma}\right) \\ = \frac{\gamma_{ij}}{4}\delta\left(\begin{Bmatrix} \delta_\delta^i + C_\delta^i \cdot f + i \cdot C^i \cdot f_{,\delta} \\ \delta_\delta^i + C_\delta^i \cdot f - i \cdot C^i \cdot f_{,\delta} \\ \delta_\delta^i - C_\delta^i \cdot f + i \cdot C^i \cdot f_{,\delta} \\ \delta_\delta^i - C_\delta^i \cdot f - i \cdot C^i \cdot f_{,\delta} \end{Bmatrix} \begin{Bmatrix} \delta_\gamma^j + C_\gamma^j \cdot f + i \cdot C^j \cdot f_{,\gamma} \\ \delta_\gamma^j + C_\gamma^j \cdot f - i \cdot C^j \cdot f_{,\gamma} \\ \delta_\gamma^j - C_\gamma^j \cdot f + i \cdot C^j \cdot f_{,\gamma} \\ \delta_\gamma^j - C_\gamma^j \cdot f - i \cdot C^j \cdot f_{,\gamma} \end{Bmatrix}\right) \\ = \delta f_{:f_{,\delta}:f_{,\gamma}}\left(\gamma_{\delta\gamma} + C_{\delta j}C_\gamma^j \cdot f^2 - C_j C^j \cdot f_{,\delta} f_{,\gamma}\right) \end{pmatrix}.$$

$$(503)$$

Now we are back at an almost classical form where we only variate with respect to the filed f and its corresponding derivatives. Unfortunately, we cannot directly apply the variation rules from Eq. (431), because Eq. (503) does not contain the square root of the metric determinant $\sqrt{\gamma}$ as we would need it to obtain complete Laplace–Beltrami operators. However, nothing hinders us from performing the following substitution in the last line of Eq. (503):

$$\delta_{f:f,\delta:f,\gamma}\left(\gamma_{\delta\gamma}+C_{\delta j}C_{\gamma}^{j}\cdot f^{2}-C_{j}C^{j}\cdot f_{,\delta}f_{,\gamma}\right)$$
$$=\delta_{F:F,\delta:F,\gamma}\left(\gamma_{\delta\gamma}+\sqrt{\gamma}\cdot C_{\delta j}C_{\gamma}^{j}\cdot F^{2}-\sqrt{\gamma}\cdot C_{j}C^{j}\cdot F_{,\delta}F_{,\gamma}\right). \quad (504)$$

Evaluation of the variation leads to

$$\delta_{f:f,\delta:f,\gamma}\left(\gamma_{\delta\gamma}+C_{\delta j}C_{\gamma}^{j}\cdot f^{2}-C_{j}C^{j}\cdot f_{,\delta}f_{,\gamma}\right)$$
$$=2\cdot\sqrt{\gamma}\cdot C_{\delta j}C_{\gamma}^{j}\cdot F\cdot\frac{\partial F}{\partial y}-\sqrt{\gamma}\cdot C_{j}C^{j}\cdot\left[F_{,\delta}\cdot\frac{\partial F_{,\gamma}}{\partial y}+F_{,\gamma}\cdot\frac{\partial F_{,\delta}}{\partial y}\right].$$

$$(505)$$

The symbol y shall denote the virtual variation parameter, which we can consider just a dummy term. Setting this into the variational integral Eq. (503) yields

$$\delta_{y}W=0=\int_{V}d^{n}x\left(R^{\delta\gamma}\left[\gamma^{\delta\gamma}\right]-\frac{1}{2}R\cdot\gamma^{\delta\gamma}+\Lambda\cdot\gamma^{\delta\gamma}+\kappa\cdot T^{\delta\gamma}\left[\gamma^{\delta\gamma}\right]\right)$$
$$\times\sqrt{\gamma}\cdot\left(2\cdot C_{\delta j}C_{\gamma}^{j}\cdot F\cdot\frac{\partial F}{\partial y}-C_{j}C^{j}\cdot\left[F_{,\delta}\cdot\frac{\partial F_{,\gamma}}{\partial y}+F_{,\gamma}\cdot\frac{\partial F_{,\delta}}{\partial y}\right]\right),$$

$$(506)$$

where for the slightly reshaped integrand, with the first factor sporting just one "governing" contravariant metric tensor, we obtain the following:

$$\delta_{y}W=0=\int_{V}d^{n}x\overbrace{\left(R^{\delta\gamma}-\frac{1}{2}R\cdot\gamma^{\delta\gamma}+\Lambda\cdot\gamma^{\delta\gamma}+\kappa\cdot T^{\delta\gamma}\right)}^{=(\ldots)^{\delta\gamma}=\mathbf{D}\cdot\gamma^{\delta\gamma}\simeq D\cdot\gamma^{\delta\gamma}}$$
$$\times\sqrt{\gamma}\cdot\left(2\cdot C_{\delta j}C_{\gamma}^{j}\cdot F\cdot\frac{\partial F}{\partial y}+C_{j}C^{j}\cdot\left[F_{,\delta}\cdot\frac{\partial F_{,\gamma}}{\partial y}+F_{,\gamma}\cdot\frac{\partial F_{,\delta}}{\partial y}\right]\right)$$
$$\xrightarrow{(\ldots)^{\delta\gamma}\simeq D\cdot\gamma^{\delta\gamma}}\int_{V}d^{n}x\cdot D\cdot\gamma^{\delta\gamma}\sqrt{\gamma}\cdot\left(2\cdot C_{\delta j}C_{\gamma}^{j}\cdot F\cdot\frac{\partial F}{\partial y}+C_{j}C^{j}\cdot\left[F_{,\delta}\cdot\frac{\partial F_{,\gamma}}{\partial y}+F_{,\gamma}\cdot\frac{\partial F_{,\delta}}{\partial y}\right]\right).$$

$$(507)$$

Integration by parts of the second addend in the last line, performed as

$$\int_{V}d^{n}x\,(\ldots)^{\delta\gamma}\sqrt{\gamma}\cdot F_{,\delta}\cdot\frac{\partial F_{,\gamma}}{\partial y}=\underbrace{\int_{V}d^{n}x\left\{[(\ldots)^{\delta\gamma}\sqrt{\gamma}\cdot F_{,\delta}]\cdot\frac{\partial F}{\partial y}\right\}_{,\gamma}}_{=S=0\text{ per definition}}$$
$$-\int_{V}d^{n}x\left[(\ldots)^{\delta\gamma}\sqrt{\gamma}\cdot F_{,\delta}\right]_{,\gamma}\cdot\frac{\partial F}{\partial y}\simeq-\int_{V}d^{n}x\cdot D\cdot\left[\gamma^{\delta\gamma}\sqrt{\gamma}\cdot F_{,\delta}\right]_{,\gamma}\cdot\frac{\partial F}{\partial y}$$

with: $\quad(\ldots)^{\delta\gamma}=\left(R^{\delta\gamma}-\frac{1}{2}R\cdot\gamma^{\delta\gamma}+\Lambda\cdot\gamma^{\delta\gamma}+\kappa\cdot T^{\delta\gamma}\right)\quad (508)$

finally gives us

$$
\begin{aligned}
0 &= \int_V d^n x \left(2 \cdot \sqrt{\gamma} \cdot (\ldots)^{\delta\gamma} C_{\delta j} C_\gamma^j \cdot F + C_j C^j \cdot \left[\left(\sqrt{\gamma} \cdot (\ldots)^{\delta\gamma} F_{,\delta} \right)_{,\gamma} + \left(\sqrt{\gamma} \cdot (\ldots)^{\delta\gamma} F_{,\gamma} \right)_{,\delta} \right] \right) \cdot \frac{\partial F}{\partial y} \\
&\simeq \int_V d^n x \left(2 \cdot \sqrt{\gamma} \cdot D \cdot \gamma^{\delta\gamma} C_{\delta j} C_\gamma^j \cdot F + C_j C^j \cdot \left[\left(\sqrt{\gamma} \cdot (\ldots)^{\delta\gamma} F_{,\delta} \right)_{,\gamma} + \left(\sqrt{\gamma} \cdot (\ldots)^{\delta\gamma} F_{,\gamma} \right)_{,\delta} \right] \right) \cdot \frac{\partial F}{\partial y} \\
&= \int_V d^n x \sqrt{\gamma} \cdot \left(2 \cdot D \cdot C_{\delta j} C^{\delta j} \cdot F + \frac{C_j C^j}{\sqrt{\gamma}} \cdot \left[\left(\sqrt{\gamma} \cdot D \cdot \gamma^{\delta\gamma} F_{,\delta} \right)_{,\gamma} + \left(\sqrt{\gamma} \cdot D \cdot \gamma^{\delta\gamma} F_{,\gamma} \right)_{,\delta} \right] \right) \cdot \frac{\partial F}{\partial y} \\
&= \int_V d^n x \cdot 2 \cdot \sqrt{\gamma} \cdot \left(D \cdot C_{\delta j} C^{\delta j} \cdot F + C_j C^j \cdot \left[\frac{F_{,\delta} \cdot \gamma^{\delta\gamma} D_{,\gamma} + F_{,\gamma} \cdot \gamma^{\delta\gamma} D_{,\delta}}{2} + D \cdot \Delta F \right] \right) \cdot \frac{\partial F}{\partial y} \\
&= \int_V d^n x \cdot 2 \cdot \sqrt{\gamma} \cdot \left(D \cdot C_{\delta j} C^{\delta j} \cdot F + C_j C^j \cdot \left[F_{,\gamma} \cdot \gamma^{\delta\gamma} D_{,\delta} + D \cdot \Delta F \right] \right) \cdot \frac{\partial F}{\partial y}. \qquad (509)
\end{aligned}
$$

Thereby S denotes the usual variation surface term, which can be set zero under the integral. Thus, we obtain the Klein–Gordon-like equation in the case of D not being a constant:

$$
D \cdot C_{\delta j} C^{\delta j} \cdot F + C_j C^j \cdot \left[F_{,\gamma} \cdot \gamma^{\delta\gamma} D_{,\delta} + D \cdot \Delta F \right] = 0 \qquad (510)
$$

and a simple Klein–Gordon equation with $D = \text{const}$:

$$
D \cdot \left(C_{\delta j} C^{\delta j} \cdot F + C_j C^j \cdot \Delta F \right) = C_{\delta j} C^{\delta j} \cdot F + C_j C^j \cdot \Delta F = 0. \qquad (511)
$$

Now we have the amazing result that the simple (mass less potential free) quantum Klein–Gordon equation can be derived from the Einstein–Hilbert action under the prerequisite that

$$
\begin{aligned}
(\ldots)^{\delta\gamma} &= \left(R^{\delta\gamma} - \frac{1}{2} R \cdot \gamma^{\delta\gamma} + \Lambda \cdot \gamma^{\delta\gamma} + \kappa \cdot T^{\delta\gamma} \right) \\
&\equiv \mathbf{D} \cdot \gamma^{\delta\gamma} \simeq D \cdot \gamma^{\delta\gamma} = \text{const} \cdot \gamma^{\delta\gamma}, \qquad (512)
\end{aligned}
$$

meaning that D would be a constant, respectively a matrix $\mathbf{D}$ of constants. However, as the term $(\ldots)^{\delta\gamma}$ is a tensor and $\gamma^{\delta\gamma}$ is a tensor, too, it is clear that in general $\mathbf{D}$ can be no tensor. In the simplest general case we could just demand that

$$
\begin{aligned}
R^{\delta\gamma} + \kappa \cdot T^{\delta\gamma} = d \cdot \gamma^{\delta\gamma} &\rightarrow \left(R^{\delta\gamma} - \frac{1}{2} R \cdot \gamma^{\delta\gamma} + \Lambda \cdot \gamma^{\delta\gamma} + \kappa \cdot T^{\delta\gamma} \right) \\
&= \left(d - \frac{R}{2} + \Lambda \right) \cdot \gamma^{\delta\gamma}, \qquad (513)
\end{aligned}
$$

which renders Eq. (510) to

$$
C_{\delta j} C^{\delta j} \cdot F + \frac{C_j C^j}{\left(d - \frac{R}{2} + \Lambda \right)} \cdot F_{,\gamma} \cdot \gamma^{\delta\gamma} \left(d - \frac{R}{2} \right)_{,\delta} + \Delta F = 0. \qquad (514)
$$

In case we want to avoid "dragging in" the term $(\ldots)^{\delta\gamma}$ into the Laplace–Beltrami operator of our final quantum equation, we could apply the following approach:

$$\delta_\gamma W = 0 = \int_V d^n x \left(R^{\delta\gamma} \left[\gamma^{\delta\gamma} \right] - \frac{1}{2} R \cdot \gamma^{\delta\gamma} + \Lambda \cdot \gamma^{\delta\gamma} + \kappa \cdot T^{\delta\gamma} \left[\gamma^{\delta\gamma} \right] \right)$$

$$\times \left(\begin{array}{l} \delta\gamma_{\delta\gamma} = \gamma_{ij}\delta \left(\frac{\partial G^i_+[x_{\nu i}]}{\partial x^\delta} \cdot \frac{\partial G^j_+[x_{\nu i}]}{\partial x^\gamma} \right) \\ = \frac{\gamma_{ij}}{4}\delta \left\{ \begin{array}{l} \delta^i_\delta + C^i_\delta \cdot f + i \cdot C_\delta \cdot \gamma^{i\alpha} f_{,\alpha} \\ \delta^i_\delta + C^i_\delta \cdot f - i \cdot C_\delta \cdot \gamma^{i\alpha} f_{,\alpha} \\ \delta^i_\delta - C^i_\delta \cdot f + i \cdot C_\delta \cdot \gamma^{i\alpha} f_{,\alpha} \\ \delta^i_\delta - C^i_\delta \cdot f - i \cdot C_\delta \cdot \gamma^{i\alpha} f_{,\alpha} \end{array} \right\} \left\{ \begin{array}{l} \delta^j_\gamma + C^j_\gamma \cdot f + i \cdot C_\gamma \cdot \gamma^{j\beta} f_{,\beta} \\ \delta^j_\gamma + C^j_\gamma \cdot f - i \cdot C_\gamma \cdot \gamma^{j\beta} f_{,\beta} \\ \delta^j_\gamma - C^j_\gamma \cdot f + i \cdot C_\gamma \cdot \gamma^{j\beta} f_{,\beta} \\ \delta^j_\gamma - C^j_\gamma \cdot f - i \cdot C_\gamma \cdot \gamma^{j\beta} f_{,\beta} \end{array} \right\} \\ = \delta_{f:f_\delta:f_\gamma} \left(\gamma_{\delta\gamma} + C_{\delta j}C^j_\gamma \cdot f^2 - \gamma_{ij}C_\gamma \cdot C_\delta \cdot \gamma^{i\alpha} f_{,\alpha}\gamma^{j\beta} f_{,\beta} \right) \end{array} \right).$$

$$(515)$$

Just as before, we perform the following substitution in the last line of Eq. (515):

$$\delta_{f:f_\delta:f_\gamma} \left(\gamma_{\delta\gamma} + C_{\delta j}C^j_\gamma \cdot f^2 - C_\gamma \cdot C_\delta \cdot \gamma_{ij}\gamma^{i\alpha} f_{,\alpha}\gamma^{j\beta} f_{,\beta} \right)$$
$$= \delta_{F:F_\delta:F_\gamma} \left(\gamma_{\delta\gamma} + \sqrt{\gamma} \cdot C_{\delta j}C^j_\gamma \cdot F^2 - \sqrt{\gamma} \cdot C_\gamma C_\delta \cdot \gamma_{ij}\gamma^{i\alpha} F_{,\alpha}\gamma^{j\beta} F_{,\beta} \right). \quad (516)$$

Evaluation of the variation leads to

$$\delta_{f:f_\delta:f_\gamma} \left(\gamma_{\delta\gamma} + C_{\delta j}C^j_\gamma \cdot f^2 - C_\gamma \cdot C_\delta \cdot \gamma_{ij}\gamma^{i\alpha} f_{,\alpha}\gamma^{j\beta} f_{,\beta} \right)$$
$$= 2 \cdot \sqrt{\gamma} \cdot C_{\delta j}C^j_\gamma \cdot F \cdot \frac{\partial F}{\partial y} - \sqrt{\gamma} \cdot C_\gamma C_\delta \cdot \gamma_{ij}\gamma^{i\alpha}\gamma^{j\beta} \left[F_{,\alpha} \cdot \frac{\partial F_{,\beta}}{\partial y} + F_{,\beta} \cdot \frac{\partial F_{,\alpha}}{\partial y} \right].$$
$$(517)$$

The symbol y shall denote the virtual variation parameter which we can consider just a dummy term again. Setting this into the variational integral (515) yields

$$\delta_\gamma W = 0 = \int_V d^n x \left(R^{\delta\gamma} \left[\gamma^{\delta\gamma} \right] - \frac{1}{2} R \cdot \gamma^{\delta\gamma} + \Lambda \cdot \gamma^{\delta\gamma} + \kappa \cdot T^{\delta\gamma} \left[\gamma^{\delta\gamma} \right] \right)$$
$$\times \sqrt{\gamma} \cdot \left(2 \cdot C_{\delta j}C^j_\gamma \cdot F \cdot \frac{\partial F}{\partial y} - C_\gamma C_\delta \cdot \gamma_{ij}\gamma^{i\alpha}\gamma^{j\beta} \left[F_{,\alpha} \cdot \frac{\partial F_{,\beta}}{\partial y} + F_{,\beta} \cdot \frac{\partial F_{,\alpha}}{\partial y} \right] \right).$$
$$(518)$$

Integration by parts of the second addend in the last line, performed as

$$\int_V d^n x \, (\ldots)^{\delta\gamma} \sqrt{\gamma} \cdot C_\gamma C_\delta \cdot \gamma_{ij} \gamma^{i\alpha} \gamma^{j\beta} F_{,\alpha} \cdot \frac{\partial F_{,\beta}}{\partial y}$$

$$= \underbrace{\int_V d^n x \left\{ [(\ldots)^{\delta\gamma} \sqrt{\gamma} \cdot C_\gamma C_\delta \cdot \gamma_{ij} \gamma^{i\alpha} \gamma^{j\beta} F_{,\alpha}] \cdot \frac{\partial F}{\partial y} \right\}_{,\beta}}_{=S=0 \text{ per definition}}$$

$$- \int_V d^n x \, [(\ldots)^{\delta\gamma} \sqrt{\gamma} \cdot C_\gamma C_\delta \cdot \gamma_{ij} \gamma^{i\alpha} \gamma^{j\beta} F_{,\alpha}]_{,\beta} \cdot \frac{\partial F}{\partial y};$$

$$(\ldots)^{\delta\gamma} = \left(R^{\delta\gamma} - \frac{1}{2} R \cdot \gamma^{\delta\gamma} + \Lambda \cdot \gamma^{\delta\gamma} + \kappa \cdot T^{\delta\gamma} \right) \tag{519}$$

finally gives us

$$0 = \int_V d^n x \left(2 \cdot \sqrt{\gamma} \cdot (\ldots)^{\delta\gamma} C_{\delta j} C_\gamma^j \cdot F + \begin{bmatrix} [(\ldots)^{\delta\gamma} \sqrt{\gamma} \cdot C_\gamma C_\delta \cdot \gamma_{ij} \gamma^{i\alpha} \gamma^{j\beta} F_{,\alpha}]_{,\beta} \\ + [(\ldots)^{\delta\gamma} \sqrt{\gamma} \cdot C_\gamma C_\delta \cdot \gamma_{ij} \gamma^{i\alpha} \gamma^{j\beta} F_{,\beta}]_{,\alpha} \end{bmatrix} \right) \cdot \frac{\partial F}{\partial y}$$

$$= \int_V d^n x \left(2 \cdot \sqrt{\gamma} \cdot (\ldots)^{\delta\gamma} C_{\delta j} C_\gamma^j \cdot F + C_\gamma C_\delta \cdot \begin{bmatrix} [(\ldots)^{\delta\gamma} \gamma^{j\beta} \gamma_{ij} \sqrt{\gamma} \cdot \gamma^{i\alpha} F_{,\alpha}]_{,\beta} \\ + [(\ldots)^{\delta\gamma} \gamma_{ij} \gamma^{i\alpha} \sqrt{\gamma} \cdot \gamma^{j\beta} F_{,\beta}]_{,\alpha} \end{bmatrix} \right) \cdot \frac{\partial F}{\partial y}$$

$$= \int_V d^n x \left(2 \cdot \sqrt{\gamma} \cdot (\ldots)^{\delta\gamma} C_{\delta j} C_\gamma^j \cdot F + C_\gamma C_\delta \cdot \begin{bmatrix} [(\ldots)^{\delta\gamma} \sqrt{\gamma} \cdot \gamma^{\beta\alpha} F_{,\alpha}]_{,\beta} \\ + [(\ldots)^{\delta\gamma} \sqrt{\gamma} \cdot \gamma^{\alpha\beta} F_{,\beta}]_{,\alpha} \end{bmatrix} \right) \cdot \frac{\partial F}{\partial y}$$

$$= \int_V d^n x \sqrt{\gamma} \cdot \left(2 \cdot (\ldots)^{\delta\gamma} C_{\delta j} C_\gamma^j \cdot F + \frac{C_\gamma C_\delta}{\sqrt{\gamma}} \cdot \begin{bmatrix} \sqrt{\gamma} \cdot \gamma^{\beta\alpha} F_{,\alpha} [(\ldots)^{\delta\gamma}]_{,\beta} + (\ldots)^{\delta\gamma} [\sqrt{\gamma} \cdot \gamma^{\beta\alpha} F_{,\alpha}]_{,\beta} \\ + \sqrt{\gamma} \cdot \gamma^{\alpha\beta} F_{,\beta} [(\ldots)^{\delta\gamma}]_{,\alpha} + (\ldots)^{\delta\gamma} [\sqrt{\gamma} \cdot \gamma^{\alpha\beta} F_{,\beta}]_{,\alpha} \end{bmatrix} \right) \cdot \frac{\partial F}{\partial y}$$

$$= \int_V d^n x \cdot \sqrt{\gamma} \cdot \left(2 \cdot (\ldots)^{\delta\gamma} C_{\delta j} C_\gamma^j \cdot F + C_\gamma C_\delta \cdot \begin{bmatrix} \gamma^{\beta\alpha} F_{,\alpha} [(\ldots)^{\delta\gamma}]_{,\beta} + \frac{(\ldots)^{\delta\gamma}}{\sqrt{\gamma}} [\sqrt{\gamma} \cdot \gamma^{\beta\alpha} F_{,\alpha}]_{,\beta} \\ + \gamma^{\alpha\beta} F_{,\beta} [(\ldots)^{\delta\gamma}]_{,\alpha} + \frac{(\ldots)^{\delta\gamma}}{\sqrt{\gamma}} [\sqrt{\gamma} \cdot \gamma^{\alpha\beta} F_{,\beta}]_{,\alpha} \end{bmatrix} \right) \cdot \frac{\partial F}{\partial y}$$

$$= \int_V d^n x \cdot 2 \cdot \sqrt{\gamma} \cdot \left((\ldots)^{\delta\gamma} C_{\delta j} C_\gamma^j \cdot F + C_\gamma C_\delta \cdot \left[\frac{\gamma^{\alpha\beta}}{2} \left([(\ldots)^{\delta\gamma}]_{,\alpha} + [(\ldots)^{\delta\gamma}]_{,\beta} \right) + (\ldots)^{\delta\gamma} \cdot \Delta F \right] \right) \cdot \frac{\partial F}{\partial y}.$$

$$\tag{520}$$

Thus, we obtain the Klein–Gordon-like equation in the case of D not being a constant:

$$(\ldots)^{\delta\gamma} C_{\delta j} C_\gamma^j \cdot F + C_\gamma C_\delta \cdot \left[\frac{\gamma^{\alpha\beta}}{2} \left([(\ldots)^{\delta\gamma}]_{,\alpha} + [(\ldots)^{\delta\gamma}]_{,\beta} \right) + (\ldots)^{\delta\gamma} \cdot \Delta F \right] = 0 \tag{521}$$

and a simple Klein–Gordon equation with $(\ldots)^{\delta\gamma} = \text{const}$, which is to say a "tensor" with only constant components:

$$(\ldots)^{\delta\gamma} \cdot (C_{\delta j} C_\gamma^j \cdot F + C_\gamma C_\delta \cdot \Delta F) = 0 \Rightarrow C_{\delta j} C_\gamma^j \cdot F + C_\gamma C_\delta \cdot \Delta F = 0. \tag{522}$$

3.12.1 Higher-Order Functional Approaches

Similar to our extensions as given above (especially cf. subsection "3.11.9 Klein–Gordon and Dirac Equations of Zero, First, Second, and n-th Order") we can seek F-vector and F-tensor approaches also for our complete extended Einstein–Hilbert action (503), that is, Eq. (506)—only this time we do not restrict ourselves to scalars in F. The vector approach like Eq. (479) gives us

$$\delta_\gamma W = 0 = \int_V d^n x \left(R^{\delta\gamma} \left[\gamma^{\delta\gamma} \right] - \frac{1}{2} R \cdot \gamma^{\delta\gamma} + \Lambda \cdot \gamma^{\delta\gamma} + \kappa \cdot T^{\delta\gamma} \left[\gamma^{\delta\gamma} \right] \right)$$

$$\times \sqrt{\gamma} \cdot \gamma_{ij} \delta \left(C^i C^j \cdot F_\delta F_\gamma + C^2 \cdot \gamma^{i\alpha} F_{\delta,\alpha} \gamma^{j\beta} F_{\gamma,\beta} \right)$$

$$= \int_V d^n x \left(R^{\delta\gamma} \left[\gamma^{\delta\gamma} \right] - \frac{1}{2} R \cdot \gamma^{\delta\gamma} + \Lambda \cdot \gamma^{\delta\gamma} + \kappa \cdot T^{\delta\gamma} \left[\gamma^{\delta\gamma} \right] \right)$$

$$\times \sqrt{\gamma} \cdot \delta \left(C_j C^j \cdot F_\delta F_\gamma + C^2 \cdot F_{\delta,j} \gamma^{j\beta} F_{\gamma,\beta} \right)$$

$$= \int_V d^n x \left(R^{\delta\gamma} \left[\gamma^{\delta\gamma} \right] - \frac{1}{2} R \cdot \gamma^{\delta\gamma} + \Lambda \cdot \gamma^{\delta\gamma} + \kappa \cdot T^{\delta\gamma} \left[\gamma^{\delta\gamma} \right] \right)$$

$$\times \sqrt{\gamma} \cdot \left(C_j C^j \cdot \left[F_\delta \cdot \frac{\partial F_\gamma}{\partial y} + F_\gamma \cdot \frac{\partial F_\delta}{\partial y} \right] \right.$$

$$\left. + C^2 \cdot \left[F_{\delta,j} \gamma^{j\beta} \cdot \frac{\partial F_{\gamma,\beta}}{\partial y} + \gamma^{j\beta} F_{\gamma,\beta} \cdot \frac{\partial F_{\delta,j}}{\partial y} \right] \right). \tag{523}$$

Integration by parts of the second addend in the last line, performed as

$$\int_V d^n x (\ldots)^{\delta\gamma} \sqrt{\gamma} \cdot F_{\delta,j} \gamma^{j\beta} \cdot \frac{\partial F_{\gamma,\beta}}{\partial y} = \underbrace{\int_V d^n x \left\{ [(\ldots)^{\delta\gamma} \sqrt{\gamma} \cdot F_{\delta,j} \gamma^{j\beta}] \cdot \frac{\partial F_\gamma}{\partial y} \right\}_{,\beta}}_{=S=0 \text{ per definition}}$$

$$- \int_V d^n x \left[(\ldots)^{\delta\gamma} \sqrt{\gamma} \cdot F_{\delta,j} \gamma^{j\beta} \right]_{,\beta} \cdot \frac{\partial F_\gamma}{\partial y};$$

$$(\ldots)^{\delta\gamma} = \left(R^{\delta\gamma} - \frac{1}{2} R \cdot \gamma^{\delta\gamma} + \Lambda \cdot \gamma^{\delta\gamma} + \kappa \cdot T^{\delta\gamma} \right) \tag{524}$$

yields

$$\delta_\gamma W = 0 = \int_V d^n x \times$$

$$\left(C_j C^j \sqrt{\gamma} \cdot (\ldots)^{\delta\gamma} \left[F_\delta \cdot \frac{\partial F_\gamma}{\partial y} + F_\gamma \cdot \frac{\partial F_\delta}{\partial y} \right] - C^2 \cdot \left[\begin{array}{l} \left[(\ldots)^{\delta\gamma} \sqrt{\gamma} \cdot F_{\delta,j} \gamma^{j\beta}\right]_{,\beta} \cdot \dfrac{\partial F_\gamma}{\partial y} \\[2mm] + \left[(\ldots)^{\delta\gamma} \sqrt{\gamma} \cdot F_{\gamma,\beta} \gamma^{j\beta}\right]_{,j} \cdot \dfrac{\partial F_\delta}{\partial y} \end{array} \right] \right)$$

$$= \int_V d^n x \times \left(\begin{array}{l} \left(C_j C^j \sqrt{\gamma} \cdot (\ldots)^{\delta\gamma} F_\delta - C^2 \cdot \left[(\ldots)^{\delta\gamma} \sqrt{\gamma} \cdot F_{\delta,j} \gamma^{j\beta}\right]_{,\beta} \right) \cdot \dfrac{\partial F_\gamma}{\partial y} \\[2mm] + \left(C_j C^j \sqrt{\gamma} \cdot (\ldots)^{\delta\gamma} F_\gamma - C^2 \cdot \left[(\ldots)^{\delta\gamma} \sqrt{\gamma} \cdot F_{\gamma,\beta} \gamma^{j\beta}\right]_{,j} \right) \cdot \dfrac{\partial F_\delta}{\partial y} \end{array} \right)$$

$$= \int_V d^n x \times \left(\begin{array}{l} \left(C_j C^j \sqrt{\gamma} \cdot (\ldots)^{\delta\gamma} F_\delta - C^2 \cdot \left[\sqrt{\gamma} \cdot F_{\delta,j} \gamma^{j\beta} (\ldots)^{\delta\gamma}_{,\beta} + (\ldots)^{\delta\gamma} \left(\sqrt{\gamma} \cdot \gamma^{j\beta} F_{\delta,j}\right)_{,\beta}\right] \right) \cdot \dfrac{\partial F_\gamma}{\partial y} \\[2mm] + \left(C_j C^j \sqrt{\gamma} \cdot (\ldots)^{\delta\gamma} F_\gamma - C^2 \cdot \left[\sqrt{\gamma} \cdot F_{\gamma,\beta} \gamma^{j\beta} (\ldots)^{\delta\gamma}_{,j} + (\ldots)^{\delta\gamma} \left(\sqrt{\gamma} \cdot \gamma^{j\beta} F_{\gamma,\beta}\right)_{,j}\right] \right) \cdot \dfrac{\partial F_\delta}{\partial y} \end{array} \right)$$

$$= \int_V d^n x \times \sqrt{\gamma} \cdot \left(\begin{array}{l} \left((\ldots)^{\delta\gamma} \left(C_j C^j F_\delta - C^2 \cdot \Delta F_\delta \right) - C^2 \cdot F_{\delta,j} \gamma^{j\beta} (\ldots)^{\delta\gamma}_{,\beta} \right) \cdot \dfrac{\partial F_\gamma}{\partial y} \\[2mm] + \left((\ldots)^{\delta\gamma} \left(C_j C^j F_\gamma - C^2 \cdot \Delta F_\gamma \right) - C^2 \cdot F_{\gamma,\beta} \gamma^{j\beta} (\ldots)^{\delta\gamma}_{,j} \right) \cdot \dfrac{\partial F_\delta}{\partial y} \end{array} \right) . \tag{525}$$

Subsequently, we obtain quantum equations of the kind

$$(\ldots)^{\delta\gamma}\left(C_j C^j F_\delta - C^2 \cdot \Delta F_\delta\right) - C^2 \cdot F_{\delta,j}\gamma^{j\beta}(\ldots)^{\delta\gamma}{}_{,\beta} = 0$$
$$(\ldots)^{\delta\gamma}\left(C_j C^j F_\gamma - C^2 \cdot \Delta F_\gamma\right) - C^2 \cdot F_{\gamma,\beta}\gamma^{j\beta}(\ldots)^{\delta\gamma}{}_{,j} = 0 \qquad (526)$$

The two equations only differ in dummy indices and are therefore identical. In the case of vanishing derivatives for the relativity term $(\ldots)^{\delta\gamma}_{,\beta} = \left(R^{\delta\gamma} - \frac{1}{2}R \cdot \gamma^{\delta\gamma} + \Lambda \cdot \gamma^{\delta\gamma} + \kappa \cdot T^{\delta\gamma}\right)_{,\beta}$ we obtain the simple Klein–Gordon equation with

$$(\ldots)^{\delta\gamma}\left(C_j C^j F_\delta - C^2 \cdot \Delta F_\delta\right) = C_j C^j F_\delta - C^2 \cdot \Delta F_\delta = 0. \qquad (527)$$

3.12.1.1 Towards "Dirac"

Even though we know that the variation of a zero does not automatically mean that the variational result also has to vanish (take the example of the Schwarzschild solution with the vanishing Ricci scalar $R = 0$), we here want to investigate the quantum factor of the integrand in Eq. (523) a bit more closely. As we see namely that this term contains first-order derivatives, one might assume to have some kind of Dirac-forms hidden there. Thus, going back to Eq. (523), we may assume that the variation could also get zero if already the term to variate is zero. Thus, by demanding

$$\delta_\gamma W = 0 = \int_V d^n x \, (\ldots)^{\delta\gamma} \times \sqrt{\gamma} \cdot \gamma_{ij}\delta \left(C^i C^j \cdot F_\delta F_\gamma + C^2 \cdot \gamma^{i\alpha}F_{\delta,\alpha}\gamma^{j\beta}F_{\gamma,\beta}\right)$$
$$\Rightarrow C^i C^j \cdot F_\delta F_\gamma + C^2 \cdot \gamma^{i\alpha}F_{\delta,\alpha}\gamma^{j\beta}F_{\gamma,\beta} = 0, \qquad (528)$$

we should be able to find suitable matrix objects $\mathbf{A}^\sigma$, $\mathbf{B}^\sigma$ allowing us the following separation:

$$C^i C^j \cdot F_\delta F_\gamma + C^2 \cdot \gamma^{i\alpha}F_{\delta,\alpha}\gamma^{j\beta}F_{\gamma,\beta} = \left(\mathbf{A}^0 C^i F_\delta + C \cdot \mathbf{A}^\alpha \gamma^{i\alpha}F_{\delta,\alpha}\right)$$
$$\times \left(\mathbf{B}^0 C^j F_\gamma + C \cdot \mathbf{B}^\beta \gamma^{j\beta}F_{\gamma,\beta}\right). \qquad (529)$$

Apart from the fact that Dirac has used the third binomial formula for his factorization, this indeed already looks Dirac-like. However, before we start searching for suitable matrix candidates (which we will not do here anyway), we want to reconsider the total integrand in Eq. (528) and try for some better symmetrization:

$$\delta_\gamma W = 0 = \int_V d^n x \, (\ldots)^{\delta\gamma} \times \sqrt{\gamma} \cdot \gamma_{ij} \delta \left(C^i C^j \cdot F_\delta F_\gamma + C^2 \cdot \gamma^{i\alpha} F_{\delta,\alpha} \gamma^{j\beta} F_{\gamma,\beta} \right)$$

$$= \int_V d^n x \times \sqrt{\gamma} \cdot \delta$$

$$\times \left(C^i \overbrace{\mathbf{e}_i \cdot \mathbf{e}_j}^{\gamma_{ij}} C^j F_\delta \overbrace{\mathbf{H}^\delta \cdot \mathbf{H}^\gamma}^{(\ldots)^{\delta\gamma}} F_\gamma + C^2 \cdot \overbrace{\mathbf{e}_i \cdot \mathbf{e}_j}^{\gamma_{ij}} \gamma^{i\alpha} F_{\delta,\alpha} \gamma^{j\beta} \overbrace{\mathbf{H}^\delta \cdot \mathbf{H}^\gamma}^{(\ldots)^{\delta\gamma}} F_{\gamma,\beta} \right)$$

$$\Rightarrow \quad C^i \overbrace{\mathbf{e}_i \cdot \mathbf{e}_j}^{\gamma_{ij}} C^j F_\delta \overbrace{\mathbf{H}^\delta \cdot \mathbf{H}^\gamma}^{(\ldots)^{\delta\gamma}} F_\gamma + C^2 \cdot \overbrace{\mathbf{e}_i \cdot \mathbf{e}_j}^{\gamma_{ij}} \gamma^{i\alpha} F_{\delta,\alpha} \gamma^{j\beta} \overbrace{\mathbf{H}^\delta \cdot \mathbf{H}^\gamma}^{(\ldots)^{\delta\gamma}} F_{\gamma,\beta} = 0.$$

$$(530)$$

Now we have completely scalar addends and seek the factorization as follows:

$$C^i \mathbf{e}_i \cdot \mathbf{e}_j C^j F_\delta \mathbf{H}^\delta \cdot \mathbf{H}^\gamma F_\gamma + C^2 \cdot \mathbf{e}_i \cdot \mathbf{e}_j \gamma^{i\alpha} F_{\delta,\alpha} \gamma^{j\beta} \mathbf{H}^\delta \cdot \mathbf{H}^\gamma F_{\gamma,\beta}$$
$$= \left(C^i \mathbf{e}_i F_\delta \mathbf{H}^\delta + C \cdot \mathbf{e}_i \gamma^{i\alpha} F_{\delta,\alpha} \mathbf{H}^\delta \right) \cdot \left(\mathbf{e}_j C^j \mathbf{H}^\gamma F_\gamma + C \cdot \mathbf{e}_j \gamma^{j\beta} \mathbf{H}^\gamma F_{\gamma,\beta} \right).$$

$$(531)$$

Now all indices have become dummy indices allocated (confined) to their very factor, and thus, the factors are perfectly equal. In the simplest case the term $(\ldots)^{\delta\gamma}$ could be proportional to the metric, and thus, $(\ldots)^{\delta\gamma} = D \cdot \gamma^{\delta\gamma} = D \cdot \mathbf{e}^\delta \cdot \mathbf{e}^\gamma$. Inserting this into Eq. (531) yields

$$\left(C^i \mathbf{e}_i F_\delta \mathbf{H}^\delta + C \cdot \mathbf{e}_i \gamma^{i\alpha} F_{\delta,\alpha} \mathbf{H}^\delta \right) \cdot \left(\mathbf{e}_j C^j \mathbf{H}^\gamma F_\gamma + C \cdot \mathbf{e}_j \gamma^{j\beta} \mathbf{H}^\gamma F_{\gamma,\beta} \right)$$
$$= \left(C^i \mathbf{e}_i F_\delta \mathbf{e}^\delta + C \cdot \mathbf{e}_i \gamma^{i\alpha} F_{\delta,\alpha} \mathbf{e}^\delta \right) \cdot \left(\mathbf{e}_j C^j \mathbf{e}^\gamma F_\gamma + C \cdot \mathbf{e}_j \gamma^{j\beta} \mathbf{e}^\gamma F_{\gamma,\beta} \right)$$
$$= \left(C^i \delta_i^\delta F_\delta + C \delta_i^\delta \gamma^{i\alpha} F_{\delta,\alpha} \right) \cdot \left(\delta_j^\gamma C^j F_\gamma + C \delta_j^\gamma \gamma^{j\beta} F_{\gamma,\beta} \right)$$
$$= \left(C^\delta F_\delta + C \gamma^{\delta\alpha} F_{\delta,\alpha} \right) \cdot \left(C^\gamma F_\gamma + C \gamma^{\gamma\beta} F_{\gamma,\beta} \right).$$

$$(532)$$

Defining the matrix S_k^σ as

$$S_k^\sigma \equiv \mathbf{e}_k \mathbf{H}^\sigma \Rightarrow K^\sigma \equiv S_k^\sigma C^k; \, H^{\sigma\alpha} \equiv C \cdot S_k^\sigma \gamma^{k\alpha} \qquad (533)$$

allows us to also give a simple equation for the general outcome of Eq. (531), namely:

$$\left(C^i \mathbf{e}_i F_\delta \mathbf{H}^\delta + C \cdot \mathbf{e}_i \gamma^{i\alpha} F_{\delta,\alpha} \mathbf{H}^\delta \right) \cdot \left(\mathbf{e}_j C^j \mathbf{H}^\gamma F_\gamma + C \cdot \mathbf{e}_j \gamma^{j\beta} \mathbf{H}^\gamma F_{\gamma,\beta} \right)$$
$$= \left(C^i S_i^\delta F_\delta + C S_i^\delta \gamma^{i\alpha} F_{\delta,\alpha} \right) \cdot \left(S_j^\gamma C^j F_\gamma + C S_j^\gamma \gamma^{j\beta} F_{\gamma,\beta} \right) \qquad . \quad (534)$$
$$= \left(K^\delta F_\delta + H^{\delta\alpha} F_{\delta,\alpha} \right) \cdot \left(K^\gamma F_\gamma + H^{\gamma\beta} F_{\gamma,\beta} \right)$$

Obviously the resulting differential equation of first order

$$K^\delta F_\delta + H^{\delta\alpha} F_{\delta,\alpha} = 0 \tag{535}$$

possesses quite some Dirac character. In contrast to the Dirac equation, however, our equation contains no quaternions and results in a scalar structure.

3.12.1.2 Second order

Second-order approaches lead us to

$$\delta_\gamma W = 0 = \int_V d^n x \left(R^{\delta\gamma} \left[\gamma^{\delta\gamma} \right] - \frac{1}{2} R \cdot \gamma^{\delta\gamma} + \Lambda \cdot \gamma^{\delta\gamma} + \kappa \cdot T^{\delta\gamma} \left[\gamma^{\delta\gamma} \right] \right)$$

$$\times \sqrt{\gamma} \cdot \delta \left(C^2 \cdot \gamma_{ij} F_\delta^i F_\gamma^j - C^\alpha C^\beta \cdot \left(\gamma_{ij} F_\delta^i \right)_{,\alpha} F^j_{\gamma,\beta} \right)$$

$$= \int_V d^n x \left(R^{\delta\gamma} \left[\gamma^{\delta\gamma} \right] - \frac{1}{2} R \cdot \gamma^{\delta\gamma} + \Lambda \cdot \gamma^{\delta\gamma} + \kappa \cdot T^{\delta\gamma} \left[\gamma^{\delta\gamma} \right] \right)$$

$$\times \sqrt{\gamma} \cdot \delta \left(C^2 \cdot F_{j\delta} F_\gamma^j - C^\alpha C^\beta \cdot F_{j\delta,\alpha} F^j_{\gamma,\beta} \right)$$

$$= \int_V d^n x \left(R^{\delta\gamma} \left[\gamma^{\delta\gamma} \right] - \frac{1}{2} R \cdot \gamma^{\delta\gamma} + \Lambda \cdot \gamma^{\delta\gamma} + \kappa \cdot T^{\delta\gamma} \left[\gamma^{\delta\gamma} \right] \right)$$

$$\times \sqrt{\gamma} \cdot \left(C^2 \cdot \left[F_{j\delta} \cdot \frac{\partial F_\gamma^j}{\partial y} + F_\gamma^j \cdot \frac{\partial F_{j\delta}}{\partial y} \right] \right.$$

$$\left. + C^\alpha C^\beta \cdot \left[F_{j\delta,\alpha} \cdot \frac{\partial F^j_{\gamma,\beta}}{\partial y} + F^j_{\gamma,\beta} \cdot \frac{\partial F_{j\delta,\alpha}}{\partial y} \right] \right). \tag{536}$$

Integration by parts of the second addend in the last line, performed as

$$\int_V d^n x \, (\ldots)^{\delta\gamma} C^\alpha C^\beta \cdot \sqrt{\gamma} \cdot \left(\gamma_{ij} \cdot F_\delta^i \right)_{,\alpha} \cdot \frac{\partial F^j_{\gamma,\beta}}{\partial y}$$

$$= \int_V d^n x \underbrace{\left\{ \lfloor (\ldots)^{\delta\gamma} C^\alpha C^\beta \cdot \sqrt{\gamma} \cdot F_{j\delta,\alpha} \rfloor \cdot \frac{\partial F_\gamma^j}{\partial y} \right\}_{,\beta}}_{=S=0 \text{ per definition}}$$

$$- \int_V d^n x \left[(\ldots)^{\delta\gamma} C^\alpha C^\beta \cdot \sqrt{\gamma} \cdot F_{j\delta,\alpha} \right]_{,\beta} \cdot \frac{\partial F_\gamma^j}{\partial y};$$

$$(\ldots)^{\delta\gamma} = \left(R^{\delta\gamma} - \frac{1}{2} R \cdot \gamma^{\delta\gamma} + \Lambda \cdot \gamma^{\delta\gamma} + \kappa \cdot T^{\delta\gamma} \right) \tag{537}$$

yields

$$
\delta_\gamma W = 0 = \int_V d^n x \left(R^{\delta\gamma}[\gamma^{\delta\gamma}] - \frac{1}{2} R \cdot \gamma^{\delta\gamma} + \Lambda \cdot \gamma^{\delta\gamma} + \kappa \cdot T^{\delta\gamma}[\gamma^{\delta\gamma}] \right) \times \sqrt{\gamma} \cdot \left(C^2 \cdot \left[F_{j\delta} \cdot \frac{\partial F_\gamma^j}{\partial y} + F_\gamma^j \cdot \frac{\partial F_{j\delta}}{\partial y} \right] \right.
$$

$$
\left. + C^\alpha C^\beta \cdot \gamma_{ij} \cdot \left[F^i_{\delta,\alpha} \cdot \frac{\partial F^j_{\gamma,\beta}}{\partial y} + F^j_{\gamma,\beta} \cdot \frac{\partial F^i_{\delta,\alpha}}{\partial y} \right] \right)
$$

$$
= \int_V d^n x \times \left(\sqrt{\gamma} \cdot C^2 \cdot (\ldots)^{\delta\gamma} \left[F_{j\delta} \cdot \frac{\partial F_\gamma^j}{\partial y} + F_\gamma^j \cdot \frac{\partial F_{j\delta}}{\partial y} \right] \right.
$$

$$
- \left[\left[(\ldots)^{\delta\gamma} C^\alpha C^\beta \cdot \sqrt{\gamma} \cdot F_{j\delta,\alpha} \right]_{,\beta} \cdot \frac{\partial F_\gamma^j}{\partial y} \right.
$$

$$
\left.\left. + \left[(\ldots)^{\delta\gamma} C^\alpha C^\beta \cdot \sqrt{\gamma} \cdot F^j_{\gamma,\beta} \right]_{,\alpha} \cdot \frac{\partial F_{j\delta}}{\partial y} \right] \right) \cdot
$$

$$
\tag{538}
$$

Now we assume that

$$
C^\alpha C^\beta = D \cdot \gamma^{\alpha\beta}
\tag{539}
$$

and are able to further simplify (539):

$$\hat{\delta}_\gamma W = 0 = \int_V d^n \mathrm{x} \times \sqrt{\gamma} \cdot \left(C^2 \cdot (\ldots)^{\delta\gamma} \cdot \left[F_{j\delta} \cdot \frac{\partial F_\gamma^j}{\partial y} + F_\gamma^j \cdot \frac{\partial F_{j\delta}}{\partial y} \right] - \frac{D}{\sqrt{\gamma}} \cdot \left[\begin{array}{l} [(\ldots)^{\delta\gamma}\gamma^{\alpha\beta} \cdot \sqrt{\gamma} \cdot F_{j\delta,\alpha}]_{,\beta} \cdot \dfrac{\partial F_\gamma^j}{\partial y} \\[2mm] + [(\ldots)^{\delta\gamma}\gamma^{\alpha\beta} \cdot \sqrt{\gamma} \cdot F_{\gamma,\beta}^j]_{,\alpha} \cdot \dfrac{\partial F_{j\delta}}{\partial y} \end{array} \right] \right)$$

$$= \int_V d^n \mathrm{x} \times \sqrt{\gamma} \cdot \left(\left(C^2 \cdot (\ldots)^{\delta\gamma} F_{j\delta} - \frac{D}{\sqrt{\gamma}} \cdot [(\ldots)^{\delta\gamma}\gamma^{\alpha\beta} \cdot \sqrt{\gamma} \cdot F_{j\delta,\alpha}]_{,\beta} \right) \cdot \frac{\partial F_\gamma^j}{\partial y} \right.$$
$$\left. + \left(C^2 \cdot (\ldots)^{\delta\gamma} F_\gamma^j - \frac{D}{\sqrt{\gamma}} \cdot [(\ldots)^{\delta\gamma}\gamma^{\alpha\beta} \cdot \sqrt{\gamma} \cdot F_{\gamma,\beta}^j]_{,\alpha} \right) \cdot \frac{\partial F_{j\delta}}{\partial y} \right)$$

$$= \int_V d^n \mathrm{x} \times \sqrt{\gamma} \cdot \left(\left(C^2 \cdot (\ldots)^{\delta\gamma} F_{j\delta} - \frac{D}{\sqrt{\gamma}} \cdot \left(\gamma^{\alpha\beta} \cdot \sqrt{\gamma} \cdot F_{j\delta,\alpha}(\ldots)^{\delta\gamma}_{,\beta} + (\ldots)^{\delta\gamma}\left[\gamma^{\alpha\beta} \cdot \sqrt{\gamma} \cdot F_{j\delta,\alpha}\right]_{,\beta} \right) \right) \cdot \frac{\partial F_\gamma^j}{\partial y} \right.$$
$$\left. + \left(C^2 \cdot (\ldots)^{\delta\gamma} F_\gamma^j - \frac{D}{\sqrt{\gamma}} \cdot \left(\gamma^{\alpha\beta} \cdot \sqrt{\gamma} \cdot F_{\gamma,\beta}^j(\ldots)^{\delta\gamma}_{,\alpha} + (\ldots)^{\delta\gamma}\left[\gamma^{\alpha\beta} \cdot \sqrt{\gamma} \cdot F_{\gamma,\beta}^j\right]_{,\alpha} \right) \right) \cdot \frac{\partial F_{j\delta}}{\partial y} \right)$$

$$= \int_V d^n \mathrm{x} \times \sqrt{\gamma} \cdot \left(\left(C^2 \cdot (\ldots)^{\delta\gamma} F_{j\delta} - D \cdot \left(\gamma^{\alpha\beta} \cdot F_{j\delta,\alpha}(\ldots)^{\delta\gamma}_{,\beta} + (\ldots)^{\delta\gamma} \Delta F_{j\delta} \right) \right) \cdot \frac{\partial F_\gamma^j}{\partial y} \right.$$
$$\left. + \left(C^2 \cdot (\ldots)^{\delta\gamma} F_\gamma^j - D \cdot \left(\gamma^{\alpha\beta} \cdot F_{\gamma,\beta}^j(\ldots)^{\delta\gamma}_{,\alpha} + (\ldots)^{\delta\gamma} \Delta F_\gamma^j \right) \right) \cdot \frac{\partial F_{j\delta}}{\partial y} \right) . \tag{540}$$

Subsequently, we obtain sets of quantum equations of the kind

$$c^2 \cdot (\ldots)^{\delta\gamma} F_{j\delta} - D \cdot \left(\gamma^{\alpha\beta} \cdot F_{j\delta\alpha}(\ldots)^{\delta\gamma}{}_{,\beta} + (\ldots)^{\delta\gamma} \Delta F_{j\delta} \right) = 0$$
$$c^2 \cdot (\ldots)^{\delta\gamma} F_{\gamma}^{j} - D \cdot \left(\gamma^{\alpha\beta} \cdot F_{\gamma\beta}^{j}(\ldots)^{\delta\gamma}{}_{,\alpha} + (\ldots)^{\delta\gamma} \Delta F_{\gamma}^{j} \right) = 0 \qquad (541)$$

Chapter 4

Various Forms of Metric x^k-Variations

We saw that with an approach of the form (388) or (389), we would have solved the problem we had with the Ricci scalar quantization, where a quantum solution did not necessarily also automatically solve the Einstein field equations. The product structure in Eq. (388) or Eq. (389) assures such a simultaneous result. However, even though we already obtained a variety of impressive outcomes, there are still a few technical issues that we intend to tackle within this chapter.

In [114, 117, 136], we considered a few options for the variation of the metric tensor. Here we repeat those leading to the Klein–Gordon equation and to the Dirac equation. Further below, when treating the connection to materials science, we will also consider the general case.

Our starting point shall be the usual tensor transformation rule for the covariant metric tensor:

$$g_{\delta\gamma} = \mathbf{g}_\delta \cdot \mathbf{g}_\gamma = \frac{\partial G^\alpha [x_k]}{\partial x^\delta} \frac{\partial G^\beta [x_k]}{\partial x^\gamma} g_{\alpha\beta}. \tag{542}$$

This already leads to quite some varieties with respect to the apparently simple term $\delta g_{\mu\nu}$ [114, 117]. The base vectors $\mathbf{g}_\delta$ to a certain metric are given as:

$$\mathbf{g}_\delta = \frac{\partial G^\alpha [x_k]}{\partial x^\delta} \mathbf{e}_\alpha, \tag{543}$$

where the functions $G^\alpha [\dots]$ denote arbitrary functions of the coordinates x_k. Here the vectors $\mathbf{e}_\alpha$ shall denote the base vectors of a fundamental coordinate

The World Formula: A Late Recognition of David Hilbert's Stroke of Genius
Norbert Schwarzer
Copyright © 2022 Jenny Stanford Publishing Pte. Ltd.
ISBN 978-981-4877-20-6 (Hardcover), 978-1-003-14644-5 (eBook)
www.jennystanford.com

system of the right (in principle arbitrary) number of dimension. Thus, we have the variation for $\delta g_{\delta\gamma}$ in (388) and (389) as follows:

$$
\delta g_{\delta\gamma} = \delta \left(\mathbf{g}_\delta \cdot \mathbf{g}_\gamma \right) = \mathbf{g}_\delta \cdot \delta\mathbf{g}_\gamma + \delta\mathbf{g}_\delta \cdot \mathbf{g}_\gamma
$$
$$
= \frac{\partial G^\alpha \, [x_k]}{\partial x^\delta} \mathbf{e}_\alpha \cdot \delta \left(\frac{\partial G^\beta \, [x_k]}{\partial x^\gamma} \mathbf{e}_\beta \right) + \delta \left(\frac{\partial G^\alpha \, [x_k]}{\partial x^\delta} \mathbf{e}_\alpha \right) \cdot \frac{\partial G^\beta \, [x_k]}{\partial x^\gamma} \mathbf{e}_\beta. \quad (544)
$$

Now we introduce two additional degrees of freedom, namely:

(a) neither the number of dimensions in which the base vectors exist and form a complete transformation (543) needs to be the same as the metric space they define,

(b) nor the variation δ is defined or fixed in any way.

In order to properly account for point (a), we shall rewrite (542) and (543) as:

$$
\delta g_{\delta\gamma} = \delta \left(\mathbf{g}_\delta \cdot \mathbf{g}_\gamma \right) = \mathbf{g}_\delta \cdot \delta\mathbf{g}_\gamma + \delta\mathbf{g}_\delta \cdot \mathbf{g}_\gamma
$$
$$
= \frac{\partial G^i \, [x_k]}{\partial x^\delta} \mathbf{e}_i \cdot \delta \left(\frac{\partial G^j \, [x_k]}{\partial x^\gamma} \mathbf{e}_j \right) + \delta \left(\frac{\partial G^i \, [x_k]}{\partial x^\delta} \mathbf{e}_i \right) \cdot \frac{\partial G^j \, [x_k]}{\partial x^\gamma} \mathbf{e}_j
$$
$$
\mathbf{g}_\delta = \frac{\partial G^j \, [x_k]}{\partial x^\delta} \mathbf{e}_j. \quad (545)
$$

Thereby the Latin indices are running to a different (potentially higher) number of dimensions N than the Greek indices, which shall be defined for a space or space-time of n dimensions.

As it is principally of no importance in what kind of fundamental coordinate system our base vectors $\mathbf{e}_j$ are defined, we here opt for the simplest starting point and demand them to be the Cartesian base vectors. Thereby we note that the ensemble of Cartesian base vectors, which is usually given as:

$$
\mathbf{e}_j = \{\mathbf{e}_0, \mathbf{e}_1, \mathbf{e}_2, \ldots, \mathbf{e}_{n-1}\} = \left\{ \begin{pmatrix} 1 \\ 0 \\ \vdots \\ 0 \end{pmatrix}, \begin{pmatrix} 0 \\ 1 \\ \vdots \\ 0 \end{pmatrix}, \ldots, \begin{pmatrix} 0 \\ 0 \\ \vdots \\ 1 \end{pmatrix} \right\}, \quad (546)
$$

has an intrinsic degree of freedom with respect to the signs of the non-zero or 1-components. Thus, we have:

$$\mathbf{e}_j = \{\mathbf{e}_0, \mathbf{e}_1, \mathbf{e}_2, \ldots, \mathbf{e}_{2 \cdot n - 1}\}$$

$$= \left\{ \begin{pmatrix} 1 \\ 0 \\ \vdots \\ 0 \end{pmatrix}, \begin{pmatrix} -1 \\ 0 \\ \vdots \\ 0 \end{pmatrix}; \begin{pmatrix} 0 \\ 1 \\ \vdots \\ 0 \end{pmatrix}, \begin{pmatrix} 0 \\ -1 \\ \vdots \\ 0 \end{pmatrix}; \ldots; \begin{pmatrix} 0 \\ 0 \\ \vdots \\ 1 \end{pmatrix}, \begin{pmatrix} 0 \\ 0 \\ \vdots \\ -1 \end{pmatrix} \right\}, \quad (547)$$

or even:

$$\mathbf{e}_j = \{\mathbf{e}_0, \mathbf{e}_1, \mathbf{e}_2, \ldots, \mathbf{e}_{4 \cdot n - 1}\}$$

$$= \left\{ \begin{pmatrix} 1 \\ 0 \\ \vdots \\ 0 \end{pmatrix}, \begin{pmatrix} -1 \\ 0 \\ \vdots \\ 0 \end{pmatrix}, \begin{pmatrix} i \\ 0 \\ \vdots \\ 0 \end{pmatrix}, \begin{pmatrix} -i \\ 0 \\ \vdots \\ 0 \end{pmatrix}; \begin{pmatrix} 0 \\ 1 \\ \vdots \\ 0 \end{pmatrix}, \begin{pmatrix} 0 \\ -1 \\ \vdots \\ 0 \end{pmatrix}, \begin{pmatrix} 0 \\ i \\ \vdots \\ 0 \end{pmatrix}, \begin{pmatrix} 0 \\ -i \\ \vdots \\ 0 \end{pmatrix}; $$

$$\ldots; \begin{pmatrix} 0 \\ 0 \\ \vdots \\ 1 \end{pmatrix}, \begin{pmatrix} 0 \\ 0 \\ \vdots \\ -1 \end{pmatrix}, \begin{pmatrix} 0 \\ 0 \\ \vdots \\ i \end{pmatrix}, \begin{pmatrix} 0 \\ 0 \\ \vdots \\ -i \end{pmatrix} \right\}. \quad (548)$$

Taking into account that all possible combinations of vector pairs (547) or quadruples (548) give Cartesian metrics, we always have 2^n, respectively 4^n permutations (c.f. [133, 134]).

In other words: The index j is not running from 0 to $n - 1$, but to $N - 1$ (as introduced above in point a) with $N = 2^n$ or $N = 4^n$.

Now using point (b) from our conditions above, we introduce a general variation vector δ_α and simply assume a variation of such a kind that we would result in the following condition:

$$\delta g_{\delta\gamma} = \delta \left(\mathbf{g}_\delta \cdot \mathbf{g}_\gamma \right) = \delta_\delta \sum_{\delta=0}^{n-1} \mathbf{g}_\delta \cdot \mathbf{g}_\gamma + \delta_\gamma \sum_{\gamma=0}^{n-1} \mathbf{g}_\gamma \cdot \mathbf{g}_\delta$$

$$= 2 \cdot \delta_\gamma \sum_{\gamma=0}^{n-1} \mathbf{g}_\gamma \cdot \mathbf{g}_\delta = 2 \cdot \delta_\gamma \sum_{\gamma=0}^{n-1} \left(\frac{\partial G^j [x_k]}{\partial x^\gamma} \mathbf{e}_j \right) \cdot \mathbf{g}_\delta = 0$$

$$\Rightarrow \sum_{\gamma=0}^{n-1} \left(\frac{\partial G^j [x_k]}{\partial x^\gamma} \mathbf{e}_j \right) = 0. \quad (549)$$

This trick leading to the desired sum in order to later produce or at least mirror the classical quantum equations had to be seen problematic, because, so far, it is not properly justified. We will therefore come back to the question of the origin of the γ-sum further below.

4.1 Matrix Option and Classical Dirac Form

The ";" in our base vector equations above shall point out that in principle we can treat all those additional base vectors as independent, which in the end shall allow us to construct matrix products of the following kind:

$$\sum_{\gamma=0}^{n-1}\left(\frac{\partial G^j\,[x_k]}{\partial x^\gamma}\gamma^\gamma\right)=0 \quad \text{with} \quad \gamma^\delta\gamma^\chi = \begin{cases} 0 & \text{for} \quad \delta \neq \chi \\ 1 & \text{for} \quad \delta = \chi \end{cases}, \tag{550}$$

where we recognize the Dirac-like character for the matrices and achieve the option for the following operator combination:

$$\sum_{\chi=0}^{n-1}\gamma^\chi\frac{\partial}{\partial x^\chi}\sum_{\gamma=0}^{n-1}\left(\gamma^\gamma\frac{\partial G^j\,[x_k]}{\partial x^\gamma}\right) = \sum_{\chi=0}^{n-1}\gamma^\chi\frac{\partial}{\partial x^\chi}\sum_{\gamma=0}^{n-1}\left(\gamma^\gamma\frac{\partial}{\partial x^\gamma}\right)G^j\,[x_k] = 0. \tag{551}$$

We recognize the Dirac equation without mass. In order to obtain such a mass term, we assume one derivative with respect to a coordinate ξ to deliver the original function times a constant as follows:

$$\frac{\partial G^j\,[x_k]}{\partial x^\xi} = i\cdot M\cdot G^j\,[x_k] \tag{552}$$

and reconstruct our factorization from above, now giving:

$$\text{with}: \quad \kappa = 0, 1, \ldots, n-2; \quad \xi = n-1$$

$$\sum_{\chi=0}^{n-1}\gamma^\chi\frac{\partial}{\partial x^\chi}\sum_{\gamma=0}^{n-1}\left(\gamma^\gamma\frac{\partial}{\partial x^\gamma}\right)G^j\,[x_k]$$

$$= \sum_{\chi=0}^{n-1}\gamma^\chi\frac{\partial}{\partial x^\chi}\left[\sum_{\kappa=0}^{n-2}\left(\gamma^\kappa\frac{\partial}{\partial x^\kappa}\right)G^j\,[x_k] + \gamma^\xi i\cdot M\cdot G^j\,[x_k]\right]$$

$$\sum_{\kappa=0}^{n-2}\left(\frac{\partial^2 G^j\,[x_k]}{(\partial x^\kappa)^2}\right) - M^2\cdot G^j\,[x_k] = 0. \tag{553}$$

Now we see that with Eq. (550) we would have a complete Dirac equation.

Please note that in contrast to the classical Dirac equation, we did the factorization without the help of the third binomial formula (c.f. [143]).

4.1.1 Dirac Equation with and without Quaternions

Because of the importance of the Dirac equation [143], we want to consider its metric origin in some more detail. Thereby we will learn that the Dirac way of applying quaternions is not the only solution.

Further in section "4.3 A Little Bit of Materials Science", we will see that a very general variation like:

$$\delta g_{\delta\gamma} = \delta\left(\mathbf{g}_\delta \cdot \mathbf{g}_\gamma\right) = \mathbf{g}_\delta \cdot \delta\mathbf{g}_\gamma + \delta\mathbf{g}_\delta \cdot \mathbf{g}_\gamma$$
$$= \frac{\partial G^i\,[x_k]}{\partial x^\delta}\mathbf{e}_i \cdot \left(\frac{\partial^2 G^j\,[x_k]}{\partial x^\alpha \partial x^\gamma}\mathbf{e}_j\right)\delta x^\alpha$$
$$+ \left(\frac{\partial^2 G^i\,[x_k]}{\partial x^\alpha \partial x^\delta}\mathbf{e}_i\right)\delta x^\alpha \cdot \frac{\partial G^j\,[x_k]}{\partial x^\gamma}\mathbf{e}_j, \tag{554}$$

would result in non-metric spaces. Thus, using point (b) from our conditions above, we simply assume a variation of such a kind that we would result in the following condition:

$$\delta g_{\delta\gamma} = \delta\left(\mathbf{g}_\delta \cdot \mathbf{g}_\gamma\right) = \mathbf{g}_\delta \cdot \delta\mathbf{g}_\gamma + \delta\mathbf{g}_\delta \cdot \mathbf{g}_\gamma = \delta g_{\delta\delta}$$
$$= 2\frac{\partial G^i\,[x_k]}{\partial x^\delta}\mathbf{e}_i \cdot \left(\frac{\partial^2 G^j\,[x_k]}{\partial x^\alpha \partial x^\delta}\mathbf{e}_j\right)\delta x^\alpha$$
$$= 2\frac{\partial G^i\,[x_k]}{\partial x^\delta}\mathbf{e}_i \cdot \left(\frac{\partial}{\partial x^\alpha}\left(g^{\alpha\beta}\frac{\partial G^j\,[x_k]}{\partial x^\delta}\mathbf{e}_j\right)\right)\delta x_\beta = 0. \tag{555}$$

We might even boldly* demand that there exists a variation rendering (554) as follows:

$$\delta g_{\delta\gamma} = \delta g_{\delta\delta} = 2\frac{\partial G^i\,[x_k]}{\partial x^\delta}\mathbf{E}_i \cdot \mathbf{E}_j\left(\frac{1}{\sqrt{g}}\frac{\partial}{\partial x^\alpha}\sqrt{g}\,g^{\alpha\beta}\frac{\partial G^j\,[x_k]}{\partial x^\beta}\right)\delta x_\delta = 0. \tag{556}$$

We recognize the complete Laplace operator:

$$\delta g_{\delta\gamma} = \delta g_{\delta\delta} = 2\frac{\partial G^i\,[x_k]}{\partial x^\delta}\mathbf{E}_i \cdot \mathbf{E}_j\left(\Delta G^j\,[x_k]\right)\delta x_\delta = 0, \tag{557}$$

which, resulting in either zero for each of the functions G^j:

$$\Delta G^j\,[x_k] = 0, \tag{558}$$

or the corresponding sum within the contraction with respect to the index j above:

$$\sum_{j=0}^{N-1}\mathbf{E}_j\left(\Delta G^j\,[x_k]\right) = \mathbf{0}, \tag{559}$$

does deliver a complete solution to the action (389).

---

*Thereby insisting on a principal degree of freedom due to condition (a) and leaving the sorting out of the final details of the procedure for later (or to the skilled and interested reader).

However, there are even more possibilities. For one, we could try and factorize Eq. (557) or Eq. (555). In the case of Eq. (555), this would lead us to:

$$\frac{\partial G^i \, [x_k]}{\partial x^\delta} \mathbf{e}_i \cdot \left(\frac{\partial}{\partial x^\alpha} \left(g^{\alpha\beta} \frac{\partial G^j \, [x_k]}{\partial x^\delta} \mathbf{e}_j \right) \right) \delta x_\beta$$

$$= \frac{\partial G^i \, [x_k]}{\partial x^\delta} \mathbf{e}_i \cdot \mathbf{e}_j \left(\frac{\partial}{\partial x^\alpha} \left(g^{\alpha\beta} \frac{\partial G^j \, [x_k]}{\partial x^\delta} \right) \right) \delta x_\beta$$

$$\left[\gamma^\delta \frac{\partial G^i \, [x_k]}{\partial x^\delta} \right] \cdot \left[\left(\gamma^\delta \frac{\partial}{\partial x^\alpha} \left(g^{\alpha\beta} \frac{\partial G^j \, [x_k]}{\partial x^\delta} \right) \right) \delta x_\beta \right] = 0, \qquad (560)$$

where we assumed the vector $\mathbf{e}_j$ not to depend on the coordinates. The objects γ^δ have to be suitable matrices in order to assure the right product outcome. We see that therefore we have to demand the conditions:

$$\gamma^\delta \gamma^\chi = \begin{cases} 0 & \text{for} \quad \delta \neq \chi \\ 1 & \text{for} \quad \delta = \chi \end{cases}, \qquad (561)$$

where we recognize the Dirac-like character of our separation. In fact, one may apply the classical Dirac gamma matrices, which are usually given as follows:

$$\gamma^\beta_{\text{Dirac}} \simeq i \cdot \gamma^\beta \quad \text{with:} \quad \gamma^0 = \begin{pmatrix} 1 & & & \\ & 1 & & \\ & & -1 & \\ & & & -1 \end{pmatrix}; \quad \gamma^1 = \begin{pmatrix} & & & 1 \\ & & 1 & \\ & -1 & & \\ -1 & & & \end{pmatrix}$$

$$\gamma^2 = \begin{pmatrix} & & & -i \\ & & i & \\ & i & & \\ -i & & & \end{pmatrix}; \quad \gamma^3 = \begin{pmatrix} & & 1 & \\ & & & -1 \\ -1 & & & \\ & 1 & & \end{pmatrix}; \quad I = \begin{pmatrix} 1 & & & \\ & 1 & & \\ & & 1 & \\ & & & 1 \end{pmatrix}$$

$$(562)$$

and find them satisfying the conditions (561). The only thing missing would be mass. Therefore, we assume one derivative with respect to a coordinate ξ to deliver the original function times a constant as follows:

$$\frac{\partial G^i \, [x_k]}{\partial x^\xi} = i \cdot M \cdot G^i \, [x_k]; \quad \frac{\partial}{\partial x^\alpha} \left(g^{\alpha\beta} \frac{\partial G^j \, [x_k]}{\partial x^\xi} \right) = -C_M \cdot M^2 \cdot G^j \, [x_k] \quad (563)$$

and construct our factorization, giving:

with : $\kappa = 0, 1, \ldots, n - 2;$ $\xi = n - 1$

$$\left[\gamma^\delta \frac{\partial G^i [x_k]}{\partial x^\delta} \right] \cdot \left[\left(\gamma^\delta \frac{\partial}{\partial x^\alpha} \left(g^{\alpha\beta} \frac{\partial G^j [x_k]}{\partial x^\delta} \right) \right) \delta x_\beta \right]$$

$$= \left[\gamma^\kappa \frac{\partial G^i [x_k]}{\partial x^\kappa} + \gamma^\xi i \cdot M \cdot G^i [x_k] \right]$$

$$\times \left[\left(\gamma^\kappa \frac{\partial}{\partial x^\alpha} \left(g^{\alpha\beta} \frac{\partial G^j [x_k]}{\partial x^\kappa} \right) \right) - C_M \cdot \gamma^\xi M^2 \cdot G^j [x_k] \right] \delta x_\beta = 0.$$

$$(564)$$

Now we see that we have two options to achieve the desired zero outcome, namely:

$$\left[\gamma^\kappa \frac{\partial G^i [x_k]}{\partial x^\kappa} + i \cdot \gamma^\xi M \cdot G^i [x_k] \right] = 0$$

$$\left[\left(\gamma^\kappa \frac{\partial}{\partial x^\alpha} \left(g^{\alpha\beta} \frac{\partial G^j [x_k]}{\partial x^\kappa} \right) \right) - C_M \cdot \gamma^\xi M^2 \cdot G^j [x_k] \right] \delta x_\beta = 0. \quad (565)$$

Please note that in contrast to the classical Dirac equation, we did the factorization without the help of the third binomial formula (c.f. [143]).

4.1.1.1 Interpretation

Classically, the Dirac equation can be evaluated as a quaternionic square root of the Klein–Gordon equation [143]. Thus, having already obtained the latter in a metric manner (e.g. [145]), we might be of the opinion that our task of also deriving the metric origin of the Dirac equation is already done.

This, however, is not correct as Dirac's introduction of the quaternions in order to obtain the square root of the Klein–Gordon equation did—de facto—neither explain why he was able to do it, nor reveal his procedure any geometric meaning.

With the Dirac equation being of total axiomatic character, this "why?" and "what does it mean?" was not to be expected anyway.

THE "WHY?":

While Dirac constructed his famous equation directly out of the Klein–Gordon equation, we here obtain a first-order differential equation like the Dirac equation as factor DIRECTLY out of the variation of the metric tensor. From Eq. (564), we can directly deduce that the Dirac (or Dirac-like) first-order differential equation is not just an operational square

root of the Klein–Gordon equation, but a real additional factor. The other factor is Klein–Gordon-like (see second line in Eq. (565)). Both such factors contain the same functions G^j and so we have a certain connection among them. Interestingly, this connection can obviously be constructed by making the Dirac factor the operational square root of the Klein–Gordon factor. However, by looking at the general starting point (389), we should conclude not only that we do have more options but also that there are probably also other ways to perform the mathematical connection among the two factors. One such other option, we named "the method of vectorial root extraction", will be considered below in the next subsection.

THE "WHAT DOES IT MEAN?":

De facto, the result of the general variation (544), using differentiation, automatically results in two differential equations, one of the first order and the other of the second order. Of course, a very simple approach for a solution would be to try and find only G^j-functions which do solve both equations. This in fact is the Dirac strategy. It should be pointed out, however, that, due to the factor-connection of the two equations, we also have the option of solutions only satisfying one of the two equations.

In other words: There seem to be more possibilities than the classical quantum equations allow and they all—apparently—are to be found in a simple (not necessarily in a mathematical sense, at least not to this author) metric variation.

4.1.1.2 Quaternion-free Dirac equation

Already in [133], by choosing a completely metric origin and introducing a new type of scalar root extraction, we were able to derive an explanation about the "why" of the Dirac equation even without resorting to an extended variation of the Einstein–Hilbert action as done here. Thereby this author is aware of the fact that already in the twenties of the last century, there have been attempts to obtain suitable extensions of Einstein's General Theory of Relativity via the introduction of metric torsion fields [167, 168]. These fields automatically resulted in spin fields and thus were seen as a way to explain the intrinsic property of spin of elementary particles (also see the nice summary in [169]). As, apart from the antimatter, spin is just the essential achievement of the Dirac equation, one felt tempted to consider twisted space-times with metric torsions as the geometric origin of the Dirac equation. Here, with the Dirac equation already emerging from a suitable variation of the metric tensor like (555), we saw, however, that we do not necessarily need such torsion fields in order to have spin (c.f. solutions (68) to (72)).

Apparently, as demonstrated above, there also seems to be an approach which extracts the Dirac operator directly out of the Einstein field equations or the Einstein–Hilbert action and thus, the General Theory of Relativity. With respect to the connection with the classical approach, we refer to [133] and present here only the new technique which does not require the use of quaternions as Dirac needed them.

It is easy to prove that the scalar product of the following vector:

$$
\mathbf{V}_\Omega = \left\{
\begin{array}{l}
a+b+c+d+e, a+b+c+d-e, a+b+c-d+e, a+b+c-d-e, \\
a+b-c+d+e, a+b-c+d-e, a+b-c-d+e, a+b-c-d-e, \\
a-b+c+d+e, a-b+c+d-e, a-b+c-d+e, a-b+c-d-e, \\
a-b-c+d+e, a-b-c+d-e, a-b-c-d+e, a-b-c-d-e
\end{array}
\right\}
$$
$$
\equiv [a \pm b \pm c \pm d \pm e]_\Omega \tag{566}
$$

gives:

$$
\mathbf{V}_\Omega \cdot \mathbf{V}_\Omega = a^2 + b^2 + c^2 + d^2 + e^2. \tag{567}
$$

Even the introduction of "virtual" parameters ε can be incorporated into vector $\mathbf{V}_\Omega$ as follows:

$$
\mathbf{V}_\Omega = \{a+b+c, a+b-c, a-b+i \cdot c, a-b-i \cdot c\}
$$
$$
\equiv [a \pm b \pm I \cdot c]_\Omega
$$
$$
\text{with}: \quad \mathbf{V}_\Omega \cdot \mathbf{V}_\Omega = a^2 + b^2. \tag{568}
$$

Now we introduce the following matrix V:

$$
V = \left[\gamma_{\beta'}^{0\alpha'} \frac{\partial}{\partial x^0} \pm \gamma_{\beta'}^{1\alpha'} \frac{\partial}{\partial x^1} \pm \gamma_{\beta'}^{2\alpha'} \frac{\partial}{\partial x^2} \pm \gamma_{\beta'}^{3\alpha'} \frac{\partial}{\partial x^3} \pm I \cdot \varepsilon^m \cdot f_m \right]. \tag{569}
$$

Please note that the "coefficient matrices" $\gamma_{\beta'}^{\mu\alpha'}$ are not connected with the so-called Dirac gamma matrices. We reveal their nature in due course. It should be pointed out that the indices α' and β' are not identical with the metric indices and could therefore also run to different dimensions. Incomplete expansion without the virtual terms leads to:

$$V_{\beta'}^{\alpha'} = \left[\gamma_{\beta'}^{\mu\alpha'} \frac{\partial}{\partial x^\mu}\right] \equiv \gamma_{\beta'}^{\mu\alpha'} \partial_\mu = \left[\gamma_{\beta'}^{0\alpha'} \frac{\partial}{\partial x^0} \pm \gamma_{\beta'}^{1\alpha'} \frac{\partial}{\partial x^1} \pm \gamma_{\beta'}^{2\alpha'} \frac{\partial}{\partial x^2} \pm \gamma_{\beta'}^{3\alpha'} \frac{\partial}{\partial x^3}\right]_\Omega$$

$$= \begin{cases} \gamma_0^{\alpha'}\left(\frac{\partial}{\partial x^0} + \frac{\partial}{\partial x^1} + \frac{\partial}{\partial x^2} + \frac{\partial}{\partial x^3}\right) \\ \gamma_1^{\alpha'}\left(\frac{\partial}{\partial x^0} + \frac{\partial}{\partial x^1} + \frac{\partial}{\partial x^2} - \frac{\partial}{\partial x^3}\right) \\ \vdots \\ \gamma_7^{\alpha'}\left(\frac{\partial}{\partial x^0} - \frac{\partial}{\partial x^1} - \frac{\partial}{\partial x^2} - \frac{\partial}{\partial x^3}\right) \end{cases} \equiv \begin{cases} \gamma_0^{\alpha'}(+++) \\ \gamma_1^{\alpha'}(++-) \\ \gamma_2^{\alpha'}(+-+) \\ \gamma_3^{\alpha'}(+--) \\ \vdots \\ \gamma_7^{\alpha'}(---) \end{cases}. \tag{570}$$

Further expansion gives us:

$$V_{\beta'}^{\alpha'} = \begin{cases} \gamma_0^0(+++) \ \gamma_0^1(+++) \ \cdots \ \gamma_0^7(+++) \\ \gamma_1^0(++-) \ \gamma_1^1(++-) \ \cdots \ \gamma_1^7(++-) \\ \gamma_2^0(+-+) \ \gamma_2^1(+-+) \ \cdots \ \gamma_2^7(+-+) \\ \gamma_3^0(+--) \ \gamma_3^1(+--) \ \cdots \ \gamma_3^7(+--) \\ \vdots \\ \gamma_7^0(---) \ \gamma_7^1(---) \ \cdots \ \gamma_7^7(---) \end{cases}. \tag{571}$$

It can be shown that with the conditions:

$$\left(\sum_{\alpha'=0}^{7} \gamma_0^{\alpha'}\right)^2 = \left(\sum_{\alpha'=0}^{7} \gamma_1^{\alpha'}\right)^2 = \left(\sum_{\alpha'=0}^{7} \gamma_2^{\alpha'}\right)^2 = \left(\sum_{\alpha'=0}^{7} \gamma_3^{\alpha'}\right)^2$$

$$= \left(\sum_{\alpha'=0}^{7} \gamma_4^{\alpha'}\right)^2 = \left(\sum_{\alpha'=0}^{7} \gamma_5^{\alpha'}\right)^2 = \left(\sum_{\alpha'=0}^{7} \gamma_6^{\alpha'}\right)^2 = \left(\sum_{\alpha'=0}^{7} \gamma_7^{\alpha'}\right)^2, \tag{572}$$

the inner product of the matrices V leads to the Klein–Gordon operator:

$$\sum_{\alpha'=0}^{7}\sum_{\beta'=0}^{7} V_{\beta'}^{\alpha'} \cdot V_{\beta'}^{\alpha'} = G\left[\gamma_{\beta'}^{\alpha'}\right] \cdot \left(\left(\frac{\partial}{\partial x^0}\right)^2 + \left(\frac{\partial}{\partial x^1}\right)^2 + \left(\frac{\partial}{\partial x^2}\right)^2 + \left(\frac{\partial}{\partial x^3}\right)^2\right). \tag{573}$$

Thereby the term $G\left[\gamma_{\beta'}^{\alpha'}\right]$ denotes a complex but scalar function of the coefficients $\gamma_{\beta'}^{\alpha'}$. Construction of the two operators:

$$V_{\beta'}^{+\alpha'} = V_{\beta'}^{\alpha'} + M \cdot \delta_{\beta'}^{\alpha'}; \quad V_{\beta'}^{-\alpha'} = V_{\beta'}^{\alpha'} - M \cdot \delta_{\beta'}^{\alpha'} \tag{574}$$

leads to the complete Klein–Gordon equation with the operator:

$$\sum_{\alpha'=0}^{7}\sum_{\beta'=0}^{7} V_{\beta'}^{+\alpha'} \cdot V_{\beta'}^{-\alpha'} = G\left[\gamma_{\beta'}^{\alpha'}\right]$$

$$\cdot \left(\left(\frac{\partial}{\partial x^0}\right)^2 + \left(\frac{\partial}{\partial x^1}\right)^2 + \left(\frac{\partial}{\partial x^2}\right)^2 + \left(\frac{\partial}{\partial x^3}\right)^2\right) - M^2. \tag{575}$$

Applying either of the operators in Eq. (574) on a function f requires the function to become a vector $\mathbf{f}_{\alpha'}$. Thus, in contrast to the Dirac form with quaternions, we not only have scalar results now when performing the matrix multiplication as an inner product, but also obtain 8 functions instead of 4. Further discussion of the new approach was given in [133].

Now we are ready to extend our approach to more general coordinates and to formally construct the whole apparatus as a true (tensor-like) transformation. But before doing this, we want to consider a simple example in four dimensions.

4.1.1.3 Cartesian example in 4D

Assuming a Cartesian coordinate system t, x, y, z, we satisfy conditions (572) via:

$$\left(\sum_{\alpha'=0}^{1} \gamma_0^{\alpha'} \right)^2 = \left(\sum_{\alpha'=0}^{1} \gamma_1^{\alpha'} \right)^2 = \dots \Rightarrow \gamma_k^k = \gamma_0^0; \quad \gamma_j^k = 0 \; \forall \; j \neq k. \quad (576)$$

This leads to partial differential equations of the kind:

$$\gamma_0^0 \cdot \left(f_0 \, [t, x, y, z]_{,t} + f_0 \, [t, x, y, z]_{,x} + f_0 \, [t, x, y, z]_{,y} + f_0 \, [t, x, y, z]_{,z} \right)$$
$$= \pm M \cdot f_0 \, [t, x, y, z]$$

$$\vdots$$

$$\gamma_0^0 \cdot \left(f_7 \, [t, x, y, z]_{,t} - f_7 \, [t, x, y, z]_{,x} - f_7 \, [t, x, y, z]_{,y} - f_7 \, [t, x, y, z]_{,z} \right)$$
$$= \pm M \cdot f_7 \, [t, x, y, z]. \quad (577)$$

Assuming a Minkowski-like metric with the trace $\{-c^2, 1,1,1\}$ changes Eq. (577) into:

$$\gamma_0^0 \cdot \left(i \cdot \frac{f_0 \, [t, x, y, z]_{,t}}{c} + f_0 \, [t, x, y, z]_{,x} + f_0 \, [t, x, y, z]_{,y} + f_0 \, [t, x, y, z]_{,z} \right)$$
$$= \pm M \cdot f_0 \, [t, x, y, z]$$

$$\vdots$$

$$\gamma_0^0 \cdot \left(i \cdot f_7 \, [t, x, y, z]_{,t} - f_7 \, [t, x, y, z]_{,x} - f_7 \, [t, x, y, z]_{,y} - f_7 \, [t, x, y, z]_{,z} \right)$$
$$= \pm M \cdot f_7 \, [t, x, y, z] \quad (578)$$

and gives us the solutions:

$$f_i \, [t, x, y, z] = e^{\pm \frac{c \cdot M}{\gamma_0^0} \cdot t} \cdot F \, (x \pm i \cdot c \cdot t, y \pm i \cdot c \cdot t, z \pm i \cdot c \cdot t). \quad (579)$$

Thereby $F(\dots)$ denotes an arbitrary function of the argument $(\dots)$. Now setting (m = mass, c = speed of light in vacuum and $\hbar$ = reduced Planck constant):

$$M = i \cdot \frac{m \cdot c^2}{\hbar}; \quad \gamma_0^0 = c \tag{580}$$

directly results in matter and antimatter solutions plus complex waves running in forward and backward direction along the coordinate x with the speed of light. As said above, solution (579) is a very simple one, resulting from condition (576). Many more solutions with different settings are possible.

Much more interesting ways to derive Dirac forms and equations without resorting to quaternions are to be found in [133, 134, 139].

4.2 A Variation Directly Leading to the Klein–Gordon Equation

Now, we move on to the following variation using partial derivatives to vary the functions $G^j[x]$:

$$\delta g_{\delta\gamma} = \delta \left(\mathbf{g}_\delta \cdot \mathbf{g}_\gamma \right) = \delta_\delta \sum_{\delta=0}^{n-1} \mathbf{g}_\delta \cdot \mathbf{g}_\gamma + \delta_\gamma \sum_{\gamma=0}^{n-1} \mathbf{g}_\gamma \cdot \mathbf{g}_\delta$$

$$= 2 \cdot \delta_\gamma \sum_{\gamma=0}^{n-1} \left(\partial^\gamma \partial_\gamma G^j[x_k] \, \mathbf{e}_j \right) \cdot \mathbf{g}_\delta 2 \cdot \delta_\gamma \sum_{\gamma=0}^{n-1} \left(\partial^\gamma \left(\frac{\partial G^j[x_k]}{\partial x^\gamma} \mathbf{e}_j \right) \right) \cdot \mathbf{g}_\delta$$

$$= 2 \cdot \delta_\gamma \sum_{\alpha,\beta=0}^{n-1} \left(\frac{\partial}{\partial x^\alpha} \left(g^{\alpha\beta} \frac{\partial G^j[x_k]}{\partial x^\beta} \mathbf{e}_j \right) \right) \cdot \mathbf{g}_\delta = 0$$

$$\Rightarrow \sum_{\alpha,\beta=0}^{n-1} \left(\frac{\partial}{\partial x^\alpha} \left(g^{\alpha\beta} \frac{\partial G^j[x_k]}{\partial x^\beta} \mathbf{e}_j \right) \right) = \Delta G^j[x_k] \, \mathbf{e}_j = 0 \tag{581}$$

and find the Klein–Gordon equation, only that again, one of the coordinates has to provide mass (and/or potential). Assuming Cartesian coordinates and choosing—as before—the coordinate ξ with:

$$\frac{\partial}{\partial x^\xi} \left(g^{\xi\xi} \frac{\partial G^j[x_k]}{\partial x^\xi} \mathbf{e}_j \right) = -M^2 \cdot G^j[x_k] \, \mathbf{e}_j, \tag{582}$$

we can rewrite Eq. (581) as

$$\xi = n - 1; \quad \delta g_{\delta\gamma} = \sum_{\alpha,\beta=0}^{n-1} \left(\frac{\partial}{\partial x^\alpha} \left(g^{\alpha\beta} \frac{\partial G^j [x_k]}{\partial x^\beta} \mathbf{e}_j \right) \right)$$

$$\sum_{\alpha,\beta=0}^{n-2} \left(\frac{\partial}{\partial x^\alpha} \left(g^{\alpha\beta} \frac{\partial G^j [x_k]}{\partial x^\beta} \mathbf{e}_j \right) \right) + \frac{\partial}{\partial x^\xi} \left(g^{\xi\xi} \frac{\partial G^j [x_k]}{\partial x^\xi} \mathbf{e}_j \right)$$

$$= \Delta G^j [x_k] \mathbf{e}_j - M^2 \cdot G^j [x_k] \mathbf{e}_j = 0. \tag{583}$$

4.2.1 The Special Case of the Schrödinger Equation—Part II

As for many people, THE quantum equation has been seen in the Schrödinger equation, our next task for this chapter should be to give a most general way to obtain the Schrödinger equation out of our metric Klein–Gordon equations (583). This time, we do not intend to obtain the classical Schrödinger equation, but its covariant analogon.

For simplicity, we assume a metric with none of the components actually being time dependent. In order to distinguish this from the general case in our considerations above, we use the usual "g" to denote the metric. The component g^{00} shall be of the form $g^{00} = C1/c^2$, with $C1$ being a constant and all other metric time components shall be $g^{0i} = 0$ ($I = 1,2,3$). We start with the Klein–Gordon form (583) (last line) but, instead of our vector function G^j, we assume (only for simplicity and brevity) a scalar function:

$$\Delta G^j [x_k] \mathbf{e}_j - M^2 \cdot G^j [x_k] \mathbf{e}_j = 0$$
$$\Rightarrow \Delta G [x_k] - M^2 \cdot G [x_k] = 0 \Rightarrow \Delta \Psi [x_k] \mathbf{e}_j - M^2 \cdot \Psi [x_k] \mathbf{e}_j = 0, \tag{584}$$

and instead of the symbol G we apply the classical Greek symbol Ψ. Then we can separate the time derivative as follows:

$$0 = \left[-M^2 + \Delta \right] \Psi = \left[-M^2 + \left(C1 \cdot \frac{\partial_t^2}{c^2} + \underbrace{\frac{1}{\sqrt{g}} \partial_\alpha \sqrt{g} \cdot g^{\alpha\beta} \partial_\beta}_{\text{3D–}\Delta\text{–Operator}} \right) \right] \Psi. \tag{585}$$

Please note that our Greek indices α and β are running only from 1 to n. Now we introduce a function $\Psi = \Phi + X$ and demand the following additional condition:

$$\partial_t \Psi = c^2 \cdot C2 \cdot (\Phi - X). \tag{586}$$

Together with Eq. (585) we obtain:

$$0 = -M^2 \left(\Phi + X \right) + \left(C1 \cdot C2 \cdot \partial_t \left(\Phi - X \right) + \frac{1}{\sqrt{g}} \partial_\alpha \sqrt{g} \cdot g^{\alpha\beta} \partial_\beta \left(\Phi + X \right) \right).$$

(587)

The following two equations summed up would result in Eq. (587):

$$0 = -M^2 \Phi + \left(C1 \cdot C2 \cdot \partial_t \Phi + \frac{1}{\sqrt{g}} \partial_\alpha \sqrt{g} \cdot g^{\alpha\beta} \partial_\beta \Phi \right)$$

$$0 = -M^2 X + \left(\frac{1}{\sqrt{g}} \partial_\alpha \sqrt{g} \cdot g^{\alpha\beta} \partial_\beta X - C1 \cdot C2 \cdot \partial_t X \right).$$

(588)

Comparing with the original Schrödinger equation as given in the form below:

$$\left[-i \cdot \hbar \cdot \partial_t - \frac{\hbar^2}{2m} \Delta_{\text{Schrödinger}} + V_{\text{Schrödinger}} \right] \Psi = 0,$$

(589)

not only gives us the matter and antimatter solutions again (c.f. [8]) but also shows us—as seen and discussed before [21]—that mass m and the potential $V_{\text{Schrödinger}}$ is now been taken on by the metric and the results of hidden (potentially compactified) dimensions due to (582). Thus, a distorted metric and additional dimensions act like an effective potential and/or mass in the Schrödinger approximation and vice versa, which is to say: what in classical physics has been described as a potential would now become a set of potentially compactified dimensions, which is providing the necessary interaction. Similarly, we have to formulate for the mass M, that what appears as (rest) mass to us is just permanently and locally curved space, whereby the curvature is the one of additional dimensions. Disregarding the antimatter solution here, setting the constants $C1$, $C2$ and reshaping the classical Schrödinger equation as:

$$C1 \cdot C2 = \frac{i \cdot 2 \cdot m}{\hbar}; \quad X = 0;$$

$$0 = \left[\frac{i \cdot 2 \cdot m}{\hbar} \cdot \partial_t + \Delta_{\text{Schrödinger}} - 2 \cdot V_{\text{Schrödinger}} \right] \Psi$$

(590)

gives us the proportionality:

$$\Rightarrow \left[\Delta_{3D} - 2 \cdot V_{\text{Schrödinger}} \right] \Psi \stackrel{\triangle}{=} -M^2 \Phi + \frac{1}{\sqrt{g}} \partial_\alpha \sqrt{g} \cdot g^{\alpha\beta} \partial_\beta \Phi$$

$$\Rightarrow \Delta_{3D} \Psi \stackrel{\triangle}{=} \Delta_{3D} \Phi; \quad \Rightarrow V_{\text{Schrödinger}} \stackrel{\triangle}{=} \frac{M^2 \Phi}{2 \cdot \Psi}.$$

(591)

This way, we can now link classical Schrödinger solutions and the corresponding Schrödinger potentials with their metric analogue.

Things are getting a bit more interesting when assuming that V was chosen such that we have $\Psi = \Psi_{\text{Schrödinger}} = \Phi_{\text{new}} = \Phi$. Then we can directly write down the equation for the determination of the potential and its connection to the metric distortion:

$$V_{\text{Schrödinger}} \triangleq \frac{M^2 \Phi}{2 \cdot \Psi} = -\frac{\frac{\partial}{\partial x^\xi} \left(g^{\xi\xi} \frac{\partial G^j [x_k]}{\partial x^\xi} \mathbf{e}_j \right)}{2 \cdot G^j [x_k] \mathbf{e}_j}. \tag{592}$$

We note: What classically is the potential V gives a metric distortion in quantum gravity.

For illustration, we apply the Schrödinger solution for the hydrogen ground state where from [115] we have:

$$\Psi_{n=1,l=0,m=0} [r, \vartheta, \varphi] = e^{i \cdot m \cdot \varphi} \cdot P_l^m [\cos \vartheta] \cdot R_{n,l} [r] = e^{-\left(\frac{r}{a_0} \right)}. \tag{593}$$

Applying the known Schrödinger hydrogen potential $\sim 1/r$ and knowing that the ground state solution is the one of a separation approach to Eq. (590) in spherical coordinates, we can use (592) and write:

$$\begin{aligned} V_{\text{Schrödinger}} \sim C_r \frac{1}{r} &= -\frac{\frac{\partial}{\partial x^\xi} \left(g^{\xi\xi} \frac{\partial G^j [x_k]}{\partial x^\xi} \mathbf{e}_j \right)}{2 \cdot G^j [x_k] \mathbf{e}_j} = -\frac{\frac{\partial}{\partial x^\xi} \left(g^{\xi\xi} \frac{\partial}{\partial x^\xi} e^{-\left(\frac{r}{a_0} \right)} \cdot f [r, x^\xi] \right)}{2 \cdot e^{-\left(\frac{r}{a_0} \right)} \cdot f [r, x^\xi]} \\ &= -\frac{\frac{\partial}{\partial x^\xi} \left(g^{\xi\xi} \frac{\partial}{\partial x^\xi} f [r, x^\xi] \right)}{2 \cdot f [r, x^\xi]}. \end{aligned} \tag{594}$$

In order to maintain the separation structure also with respect to the function $f[\dots]$, we have to get rid of the r-functionality of this function and get:

$$\begin{aligned} V_{\text{Schrödinger}} \sim C_r \frac{1}{r} &= -\frac{\frac{\partial}{\partial x^\xi} \left(g^{\xi\xi} [r, \dots] \frac{\partial}{\partial x^\xi} f [x^\xi] \right)}{2 \cdot f [x^\xi]} \\ &= -\frac{\frac{\partial}{\partial x^\xi} g^{\xi\xi} [r, \dots] \frac{\partial}{\partial x^\xi} f [x^\xi] + g^{\xi\xi} [r, \dots] \frac{\partial^2}{(\partial x^\xi)^2} f [x^\xi]}{2 \cdot f [x^\xi]}. \end{aligned} \tag{595}$$

A very simple solution to Eq. (595) could be found via the approach $\frac{\partial^2}{(\partial x^\xi)^2} f [x^\xi] = 0$, which reduces the equation to:

$$C_r \frac{1}{r} = -\frac{\frac{\partial}{\partial x^\xi} g^{\xi\xi} \left[r, \ldots\right] \frac{\partial}{\partial x^\xi} f\left[x^\xi\right]}{2 \cdot f\left[x^\xi\right]}$$

$$\xrightarrow{f\left[x^\xi\right]=C_\xi \cdot x^\xi} \frac{\frac{\partial}{\partial x^\xi} g^{\xi\xi}\left[r, x^\xi\right] C_\xi}{2 \cdot f\left[x^\xi\right]} = \frac{\frac{\partial}{\partial x^\xi} g^{\xi\xi}\left[r, x^\xi\right]}{2 \cdot x^\xi} = -C_r \frac{1}{r}. \quad (596)$$

Now we only have to solve for the metric component $g^{\xi\xi}\left[r, x^\xi\right]$, which gives:

$$\frac{\frac{\partial}{\partial x^\xi} g^{\xi\xi}\left[r, x^\xi\right]}{2 \cdot x^\xi} = \frac{\frac{\partial}{\partial x^\xi} G^{\xi\xi}\left[r\right] \cdot \left(x^\xi\right)^2}{2 \cdot x^\xi} = G^{\xi\xi}\left[r\right]$$

$$= -C_r \frac{1}{r} \Rightarrow g^{\xi\xi}\left[r, x^\xi\right] = -C_r \frac{\left(x^\xi\right)^2}{r}. \quad (597)$$

We note that in contrast to the base vector variation as a sum of function as applied in section "3.11 The Additive Variation of the Metric Tensor: Brief Introduction", this time we have no direct metric analog for mass and classical potential but have to resort to the assumption of suitable additional coordinates, which is to say potentially compactified dimensions (c.f. (471)).

4.3 A Little Bit of Materials Science

It was shown in a variety of papers by this author (e.g. [148] to [151]) that one major problem of mathematically dealing with time- and stress-dependent material properties results from the space-time dependency of these very properties. It was even shown with first-principles methods [149] that the usual time-functional approach, which expands the ordinary linear theory of elasticity to an effective (global) time dependency, simply cannot work in a general manner because its basic assumption of material homogeneity does not hold. The integral (space or time integral) effective laws applied to overcome this problem, on the other hand, are always only valid in connection with a certain group of experiments, respectively deformations and stress situations or processes.

A very general variation of Eq. (668) could be as follows:

$$\delta g_{\delta\gamma} = \delta\left(\mathbf{g}_\delta \cdot \mathbf{g}_\gamma\right) = \mathbf{g}_\delta \cdot \delta\mathbf{g}_\gamma + \delta\mathbf{g}_\delta \cdot \mathbf{g}_\gamma$$

$$= \frac{\partial G^i\left[x_k\right]}{\partial x^\delta}\mathbf{e}_i \cdot \left(\frac{\partial^2 G^j\left[x_k\right]}{\partial x^\alpha \partial x^\gamma}\mathbf{e}_j\right)\delta x^\alpha + \left(\frac{\partial^2 G^i\left[x_k\right]}{\partial x^\alpha \partial x^\delta}\mathbf{e}_i\right)\delta x^\alpha \cdot \frac{\partial G^j\left[x_k\right]}{\partial x^\gamma}\mathbf{e}_j.$$

$$(598)$$

We note that setting (598) equal to zero automatically fulfills condition (86) (or the classical form (87)). A non-trivial solution could be found with an approach of the kind:

$$G^j [x_k] = C^j \cdot G [x_k], \tag{599}$$

with the C^j denoting constants, satisfying the condition:

$$\sum_{j=0}^{N-1} \left(C^j \right)^2 = 0. \tag{600}$$

In result, however, we obtain non-metric spaces as the subsequent metric tensor for such base-vectors would be identical zero in all components (see also the corresponding discussion in [135]). So, as we assumed that the general form does not give us other than trivial or non-metric solutions, we asked what kind of symmetries or summed up variations could be more successful. In [114, 117] and above in section "3.10 The Variation of the Metric Tensor", we already introduced such options and found the classical quantum equations. Here now we want to go back to the general variation (598), assume summation over one index as follows:

$$
\begin{aligned}
\delta g_{\delta\gamma} = \delta \left(\mathbf{g}_\delta \cdot \mathbf{g}_\gamma \right) &= \mathbf{g}_\delta \cdot \delta \mathbf{g}_\gamma + \delta \mathbf{g}_\delta \cdot \mathbf{g}_\gamma \\
&= \frac{\partial G^i [x_k]}{\partial x^\delta} \mathbf{e}_i \cdot \left(\frac{\partial^2 G^j [x_k]}{\partial x^\alpha \partial x^\gamma} \mathbf{e}_j \right) \delta x^\alpha + \left(\frac{\partial^2 G^i [x_k]}{\partial x^\alpha \partial x^\delta} \mathbf{e}_i \right) \delta x^\alpha \cdot \frac{\partial G^j [x_k]}{\partial x^\gamma} \mathbf{e}_j \\
&= \frac{\partial G^i [x_k]}{\partial x^\delta} \mathbf{e}_i \cdot \sum_{\alpha=0}^{n-1} \left(\frac{\partial^2 G^j [x_k]}{\partial x^\alpha \partial x^\gamma} \mathbf{e}_j \right) \delta x^\alpha + \sum_{\alpha=0}^{n-1} \left(\frac{\partial^2 G^i [x_k]}{\partial x^\alpha \partial x^\delta} \mathbf{e}_i \right) \delta x^\alpha \cdot \frac{\partial G^j [x_k]}{\partial x^\gamma} \mathbf{e}_j \\
&= \frac{\partial G^i [x_k]}{\partial x^\delta} \mathbf{e}_i \cdot \sum_{\alpha=0}^{n-1} \left(\frac{\partial^2 G^j [x_k]}{\partial x^\alpha \partial x^\gamma} \mathbf{e}_j \right) \delta^\alpha + \sum_{\alpha=0}^{n-1} \left(\frac{\partial^2 G^i [x_k]}{\partial x^\alpha \partial x^\delta} \mathbf{e}_i \right) \delta^\alpha \cdot \frac{\partial G^j [x_k]}{\partial x^\gamma} \mathbf{e}_j \\
\Rightarrow \sum_{\alpha=0}^{n-1} & \left(\frac{\partial^2 G^j [x_k]}{\partial x^\alpha \partial x^\gamma} \mathbf{e}_j \right) \delta^\alpha = 0, \tag{601}
\end{aligned}
$$

and try techniques known from the theory of elasticity [152]. These techniques, known there as the 3-function-ansatz and the Galerkin approach, were introduced in the thirties of the last century by Heinz Neuber [153]. It appears advisable to extend this ansatz to an n-function approach with n giving the number of dimensions of the space-time under consideration.

Please note that by assuming non-equal variation constants δ^α for the various α, we could even derive anisotropic equations.

For simplicity we only give the ansatz in Cartesian coordinates:

$$G^j[x_k] = g^{jl}\partial_l G[x_k]; \quad \Delta G[x_k] = 0$$

$$G^j[x_k] = x_\xi \cdot g^{jl}\partial_l G[x_k] + \alpha \cdot G[x_k]; \quad \xi \text{ any of } 0, 1, \ldots, n-1$$

$$G^j[x_k] = \left\{ \ldots, \overbrace{\frac{\partial G[x_k]}{\partial x_\zeta}}^{pos\ \xi}, \ldots, -\overbrace{\frac{\partial G[x_k]}{\partial x_\xi}}^{pos\ \zeta}, \ldots \right\}; \quad \forall(\ldots, \ldots) = 0$$

$$G^j[x_k] = \left\{ \ldots, -\overbrace{\frac{\partial G[x_k]}{\partial x_\zeta}}^{pos\ \xi}, \ldots, \overbrace{\frac{\partial G[x_k]}{\partial x_\xi}}^{pos\ \zeta}, \ldots \right\}; \quad \forall(\ldots, \ldots) = 0. \quad (602)$$

Any combination of the basic solutions (602) is also a solution. We see that we can construct non-trivial solutions to (598), (601) out of combinations of harmonic functions and their derivatives. In order to satisfy (601) in its currently given form, we have to set $\alpha = -1$.

We also find spin-like solutions as given above (see Eqs. (69) to (72)).

4.3.1 Deriving the Equations of Elasticity from the Metric Origin Out of the Einstein–Hilbert Action

With respect to materials science applications, it is probably of somewhat more interest to consider variations of the kind:

$$\delta g_{\delta\gamma} = \delta\left(\mathbf{g}_\delta \cdot \mathbf{g}_\gamma\right) = \mathbf{g}_\delta \cdot \delta\mathbf{g}_\gamma + \delta\mathbf{g}_\delta \cdot \mathbf{g}_\gamma$$

$$= \frac{\partial G^i[x_k]}{\partial x^\delta}\mathbf{e}_i \cdot \left(\frac{\partial^2 G^j[x_k]}{\partial x^\beta \partial x^\gamma}\mathbf{e}_j\right)\delta x^\beta + \left(\frac{\partial^2 G^i[x_k]}{\partial x^\chi \partial x^\delta}\mathbf{e}_i\right)\delta x^\chi \cdot \frac{\partial G^j[x_k]}{\partial x^\gamma}\mathbf{e}_j$$

$$= \frac{\partial G^i[x_k]}{\partial x^\delta}\mathbf{e}_i \cdot \left(\frac{\partial^2 G^j[x_k]}{\partial x^\gamma \partial x^\gamma}\mathbf{e}_j\right)C_\beta^\gamma\delta x^\beta + \left(\frac{\partial^2 G^i[x_k]}{\partial x^\chi \partial x^\delta}\mathbf{e}_i\right)\delta x^\chi \cdot \frac{\partial G^j[x_k]}{\partial x^\gamma}\mathbf{e}_j$$

$$= \frac{\partial G^i[x_k]}{\partial x^\delta}\mathbf{e}_i \cdot \left(\frac{\partial^2 G^j[x_k]}{(\partial x^\gamma)^2}\mathbf{e}_j\right)C_\beta^\gamma\delta x^\beta + \left(\frac{\partial^2 G^i[x_k]}{\partial x^\chi \partial x^\delta}\mathbf{e}_i\right)\delta x^\beta C_\beta^\chi \cdot C_\gamma^\delta\frac{\partial G^j[x_k]}{\partial x^\delta}\mathbf{e}_j.$$

$$(603)$$

The experienced reader recognizes the structural elements of the basic equation of elasticity (e.g. [152], pp. 166), which can be given for an isotropic material with the Poisson's ratio v as:

$$\left(\overbrace{(1 - 2 \cdot v)}^{\stackrel{\wedge}{=}a} \cdot \Delta G^j \left[x_k\right] + \left(\frac{\partial^2 G^j \left[x_k\right]}{\partial x^\gamma \partial x^\delta} \right) \right) \mathbf{e}_j = 0. \tag{604}$$

A simple exchange of the dummy indices in (603) leads us to:

$$\delta g_{\delta \gamma} = \delta \left(\mathbf{g}_\delta \cdot \mathbf{g}_\gamma \right) = \mathbf{g}_\delta \cdot \delta \mathbf{g}_\gamma + \delta \mathbf{g}_\delta \cdot \mathbf{g}_\gamma$$

$$= \frac{\partial G^i \left[x_k\right]}{\partial x^\delta} \mathbf{e}_i \cdot \left(\frac{\partial^2 G^j \left[x_k\right]}{(\partial x^\gamma)^2} \mathbf{e}_j \right) C_\beta^\gamma \delta x^\beta + \left(\frac{\partial^2 G^i \left[x_k\right]}{\partial x^\chi \partial x^\delta} \mathbf{e}_i \right) \delta x^\beta C_\beta^\chi \cdot C_\gamma^\delta \frac{\partial G^j \left[x_k\right]}{\partial x^\delta} \mathbf{e}_j$$

$$= \frac{\partial G^j \left[x_k\right]}{\partial x^\delta} \mathbf{e}_j \cdot \left(\frac{\partial^2 G^i \left[x_k\right]}{(\partial x^\gamma)^2} \mathbf{e}_i \right) C_\beta^\gamma \delta x^\beta + \left(\frac{\partial^2 G^i \left[x_k\right]}{\partial x^\chi \partial x^\delta} \mathbf{e}_i \right) \delta x^\beta C_\beta^\chi \cdot C_\gamma^\delta \frac{\partial G^j \left[x_k\right]}{\partial x^\delta} \mathbf{e}_j$$

$$= \left(\left(\frac{\partial^2 G^i \left[x_k\right]}{(\partial x^\gamma)^2} \mathbf{e}_i \right) C_\beta^\gamma + \left(\frac{\partial^2 G^i \left[x_k\right]}{\partial x^\chi \partial x^\delta} \mathbf{e}_i \right) C_\beta^\chi C_\gamma^\delta \right) \cdot \delta x^\beta \frac{\partial G^j \left[x_k\right]}{\partial x^\delta} \mathbf{e}_j, \tag{605}$$

and assuming $C_\beta^\gamma = b \cdot \delta_\beta^\gamma$, $C_\beta^\chi C_\gamma^\delta = b^2 \cdot \delta_\beta^\chi \delta_\gamma^\delta$, $a = 1/b$ yields the isotropic equation:

$$\delta g_{\delta \gamma} = \delta \left(\mathbf{g}_\delta \cdot \mathbf{g}_\gamma \right) = \mathbf{g}_\delta \cdot \delta \mathbf{g}_\gamma + \delta \mathbf{g}_\delta \cdot \mathbf{g}_\gamma$$

$$= \left(\left(\frac{\partial^2 G^i \left[x_k\right]}{(\partial x^\gamma)^2} \mathbf{e}_i \right) C_\beta^\gamma + \left(\frac{\partial^2 G^i \left[x_k\right]}{\partial x^\chi \partial x^\delta} \mathbf{e}_i \right) C_\beta^\chi C_\gamma^\delta \right) \cdot \delta x^\beta \frac{\partial G^j \left[x_k\right]}{\partial x^\delta} \mathbf{e}_j$$

$$= b^2 \cdot \left(\frac{\Delta G^i \left[x_k\right] \mathbf{e}_i}{b} + \left(\frac{\partial^2 G^i \left[x_k\right]}{\partial x^\gamma \partial x^\delta} \mathbf{e}_i \right) \right) \cdot \delta x^\beta \frac{\partial G^j \left[x_k\right]}{\partial x^\delta} \mathbf{e}_j$$

$$= b^2 \cdot \left(a \cdot \Delta G^i \left[x_k\right] \mathbf{e}_i + \left(\frac{\partial^2 G^i \left[x_k\right]}{\partial x^\gamma \partial x^\delta} \mathbf{e}_i \right) \right) \cdot \delta x^\beta \frac{\partial G^j \left[x_k\right]}{\partial x^\delta} \mathbf{e}_j. \tag{606}$$

With the n-function-ansatz from (602) we can solve:

$$a \cdot \Delta G^j \left[x_k\right] \mathbf{e}_j + \left(\frac{\partial^2 G^j \left[x_k\right]}{\partial x^\gamma \partial x^\delta} \mathbf{e}_j \right)$$

$$= \left(\overbrace{(1 - 2 \cdot v)}^{\stackrel{\wedge}{=}a} \cdot \Delta G^j \left[x_k\right] + \left(\frac{\partial^2 G^j \left[x_k\right]}{\partial x^\gamma \partial x^\delta} \right) \right) \mathbf{e}_j = 0, \tag{607}$$

with $\alpha = -1 - 2 {*} a$.

As in connection with the general Eq. (601) it should be pointed out that with non-equal δx^α or δ^α for the various α, we could even derive anisotropic equations, respectively equations for anisotropic space-times.

It should be noted that, as shown before in section "4.1 Matrix Option and Classical Dirac Form", one could also extract a Dirac-like equation from Eq. (606). This could be obtained via $\delta x^\beta \to \delta x^\delta$ leading to:

$$
\begin{aligned}
\delta g_{\delta\gamma} &= \delta \left(\mathbf{g}_\delta \cdot \mathbf{g}_\gamma \right) = \mathbf{g}_\delta \cdot \delta\mathbf{g}_\gamma + \delta\mathbf{g}_\delta \cdot \mathbf{g}_\gamma \\
&= b^2 \cdot \left(a \cdot \Delta G^i \left[x_k \right] \mathbf{e}_i + \left(\frac{\partial^2 G^i \left[x_k \right]}{\partial x^\gamma \partial x^\delta} \mathbf{e}_i \right) \right) \cdot \delta x^\beta \frac{\partial G^j \left[x_k \right]}{\partial x^\delta} \mathbf{e}_j \\
&= b^2 \cdot \left(a \cdot \Delta G^i \left[x_k \right] \mathbf{e}_i + \left(\frac{\partial^2 G^i \left[x_k \right]}{\partial x^\gamma \partial x^\delta} \mathbf{e}_i \right) \right) \cdot \delta x^\delta \frac{\partial G^j \left[x_k \right]}{\partial x^\delta} \mathbf{e}_j \\
&\Rightarrow \sum_{\delta=0}^{n-1} \delta x^\delta \frac{\partial G^j \left[x_k \right]}{\partial x^\delta} \mathbf{e}_j = 0.
\end{aligned}
\tag{608}
$$

We have to point out that even though these Dirac-like equations are anything but close to classical elasticity equations, they still give displacement distributions of spaces and might automatically include defects (like dislocations) which require quite some math-construction in the classical technical mechanics.

4.3.2 Realization and Application in Materials Science

It should be pointed out that the *n*-function-ansatz was already widely used in connection with 3D elastic problems of inhomogeneous materials (e.g. [154–157]). We also developed a very sophisticated and comprehensive software package [158]. Newer prototypes of this package meanwhile even allow the consideration of higher dimensions, especially time. Subsequently, we have the possibility to better simulate and understand time-dependent material behavior, defects, degradation, failure and other non-linear effects. Another advantage can be seen in the fact that our solutions are automatically quantized allowing us to directly consider complex uncertainty budgets without further ado.

4.3.3 Interpretation

It is important to note that the backbone of our approach are harmonic functions (c.f. the first line in our very general *n*-function-ansatz in Eq. (602)).

Thus, having obtained the general formula (598) as a form of very general quantum equations directly out of the Einstein–Hilbert action,

it is of little wonder why the classical quantum equations are so dominated by harmonics. After all, the Laplace and Poisson equations are just a part of the total set of equations. Their solutions, which are classically seen as wave functions, reveal themselves as space-time displacements.

4.4 But Where Is Thermodynamics?

Having worked out the similarity of quantum gravity and the theory of elasticity now allows us to try and find other fundamental fields inside the already derived general equation(s). Here we start with thermodynamics.

It is well known from the thermodynamics of deformed bodies that the work of deformation can be described as the inner product of the stress tensor σ^{jk} with the deformation tensor u_{jk}:

$$W_D = \sigma^{jk} u_{jk}. \tag{609}$$

Knowing that the functions G^j are displacement fields and closely investigating them, we conclude that the variation of the metric tensor (608) is nothing else than the holistic equivalent of a deformation tensor and thus, that the integrand in (87), (389) is giving us nothing else but a very general form of deformation work.

If this is the case, however, and especially as we assume the Einstein–Hilbert action to "contain it all", we have to ask where we could find the equivalent to thermal energy? After all, sticking to the simpler classical action form (87):

$$\left(R^{\delta\gamma} - \frac{1}{2} R g^{\delta\gamma} + \Lambda g^{\delta\gamma} + \kappa T^{\delta\gamma} \right) \delta g_{\delta\gamma}, \tag{610}$$

none of the addends in parenthesis has the makings of a term $T*dS$ or $S*dT$ with S denoting the entropy and T denoting temperature.

Obviously we are missing something ... at least as long as we consider thermodynamics as so fundamental that it has to show up in a truly fundamental theory.

Thus, with thermodynamics missing inside the classical metric and the newer base-vector variation of the Einstein–Hilbert action, we have to look for a different way to obtain what we need.

4.5 The Variation with Respect to Ensemble Parameters

Reconsideration of Eq. (389) and thereby incorporating our degrees of freedom (545) leads to:

$$\delta_g W = 0$$

$$= \int_V d^n x \left(\overbrace{R^{\delta\gamma} - \frac{1}{2} R g^{\delta\gamma} + \Lambda g^{\delta\gamma} + \kappa T^{\delta\gamma}}^{\text{Relativity}} \right)$$

$$\times \left(\underbrace{\frac{\partial G^i [x_k]}{\partial x^\delta} \mathbf{e}_i \cdot \delta \left(\frac{\partial G^j [x_k]}{\partial x^\gamma} \mathbf{e}_j \right) + \delta \left(\frac{\partial G^i [x_k]}{\partial x^\delta} \mathbf{e}_i \right) \cdot \frac{\partial G^j [x_k]}{\partial x^\gamma} \mathbf{e}_j}_{\text{Quantum}} + \underbrace{\frac{\partial g_{\delta\gamma}}{\partial n} \delta n}_{\text{ThD}} \right) .$$

$$(611)$$

We have assumed that we have a closed system and that therefore the amount of information, respectively the number of dimensions should not change, meaning $\frac{\partial g_{\alpha\beta}}{\partial n} \delta n = 0$ (c.f. [163, 171]). What still could change, respectively be variated, however, would be the number of $i-j$ ensembles and the corresponding base-vectors. For instance, we could understand these ensembles as objects with different centers of gravity, like:

$$\frac{\partial G^j [x_k]}{\partial x^\gamma} \mathbf{e}_j = \frac{\partial G^j [x_k - \xi_{kj}]}{\partial x^\gamma} \mathbf{e}_j, \qquad (612)$$

with the ξ_{kj} denoting the various centers of gravity. Now one could perform the variation with respect to exactly these coordinates and find (thereby simplifying via our condition of the closed system $\frac{\partial g_{\alpha\beta}}{\partial n} \delta n = 0$):

$$X_k \equiv x_k - \xi_k; \delta_g W = \int_V d^n x \left(\overbrace{R^{\delta\gamma} - \frac{1}{2} R g^{\delta\gamma} + \Lambda g^{\delta\gamma} + \kappa T^{\delta\gamma}}^{\text{Relativity}} \right)$$

$$\times \left(\underbrace{\frac{\partial G^i [X_k]}{\partial x^\delta} \mathbf{e}_i \cdot \delta_x \left(\frac{\partial G^j [X_k]}{\partial x^\gamma} \mathbf{e}_j \right) + \delta_x \left(\frac{\partial G^i [X_k]}{\partial x^\delta} \mathbf{e}_i \right) \cdot \frac{\partial G^j [X_k]}{\partial x^\gamma} \mathbf{e}_j}_{\text{Quantum}} \right.$$

$$\left. + \underbrace{\frac{\partial G^i [X_k]}{\partial x^\delta} \mathbf{e}_i \cdot \delta_\xi \left(\frac{\partial G^j [X_k]}{\partial x^\gamma} \mathbf{e}_j \right) + \delta_\xi \left(\frac{\partial G^i [X_k]}{\partial x^\delta} \mathbf{e}_i \right) \cdot \frac{\partial G^j [X_k]}{\partial x^\gamma} \mathbf{e}_j}_{\text{"?"} \Rightarrow \text{2nd law of Thermodynamics}} \right) .$$

$$(613)$$

We realize that the additional variation is just the positioning of the first derivatives of metric displacements and their vectors. In other words: This variation is about positioning the individual centers of gravity within the metric space-time and their vectors of changes against the various coordinates x. In our "Science Riddle 20" [172], we discussed the effect of chaotic distributions for huge ensembles. We came to the conclusion that states are statistically favored where the various displacements cancel each other out. This then leads to a disappearance of the term "?". Interestingly, however, such equal distributions of vector orientations and magnitudes also coincides with states with maximum classical entropy, where we exactly obtain $? = 0$ (c.f. Fig. 4.5.1). Thus, it is clear: The "?" must stand for thermodynamics and that it is a driving force for the second law of thermodynamics.

However, there is also the chance to make this variational term to zero with suitable symmetrical structures of the ensembles also canceling each other out (Figs. 4.5.2a and 4.5.2b). Then the term "?" would stand for a force of "self-organization" and/or self-structuring of the very system (or parts of it). The whole could also be achieved as a proper combination of the "?"- and the "Quantum"-term which then has to give a resulting zero in sum.

In this whole context, we also make out the parameter N (the number to which the indices i and j can run) as yet another quantity defining upon the form of enclosure of a given system. With n and N being fixed, we should speak about a truly closed system in the classical meaning.

From all what has been elaborated above, we conclude that the term "?" not only presents the metric manifestation of the second law of thermodynamics, but also gives the reason for the self-organization vector and evolution we observe within our universe as a rather omnipresent property.

With this, we might want to rewrite Eq. (613) as follows:

$$X_k \equiv x_k - \xi_k; \quad \delta_g W = \int_V d^n x \overbrace{\left(R^{\delta\gamma} - \frac{1}{2} R g^{\delta\gamma} + \Lambda g^{\delta\gamma} + \kappa T^{\delta\gamma} \right)}^{\text{Relativity}}$$

$$\times \left(\underbrace{\frac{\partial G^i [X_k]}{\partial x^\delta} \mathbf{e}_i \cdot \delta_x \left(\frac{\partial G^j [X_k]}{\partial x^\gamma} \mathbf{e}_j \right) + \delta_x \left(\frac{\partial G^i [X_k]}{\partial x^\delta} \mathbf{e}_i \right) \cdot \frac{\partial G^j [X_k]}{\partial x^\gamma} \mathbf{e}_j}_{\text{Quantum}} \right.$$

$$\left. + \underbrace{\frac{\partial G^i [X_k]}{\partial x^\delta} \mathbf{e}_i \cdot \delta_\xi \left(\frac{\partial G^j [X_k]}{\partial x^\gamma} \mathbf{e}_j \right) + \delta_\xi \left(\frac{\partial G^i [X_k]}{\partial x^\delta} \mathbf{e}_i \right) \cdot \frac{\partial G^j [X_k]}{\partial x^\gamma} \mathbf{e}_j}_{\text{Thermodynamics \& Evolution}} \right) .$$

$$(614)$$

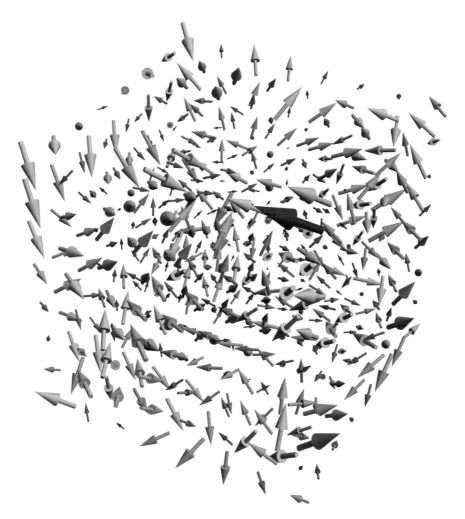

Figure 4.5.1 Random distribution of displacement vector field for G^j in order to achieve a structural solution to Eq. (613). The solution also results in maximum entropy.

Figure 4.5.2a Symmetric distribution of a displacement vector field for G^j in order to achieve a structural solution to Eq. (613). The solution corresponds to a non-extremal entropy.

Figure 4.5.2b Symmetric distribution of a displacement vector field for G^j in order to achieve a structural solution to Eq. (613). The solution corresponds to a non-extremal entropy.

It should be pointed out that the structure also allows an easy incorporation of interaction among the various centers of gravity (e.g. [172]).

4.6 Ordinary Derivative Variation and the Ideal Gas

It should be noted that performing the variation as ordinary derivation and with the simple setting for X_k as set in Eq. (614), the variation of the quantum and the thermo-evolution term can cancel each other out, because we could perform it as follows:

$$\delta_x \left(\frac{\partial G^j \ [X_k]}{\partial x^\gamma} \mathbf{e}_j \right) = \frac{\partial}{\partial x^\beta} \left(\frac{\partial G^j \ [X_k]}{\partial x^\gamma} \mathbf{e}_j \right) \delta x^\beta$$

$$\delta_\xi \left(\frac{\partial G^j \ [X_k]}{\partial x^\gamma} \mathbf{e}_j \right) = \frac{\partial}{\partial \xi^\beta} \left(\frac{\partial G^j \ [X_k]}{\partial x^\gamma} \mathbf{e}_j \right) \delta \xi^\beta = \frac{\partial}{\partial x^\gamma} \left(\frac{\partial G^j \ [X_k]}{\partial \xi^\beta} \mathbf{e}_j \right) \delta \xi^\beta.$$

$$(615)$$

Now we assume that we can perform the following simplification in the second line:

$$\frac{\partial}{\partial \xi^\beta} \left(\frac{\partial G^j \ [X_k]}{\partial x^\gamma} \mathbf{e}_j \right) \delta \xi^\beta = \left(\frac{\partial^2 G^j \ [X_k]}{\partial x^\gamma \partial \xi^\beta} \mathbf{e}_j \right) \delta \xi^\beta = \left(-\frac{\partial^2 G^j \ [X_k]}{\partial x^\gamma \partial x^\beta} \mathbf{e}_j \right) \delta x^\beta.$$

$$(616)$$

It has to be pointed out that this is not possible in cases of position dependent base vectors, for instance. However, in Cartesian coordinates, we could also simplify the first line of Eq. (615) and then obtain in sum:

$$\left(\frac{\partial^2 G^j \ [X_k]}{\partial x^\beta \partial x^\gamma} \mathbf{e}_j \right) \delta x^\beta \Rightarrow \delta_x \left(\frac{\partial G^j \ [X_k]}{\partial x^\gamma} \mathbf{e}_j \right) + \delta_\xi \left(\frac{\partial G^j \ [X_k]}{\partial x^\gamma} \mathbf{e}_j \right) = 0, \quad (617)$$

which automatically gives us:

$$0 = \left(\underbrace{\frac{\partial G^i \ [X_k]}{\partial x^\delta} \mathbf{e}_i \cdot \delta_x \left(\frac{\partial G^j \ [X_k]}{\partial x^\gamma} \mathbf{e}_j \right) + \delta_x \left(\frac{\partial G^i \ [X_k]}{\partial x^\delta} \mathbf{e}_i \right) \cdot \frac{\partial G^j \ [X_k]}{\partial x^\gamma} \mathbf{e}_j}_{\text{Quantum}} + \underbrace{\frac{\partial G^i \ [X_k]}{\partial x^\delta} \mathbf{e}_i \cdot \delta_\xi \left(\frac{\partial G^j \ [X_k]}{\partial x^\gamma} \mathbf{e}_j \right) + \delta_\xi \left(\frac{\partial G^i \ [X_k]}{\partial x^\delta} \mathbf{e}_i \right) \cdot \frac{\partial G^j \ [X_k]}{\partial x^\gamma} \mathbf{e}_j}_{\text{Thermodynamics \& Evolution}} \right).$$

$$(618)$$

In order to avoid such a triviality, either we shall demand the variation with respect to ξ_k to be different or we set X_k as follows:

$$X_k \equiv x_k - f_k \ [\xi_k]. \tag{619}$$

Obviously already the simplest linear form like $f[\xi_k] = m_k^* \xi_k$ suffices to force both terms in Eq. (615) to give zero independently, which, having assumed the same structural variation with respect to ξ_k and x_k, only requires the solution of one. This only leaves statistical laws to take care about the distribution of the various i-j-gravity centers. There is no interaction.

As an example, we want to consider solutions to the following variational outcome (from [162]):

$$\sum_{\alpha=0}^{n-1} \frac{\partial^2 G^j [x_k]}{\partial x^\alpha \partial x^\gamma} \mathbf{e}_j = 0 = \sum_{\alpha=0}^{n-1} \frac{\partial^2 G^i [x_k]}{\partial x^\alpha \partial x^\delta} \mathbf{e}_i. \qquad (620)$$

Even when introducing point solutions from [162], reading for a single center at $\xi_k = 0$:

$$G^j [x_k] = \pm C_s \cdot \left(\frac{\dfrac{t}{\left(x^2 + y^2 + z^2\right)^2 \left(1 + \frac{t^2}{x^2+y^2+z^2}\right)}}{+ \dfrac{\arctan\left[\frac{t}{\sqrt{x^2+y^2+z^2}}\right]}{\left(x^2 + y^2 + z^2\right)^{3/2}}} \right) \cdot \{t, x, y, z\}, \qquad (621)$$

we still do not get anything but an identical solution already to the quantum term (regarding the evaluation in connection with the variation we have to refer to [162]). Thereby C_s stands for a suitable constant. We recognize the potential of a point charge as limit for $t \to \infty$. The resulting $t \to \infty$-limit would then read:

$$\lim_{t \to \infty} G^j [x_k] = \mp C_s \cdot \frac{\pi}{2} \cdot \frac{1}{\left(x^2 + y^2 + z^2\right)^{3/2}} \cdot \{0, x, y, z\}. \qquad (622)$$

Assuming two equal centers placed at $\pm x_0$ would give us:

$$\lim_{t \to \infty} G^j [x_k] = \mp C_s \cdot \frac{\pi}{2} \cdot \frac{1}{\left((x + x_0)^2 + y^2 + z^2\right)^{3/2}} \cdot \{0, x \pm x_0, y, z\}, \qquad (623)$$

but still already the fulfillment of the quantum term, which is guaranteed due to our solution (621) to (620) and thus, both terms in Eq. (615). This means that even the presence of "point charges" would not force our system to follow anything else but statistics, which leads to the second law of thermodynamics and provides neither any interaction nor evolutionary driving forces.

We might see this approach above therefore as a quantum gravity model for an ideal gas.

4.7 Combined Successive Variation

Things are getting significantly different the moment we allow the variation to be performed somehow simultaneously or in a successive manner, but then we have variations of second order and these are considered small of second order. So, here is an example:

$$
\delta_\xi \left(\frac{\partial G^i \, [X_k]}{\partial x^\delta} \mathbf{e}_i \cdot \delta_x \left(\frac{\partial G^j \, [X_k]}{\partial x^\gamma} \mathbf{e}_j \right) + \delta_x \left(\frac{\partial G^i \, [X_k]}{\partial x^\delta} \mathbf{e}_i \right) \cdot \frac{\partial G^j \, [X_k]}{\partial x^\gamma} \mathbf{e}_j \right)
$$
$$
= \delta_\xi \left(\frac{\partial G^i \, [X_k]}{\partial x^\delta} \mathbf{e}_i \right) \cdot \delta_x \left(\frac{\partial G^j \, [X_k]}{\partial x^\gamma} \mathbf{e}_j \right) + \delta_\xi \delta_x \left(\frac{\partial G^i \, [X_k]}{\partial x^\delta} \mathbf{e}_i \right) \cdot \frac{\partial G^j \, [X_k]}{\partial x^\gamma} \mathbf{e}_j
$$
$$
+ \frac{\partial G^i \, [X_k]}{\partial x^\delta} \mathbf{e}_i \cdot \delta_\xi \delta_x \left(\frac{\partial G^j \, [X_k]}{\partial x^\gamma} \mathbf{e}_j \right) + \delta_x \left(\frac{\partial G^i \, [X_k]}{\partial x^\delta} \mathbf{e}_i \right) \cdot \delta_\xi \left(\frac{\partial G^j \, [X_k]}{\partial x^\gamma} \mathbf{e}_j \right).
$$

$$(624)$$

Assuming all double variations to be small of higher order, we could simplify:

$$
\delta_\xi \left(\frac{\partial G^i \, [X_k]}{\partial x^\delta} \mathbf{e}_i \cdot \delta_x \left(\frac{\partial G^j \, [X_k]}{\partial x^\gamma} \mathbf{e}_j \right) + \delta_x \left(\frac{\partial G^i \, [X_k]}{\partial x^\delta} \mathbf{e}_i \right) \cdot \frac{\partial G^j \, [X_k]}{\partial x^\gamma} \mathbf{e}_j \right)
$$
$$
= \delta_\xi \left(\frac{\partial G^i \, [X_k]}{\partial x^\delta} \mathbf{e}_i \right) \cdot \delta_x \left(\frac{\partial G^j \, [X_k]}{\partial x^\gamma} \mathbf{e}_j \right) + \delta_x \left(\frac{\partial G^i \, [X_k]}{\partial x^\delta} \mathbf{e}_i \right) \cdot \delta_\xi \left(\frac{\partial G^j \, [X_k]}{\partial x^\gamma} \mathbf{e}_j \right).
$$

$$(625)$$

Still we have obtained products of variated terms we might also consider small of second order, and thus, for here and now, we are not going to consider this path any further.

Thus, we conclude that our form of rather simple (even though extended) variation by the means of ordinary derivatives does not do the job and we have to investigate more general options.

4.8 Lie and Covariant Variation

4.8.1 Covariant Variation

Even though we already have considered covariant variation forms in connection with the Einstein–Hilbert surface term in [51], we here want to revisit the problem with respect to the simple variation of base vectors. The informed reader knows, of course, that covariant variations of the type:

$$\left(R^{\delta\gamma} - \frac{1}{2} R g^{\delta\gamma} + \Lambda g^{\delta\gamma} + \kappa T^{\delta\gamma} \right) \delta g_{\delta\gamma}$$

$$= \left(R^{\delta\gamma} - \frac{1}{2} R g^{\delta\gamma} + \Lambda g^{\delta\gamma} + \kappa T^{\delta\gamma} \right) g_{\delta\gamma;\beta} \cdot \delta\left(x^\beta \right) \qquad (626)$$

directly give zero, because it always holds that:

$$g_{\delta\gamma;\beta} = g^{\delta\gamma}{}_{;\beta} = 0. \qquad (627)$$

But what about the covariant variation of the base vectors, thereby looking for intrinsic solutions just as we saw them with the one dimension solutions to the Einstein field equations or "infinity stones" (see [93, 96])? We know that we can write Eq. (627) in its base vector decomposed form, reading:

$$g_{\alpha\beta;\gamma} = \mathbf{g}_\alpha \cdot \left(\mathbf{g}_{\beta,\gamma} - \mathbf{g}_\sigma \Gamma^\sigma_{\gamma\beta} \right) + \mathbf{g}_\beta \cdot \left(\mathbf{g}_{\alpha,\gamma} - \mathbf{g}_\sigma \Gamma^\sigma_{\gamma\alpha} \right). \qquad (628)$$

Now we expand just one of the addends and find:

$$\mathbf{g}_\alpha \cdot \left(\mathbf{g}_{\beta,\gamma} - \mathbf{g}_\sigma \Gamma^\sigma_{\gamma\beta} \right) = -\frac{1}{2} \left(\mathbf{g}_\alpha \cdot \left(\mathbf{g}_{\gamma,\beta} - \mathbf{g}_{\beta,\gamma} \right) + \mathbf{g}_\beta \cdot \left(\mathbf{g}_{\alpha,\gamma} - \mathbf{g}_{\gamma,\alpha} \right) \right.$$
$$\left. + \mathbf{g}_\gamma \cdot \left(\mathbf{g}_{\alpha,\beta} - \mathbf{g}_{\beta,\alpha} \right) \right). \qquad (629)$$

We demand it to give an intrinsic zero if combined with its partnering factor from Eq. (626). In order to keep things simple, we only pick the terms with the contravariant metric tensor $\left(-\frac{1}{2} R g^{\delta\gamma} + \Lambda g^{\delta\gamma} \right)$ leading to:

$$-\frac{1}{2} \left(\Lambda - \frac{1}{2} R \right) g^{\alpha\beta} \left(\mathbf{g}_\alpha \cdot \left(\mathbf{g}_{\gamma,\beta} - \mathbf{g}_{\beta,\gamma} \right) + \mathbf{g}_\beta \cdot \left(\mathbf{g}_{\alpha,\gamma} - \mathbf{g}_{\gamma,\alpha} \right) \right.$$
$$\left. + \mathbf{g}_\gamma \cdot \left(\mathbf{g}_{\alpha,\beta} - \mathbf{g}_{\beta,\alpha} \right) \right) = 0$$

$$\Rightarrow 0 = -\frac{1}{2} \mathbf{g}^\alpha \cdot \mathbf{g}^\beta \cdot \left(\mathbf{g}_\alpha \cdot \left(\mathbf{g}_{\gamma,\beta} - \mathbf{g}_{\beta,\gamma} \right) + \mathbf{g}_\beta \cdot \left(\mathbf{g}_{\alpha,\gamma} - \mathbf{g}_{\gamma,\alpha} \right) \right.$$
$$\left. + \mathbf{g}_\gamma \cdot \left(\mathbf{g}_{\alpha,\beta} - \mathbf{g}_{\beta,\alpha} \right) \right)$$

$$= -\frac{1}{2} \left(\mathbf{g}^\alpha \cdot \delta^\beta_\alpha \left(\mathbf{g}_{\gamma,\beta} - \mathbf{g}_{\beta,\gamma} \right) + \mathbf{g}^\beta \cdot \delta^\alpha_\beta \left(\mathbf{g}_{\alpha,\gamma} - \mathbf{g}_{\gamma,\alpha} \right) \right.$$
$$\left. + \mathbf{g}^\alpha \cdot \mathbf{g}^\beta \cdot \mathbf{g}_\gamma \cdot \left(\mathbf{g}_{\alpha,\beta} - \mathbf{g}_{\beta,\alpha} \right) \right)$$

$$= -\frac{1}{2} \left(\mathbf{g}^\beta \cdot \left(\mathbf{g}_{\gamma,\beta} - \mathbf{g}_{\beta,\gamma} \right) + \mathbf{g}^\alpha \cdot \left(\mathbf{g}_{\alpha,\gamma} - \mathbf{g}_{\gamma,\alpha} \right) + \mathbf{g}^\alpha \cdot \delta^\beta_\gamma \cdot \left(\mathbf{g}_{\alpha,\beta} - \mathbf{g}_{\beta,\alpha} \right) \right)$$

$$= -\frac{1}{2} \left(\mathbf{g}^\beta \cdot \left(\mathbf{g}_{\gamma,\beta} - \mathbf{g}_{\beta,\gamma} \right) + \mathbf{g}^\alpha \cdot \left(\mathbf{g}_{\alpha,\gamma} - \mathbf{g}_{\gamma,\alpha} \right) + \mathbf{g}^\alpha \cdot \left(\mathbf{g}_{\alpha,\gamma} - \mathbf{g}_{\gamma,\alpha} \right) \right)$$

$$- -\frac{1}{2} \left(\mathbf{g}^\beta \cdot \left(\mathbf{g}_{\gamma,\beta} - \mathbf{g}_{\beta,\gamma} \right) + 2 \cdot \mathbf{g}^\alpha \cdot \left(\mathbf{g}_{\alpha,\gamma} - \mathbf{g}_{\gamma,\alpha} \right) \right). \qquad (630)$$

Considering β a dummy index allows us to simplify our intrinsic zero demand to:

$$\Rightarrow 0 = \mathbf{g}^\alpha \cdot \left(\mathbf{g}_{\alpha,\gamma} - \mathbf{g}_{\gamma,\alpha}\right). \tag{631}$$

That this is no unjustified assumption can be seen when reconsidering the simplification in the third line of (630) and performing it as follows:

$$= -\frac{1}{2}\left(\mathbf{g}^\alpha \cdot \delta_\alpha^\beta \left(\mathbf{g}_{\gamma,\beta} - \mathbf{g}_{\beta,\gamma}\right) + \mathbf{g}^\beta \cdot \delta_\beta^\alpha \left(\mathbf{g}_{\alpha,\gamma} - \mathbf{g}_{\gamma,\alpha}\right)\right.$$
$$\left. + \mathbf{g}^\alpha \cdot \mathbf{g}^\beta \cdot \mathbf{g}_\gamma \cdot \left(\mathbf{g}_{\alpha,\beta} - \mathbf{g}_{\beta,\alpha}\right)\right)$$
$$= -\frac{1}{2}\left(\mathbf{g}^\alpha \cdot \left(\mathbf{g}_{\alpha,\gamma} - \mathbf{g}_{\gamma,\alpha}\right) + \mathbf{g}^\alpha \cdot \left(\mathbf{g}_{\alpha,\gamma} - \mathbf{g}_{\gamma,\alpha}\right) + \mathbf{g}^\alpha \cdot \delta_\gamma^\beta \cdot \left(\mathbf{g}_{\alpha,\beta} - \mathbf{g}_{\beta,\alpha}\right)\right)$$
$$= -\frac{1}{2}\left(2 \cdot \mathbf{g}^\alpha \cdot \left(\mathbf{g}_{\alpha,\gamma} - \mathbf{g}_{\gamma,\alpha}\right) + \mathbf{g}^\alpha \cdot \left(\mathbf{g}_{\alpha,\gamma} - \mathbf{g}_{\gamma,\alpha}\right)\right)$$
$$= -\frac{3}{2} \cdot \mathbf{g}^\alpha \cdot \left(\mathbf{g}_{\alpha,\gamma} - \mathbf{g}_{\gamma,\alpha}\right) \quad \Rightarrow \quad 0 = \mathbf{g}^\alpha \cdot \left(\mathbf{g}_{\alpha,\gamma} - \mathbf{g}_{\gamma,\alpha}\right). \tag{632}$$

Incorporation of the transformation rule for the base vectors gives us:

$$0 = \mathbf{g}^\alpha \cdot \left(\left(\frac{\partial G^j [X_k]}{\partial x^\alpha}\mathbf{e}_j\right)_{,\gamma} - \left(\frac{\partial G^j [X_k]}{\partial x^\gamma}\mathbf{e}_j\right)_{,\alpha}\right)$$

$$= \left(\frac{\partial G^\alpha [X_k]}{\partial x^i}\mathbf{e}^i\right) \cdot \left(\left(\frac{\partial G^j [X_k]}{\partial x^\alpha}\mathbf{e}_j\right)_{,\gamma} - \left(\frac{\partial G^j [X_k]}{\partial x^\gamma}\mathbf{e}_j\right)_{,\alpha}\right)$$

$$= \left(\frac{\partial G^\alpha [X_k]}{\partial x^i}\mathbf{e}^i\right) \cdot \mathbf{e}_j \left(\left(\frac{\partial G^j [X_k]}{\partial x^\alpha}\right)_{,\gamma} - \left(\frac{\partial G^j [X_k]}{\partial x^\gamma}\right)_{,\alpha}\right), \tag{633}$$

where in the last line we have used the fact that the derivative of the base vectors $\mathbf{e}_j$ has already been taking into account by the means of the Christoffel symbols in (628). The same is assured if we consider the basis $\mathbf{e}_j$ everywhere as the one of Cartesian or affine coordinates.

We realize that this automatically solves our summation problem as addressed in previous works and—more or less halfheartedly—discussed in [171]. It was shown in previous papers (e.g. [171] and see references therein) and earlier in this book that (633) can be solved with the following ansatz for $G^j[\dots]$ consisting of harmonic functions $G[\dots]$:

$$G^j [x_k] = g^{jl}\partial_l G [x_k]; \quad \Delta G [x_k] = 0, \tag{634}$$

$$G^j [x_k] = \pm \left\{ \dots, \overbrace{\frac{\partial G [x_k]}{\partial x_\zeta}}^{\text{pos } \xi}, \dots, -\overbrace{\frac{\partial G [x_k]}{\partial x_\xi}}^{\text{pos } \zeta}, \dots \right\};$$

$$\Delta G [x_k] = 0 \quad \& \quad \forall (\dots, \dots) = 0. \tag{635}$$

Thereby we considered the second type of solutions (635) spin-like.

Please note that even though we can easily see that (627) is fulfilled because we have

$$g_{\alpha\beta;\gamma} = \mathbf{g}_\alpha \cdot \left(\mathbf{g}_{\beta,\gamma} - \mathbf{g}_\sigma \Gamma^\sigma_{\gamma\beta}\right) + \mathbf{g}_\beta \cdot \left(\mathbf{g}_{\alpha,\gamma} - \mathbf{g}_\sigma \Gamma^\sigma_{\gamma\alpha}\right)$$

$$= -\frac{1}{2}\left(\mathbf{g}_\alpha \cdot \left(\mathbf{g}_{\gamma,\beta} - \mathbf{g}_{\beta,\gamma}\right) + \mathbf{g}_\beta \cdot \left(\mathbf{g}_{\alpha,\gamma} - \mathbf{g}_{\gamma,\alpha}\right) + \mathbf{g}_\gamma \cdot \left(\mathbf{g}_{\alpha,\beta} - \mathbf{g}_{\beta,\alpha}\right)\right)$$

$$-\frac{1}{2}\left(\mathbf{g}_\beta \cdot \left(\mathbf{g}_{\gamma,\alpha} - \mathbf{g}_{\alpha,\gamma}\right) + \mathbf{g}_\alpha \cdot \left(\mathbf{g}_{\beta,\gamma} - \mathbf{g}_{\gamma,\beta}\right) + \mathbf{g}_\gamma \cdot \left(\mathbf{g}_{\beta,\alpha} - \mathbf{g}_{\alpha,\beta}\right)\right)$$

$$= -\frac{1}{2}\left(\begin{array}{c}\mathbf{g}_\alpha \cdot \left(\mathbf{g}_{\gamma,\beta} - \mathbf{g}_{\beta,\gamma}\right) + \mathbf{g}_\alpha \cdot \left(\mathbf{g}_{\beta,\gamma} - \mathbf{g}_{\gamma,\beta}\right) \\ +\mathbf{g}_\beta \cdot \left(\mathbf{g}_{\alpha,\gamma} - \mathbf{g}_{\gamma,\alpha}\right) + \mathbf{g}_\beta \cdot \left(\mathbf{g}_{\gamma,\alpha} - \mathbf{g}_{\alpha,\gamma}\right) \\ +\mathbf{g}_\gamma \cdot \left(\mathbf{g}_{\alpha,\beta} - \mathbf{g}_{\beta,\alpha}\right) + \mathbf{g}_\gamma \cdot \left(\mathbf{g}_{\beta,\alpha} - \mathbf{g}_{\alpha,\beta}\right)\end{array}\right)$$

$$= -\frac{1}{2}\left(\begin{array}{c}\mathbf{g}_\alpha \cdot \left(\mathbf{g}_{\gamma,\beta} - \mathbf{g}_{\beta,\gamma}\right) - \mathbf{g}_\alpha \cdot \left(\mathbf{g}_{\gamma,\beta} - \mathbf{g}_{\beta,\gamma}\right) \\ +\mathbf{g}_\beta \cdot \left(\mathbf{g}_{\alpha,\gamma} - \mathbf{g}_{\gamma,\alpha}\right) - \mathbf{g}_\beta \cdot \left(\mathbf{g}_{\alpha,\gamma} - \mathbf{g}_{\gamma,\alpha}\right) \\ +\mathbf{g}_\gamma \cdot \left(\mathbf{g}_{\alpha,\beta} - \mathbf{g}_{\beta,\alpha}\right) - \mathbf{g}_\gamma \cdot \left(\mathbf{g}_{\alpha,\beta} - \mathbf{g}_{\beta,\alpha}\right)\end{array}\right) = 0, \tag{636}$$

this does not mean that the solution to Eq. (631) or (633) would not be realized.

After all, it is not up to us to decide what type of effective zero the universe realized in order to satisfy the extremal condition (87) in the end.

Thus, all options for possible solutions of the total variation of the kind:

$$\delta_g W = 0$$

$$= \int_V d^n x \left(\begin{array}{c}R^{\alpha\beta} - \frac{1}{2}Rg^{\alpha\beta} \\ +\Lambda g^{\alpha\beta} + \kappa T^{\alpha\beta}\end{array}\right) \cdot \left(\mathbf{g}_\alpha \cdot \left(\mathbf{g}_{\beta,\gamma} - \mathbf{g}_\sigma \Gamma^\sigma_{\gamma\beta}\right) + \mathbf{g}_\beta \cdot \left(\mathbf{g}_{\alpha,\gamma} - \mathbf{g}_\sigma \Gamma^\sigma_{\gamma\alpha}\right)\right)$$

$$= -\int_V d^n x \frac{\left(R^{\alpha\beta} + \kappa T^{\alpha\beta}\right)}{2}$$

$$\cdot \left(\begin{array}{c}\left(\mathbf{g}_\alpha \cdot \left(\mathbf{g}_{\gamma,\beta} - \mathbf{g}_{\beta,\gamma}\right) + \mathbf{g}_\beta \cdot \left(\mathbf{g}_{\alpha,\gamma} - \mathbf{g}_{\gamma,\alpha}\right) + \mathbf{g}_\gamma \cdot \left(\mathbf{g}_{\alpha,\beta} - \mathbf{g}_{\beta,\alpha}\right)\right) \\ + \left(\mathbf{g}_\beta \cdot \left(\mathbf{g}_{\gamma,\alpha} - \mathbf{g}_{\alpha,\gamma}\right) + \mathbf{g}_\alpha \cdot \left(\mathbf{g}_{\beta,\gamma} - \mathbf{g}_{\gamma,\beta}\right) + \mathbf{g}_\gamma \cdot \left(\mathbf{g}_{\beta,\alpha} - \mathbf{g}_{\alpha,\beta}\right)\right)\end{array}\right)$$

$$-\frac{3}{2} \cdot \int_V d^n x \left(\Lambda - \frac{1}{2}R\right) \cdot \left(\mathbf{g}^\alpha \cdot \left(\mathbf{g}_{\alpha,\gamma} - \mathbf{g}_{\gamma,\alpha}\right) + \mathbf{g}^\beta \cdot \left(\mathbf{g}_{\beta,\gamma} - \mathbf{g}_{\gamma,\beta}\right)\right)$$

$$\tag{637}$$

have to be taken into account. Using the result from Eq. (633) and taking into account that the bases $\mathbf{e}_j$ have to be considered bases of affine spaces, we can simplify the last line of Eq. (637):

$$0 = - \int_V d^n x \frac{\left(R^{\alpha\beta} + \kappa T^{\alpha\beta} \right)}{2}$$

$$\cdot \left(\begin{array}{c} \left(\mathbf{g}_\alpha \cdot \left(\mathbf{g}_{\gamma,\beta} - \mathbf{g}_{\beta,\gamma} \right) + \mathbf{g}_\beta \cdot \left(\mathbf{g}_{\alpha,\gamma} - \mathbf{g}_{\gamma,\alpha} \right) + \mathbf{g}_\gamma \cdot \left(\mathbf{g}_{\alpha,\beta} - \mathbf{g}_{\beta,\alpha} \right) \right) \\ + \left(\mathbf{g}_\beta \cdot \left(\mathbf{g}_{\gamma,\alpha} - \mathbf{g}_{\alpha,\gamma} \right) + \mathbf{g}_\alpha \cdot \left(\mathbf{g}_{\beta,\gamma} - \mathbf{g}_{\gamma,\beta} \right) + \mathbf{g}_\gamma \cdot \left(\mathbf{g}_{\beta,\alpha} - \mathbf{g}_{\alpha,\beta} \right) \right) \end{array} \right)$$

$$- \frac{3}{2} \cdot \int_V d^n x \left(\Lambda - \frac{1}{2} R \right)$$

$$\cdot \left(\begin{array}{c} \left(\frac{\partial G^\alpha [X_k]}{\partial x^i} \mathbf{e}^i \right) \cdot \mathbf{e}_j \left(\left(\frac{\partial G^j [X_k]}{\partial x^\alpha} \right)_{,\gamma} - \left(\frac{\partial G^j [X_k]}{\partial x^\gamma} \right)_{,\alpha} \right) \\ + \left(\frac{\partial G^\beta [X_k]}{\partial x^i} \mathbf{e}^i \right) \cdot \mathbf{e}_j \left(\left(\frac{\partial G^j [X_k]}{\partial x^\beta} \right)_{,\gamma} - \left(\frac{\partial G^j [X_k]}{\partial x^\beta} \right)_{,\alpha} \right) \end{array} \right). \tag{638}$$

Exchanging one of the dummy indices results in:

$$0 = - \int_V d^n x \frac{\left(R^{\alpha\beta} + \kappa T^{\alpha\beta} \right)}{2}$$

$$\cdot \left(\begin{array}{c} \left(\mathbf{g}_\alpha \cdot \left(\mathbf{g}_{\gamma,\beta} - \mathbf{g}_{\beta,\gamma} \right) + \mathbf{g}_\beta \cdot \left(\mathbf{g}_{\alpha,\gamma} - \mathbf{g}_{\gamma,\alpha} \right) + \mathbf{g}_\gamma \cdot \left(\mathbf{g}_{\alpha,\beta} - \mathbf{g}_{\beta,\alpha} \right) \right) \\ + \left(\mathbf{g}_\beta \cdot \left(\mathbf{g}_{\gamma,\alpha} - \mathbf{g}_{\alpha,\gamma} \right) + \mathbf{g}_\alpha \cdot \left(\mathbf{g}_{\beta,\gamma} - \mathbf{g}_{\gamma,\beta} \right) + \mathbf{g}_\gamma \cdot \left(\mathbf{g}_{\beta,\alpha} - \mathbf{g}_{\alpha,\beta} \right) \right) \end{array} \right)$$

$$- 3 \cdot \int_V d^n x \left(\Lambda - \frac{1}{2} R \right) \cdot \left(\left(\frac{\partial G^\alpha [X_k]}{\partial x^i} \mathbf{e}^i \right) \right.$$

$$\left. \cdot \mathbf{e}_j \left(\left(\frac{\partial G^j [X_k]}{\partial x^\alpha} \right)_{,\gamma} - \left(\frac{\partial G^j [X_k]}{\partial x^\gamma} \right)_{,\alpha} \right) \right). \tag{639}$$

We see that there are many possibilities to form the resulting zero and assuming a universe applying all these options, it is of little wonder why in quantum electrodynamics (or all similar classical field theories) we always have to sum over so many (infinitely many) paths in order to get the total result.

4.8.2 "Covariant" Variation with Respect to the Gravity Centers ξ^j

Similar to the evaluation above, we might also consider a variation with respect to the space vectors of the gravity centers. Thereby we have to take into account that only the function $G[\dots]$ does depend on the gravity center position, but not the purely coordinate dependent base vectors. This would give us:

$$g_{\alpha\beta;\xi^{ij}} = (\mathbf{g}_\alpha \cdot \mathbf{g}_\beta)_{;\xi^{ij}} = \left(\frac{\partial G^j[X_k]}{\partial x^\alpha} \mathbf{e}_j \cdot \frac{\partial G^i[X_k]}{\partial x^\beta} \mathbf{e}_i \right)_{;\xi^{ij}}$$

$$= \mathbf{e}_j \cdot \mathbf{e}_i \left(\frac{\partial G^j[X_k]}{\partial x^\alpha} \frac{\partial G^i[X_k]}{\partial x^\beta} \right)_{,\xi^{ij}}$$

$$= \mathbf{e}_j \cdot \mathbf{e}_i \left(\frac{\partial^2 G^j}{\partial \xi^j \partial x^\alpha} \frac{\partial G^i}{\partial x^\beta} + \frac{\partial^2 G^i}{\partial \xi^i \partial x^\beta} \frac{\partial G^j}{\partial x^\alpha} \right). \tag{640}$$

As we have to take the centers of gravity as coordinate vectors with respect to all dimensions, we should write ξ^j and the corresponding $G^j[\dots]$ as:

$$\xi^j = \xi^{\gamma j}; \quad G^j[X_k] = G^j\left[x^\gamma - \xi^{\gamma j}\right]. \tag{641}$$

This allows us to rearrange Eq. (640) in the following way:

$$g_{\alpha\beta;\xi^{ij}} = \mathbf{e}_j \cdot \mathbf{e}_i \left(\frac{\partial^2 G^j\left[x^\gamma - \xi^{\gamma j}\right]}{\partial \xi^{\gamma j} \partial x^\alpha} \frac{\partial G^i\left[x^\gamma - \xi^{\gamma j}\right]}{\partial x^\beta} \right.$$

$$\left. + \frac{\partial^2 G^i\left[x^\gamma - \xi^{\gamma j}\right]}{\partial \xi^{\gamma i} \partial x^\beta} \frac{\partial G^j\left[x^\gamma - \xi^{\gamma j}\right]}{\partial x^\alpha} \right)$$

$$= \mathbf{e}_j \cdot \mathbf{e}_i \left(-\frac{\partial^2 G^j\left[x^\gamma - \xi^{\gamma j}\right]}{\partial x^\gamma \partial x^\alpha} \frac{\partial G^i\left[x^\gamma - \xi^{\gamma j}\right]}{\partial x^\beta} \right.$$

$$\left. - \frac{\partial^2 G^i\left[x^\gamma - \xi^{\gamma j}\right]}{\partial x^\gamma \partial x^\beta} \frac{\partial G^j\left[x^\gamma - \xi^{\gamma j}\right]}{\partial x^\alpha} \right). \tag{642}$$

Combination with Eq. (626), where we only consider the simpler factors $\left(\frac{1}{2} R g^{\delta\gamma} + \Lambda g^{\delta\gamma} \right)$ as done in Eq. (630), yields:

$$\left(\Lambda - \frac{1}{2} R \right) g^{\alpha\beta} g_{\alpha\beta;\xi^{ij}} = \left(\Lambda - \frac{1}{2} R \right) g^{\alpha\beta} \mathbf{e}_j \cdot \mathbf{e}_i \left(\begin{array}{c} -\frac{\partial^2 G^j\left[x^\gamma - \xi^{\gamma j}\right]}{\partial x^\gamma \partial x^\alpha} \frac{\partial G^i\left[x^\gamma - \xi^{\gamma j}\right]}{\partial x^\beta} \\ -\frac{\partial^2 G^i\left[x^\gamma - \xi^{\gamma j}\right]}{\partial x^\gamma \partial x^\beta} \frac{\partial G^j\left[x^\gamma - \xi^{\gamma j}\right]}{\partial x^\alpha} \end{array} \right)$$

$$\Rightarrow \mathbf{g}^\alpha \cdot \mathbf{g}^\beta \cdot \left(-\mathbf{e}_j \frac{\partial^2 G^j\left[x^\gamma - \xi^{\gamma j}\right]}{\partial x^\gamma \partial x^\alpha} \mathbf{g}_\beta - \mathbf{e}_i \frac{\partial^2 G^i\left[x^\gamma - \xi^{\gamma j}\right]}{\partial x^\gamma \partial x^\beta} \mathbf{g}_\alpha \right)$$

$$= \left(-\mathbf{g}^\beta \cdot \mathbf{e}_j \frac{\partial^2 G^j\left[x^\gamma - \xi^{\gamma j}\right]}{\partial x^\gamma \partial x^\alpha} \delta^\alpha_\beta - \mathbf{g}^\alpha \cdot \mathbf{e}_i \frac{\partial^2 G^i\left[x^\gamma - \xi^{\gamma j}\right]}{\partial x^\gamma \partial x^\beta} \delta^\beta_\alpha \right)$$

$$= \left(-\mathbf{g}^\alpha \cdot \mathbf{e}_j \frac{\partial^2 G^j\left[x^\gamma - \xi^{\gamma j}\right]}{\partial x^\gamma \partial x^\alpha} - \mathbf{g}^\alpha \cdot \mathbf{e}_i \frac{\partial^2 G^i\left[x^\gamma - \xi^{\gamma j}\right]}{\partial x^\gamma \partial x^\alpha} \right)$$

$$= -2 \cdot \mathbf{g}^\alpha \cdot \mathbf{e}_j \frac{\partial^2 G^j\left[x^\gamma - \xi^{\gamma j}\right]}{\partial x^\gamma \partial x^\alpha}. \tag{643}$$

Please note that we have used the freedom of considering i and j as dummy indices.

4.8.3 Lie-Variation

It has to be pointed out that we do know that in addition to the classical Lie-derivative, which is performed as a derivative along a given vector, one can also construct such a derivative along higher order tensors. However, for brevity, we will not consider such generalized cases here as it totally suffices to use simple vectors in order to see that covariant and Lie-derivatives provide the driving forces for evolution (please see section "4.9 Evolution as an Inevitable Result").

One of the already well-known forms of Lie-variation of the metric results is the Killing equation.

This variation keeps the metric constant, which holds for ordinary coordinate transformations and all metric transformations along Killing vector fields. The equation to determine the corresponding vector fields v_α is the so-called Killing equation and reads (e.g. [173]):

$$0 = v_{\alpha;\beta} + v_{\beta;\alpha}. \tag{644}$$

Thereby the ";" denotes the covariant derivative. With the metric unchanged along transformations of the kind

$$g^{ij}\left[x^i\right] \rightarrow g^{ij}\left[x^i + \varepsilon \cdot v^i\right], \tag{645}$$

of course, also the Ricci scalar is conserved. Some discussion in connection with this equation and its deeper connection with the Dirac equation was already given in [113, 114].

Here now we want to study a variety of Lie-variations and start with the Lie-variation of the base vector along a certain vector $\mathbf{x}^\sigma$:

$$L_x \mathbf{g}_\alpha = x^\chi \partial_\chi \mathbf{g}_\alpha + \mathbf{g}_\chi \partial_\alpha x^\chi. \tag{646}$$

Assuming this very vector $\mathbf{x}^\sigma$ to be the contravariant position vector, we can write:

$$L_x \mathbf{g}_\alpha = x^\chi \partial_\chi \mathbf{g}_\alpha + \mathbf{g}_\chi \frac{\partial x^\chi}{\partial x^\alpha} = x^\chi \partial_\chi \mathbf{g}_\alpha + \mathbf{g}_\chi b \cdot \delta_\alpha^\chi = x^\chi \partial_\chi \mathbf{g}_\alpha + b \cdot \mathbf{g}_\alpha = \left(x^\chi \partial_\chi + b\right) \cdot \mathbf{g}_\alpha. \tag{647}$$

Please note that we usually set $b = 1$. The resulting equation has quite some similarity with the Dirac equation [143].

Another possibility would be the Lie-variation with respect to the positions of the various centers of gravity, which we name ξ^j here:

$$L_\xi \mathbf{g}_\alpha = \xi^j \partial_j \mathbf{g}_\alpha + \mathbf{g}_j \frac{\partial \xi^j}{\partial x^\alpha} = \xi^j \partial_j \mathbf{g}_\alpha + \mathbf{g}_j \cdot 0 = \xi^j \partial_j \mathbf{g}_\alpha, \tag{648}$$

or with respect to the difference vector $x^\sigma - \xi_j^\sigma$:

$$L_{(x^\sigma - \xi_j^\sigma)} \mathbf{g}_\alpha = \left(x^\sigma - \xi_j^\sigma\right) \partial_\sigma \mathbf{g}_\alpha + \mathbf{g}_\sigma \frac{\partial \left(x^\sigma - \xi_j^\sigma\right)}{\partial x^\alpha} = \left(x^\sigma - \xi_j^\sigma\right) \partial_\sigma \mathbf{g}_\alpha + \mathbf{g}_\sigma b \cdot \delta_\alpha^\sigma$$

$$= \left(x^\sigma - \xi_j^\sigma\right) \partial_\sigma \mathbf{g}_\alpha + b \cdot \mathbf{g}_\alpha = \left(\left(x^\sigma - \xi_j^\sigma\right) \partial_\sigma + b\right) \cdot \mathbf{g}_\alpha. \tag{649}$$

In connection with the Einstein–Hilbert factor from Eq. (626), the corresponding variations then read as follows:

to Eq. (647):

$$\left(\Lambda - \frac{1}{2}R\right) g^{\alpha\beta} \left(\mathbf{g}_\beta \cdot L_x \mathbf{g}_\alpha + \mathbf{g}_\alpha \cdot L_x \mathbf{g}_\beta\right)$$

$$= \left(\Lambda - \frac{1}{2}R\right) \mathbf{g}^\alpha \cdot \mathbf{g}^\beta \cdot \left(\mathbf{g}_\beta \cdot \left(x^\chi \partial_\chi + b\right) \cdot \mathbf{g}_\alpha + \mathbf{g}_\alpha \cdot \left(x^\chi \partial_\chi + b\right) \cdot \mathbf{g}_\beta\right)$$

$$= \left(\Lambda - \frac{1}{2}R\right) \left(\mathbf{g}^\beta \cdot \delta_\beta^\alpha \left(x^\chi \partial_\chi + b\right) \cdot \mathbf{g}_\alpha + \mathbf{g}^\alpha \cdot \delta_\alpha^\beta \left(x^\chi \partial_\chi + b\right) \cdot \mathbf{g}_\beta\right)$$

$$= \left(\Lambda - \frac{1}{2}R\right) \left(\mathbf{g}^\beta \cdot \left(x^\chi \partial_\chi + b\right) \cdot \mathbf{g}_\beta + \mathbf{g}^\alpha \cdot \left(x^\chi \partial_\chi + b\right) \cdot \mathbf{g}_\alpha\right), \tag{650}$$

to Eq. (648):

$$\left(\Lambda - \frac{1}{2}R\right) g^{\alpha\beta} \left(\mathbf{g}_\beta \cdot L_\xi \mathbf{g}_\alpha + \mathbf{g}_\alpha \cdot L_\xi \mathbf{g}_\beta\right)$$

$$= \left(\Lambda - \frac{1}{2}R\right) \left(\mathbf{g}^\beta \cdot \xi^j \partial_j \cdot \mathbf{g}_\beta + \mathbf{g}^\alpha \cdot \xi^j \partial_j \cdot \mathbf{g}_\alpha\right)$$

$$\tag{651}$$

and to Eq. (649):

$$\left(\Lambda - \frac{1}{2}R\right) g^{\alpha\beta} \left(\mathbf{g}_\beta \cdot l_{(x^\sigma - \xi_j^\sigma)} \mathbf{g}_\alpha + \mathbf{g}_\alpha \cdot L_{(x^\sigma - \xi_j^\sigma)} \mathbf{g}_\beta\right)$$

$$= \left(\Lambda - \frac{1}{2}R\right) \left(\mathbf{g}^\beta \cdot \left(\left(x^\sigma - \xi_j^\sigma\right) \partial_\sigma + b\right) \cdot \mathbf{g}_\beta + \mathbf{g}^\alpha \cdot \left(\left(x^\sigma - \xi_j^\sigma\right) \partial_\sigma + b\right) \cdot \mathbf{g}_\alpha\right). \tag{652}$$

Demanding all these variations to deliver zero outcomes and incorporating the base vector transformation, thereby remembering that we must not perform the derivation over the bases $\mathbf{e}_j$ (c.f. end of section "4.8.1 Covariant Variation") leads to the following equations:

to Eq. (647):

$$
\begin{aligned}
0 &= \left(\Lambda - \frac{1}{2}R\right) g^{\alpha\beta} \left(\mathbf{g}_\beta \cdot L_x \mathbf{g}_\alpha + \mathbf{g}_\alpha \cdot L_x \mathbf{g}_\beta\right) \\
&= \left(\Lambda - \frac{1}{2}R\right) \left(\mathbf{g}^\beta \cdot (x^\chi \partial_\chi + b) \cdot \left(\frac{\partial G^j}{\partial x^\beta}\mathbf{e}_j\right) + \mathbf{g}^\alpha \cdot (x^\chi \partial_\chi + b) \cdot \left(\frac{\partial G^j}{\partial x^\alpha}\mathbf{e}_j\right)\right) \\
&\Rightarrow 0 = 2 \cdot \left(\Lambda - \frac{1}{2}R\right) \left(\left(\frac{\partial G^\alpha}{\partial x^i}\mathbf{e}^i\right) \cdot (x^\chi \partial_\chi + b) \cdot \left(\frac{\partial G^j}{\partial x^\alpha}\mathbf{e}_j\right)\right) \\
&= 2 \cdot \left(\Lambda - \frac{1}{2}R\right) \left(\left(\frac{\partial G^\alpha}{\partial x^i}\mathbf{e}^i\right) \cdot \mathbf{e}_j \left(x^\chi \partial_\chi + b\right) \cdot \left(\frac{\partial G^j}{\partial x^\alpha}\right)\right),
\end{aligned} \tag{653}
$$

to Eq. (648):

$$
\begin{aligned}
&\left(\Lambda - \frac{1}{2}R\right) g^{\alpha\beta} \left(\mathbf{g}_\beta \cdot L_\xi \mathbf{g}_\alpha + \mathbf{g}_\alpha \cdot L_\xi \mathbf{g}_\beta\right) \\
&\Rightarrow 0 = 2 \cdot \left(\Lambda - \frac{1}{2}R\right) \left(\left(\frac{\partial G^\alpha}{\partial x^i}\mathbf{e}^i\right) \cdot \xi^j \partial_j \cdot \left(\frac{\partial G^j}{\partial x^\alpha}\mathbf{e}_j\right)\right) \\
&= 2 \cdot \left(\Lambda - \frac{1}{2}R\right) \left(\left(\frac{\partial G^\alpha}{\partial x^i}\mathbf{e}^i\right) \cdot \mathbf{e}_j \xi^j \partial_j \cdot \left(\frac{\partial G^j}{\partial x^\alpha}\right)\right)
\end{aligned} \tag{654}
$$

and to Eq. (649):

$$
\begin{aligned}
&\left(\Lambda - \frac{1}{2}R\right) g^{\alpha\beta} \left(\mathbf{g}_\beta \cdot L_{(x^\sigma - \xi_j^\sigma)} \mathbf{g}_\alpha + \mathbf{g}_\alpha \cdot L_{(x^\sigma - \xi_j^\sigma)} \mathbf{g}_\beta\right) \\
&\Rightarrow 0 = 2 \cdot \left(\Lambda - \frac{1}{2}R\right) \left(\left(\frac{\partial G^\alpha}{\partial x^i}\mathbf{e}^i\right) \cdot \left((x^\sigma - \xi_j^\sigma) \partial_\sigma + b\right) \cdot \left(\frac{\partial G^j}{\partial x^\alpha}\mathbf{e}_j\right)\right) \\
&= 2 \cdot \left(\Lambda - \frac{1}{2}R\right) \left(\left(\frac{\partial G^\alpha}{\partial x^i}\mathbf{e}^i\right) \cdot \mathbf{e}_j \cdot \left((x^\sigma - \xi_j^\sigma) \partial_\sigma + b\right) \cdot \left(\frac{\partial G^j}{\partial x^\alpha}\right)\right).
\end{aligned} \tag{655}
$$

4.9 Evolution as an Inevitable Result

In difference to the purely ordinary variation derivative forms as discussed in sections "4.6 Ordinary Derivative Variation and the Ideal Gas" and "4.7 Combined Successive Variation" and as applied in some of our earlier papers (see [171] and references therein), the Lie- and the covariant variation forms as introduced in the last section above are clearly providing significant differences for the variation with respect to the main coordinates x^α and the variation with respect to the position vectors of the various gravity centers ξ^α. While it was shown with respect to the ordinary derivative based variations that there is no additional force coming from the various gravity centers, which is to say that they do not produce any interaction with each other, we clearly recognize significant differences regarding the variation of the two for the Lie- and the covariant forms.

This means that while we know from [171], subsection "Incorporating Interaction", that we can always actively (by hand) force some kind of gravity center interaction simply by defining the functions $G^j[\ldots]$ as complex objects like:

$$G^j[\ldots] = G^j\left[X_{kj}\right] \equiv G^j\left[X_{k0}, X_{k1}, \ldots, X_{ki}, \ldots, X_{k(N-1)}\right]; \quad X_{ki} \equiv x_k - \xi_{ki},$$
(656)

we do obtain this interaction also in cases of non-cross interaction as postulated in Eq. (656), but simply get it as an automatic outcome due to the simple existence of more than one center of gravity, because of the difference of variations with respect to x^α and ξ^α. Even though the functions $G^j[\ldots]$ stay as simple as:

$$G^j[\ldots] = G^j\left[X_{kj}\right]; \quad X_k \equiv x_k - \xi_{kj}$$
(657)

and thus do explicitly not show any sign of interaction (not directly, anyway), still the interaction between the centers of gravity comes into existence the moment the system is allowed to vary. The only prerequisite is that this variation is with respect to both x^α and ξ^α and that this variation is not just based on ordinary derivatives.

As an example, we can consider the covariant derivative, where due to Eq. (627) we are automatically only left with Eq. (643), which is to say:

$$0 = -2 \cdot \mathbf{g}^\alpha \cdot \mathbf{e}_j \frac{\partial^2 G^j\left[x^\gamma - \xi^{\gamma j}\right]}{\partial x^\gamma \partial x^\alpha}.$$
(658)

But this is not the whole variation condition, because now we miss the volume integral. Completed with Eq. (87), we have to assure that:

$$
\begin{aligned}
\delta_g W = 0 &= \int_V d^n x \left(\begin{array}{c} R^{\alpha\beta} - \tfrac{1}{2} Rg^{\alpha\beta} \\ +\Lambda g^{\alpha\beta} + \kappa T^{\alpha\beta} \end{array} \right) \cdot \mathbf{e}_j \cdot \mathbf{e}_i \left(\begin{array}{c} -\frac{\partial^2 G^j\left[x^\gamma - \xi^{\gamma j}\right]}{\partial x^\gamma \partial x^\alpha} \frac{\partial G^i\left[x^\gamma - \xi^{\gamma i}\right]}{\partial x^\beta} \\ -\frac{\partial^2 G^i\left[x^\gamma - \xi^{\gamma i}\right]}{\partial x^\gamma \partial x^\beta} \frac{\partial G^j\left[x^\gamma - \xi^{\gamma j}\right]}{\partial x^\alpha} \end{array} \right) \\
&= \int_V d^n x \left(R^{\alpha\beta} + \kappa T^{\alpha\beta} \right) \cdot \mathbf{e}_j \cdot \mathbf{e}_i \left(\begin{array}{c} -\frac{\partial^2 G^j\left[x^\gamma - \xi^{\gamma j}\right]}{\partial x^\gamma \partial x^\alpha} \frac{\partial G^i\left[x^\gamma - \xi^{\gamma i}\right]}{\partial x^\beta} \\ -\frac{\partial^2 G^i\left[x^\gamma - \xi^{\gamma i}\right]}{\partial x^\gamma \partial x^\beta} \frac{\partial G^j\left[x^\gamma - \xi^{\gamma j}\right]}{\partial x^\alpha} \end{array} \right) \\
&\quad -2 \cdot \int_V d^n x \left(\Lambda - \frac{1}{2} R \right) \cdot \mathbf{g}^\alpha \cdot \mathbf{e}_j \frac{\partial^2 G^j\left[x^\gamma - \xi^{\gamma j}\right]}{\partial x^\gamma \partial x^\alpha}
\end{aligned}
$$
(659)

is satisfied. Here we have a variety of options; for instance, we could solve Eq. (658) or the Einstein field equations or try for positioning the various centers of gravity in order to achieve the right zero outcome. Also, mixtures of these options are thinkable as there does not seem to be a rule forcing us (or the universe) into a certain strategy and subsequent outcome.

Things are getting a bit more interesting if we do not directly apply the condition (627) but look for intrinsic solutions to the whole as shown above in section "4.8.1 Covariant Variation". Adding up the results from the covariant variation (639) and (643) yields:

$$
3 \cdot \mathbf{g}^\alpha \cdot \left(\mathbf{g}_{\gamma,\alpha} - \mathbf{g}_{\alpha,\gamma} \right) - 2 \cdot \mathbf{g}^\alpha \cdot \mathbf{e}_j \frac{\partial^2 G^j \left[x^\gamma - \xi^{\gamma j} \right]}{\partial x^\gamma \partial x^\alpha}
$$

$$
= \mathbf{g}^\alpha \cdot \left(3 \cdot \left(\mathbf{g}_{\gamma,\alpha} - \mathbf{g}_{\alpha,\gamma} \right) - 2 \cdot \mathbf{e}_j \frac{\partial^2 G^j \left[x^\gamma - \xi^{\gamma j} \right]}{\partial x^\gamma \partial x^\alpha} \right)
$$

$$
= \mathbf{g}^\alpha \cdot \mathbf{e}_j \left(3 \cdot \left(\frac{\partial^2 G^j \left[x^\gamma - \xi^{\gamma j} \right]}{\partial x^\alpha \partial x^\gamma} - \frac{\partial^2 G^j \left[x^\gamma - \xi^{\gamma j} \right]}{\partial x^\gamma \partial x^\alpha} \right) \right.
$$

$$
\left. -2 \cdot \frac{\partial^2 G^j \left[x^\gamma - \xi^{\gamma j} \right]}{\partial x^\gamma \partial x^\alpha} \right)
$$

$$
= \mathbf{g}^\alpha \cdot \mathbf{e}_j \left(3 \cdot \frac{\partial^2 G^j \left[x^\gamma - \xi^{\gamma j} \right]}{\partial x^\alpha \partial x^\gamma} - 5 \cdot \frac{\partial^2 G^j \left[x^\gamma - \xi^{\gamma j} \right]}{\partial x^\gamma \partial x^\alpha} \right). \tag{660}
$$

It does not require intensive math to recognize that a solution to Eq. (660) can be found by a combination of Eqs. (602) and (635) and a suitable positioning of the various centers of gravity. So, as a very simple example, we might just assume to have the inner quantum solution with the condition:

$$
0 = \mathbf{g}^\alpha \cdot \left(\left(\frac{\partial G^j \left[x^\gamma - \xi^{\gamma j} \right]}{\partial x^\alpha} \mathbf{e}_j \right)_{,\gamma} - \left(\frac{\partial G^j \left[x^\gamma - \xi^{\gamma j} \right]}{\partial x^\gamma} \mathbf{e}_j \right)_{,\alpha} \right), \tag{661}
$$

with a suitable selection of functions $G^j[\dots]$. Now, even though usually the volume integral of the variational task (87) has been ignored in a sense that one simply demands the fulfillment with respect to the whole integrand, this does not mean that the integral could still not serve a purpose. In our case here, we have the terms

$$
-\frac{3}{2} \cdot \mathbf{g}^\alpha \cdot \left(\mathbf{g}_{\alpha,\gamma} - \mathbf{g}_{\gamma,\alpha} \right) - 2 \cdot \mathbf{g}^\alpha \cdot \mathbf{e}_j \frac{\partial^2 G^j \left[x^\gamma - \xi^{\gamma j} \right]}{\partial x^\gamma \partial x^\alpha} = -\mathbf{g}^\alpha \cdot 2 \cdot \mathbf{e}_j \frac{\partial^2 G^j \left[x^\gamma - \xi^{\gamma j} \right]}{\partial x^\gamma \partial x^\alpha}
$$

$$\tag{662}$$

as leftovers from Eq. (660) with our $G^j[\ldots]$-solutions from Eq. (661). Just for the sake of simplicity, assuming:

$$R - 2\kappa\, L_M = 0 \tag{663}$$

and inserting the leftovers from Eq. (662) into Eq. (87), resulting in:

$$\delta_g W = 0 = -\int_V d^n x \wedge \left(\mathbf{g}^\alpha \cdot 2 \cdot \mathbf{e}_j\, \frac{\partial^2 G^j\left[x^\gamma - \xi^{\gamma j}\right]}{\partial x^\gamma \partial x^\alpha} \right) \tag{664}$$

gives us a positioning condition for the various centers of gravity j within the volume V.

Thus, we have a driving force towards a certain structuring or ordering without any actively proposed interaction but simply because we allow for the simultaneous variation of x^α and ξ^α.

For completeness, we want to give the variation integral for both variations of x^α and ξ^α:

$$0 = \int_V d^n x\, \frac{\left(R^{\alpha\beta} + \kappa\, T^{\alpha\beta}\right)}{2}$$

$$\cdot \left(\begin{array}{c} \left(\mathbf{g}_\alpha \cdot \left(\mathbf{g}_{\gamma,\beta} - \mathbf{g}_{\beta,\gamma}\right) + \mathbf{g}_\beta \cdot \left(\mathbf{g}_{\alpha,\gamma} - \mathbf{g}_{\gamma,\alpha}\right) + \mathbf{g}_\gamma \cdot \left(\mathbf{g}_{\alpha,\beta} - \mathbf{g}_{\beta,\alpha}\right)\right) \\ + \left(\mathbf{g}_\beta \cdot \left(\mathbf{g}_{\gamma,\alpha} - \mathbf{g}_{\alpha,\gamma}\right) + \mathbf{g}_\alpha \cdot \left(\mathbf{g}_{\beta,\gamma} - \mathbf{g}_{\gamma,\beta}\right) + \mathbf{g}_\gamma \cdot \left(\mathbf{g}_{\beta,\alpha} - \mathbf{g}_{\alpha,\beta}\right)\right) \end{array} \right)$$

$$+ 3 \cdot \int_V d^n x \left(\Lambda - \frac{1}{2} R \right) \cdot \left(\left(\frac{\partial G^\alpha\left[X_k\right]}{\partial x^i} \mathbf{e}^i \right) \right.$$

$$\left. \cdot \mathbf{e}_j \left(\left(\frac{\partial G^j\left[X_k\right]}{\partial x^\alpha} \right)_{,\gamma} - \left(\frac{\partial G^j\left[X_k\right]}{\partial x^\gamma} \right)_{,\alpha} \right) \right)$$

$$+ \int_V d^n x \left(R^{\alpha\beta} + \kappa\, T^{\alpha\beta} \right) \cdot \mathbf{e}_j \cdot \mathbf{e}_i \left(\begin{array}{c} \frac{\partial^2 G^j\left[x^\gamma - \xi^{\gamma j}\right]}{\partial x^\gamma \partial x^\alpha} \frac{\partial G^i\left[x^\gamma - \xi^{\gamma j}\right]}{\partial x^\beta} \\ + \frac{\partial^2 G^i\left[x^\gamma - \xi^{\gamma j}\right]}{\partial x^\gamma \partial x^\beta} \frac{\partial G^j\left[x^\gamma - \xi^{\gamma j}\right]}{\partial x^\alpha} \end{array} \right)$$

$$+ 2 \cdot \int_V d^n x \left(\Lambda - \frac{1}{2} R \right) \cdot \mathbf{g}^\alpha \cdot \mathbf{e}_j\, \frac{\partial^2 G^j\left[x^\gamma - \xi^{\gamma j}\right]}{\partial x^\gamma \partial x^\alpha}. \tag{665}$$

Thus, including the positioning possibilities for the gravity centers, the universe has a huge number of options to make this equation come true.

Similar gravity center position optimization (or extremalization) can be observed from the Lie-forms (654) and (655).

However, if understanding intrinsic, which is to say system-immanent, forces, leading to ordering or structuring, as a form of self-organization, we have what is of need to explain an inner desire for evolution of each and every system where there would be more than one center of gravity. Thereby, and this is important, the interaction is not restricted to gravity. We only used the name "centers of gravity" because we here are still dealing with general metrical objects.

4.10 Consequences

4.10.1 A World Formula (?)

Interpreting the first factor under the variation integral in Eq. (388):

$$\left(\begin{array}{c} f'\,[R]\,R^{\delta\gamma} - \frac{1}{2} f\,[R]\,g^{\delta\gamma} + f''\,[R]\left[R^{;\delta\gamma} - \Delta_g R \cdot g^{\delta\gamma} \right] \\ + f'''\,[R]\left[R^{;\delta}\,R^{;\gamma} - R^{;\sigma}\,R_{;\sigma} \cdot g^{\delta\gamma} \right] + \underbrace{8\pi\,G}_{\kappa}\,T^{\delta\gamma} \end{array} \right) = 0 \qquad (666)$$

or its classical equivalent (389):

$$R^{\delta\gamma} - \frac{1}{2} R g^{\delta\gamma} + \Lambda g^{\delta\gamma} + \kappa\,T^{\delta\gamma} = 0 \qquad (667)$$

as the part of the General Theory of Relativity and the second factor:

$$\delta g_{\delta\gamma} = \delta\left(\mathbf{g}_\delta \cdot \mathbf{g}_\gamma \right) = \mathbf{g}_\delta \cdot \delta\mathbf{g}_\gamma + \delta\mathbf{g}_\delta \cdot \mathbf{g}_\gamma$$

$$= \frac{\partial G^\alpha\,[x_k]}{\partial x^\delta}\mathbf{e}_\alpha \cdot \delta\left(\frac{\partial G^\beta\,[x_k]}{\partial x^\gamma}\mathbf{e}_\beta \right) + \delta\left(\frac{\partial G^\alpha\,[x_k]}{\partial x^\delta}\mathbf{e}_\alpha \right) \cdot \frac{\partial G^\beta\,[x_k]}{\partial x^\gamma}\mathbf{e}_\beta = 0$$

$$(668)$$

as the quantum theoretical part, we see that the solution of one automatically fulfills the whole variation condition (388) (or (389)) and thus, both theories.

One might therefore see Eq. (388) (or Eq. (389)) as a world formula.

What we still have to solve, however, is the riddle of how to obtain suitable sums for the derivation of classical quantum equations. We will once more come to that point further below in this chapter.

4.10.2 What Is the Nature of the Classical Quantum Theoretical Wave Function?

As a by-product, we also saw that the nature of the quantum theoretical wave function is just the set of transformation functions $G^j\,[x_k]$ allowing the transformation from a coordinate system A with coordinates x_k to B with coordinates $G^j[x_k]$. In the following, we shall name the quantum functions "G-fluctuations of the metric base".

More discussion about the consequences of our result has been presented in [136].

In section "4.3.1 Deriving the Equations of Elasticity" above, we also saw that the $G^j[x_k]*\mathbf{e}_j$ can be seen as space-time displacement vectors.

4.11 An Important Question!

Question: Which of the many options given above are used in our universe?

Somebody like Dirac might probably have given the answer:

"The most beautiful equation must be the right one!"

However, this author would rather not go for any such a restriction or selection and wants to put it as follows:

"Our universe probably uses all degrees of freedom (and more). The simpler ones are thereby—perhaps—used more frequently than the more complicated ones; respectively, the ones with the highest probability are most likely also the more frequently realized ones."

And then this author would also like to add:

"Obviously, it does not lie in the nature of nature to not use all possibilities as long as they are 'allowed', which is to say, possible."

4.12 The Fundamental Connection to Materials Science

It was shown in a variety of papers by this author (e.g. [147–150]) that one major problem of mathematically dealing with time- and stress-dependent material properties results from the space-time dependency of these very properties. It was even shown with first-principles methods [148] that the usual time-functional approach, which expands the ordinary linear theory of elasticity to an effective (global) time dependency, simply cannot work in a general manner because its basic assumption of material homogeneity does not hold. The integral (space or time integral) effective laws applied to overcome this problem, on the other hand, are always only valid in connection with a certain group of experiments, respectively deformations and stress situations or processes.

The problem, so it was shown in [148], seems to result from the restriction of the material constitutive laws to linearity, homogeneity and the time-dependency only being functionality (add-on like or grafted) incorporated.

This author is of the opinion that the incorporation of time and non-linearity as true additional degrees of freedom provides quite some potential for mirroring the natural behavior much better than the current approaches can ever do. Degrees of freedom, however, are just dimensions and so we end up with elastic material space-times instead of the usual linear elastic spaces being (time-functionally) made time dependent.

Now it was shown by the author that any kind of process a certain space-time is subjected to, including those which add energy, mass or any other kind of "stress", can be described by the transformation of the kind (84) (e.g. [82]).

Naturally, this can also be done to a material space-time, and the learned materials scientist will immediately realize the connection, because we have the following analogy:

$$G_{\alpha\beta} = F \left[f \left[t, x, y, z \right] \right]_{\alpha\beta}^{ij} g_{ij} \xrightarrow{\text{simplest}} G_{\alpha\beta} = F \left[f \left[t, x, y, z \right] \right] \cdot \delta_\alpha^i \delta_\beta^j g_{ij}$$

$$\Updownarrow$$

$$\varepsilon_{\alpha\beta} = E_{\alpha\beta}^{ij} \sigma_{ij} \xrightarrow{\text{simplest}} \varepsilon_{\alpha\beta} = E \cdot \delta_\alpha^i \delta_\beta^j \sigma_{ij}. \tag{669}$$

This connection not only provides us with quite some interesting analogy between the cosmologic space-time and elasticity but also gives us access to the mathematical apparatus on both sides of the wall. For instance, while there is still a great struggle in dealing with non-homogeneous Young's moduli, there are methods to deal with general $F[\ldots]$ in Eq. (84). As mathematically, however, there is no difference to Eq. (669), we immediately see one advantage in forming this connection. The explicit time-dependency in F of Eq. (84) and the classical problems in treating creep in Eq. (669) are adding another point.

Just in order to give an example that there are also situations where materials science can help to dramatically simplify certain problems in General Theory of Relativity, we want to consider a complex diagonal metric g_{ij}. It can be shown that there always exists a transformation:

$$G_{\alpha\beta} = F_{\alpha\beta}^{ij} g_{ij} \xrightarrow{\text{isotropic}} G_{\alpha\beta} = \delta_j^i \cdot \Psi\,[t, x, y, z], \tag{670}$$

which renders the resulting metric completely isotropic. The skilled materials scientist will have no difficulties in finding that the transformation necessary to achieve this does have the following structure:

$$F_{\alpha\beta}^{ij} = \begin{pmatrix} 1 & \nu & \nu & \nu & & & & & \\ \nu & 1 & \nu & \nu & & & \vdots & & \\ \nu & \nu & 1 & \nu & & \cdots & 0 & \cdots & \\ \nu & \nu & \nu & 1 & & & \vdots & & \\ & & & & 1+\nu & & & & \\ & & & & & 1+\nu & & & \\ & \vdots & & & & & 1+\nu & & \\ & \cdots\,0\,\cdots & & & & & & 1+\nu & \\ & \vdots & & & & & & & 1+\nu \\ & & & & & & & & & 1+\nu \end{pmatrix}. \tag{671}$$

Thereby we have used the symmetries of the transformation (being the same for F and the Young's modulus tensor) and applied the Klein–Voigt presentation for shortage. The reader will already have recognized the typical form of a Young's modulus tensor, only that it is not in three, but in four material dimensions. Thus, it automatically includes different time behavior at each position in space, giving a much greater flexibility to describe real material problems.

Now, with the connection established, we might like to ask how to model and understand processes (transformations) our 4-dimensional material space-time is subjected to. Well, the answer is not too difficult if we just turn our attention to the energy momentum tensor and observe what happens to it when something has been done to our material. The corresponding mathematical apparatus was given in [113].

Thereby it was worked out how one has to connect solutions from the general approach (8) (or (669) if we are already thinking in material space-times) to the resulting energy momentum tensor and its changes. As shown in [113] and chapter "2 An Unusual Introduction", section "2.1 Derivation of Electro-Magnetic Interaction (and Matter) via a Set of Creative Transformations", we will not need anything else but the field tensor of the electromagnetic interaction to close the circle. The electrostatic and magnetic vector fields $\mathbf{E}$ and $\mathbf{B}$ with components E_i and B_i, on the other hand, are forming the energy momentum tensor in contravariant form, which was already given above as follows:

$$\left(T^{\alpha\beta}\right) = \begin{pmatrix} \dfrac{1}{2}\cdot(\mathbf{E}\cdot\mathbf{E}+\mathbf{B}\cdot\mathbf{B}) & (\mathbf{E}\times\mathbf{B})^{T} \\ \mathbf{E}\times\mathbf{B} & \dfrac{1}{2}\cdot(\mathbf{E}\cdot\mathbf{E}+\mathbf{B}\cdot\mathbf{B})\cdot\delta_{ik} - E_i E_k - B_i B_k \end{pmatrix}$$

$$\delta_{ik} = \begin{pmatrix} 1 & 0 & 0 \\ 0 & 1 & 0 \\ 0 & 0 & 1 \end{pmatrix}. \tag{672}$$

Now some critical readers might ask what—on earth—the electromagnetic field tensors have to do with the strains and stresses in general materials. After all, materials are consisting of atoms.

This author answers with a counter question, namely:

Well, and what holds these atoms together? What is the field acting between them?

Yes, it is a field completely bases on electromagnetic interaction and as such, the E–B form to better and more holistically describe, interpret and understand the inner workings of materials just seems to fit perfectly.

Thereby it might come in very handy, especially in connection with questions regarding spin-glasses, spin-bio-molecules and all sorts of constructive solutions regarding quantum computer concepts, that we also already have a Dirac-based approach.

It should explicitly be noted, however, that the restriction to the E–B–energy–momentum tensor only applies for "ordinary", which is to say earthly materials as been usually treated by classical materials science. Pure photonic matter, strong or weak force dominated matter (like atom cores), degenerated matter due to huge gravity fields or even black hole material cannot be treated that way.

4.13 Something about Metric Temperature—Only Another Option

Even though we already established a thermodynamic concept on the basis of ensembles of centers of gravity in the sections above, we here want to discuss an alternative approach. Temperature is a phenomenological quantity. In order to establish a suitable metric equivalent, we start with our generalized definition of the base vectors (545). We simply assume that there are vector components for the 0-component as follows:

$$
\mathbf{g}_{\delta=0} = \frac{\partial G^j}{\partial x^0}\mathbf{e}_j = \frac{\partial G^j}{\partial t}\mathbf{e}_j = \frac{\partial G^j}{\partial t}\left\{\begin{pmatrix}1\\0\\0\\0\end{pmatrix},\begin{pmatrix}0\\1\\0\\0\end{pmatrix},\begin{pmatrix}0\\0\\1\\0\end{pmatrix},\begin{pmatrix}0\\0\\0\\1\end{pmatrix}\right\}_j . \quad (673)
$$

In order to later discuss the question whether with respect to a metric temperature equivalent we should in- or exclude the 0-component itself, here we separate the vector:

$$
\mathbf{g}_0 \equiv G_0 + V_0 = \frac{\partial G^j}{\partial t}\left\{\begin{pmatrix}1\\0\\0\\0\end{pmatrix}\right\}_j + \frac{\partial G^j}{\partial t}\left\{\begin{pmatrix}0\\1\\0\\0\end{pmatrix},\begin{pmatrix}0\\0\\1\\0\end{pmatrix},\begin{pmatrix}0\\0\\0\\1\end{pmatrix}\right\}_j . \quad (674)
$$

We shall start with the time derivative of the spatial components only.

For illustration, one might simply take spatial "points" with certain positions $\mathbf{x}_j$ at a time $t = 0$ moving with a certain velocity $\mathbf{v}_j$:

$$
V_0 = \sum_{j=0}^{N-1}\frac{\partial X^j}{\partial t}\left\{\begin{pmatrix}0\\1\\0\\0\end{pmatrix},\begin{pmatrix}0\\0\\1\\0\end{pmatrix},\begin{pmatrix}0\\0\\0\\1\end{pmatrix}\right\}_j
$$

$$
= \sum_{j=0}^{N-1}\frac{\partial X\left[\mathbf{x}_j + \mathbf{v}_j \cdot t\right]^j}{\partial t}\left\{\begin{pmatrix}0\\1\\0\\0\end{pmatrix},\begin{pmatrix}0\\0\\1\\0\end{pmatrix},\begin{pmatrix}0\\0\\0\\1\end{pmatrix}\right\}_j
$$

$$
= \sum_{j=0}^{N-1}\left\{\begin{pmatrix}0\\v_x\\0\\0\end{pmatrix}+\begin{pmatrix}0\\0\\v_y\\0\end{pmatrix}+\begin{pmatrix}0\\0\\0\\v_z\end{pmatrix}\right\}_j . \quad (675)
$$

Now we can evaluate the variance of the velocity distribution:

$$
(\Delta V)^2 = \sum_{j=0}^{N-1} \left(\frac{\partial X^j}{\partial t} \left\{ \begin{pmatrix} 0 \\ 1 \\ 0 \\ 0 \end{pmatrix}, \begin{pmatrix} 0 \\ 0 \\ 1 \\ 0 \end{pmatrix}, \begin{pmatrix} 0 \\ 0 \\ 0 \\ 1 \end{pmatrix} \right\}_j - V_0 \right)
$$

$$
\cdot \left(\frac{\partial X^j}{\partial t} \left\{ \begin{pmatrix} 0 \\ 1 \\ 0 \\ 0 \end{pmatrix}, \begin{pmatrix} 0 \\ 0 \\ 1 \\ 0 \end{pmatrix}, \begin{pmatrix} 0 \\ 0 \\ 0 \\ 1 \end{pmatrix} \right\}_j - V_0 \right)
$$

$$
= \sum_{j=0}^{N-1} \left(\left\{ \begin{pmatrix} 0 \\ v_x \\ 0 \\ 0 \end{pmatrix} + \begin{pmatrix} 0 \\ 0 \\ v_y \\ 0 \end{pmatrix} + \begin{pmatrix} 0 \\ 0 \\ 0 \\ v_z \end{pmatrix} \right\}_j - V_0 \right)
$$

$$
\cdot \left(\left\{ \begin{pmatrix} 0 \\ v_x \\ 0 \\ 0 \end{pmatrix} + \begin{pmatrix} 0 \\ 0 \\ v_y \\ 0 \end{pmatrix} + \begin{pmatrix} 0 \\ 0 \\ 0 \\ v_z \end{pmatrix} \right\}_j - V_0 \right). \tag{676}
$$

This variance could be considered a suitable proportionality to temperature. However, as in classical thermodynamics the average kinetic energy gives the temperature, we still need an expression for something metrically equivalent to the mass m.

We simply assume that the complete vector $\mathbf{g}_0$ should provide the whole and thus, define a metric-T-dependency as follows:

$$
T \sim (\Delta V)^2 = \sum_{j=0}^{N-1} \left(\frac{\partial X^j}{\partial t} \left\{ \begin{pmatrix} 0 \\ 1 \\ 0 \\ 0 \end{pmatrix}, \begin{pmatrix} 0 \\ 0 \\ 1 \\ 0 \end{pmatrix}, \begin{pmatrix} 0 \\ 0 \\ 0 \\ 1 \end{pmatrix} \right\}_j - V_0 \right)^2
$$

$$
= \sum_{j=0}^{N-1} \left(\left\{ \begin{pmatrix} 0 \\ \sqrt{\frac{m}{2}} \cdot v_x \\ 0 \\ 0 \end{pmatrix} + \begin{pmatrix} 0 \\ 0 \\ \sqrt{\frac{m}{2}} \cdot v_y \\ 0 \end{pmatrix} + \begin{pmatrix} 0 \\ 0 \\ 0 \\ \sqrt{\frac{m}{2}} \cdot v_z \end{pmatrix} \right\}_j - V_0 \right)^2, \tag{677}
$$

where we have to change the definition for V_0 accordingly to:

$$V_0 = \sum_{j=0}^{N-1} \left\{ \begin{pmatrix} 0 \\ \sqrt{\frac{m}{2}} \cdot v_x \\ 0 \\ 0 \end{pmatrix} + \begin{pmatrix} 0 \\ 0 \\ \sqrt{\frac{m}{2}} \cdot v_y \\ 0 \end{pmatrix} + \begin{pmatrix} 0 \\ 0 \\ 0 \\ \sqrt{\frac{m}{2}} \cdot v_z \end{pmatrix} \right\}_j . \tag{678}$$

It should explicitly be noted that the approach above still allows us the construction of a flat space metric . . . only, that it would be one with some kind of intrinsic temperature.

The author is aware of the fact that the simple illustration used above could be seen as problematic with respect to the restriction of the speed of light as maximum velocity. However, as this

(a) is only an illustration for the general—and correct—form (673) and
(b) we still are only talking about space points and not necessarily massive objects,

we will not consider this issue here.

Going back to the general form (673) we started with, we would have a metric temperature equivalent as follows:

$$T \sim (\Delta V)^2 = \sum_{j=0}^{N-1} \left(\frac{\partial G^j}{\partial t} \left\{ \begin{pmatrix} 0 \\ 1 \\ 0 \\ 0 \end{pmatrix}, \begin{pmatrix} 0 \\ 0 \\ 1 \\ 0 \end{pmatrix}, \begin{pmatrix} 0 \\ 0 \\ 0 \\ 1 \end{pmatrix} \right\}_j \right.$$
$$\left. - \sum_{j=0}^{N-1} \frac{\partial G^j}{\partial t} \left\{ \begin{pmatrix} 0 \\ 1 \\ 0 \\ 0 \end{pmatrix}, \begin{pmatrix} 0 \\ 0 \\ 1 \\ 0 \end{pmatrix}, \begin{pmatrix} 0 \\ 0 \\ 0 \\ 1 \end{pmatrix} \right\}_j \right)^2 . \tag{679}$$

Also taking the first (time) component into account extends the expression above to:

$$T \sim (\Delta V)^2 = \sum_{j=0}^{N-1} \left(\frac{\partial G^j}{\partial t} \left\{ \begin{pmatrix} 1 \\ 0 \\ 0 \\ 0 \end{pmatrix}, \begin{pmatrix} 0 \\ 1 \\ 0 \\ 0 \end{pmatrix}, \begin{pmatrix} 0 \\ 0 \\ 1 \\ 0 \end{pmatrix}, \begin{pmatrix} 0 \\ 0 \\ 0 \\ 1 \end{pmatrix} \right\}_j \right.$$
$$\left. - \sum_{j=0}^{N-1} \frac{\partial G^j}{\partial t} \left\{ \begin{pmatrix} 1 \\ 0 \\ 0 \\ 0 \end{pmatrix}, \begin{pmatrix} 0 \\ 1 \\ 0 \\ 0 \end{pmatrix}, \begin{pmatrix} 0 \\ 0 \\ 1 \\ 0 \end{pmatrix}, \begin{pmatrix} 0 \\ 0 \\ 0 \\ 1 \end{pmatrix} \right\}_j \right)^2 . \tag{680}$$

Extension to arbitrary dimensions n should look as follows:

$$T \sim (\Delta V)^2 = \sum_{j=0}^{N-1} \left(\frac{\partial G^j}{\partial t} \left\{ \begin{pmatrix} 0 \\ 1 \\ \vdots \\ 0 \end{pmatrix}, \cdots, \begin{pmatrix} 0 \\ 0 \\ \vdots \\ 1 \end{pmatrix} \right\}_j \right.$$

$$\left. - \sum_{j=0}^{N-1} \frac{\partial G^j}{\partial t} \left\{ \begin{pmatrix} 0 \\ 1 \\ \vdots \\ 0 \end{pmatrix}, \cdots, \begin{pmatrix} 0 \\ 0 \\ \vdots \\ 1 \end{pmatrix} \right\}_j \right)^2, \qquad (681)$$

respectively:

$$T \sim (\Delta V)^2 = \sum_{j=0}^{N-1} \left(\frac{\partial G^j}{\partial t} \left\{ \begin{pmatrix} 1 \\ 0 \\ \vdots \\ 0 \end{pmatrix}, \begin{pmatrix} 0 \\ 1 \\ \vdots \\ 0 \end{pmatrix}, \cdots, \begin{pmatrix} 0 \\ 0 \\ \vdots \\ 1 \end{pmatrix} \right\}_j \right.$$

$$\left. - \sum_{j=0}^{N-1} \frac{\partial G^j}{\partial t} \left\{ \begin{pmatrix} 1 \\ 0 \\ \vdots \\ 0 \end{pmatrix}, \begin{pmatrix} 0 \\ 1 \\ \vdots \\ 0 \end{pmatrix}, \cdots, \begin{pmatrix} 0 \\ 0 \\ \vdots \\ 1 \end{pmatrix} \right\}_j \right)^2. \qquad (682)$$

4.13.1 Example

It is obvious that static quantum Eigen state solutions, like the Schrödinger hydrogen or the quantum harmonic oscillator, would not have a temperature, as their base vector solutions do not show any time dependency. Things are different with solutions of the type (579). Assuming that we have an ensemble of equally distributed forward and backward waves of matter character (negative sign before M), with velocities $\mathbf{v}_j$ in all three directions x, y, z, the derivative terms for $F[\ldots]$ would cancel each other and we would end up with an effective non-zero total velocity:

$$G^j \left[t, x, y, z \right] = e^{-i \cdot M \cdot t} \cdot F_j \left(x \pm i \cdot v_{xj} \cdot t, \, y \pm i \cdot v_{yj} \cdot t, \, z \pm i \cdot v_{zj} \cdot t \right)$$

$$\equiv e^{-i \cdot M \cdot t} \cdot F_j$$

$$\frac{\partial G^j}{\partial t} = -i \cdot M \cdot e^{-i \cdot M \cdot t} \cdot F_j + i \cdot e^{-i \cdot M \cdot t}$$

$$\cdot \left(\pm v_{xj} \cdot F_j^{(1,0,0)} \pm v_{yj} \cdot F_j^{(0,1,0)} \pm v_{zj} \cdot F_j^{(0,0,1)} \right)$$

$$V_0 = -i \cdot M \cdot e^{-c \cdot i \cdot M \cdot t} \cdot \sum_{j=0}^{N-1} F_j \cdot \left\{ \begin{pmatrix} 0 \\ 1 \\ 0 \\ 0 \end{pmatrix}, \begin{pmatrix} 0 \\ 0 \\ 1 \\ 0 \end{pmatrix}, \begin{pmatrix} 0 \\ 0 \\ 0 \\ 1 \end{pmatrix} \right\}_j. \tag{683}$$

For no particular reason, simply out of curiosity, we have started with the purely spatial form (679). For simplicity we assume $V_0 = 0$ and obtain for the metric temperature equivalent:

$$V_0 = \sum_{j=0}^{N-1} \frac{\partial G^j}{\partial t} \left\{ \begin{pmatrix} 0 \\ 1 \\ 0 \\ 0 \end{pmatrix}, \begin{pmatrix} 0 \\ 0 \\ 1 \\ 0 \end{pmatrix}, \begin{pmatrix} 0 \\ 0 \\ 0 \\ 1 \end{pmatrix} \right\}_j$$

$$\equiv 0; T \sim \sum_{j=0}^{N-1} \left(\frac{\partial G^j}{\partial t} \left\{ \begin{pmatrix} 0 \\ 1 \\ 0 \\ 0 \end{pmatrix}, \begin{pmatrix} 0 \\ 0 \\ 1 \\ 0 \end{pmatrix}, \begin{pmatrix} 0 \\ 0 \\ 0 \\ 1 \end{pmatrix} \right\}_j - 0 \right)^2, \tag{684}$$

$$T \sim -e^{-2 \cdot i \cdot M \cdot t} \cdot \sum_{j=0}^{N-1} \left(\left(v_{xj} \cdot F_j^{(1,0,0)} \right)^2 + \left(v_{yj} \cdot F_j^{(1,0,0)} \right)^2 + \left(v_{zj} \cdot F_j^{(1,0,0)} \right)^2 \right).$$
$$\tag{685}$$

Consideration of all metric, respectively base vector components on the other hand yields:

$$V_0 = \sum_{j=0}^{N-1} \frac{\partial G^j}{\partial t} \left\{ \begin{pmatrix} 1 \\ 0 \\ 0 \\ 0 \end{pmatrix}, \begin{pmatrix} 0 \\ 1 \\ 0 \\ 0 \end{pmatrix}, \begin{pmatrix} 0 \\ 0 \\ 1 \\ 0 \end{pmatrix}, \begin{pmatrix} 0 \\ 0 \\ 0 \\ 1 \end{pmatrix} \right\}_j$$

$$\equiv 0; T \sim \sum_{j=0}^{N-1} \left(\frac{\partial G^j}{\partial t} \left\{ \begin{pmatrix} 1 \\ 0 \\ 0 \\ 0 \end{pmatrix}, \begin{pmatrix} 0 \\ 1 \\ 0 \\ 0 \end{pmatrix}, \begin{pmatrix} 0 \\ 0 \\ 1 \\ 0 \end{pmatrix}, \begin{pmatrix} 0 \\ 0 \\ 0 \\ 1 \end{pmatrix} \right\}_j - 0 \right)^2, \tag{686}$$

$$T \sim -e^{-2 \cdot i \cdot M \cdot t} \cdot \sum_{j=0}^{N-1} \left(M^2 \cdot \left(F_j \right)^2 + \left(v_{xj} \cdot F_j^{(1,0,0)} \right)^2 \right.$$
$$\left. + \left(v_{yj} \cdot F_j^{(1,0,0)} \right)^2 + \left(v_{zj} \cdot F_j^{(1,0,0)} \right)^2 \right). \tag{687}$$

As for here and now we only intended to work out a way to consider temperature also within a metric mathematical environment without the introduction of centers of gravity, we will not give any discussion to the approach presented in this section. Apparently, however, the center of gravity approach seems to be much more promising than the method used here (e.g. see [159, 163]).

4.14 Incorporating Interaction

It is evident that with the introduction of centers of gravity, we should also consider the option for interaction among those various centers. This would simply be achieved by allowing the G^j to not only depend on the differential coordinates X_k as given in (613), but also to have dependencies as:

$$G^j [\ldots] = G^j \left[X_{kj} \right] \equiv G^j \left[X_{k0}, X_{k1}, \ldots, X_{ki}, \ldots, X_{k(N-1)} \right]; \quad X_{ki} \equiv x_k - \xi_{ki}.$$
$$(688)$$

This immediately renders the variation with respect to the relative or position coordinates extremely complex even in the smallest ensembles. But this is not so different from the usual complexity known from statistical mechanics. Here now we just face a quantum gravity interactive statistics, where the complete variation should read:

$$\delta_g W = \int\limits_V d^n x \overbrace{\left(R^{\delta\gamma} - \frac{1}{2} R g^{\delta\gamma} + \Lambda g^{\delta\gamma} + \kappa T^{\delta\gamma} \right)}^{\text{Relativity}}$$

$$\times \left(\underbrace{\frac{\partial G^i \left[X_{ki} \right]}{\partial x^\delta} \mathbf{e}_i \cdot \delta_x \left(\frac{\partial G^j \left[X_{kj} \right]}{\partial x^\gamma} \mathbf{e}_j \right) + \delta_x \left(\frac{\partial G^i \left[X_{ki} \right]}{\partial x^\delta} \mathbf{e}_i \right) \cdot \frac{\partial G^j \left[X_{kj} \right]}{\partial x^\gamma} \mathbf{e}_j}_{\text{Quantum}} \right.$$

$$\left. + \underbrace{\sum_{i,j=0}^{N-1} \left(\frac{\partial G^i \left[X_{ki} \right]}{\partial x^\delta} \mathbf{e}_i \cdot \delta_{\xi j} \left(\frac{\partial G^j \left[X_{kj} \right]}{\partial x^\gamma} \mathbf{e}_j \right) + \delta_{\xi i} \left(\frac{\partial G^i \left[X_{ki} \right]}{\partial x^\delta} \mathbf{e}_i \right) \cdot \frac{\partial G^j \left[X_{kj} \right]}{\partial x^\gamma} \mathbf{e}_j \right)}_{\text{2nd law of Thermodynamics \& Interaction}} \right).$$
$$(689)$$

Summing up all variations considered here in just one world formula also for N- and n-variable (perhaps open) systems would now give us the following two forms:

$$\delta_g W = \left(\int_V d^n x \left(\overbrace{[R - 2\kappa L_M - 2\Lambda] \cdot \frac{\partial \sqrt{-g}}{\partial n} + \sqrt{-g} \cdot \frac{\partial [R - 2\kappa L_M - 2\Lambda]}{\partial n}}^{\text{Thermodynamics / Exchange / Open Systems}} \right) \right) \delta n$$

$$+ \int_V d^n x \left(\overbrace{R^{\delta\gamma} - \frac{1}{2} R g^{\delta\gamma} + \Lambda g^{\delta\gamma} + \kappa T^{\delta\gamma}}^{\text{Relativity}} \right)$$

$$\times \left(\underbrace{\frac{\partial G^i [X_{ki}]}{\partial x^\delta} \mathbf{e}_i \cdot \delta_x \left(\frac{\partial G^j [X_{kj}]}{\partial x^\gamma} \mathbf{e}_j \right) + \delta_x \left(\frac{\partial G^i [X_{ki}]}{\partial x^\delta} \mathbf{e}_i \right) \cdot \frac{\partial G^j [X_{kj}]}{\partial x^\gamma} \mathbf{e}_j + \underbrace{\frac{\partial g_{\delta\gamma}}{\partial n} \delta n}_{\text{ThD}}}_{\text{Quantum}} \right.$$
$$\left. + \underbrace{\delta_N \left(\sum_{i,j=0}^{N-1} \frac{\partial G^i [X_{ki}]}{\partial x^\delta} \mathbf{e}_i \cdot \delta_{\xi j} \left(\frac{\partial G^j [X_{kj}]}{\partial x^\gamma} \mathbf{e}_j \right) + \delta_{\xi i} \left(\frac{\partial G^i [X_{ki}]}{\partial x^\delta} \mathbf{e}_i \right) \cdot \frac{\partial G^j [X_{kj}]}{\partial x^\gamma} \mathbf{e}_j \right)}_{\text{2}^{\text{nd}}\text{ law of Thermodynamics \& Interaction}} \right)$$

$$(690)$$

or somewhat simpler with the *N*-variation as separate addend:

$$\delta_g W = \left(\int_V d^n x \left(\overbrace{[R - 2\kappa L_M - 2\Lambda] \cdot \frac{\partial \sqrt{-g}}{\partial n} + \sqrt{-g} \cdot \frac{\partial [R - 2\kappa L_M - 2\Lambda]}{\partial n}}^{\text{Thermodynamics / Exchange / Open Systems}} \right) \right) \delta n$$

$$+ \int_V d^n x \left(\overbrace{R^{\delta\gamma} - \frac{1}{2} R g^{\delta\gamma} + \Lambda g^{\delta\gamma} + \kappa T^{\delta\gamma}}^{\text{Relativity}} \right)$$

$$\times \left(\underbrace{\frac{\partial G^i [X_{ki}]}{\partial x^\delta} \mathbf{e}_i \cdot \delta_x \left(\frac{\partial G^j [X_{kj}]}{\partial x^\gamma} \mathbf{e}_j \right) + \delta_x \left(\frac{\partial G^i [X_{ki}]}{\partial x^\delta} \mathbf{e}_i \right) \cdot \frac{\partial G^j [X_{kj}]}{\partial x^\gamma} \mathbf{e}_j + \underbrace{\frac{\partial g_{\delta\gamma}}{\partial n} \delta n}_{\text{ThD}}}_{\text{Quantum}} \right.$$
$$\left. + \underbrace{\frac{\partial g_{\alpha\beta}}{\partial N} \delta_N + \sum_{i,j=0}^{N-1} \left(\frac{\partial G^i [X_{ki}]}{\partial x^\delta} \mathbf{e}_i \cdot \delta_{\xi j} \left(\frac{\partial G^j [X_{kj}]}{\partial x^\gamma} \mathbf{e}_j \right) + \delta_{\xi i} \left(\frac{\partial G^i [X_{ki}]}{\partial x^\delta} \mathbf{e}_i \right) \cdot \frac{\partial G^j [X_{kj}]}{\partial x^\gamma} \mathbf{e}_j \right)}_{\text{2}^{\text{nd}}\text{ law of Thermodynamics \& Interaction}} \right).$$

$$(691)$$

4.15 Derivation of the Diffusion Equation

As a very simple example, here we want to repeat the essentials of the derivation of the diffusion equation from the fundamental Eq. (691) and demand the following boundary conditions to be fulfilled:

$$\int_V d^n x \left([R - 2\kappa L_M - 2\Lambda] \cdot \frac{\partial \sqrt{-g}}{\partial n} + \sqrt{-g} \cdot \frac{\partial [R - 2\kappa L_M - 2\Lambda]}{\partial n} \right) \cdot \delta n = 0$$

$$\frac{\partial g_{\alpha\beta}}{\partial n} \delta n = 0$$

$$\frac{\partial G^i [X_{ki}]}{\partial x^\delta} \mathbf{e}_i \cdot \delta_x \left(\frac{\partial G^j [X_{kj}]}{\partial x^\gamma} \mathbf{e}_j \right) + \delta_x \left(\frac{\partial G^i [X_{ki}]}{\partial x^\delta} \mathbf{e}_i \right) \cdot \frac{\partial G^j [X_{kj}]}{\partial x^\gamma} \mathbf{e}_j = 0,$$

$$(692)$$

which results in the remaining equation:

$$\frac{\partial g_{\delta\gamma}}{\partial N} \delta_N + \sum_{i,j=0}^{N-1} \left(\frac{\partial G^i [X_{ki}]}{\partial x^\delta} \mathbf{e}_i \cdot \delta_{\xi j} \left(\frac{\partial G^j [X_{kj}]}{\partial x^\gamma} \mathbf{e}_j \right) \right.$$

$$\left. + \delta_{\xi i} \left(\frac{\partial G^i [X_{ki}]}{\partial x^\delta} \mathbf{e}_i \right) \cdot \frac{\partial G^j [X_{kj}]}{\partial x^\gamma} \mathbf{e}_j \right) = 0. \qquad (693)$$

For the sake of simplicity, we ignore any particle interaction and reduce (693) to:

$$\frac{\partial g_{\delta\gamma}}{\partial N} \delta_N + \frac{\partial G^i [X_k]}{\partial x^\delta} \mathbf{e}_i \cdot \delta_\xi \left(\frac{\partial G^j [X_k]}{\partial x^\gamma} \mathbf{e}_j \right) + \delta_\xi \left(\frac{\partial G^i [X_k]}{\partial x^\delta} \mathbf{e}_i \right) \cdot \frac{\partial G^j [X_k]}{\partial x^\gamma} \mathbf{e}_j = 0,$$

$$(694)$$

with the definition for X_k as given in Eq. (613). Similar to our evaluation from section "Motivation" in [159] (see also section "2.2 Intelligent Zero Approaches—Just one Example" in this book), we perform the variation of the second and third addends in (694) as follows:

$$G^j [X_k] \equiv G^j; \quad \frac{\partial G^i}{\partial x^\delta} \mathbf{e}_i \cdot \delta_\xi \left(\frac{\partial G^j}{\partial x^\gamma} \mathbf{e}_j \right) + \delta_\xi \left(\frac{\partial G^i}{\partial x^\delta} \mathbf{e}_i \right) \cdot \frac{\partial G^j}{\partial x^\gamma} \mathbf{e}_j$$

$$= -C_\delta^\gamma \frac{\partial G^i}{\partial x^\gamma} \mathbf{e}_i \cdot \left(\frac{\partial^2 G^j}{\partial x^\sigma \partial x^\gamma} \mathbf{e}_j \right) \delta^\sigma - \left(\frac{\partial^2 G^i}{\partial x^\sigma \partial x^\delta} \mathbf{e}_i \right) \delta^\sigma \cdot \frac{\partial G^j}{\partial x^\gamma} \mathbf{e}_j.$$

$$(695)$$

Exchanging the dummy indices and assuming $C_\delta^\gamma = b \cdot \delta_\delta^\gamma$ leads to:

$$= -\left(C_\delta^\gamma \left(\frac{\partial^2 G^i}{\partial x^\sigma \partial x^\gamma} \mathbf{e}_i \right) + \left(\frac{\partial^2 G^i}{\partial x^\sigma \partial x^\delta} \mathbf{e}_i \right) \right) \delta^\sigma \cdot \frac{\partial G^j}{\partial x^\gamma} \mathbf{e}_j$$

$$= -(b+1) \left(\frac{\partial^2 G^i}{\partial x^\sigma \partial x^\delta} \mathbf{e}_i \right) \delta^\sigma \cdot \frac{\partial G^j}{\partial x^\gamma} \mathbf{e}_j$$

$$= -(b+1) C_\sigma^\delta \left(\frac{\partial^2 G^i}{\partial x^\sigma \partial x^\delta} \mathbf{e}_i \right) \delta^\sigma \cdot \frac{\partial G^j}{\partial x^\gamma} \mathbf{e}_j$$

$$= -(b+1) \cdot b \cdot \delta_\sigma^\delta \left(\frac{\partial^2 G^i}{\partial x^\sigma \partial x^\delta} \mathbf{e}_i \right) \delta^\sigma \cdot \frac{\partial G^j}{\partial x^\gamma} \mathbf{e}_j$$

$$= -(b+1) \cdot b \left(\frac{\partial^2 G^i}{(\partial x^\sigma)^2} \mathbf{e}_i \right) \delta^\sigma \cdot \frac{\partial G^j}{\partial x^\gamma} \mathbf{e}_j$$

$$= -(b+1) \cdot b \left(\frac{\partial^2 G^i \mathbf{e}_i}{(\partial x^\sigma)^2} \right) \delta^\sigma \cdot \frac{\partial G^j}{\partial x^\gamma} \mathbf{e}_j$$

$$= -(b+1) \cdot b \cdot \Delta \left(G^i \mathbf{e}_i \right) \delta^\sigma \cdot \frac{\partial G^j}{\partial x^\gamma} \mathbf{e}_j. \tag{696}$$

Now we assume the variation regarding N to result in zeros with respect to all spatial derivatives. The justification for this is the typical continuity equation approach where we demand that under the volume considered in our variational integral (691) any change of N does not change the derivatives with respect to the spatial coordinates, meaning:

$$\frac{\partial g_{\delta\gamma}}{\partial N} \delta_N = \frac{\partial \left(\frac{\partial G^j}{\partial x^\gamma} \mathbf{e}_j \frac{\partial G^i}{\partial x^\delta} \mathbf{e}_i \right)}{\partial N} \delta_N = 0; \quad \delta, \gamma \neq 0. \tag{697}$$

With respect to the time-coordinate, however, we can have very complex dependencies. The simplest would be a linear one like:

$$\frac{\partial \left(\frac{\partial G^j}{\partial x^0} \mathbf{e}_j \right)}{\partial N} \delta_N = D^{-1} \cdot \frac{\partial G^j}{\partial x^0} \mathbf{e}_j \cdot \delta^\delta. \tag{698}$$

This gives us the following for the first addend:

$$\frac{\partial g_{\delta\gamma}}{\partial N} \delta_N = D^{-1} \cdot \left(\frac{\partial G^i}{\partial x^\delta} \mathbf{e}_i \cdot \frac{\partial G^j}{\partial x^0} \mathbf{e}_j + \frac{\partial G^i}{\partial x^0} \mathbf{e}_i \cdot \frac{\partial G^j}{\partial x^\gamma} \mathbf{e}_j \right) \cdot \delta^\delta. \tag{699}$$

In connection with the result from Eq. (696), again using $C_\delta^\gamma = b \cdot \delta_\delta^\gamma$ and exchanging the dummy indices, we obtain:

$$0 = D^{-1} \cdot \left(\frac{\partial G^i}{\partial x^\delta} \mathbf{e}_i \cdot \frac{\partial G^j}{\partial x^0} \mathbf{e}_j + \frac{\partial G^i}{\partial x^0} \mathbf{e}_i \cdot \frac{\partial G^j}{\partial x^\gamma} \mathbf{e}_j \right) \cdot \delta^\sigma$$

$$- (b+1) \cdot b \cdot \Delta \left(G^i \mathbf{e}_i \right) \delta^\sigma \cdot \frac{\partial G^j}{\partial x^\gamma} \mathbf{e}_j$$

$$= D^{-1} \cdot (b+1) \cdot \left(\frac{\partial G^i}{\partial x^0} \mathbf{e}_i \right) \cdot \frac{\partial G^j}{\partial x^\gamma} \mathbf{e}_j \cdot \delta^\sigma - (b+1) \cdot b \cdot \Delta \left(G^i \mathbf{e}_i \right) \delta^\sigma \cdot \frac{\partial G^j}{\partial x^\gamma} \mathbf{e}_j$$

$$= (b+1) \cdot \left(\left(\frac{\partial G^i}{\partial x^0} \mathbf{e}_i \right) - D \cdot b \cdot \Delta \left(G^i \mathbf{e}_i \right) \right) \delta^\sigma \cdot \frac{\partial G^j}{\partial x^\gamma} \mathbf{e}_j$$

$$\Rightarrow \quad \left(\frac{\partial G^i}{\partial x^0} \mathbf{e}_i \right) = D \cdot b \cdot \Delta \left(G^i \mathbf{e}_i \right), \tag{700}$$

where we recognize the diffusion equation in its simplest (homogeneous) form. Along the way of our derivation, we can easily make out passages where we can incorporate anisotropy and inhomogeneity in very general manners.

REPETITION

Chapter 5

Reconsideration of the Ordinary Derivative x^k Variation

5.1 Scalar Approach: Pre-Considerations

Similar to the procedure applied in connection with the additive metric variation above, we rewrite Eq. (389) as follows:

$$\delta_\gamma W = 0 = \int_V d^n x \left(\overbrace{R^{\delta\gamma} \left[\gamma^{\delta\gamma}\right] - \frac{1}{2} R \cdot \gamma^{\delta\gamma} + \Lambda \cdot \gamma^{\delta\gamma} + \kappa \cdot T^{\delta\gamma} \left[\gamma^{\delta\gamma}\right]}^{\text{Relativity}} \right)$$

$$\times \underbrace{\left(\begin{array}{l} \delta g_{\delta\gamma} = \delta \left(\frac{\partial G^i \left[x_{\text{vi}}\right]}{\partial x^\delta} \cdot \frac{\partial G^j \left[x_{\text{vi}}\right]}{\partial x^\gamma} \gamma_{ij} \right) = \gamma_{ij} \cdot \delta \left(\frac{\partial G^i \left[x_{\text{vi}}\right]}{\partial x^\delta} \cdot \frac{\partial G^j \left[x_{\text{vi}}\right]}{\partial x^\gamma} \right) \\ = \gamma_{ij} \cdot \left(\frac{\partial^2 G^i \left[x_{\text{vi}}\right]}{\partial x^\sigma \partial x^\delta} \cdot \frac{\partial G^j \left[x_{\text{vi}}\right]}{\partial x^\gamma} + \frac{\partial G^i \left[x_{\text{vi}}\right]}{\partial x^\delta} \cdot \frac{\partial^2 G^j \left[x_{\text{vi}}\right]}{\partial x^\sigma \partial x^\gamma} \right) \delta^\sigma \end{array} \right)}_{\text{Quantum}}.$$

(701)

Thereby, just as done in the section "3.12 Connection with the Extended Einstein–Hilbert Variation (389)", we used the transformed coordinates only in connection with the variated metric, because this leads to simple products of the untransformed metric $\gamma_{\delta\gamma}$. Applying the same abbreviations as used in

The World Formula: A Late Recognition of David Hilbert's Stroke of Genius
Norbert Schwarzer
Copyright © 2022 Jenny Stanford Publishing Pte. Ltd.
ISBN 978-981-4877-20-6 (Hardcover), 978-1-003-14644-5 (eBook)
www.jennystanford.com

this book before, we can evolve the Einstein–Hilbert action as follows:

$$\delta_\gamma W = 0$$

$$= \int_V d^n x \, (\ldots)^{\delta\gamma} \gamma_{ij} \cdot \left(\frac{\partial^2 G^i \, [x_{\nabla i}]}{\partial x^\sigma \partial x^\delta} \cdot \frac{\partial G^j \, [x_{\nabla i}]}{\partial x^\gamma} + \frac{\partial G^i \, [x_{\nabla i}]}{\partial x^\delta} \cdot \frac{\partial^2 G^j \, [x_{\nabla i}]}{\partial x^\sigma \partial x^\gamma} \right) \delta^\sigma$$

$$= \int_V d^n x \, (\ldots)^{\delta\gamma} \gamma_{ij} \cdot \left(\frac{\partial^2 G^i}{\partial x^\sigma \partial x^\delta} \cdot \frac{\partial G^j}{\partial x^\gamma} + \frac{\partial G^i}{\partial x^\delta} \cdot \frac{\partial^2 G^j}{\partial x^\sigma \partial x^\gamma} \right) \delta^\sigma$$

$$= \int_V d^n x \cdot \mathbf{H}^\delta \cdot \mathbf{H}^\gamma \cdot \mathbf{e}_i \cdot \mathbf{e}_j \cdot \left(\frac{\partial^2 G^i}{\partial x^\sigma \partial x^\delta} \cdot \frac{\partial G^j}{\partial x^\gamma} + \frac{\partial G^i}{\partial x^\delta} \cdot \frac{\partial^2 G^j}{\partial x^\sigma \partial x^\gamma} \right) \delta^\sigma$$

$$= \int_V d^n x \cdot \left(A \cdot S_i^\delta S_j^\gamma + B \cdot S_i^\gamma S_j^\delta \right) \left(\frac{\partial^2 G^i}{\partial x^\sigma \partial x^\delta} \cdot \frac{\partial G^j}{\partial x^\gamma} + \frac{\partial G^i}{\partial x^\delta} \cdot \frac{\partial^2 G^j}{\partial x^\sigma \partial x^\gamma} \right) \delta^\sigma \, ;$$

$$A + B = 1. \tag{702}$$

For illustration, we consider the simplest case, where we set $(\ldots)^{\delta\gamma} = D \cdot \gamma^{\delta\gamma}$ and obtain:

$$\delta_\gamma W = 0$$

$$= \int_V d^n x \cdot D \left(A \cdot \delta_i^\delta \delta_j^\gamma + B \cdot \delta_i^\gamma \delta_j^\delta \right) \left(\frac{\partial^2 G^i}{\partial x^\sigma \partial x^\delta} \cdot \frac{\partial G^j}{\partial x^\gamma} + \frac{\partial G^i}{\partial x^\delta} \cdot \frac{\partial^2 G^j}{\partial x^\sigma \partial x^\gamma} \right) \delta^\sigma$$

$$= \int_V d^n x \cdot D \left(A \left(\frac{\partial^2 G^\delta}{\partial x^\sigma \partial x^\delta} \cdot \frac{\partial G^\gamma}{\partial x^\gamma} + \frac{\partial G^\delta}{\partial x^\delta} \cdot \frac{\partial^2 G^\gamma}{\partial x^\sigma \partial x^\gamma} \right) \right.$$

$$\left. + B \left(\frac{\partial^2 G^\gamma}{\partial x^\sigma \partial x^\delta} \cdot \frac{\partial G^\delta}{\partial x^\gamma} + \frac{\partial G^\gamma}{\partial x^\delta} \cdot \frac{\partial^2 G^\delta}{\partial x^\sigma \partial x^\gamma} \right) \right) \delta^\sigma. \tag{703}$$

Now we substitute $\frac{\partial G^\gamma}{\partial x^{\gamma:\delta}} = \frac{\partial \sqrt{\gamma} \cdot \gamma^{\gamma\alpha} \cdot F_{,\alpha}}{\partial x^{\gamma:\delta}}$; $\frac{\partial G^\delta}{\partial x^{\gamma:\delta}} = \frac{\partial \sqrt{\gamma} \cdot \gamma^{\delta\alpha} \cdot F_{,\alpha}}{\partial x^{\gamma:\delta}}$ and inserting in Eq. (703) gives us:

$$\delta_\gamma W = 0$$

$$= \int_V d^n x \cdot \left(2A \left(\sqrt{\gamma} \Delta F \right)_{,\sigma} \cdot \sqrt{\gamma} \Delta F + B \left(\begin{array}{c} \frac{\partial^2 \left(\sqrt{\gamma} \cdot \gamma^{\gamma\alpha} \cdot F_{,\alpha} \right)}{\partial x^\sigma \partial x^\delta} \cdot \frac{\partial \sqrt{\gamma} \cdot \gamma^{\delta\alpha} \cdot F_{,\alpha}}{\partial x^\gamma} \\ + \frac{\partial \sqrt{\gamma} \cdot \gamma^{\gamma\alpha} \cdot F_{,\alpha}}{\partial x^\delta} \cdot \frac{\partial^2 \left(\sqrt{\gamma} \cdot \gamma^{\delta\alpha} \cdot F_{,\alpha} \right)}{\partial x^\sigma \partial x^\gamma} \end{array} \right) \right) \delta^\sigma$$

$$= \int_V d^n x \cdot \left(A\gamma_{,\sigma} \cdot \Delta F^2 + 2A \cdot \gamma \cdot \Delta F_{,\sigma} \Delta F + B \left(\begin{array}{c} \frac{\partial^2 \left(\sqrt{\gamma} \cdot \gamma^{\gamma\alpha} \cdot F_{,\alpha} \right)}{\partial x^\sigma \partial x^\delta} \cdot \frac{\partial \sqrt{\gamma} \cdot \gamma^{\delta\alpha} \cdot F_{,\alpha}}{\partial x^\gamma} \\ + \frac{\partial \sqrt{\gamma} \cdot \gamma^{\gamma\alpha} \cdot F_{,\alpha}}{\partial x^\delta} \cdot \frac{\partial^2 \left(\sqrt{\gamma} \cdot \gamma^{\delta\alpha} \cdot F_{,\alpha} \right)}{\partial x^\sigma \partial x^\gamma} \end{array} \right) \right) \delta^\sigma. \tag{704}$$

Without further ado, we can obtain a "Klein–Gordon like" equation as follows:

$$0 = \Delta F + \frac{2 \cdot \gamma \cdot \Delta F_{,\sigma}}{\gamma_{,\sigma}} + \frac{B}{A \cdot \gamma_{,\sigma} \cdot \Delta F} \left(\frac{\frac{\partial^2 \left(\sqrt{\gamma} \cdot \gamma^{\gamma\alpha} \cdot F_{,\alpha} \right)}{\partial x^{\sigma} \partial x^{\delta}} \cdot \frac{\partial \sqrt{\gamma} \cdot \gamma^{\delta\alpha} \cdot F_{,\alpha}}{\partial x^{\gamma}}}{+ \frac{\partial \sqrt{\gamma} \cdot \gamma^{\gamma\alpha} \cdot F_{,\alpha}}{\partial x^{\delta}} \cdot \frac{\partial^2 \left(\sqrt{\gamma} \cdot \gamma^{\delta\alpha} \cdot F_{,\alpha} \right)}{\partial x^{\sigma} \partial x^{\gamma}}} \right). \quad (705)$$

There we used the "..." signs, because we are missing the absolute F, which would assure the Eigen-character of the true Klein–Gordon equation. We therefore now apply the substitution:

$$\frac{\partial G^{\gamma}}{\partial x^{\gamma:\delta}} = \frac{\partial \left(C^{\gamma} \cdot F + \sqrt{\gamma} \cdot \gamma^{\gamma\alpha} \cdot F_{,\alpha} \right)}{\partial x^{\gamma:\delta}}; \quad \frac{\partial G^{\delta}}{\partial x^{\gamma:\delta}} = \frac{\partial \left(C^{\delta} \cdot F + \sqrt{\gamma} \cdot \gamma^{\delta\alpha} \cdot F_{,\alpha} \right)}{\partial x^{\gamma:\delta}}. \quad (706)$$

Substitution into Eq. (703) yields:

$$\delta_{\gamma} W = 0 = \int_{V} d^n x \cdot \left(\begin{array}{c} A \left(\frac{\frac{\partial^2 \left(F \cdot C^{\delta} + \sqrt{\gamma} \cdot \gamma^{\delta\alpha} \cdot F_{,\alpha} \right)}{\partial x^{\sigma} \partial x^{\delta}} \cdot \frac{\partial \left(F \cdot C^{\gamma} + \sqrt{\gamma} \cdot \gamma^{\gamma\alpha} \cdot F_{,\alpha} \right)}{\partial x^{\gamma}}}{+ \frac{\partial \left(F \cdot C^{\delta} + \sqrt{\gamma} \cdot \gamma^{\delta\alpha} \cdot F_{,\alpha} \right)}{\partial x^{\delta}} \cdot \frac{\partial^2 \left(F \cdot C^{\gamma} + \sqrt{\gamma} \cdot \gamma^{\gamma\alpha} \cdot F_{,\alpha} \right)}{\partial x^{\sigma} \partial x^{\gamma}}} \right) \\ + B \left(\frac{\frac{\partial^2 \left(F \cdot C^{\gamma} + \sqrt{\gamma} \cdot \gamma^{\gamma\alpha} \cdot F_{,\alpha} \right)}{\partial x^{\sigma} \partial x^{\delta}} \cdot \frac{\partial \left(F \cdot C^{\delta} + \sqrt{\gamma} \cdot \gamma^{\delta\alpha} \cdot F_{,\alpha} \right)}{\partial x^{\gamma}}}{+ \frac{\partial \left(F \cdot C^{\gamma} + \sqrt{\gamma} \cdot \gamma^{\gamma\alpha} \cdot F_{,\alpha} \right)}{\partial x^{\delta}} \cdot \frac{\partial^2 \left(F \cdot C^{\delta} + \sqrt{\gamma} \cdot \gamma^{\delta\alpha} \cdot F_{,\alpha} \right)}{\partial x^{\sigma} \partial x^{\gamma}}} \right) \end{array} \right) \delta^{\sigma}. \quad (707)$$

As before, we can dramatically simplify the first addend:

$$\delta_{\gamma} W = 0 = \int_{V} d^n x \cdot \left(\begin{array}{c} 2A \left(C^{\alpha} F_{,\alpha} \cdot \left(C^{\alpha} F_{,\alpha} \right)_{,\sigma} + \left(\sqrt{\gamma} \Delta F \right)_{,\sigma} \cdot \sqrt{\gamma} \Delta F \right) \\ + B \left(\frac{\frac{\partial^2 \left(F \cdot C^{\gamma} + \sqrt{\gamma} \cdot \gamma^{\gamma\alpha} \cdot F_{,\alpha} \right)}{\partial x^{\sigma} \partial x^{\delta}} \cdot \frac{\partial \left(F \cdot C^{\delta} + \sqrt{\gamma} \cdot \gamma^{\delta\alpha} \cdot F_{,\alpha} \right)}{\partial x^{\gamma}}}{+ \frac{\partial \left(F \cdot C^{\gamma} + \sqrt{\gamma} \cdot \gamma^{\gamma\alpha} \cdot F_{,\alpha} \right)}{\partial x^{\delta}} \cdot \frac{\partial^2 \left(F \cdot C^{\delta} + \sqrt{\gamma} \cdot \gamma^{\delta\alpha} \cdot F_{,\alpha} \right)}{\partial x^{\sigma} \partial x^{\gamma}}} \right) \end{array} \right) \delta^{\sigma}. \quad (708)$$

We still do not see how we could obtain a "classical" Klein–Gordon (Eigen-) equation from this structure. If, however, assuming the constant vector $\mathbf{C}^{\alpha}$ to be $\mathbf{C}^{\alpha} = \mathbf{C}^* x^{\alpha}$, then things change as follows:

$$\delta_{\gamma} W = 0 = \int_{V} d^n x \cdot \left(\begin{array}{c} 2A \left(C \left(5 \cdot F_{,\sigma} + x^{\alpha} F_{,\alpha\sigma} \right) + \left(\sqrt{\gamma} \Delta F \right)_{,\sigma} \right) \cdot \left(C \left(4 \cdot F + x^{\alpha} F_{,\alpha} \right) + \sqrt{\gamma} \Delta F \right) \\ + B \left(\frac{\frac{\partial^2 \left(F \cdot C \cdot x^{\gamma} + \sqrt{\gamma} \cdot \gamma^{\gamma\alpha} \cdot F_{,\alpha} \right)}{\partial x^{\sigma} \partial x^{\delta}} \cdot \frac{\partial \left(F \cdot C \cdot x^{\delta} + \sqrt{\gamma} \cdot \gamma^{\delta\alpha} \cdot F_{,\alpha} \right)}{\partial x^{\gamma}}}{+ \frac{\partial \left(F \cdot C \cdot x^{\gamma} + \sqrt{\gamma} \cdot \gamma^{\gamma\alpha} \cdot F_{,\alpha} \right)}{\partial x^{\delta}} \cdot \frac{\partial^2 \left(F \cdot C \cdot x^{\delta} + \sqrt{\gamma} \cdot \gamma^{\delta\alpha} \cdot F_{,\alpha} \right)}{\partial x^{\sigma} \partial x^{\gamma}}} \right) \end{array} \right) \delta^{\sigma}$$

$$= \int_{V} d^n x \cdot \left(\begin{array}{c} 2A \left(C \left(5 \cdot F_{,\sigma} + x^{\alpha} F_{,\alpha\sigma} \right) + \left(\sqrt{\gamma} \Delta F \right)_{,\sigma} \right) \cdot \left(C \left(4 \cdot F + x^{\alpha} F_{,\alpha} \right) + \sqrt{\gamma} \Delta F \right) \\ + B \left(\frac{\left(C \cdot \left(F \cdot \delta_{\delta}^{\gamma} + F_{,\delta} \cdot x^{\gamma} \right) + \frac{\partial \left(\sqrt{\gamma} \cdot \gamma^{\gamma\alpha} \cdot F_{,\alpha} \right)}{\partial x^{\delta}} \right)_{,\sigma} \cdot \left(C \cdot \left(F \cdot \delta_{\gamma}^{\delta} + F_{,\gamma} \cdot x^{\delta} \right) + \frac{\partial \left(\sqrt{\gamma} \cdot \gamma^{\delta\alpha} \cdot F_{,\alpha} \right)}{\partial x^{\gamma}} \right)}{+ \left(C \cdot \left(F \cdot \delta_{\delta}^{\gamma} + F_{,\delta} \cdot x^{\gamma} \right) + \frac{\partial \left(\sqrt{\gamma} \cdot \gamma^{\gamma\alpha} \cdot F_{,\alpha} \right)}{\partial x^{\delta}} \right) \cdot \left(C \cdot \left(F \cdot \delta_{\gamma}^{\delta} + F_{,\gamma} \cdot x^{\delta} \right) + \frac{\partial \left(\sqrt{\gamma} \cdot \gamma^{\delta\alpha} \cdot F_{,\alpha} \right)}{\partial x^{\gamma}} \right)_{,\sigma}} \right) \end{array} \right) \delta^{\sigma}. \quad (709)$$

In order to get a better overview, we assume the first derivatives to vanish and can simplify as follows:

$$\delta_\gamma W = 0 = \int_V d^n x \cdot C^2 \left(\begin{array}{l} 2A\,(5 \cdot F_{,\sigma} + x^\alpha F_{,\alpha\sigma}) \cdot (4 \cdot F + x^\alpha F_{,\alpha}) \\ +B \left(\begin{array}{l} \left(F \cdot \delta_\delta^\gamma + F_{,\delta} \cdot x^\gamma \right)_{,\sigma} \cdot \left(F \cdot \delta_\gamma^\delta + F_{,\gamma} \cdot x^\delta \right) \\ + \left(F \cdot \delta_\delta^\gamma + F_{,\delta} \cdot x^\gamma \right) \cdot \left(F \cdot \delta_\gamma^\delta + F_{,\gamma} \cdot x^\delta \right)_{,\sigma} \end{array} \right) \end{array} \right) \delta^\sigma$$

$$= \int_V d^n x \cdot C^2 \left(\begin{array}{l} 2A\,(5 \cdot F_{,\sigma} + x^\alpha F_{,\alpha\sigma}) \cdot (4 \cdot F + x^\alpha F_{,\alpha}) \\ +B \left(\begin{array}{l} \left(F_{,\sigma} \cdot \left(\delta_\delta^\gamma + \delta_\sigma^\gamma \right) + F_{,\delta\sigma} \cdot x^\gamma \right) \cdot \left(F \cdot \delta_\gamma^\delta + F_{,\gamma} \cdot x^\delta \right) \\ + \left(F \cdot \delta_\delta^\gamma + F_{,\delta} \cdot x^\gamma \right) \cdot \left(F_{,\sigma} \cdot \left(\delta_\gamma^\delta + \delta_\sigma^\delta \right) + F_{,\gamma\sigma} \cdot x^\delta \right) \end{array} \right) \end{array} \right) \delta^\sigma$$

$$= \int_V d^n x \cdot C^2 \left(2A\,(x^\alpha F_{,\alpha\sigma}) \cdot (4 \cdot F) + B \left(\begin{array}{l} \left(F_{,\delta\sigma} \cdot x^\gamma \right) \cdot \left(F \cdot \delta_\gamma^\delta \right) \\ + \left(F \cdot \delta_\delta^\gamma \right) \cdot \left(F_{,\gamma\sigma} \cdot x^\delta \right) \end{array} \right) \right) \delta^\sigma$$

$$= \int_V d^n x \cdot C^2 \left(2A\,(x^\alpha F_{,\alpha\sigma}) \cdot (4 \cdot F) + B\,(2x^\alpha F_{,\alpha\sigma} \cdot F) \right) \delta^\sigma. \tag{710}$$

We have to realize that something seems to be wrong with our obviously too simple approach, as we are unable to reproduce the classical Klein–Gordon equation in a suitable form as we were so easily able to with the additive base vector approach in section "3.12 Connection with the Extended Einstein–Hilbert Variation (389)".

5.1.1 Scalar Approach: A Trial

So, what could we do?

So far, we have not used the degree of freedom offering itself in the definition of the structure of the functions G^j. Thus, going back to Eq. (701) and applying the method of vectorial root extraction as introduced in [8], for instance, we can write:

$$\delta_\gamma W = 0 = \int_V d^n x \sqrt{\gamma} \cdot \left(R^{\delta\gamma} \left[\gamma^{\delta\gamma} \right] - \frac{1}{2} R \cdot \gamma^{\delta\gamma} + \Lambda \cdot \gamma^{\delta\gamma} + \kappa \cdot T^{\delta\gamma} \left[\gamma^{\delta\gamma} \right] \right)$$

$$\times \left(\begin{array}{l} \delta g_{\delta\gamma} = \delta \left(\frac{\partial G^i \,[x_{\forall j}]}{\partial x^\delta} \cdot \frac{\partial G^j \,[x_{\forall j}]}{\partial x^\gamma} \gamma_{ij} \right) = \gamma_{ij} \cdot \delta \left(\frac{\partial G^i \,[x_{\forall j}]}{\partial x^\delta} \cdot \frac{\partial G^j \,[x_{\forall j}]}{\partial x^\gamma} \right) \\[2mm] = \frac{\gamma_{ij}}{4} \delta \left(\left\{ \begin{array}{l} \delta_\delta^i + C_\delta^i \cdot f + i \cdot C^i \cdot f_{,\delta} \\ \delta_\delta^i + C_\delta^i \cdot f - i \cdot C^i \cdot f_{,\delta} \\ \delta_\delta^i - C_\delta^i \cdot f + i \cdot C^i \cdot f_{,\delta} \\ \delta_\delta^i - C_\delta^i \cdot f - i \cdot C^i \cdot f_{,\delta} \end{array} \right\} \left\{ \begin{array}{l} \delta_\gamma^j + C_\gamma^j \cdot f + i \cdot C^j \cdot f_{,\gamma} \\ \delta_\gamma^j + C_\gamma^j \cdot f - i \cdot C^j \cdot f_{,\gamma} \\ \delta_\gamma^j - C_\gamma^j \cdot f + i \cdot C^j \cdot f_{,\gamma} \\ \delta_\gamma^j - C_\gamma^j \cdot f - i \cdot C^j \cdot f_{,\gamma} \end{array} \right\} \right) \\[2mm] = \delta \left(\delta_\delta^i \delta_\gamma^j + C_\delta^i C_\gamma^j \cdot f^2 - C^i C^j \cdot f_{,\delta} f_{,\gamma} \right) \gamma_{ij} \\[2mm] = \left(2 C_\delta^i C_\gamma^j \cdot f \cdot f_{,\sigma} - C^i C^j \cdot \left(f_{,\delta\sigma} f_{,\gamma} + f_{,\delta} f_{,\gamma\sigma} \right) \right) \gamma_{ij} \delta^\sigma \end{array} \right). \tag{711}$$

Now, comparing with Eq. (498), we see the connection to our additive base vectors and the variation with respect to the arbitrary or even virtual parameter y (c.f. section "3.12 Connection with the Extended Einstein–Hilbert Variation (389)"). The only significant difference is that now we did not need the introduction of the virtual parameter, but directly variated with respect to the coordinates respectively dimensions. Our problem was that we could not factor out the derivatives with respect to the variational index σ. However, after performing the integration by parts for the integrand with the terms $C^i C^j \cdot \left(f_{,\delta\sigma} f_{,\gamma} + f_{,\delta} f_{,\gamma\sigma} \right)$ as follows:

$$\int_V d^n x \, (\ldots)^{\delta\gamma} \sqrt{\gamma} \cdot C^i C^j \cdot \gamma_{ij} f_{,\delta} f_{,\gamma\sigma}$$

$$= \int_V d^n x \left\{ \left[(\ldots)^{\delta\gamma} \sqrt{\gamma} \cdot C^i C^j \cdot \gamma_{ij} f_{,\delta} \right] \cdot f_{,\sigma} \right\}_{,\gamma}$$

$$\underbrace{\phantom{= \int_V d^n x \left\{ \left[(\ldots)^{\delta\gamma} \sqrt{\gamma} \cdot C^i C^j \cdot \gamma_{ij} f_{,\delta} \right] \cdot f_{,\sigma} \right\}_{,\gamma}}}_{=S=0 \text{ per definition}}$$

$$- \int_V d^n x \left[(\ldots)^{\delta\gamma} \sqrt{\gamma} \cdot C^i C^j \cdot \gamma_{ij} f_{,\delta} \right]_{,\gamma} \cdot f_{,\sigma}$$

$$\simeq - \int_V d^n x \cdot \left[D \cdot \gamma^{\delta\gamma} \sqrt{\gamma} \cdot C^i C^j \cdot \gamma_{ij} f_{,\delta} \right]_{,\gamma} \cdot f_{,\sigma}$$

with : $\quad (\ldots)^{\delta\gamma} = \left(R^{\delta\gamma} - \frac{1}{2} R \cdot \gamma^{\delta\gamma} + \Lambda \cdot \gamma^{\delta\gamma} + \kappa \cdot T^{\delta\gamma} \right) \simeq D \cdot \gamma^{\delta\gamma},$ (712)

we can further evaluate:

$$0 = \int_V d^n x \sqrt{\gamma} \cdot \left(2 \cdot (\ldots)^{\delta\gamma} C_{\delta j} C_{\gamma}^j \cdot f + \left[(\ldots)^{\delta\gamma} C^i C^j \cdot \gamma_{ij} f_{,\delta} \right]_{,\gamma} \right.$$

$$\left. + \left[(\ldots)^{\delta\gamma} C^i C^j \cdot \gamma_{ij} f_{,\gamma} \right]_{,\delta} \right) \cdot f_{,\sigma}$$

$$= \int_V d^n x \sqrt{\gamma} \cdot \left(2 \cdot (\ldots)^{\delta\gamma} C_{\delta j} C_{\gamma}^j \cdot f + \left[(\ldots)^{\delta\gamma} C_j C^j f_{,\delta} \right]_{,\gamma} \right.$$

$$\left. + \left[(\ldots)^{\delta\gamma} C_j C^j f_{,\gamma} \right]_{,\delta} \right) \cdot f_{,\sigma}$$

$$\simeq \int_V d^n x \left(2 \cdot D \cdot \sqrt{\gamma} \cdot \gamma^{\delta\gamma} C_{\delta j} C_{\gamma}^j \cdot f + \left[\sqrt{\gamma} \cdot D \cdot \gamma^{\delta\gamma} C_j C^j f_{,\delta} \right]_{,\gamma} \right.$$

$$\left. + \left[\sqrt{\gamma} \cdot D \cdot \gamma^{\delta\gamma} C_j C^j f_{,\gamma} \right]_{,\delta} \right) \cdot f_{,\sigma}$$

$$= \int_V d^n x \cdot 2 \cdot \left(D \cdot \sqrt{\gamma} \cdot \gamma^{\delta\gamma} C_{\delta j} C_{\gamma}^j \cdot f + \sqrt{\gamma} \cdot \gamma^{\delta\gamma} f_{,\delta} \left[D \cdot C_j C^j \right]_{,\gamma} \right.$$

$$+ D \cdot C_j C^j \Delta f \Big) \cdot f_{,\sigma}.$$ (713)

Thereby S denotes the usual variation surface term, which can be set to zero under the integral. Thus, we obtain the Klein–Gordon-like equation in the case of $\left[D \cdot C_j C^j\right]_{,\gamma}$ not giving zero:

$$D \cdot \sqrt{\gamma} \cdot \gamma^{\delta\gamma} C_{\delta j} C_\gamma^j \cdot f + \sqrt{\gamma} \cdot \gamma^{\delta\gamma} f_{,\delta} \left[D \cdot C_j C^j\right]_{,\gamma} + D \cdot C_j C^j \Delta f = 0. \quad (714)$$

Comparison with the classical Klein–Gordon equation with potential given as:

$$\frac{M^2 \cdot c^2}{\hbar^2} \cdot f + V \cdot f + \Delta f = 0 \quad (715)$$

reveals the metric nature of mass and potential as follows:

$$\frac{M^2 \cdot c^2}{\hbar^2} + V = \frac{D \cdot \sqrt{\gamma} \cdot \gamma^{\delta\gamma} C_{\delta j} C_\gamma^j \cdot f + \sqrt{\gamma} \cdot \gamma^{\delta\gamma} f_{,\delta} \left[D \cdot C_j C^j\right]_{,\gamma}}{D \cdot C_j C^j \cdot f}$$

$$= \frac{\sqrt{\gamma} \cdot \gamma^{\delta\gamma} C_{\delta j} C_\gamma^j}{C_j C^j} + \frac{\sqrt{\gamma} \cdot \gamma^{\delta\gamma} f_{,\delta} \left[D \cdot C_j C^j\right]_{,\gamma}}{D \cdot C_j C^j \cdot f}. \quad (716)$$

As an assumption, one might just set:

$$M^2 = \frac{\hbar^2}{c^2} \cdot \frac{\sqrt{\gamma} \cdot \gamma^{\delta\gamma} C_{\delta j} C_\gamma^j}{C_j C^j}; \quad V = \frac{\sqrt{\gamma} \cdot \gamma^{\delta\gamma} f_{,\delta} \left[D \cdot C_j C^j\right]_{,\gamma}}{D \cdot C_j C^j \cdot f}. \quad (717)$$

5.1.2 Scalar Approach: Some Refinements

Not being satisfied with the approximation:

$$(\ldots)^{\delta\gamma} = \left(R^{\delta\gamma} - \frac{1}{2}R \cdot \gamma^{\delta\gamma} + \Lambda \cdot \gamma^{\delta\gamma} + \kappa \cdot T^{\delta\gamma}\right) \simeq D \cdot \sqrt{\gamma} \cdot \gamma^{\delta\gamma} \quad (718)$$

that we used above, we apply another scalar function approach as follows:

$$\delta_\gamma W - 0 = \int_V d^n x \sqrt{\gamma} \cdot \left(R^{\delta\gamma} - \frac{1}{2}R \cdot \gamma^{\delta\gamma} + \Lambda \cdot \gamma^{\delta\gamma} + \kappa \cdot T^{\delta\gamma}\right)$$

$$\times \left(\begin{array}{l} \delta g_{\delta\gamma} = \delta\left(\dfrac{\partial G^i\,[x_{\gamma i}]}{\partial x^\delta} \cdot \dfrac{\partial G^j\,[x_{\gamma i}]}{\partial x^\gamma}\gamma_{ij}\right) = \gamma_{ij} \cdot \delta\left(\dfrac{\partial G^i\,[x_{\gamma i}]}{\partial x^\delta} \cdot \dfrac{\partial G^j\,[x_{\gamma i}]}{\partial x^\gamma}\right) \\[4pt] = \dfrac{\gamma_{ij}}{4}\delta\left(\left\{\begin{array}{l} \delta_\delta^i + M \cdot \delta_\delta^i \cdot f + \delta_\delta^i \cdot C^\alpha f_{,\alpha} \\ \delta_\delta^i + M \cdot \delta_\delta^i \cdot f - \delta_\delta^i \cdot C^\alpha f_{,\alpha} \\ \delta_\delta^i - M \cdot \delta_\delta^i \cdot f + \delta_\delta^i \cdot C^\alpha f_{,\alpha} \\ \delta_\delta^i - M \cdot \delta_\delta^i \cdot f - \delta_\delta^i \cdot C^\alpha f_{,\alpha} \end{array}\right\}\left\{\begin{array}{l} \delta_\gamma^j \mid M \cdot \delta_\gamma^j \cdot f + \delta_\gamma^j \cdot C^\beta f_{,\beta} \\ \delta_\gamma^j + M \cdot \delta_\gamma^j \cdot f - \delta_\gamma^j \cdot C^\beta f_{,\beta} \\ \delta_\gamma^j - M \cdot \delta_\gamma^j \cdot f + \delta_\gamma^j \cdot C^\beta f_{,\beta} \\ \delta_\gamma^j - M \cdot \delta_\gamma^j \cdot f - \delta_\gamma^j \cdot C^\beta f_{,\beta} \end{array}\right\}\right) \\[4pt] = \delta\left(\delta_\delta^i \delta_\gamma^j + M^2 \delta_\delta^i \delta_\gamma^j \cdot f^2 + \delta_\delta^i \delta_\gamma^j \cdot C^\alpha C^\beta \cdot f_{,\alpha} f_{,\beta}\right)\gamma_{ij} \\[4pt] = \left(2 \cdot M^2 \delta_\delta^i \delta_\gamma^j \cdot f \cdot f_{,\alpha} + \delta_\delta^i \delta_\gamma^j \cdot C^\alpha C^\beta \cdot \left(f_{,\alpha\sigma} f_{,\beta} + f_{,\alpha} f_{,\beta\sigma}\right)\right)\gamma_{ij}\delta^v \end{array}\right).$$

$$(719)$$

Thereby we introduced the symbols γ^{ab} and γ_{ab} to distinguish between the variated g^{ab} and g_{ab} and the non-variated metric γ^{ab} and γ_{ab}. As before, we seek to perform the variation under the integral in scalar functional form (applying a scalar function f and its derivatives) and align it close to the usual tensor transformations.

The special vectors $\{\ldots\}$ simply give all possible arrangements of the f-terms and their derivatives. After performing the integration by parts for the integrand with the terms $\mathbf{C}^{\alpha}\mathbf{C}^{\beta} \cdot \left(f_{,\alpha\sigma}\, f_{,\beta} + f_{,\alpha}\, f_{,\beta\sigma} \right)$ as follows:

$$
\int_V d^n x \, (\ldots)^{\delta\gamma} \, \sqrt{\gamma} \cdot \mathbf{C}^{\alpha}\mathbf{C}^{\beta} \cdot \gamma_{ij}\, f_{,\alpha}\, f_{,\beta\sigma}
$$

$$
= \int_V d^n x \, \left\{ \left[(\ldots)^{\delta\gamma}\, \sqrt{\gamma} \cdot \mathbf{C}^{\alpha}\mathbf{C}^{\beta} \cdot \gamma_{ij}\, f_{,\alpha} \right] \cdot f_{,\sigma} \right\}_{,\beta}
$$

$$
\underbrace{}_{=S=0 \text{ per definition}}
$$

$$
- \int_V d^n x \, \left[(\ldots)^{\delta\gamma}\, \sqrt{\gamma} \cdot \mathbf{C}^{\alpha}\mathbf{C}^{\beta} \cdot \gamma_{ij}\, f_{,\alpha} \right]_{,\beta} \cdot f_{,\sigma}; \quad (\ldots)^{\delta\gamma}
$$

$$
= \left(R^{\delta\gamma} - \frac{1}{2} R \cdot \gamma^{\delta\gamma} + \Lambda \cdot \gamma^{\delta\gamma} + \kappa \cdot T^{\delta\gamma} \right), \tag{720}
$$

we can further evaluate Eq. (719):

$$
0 = \int_V d^n x \left(\begin{array}{l} 2 \cdot \sqrt{\gamma} \cdot (\ldots)^{\delta\gamma}\, \gamma_{\delta\gamma}\, M^2 \cdot f \\ - \left[\sqrt{\gamma} \cdot (\ldots)^{\delta\gamma}\, \mathbf{C}^{\alpha}\mathbf{C}^{\beta}\, \gamma_{\delta\gamma}\, f_{,\alpha} \right]_{,\beta} - \left[\sqrt{\gamma} \cdot (\ldots)^{\delta\gamma}\, \mathbf{C}^{\alpha}\mathbf{C}^{\beta}\, \gamma_{\delta\gamma}\, f_{,\beta} \right]_{,\alpha} \end{array} \right) \cdot f_{,\sigma}
$$

$$
= \int_V d^n x \left(\begin{array}{l} 2 \cdot \sqrt{\gamma} \cdot (\ldots)^{\delta\gamma}\, \gamma_{\delta\gamma}\, M^2 \cdot f \\ - \left[(\ldots)^{\delta\gamma}\, \gamma_{\delta\gamma} \right]_{,\beta} \sqrt{\gamma} \cdot \mathbf{C}^{\alpha}\mathbf{C}^{\beta}\, f_{,\alpha} - (\ldots)^{\delta\gamma}\, \gamma_{\delta\gamma} \left[\sqrt{\gamma} \cdot \mathbf{C}^{\alpha}\mathbf{C}^{\beta}\, f_{,\alpha} \right]_{,\beta} \\ - \left[(\ldots)^{\delta\gamma}\, \gamma_{\delta\gamma} \right]_{,\alpha} \sqrt{\gamma} \cdot \mathbf{C}^{\alpha}\mathbf{C}^{\beta}\, f_{,\beta} - (\ldots)^{\delta\gamma}\, \gamma_{\delta\gamma} \left[\sqrt{\gamma} \cdot \mathbf{C}^{\alpha}\mathbf{C}^{\beta}\, f_{,\beta} \right]_{,\alpha} \end{array} \right) \cdot f_{,\sigma}
$$

$$
\overset{\mathbf{C}^{\alpha}\mathbf{C}^{\beta} \xrightarrow{\equiv} d \cdot \gamma^{\alpha\beta}}{=} \int_V d^n x \times 2 \cdot \sqrt{\gamma} \cdot \left(\begin{array}{l} (\ldots)^{\delta\gamma}\, \gamma_{\delta\gamma}\, M^2 \cdot f - d \cdot (\ldots)^{\delta\gamma}\, \gamma_{\delta\gamma}\, \Delta f \\ - \frac{d}{2} \cdot \left(\left[(\ldots)^{\delta\gamma}\, \gamma_{\delta\gamma} \right]_{,\beta}\, \gamma^{\alpha\beta}\, f_{,\alpha} + d \cdot \left[(\ldots)^{\delta\gamma}\, \gamma_{\delta\gamma} \right]_{,\alpha}\, \gamma^{\alpha\beta}\, f_{,\beta} \right) \end{array} \right) \cdot f_{,\sigma}
$$

$$
= \int_V d^n x \times 2 \cdot \sqrt{\gamma} \cdot \left((\ldots)^{\delta\gamma}\, \gamma_{\delta\gamma}\, M^2 \cdot f - d \cdot \left((\ldots)^{\delta\gamma}\, \gamma_{\delta\gamma}\, \Delta f + \left[(\ldots)^{\delta\gamma}\, \gamma_{\delta\gamma} \right]_{,\alpha}\, f^{,\alpha} \right) \right) \cdot f_{,\sigma}.
$$

$$
\tag{721}
$$

From the last line, we extract the integrand and set it to zero, and factorizing out the scalar term $(\ldots)^{\delta\gamma}\, \gamma_{\delta\gamma}$ finally gives us

$$
M^2 \cdot f - d \cdot \left(\Delta f + \frac{\left[(\ldots)^{\delta\gamma}\, \gamma_{\delta\gamma} \right]_{,\alpha}\, f^{,\alpha}}{(\ldots)^{\delta\gamma}\, \gamma_{\delta\gamma}} \right) = 0. \tag{722}
$$

Comparison with the classical Klein–Gordon equation with mass m and potential V, here now given as:

$$\frac{m^2 \cdot c^2}{\hbar^2} \cdot f - \Delta f - V \cdot f = 0, \tag{723}$$

reveals the metric apparent nature of mass* and potential as follows:

$$\frac{M^2}{d} = \frac{m^2 \cdot c^2}{\hbar^2}; \quad V = \frac{\left[(\ldots)^{\delta\gamma} \, \gamma_{\delta\gamma} \right]_{,\alpha} f^{,\alpha}}{d \cdot (\ldots)^{\delta\gamma} \, \gamma_{\delta\gamma} \cdot f}. \tag{724}$$

In addition, we also obtain the wave function f via our Einstein–Hilbert starting point as a quasi-harmonic on the space-time dimensions.

5.1.3 Scalar Approach: Avoiding the Introduction of γ^{ab}

Not satisfied with the inconsistency of using the non-variated metric γ^{ab} and γ_{ab} within our approach (Eq. (719)), which we might take as a kind of approximation, we apply another scalar function approach as follows:

$$\delta_\gamma W = 0 = \int_V d^n x \sqrt{g} \cdot \left(R^{\delta\gamma} - \frac{1}{2} R \cdot g^{\delta\gamma} + \Lambda \cdot g^{\delta\gamma} + \kappa \cdot T^{\delta\gamma} \right)$$

$$\times \begin{pmatrix} \delta g_{\delta\gamma} = \delta \left(\frac{\partial G^i\,[x_{\forall i}]}{\partial x^\delta} \cdot \frac{\partial G^j\,[x_{\forall i}]}{\partial x^\gamma} \gamma_{ij} \right) = g_{ij} \cdot \delta \left(\frac{\partial G^i\,[x_{\forall i}]}{\partial x^\delta} \cdot \frac{\partial G^j\,[x_{\forall i}]}{\partial x^\gamma} \right) \\[4pt] = \frac{g_{ij}}{4} \delta \left(\left\{ \begin{matrix} \delta^i_\delta + M \cdot \delta^i_\delta \cdot f + \delta^i_\delta \cdot \mathbf{C}^\alpha f_{,\alpha} \\ \delta^i_\delta + M \cdot \delta^i_\delta \cdot f - \delta^i_\delta \cdot \mathbf{C}^\alpha f_{,\alpha} \\ \delta^i_\delta - M \cdot \delta^i_\delta \cdot f + \delta^i_\delta \cdot \mathbf{C}^\alpha f_{,\alpha} \\ \delta^i_\delta - M \cdot \delta^i_\delta \cdot f - \delta^i_\delta \cdot \mathbf{C}^\alpha f_{,\alpha} \end{matrix} \right\} \left\{ \begin{matrix} \delta^j_\gamma + M \cdot \delta^j_\gamma \cdot f + \delta^j_\gamma \cdot \mathbf{C}^\beta f_{,\beta} \\ \delta^j_\gamma + M \cdot \delta^j_\gamma \cdot f - \delta^j_\gamma \cdot \mathbf{C}^\beta f_{,\beta} \\ \delta^j_\gamma - M \cdot \delta^j_\gamma \cdot f + \delta^j_\gamma \cdot \mathbf{C}^\beta f_{,\beta} \\ \delta^j_\gamma - M \cdot \delta^j_\gamma \cdot f - \delta^j_\gamma \cdot \mathbf{C}^\beta f_{,\beta} \end{matrix} \right\} \right) \\[4pt] = g_{ij} \delta \left(\delta^i_\delta \delta^j_\gamma + M^2 \delta^i_\delta \delta^j_\gamma \cdot f^2 + \delta^i_\delta \delta^j_\gamma \cdot \mathbf{C}^\alpha \mathbf{C}^\beta \cdot f_{,\alpha} f_{,\beta} \right) \\[4pt] = g_{\delta\gamma} \delta \left(1 + M^2 \cdot f^2 + \mathbf{C}^\alpha \mathbf{C}^\beta \cdot f_{,\alpha} f_{,\beta} \right) \\[4pt] = g_{\delta\gamma} \left(2 \cdot M^2 \cdot f \cdot f_{,\sigma} + \mathbf{C}^\alpha \mathbf{C}^\beta \cdot \left(f_{,\alpha\sigma} f_{,\beta} + f_{,\alpha} f_{,\beta\sigma} \right) \right) \delta^\sigma \end{pmatrix}$$

$$\tag{725}$$

The special vectors $\{\ldots\}$ simply give all possible arrangements of the f-terms and their derivatives.

Please note that the step in the second line of Eq. (725) requires some discussion which we are going to present here. The simplest "justification" could be given by stating that we applied "a special form of variation", where the outcome in the end gives proof that what we did could not have been

---

*So far, the parameter M was simply introduced as a mere parameter for the linear terms of f within our variational approach (Eq. (719)). Thus, we cannot really consider Eq. (724) a revelation of the nature of mass. But we will see in the next subsection entitled "Scalar Approach: Avoiding the Introduction of γ^{ab}", to which we, in fact, are already very close.

"completely meaningless". However, as the problematic part is just the second "=" sign, meaning:

$$\delta g_{\delta\gamma} = \delta\left(\frac{\partial G^i\,[x_{\forall i}]}{\partial x^\delta} \cdot \frac{\partial G^j\,[x_{\forall i}]}{\partial x^\gamma}\gamma_{ij}\right) \xrightarrow{\text{problematic}} g_{ij}\cdot\delta\left(\frac{\partial G^i\,[x_{\forall i}]}{\partial x^\delta}\cdot\frac{\partial G^j\,[x_{\forall i}]}{\partial x^\gamma}\right),$$

(726)

we could simply argue that, as the variation is arbitrary until we have fixed its structure completely, we have simply forced all variational changes into the Jacobi matrices $\left(\frac{\partial G^i\,[x_{\forall i}]}{\partial x^\delta} \cdot \frac{\partial G^j\,[x_{\forall i}]}{\partial x^\gamma}\right)$, and thus, whatever the variation does, it has been taken on by the structural content of these matrices. In the case here (third to fifth lines in Eq. (725)), the variation acts on the functions f, its derivatives and the base vector elements $\mathbf{C}^\alpha\mathbf{C}^\beta$. In other words, we would not violate the principal idea of the variation by restricting its action onto certain structures and using degrees of freedom for simplification and/or compactification. Another, more classical, justification was presented above in section "2.2 Intelligent Zero Approaches: Just one Example" and considers the infinitesimal character of the variation with respect to the contravariant metric tensor. However, as elaborated above, this is not really of need here. Things are very simply in connection with the wrapper or wrapper-like approaches where we usually have structures of the form:

$$g_{\delta\gamma} = (1 + h\,[x_{\forall i}])^{\rho[n]}\,\gamma_{\delta\gamma}$$
$$\Rightarrow \delta g_{\delta\gamma} = \gamma_{\delta\gamma}\rho\,[n]\cdot(1 + h\,[x_{\forall i}])^{\rho[n]-1}\,\delta h\,[x_{\forall i}]$$
$$= \frac{g_{\delta\gamma}}{(1 + h\,[x_{\forall i}])}\rho\,[n]\cdot\delta h\,[x_{\forall i}].$$

(727)

Thereby the exponent $\rho[n]$ is a function of the dimension n. We see that in cases of $h \ll 1$, we can always approximate:

$$\delta g_{\delta\gamma} \simeq g_{\delta\gamma}\rho\,[n]\cdot\delta h\,[x_{\forall i}],$$

(728)

which we will frequently use below without further notice.

After performing the integration by parts for the integrand with the terms $\mathbf{C}^\alpha\mathbf{C}^\beta\cdot\left(f_{,\alpha\sigma}\,f_{,\beta} + f_{,\alpha}\,f_{,\beta\sigma}\right)$ as follows:

$$\int_V d^n x\,(\ldots)^{\delta\gamma}\,\sqrt{g}\cdot\mathbf{C}^\alpha\mathbf{C}^\beta\cdot g_{ij}\,f_{,\alpha}\,f_{,\beta\sigma}$$

$$= \int_V d^n x\,\left\{\left[(\ldots)^{\delta\gamma}\,\sqrt{g}\cdot\mathbf{C}^\alpha\mathbf{C}^\beta\cdot g_{ij}\,f_{,\alpha}\right]\cdot f_{,\sigma}\right\}_{,\beta}$$

$$\underbrace{\phantom{= \int_V d^n x\,\left\{\left[(\ldots)^{\delta\gamma}\,\sqrt{g}\cdot\mathbf{C}^\alpha\mathbf{C}^\beta\cdot g_{ij}\,f_{,\alpha}\right]\cdot f_{,\sigma}\right\}_{,\beta}}}_{=S=0\text{ per definition}}$$

$$-\int_V d^n x \left[(\ldots)^{\delta\gamma} \sqrt{g} \cdot C^\alpha C^\beta \cdot g_{ij} f_{,\alpha}\right]_{,\beta} \cdot f_{,\sigma}; \quad (\ldots)^{\delta\gamma}$$

$$= \left(R^{\delta\gamma} - \frac{1}{2} R \cdot g^{\delta\gamma} + \Lambda \cdot g^{\delta\gamma} + \kappa \cdot T^{\delta\gamma}\right), \qquad (729)$$

we can further evaluate Eq. (719):

$$0 = \int_V d^n x \begin{pmatrix} 2 \cdot \sqrt{g} \cdot (\ldots)^{\delta\gamma} g_{\delta\gamma} M^2 \cdot f \\ -\left[\sqrt{g} \cdot (\ldots)^{\delta\gamma} C^\alpha C^\beta g_{\delta\gamma} f_{,\alpha}\right]_{,\beta} - \left[\sqrt{g} \cdot (\ldots)^{\delta\gamma} C^\alpha C^\beta g_{\delta\gamma} f_{,\beta}\right]_{,\alpha} \end{pmatrix} \cdot f_{,\sigma}$$

$$= \int_V d^n x \begin{pmatrix} 2 \cdot \sqrt{g} \cdot (\ldots)^{\delta\gamma} g_{\delta\gamma} M^2 \cdot f \\ -\left[(\ldots)^{\delta\gamma} g_{\delta\gamma}\right]_{,\beta} \sqrt{g} \cdot C^\alpha C^\beta f_{,\alpha} - (\ldots)^{\delta\gamma} g_{\delta\gamma} \left[\sqrt{g} \cdot C^\alpha C^\beta f_{,\alpha}\right]_{,\beta} \\ -\left[(\ldots)^{\delta\gamma} g_{\delta\gamma}\right]_{,\alpha} \sqrt{g} \cdot C^\alpha C^\beta f_{,\beta} - (\ldots)^{\delta\gamma} g_{\delta\gamma} \left[\sqrt{g} \cdot C^\alpha C^\beta f_{,\beta}\right]_{,\alpha} \end{pmatrix} \cdot f_{,\sigma}$$

$$\xrightarrow{C^\alpha C^\beta \equiv d \cdot g^{\alpha\beta}} = \int_V d^n x \times 2 \cdot \sqrt{g} \cdot \begin{pmatrix} (\ldots)^{\delta\gamma} g_{\delta\gamma} M^2 \cdot f - d \cdot (\ldots)^{\delta\gamma} g_{\delta\gamma} \Delta f \\ -\frac{d}{2} \cdot \left(\left[(\ldots)^{\delta\gamma} g_{\delta\gamma}\right]_{,\beta} g^{\alpha\beta} f_{,\alpha} + d \cdot \left[(\ldots)^{\delta\gamma} g_{\delta\gamma}\right]_{,\alpha} g^{\alpha\beta} f_{,\beta}\right) \end{pmatrix} \cdot f_{,\sigma}$$

$$= \int_V d^n x \times 2 \cdot \sqrt{g} \cdot \left((\ldots)^{\delta\gamma} g_{\delta\gamma} M^2 \cdot f - d \cdot \left((\ldots)^{\delta\gamma} g_{\delta\gamma} \Delta f + \left[(\ldots)^{\delta\gamma} g_{\delta\gamma}\right]_{,\alpha} f^{,\alpha}\right)\right) \cdot f_{,\sigma}.$$

$$(730)$$

From the last line, we extract the integrand and set it to zero, and factorizing out the scalar term $(\ldots)^{\delta\gamma} g_{\delta\gamma}$ finally gives us:

$$M^2 \cdot f - d \cdot \left(\Delta f + \frac{\left[(\ldots)^{\delta\gamma} g_{\delta\gamma}\right]_{,\alpha} f^{,\alpha}}{(\ldots)^{\delta\gamma} g_{\delta\gamma}}\right) = 0. \qquad (731)$$

A comparison with our result above, namely Eq. (722), shows us that in the case of a scalar function approach, there is no problem in directly applying the complete metric and dimension variation. We need not distinguish between the variated (or rather variationally transformed) metric and the non-variated one.

Evaluation of $(\ldots)^{\delta\gamma} g_{\delta\gamma}$ gives us:

$$\left(R^{\delta\gamma} - \frac{1}{2} R \cdot g^{\delta\gamma} + \Lambda \cdot g^{\delta\gamma} + \kappa \cdot T^{\delta\gamma}\right) g_{\delta\gamma} = R - n \cdot \frac{R}{2} + n \cdot \Lambda + \kappa \cdot T^{\delta\gamma} g_{\delta\gamma}$$

$$\xrightarrow{n=4} = 4 \cdot \Lambda + \kappa \cdot T^{\delta\gamma} g_{\delta\gamma} - R. \qquad (732)$$

Assuming a space-time of 4 dimensions and setting the result into Eq. (731) yields:

$$M^2 \cdot f - d \cdot \left(\Delta f + \frac{\left[R - n \cdot \frac{R}{2} + n \cdot \Lambda + \kappa \cdot T^{\delta\gamma} g_{\delta\gamma}\right]_{,\alpha} f^{,\alpha}}{R - n \cdot \frac{R}{2} + n \cdot \Lambda + \kappa \cdot T^{\delta\gamma} g_{\delta\gamma}}\right) = 0$$

$$\xrightarrow{n=4 \quad \& \quad 4 \cdot \Lambda + \kappa \cdot T^{\delta\gamma} g_{\delta\gamma} = 0} M^2 \cdot f - d \cdot \left(\Delta f + \frac{R_{,\alpha} f^{,\alpha}}{R}\right) = 0. \qquad (733)$$

5.1.3.1 A first and rather timid trial to interpret (733)

Now, in using the last line of Eq. (733) and assuming f to be a harmonic in an $n = 4$-space-time, we could finally obtain the parameter mass M or the true physical mass m as:

$$M = \sqrt{\frac{d \cdot R_{,\alpha} f^{,\alpha}}{R \cdot f}} \quad \Rightarrow \quad m = \sqrt{\frac{d \cdot \hbar \cdot R_{,\alpha} f^{,\alpha}}{c \cdot R \cdot f}}. \tag{734}$$

This should give the rest mass when the time-dependencies of R and f cancel each other out in such a way that:

$$\frac{\partial \left(\frac{R_{,\alpha} f^{,\alpha}}{R \cdot f} \right)}{\partial t} = 0. \tag{735}$$

It needs to be pointed out that so far we only dragged the energy momentum tensor and the corresponding Lagrangian L_M with us for the reason of completeness. However, Einstein and Hilbert had postulated the term and with our finding here, namely that mass and potential do not need such a postulation, we might tend to conclude that its introduction was not such a good idea in the first place.

In other words: It has to be stated that with the more complete variation (in comparison to the Hilbert variation from [137]) as shown here, the postulation of the matter term L_M as introduced by Hilbert and Einstein was not needed at all. In fact and looking back, this seems to have been the greater "blunder" rather than the introduction of the cosmological constant Λ by Einstein, because it suggested a finish line where there should not have been one. Without the early fixation of the variation of the Einstein–Hilbert action solely with respect to the metric (and ending there), things may have been easier later on for others. This should not be understood as a criticism on Einstein or Hilbert but merely as an explanation why it was so difficult to see—in retrospect of our own evaluations here—the relatively simple path we walked here . . . thereby simply extending the work of Einstein and Hilbert in not very innovative manner.

Still, for the time being we are going to further drag the term of the energy momentum tensor or the matter Lagrangian through our further evaluations in order to have it ready in case we might need it—and be it just for discussion about its obsolete character.

As a nice by-product, we now have the means to compare the traditional variation under the integral with our simple wrapper approach from Eq. (84). In four dimensions, Eq. (84) requires the setting of $F[f] = f^2$ in order to make the Ricci scalar R a pure Laplace operator of the function $f = f[x^k]$. On the other hand, we take it from Eqs. (725) (last line) and (730) that with $d = 0$, we obtain the metric variation of:

$$G_{\delta\gamma} = \delta g_{\delta\gamma} = 2 \cdot M^2 \delta^i_\delta \delta^j_\gamma \cdot f \cdot f_{,\sigma} g_{ij} \delta^\sigma. \tag{736}$$

Combining with Eq. (84) and demanding that the two functions $f[\ldots]$ from Eqs. (84) and (725) should be the same, we immediately extract:

$$G_{\delta\gamma} = \delta g_{\delta\gamma} = 2 \cdot M^2 \delta^i_\delta \delta^j_\gamma \cdot f \cdot f_{,\sigma} g_{ij} \delta^\sigma = A \cdot \delta^i_\delta \delta^j_\gamma \cdot f^2; \quad A = \text{const} = 2 \cdot M^2$$

$$\Rightarrow f = C_f \cdot e^{\sum\limits_{\sigma=0}^{n-1} x^\sigma}. \tag{737}$$

5.1.3.2 Towards "Dirac"

Realizing that the constant terms in the $\{\ldots\}$-vectors in Eq. (725) are vanishing anyway during the variation, we could also already start with a variation of the form:

$$\delta_\gamma W = 0 = \int_V d^n x \sqrt{g} \cdot \left(R^{\delta\gamma} - \frac{1}{2} R \cdot g^{\delta\gamma} + \Lambda \cdot g^{\delta\gamma} + \kappa \cdot T^{\delta\gamma} \right)$$

$$\times \left(\begin{array}{c} \delta g_{\delta\gamma} = \delta \left(\frac{\partial G^i[x_{\forall i}]}{\partial x^\delta} \cdot \frac{\partial G^j[x_{\forall i}]}{\partial x^\gamma} \gamma_{ij} \right) = g_{ij} \cdot \delta \left(\frac{\partial G^i[x_{\forall i}]}{\partial x^\delta} \cdot \frac{\partial G^j[x_{\forall i}]}{\partial x^\gamma} \right) \\ = \frac{g_{ij}}{4} \delta \left(\left\{ \begin{array}{c} M \cdot \delta^i_\delta \cdot f + \delta^i_\delta \cdot \mathbf{C}^\alpha f_{,\alpha} \\ M \cdot \delta^i_\delta \cdot f - \delta^i_\delta \cdot \mathbf{C}^\alpha f_{,\alpha} \end{array} \right\} \left\{ \begin{array}{c} M \cdot \delta^j_\gamma \cdot f + \delta^j_\gamma \cdot \mathbf{C}^\beta f_{,\beta} \\ M \cdot \delta^j_\gamma \cdot f - \delta^j_\gamma \cdot \mathbf{C}^\beta f_{,\beta} \end{array} \right\} \right) \end{array} \right) . \tag{738}$$

Instead of proceeding with:

$$\delta_\gamma W = 0 = \int_V d^n x \sqrt{g} \cdot \left(R^{\delta\gamma} - \frac{1}{2} R \cdot g^{\delta\gamma} + \Lambda \cdot g^{\delta\gamma} + \kappa \cdot T^{\delta\gamma} \right)$$

$$\times \left(\begin{array}{c} \delta g_{\delta\gamma} = \ldots = \delta \left(M^2 \delta^i_\delta \delta^j_\gamma \cdot f^2 + \delta^i_\delta \delta^j_\gamma \cdot \mathbf{C}^\alpha \mathbf{C}^\beta \cdot f_{,\alpha} f_{,\beta} \right) g_{ij} \\ = \left(2 \cdot M^2 \delta^i_\delta \delta^j_\gamma \cdot f \cdot f_{,\sigma} + \delta^i_\delta \delta^j_\gamma \cdot \mathbf{C}^\alpha \mathbf{C}^\beta \cdot \left(f_{,\alpha\sigma} f_{,\beta} + f_{,\alpha} f_{,\beta\sigma} \right) \right) g_{ij} \delta^\sigma \end{array} \right), \tag{739}$$

this time, we apply the Leibnitz rule and end up with:

$$\delta_\gamma W = 0 = \int_V d^n x \sqrt{g} \cdot \left(R^{\delta\gamma} - \frac{1}{2} R \cdot g^{\delta\gamma} + \Lambda \cdot g^{\delta\gamma} + \kappa \cdot T^{\delta\gamma} \right)$$

$$
\times
\left(
\begin{array}{l}
\delta g_{\delta\gamma} = \frac{g_{ij}}{4}
\left(
\begin{array}{l}
\left\{ \begin{array}{l} M \cdot \delta_\gamma^j \cdot f + \delta_\gamma^j \cdot \mathbf{C}^\beta f_{,\beta} \\ M \cdot \delta_\gamma^j \cdot f - \delta_\gamma^j \cdot \mathbf{C}^\beta f_{,\beta} \end{array} \right\}
\delta
\left\{ \begin{array}{l} M \cdot \delta_\delta^i \cdot f + \delta_\delta^i \cdot \mathbf{C}^\alpha f_{,\alpha} \\ M \cdot \delta_\delta^i \cdot f - \delta_\delta^i \cdot \mathbf{C}^\alpha f_{,\alpha} \end{array} \right\} \\
+ \left\{ \begin{array}{l} M \cdot \delta_\delta^i \cdot f + \delta_\delta^i \cdot \mathbf{C}^\alpha f_{,\alpha} \\ M \cdot \delta_\delta^i \cdot f - \delta_\delta^i \cdot \mathbf{C}^\alpha f_{,\alpha} \end{array} \right\}
\delta
\left\{ \begin{array}{l} M \cdot \delta_\gamma^j \cdot f + \delta_\gamma^j \cdot \mathbf{C}^\beta f_{,\beta} \\ M \cdot \delta_\gamma^j \cdot f - \delta_\gamma^j \cdot \mathbf{C}^\beta f_{,\beta} \end{array} \right\}
\end{array}
\right) \\
= \frac{g_{ij}}{4}
\left(
\begin{array}{l}
\left\{ \begin{array}{l} M \cdot \delta_\gamma^j \cdot f + \delta_\gamma^j \cdot \mathbf{C}^\beta f_{,\beta} \\ M \cdot \delta_\gamma^j \cdot f - \delta_\gamma^j \cdot \mathbf{C}^\beta f_{,\beta} \end{array} \right\}
\left\{ \begin{array}{l} M \cdot \delta_\delta^i \cdot f_{,\sigma} + \delta_\delta^i \cdot \mathbf{C}^\alpha f_{,\alpha\sigma} \\ M \cdot \delta_\delta^i \cdot f_{,\sigma} - \delta_\delta^i \cdot \mathbf{C}^\alpha f_{,\alpha\sigma} \end{array} \right\} \delta^\sigma \\
+ \left\{ \begin{array}{l} M \cdot \delta_\delta^i \cdot f + \delta_\delta^i \cdot \mathbf{C}^\alpha f_{,\alpha} \\ M \cdot \delta_\delta^i \cdot f - \delta_\delta^i \cdot \mathbf{C}^\alpha f_{,\alpha} \end{array} \right\}
\left\{ \begin{array}{l} M \cdot \delta_\gamma^j \cdot f_{,\sigma} + \delta_\gamma^j \cdot \mathbf{C}^\beta f_{,\beta\sigma} \\ M \cdot \delta_\gamma^j \cdot f_{,\sigma} - \delta_\gamma^j \cdot \mathbf{C}^\beta f_{,\beta\sigma} \end{array} \right\} \delta^\sigma
\end{array}
\right)
\end{array}
\right).
$$

$$(740)$$

Easily we could now perform the multiplication and finish the evaluation just as before leading us to Eq. (731). However, there is also the possibility to demand the factor(s) $\left\{ \begin{array}{l} M \cdot \delta_\gamma^j \cdot f + \delta_\gamma^j \cdot \mathbf{C}^\beta f_{,\beta} \\ M \cdot \delta_\gamma^j \cdot f - \delta_\gamma^j \cdot \mathbf{C}^\beta f_{,\beta} \end{array} \right\}$ to either give zero or result in a zero in connection with the subsequent product $\left\{ \begin{array}{l} M \cdot \delta_\gamma^j \cdot f + \delta_\gamma^j \cdot \mathbf{C}^\beta f_{,\beta} \\ M \cdot \delta_\gamma^j \cdot f - \delta_\gamma^j \cdot \mathbf{C}^\beta f_{,\beta} \end{array} \right\} \left\{ \begin{array}{l} M \cdot \delta_\delta^i \cdot f_{,\sigma} + \delta_\delta^i \cdot \mathbf{C}^\alpha f_{,\alpha\sigma} \\ M \cdot \delta_\delta^i \cdot f_{,\sigma} - \delta_\delta^i \cdot \mathbf{C}^\alpha f_{,\alpha\sigma} \end{array} \right\}$. At first, it may appear a bit strange that with $\left\{ \begin{array}{l} M \cdot \delta_\gamma^j \cdot f + \delta_\gamma^j \cdot \mathbf{C}^\beta f_{,\beta} \\ M \cdot \delta_\gamma^j \cdot f - \delta_\gamma^j \cdot \mathbf{C}^\beta f_{,\beta} \end{array} \right\} = \left(\begin{array}{c} 0 \\ 0 \end{array} \right)$ or $\left\{ \begin{array}{l} M \cdot \delta_\gamma^j \cdot f + \delta_\gamma^j \cdot \mathbf{C}^\beta f_{,\beta} \\ M \cdot \delta_\gamma^j \cdot f - \delta_\gamma^j \cdot \mathbf{C}^\beta f_{,\beta} \end{array} \right\} \left\{ \begin{array}{l} M \cdot \delta_\delta^i \cdot f_{,\sigma} + \delta_\delta^i \cdot \mathbf{C}^\alpha f_{,\alpha\sigma} \\ M \cdot \delta_\delta^i \cdot f_{,\sigma} - \delta_\delta^i \cdot \mathbf{C}^\alpha f_{,\alpha\sigma} \end{array} \right\} = 0$, one has two conditions for just one function f, but the reader may prove that the following is correct:

$$\left\{ \begin{array}{l} M \cdot f + i \cdot \mathbf{C}^\beta f_{,\beta} \\ M \cdot f - i \cdot \mathbf{C}^\beta f_{,\beta} \end{array} \right\} \left\{ \begin{array}{l} M \cdot f_{,\sigma} - i \cdot \mathbf{C}^\alpha f_{,\alpha\sigma} \\ M \cdot f_{,\sigma} + i \cdot \mathbf{C}^\alpha f_{,\alpha\sigma} \end{array} \right\} = M^2 \cdot f \cdot f_{,\sigma} + \mathbf{C}^\beta f_{,\beta} \mathbf{C}^\alpha f_{,\alpha\sigma}.$$

$$(741)$$

The corresponding form for Eq. (738) (before the variation) would simply be:

$$\left\{ \begin{array}{l} M \cdot f + i \cdot \mathbf{C}^\beta f_{,\beta} \\ M \cdot f - i \cdot \mathbf{C}^\beta f_{,\beta} \end{array} \right\} \left\{ \begin{array}{l} M \cdot f - i \cdot \mathbf{C}^\alpha f_{,\alpha} \\ M \cdot f + i \cdot \mathbf{C}^\alpha f_{,\alpha} \end{array} \right\} = M^2 \cdot f \cdot f + \mathbf{C}^\beta f_{,\beta} \mathbf{C}^\alpha f_{,\alpha}. \quad (742)$$

Assuming the situation in 4 dimensions, applying again $\mathbf{C}^\beta \mathbf{C}^\alpha = d \cdot g^{\alpha\beta}$ and taking the Minkowski metric:

$$\eta_{\alpha\beta} = \begin{pmatrix} -c^2 & 0 & 0 & 0 \\ 0 & 1 & 0 & 0 \\ 0 & 0 & 1 & 0 \\ 0 & 0 & 0 & 1 \end{pmatrix}$$

$$(743)$$

makes the apparently two vector component products prod1 and prod2 resulting from Eq. (742), yielding

$$\begin{Bmatrix} M \cdot f + i \cdot \mathbf{C}^\beta f_{,\beta} \\ M \cdot f - i \cdot \mathbf{C}^\beta f_{,\beta} \end{Bmatrix} \underbrace{\begin{Bmatrix} M \cdot f - i \cdot \mathbf{C}^\alpha f_{,\alpha} \\ M \cdot f + i \cdot \mathbf{C}^\alpha f_{,\alpha} \end{Bmatrix}}_{\text{prod1}}$$

$$= \left(\underbrace{\begin{aligned} & \left(M \cdot f + i \cdot \mathbf{C}^\beta f_{,\beta} \right) \cdot \left(M \cdot f - i \cdot \mathbf{C}^\alpha f_{,\alpha} \right) \\ & + \left(M \cdot f - i \cdot \mathbf{C}^\beta f_{,\beta} \right) \cdot \left(M \cdot f + i \cdot \mathbf{C}^\alpha f_{,\alpha} \right) \end{aligned}}_{\text{prod2}} \right), \qquad (744)$$

identical.

Thus, just one of the four factors:

$$\left(M \cdot f + i \cdot \mathbf{C}^\beta f_{,\beta} \right), \left(M \cdot f - i \cdot \mathbf{C}^\alpha f_{,\alpha} \right), \left(M \cdot f - i \cdot \mathbf{C}^\beta f_{,\beta} \right), \left(M \cdot f + i \cdot \mathbf{C}^\alpha f_{,\alpha} \right), \tag{745}$$

needs to give zero in order to have the whole scalar product (Eq. (742)) to vanish.

Now we insist that instead of a vector $\{\ldots\}$-structure, we intend to obtain the result from Eq. (742) via a quaternion structure as applied by Dirac [143]. Easily we realize that under the additional assumption that f becomes some kind of function vector $\mathbf{f}^*$, this can be achieved by the Dirac matrices as given in Eqs. (561) and (562) and yielding the almost[†] classical Dirac equation:

$$M \cdot \mathbf{f} + \gamma^\beta \mathbf{f}_{,\beta} = 0. \tag{746}$$

However, even without the assumption of the Minkowski flat space-time, we can easily check that Eq. (746) would also be the correct root extraction for any:

$$M^2 \cdot f \cdot f + \mathbf{C}^\beta f_{,\beta} \mathbf{C}^\alpha f_{,\alpha} = 0, \tag{747}$$

being just the right hand side of Eq. (742).

Thus, the interesting finding here is not the fact that we have extracted the Dirac equation out of the Einstein–Hilbert action, but that it appeared as a true square root. The original Dirac approach, namely, started with the Klein–Gordon equation (Eq. (723)) ($V = 0$) and by introducing the quaternions

---

*In the classical Dirac theory, we here have antimatter and the spin leading to 4-vectors, but in the vector approach in the section below, we will see that the derivation of the metric origin has some deeper meaning within the structure of our extended Einstein–Hilbert action variation (740).

†The reader may prove that one can easily make it the classical from, simply by adapting signs and factors.

via his gamma matrices, Dirac was able to factorize this equation. Here now we did not extract the root out of the Klein–Gordon equation, but out of the right hand side of Eq. (742), which itself results from an extension of the Einstein–Hilbert action. Still we ended up with the same Dirac-type equation (Eq. (746)), with the little snag that, just as Dirac had to do, we are forced to make f a vector in order to "squeeze" it into the quaternionic structure of the Dirac theory. We see this as a drawback, because so far we saw no reason why, in our attempt to have THE simplest approach, f should not be "allowed" to just stay a scalar (see also discussion below in this book). After all, from the metrical standpoint, there is no need to actually assume anything but a scalar . . . at least this author does not see any. The square root extraction of the right hand side of Eq. (742) is just a mathematical operation. Performing it in one (scalars or vectors) or the other (quaternions) way cannot count as justification for the sudden appearance of **f**-vectors . . . not in the purely metrical picture anyway, where nothing has forced us to introduce the quaternions except our very own desire to compare our extraction from the extended Einstein–Hilbert action with the classical Dirac theory. Thus, we conclude that the Dirac equation, even though working well, cannot be the only description or mirror of the square root of our metric Eq. (742). Within the Dirac theory (e.g. [170]), it is the introduction of spinors which makes the whole thing work, but as—so far—we see no need for spinor within our metric picture, we refrain from so simply taking on this recipe but rather move on seeking for an alternative. We will find it in realizing that the right hand side of Eq. (742) does not need to vanish but could also just deliver a constant.

Before we now try to evaluate metric Dirac equations from our scalar starting point (Eqs. (719) and (738)) or the corresponding previous forms (Eqs. (711) and (719)), we first want to investigate function vectors $\mathbf{f}_\delta$ instead of scalars under the variational integral.

5.1.4 Scalar Approach: Only to Be Seen as an Additional Trial Using σ

Now, more or less for completeness and again also because we are not satisfied with the inconsistency of using the non-variated metric γ^{ab} and γ_{ab} within our approach (Eqs. (719)), which we might take as a kind of approximation, we apply another scalar function approach as follows:

$$\delta_\gamma W = 0 = \int_V d^n x \sqrt{g} \cdot \left(R^{\delta\gamma} - \frac{1}{2} R \cdot g^{\delta\gamma} + \Lambda \cdot g^{\delta\gamma} + \kappa \cdot T^{\delta\gamma} \right)$$

$$\left(\begin{array}{l} \delta g_{\delta\gamma} = \delta \left(\frac{\partial G^i [x_{\forall i}]}{\partial x^\delta} \cdot \frac{\partial G^j [x_{\forall i}]}{\partial x^\gamma} \gamma_{ij} \right) = g_{ij} \cdot \delta \left(\frac{\partial G^i [x_{\forall i}]}{\partial x^\delta} \cdot \frac{\partial G^j [x_{\forall i}]}{\partial x^\gamma} \right) \\[4mm] \qquad = \frac{g_{\delta\gamma}}{4} \delta_\sigma \left(\left\{ \begin{array}{l} 1 + h^\sigma \cdot f + C^{\sigma\alpha} f_{,\alpha} \\ 1 + h^\sigma \cdot f + C^{\sigma\alpha} f_{,\alpha} \\ 1 + h^\sigma \cdot f + C^{\sigma\alpha} f_{,\alpha} \\ 1 + h^\sigma \cdot f + C^{\sigma\alpha} f_{,\alpha} \end{array} \right\} \left\{ \begin{array}{l} 1 + h^\sigma \cdot f + C^{\sigma\beta} f_{,\beta} \\ 1 + h^\sigma \cdot f + C^{\sigma\beta} f_{,\beta} \\ 1 + h^\sigma \cdot f + C^{\sigma\beta} f_{,\beta} \\ 1 + h^\sigma \cdot f + C^{\sigma\beta} f_{,\beta} \end{array} \right\} \right) \\[4mm] \qquad = g_{\delta\gamma} \delta_\sigma \left(1 + M^\sigma f \cdot M^\sigma f + C^{\sigma\alpha} C^{\sigma\beta} f_{,\alpha} f_{,\beta} \right) \\[2mm] \qquad = \left(2 \cdot M^\sigma f \cdot M^\sigma f_{,\sigma} + (C^{\sigma\alpha} f_{,\alpha})_{,\sigma} C^{\sigma\beta} f_{,\beta} + C^{\sigma\alpha} f_{,\alpha} \left(C^{\sigma\beta} f_{,\beta} \right)_{,\sigma} \right) \delta x_\sigma \end{array} \right) .$$

$$(748)$$

The problematic part lies in the second line of Eq. (748), but it will be shown later on in a bit more detail that we may see the corresponding operation as follows:

(A) We could start with the following definition and subsequent approximation for the tensor $g_{\delta\gamma}$:

$$g_{\delta\gamma} = \frac{\partial G^i [x_{\forall i}]}{\partial x^\delta} \cdot \frac{\partial G^j [x_{\forall i}]}{\partial x^\gamma} \gamma_{ij}$$

$$\Rightarrow$$

$$\gamma_{ij} = g_{\delta\gamma} \left(\frac{\partial G^i [x_{\forall i}]}{\partial x^\delta} \right)^{-1} \left(\frac{\partial G^j [x_{\forall i}]}{\partial x^\gamma} \right)^{-1}$$

$$= g_{\delta\gamma} \delta^\delta_i \delta^\gamma_j \cdot \left(1 + M^\sigma f \cdot M^\sigma f + C^{\sigma\alpha} C^{\sigma\beta} f_{,\alpha} f_{,\beta} \right)^{-1}$$

$$\underbrace{\phantom{1 \gg M^\sigma f \cdot M^\sigma f + C^{\sigma\alpha} C^{\sigma\beta} f_{,\alpha} f_{,\beta}}}_{1 \gg M^\sigma f \cdot M^\sigma f + C^{\sigma\alpha} C^{\sigma\beta} f_{,\alpha} f_{,\beta}} \simeq g_{\delta\gamma} \delta^\delta_i \delta^\gamma_j .$$

$$(749)$$

(B) Now, we assume the variation only to act on the special transformation matrices (these are no ordinary Jacobi matrices, which would not change the physical tensor properties), respectively, taking the approach from Eq. (748), the variation only acts on the function vector **f** and its derivatives, which gives us:

$$\delta g_{\delta\gamma} = \gamma_{\delta\gamma} \delta_\sigma \left(1 + M^\sigma f \cdot M^\sigma f + C^{\sigma\alpha} C^{\sigma\beta} f_{,\alpha} f_{,\beta} \right)$$

$$= g_{\delta\gamma} \cdot \left(1 + M^\sigma f \cdot M^\sigma f + C^{\sigma\alpha} C^{\sigma\beta} f_{,\alpha} f_{,\beta} \right)^{-1} \delta_\sigma$$

$$\left(1 + M^\sigma f \cdot M^\sigma f + C^{\sigma\alpha} C^{\sigma\beta} f_{,\alpha} f_{,\beta} \right)$$

$$\underbrace{\phantom{1 \gg M^\sigma f \cdot M^\sigma f + C^{\sigma\alpha} C^{\sigma\beta} f_{,\alpha} f_{,\beta}}}_{1 \gg M^\sigma f \cdot M^\sigma f + C^{\sigma\alpha} C^{\sigma\beta} f_{,\alpha} f_{,\beta}} \simeq g_{\delta\gamma} \delta_\sigma \left(1 + M^\sigma f \cdot M^\sigma f + C^{\sigma\alpha} C^{\sigma\beta} f_{,\alpha} f_{,\beta} \right) .$$

$$(750)$$

Due to the approximation in the last line, we need not bother about the potentially problematic division by a vector at the moment, but in order to potentially reuse Eq. (750) also in the non-approximated form, we may just demand that there is a vector $\mathbf{M}_\sigma$ as such that we could write:

$$
\begin{aligned}
\delta g_{\delta\gamma} &= \gamma_{\delta\gamma}\delta_\sigma \left(1 + M^\sigma f \cdot M^\sigma f + C^{\sigma\alpha}C^{\sigma\beta} f_{,\alpha} f_{,\beta}\right) \\
&= g_{\delta\gamma} \cdot \left(1 + h_\sigma M^\sigma f \cdot h_\sigma M^\sigma f + h_\sigma C^{\sigma\alpha} h_\sigma C^{\sigma\beta} f_{,\alpha} f_{,\beta}\right)^{-1} \delta_\sigma \\
&\quad \left(1 + M^\sigma f \cdot M^\sigma f + C^{\sigma\alpha}C^{\sigma\beta} f_{,\alpha} f_{,\beta}\right).
\end{aligned} \tag{751}
$$

Now we have a variety of options to move on with the variational integral and could reshape Eq. (748) as follows:

$$
\begin{aligned}
0 &= \int_V d^n x \sqrt{g} \cdot (\ldots)^{\delta\gamma} \left(2 \cdot M^\sigma f \cdot M^\sigma f_{,\sigma} + (C^{\sigma\alpha} f_{,\alpha})_{,\sigma} C^{\sigma\beta} f_{,\beta} + C^{\sigma\alpha} f_{,\alpha} \left(C^{\sigma\beta} f_{,\beta}\right)_{,\sigma}\right) \delta x_\sigma \\
&\quad 2 \cdot \int_V d^n x \sqrt{g} \cdot (\ldots)^{\delta\gamma} \left(M^\sigma f \cdot M^\sigma f_{,\sigma} + d \cdot \Delta f \cdot C^{\sigma\beta} f_{,\beta}\right) \delta x_\sigma \\
&\quad 2 \cdot \int_V d^n x \left(\sqrt{g} \cdot (\ldots)^{\delta\gamma} M^\sigma f \cdot M^\sigma f_{,\sigma} - \left[\sqrt{g} \cdot (\ldots)^{\delta\gamma} d \cdot \Delta f \cdot C^{\sigma\beta}\right]_{,\beta} f\right) \delta x_\sigma \\
\text{or} \\
&= \int_V d^n x \sqrt{g} \cdot (\ldots)^{\delta\gamma} \left(2 \cdot M^\sigma f \cdot M^\sigma f_{,\sigma} + 2 \left(C^{\sigma\alpha} f_{,\alpha}\right)_{,\sigma} C^{\sigma\beta} f_{,\beta}\right) \delta x_\sigma \\
&= \int_V d^n x \sqrt{g} \cdot (\ldots)^{\delta\gamma} \left(2 \cdot M^\sigma f \cdot M^\sigma f_{,\sigma} + 2 \left(C^{\sigma\alpha}{}_{,\sigma} f_{,\alpha} + C^{\sigma\alpha} f_{,\alpha\sigma}\right) C^{\sigma\beta} f_{,\beta}\right) \delta x_\sigma \\
&= 2 \cdot \int_V d^n x \left(\sqrt{g} \cdot (\ldots)^{\delta\gamma} M^\sigma f \cdot M^\sigma f_{,\sigma} + \left(\begin{array}{c} \sqrt{g} \cdot (\ldots)^{\delta\gamma} C^{\sigma\beta} f_{,\beta} \overbrace{C^{\sigma\alpha}{}_{,\sigma} f_{,\alpha}}^{=C^{\sigma\alpha}{}_{,\alpha} f_{,\sigma}} \\ - \left[\sqrt{g} \cdot (\ldots)^{\delta\gamma} C^{\sigma\beta} f_{,\beta} C^{\sigma\alpha}\right]_{,\alpha} f_{,\sigma} \end{array}\right)\right) \delta x_\sigma \\
&= 2 \cdot \int_V d^n x \left(\sqrt{g} \cdot (\ldots)^{\delta\gamma} M^\sigma f \cdot M^\sigma + \left(\begin{array}{c} \sqrt{g} \cdot (\ldots)^{\delta\gamma} C^{\sigma\beta} f_{,\beta} C^{\sigma\alpha}{}_{,\alpha} \\ - \left[\sqrt{g} \cdot (\ldots)^{\delta\gamma} C^{\sigma\beta} f_{,\beta} C^{\sigma\alpha}\right]_{,\alpha} \end{array}\right)\right) f_{,\sigma} \delta x_\sigma \\
\text{or} \\
&= 2 \cdot \int_V d^n x \left(\sqrt{g} \cdot (\ldots)^{\delta\gamma} M^\sigma f \cdot M^\sigma f_{,\sigma} - \left(C^{\sigma\alpha} f_{,\alpha}\right) \left[\sqrt{g} \cdot (\ldots)^{\delta\gamma} C^{\sigma\beta} f_{,\beta}\right]_{,\sigma}\right) \delta x_\sigma,
\end{aligned} \tag{752}
$$

but as none of these attempts gives us any more information than we already have achieved with the other scalar approaches above, we leave it to the interested reader to further play with this approach and instead move on to vector approaches for the functions f.

5.2 Vector Approach

5.2.1 Vector Approach: First Trial

Another way to avoid the approximation:

$$(\ldots)^{\delta\gamma} = \left(R^{\delta\gamma} - \frac{1}{2} R \cdot \gamma^{\delta\gamma} + \Lambda \cdot \gamma^{\delta\gamma} + \kappa \cdot T^{\delta\gamma} \right) \simeq D \cdot \sqrt{\gamma} \cdot \gamma^{\delta\gamma}, \quad (753)$$

which we used above, could perhaps be that we here apply a vector function approach as follows:

$$\delta_\gamma W = 0 = \int_V d^n x \left(R^{\delta\gamma} \left[\gamma^{\delta\gamma} \right] - \frac{1}{2} R \cdot \gamma^{\delta\gamma} + \Lambda \cdot \gamma^{\delta\gamma} + \kappa \cdot T^{\delta\gamma} \left[\gamma^{\delta\gamma} \right] \right)$$

$$\times \sqrt{\gamma} \cdot \gamma_{ij} \delta \left(C^i C^j \cdot f_\delta f_\gamma + C^2 \cdot \gamma^{i\alpha} f_{\delta,\alpha} \gamma^{j\beta} f_{\gamma,\beta} \right)$$

$$= \int_V d^n x \left(R^{\delta\gamma} \left[\gamma^{\delta\gamma} \right] - \frac{1}{2} R \cdot \gamma^{\delta\gamma} + \Lambda \cdot \gamma^{\delta\gamma} + \kappa \cdot T^{\delta\gamma} \left[\gamma^{\delta\gamma} \right] \right)$$

$$\times \sqrt{\gamma} \cdot \delta \left(C_j C^j \cdot f_\delta f_\gamma + C^2 \cdot f_{\delta,j} \gamma^{j\beta} f_{\gamma,\beta} \right)$$

$$= \int_V d^n x \left(R^{\delta\gamma} \left[\gamma^{\delta\gamma} \right] - \frac{1}{2} R \cdot \gamma^{\delta\gamma} + \Lambda \cdot \gamma^{\delta\gamma} + \kappa \cdot T^{\delta\gamma} \left[\gamma^{\delta\gamma} \right] \right)$$

$$\times \sqrt{\gamma} \cdot \left(C_j C^j \cdot \left[f_\delta \cdot f_{\gamma,\sigma} + f_\gamma \cdot f_{\delta,\sigma} \right] + C^2 \right.$$

$$\left. \cdot \left[f_{\delta,j} \gamma^{j\beta} \cdot f_{\gamma,\beta\sigma} + \gamma^{j\beta} f_{\gamma,\beta} \cdot f_{\delta,j\sigma} \right] \right). \quad (754)$$

Integration by parts of the second addend in the last line, performed as:

$$\int_V d^n x \, (\ldots)^{\delta\gamma} \sqrt{\gamma} \cdot C^2 \cdot f_{\delta,j} \gamma^{j\beta} \cdot f_{\gamma,\beta\sigma}$$

$$= \underbrace{\int_V d^n x \left\{ \left[(\ldots)^{\delta\gamma} C^2 \cdot \sqrt{\gamma} \cdot f_{\delta,j} \gamma^{j\beta} \right] \cdot f_{\gamma,\sigma} \right\}_{,\beta}}_{=S=0 \text{ per definition}}$$

$$- \int_V d^n x \left[(\ldots)^{\delta\gamma} C^2 \cdot \sqrt{\gamma} \cdot f_{\delta,j} \gamma^{j\beta} \right]_{,\beta} \cdot f_{\gamma,\sigma}; \quad (\ldots)^{\delta\gamma}$$

$$= \left(R^{\delta\gamma} - \frac{1}{2} R \cdot \gamma^{\delta\gamma} + \Lambda \cdot \gamma^{\delta\gamma} + \kappa \cdot T^{\delta\gamma} \right)$$

$$(755)$$

yields:

$$\delta_\gamma W = 0 = \int_V d^n x$$

$$\times \left(c_j C^j \sqrt{\gamma} \cdot (\dots)^{\delta\gamma} \left[f_\delta \cdot f_{\gamma,\sigma} + f_\gamma \cdot f_{\delta,\sigma} \right] - C^2 \cdot \left[\begin{array}{l} \left[(\dots)^{\delta\gamma} \sqrt{\gamma} \cdot f_{\delta,j} \gamma^{j\beta} \right]_{,\beta} \cdot f_{\gamma,\sigma} \\ + \left[(\dots)^{\delta\gamma} \sqrt{\gamma} \cdot f_{\gamma,\beta} \gamma^{j\beta} \right]_{,j} \cdot f_{\delta,\sigma} \end{array} \right] \right)$$

$$= \int_V d^n x \times \left(\begin{array}{l} \left(c_j C^j \sqrt{\gamma} \cdot (\dots)^{\delta\gamma} f_\delta - C^2 \cdot \left[(\dots)^{\delta\gamma} \sqrt{\gamma} \cdot f_{\delta,j} \gamma^{j\beta} \right]_{,\beta} \right) \cdot f_{\gamma,\sigma} \\ + \left(c_j C^j \sqrt{\gamma} \cdot (\dots)^{\delta\gamma} f_\gamma - C^2 \cdot \left[(\dots)^{\delta\gamma} \sqrt{\gamma} \cdot f_{\gamma,\beta} \gamma^{j\beta} \right]_{,j} \right) \cdot f_{\delta,\sigma} \end{array} \right)$$

$$= \int_V d^n x \times \left(\begin{array}{l} \left(c_j C^j \sqrt{\gamma} \cdot (\dots)^{\delta\gamma} f_\delta - C^2 \cdot \left[\sqrt{\gamma} \cdot f_{\delta,j} \gamma^{j\beta} (\dots)^{\delta\gamma}_{,\beta} + (\dots)^{\delta\gamma} \left(\sqrt{\gamma} \cdot \gamma^{j\beta} f_{\delta,j} \right)_{,\beta} \right] \right) \cdot f_{\gamma,\sigma} \\ + \left(c_j C^j \sqrt{\gamma} \cdot (\dots)^{\delta\gamma} f_\gamma - C^2 \cdot \left[\sqrt{\gamma} \cdot f_{\gamma,\beta} \gamma^{j\beta} (\dots)^{\delta\gamma}_{,j} + (\dots)^{\delta\gamma} \left(\sqrt{\gamma} \cdot \gamma^{j\beta} f_{\gamma,\beta} \right)_{,j} \right] \right) \cdot f_{\delta,\sigma} \end{array} \right)$$

$$= \int_V d^n x \times \sqrt{\gamma} \cdot \left(\begin{array}{l} \left((\dots)^{\delta\gamma} \left(C_j C^j f_\delta - C^2 \cdot f_{\delta,j} \gamma^{j\beta} (\dots)^{\delta\gamma}_{,\beta} \right) \cdot f_{\gamma,\sigma} \\ + \left((\dots)^{\delta\gamma} \left(C_j C^j f_\gamma - C^2 \cdot f_{\gamma,\beta} \gamma^{j\beta} (\dots)^{\delta\gamma}_{,j} \right) \cdot f_{\delta,\sigma} \end{array} \right).$$

$$(756)$$

Please note that in contrast to the scalar f-function approach above and without loss of generality, we can assume C to be a constant. Subsequently, we obtain quantum equations of the kind:

$$(\ldots)^{\delta\gamma}\left(C_j C^j f_\delta - C^2 \cdot \Delta f_\delta\right) - C^2 \cdot f_{\delta,j}\gamma^{j\beta}(\ldots)^{\delta\gamma}{}_{,\beta} = 0$$
$$(\ldots)^{\delta\gamma}\left(C_j C^j f_\gamma - C^2 \cdot \Delta f_\gamma\right) - C^2 \cdot f_{\gamma,\beta}\gamma^{j\beta}(\ldots)^{\delta\gamma}{}_{,j} = 0. \tag{757}$$

The two sets of equations only differ in dummy indices and are therefore identical. In the case of vanishing derivatives for the relativity term like $(\ldots)^{\delta\gamma}{}_{,\beta} = \left(R^{\delta\gamma} - \frac{1}{2}R \cdot \gamma^{\delta\gamma} + \Lambda \cdot \gamma^{\delta\gamma} + \kappa \cdot T^{\delta\gamma}\right)_{,\beta} = 0$, we obtain the simple Klein–Gordon equation with:

$$(\ldots)^{\delta\gamma}\left(C_j C^j f_\delta - C^2 \cdot \Delta f_\delta\right) = C_j C^j f_\delta - C^2 \cdot \Delta f_\delta = 0. \tag{758}$$

We note that the function vector $\mathbf{f}_\delta$ or $\mathbf{f}_\gamma$ in Eq. (757), in contrast to the classical Klein–Gordon equation, if constructed out of the Dirac equation, does not contain all vector components in a decoupled manner, but provides—in principle—independent equations for each vector component. The last term in both equations of Eq. (757) assures this. Thus, in situations where we have $(\ldots)^{\delta\gamma}{}_{,\beta} = \left(R^{\delta\gamma} - \frac{1}{2}R \cdot \gamma^{\delta\gamma} + \Lambda \cdot \gamma^{\delta\gamma} + \kappa \cdot T^{\delta\gamma}\right)_{,\beta} = 0$, we would have the classical decoupling of the wave function components.

5.2.1.1 Towards inner covariance

The final result of our variation (Eq. (754)) is a scalar (Eq. (756)). It only becomes a vector (respectively two vectors), because we factored the variational terms $f_{\gamma,\sigma}$, $f_{\delta,\sigma}$ in Eq. (757) out. Leaving them in there and obtaining as final equation:

$$\begin{pmatrix}(\ldots)^{\delta\gamma}\left(C_j C^j f_\delta - C^2 \cdot \Delta f_\delta\right) \\ -C^2 \cdot f_{\delta,j}\gamma^{j\beta}(\ldots)^{\delta\gamma}{}_{,\beta}\end{pmatrix} \cdot f_{\gamma,\sigma} + \begin{pmatrix}(\ldots)^{\delta\gamma}\left(C_j C^j f_\gamma - C^2 \cdot \Delta f_\gamma\right) \\ -C^2 \cdot f_{\gamma,\beta}\gamma^{j\beta}(\ldots)^{\delta\gamma}{}_{,j}\end{pmatrix} \cdot f_{\delta,\sigma} = 0, \tag{759}$$

not only allows for more general solutions, as we now have a sum of more terms at hand, but also gives just one equation instead of two (thereby noting that the two equations of Eq. (757) are structurally equal). For a scalar, however, we would not need to bother about covariance. Thus, the question would be to put Eq. (757) into a scalar form. Realizing that Eq. (759) still is the integrand of Eq. (756), we substitute $f_\alpha = F_{,\alpha}$ and perform integration by parts as follows:

$$\delta_\gamma W = 0 = \int_V d^n x \times \sqrt{\gamma} \cdot \left(\left(\begin{matrix} (\ldots)^{\delta\gamma} \left(C_j C^j F_{,\delta} - C^2 \cdot \Delta F_{,\delta} \right) \\ -C^2 \cdot F_{,\delta j} \gamma^{j\beta} (\ldots)^{\delta\gamma}_{,\beta} \end{matrix} \right) \right.$$

$$\left. \cdot F_{,\gamma\sigma} + \left(\begin{matrix} (\ldots)^{\delta\gamma} \left(C_j C^j F_{\gamma} - C^2 \cdot \Delta F_{,\gamma} \right) \\ -C^2 \cdot F_{,\gamma\beta} \gamma^{j\beta} (\ldots)^{\delta\gamma}_{,j} \end{matrix} \right) \cdot F_{,\delta\sigma} \right)$$

$$= - \int_V d^n x \times \left(\left(\sqrt{\gamma} \cdot \left(\begin{matrix} (\ldots)^{\delta\gamma} \left(C_j C^j F_{,\delta} - C^2 \cdot \Delta F_{,\delta} \right) \\ -C^2 \cdot F_{,\delta j} \gamma^{j\beta} (\ldots)^{\delta\gamma}_{,\beta} \end{matrix} \right) \right) \right)_{,\gamma}$$

$$+ \left(\sqrt{\gamma} \cdot \left(\begin{matrix} (\ldots)^{\delta\gamma} \left(C_j C^j F_{\gamma} - C^2 \cdot \Delta F_{,\gamma} \right) \\ -C^2 \cdot F_{,\gamma\beta} \gamma^{j\beta} (\ldots)^{\delta\gamma}_{,j} \end{matrix} \right) \right)_{,\delta} \right) \cdot F_{,\sigma}. \quad (760)$$

Once again assuming a situation where we could set (potentially approximate) $(\ldots)^{\delta\gamma} \simeq D \cdot \gamma^{\delta\gamma}$, we immediately recognize complete Laplace operators and simplify as follows:

$$\delta_\gamma W = 0 = \simeq - \int_V d^n x \cdot D \cdot \left(\left(\sqrt{\gamma} \cdot \left(\begin{matrix} \gamma^{\delta\gamma} \left(C_j C^j F_{,\delta} - C^2 \cdot \Delta F_{,\delta} \right) \\ -C^2 \cdot F_{,\delta j} \gamma^{j\beta} \gamma^{\delta\gamma}_{,\beta} \end{matrix} \right) \right) \right)_{,\gamma}$$

$$+ \left(\sqrt{\gamma} \cdot \left(\begin{matrix} \gamma^{\delta\gamma} \left(C_j C^j F_{\gamma} - C^2 \cdot \Delta F_{,\gamma} \right) \\ -C^2 \cdot F_{,\gamma\beta} \gamma^{j\beta} \gamma^{\delta\gamma}_{,j} \end{matrix} \right) \right)_{,\delta} \right) \cdot F_{,\sigma}$$

$$= \int_V d^n x \cdot D \cdot \left(\begin{matrix} \sqrt{\gamma} \cdot \left(C^2 \cdot \Delta^2 F - C_j C^j \Delta F \right) + C^2 \cdot \left(\sqrt{\gamma} \cdot F_{,\delta j} \gamma^{j\beta} \gamma^{\delta\gamma}_{,\beta} \right)_{,\gamma} \\ + \sqrt{\gamma} \cdot \left(C^2 \cdot \Delta^2 F - C_j C^j \Delta F \right) + C^2 \cdot \left(\sqrt{\gamma} \cdot F_{,\gamma\beta} \gamma^{j\beta} \gamma^{\delta\gamma}_{,j} \right)_{,\delta} \end{matrix} \right) \cdot F_{,\sigma}$$

$$= \int_V d^n x \cdot D \cdot \left(2 \cdot \sqrt{\gamma} \cdot \left(C^2 \cdot \Delta^2 F - C_j C^j \Delta F \right) + C^2 \cdot \left[\begin{matrix} \left(\sqrt{\gamma} \cdot F_{,\delta\alpha} \gamma^{\alpha\beta} \gamma^{\delta\gamma}_{,\beta} \right)_{,\gamma} \\ + \left(\sqrt{\gamma} \cdot F_{,\gamma\beta} \gamma^{j\beta} \gamma^{\delta\gamma}_{,j} \right)_{,\delta} \end{matrix} \right] \right) \cdot F_{,\sigma}.$$

$$(761)$$

We obtained just one inner scalar equation with apparently bi- and unoharmonic operators, because without further ado the integrand looks as follows:

$$0 = 2 \cdot \sqrt{\gamma} \cdot \left(C^2 \cdot \Delta^2 F - C_j C^j \Delta F \right)$$

$$+ C^2 \cdot \left[\left(\sqrt{\gamma} \cdot F_{,\delta\alpha} \gamma^{\alpha\beta} \gamma^{\delta\gamma}_{,\beta} \right)_{,\gamma} + \left(\sqrt{\gamma} \cdot F_{,\gamma\beta} \gamma^{j\beta} \gamma^{\delta\gamma}_{,j} \right)_{,\delta} \right]$$

$$\Rightarrow 2 \cdot \sqrt{\gamma} \cdot \left(C^2 \cdot \Delta^2 F - C_j C^j \Delta F \right)$$

$$+ C^2 \cdot \left[\left(\sqrt{\gamma} \cdot F_{,\delta}^{\;\beta} \gamma^{\delta\gamma}_{,\beta} \right)_{,\gamma} + \left(\sqrt{\gamma} \cdot F_{,\gamma}^{\;\beta} \gamma^{\delta\gamma}_{,\beta} \right)_{,\delta} \right] = 0. \quad (762)$$

However, closer evaluation reveals that the biharmonic operator can be made to vanish:

$$2 \cdot \sqrt{\gamma} \cdot \left(C^2 \cdot \Delta^2 F - C_j C^j \Delta F\right) + C^2 \cdot \left[\left(\sqrt{\gamma} \cdot F_{,\delta}{}^\beta \gamma^{\delta\gamma}{}_{,\beta}\right)_{,\gamma} + \left(\sqrt{\gamma} \cdot F_{,\gamma}{}^\beta \gamma^{\delta\gamma}{}_{,\beta}\right)_{,\delta}\right]$$

$$= 2 \cdot \sqrt{\gamma} \cdot \left(C^2 \cdot \Delta^2 F - C_j C^j \Delta F\right)$$

$$+ C^2 \cdot \left[\begin{array}{c} \left(\left(\sqrt{\gamma} \cdot F_{,\delta}{}^\beta \gamma^{\delta\gamma}\right)_{,\beta} - \overbrace{\left(\sqrt{\gamma} \cdot F_{,\delta}{}^\beta\right)_{,\beta}}^{\sqrt{\gamma} \cdot \Delta F_{,\delta}} \gamma^{\delta\gamma}\right)_{,\gamma} \\ + \left(\left(\sqrt{\gamma} \cdot F_{,\gamma}{}^\beta \gamma^{\delta\gamma}\right)_{,\beta} - \underbrace{\left(\sqrt{\gamma} \cdot F_{,\gamma}{}^\beta\right)_{,\beta}}_{\sqrt{\gamma} \cdot \Delta F_{,\gamma}} \gamma^{\delta\gamma}\right)_{,\delta} \end{array}\right]$$

$$= 2 \cdot \sqrt{\gamma} \cdot \left(C^2 \cdot \Delta^2 F - C_j C^j \Delta F\right)$$

$$+ C^2 \cdot \left[\begin{array}{c} \left(\sqrt{\gamma} \cdot F^{,\gamma\beta}\right)_{,\beta\gamma} - \overbrace{\left(\sqrt{\gamma} \cdot \Delta F_{,\delta} \gamma^{\delta\gamma}\right)_{,\gamma}}^{\sqrt{\gamma} \cdot \Delta^2 F} \\ + \left(\sqrt{\gamma} \cdot F^{,\delta\beta}\right)_{,\beta\delta} - \underbrace{\left(\sqrt{\gamma} \cdot \Delta F_{,\gamma} \gamma^{\delta\gamma}\right)_{,\delta}}_{\sqrt{\gamma} \cdot \Delta^2 F} \end{array}\right]$$

$$= 2 \cdot \sqrt{\gamma} \cdot \left(C^2 \cdot \Delta^2 F - C_j C^j \Delta F\right)$$

$$+ C^2 \cdot \left[\left(\sqrt{\gamma} \cdot F^{,\gamma\beta}\right)_{,\beta\gamma} + \left(\sqrt{\gamma} \cdot F^{,\delta\beta}\right)_{,\beta\delta} - 2 \cdot \sqrt{\gamma} \cdot \Delta^2 F\right]$$

$$= C^2 \cdot \left[\left(\sqrt{\gamma} \cdot F^{,\gamma\beta}\right)_{,\beta\gamma} + \left(\sqrt{\gamma} \cdot F^{,\delta\beta}\right)_{,\beta\delta}\right] - 2 \cdot \sqrt{\gamma} \cdot C_j C^j \Delta F = 0. \quad (763)$$

This leaves us with a fascinatingly simple equation of Klein–Gordon type, namely:

$$C^2 \cdot \left[\left(\sqrt{\gamma} \cdot F^{,\gamma\beta}\right)_{,\beta\gamma} + \left(\sqrt{\gamma} \cdot F^{,\delta\beta}\right)_{,\beta\delta}\right] - 2 \cdot \sqrt{\gamma} \cdot C_j C^j \Delta F$$

$$= 2 \cdot \left(C^2 \cdot \left(\sqrt{\gamma} \cdot F^{,\gamma\beta}\right)_{,\beta\gamma} - \sqrt{\gamma} \cdot C_j C^j \Delta F\right)$$

$$= \sqrt{\gamma} \cdot \left[\begin{array}{c} C^2 \cdot \left(\left(\dfrac{\gamma^{\lambda\kappa}\gamma_{\lambda\kappa,\gamma}\gamma^{\lambda\kappa}\gamma_{\lambda\kappa,\beta}}{4} + \dfrac{\left(\gamma^{\lambda\kappa}\gamma_{\lambda\kappa,\beta}\right)_{,\gamma}}{2}\right) F^{,\gamma\beta} + F^{,\gamma\beta}{}_{,\beta\gamma} + \dfrac{\gamma^{\lambda\kappa}\gamma_{\lambda\kappa,\gamma}}{2} F^{,\gamma\beta}{}_{,\beta} + \dfrac{\gamma^{\lambda\kappa}\gamma_{\lambda\kappa,\beta}}{2} F^{,\gamma\beta}{}_{,\gamma} \\ \left(\dfrac{\gamma^{\lambda\kappa}\gamma_{\lambda\kappa,\delta}\gamma^{\lambda\kappa}\gamma_{\lambda\kappa,\beta}}{4} + \dfrac{\left(\gamma^{\lambda\kappa}\gamma_{\lambda\kappa,\beta}\right)_{,\delta}}{2}\right) F^{,\delta\beta} + F^{,\delta\beta}{}_{,\beta\delta} + \dfrac{\gamma^{\lambda\kappa}\gamma_{\lambda\kappa,\delta}}{2} F^{,\delta\beta}{}_{,\beta} + \dfrac{\gamma^{\lambda\kappa}\gamma_{\lambda\kappa,\beta}}{2} F^{,\delta\beta}{}_{,\delta} \\ -2 \cdot C_j C^j \Delta F \end{array}\right] = 0$$

$$\Rightarrow C^2 \cdot \left(\left(\dfrac{\gamma^{\lambda\kappa}\gamma_{\lambda\kappa,\gamma}\gamma^{\lambda\kappa}\gamma_{\lambda\kappa,\beta}}{4} + \dfrac{\left(\gamma^{\lambda\kappa}\gamma_{\lambda\kappa,\beta}\right)_{,\gamma}}{2}\right) F^{,\gamma\beta} + F^{,\gamma\beta}{}_{,\beta\gamma}\right.$$

$$\left. + \dfrac{\gamma^{\lambda\kappa}\gamma_{\lambda\kappa,\gamma}}{2} F^{,\gamma\beta}{}_{,\beta} + \dfrac{\gamma^{\lambda\kappa}\gamma_{\lambda\kappa,\beta}}{2} F^{,\gamma\beta}{}_{,\gamma}\right) - C_j C^j \Delta F$$

$$= C^2 \cdot \left(\left(\dfrac{\gamma^{\lambda\kappa}\gamma_{\lambda\kappa,\gamma}\gamma^{\lambda\kappa}\gamma_{\lambda\kappa,\beta}}{4} + \dfrac{\left(\gamma^{\lambda\kappa}\gamma_{\lambda\kappa,\beta}\right)_{,\gamma}}{2}\right) F^{,\gamma\beta} + F^{,\gamma\beta}{}_{,\beta\gamma} + 2 \cdot \gamma^{\lambda\kappa}\gamma_{\lambda\kappa,\gamma} F^{,\gamma\beta}{}_{,\beta}\right) - C_j C^j \Delta F = 0.$$

$$(764)$$

It should be pointed out that we also have another way to present Eq. (764) in the short, unexpanded form of the first two lines, namely as:

$$0 = 2 \cdot \left(C^2 \cdot \left(\sqrt{\gamma} \cdot F^{,\gamma\beta} \right)_{,\beta\gamma} - \sqrt{\gamma} \cdot C_j C^j \Delta F \right)$$

$$= 2 \cdot \left(C^2 \cdot \left(\sqrt{\gamma} \cdot \gamma^{\beta\kappa} F^{,\gamma}{}_{\kappa} \right)_{,\beta\gamma} - \sqrt{\gamma} \cdot C_j C^j \Delta F \right)$$

$$= 2 \cdot \left(C^2 \cdot \left(\sqrt{\gamma} \cdot \Delta F^{,\gamma} \right)_{,\gamma} - \sqrt{\gamma} \cdot C_j C^j \Delta F \right)$$

$$= 2 \cdot \left(C^2 \cdot \left(\sqrt{\gamma} \cdot \Delta \left(\gamma^{\gamma\kappa} F_{,\kappa} \right) \right)_{,\gamma} - \sqrt{\gamma} \cdot C_j C^j \Delta F \right). \tag{765}$$

We recognize the typical structural elements of the equations of elasticity and will come back to these results in chapter "10 An Elastic World Formula".

It may still be seen as a disadvantage that we have applied an ordinary derivative and not something which guaranties also an intrinsic tensor outcome also without the setting of our vector function f_α to a covariant derivative vector via $f_\alpha = F_{,\alpha}$. Considering the covariant derivative of a metric from Eq. (77) tells us that we require just two other types of variation, namely:

$$\delta_\gamma W = 0 = \int_V d^n x \left(R^{\delta\gamma} \left[\gamma^{\delta\gamma} \right] - \frac{1}{2} R \cdot \gamma^{\delta\gamma} + \Lambda \cdot \gamma^{\delta\gamma} + \kappa \cdot T^{\delta\gamma} \left[\gamma^{\delta\gamma} \right] \right)$$

$$\times \sqrt{\gamma} \cdot \gamma_{ij} \delta_\delta \left(C^i C^j \cdot f_\sigma f_\gamma + C^2 \cdot \gamma^{i\alpha} f_{\sigma,\alpha} \gamma^{j\beta} f_{\gamma,\beta} \right) \tag{766}$$

and:

$$\delta_\gamma W = 0 = \int_V d^n x \left(R^{\delta\gamma} \left[\gamma^{\delta\gamma} \right] - \frac{1}{2} R \cdot \gamma^{\delta\gamma} + \Lambda \cdot \gamma^{\delta\gamma} + \kappa \cdot T^{\delta\gamma} \left[\gamma^{\delta\gamma} \right] \right)$$

$$\times \sqrt{\gamma} \cdot \gamma_{ij} \delta_\gamma \left(C^i C^j \cdot f_\delta f_\sigma + C^2 \cdot \gamma^{i\alpha} f_{\delta,\alpha} \gamma^{j\beta} f_{\sigma,\beta} \right). \tag{767}$$

Now performing the variations and integration by parts yields:

$$\delta_\gamma W = 0$$

$$= \int_V d^n x \times \sqrt{\gamma} \cdot \left(\begin{array}{c} ((\ldots)^{\delta\gamma} \left(C_j C^j f_\sigma - C^2 \cdot \Delta f_\sigma \right) - C^2 \cdot f_{\sigma,j} \gamma^{j\beta} (\ldots)^{\delta\gamma}{}_{,\beta}) \cdot f_{\gamma,\delta} \\ + ((\ldots)^{\delta\gamma} \left(C_j C^j f_\gamma - C^2 \cdot \Delta f_\gamma \right) - C^2 \cdot f_{\gamma,\beta} \gamma^{j\beta} (\ldots)^{\delta\gamma}{}_{,j}) \cdot f_{\sigma,\delta} \end{array} \right) \tag{768}$$

for Eq. (766) and:

$$\delta_\gamma W = 0$$

$$= \int_V d^n x \times \sqrt{\gamma} \cdot \left(\begin{array}{l} ((\ldots)^{\delta\gamma} \left(C_j C^j f_\delta - C^2 \cdot \Delta f_\delta \right) - C^2 \cdot f_{\delta,j} \gamma^{j\beta} (\ldots)^{\delta\gamma}{}_{,\beta}) \cdot f_{\sigma,\gamma} \\ + ((\ldots)^{\delta\gamma} \left(C_j C^j f_\sigma - C^2 \cdot \Delta f_\sigma \right) - C^2 \cdot f_{\sigma,\beta} \gamma^{j\beta} (\ldots)^{\delta\gamma}{}_{,j}) \cdot f_{\delta,\gamma} \end{array} \right)$$

$$(769)$$

with respect to Eq. (767). Please note that according to Eq. (77), the covariant sum of all these variations in combination with Eq. (756) would give zero, but we could still seek for intrinsic solutions. Observing Eqs. (768) and (769), we see that in contrast to Eq. (756), where with Eq. (757), we have obtained two identical equations, we now apparently have a somewhat more difficult situation. We can easily see this when introducing a few abbreviations. As an example, we use Eq. (768) (the results are just similar for Eq. (769)):

$$((\ldots)^{\delta\gamma} \left(C_j C^j f_\sigma - C^2 \cdot \Delta f_\sigma \right) - C^2 \cdot f_{\sigma,j} \gamma^{j\beta} (\ldots)^{\delta\gamma}{}_{,\beta}) \cdot f_{\gamma,\delta}$$
$$+ ((\ldots)^{\delta\gamma} \left(C_j C^j f_\gamma - C^2 \cdot \Delta f_\gamma \right) - C^2 \cdot f_{\gamma,\beta} \gamma^{j\beta} (\ldots)^{\delta\gamma}{}_{,j}) \cdot f_{\sigma,\delta}$$
$$= \left((\ldots)^{\delta\gamma} \begin{pmatrix} : \\ : \end{pmatrix}_\sigma - \begin{pmatrix} \cdot \cdot \end{pmatrix}_\sigma^\beta (\ldots)^{\delta\gamma}{}_{,\beta} \right) \cdot f_{\gamma,\delta}$$
$$+ \left((\ldots)^{\delta\gamma} \begin{pmatrix} : \\ : \end{pmatrix}_\gamma - \begin{pmatrix} \cdot \cdot \end{pmatrix}_\gamma^j (\ldots)^{\delta\gamma}{}_{,j} \right) \cdot f_{\sigma,\delta}. \qquad (770)$$

Now we simply substitute $f_\gamma = F_{,\gamma}$, and after integration by parts we have (under the integral Eq. (768)):

$$= -\left(\left((\ldots)^{\delta\gamma} \begin{pmatrix} : \\ : \end{pmatrix}_\sigma - \begin{pmatrix} \cdot \cdot \end{pmatrix}_\sigma^\beta (\ldots)^{\delta\gamma}{}_{,\beta} \right)_{,\gamma} + \left((\ldots)^{\delta\gamma} \begin{pmatrix} : \\ : \end{pmatrix}_\gamma - \begin{pmatrix} \cdot \cdot \end{pmatrix}_\gamma^j (\ldots)^{\delta\gamma}{}_{,j} \right)_{,\sigma} \right) \cdot F_{,\delta}$$

$$= -\left((\ldots)^{\delta\gamma} \left(C_j C^j F_{,\sigma\gamma} - C^2 \cdot \Delta F_{,\sigma\gamma} \right) + (\ldots)^{\delta\gamma}{}_{,\gamma} \begin{pmatrix} : \\ : \end{pmatrix}_\sigma - C^2 \cdot \left(F_{,\sigma j} \gamma^{j\beta} (\ldots)^{\delta\gamma}{}_{,\beta} \right)_{,\gamma} \right) \cdot F_{,\delta}$$

$$- \left((\ldots)^{\delta\gamma} \left(C_j C^j F_{,\gamma\sigma} - C^2 \cdot \Delta F_{,\gamma\sigma} \right) + (\ldots)^{\delta\gamma}{}_{,\sigma} \begin{pmatrix} : \\ : \end{pmatrix}_\gamma - C^2 \cdot \left(F_{,\gamma\beta} \gamma^{j\beta} (\ldots)^{\delta\gamma}{}_{,j} \right)_{,\sigma} \right) \cdot F_{,\delta}$$

$$= -\left(\begin{array}{l} 2 \cdot (\ldots)^{\delta\gamma} \left(C_j C^j F_{,\sigma\gamma} - C^2 \cdot \Delta F_{,\sigma\gamma} \right) + (\ldots)^{\delta\gamma}{}_{,\gamma} \begin{pmatrix} : \\ : \end{pmatrix}_\sigma + (\ldots)^{\delta\gamma}{}_{,\sigma} \begin{pmatrix} : \\ : \end{pmatrix}_\gamma \\ -C^2 \cdot \left[\left(F_{,\gamma\beta} \gamma^{j\beta} (\ldots)^{\delta\gamma}{}_{,j} \right)_{,\sigma} + \left(F_{,\sigma j} \gamma^{j\beta} (\ldots)^{\delta\gamma}{}_{,\beta} \right)_{,\gamma} \right] \end{array} \right) \cdot F_{,\delta}.$$

$$(771)$$

Please note that, in contrast to the scalar f-function approach above and without loss of generality, we can assume C to be a constant. Subsequently, we obtain quantum equations of the kind:

$$(\ldots)^{\delta\gamma}\left(C_jC^j f_\delta - C^2 \cdot \Delta f_\delta\right) - C^2 \cdot f_{\delta,j}\gamma^{j\beta}(\ldots)^{\delta\gamma}{}_{,\beta} = 0$$

$$(\ldots)^{\delta\gamma}\left(C_jC^j f_\gamma - C^2 \cdot \Delta f_\gamma\right) - C^2 \cdot f_{\gamma,\beta}\gamma^{j\beta}(\ldots)^{\delta\gamma}{}_{,j} = 0. \qquad (772)$$

Alternatively, and following the recipe which has led us to Eq. (764) in an only slightly adapted form, we can also apply the integration by parts a second time onto Eqs. (768) and (769) with respect to the expressions $f_{\sigma,\gamma}$, $f_{\delta,\gamma}$, $f_{\sigma,\delta}$ and always obtain equations of the type:

$$0 = C^2 \cdot \left(\sqrt{\gamma} \cdot F^{,\gamma\beta}\right)_{,\beta\gamma} - \sqrt{\gamma} \cdot C_jC^j \Delta F$$

$$\Rightarrow C^2 \cdot \left(\left(\frac{\gamma^{\lambda\kappa}\gamma_{\lambda\kappa,\gamma}\gamma^{\lambda\kappa}\gamma_{\lambda\kappa,\beta}}{4} + \frac{\left(\gamma^{\lambda\kappa}\gamma_{\lambda\kappa,\beta}\right)_{,\gamma}}{2}\right)F^{,\gamma\beta} + F^{,\gamma\beta}{}_{,\beta\gamma} + 2 \cdot \gamma^{\lambda\kappa}\gamma_{\lambda\kappa,\gamma}F^{,\gamma\beta}{}_{,\beta}\right)$$

$$-C_jC^j \Delta F = 0$$

$$0 = C^2 \cdot \left(\sqrt{\gamma} \cdot h_{,\sigma}{}^{\gamma\beta}\right)_{,\beta\gamma} - \sqrt{\gamma} \cdot C_jC^j \Delta h_{,\sigma}; \quad \text{with:} \quad h_{,\sigma} = F. \qquad (773)$$

Thereby the last equation resulted from the bold move of expanding the first line in Eq. (768) and the second line in Eq. (769) as follows:

$$\delta_\gamma W = 0$$

$$= \int_V d^n x \times \sqrt{\gamma} \cdot \left(\begin{array}{l}((\ldots)^{\delta\gamma}\left(C_jC^j h_{,\gamma\sigma} - C^2 \cdot \Delta h_{,\gamma\sigma}\right) - C^2 \cdot h_{,\gamma\sigma j}\gamma^{j\beta}(\ldots)^{\delta\gamma}{}_{,\beta}) \cdot h_{,\gamma\gamma\delta} \\ +\ldots \end{array}\right),$$

$$(774)$$

$$\delta_\gamma W = 0$$

$$= \int_V d^n x \times \sqrt{\gamma} \cdot \left(\begin{array}{l}\ldots \\ + ((\ldots)^{\delta\gamma}\left(C_jC^j h_{,\delta\sigma} - C^2 \cdot \Delta h_{,\delta\sigma}\right) - C^2 \cdot h_{\delta\sigma\beta}\gamma^{j\beta}(\ldots)^{\delta\gamma}{}_{,j}) \cdot h_{,\delta\delta\gamma}\end{array}\right).$$

$$(775)$$

Now integration by parts (only of the corresponding parts being considered) yields:

$$-\int_V d^n x \times \left(\sqrt{\gamma} \cdot ((\ldots)^{\delta\gamma}\left(C_jC^j h_{,\gamma\sigma} - C^2 \cdot \Delta h_{,\gamma\sigma}\right) - C^2 \cdot h_{,\gamma\sigma j}\gamma^{j\beta}(\ldots)^{\delta\gamma}{}_{,\beta}))_{,\delta} \cdot h_{,\gamma\gamma},$$

$$(776)$$

$$-\int_V d^n x \times \left(\sqrt{\gamma} \cdot ((\ldots)^{\delta\gamma}\left(C_jC^j h_{,\delta\sigma} - C^2 \cdot \Delta h_{,\delta\sigma}\right) - C^2 \cdot h_{\delta\sigma\beta}\gamma^{j\beta}(\ldots)^{\delta\gamma}{}_{,j}))_{,\gamma} \cdot h_{,\delta\delta}.$$

$$(777)$$

This can be brought into the form given in the last line of Eq. (773).

We realized that we had to introduce a vector and substitute it for the function F in order to allow us to bring all parts of the variations (Eqs. (768) and (769)), which results from the Christoffel symbols of the covariant variation of the metric tensor into the same form to have an consistent intrinsic solution in the end. However, as our starting point already was a function vector $\mathbf{f}_\alpha$, we wonder whether not a second-order approach as introduced in section "3.12.1 Higher-Order Functional Approaches" would be more suitable.

5.2.1.2 Towards "Dirac"

Starting with one of the two equations (Eq. (757)), we will now again seek for a metric Dirac-like equation

$$
\begin{aligned}
(\ldots)^{\delta\gamma} &\left(C_j C^j f_\delta - C^2 \cdot \Delta f_\delta \right) - C^2 \cdot f_{\delta,j} \gamma^{j\beta} (\ldots)^{\delta\gamma}{}_{,\beta} = 0 \\
&= \mathbf{H}^\delta \cdot \mathbf{H}^\gamma \left(C_j C^j f_\delta - C^2 \cdot \Delta f_\delta \right) - C^2 \cdot f_{\delta,j} \gamma^{j\beta} \left(\mathbf{H}^\delta \cdot \mathbf{H}^\gamma \right)_{,\beta} \\
&= \mathbf{H}^\delta \cdot \mathbf{H}^\gamma \left(C_j C^j f_\delta - C^2 \cdot \Delta f_\delta \right) - C^2 \cdot f_{\delta,j} \gamma^{j\beta} \left(\mathbf{H}^\delta \right)_{,\beta} \cdot \mathbf{H}^\gamma - C^2 \cdot f_{\delta,j} \gamma^{j\beta} \mathbf{H}^\delta \cdot \left(\mathbf{H}^\gamma \right)_{,\beta} \\
&= \left[\frac{\mathbf{H}^\delta}{2} \cdot \left(C_j C^j f_\delta - C^2 \cdot \Delta f_\delta \right) - C^2 \cdot f_{\delta,j} \gamma^{j\beta} \left(\mathbf{H}^\delta \right)_{,\beta} \right] \cdot \mathbf{H}^\gamma \\
&+ \mathbf{H}^\delta \cdot \left[\frac{\mathbf{H}^\gamma}{2} \cdot \left(C_j C^j f_\delta - C^2 \cdot \Delta f_\delta \right) - C^2 \cdot f_{\delta,j} \gamma^{j\beta} \left(\mathbf{H}^\gamma \right)_{,\beta} \right].
\end{aligned}
\tag{778}
$$

As we recognize the principal similarity with our Dirac-like evaluations for the result with the quantum factor of the integrand in Eq. (523), we will try to find a better "Dirac-process" than we did there. So, we will repeat the evaluation Eqs. (528) to (535), but build in some technical changes. Thus, going back to Eq. (523), we may assume that the variation could also get zero if already the term to variate is zero. Thus, by demanding

$$
\delta_\gamma W = 0 = \int_V d^n x \, (\ldots)^{\delta\gamma} \times \sqrt{\gamma} \cdot \gamma_{ij} \delta \left(C^i C^j \cdot f_\delta f_\gamma + C^2 \cdot \gamma^{i\alpha} f_{\delta,\alpha} \gamma^{j\beta} f_{\gamma,\beta} \right)
$$

$$
\Rightarrow C^i C^j \cdot f_\delta f_\gamma + C^2 \cdot \gamma^{i\alpha} f_{\delta,\alpha} \gamma^{j\beta} f_{\gamma,\beta} = 0
\tag{779}
$$

we should be able to find suitable matrix objects $\mathbf{A}^\sigma$, $\mathbf{B}^\sigma$ allowing us the following separation:

$$
C^i C^j \cdot f_\delta f_\gamma + C^2 \cdot \gamma^{i\alpha} f_{\delta,\alpha} \gamma^{j\beta} f_{\gamma,\beta} = \left(\mathbf{A}^0 C^i f_\delta + C \cdot \mathbf{A}^\alpha \gamma^{i\alpha} f_{\delta,\alpha} \right) \left(\mathbf{B}^0 C^j f_\gamma + C \cdot \mathbf{B}^\beta \gamma^{j\beta} f_{\gamma,\beta} \right).
\tag{780}
$$

Apart from the fact that Dirac has used the third binomial formula for his factorization, this indeed already looks Dirac-like. However, before we start searching for suitable matrix candidates (which we will not do here anyway), we want to reconsider the total integrand in Eq. (528) and try for some better symmetrization:

$$\delta_\gamma W = 0 = \int_V d^n x \, (\dots)^{\delta\gamma} \times \sqrt{\gamma} \cdot \gamma_{ij} \delta \left(C^i C^j \cdot f_\delta f_\gamma + C^2 \cdot \gamma^{i\alpha} f_{\delta,\alpha} \gamma^{j\beta} f_{\gamma,\beta} \right)$$

$$= \int_V d^n x \times \sqrt{\gamma} \cdot \delta \left(C^i \overbrace{\mathbf{e}_i \cdot \mathbf{e}_j}^{\gamma_{ij}} C^j f_\delta \overbrace{\mathbf{H}^\delta \cdot \mathbf{H}^\gamma}^{(\dots)^{\delta\gamma}} f_\gamma + C^2 \cdot \overbrace{\mathbf{e}_i \cdot \mathbf{e}_j}^{\gamma_{ij}} \gamma^{i\alpha} f_{\delta,\alpha} \gamma^{j\beta} \overbrace{\mathbf{H}^\delta \cdot \mathbf{H}^\gamma}^{(\dots)^{\delta\gamma}} f_{\gamma,\beta} \right)$$

$$\Rightarrow C^i \overbrace{\mathbf{e}_i \cdot \mathbf{e}_j}^{\gamma_{ij}} C^j f_\delta \overbrace{\mathbf{H}^\delta \cdot \mathbf{H}^\gamma}^{(\dots)^{\delta\gamma}} f_\gamma + C^2 \cdot \overbrace{\mathbf{e}_i \cdot \mathbf{e}_j}^{\gamma_{ij}} \gamma^{i\alpha} f_{\delta,\alpha} \gamma^{j\beta} \overbrace{\mathbf{H}^\delta \cdot \mathbf{H}^\gamma}^{(\dots)^{\delta\gamma}} f_{\gamma,\beta} = 0. \tag{781}$$

This can be grouped in two different ways:

$$\delta_\gamma W = 0 = \int_V d^n x \cdot \mathbf{H}^\delta \cdot \mathbf{H}^\gamma \cdot \mathbf{e}_i \cdot \mathbf{e}_j \sqrt{\gamma} \cdot \delta_\sigma \left(C^i C^j \cdot f_\delta f_\gamma + C^2 \cdot \gamma^{i\alpha} f_{\delta,\alpha} \gamma^{j\beta} f_{\gamma,\beta} \right)$$

$$= \int_V d^n x \cdot \left(A^2 \cdot S_i^\delta S_j^\gamma + B^2 \cdot S_i^\gamma S_j^\delta \right) \sqrt{\gamma} \cdot \delta_\sigma \left(C^i C^j \cdot f_\delta f_\gamma + C^2 \cdot \gamma^{i\alpha} f_{\delta,\alpha} \gamma^{j\beta} f_{\gamma,\beta} \right);$$

$$A^2 + B^2 = 1. \tag{782}$$

For illustration, we consider the simplest case, where we set $(\dots)^{\delta\gamma} = D \cdot \gamma^{\delta\gamma}$ and obtain:

$$\delta_\gamma W = 0 = \int_V d^n x \cdot D \left(A^2 \cdot \delta_i^\delta \delta_j^\gamma + B^2 \cdot \delta_i^\gamma \delta_j^\delta \right) \sqrt{\gamma} \cdot \delta_\sigma \left(C^i C^j \cdot f_\delta f_\gamma + C^2 \cdot \gamma^{i\alpha} f_{\delta,\alpha} \gamma^{j\beta} f_{\gamma,\beta} \right)$$

$$= \int_V d^n x \cdot D \sqrt{\gamma} \cdot \delta_\sigma \left(A^2 \left(C^\delta C^\gamma \cdot f_\delta f_\gamma + C^2 \cdot \gamma^{\delta\alpha} f_{\delta,\alpha} \gamma^{\gamma\beta} f_{\gamma,\beta} \right) \right.$$

$$\left. + B^2 \left(C^\gamma C^\delta \cdot f_\delta f_\gamma + C^2 \cdot \gamma^{\gamma\alpha} f_{\delta,\alpha} \gamma^{\delta\beta} f_{\gamma,\beta} \right) \right)$$

$$= \int_V d^n x \cdot D \sqrt{\gamma} \cdot \delta_\sigma \left((A^2 + B^2) C^\delta C^\gamma \cdot f_\delta f_\gamma \right.$$

$$\left. + C^2 \cdot \left(A^2 \cdot \gamma^{\delta\alpha} f_{\delta,\alpha} \gamma^{\gamma\beta} f_{\gamma,\beta} + B^2 \cdot \gamma^{\gamma\alpha} f_{\delta,\alpha} \gamma^{\delta\beta} f_{\gamma,\beta} \right) \right)$$

$$= \int_V d^n x \cdot D \sqrt{\gamma} \cdot \delta_\sigma \left(\overbrace{A \cdot C^\gamma f_\delta}^{a_1} \cdot \overbrace{A \cdot C^\delta f_\gamma}^{a_2} + \overbrace{B \cdot C^\delta f_\delta}^{a_3} \cdot \overbrace{B \cdot C^\gamma f_\gamma}^{a_4} \right.$$

$$\left. + \underbrace{C \cdot A \cdot f_{\delta,}{}^\delta}_{b_1} \cdot \underbrace{C \cdot A \cdot f_{\gamma,}{}^\gamma}_{b_2} + \underbrace{C \cdot B \cdot f_{\delta,}{}^\gamma}_{\Theta_\delta^\gamma} \cdot \underbrace{C \cdot B \cdot f_{\gamma,}{}^\delta}_{\Theta_\gamma^\delta} \right).$$

$$\tag{783}$$

We already see that factorization does lead to a true matrix of equations, while in the Dirac case the result is always a vector of equations. Thus, we cannot hope to perfectly reproduce the Dirac equation or something similar from the expression in parenthesis in the last line (:::) of Eq. (783). Still intending to obtain differential equations of first order, the apparently simplest way seems to be via factorization of (::::).

Thus, we can apply the method of vectorial root extraction or factorization (e.g. [8]) and write:

$$
\left(
\overbrace{A \cdot C^\gamma f_\delta}^{a_1} \cdot \overbrace{A \cdot C^\delta f_\gamma}^{a_2} + \overbrace{B \cdot C^\delta f_\delta}^{a_3} \cdot \overbrace{B \cdot C^\gamma f_\gamma}^{a_4} \\
+ \overbrace{C \cdot A \cdot f_{\delta,}{}^\delta}^{b_1} \cdot \overbrace{C \cdot A \cdot f_{\gamma,}{}^\gamma}^{b_2} + \overbrace{C \cdot B \cdot f_{\delta,}{}^\gamma}^{\Theta_\delta^\gamma} \cdot \overbrace{C \cdot B \cdot f_{\gamma,}{}^\delta}^{\Theta_\gamma^\delta}
\right)
$$

$$
= \overbrace{a_1 \cdot a_2 + a_3 \cdot a_4}^{a^2} + \overbrace{b_1 \cdot b_2}^{b^2} + \Theta_\delta^\gamma \Theta_\gamma^\delta = a^2 + b^2 + \Theta_\delta^\gamma \Theta_\gamma^\delta
$$

$$
= \frac{1}{8 \cdot n} \cdot
\begin{pmatrix}
\sqrt{n} \cdot a_1 + a_3 \cdot \delta_\delta^\gamma + b_1 \cdot \delta_\delta^\gamma + \sqrt{n} \cdot \Theta_\delta^\gamma \\
\sqrt{n} \cdot a_1 + a_3 \cdot \delta_\delta^\gamma + b_1 \cdot \delta_\delta^\gamma - \sqrt{n} \cdot \Theta_\delta^\gamma \\
\sqrt{n} \cdot a_1 + a_3 \cdot \delta_\delta^\gamma - b_1 \cdot \delta_\delta^\gamma + \sqrt{n} \cdot \Theta_\delta^\gamma \\
\sqrt{n} \cdot a_1 + a_3 \cdot \delta_\delta^\gamma - b_1 \cdot \delta_\delta^\gamma - \sqrt{n} \cdot \Theta_\delta^\gamma \\
\sqrt{n} \cdot a_1 - a_3 \cdot \delta_\delta^\gamma + b_1 \cdot \delta_\delta^\gamma + \sqrt{n} \cdot \Theta_\delta^\gamma \\
\sqrt{n} \cdot a_1 - a_3 \cdot \delta_\delta^\gamma + b_1 \cdot \delta_\delta^\gamma - \sqrt{n} \cdot \Theta_\delta^\gamma \\
\sqrt{n} \cdot a_1 - a_3 \cdot \delta_\delta^\gamma - b_1 \cdot \delta_\delta^\gamma + \sqrt{n} \cdot \Theta_\delta^\gamma \\
\sqrt{n} \cdot a_1 - a_3 \cdot \delta_\delta^\gamma - b_1 \cdot \delta_\delta^\gamma - \sqrt{n} \cdot \Theta_\delta^\gamma
\end{pmatrix}
$$

$$
\cdot
\begin{pmatrix}
\sqrt{n} \cdot a_2 + a_4 \cdot \delta_\gamma^\delta + b_2 \cdot \delta_\gamma^\delta + \sqrt{n} \cdot \Theta_\gamma^\delta \\
\sqrt{n} \cdot a_2 + a_4 \cdot \delta_\gamma^\delta + b_2 \cdot \delta_\gamma^\delta - \sqrt{n} \cdot \Theta_\gamma^\delta \\
\sqrt{n} \cdot a_2 + a_4 \cdot \delta_\gamma^\delta - b_2 \cdot \delta_\gamma^\delta + \sqrt{n} \cdot \Theta_\gamma^\delta \\
\sqrt{n} \cdot a_2 + a_4 \cdot \delta_\gamma^\delta - b_2 \cdot \delta_\gamma^\delta - \sqrt{n} \cdot \Theta_\gamma^\delta \\
\sqrt{n} \cdot a_2 - a_4 \cdot \delta_\gamma^\delta + b_2 \cdot \delta_\gamma^\delta + \sqrt{n} \cdot \Theta_\gamma^\delta \\
\sqrt{n} \cdot a_2 - a_4 \cdot \delta_\gamma^\delta + b_2 \cdot \delta_\gamma^\delta - \sqrt{n} \cdot \Theta_\gamma^\delta \\
\sqrt{n} \cdot a_2 - a_4 \cdot \delta_\gamma^\delta - b_2 \cdot \delta_\gamma^\delta + \sqrt{n} \cdot \Theta_\gamma^\delta \\
\sqrt{n} \cdot a_2 - a_4 \cdot \delta_\gamma^\delta - b_2 \cdot \delta_\gamma^\delta - \sqrt{n} \cdot \Theta_\gamma^\delta
\end{pmatrix}
= K^2. \quad (784)
$$

Thereby n denotes the number of dimensions and K shall be a constant, where we take it, that the variation of a constant vanishes. Obviously, the easiest way to fulfill Eq. (784) would be to demand that each one of the factors gives a vector of constants, which means:

$$0 = a^2 + b^2 + \Theta_\delta^\gamma \Theta_\gamma^\delta; \quad \sum_{i=1}^{8} K_{i\delta}^\gamma k_{i\gamma}^\delta = 8 \cdot n \cdot K^2$$

$$\Rightarrow \begin{pmatrix} \sqrt{n} \cdot a_1 + a_3 \cdot \delta_\delta^\gamma + b_1 \cdot \delta_\delta^\gamma + \sqrt{n} \cdot \Theta_\delta^\gamma \\ \sqrt{n} \cdot a_1 + a_3 \cdot \delta_\delta^\gamma + b_1 \cdot \delta_\delta^\gamma - \sqrt{n} \cdot \Theta_\delta^\gamma \\ \sqrt{n} \cdot a_1 + a_3 \cdot \delta_\delta^\gamma - b_1 \cdot \delta_\delta^\gamma + \sqrt{n} \cdot \Theta_\delta^\gamma \\ \sqrt{n} \cdot a_1 + a_3 \cdot \delta_\delta^\gamma - b_1 \cdot \delta_\delta^\gamma - \sqrt{n} \cdot \Theta_\delta^\gamma \\ \sqrt{n} \cdot a_1 - a_3 \cdot \delta_\delta^\gamma + b_1 \cdot \delta_\delta^\gamma + \sqrt{n} \cdot \Theta_\delta^\gamma \\ \sqrt{n} \cdot a_1 - a_3 \cdot \delta_\delta^\gamma + b_1 \cdot \delta_\delta^\gamma - \sqrt{n} \cdot \Theta_\delta^\gamma \\ \sqrt{n} \cdot a_1 - a_3 \cdot \delta_\delta^\gamma - b_1 \cdot \delta_\delta^\gamma + \sqrt{n} \cdot \Theta_\delta^\gamma \\ \sqrt{n} \cdot a_1 - a_3 \cdot \delta_\delta^\gamma - b_1 \cdot \delta_\delta^\gamma - \sqrt{n} \cdot \Theta_\delta^\gamma \end{pmatrix}$$

$$= \begin{pmatrix} K_{1\delta}^\gamma \\ K_{2\delta}^\gamma \\ K_{3\delta}^\gamma \\ K_{4\delta}^\gamma \\ K_{5\delta}^\gamma \\ K_{6\delta}^\gamma \\ K_{7\delta}^\gamma \\ K_{8\delta}^\gamma \end{pmatrix} \cup \begin{pmatrix} \sqrt{n} \cdot a_2 + a_4 \cdot \delta_\gamma^\delta + b_2 \cdot \delta_\gamma^\delta + \sqrt{n} \cdot \Theta_\gamma^\delta \\ \sqrt{n} \cdot a_2 + a_4 \cdot \delta_\gamma^\delta + b_2 \cdot \delta_\gamma^\delta - \sqrt{n} \cdot \Theta_\gamma^\delta \\ \sqrt{n} \cdot a_2 + a_4 \cdot \delta_\gamma^\delta - b_2 \cdot \delta_\gamma^\delta + \sqrt{n} \cdot \Theta_\gamma^\delta \\ \sqrt{n} \cdot a_2 + a_4 \cdot \delta_\gamma^\delta - b_2 \cdot \delta_\gamma^\delta - \sqrt{n} \cdot \Theta_\gamma^\delta \\ \sqrt{n} \cdot a_2 - a_4 \cdot \delta_\gamma^\delta + b_2 \cdot \delta_\gamma^\delta + \sqrt{n} \cdot \Theta_\gamma^\delta \\ \sqrt{n} \cdot a_2 - a_4 \cdot \delta_\gamma^\delta + b_2 \cdot \delta_\gamma^\delta - \sqrt{n} \cdot \Theta_\gamma^\delta \\ \sqrt{n} \cdot a_2 - a_4 \cdot \delta_\gamma^\delta - b_2 \cdot \delta_\gamma^\delta + \sqrt{n} \cdot \Theta_\gamma^\delta \\ \sqrt{n} \cdot a_2 - a_4 \cdot \delta_\gamma^\delta - b_2 \cdot \delta_\gamma^\delta - \sqrt{n} \cdot \Theta_\gamma^\delta \end{pmatrix} = \begin{pmatrix} k_{1\gamma}^\delta \\ k_{2\gamma}^\delta \\ k_{3\gamma}^\delta \\ k_{4\gamma}^\delta \\ \vdots \\ k_{8\gamma}^\delta \end{pmatrix}.$$

$$(785)$$

Just taking the first factor and write it all out, leads to the following set of differential equations:

$$\begin{pmatrix} \sqrt{n} \cdot A \cdot C^\gamma f_\delta + B \cdot C^\alpha f_\alpha \cdot \delta_\delta^\gamma + C \cdot A \cdot f_{\alpha,}{}^\alpha \cdot \delta_\delta^\gamma + \sqrt{n} \cdot B \cdot f_{\delta,}{}^\gamma \\ \sqrt{n} \cdot A \cdot C^\gamma f_\delta + B \cdot C^\alpha f_\alpha \cdot \delta_\delta^\gamma + C \cdot A \cdot f_{\alpha,}{}^\alpha \cdot \delta_\delta^\gamma - \sqrt{n} \cdot B \cdot f_{\delta,}{}^\gamma \\ \vdots \\ \sqrt{n} \cdot A \cdot C^\gamma f_\delta - B \cdot C^\alpha f_\alpha \cdot \delta_\delta^\gamma - C \cdot A \cdot f_{\alpha,}{}^\alpha \cdot \delta_\delta^\gamma - \sqrt{n} \cdot B \cdot f_{\delta,}{}^\gamma \end{pmatrix} = \begin{pmatrix} K_{1\delta}^\gamma \\ K_{2\delta}^\gamma \\ \vdots \\ K_{8\delta}^\gamma \end{pmatrix},$$

$$(786)$$

where we have exchanged the dummy indices for the a and b terms in order to avoid confusion with the running indices of the matrix terms δ_δ^γ, $C^\gamma f_\delta$ and $f_{\delta,}{}^\gamma$. It needs to be pointed out that the complex factorization performed above is not the only option. One could also group as follows:

$$K^2 = \frac{1}{2 \cdot n} \cdot \left(\overbrace{A \cdot C^\gamma f_\delta}^{a_1} \cdot \overbrace{A \cdot C^\delta f_\gamma}^{a_2} + C \cdot A \cdot f_{\delta,}{}^\delta \cdot \overbrace{C \cdot A \cdot f_{\gamma,}{}^\gamma}^{b_2} \right.$$
$$\left. + \underbrace{B \cdot C^\delta f_\delta}_{a_3} \cdot \underbrace{B \cdot C^\gamma f_\gamma}_{a_4} + C \cdot B \cdot f_{\delta,}{}^\gamma \cdot \underbrace{C \cdot B \cdot f_{\gamma,}{}^\delta}_{\Theta_\gamma^\delta} \right)$$

$$= \frac{1}{2 \cdot n} \cdot \left(\overbrace{a_1 \cdot a_2}^{a1^2} + \overbrace{a_3 \cdot a_4}^{a2^2} + \overbrace{b_1 \cdot b_2}^{b^2} + \Theta_\delta^\gamma \Theta_\gamma^\delta \right) = \frac{1}{2 \cdot n} \cdot \left(\overbrace{a1^2 + b^2}^{K1^2} + \overbrace{a2^2 + \Theta_\delta^\gamma \Theta_\gamma^\delta}^{K2^2} \right)$$

$$= \frac{1}{2 \cdot n} \cdot \left(\underbrace{\left(\begin{array}{c} \sqrt{n} \cdot a_1 + b_1 \cdot \delta_\delta^\gamma \\ \sqrt{n} \cdot a_1 - b_1 \cdot \delta_\delta^\gamma \end{array} \right) \cdot \left(\begin{array}{c} \sqrt{n} \cdot a_2 + b_2 \cdot \delta_\gamma^\delta \\ \sqrt{n} \cdot a_2 - b_2 \cdot \delta_\gamma^\delta \end{array} \right)}_{K1^2} \right.$$

$$\left. + \underbrace{\left(\begin{array}{c} a_3 \cdot \delta_\delta^\gamma + \sqrt{n} \cdot \Theta_\delta^\gamma \\ a_3 \cdot \delta_\delta^\gamma - \sqrt{n} \cdot \Theta_\delta^\gamma \end{array} \right) \cdot \left(\begin{array}{c} a_4 \cdot \delta_\gamma^\delta + \sqrt{n} \cdot \Theta_\gamma^\delta \\ a_4 \cdot \delta_\gamma^\delta - \sqrt{n} \cdot \Theta_\gamma^\delta \end{array} \right)}_{K2^2} \right). \tag{787}$$

The corresponding first-order differential equations subsequently read:

$$\left(\begin{array}{c} \sqrt{n} \cdot a_1 + b_1 \cdot \delta_\delta^\gamma \\ \sqrt{n} \cdot a_1 - b_1 \cdot \delta_\delta^\gamma \end{array} \right) = \left(\begin{array}{c} K1_{1\delta}^\gamma \\ K1_{2\delta}^\gamma \end{array} \right); \quad \left(\begin{array}{c} \sqrt{n} \cdot a_2 + b_2 \cdot \delta_\gamma^\delta \\ \sqrt{n} \cdot a_2 - b_2 \cdot \delta_\gamma^\delta \end{array} \right) = \left(\begin{array}{c} k1_{1\gamma}^\delta \\ k1_{2\gamma}^\delta \end{array} \right)$$

$$\left(\begin{array}{c} a_3 \cdot \delta_\delta^\gamma + \sqrt{n} \cdot \Theta_\delta^\gamma \\ a_3 \cdot \delta_\delta^\gamma - \sqrt{n} \cdot \Theta_\delta^\gamma \end{array} \right) = \left(\begin{array}{c} K2_{1\delta}^\gamma \\ K2_{2\delta}^\gamma \end{array} \right); \quad \left(\begin{array}{c} a_4 \cdot \delta_\gamma^\delta + \sqrt{n} \cdot \Theta_\gamma^\delta \\ a_4 \cdot \delta_\gamma^\delta - \sqrt{n} \cdot \Theta_\gamma^\delta \end{array} \right) = \left(\begin{array}{c} k2_{1\gamma}^\delta \\ k2_{2\gamma}^\delta \end{array} \right)$$

$$\sum_{i=1}^{2} K1_{i\delta}^\gamma k1_{i\gamma}^\delta = K1^2; \quad \sum_{i=1}^{2} K2_{i\delta}^\gamma k2_{i\gamma}^\delta = K2^2. \tag{788}$$

Now we have the more or less philosophical problem that there are four different equations for all function components of which all must be fulfilled. We therefore consider this only a temporary result where more technical honing and discussion will be needed.

Before coming to this, however, we are going to consider more functional vector approaches for our extended Einstein–Hilbert action.

5.2.2 Vector Approach: Some Refinements

Reusing the refinements from our scalar approach (Eq. (719)), we here apply a new vector function approach as follows:

$$\delta_\gamma W = 0 = \int_V d^n x \sqrt{\gamma} \cdot \left(R^{\delta\gamma} - \frac{1}{2} R \cdot \gamma^{\delta\gamma} + \Lambda \cdot \gamma^{\delta\gamma} + \kappa \cdot T^{\delta\gamma} \right)$$

$$\times \begin{pmatrix} \delta g_{\delta\gamma} = \delta \left(\frac{\partial G^i [x_{\nu i}]}{\partial x^\delta} \cdot \frac{\partial G^j [x_{\nu i}]}{\partial x^\gamma} \gamma_{ij} \right) = \gamma_{ij} \cdot \delta \left(\frac{\partial G^i [x_{\nu i}]}{\partial x^\delta} \cdot \frac{\partial G^j [x_{\nu i}]}{\partial x^\gamma} \right) \\ = \frac{\gamma_{ij}}{4} \delta \left(\delta^i_\delta \delta^j_\gamma \begin{Bmatrix} 1 + M \cdot C_k f^k + \mathbf{C}^\alpha C_k f^k_{,\alpha} \\ 1 + M \cdot C_k f^k - \mathbf{C}^\alpha C_k f^k_{,\alpha} \\ 1 - M \cdot C_k f^k + \mathbf{C}^\alpha C_k f^k_{,\alpha} \\ 1 - M \cdot C_k f^k - \mathbf{C}^\alpha C_k f^k_{,\alpha} \end{Bmatrix} \begin{Bmatrix} 1 + M \cdot C_l f^l + \mathbf{C}^\beta C_l f^l_{,\beta} \\ 1 + M \cdot C_l f^l - \mathbf{C}^\beta C_l f^l_{,\beta} \\ 1 - M \cdot C_l f^l + \mathbf{C}^\beta C_l f^l_{,\beta} \\ 1 - M \cdot C_l f^l - \mathbf{C}^\beta C_l f^l_{,\beta} \end{Bmatrix} \right) \\ = \delta \left[C_k C_l \delta^i_\delta \delta^j_\gamma \left(1 + M^2 \cdot f^k f^l + \mathbf{C}^\alpha \mathbf{C}^\beta \cdot f^k_{,\alpha} f^l_{,\beta} \right) \right] \gamma_{ij} \\ = C_k C_l \delta^i_\delta \delta^j_\gamma \left(M^2 \cdot \left(f^k f^l_{,\sigma} + f^k_{,\sigma} f^l \right) + \mathbf{C}^\alpha \mathbf{C}^\beta \cdot \left(f^k_{,\alpha\sigma} f^l_{,\beta} + f^k_{,\alpha} f^l_{,\beta\sigma} \right) \right) \gamma_{ij} \delta^\sigma \end{pmatrix}$$

$$(789)$$

Integration by parts of the second addend in the last line, performed as above, yields:

$$\delta_\gamma W = 0 = \int_V d^n x \times \begin{pmatrix} \sqrt{\gamma} \cdot (\ldots)^{\delta\gamma} \gamma_{\delta\gamma} C_k C_l M^2 \cdot \left(f^k f^l_{,\sigma} + f^k_{,\sigma} f^l \right) \\ - \begin{bmatrix} [(\ldots)^{\delta\gamma} \gamma_{\delta\gamma} \mathbf{C}^\alpha \mathbf{C}^\beta C_k C_l \sqrt{\gamma} \cdot f^l_{,\beta}]_{,\alpha} \cdot f^k_{,\sigma} \\ + [(\ldots)^{\delta\gamma} \gamma_{\delta\gamma} \mathbf{C}^\alpha \mathbf{C}^\beta C_k C_l \sqrt{\gamma} \cdot f^k_{,\alpha}]_{,\beta} \cdot f^l_{,\sigma} \end{bmatrix} \end{pmatrix}$$

$$= \int_V d^n x \times \begin{pmatrix} \left(\sqrt{\gamma} \cdot (\ldots)^{\delta\gamma} \gamma_{\delta\gamma} C_k C_l M^2 f^l - \left[(\ldots)^{\delta\gamma} \gamma_{\delta\gamma} \overbrace{\mathbf{C}^\alpha \mathbf{C}^\beta}^{=d \cdot \gamma^{\alpha\beta}} C_k C_l \sqrt{\gamma} \cdot f^l_{,\beta} \right]_{,\alpha} \right) \cdot f^k_{,\sigma} \\ + \left(\sqrt{\gamma} \cdot (\ldots)^{\delta\gamma} \gamma_{\delta\gamma} C_k C_l M^2 f^k - \left[(\ldots)^{\delta\gamma} \gamma_{\delta\gamma} \mathbf{C}^\alpha \mathbf{C}^\beta C_k C_l \sqrt{\gamma} \cdot f^k_{,\alpha} \right]_{,\beta} \right) \cdot f^l_{,\sigma} \end{pmatrix}$$

$$= \int_V d^n x \times \begin{pmatrix} \left(\sqrt{\gamma} \cdot (\ldots)^{\delta\gamma} \gamma_{\delta\gamma} C_k C_l M^2 f^l - d \cdot \begin{pmatrix} [(\ldots)^{\delta\gamma} \gamma_{\delta\gamma} C_k C_l]_{,\alpha} \gamma^{\alpha\beta} \sqrt{\gamma} \cdot f^l_{,\beta} \\ + (\ldots)^{\delta\gamma} \gamma_{\delta\gamma} C_k C_l \sqrt{\gamma} \cdot \Delta f^l \end{pmatrix} \right) \cdot f^k_{,\sigma} \\ + \left(\sqrt{\gamma} \cdot (\ldots)^{\delta\gamma} \gamma_{\delta\gamma} C_k C_l M^2 f^k - d \cdot \begin{pmatrix} [(\ldots)^{\delta\gamma} \gamma_{\delta\gamma} C_k C_l]_{,\beta} \gamma^{\alpha\beta} \sqrt{\gamma} \cdot f^k_{,\alpha} \\ + (\ldots)^{\delta\gamma} \gamma_{\delta\gamma} C_k C_l \sqrt{\gamma} \cdot \Delta f^k \end{pmatrix} \right) \cdot f^l_{,\sigma} \end{pmatrix}$$

$$(790)$$

Subsequently, we obtain quantum equations of the kind:

$$\left(\sqrt{\gamma} \cdot (\ldots)^{\delta\gamma} \gamma_{\delta\gamma} C_k C_l M^2 f^l - d \cdot \begin{pmatrix} [(\ldots)^{\delta\gamma} \gamma_{\delta\gamma} C_k C_l]_{,\alpha} \gamma^{\alpha\beta} \sqrt{\gamma} \cdot f^l_{,\beta} \\ + (\ldots)^{\delta\gamma} \gamma_{\delta\gamma} C_k C_l \sqrt{\gamma} \cdot \Delta f^l \end{pmatrix} \right) \cdot f^k_{,\sigma} = 0$$

$$\Rightarrow (\ldots)^{\delta\gamma} \gamma_{\delta\gamma} C_k C_l M^2 f^l - d \cdot \left([(\ldots)^{\delta\gamma} \gamma_{\delta\gamma} C_k C_l]_{,\alpha} \gamma^{\alpha\beta} f^l_{,\beta} + (\ldots)^{\delta\gamma} \gamma_{\delta\gamma} C_k C_l \Delta f^l \right) = 0.$$

$$(791)$$

In the case of the vanishing of the derivative terms $[(\ldots)^{\delta\gamma} \gamma_{\delta\gamma} C_k C_l]_{,\alpha} = 0$ we obtain the simple Klein–Gordon equation with:

$$(\ldots)^{\delta\gamma} \gamma_{\delta\gamma} C_k C_l M^2 f^l - d \cdot \left(0 \cdot \gamma^{\alpha\beta} f^l_{,\beta} + (\ldots)^{\delta\gamma} \gamma_{\delta\gamma} C_k C_l \Delta f^l \right) = 0$$

$$\Rightarrow M^2 f^l - d \cdot \Delta f^l = 0.$$

$$(792)$$

As before, we note that the function vector $\mathbf{f}^k$ or $\mathbf{f}^l$ in Eq. (791), in contrast to the classical Klein–Gordon equation, if constructed out of the classical Dirac equation, does not contain all vector components in a decoupled or degenerated manner, but provides—in principle—independent equations for each vector component. It is the term $\left[(\ldots)^{\delta\gamma} \, \gamma_{\delta\gamma} C_k C_l \right]_{,\alpha} \gamma^{\alpha\beta} f^l_{,\beta}$ in Eq. (791) (last line) assuring this. Thus, in situations where we have $\left[(\ldots)^{\delta\gamma} \, \gamma_{\delta\gamma} C_k C_l \right]_{,\alpha} = 0$, we would have the classical decoupling of the wave function components. So far, we have not fixed the matrix $C_k C_l$ yet.

5.2.2.1 Towards "Dirac" again and about a potential metric Pauli exclusion principle

Realizing that the constant terms in the $\{\ldots\}$-vectors in Eq. (725) are vanishing anyway during the variation, we could also already start with a variation of the form:

$$\delta_\gamma W = 0 = \int_V d^n x \sqrt{g} \cdot \left(R^{\delta\gamma} - \frac{1}{2} R \cdot g^{\delta\gamma} + \Lambda \cdot g^{\delta\gamma} + \kappa \cdot T^{\delta\gamma} \right)$$

$$\times \left(\begin{array}{c} \delta g_{\delta\gamma} = \delta \left(\frac{\partial G^i [x_{\forall i}]}{\partial x^\delta} \cdot \frac{\partial G^j [x_{\forall i}]}{\partial x^\gamma} \gamma_{ij} \right) = g_{ij} \cdot \delta \left(\frac{\partial G^i [x_{\forall i}]}{\partial x^\delta} \cdot \frac{\partial G^j [x_{\forall i}]}{\partial x^\gamma} \right) \\ = \frac{g_{ij}}{4} \delta \left(\delta^i_\delta \delta^j_\gamma \left\{ \begin{array}{c} M \cdot C^k f_k + \mathbf{C}^\alpha C^k f_{k,\alpha} \\ M \cdot C^k f_k - \mathbf{C}^\alpha C^k f_{k,\alpha} \end{array} \right\} \left\{ \begin{array}{c} M \cdot C^l f_l + \mathbf{C}^\beta C^l f_{l,\beta} \\ M \cdot C^l f_l - \mathbf{C}^\beta C^l f_{l,\beta} \end{array} \right\} \right) \end{array} \right) \cdot$$

$$\tag{793}$$

As done in the scalar case above, we now symmetrize our $\{\ldots\}$-vector as follows:

$$\left\{ \begin{array}{c} M \cdot C^k f_k + i \cdot \mathbf{C}^\alpha C^k f_{k,\alpha} \\ M \cdot C^k f_k - i \cdot \mathbf{C}^\alpha C^k f_{k,\alpha} \end{array} \right\} \left\{ \begin{array}{c} M \cdot C^l f_l - i \cdot \mathbf{C}^\beta C^l f_{l,\beta} \\ M \cdot C^l f_l + i \cdot \mathbf{C}^\beta C^l f_{l,\beta} \end{array} \right\}$$

$$= M^2 \cdot C^k f_k \cdot C^l f_l + \mathbf{C}^\beta C^l f_{l,\beta} \mathbf{C}^\alpha C^k f_{k,\alpha}. \tag{794}$$

Assuming the situation in 4 dimensions, applying again $\mathbf{C}^\beta \mathbf{C}^\alpha = d \cdot g^{\alpha\beta}$ and taking the Minkowski metric:

$$\eta_{\alpha\beta} = \begin{pmatrix} -c^2 & 0 & 0 & 0 \\ 0 & 1 & 0 & 0 \\ 0 & 0 & 1 & 0 \\ 0 & 0 & 0 & 1 \end{pmatrix} \tag{795}$$

makes the apparently two vector component products prod1 and prod2 resulting from Eq. (742), yielding

$$\left\{ \begin{array}{l} M \cdot C^k f_k + i \cdot \mathbf{C}^\alpha C^k f_{k,\alpha} \\ M \cdot C^k f_k - i \cdot \mathbf{C}^\alpha C^k f_{k,\alpha} \end{array} \right\} \left\{ \begin{array}{l} M \cdot C^l f_l - i \cdot \mathbf{C}^\beta C^l f_{l,\beta} \\ M \cdot C^l f_l + i \cdot \mathbf{C}^\beta C^l f_{l,\beta} \end{array} \right\}$$

$$= \left(\overbrace{ \begin{array}{l} \left(M \cdot C^l f_l - i \cdot \mathbf{C}^\beta C^l f_{l,\beta} \right) \cdot \left(M \cdot C^k f_k + i \cdot \mathbf{C}^\alpha C^k f_{k,\alpha} \right) \\ + \left(M \cdot C^l f_l + i \cdot \mathbf{C}^\beta C^l f_{l,\beta} \right) \cdot \left(M \cdot C^k f_k - i \cdot \mathbf{C}^\alpha C^k f_{k,\alpha} \right) \end{array} }^{\text{prod1}} \right)_{\underbrace{\qquad\qquad\qquad}_{\text{prod2}}}, \quad (796)$$

identical.

Thus, just one of the four factors:

$$M \cdot C^k f_k + i \cdot \mathbf{C}^\alpha C^k f_{k,\alpha}, \; M \cdot C^k f_k - i \cdot \mathbf{C}^\alpha C^k f_{k,\alpha},$$

$$M \cdot C^l f_l - i \cdot \mathbf{C}^\beta C^l f_{l,\beta}, \; M \cdot C^l f_l + i \cdot \mathbf{C}^\beta C^l f_{l,\beta}, \qquad (797)$$

needs to give zero in order to have the whole scalar product (Eq. (794)) to vanish.

Now, as before at the end of section "5.1.3 Scalar Approach: Avoiding the Introduction of γ^{ab}/Towards 'Dirac", we insist that instead of a vector $\{\ldots\}$-structure, we intend to obtain the result from Eq. (794) via a quaternion structure as applied by Dirac [143]. And in fact we can easily see that this can be achieved by the Dirac matrices as given in Eqs. (561) and (562) and yield a "Dirac-like" equation:

$$M \cdot C^k f_k + \gamma^\beta C^k f_{k,\beta} = 0. \qquad (798)$$

But for some reason the vector structure of $\mathbf{f}_k$ is not right, which is to say it is not truly Dirac-compatible. But why, so we ask ourselves, should the simple switch from our $\{\ldots\}$-vector form (where the $\mathbf{f}_k$ are perfectly ok) to the quaternions demand a complete shift of the vector structure of the $\mathbf{f}_k$?

Something, obviously, is missing!

We try to solve the riddle by considering the particle at rest problem with just a scalar $f = f[t]$. From Eq. (745), we extract the two solutions:

$$f = C_t \cdot e^{\pm i \cdot \frac{M}{c0} \cdot t}, \qquad (799)$$

where we clearly recognize the matter and antimatter solution. The same solution is directly been obtained from the right hand side of Eq. (742) with:

$$M^2 \cdot f \cdot f + \mathbf{C}^\beta f_{,\beta} \mathbf{C}^\alpha f_{,\alpha} = 0$$

$$= \left\{ \begin{array}{l} M \cdot f + i \cdot \mathbf{C}^\beta f_{,\beta} \\ M \cdot f - i \cdot \mathbf{C}^\beta f_{,\beta} \end{array} \right\} \left\{ \begin{array}{l} M \cdot f - i \cdot \mathbf{C}^\alpha f_{,\alpha} \\ M \cdot f + i \cdot \mathbf{C}^\alpha f_{,\alpha} \end{array} \right\} = \left\{ \begin{array}{l} M \cdot f + \mathbf{C}^\beta f_{,\beta} \\ M \cdot f - \mathbf{C}^\beta f_{,\beta} \end{array} \right\} \left\{ \begin{array}{l} M \cdot f + \mathbf{C}^\alpha f_{,\alpha} \\ M \cdot f - \mathbf{C}^\alpha f_{,\alpha} \end{array} \right\},$$

$$(800)$$

where it does not matter which form of the $\{\ldots\}$-vectors (symmetric or not) we had. As this is just the classical Dirac particle at rest solution with matter and antimatter, we simply wonder where the spin may be hiding within our approach.

In order to find it, we are now going back to our starting approach for the scalar f from Eq. (725) and simply restructure the $\{\ldots\}$-vectors according to what we did in Eqs. (741), (742) and (794):

$$\delta_\gamma W = 0 = \int_V d^n x \sqrt{g} \cdot \left(R^{\delta\gamma} - \frac{1}{2} R \cdot g^{\delta\gamma} + \Lambda \cdot g^{\delta\gamma} + \kappa \cdot T^{\delta\gamma} \right)$$

$$
\begin{pmatrix}
\delta g_{\delta\gamma} = \delta \left(\dfrac{\partial G^i [x_{vi}]}{\partial x^\delta} \cdot \dfrac{\partial G^j [x_{vi}]}{\partial x^\gamma} \gamma_{ij} \right) = g_{ij} \cdot \delta \left(\dfrac{\partial G^i [x_{vi}]}{\partial x^\delta} \cdot \dfrac{\partial G^j [x_{vi}]}{\partial x^\gamma} \right) \\
= \dfrac{g_{ij}}{4} \delta_\delta^i \delta_\gamma^j \delta \begin{Bmatrix} 1 + M \cdot f + \mathbf{C}^\alpha f_{,\alpha} \\ 1 + M \cdot f - \mathbf{C}^\alpha f_{,\alpha} \\ 1 - M \cdot f + \mathbf{C}^\alpha f_{,\alpha} \\ 1 - M \cdot f - \mathbf{C}^\alpha f_{,\alpha} \end{Bmatrix} \begin{Bmatrix} 1 + M \cdot f + \mathbf{C}^\beta f_{,\beta} \\ 1 + M \cdot f - \mathbf{C}^\beta f_{,\beta} \\ 1 - M \cdot f + \mathbf{C}^\beta f_{,\beta} \\ 1 - M \cdot f - \mathbf{C}^\beta f_{,\beta} \end{Bmatrix} \\
\times \\
= \dfrac{g_{\delta\gamma}}{4} \delta \begin{Bmatrix} 1 + M \cdot f + \mathbf{C}^\alpha f_{,\alpha} \\ 1 + M \cdot f - \mathbf{C}^\alpha f_{,\alpha} \\ 1 - M \cdot f + \mathbf{C}^\alpha f_{,\alpha} \\ 1 - M \cdot f - \mathbf{C}^\alpha f_{,\alpha} \end{Bmatrix} \begin{Bmatrix} 1 + M \cdot f + \mathbf{C}^\beta f_{,\beta} \\ 1 + M \cdot f - \mathbf{C}^\beta f_{,\beta} \\ 1 - M \cdot f + \mathbf{C}^\beta f_{,\beta} \\ 1 - M \cdot f - \mathbf{C}^\beta f_{,\beta} \end{Bmatrix}
\end{pmatrix} . \quad (801)
$$

The reader may prove that the same result of the scalar product (as we had in Eq. (725)) will still be obtained from the symmetrized form, namely:

$$
\delta \begin{Bmatrix} 1 + i \cdot M \cdot f + i \cdot \mathbf{C}^\alpha f_{,\alpha} \\ 1 + i \cdot M \cdot f - i \cdot \mathbf{C}^\alpha f_{,\alpha} \\ 1 - i \cdot M \cdot f + i \cdot \mathbf{C}^\alpha f_{,\alpha} \\ 1 - i \cdot M \cdot f - i \cdot \mathbf{C}^\alpha f_{,\alpha} \end{Bmatrix} \begin{Bmatrix} 1 - i \cdot M \cdot f - i \cdot \mathbf{C}^\beta f_{,\beta} \\ 1 - i \cdot M \cdot f + i \cdot \mathbf{C}^\beta f_{,\beta} \\ 1 + i \cdot M \cdot f - i \cdot \mathbf{C}^\beta f_{,\beta} \\ 1 + i \cdot M \cdot f + i \cdot \mathbf{C}^\beta f_{,\beta} \end{Bmatrix}
$$

$$= \delta \left(1 + M^2 \cdot f^2 + \mathbf{C}^\alpha \mathbf{C}^\beta \cdot f_{,\alpha} f_{,\beta} \right). \quad (802)$$

We demand the outcome of Eq. (802) to be zero after the variation, respectively, a constant before, which is to say:

$$M^2 \cdot f^2 + \mathbf{C}^\alpha \mathbf{C}^\beta \cdot f_{,\alpha} f_{,\beta} = Q^2. \quad (803)$$

The particle at rest solutions would now read:

$$f[t] = \frac{\pm Q \cdot \tan\left[M \cdot \left(\pm \frac{t}{C^0} + C_t \right) \right]}{M \cdot \sqrt{\sec\left[M \cdot \left(\pm \frac{t}{C^0} + C_t \right) \right]^2}}$$

$$= +\frac{Q}{M} \cdot \tan\left[M \cdot \left(\pm \frac{t}{C^0} + C_t \right) \right] \cdot \sqrt{\cos\left[M \cdot \left(\pm \frac{t}{C^0} + C_t \right) \right]^2}. \quad (804)$$

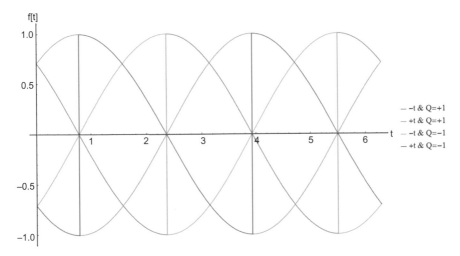

Figure 5.2.1 "Ripped photon" (see text) solutions to particles obtained from Eq. (803).

Setting $M = C^0 = 1$ and $C_t = \pi/2$, we can distinguish four solutions (see Fig. 5.2.1).

We see that we obtain some saw-tooth solutions which give the impression of photons (complete oscillations) being ripped apart in order to bring about two particles (matter and antimatter). In this case, the spin should have something to do with the shift parameter C_1. Setting $C_1 = 0$, the $++\&--$ and the $+-\&-+$ solutions, thereby the signs standing for Q and the sign before t, respectively, degenerate (c.f. Fig. 5.2.2). We conclude that the Q-parameter is clearly connected with the sign before t and thus, the matter and antimatter character of the particle. In other words, the solution Eq. (804) can only have negative Q for matter $(-t)$ and positive Q for antimatter $(+t)$ solutions. This agrees perfectly with the charged leptons (electron, positron, muon, anti-muon, tauon and anti-tauon).

We see that we can completely avoid the sign in front of the time-t-dimension and take on the coding of matter and antimatter character just completely via the Q, obviously being the elementary charge. In order to still obtain four non-degenerate* solutions, we apply the parameter C_1 in $\pm$ as shown in Fig. 5.2.3.

---

*Please note that in contrast to the classical quantum theory, here we use the expression "degenerate" not for undistinguishable energy states, but for undistinguishable f-solutions.

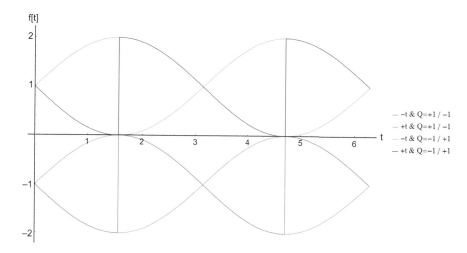

Figure 5.2.2 "Ripped photon" solutions to particles obtained from Eq. (803) for $C_1 = 0$. In order to still see the difference of the degenerated states, we have used an offset (see legend "/± 1").

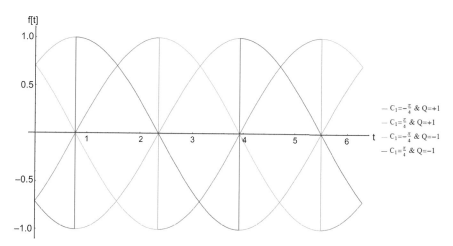

Figure 5.2.3 "Ripped photon" solutions to particles obtained from Eq. (803) only for $t = +t$, but $C_1 = \pm \pi/4$.

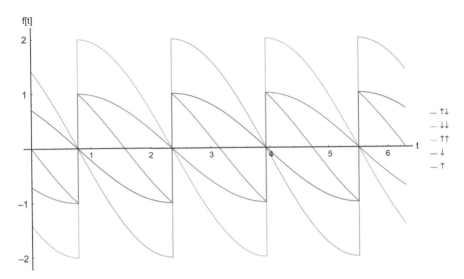

Figure 5.2.4 "Ripped photon" solutions to particles obtained from Eq. (803) only for $t = +t$, but $C_1 = \pm \pi/4$ in various spin-pairings. For comparison, we also show the single up and down states again.

As it is clear from the figures above that matter and antimatter solutions cancel each other out, we would be interested in combining matter solutions with different C_1-states. This is shown in Fig. 5.2.4, where we clearly see that anti-parallel mixtures (blue) are leading to significantly lower overall (integrated) oscillations than the parallel pairings (yellow and green) and even the singular states (red and lilac in Fig. 5.2.4). Associating, as one classically does, the integrated oscillations with energy, the anti-parallel pairings are significantly favored against parallel ones. This clearly is spin-behavior of elementary particles and following the Pauli exclusion principle [175], we might conclude to have found C_1 as an elementary spin-parameter.

5.2.3 An Intermediate Metric Interpretation

Inserting our solution Eq. (804) into the variated metric being given as follows:

$$G_{\delta\gamma} = g_{\delta\gamma}\left(1 + M^2 \cdot f^2 + \mathbf{C}^\alpha \mathbf{C}^\beta \cdot f_{,\alpha} f_{,\beta}\right) = g_{\delta\gamma}\left(1 - Q^2\right), \qquad (805)$$

we see that this Dirac-like variation, which we have considered here in the chapter above, only provides a constant scaling of the metric, but at the same time it gives significant changes to the coordinate transformations via:

$$\frac{\partial G^i\,[x_{\forall i}]}{\partial x^\delta} = \delta^i_\delta \left\{ \begin{array}{l} 1 + M \cdot f + \mathbf{C}^\alpha\,f_{,\alpha} \\ 1 + M \cdot f - \mathbf{C}^\alpha\,f_{,\alpha} \\ 1 - M \cdot f + \mathbf{C}^\alpha\,f_{,\alpha} \\ 1 - M \cdot f - \mathbf{C}^\alpha\,f_{,\alpha} \end{array} \right\} \equiv \delta^i_\delta \left\{ 1 \pm M \cdot f \pm \mathbf{C}^\alpha\,f_{,\alpha} \right\}$$

$$\xrightarrow{\text{at rest}} = \delta^i_\delta \left\{ 1 \pm \frac{(\pm Q) \cdot \tan\left[M \cdot \left(\frac{t}{C^0} \pm C_1\right)\right]}{\sqrt{\sec\left[M \cdot \left(\frac{t}{C^0} \pm C_1\right)\right]^2}} \pm \frac{(\pm Q)}{\sqrt{\sec\left[M \cdot \left(\frac{t}{C^0} \pm C_1\right)\right]^2}} \right\}.$$

$$(806)$$

We recognize truncated sin- and cos-oscillations in four possible constellations which are metrically defining, respectively describing a leptonic particle. The frequency of these oscillations has been determined by the mass-parameter M. Knowing that in a Minkowski metric the metric base vector component reads $C^0 = 1/c$ (still we assume $d = 1$), we may like to rewrite Eq. (806) as follows:

$$\frac{\partial G^i\,[x_{\forall i}]}{\partial x^\delta} \xrightarrow{\text{at rest}} = \delta^i_\delta \left\{ 1 \pm \frac{(\pm Q) \cdot \tan\left[M \cdot (ct \pm C_1)\right]}{\sqrt{\sec\left[M \cdot (ct \pm C_1)\right]^2}} \pm \frac{(\pm Q)}{\sqrt{\sec\left[M \cdot (ct \pm C_1)\right]^2}} \right\}$$

$$\xrightarrow{(\pm Q)=q} = \delta^i_\delta \left\{ 1 \pm q \cdot \sqrt{\sin\left[M \cdot (ct \pm C_1)\right]^2} \pm q \cdot \sqrt{\cos\left[M \cdot (ct \pm C_1)\right]^2} \right\}$$

$$= \delta^i_\delta \left\{ \begin{array}{l} 1 + q \cdot \sqrt{\sin\left[M \cdot (ct \pm C_1)\right]^2} + q \cdot \sqrt{\cos\left[M \cdot (ct \pm C_1)\right]^2} \\ 1 + q \cdot \sqrt{\sin\left[M \cdot (ct \pm C_1)\right]^2} - q \cdot \sqrt{\cos\left[M \cdot (ct \pm C_1)\right]^2} \\ 1 - q \cdot \sqrt{\sin\left[M \cdot (ct \pm C_1)\right]^2} + q \cdot \sqrt{\cos\left[M \cdot (ct \pm C_1)\right]^2} \\ 1 - q \cdot \sqrt{\sin\left[M \cdot (ct \pm C_1)\right]^2} - q \cdot \sqrt{\cos\left[M \cdot (ct \pm C_1)\right]^2} \end{array} \right\}.$$

$$(807)$$

Ignoring the constant 1, the various $\{\ldots\}_k$-components can be illustrated as shown in Fig. 5.2.5.

Please note that, interestingly, we could also have obtained these solutions for any other Cartesian coordinate, because each form like (from Eq. (803)):

$$M^2 \cdot f^2 + D^2 \cdot g^{\alpha\alpha} \cdot (f_{,\alpha})^2 = Q^2; \quad \alpha = 1, 2, 3; \quad g^{\alpha\alpha} = 1 \qquad (808)$$

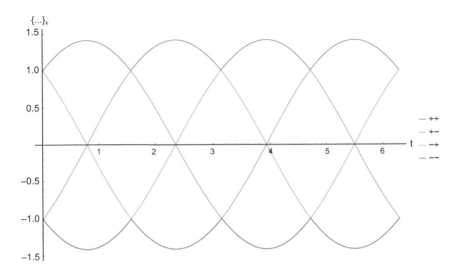

Figure 5.2.5 {...}-Components of the base vectors according to the leptonic particle at rest solution (Eq. (807)). For the reason of simplicity, we have set $q = Q = 1$ and $C_1 = 0$.

ends up with solutions of the kind:

$$f [x^\alpha] = \frac{\pm Q \cdot \tan \left[M \cdot \left(\pm \frac{x^\alpha}{D} + C_\alpha \right) \right]}{M \cdot \sqrt{\sec \left[M \cdot \left(\pm \frac{x^\alpha}{D} + C_\alpha \right) \right]^2}}$$

$$= \pm \frac{Q}{M} \cdot \tan \left[M \cdot \left(\pm \frac{x^\alpha}{D} + C_\alpha \right) \right] \cdot \sqrt{\cos \left[M \cdot \left(\pm \frac{x^\alpha}{D} + C_\alpha \right) \right]^2}$$

$$(809)$$

in the case of positive D^2 and just the two solutions:

$$f [x^\alpha] = \frac{1}{2} \cdot \left(Q^2 e^{M \cdot \left(\pm \frac{x^\alpha}{D} - C_\alpha \right)} + \frac{e^{-M \cdot \left(\pm \frac{x^\alpha}{D} - C_\alpha \right)}}{M^2} \right) \qquad (810)$$

in the case of negative D^2. In the case of $D^2 < 0$, we have to distinguish between the case that D is imaginary, which does not create any problems regarding the behavior in infinity, because the solutions are just endless oscillations and D being real, which automatically (in order to avoid infinitely growing solutions) excludes all solutions with not vanishing Q. In this case, we would end up with just one possible (finite) solution, namely:

$$f[x^\alpha] = \frac{1}{2} \cdot \begin{cases} \dfrac{e^{M \cdot \left(\frac{x^\alpha}{D} - C_\alpha\right)}}{M^2} & \text{for} \quad x^\alpha < 0 \\[4mm] \dfrac{e^{-M \cdot \left(+\frac{x^\alpha}{D} - C_\alpha\right)}}{M^2} & \text{for} \quad x^\alpha \geq 0 \end{cases}. \tag{811}$$

This automatically leads to questions about the correct form of the sign convention, because obviously it does matter now which dimensions (the spatial or the timely ones) get the negative and which the positive sign. We are going to discuss this aspect further in section "10.2 The 'Elastic' Dirac Equation".

Even though we obtained some obviously not completely meaningless solutions, we still wonder why we did not need to actually perform the variation under the integrals and just extracted the Dirac forms from our original approach, thereby applying a special base vector structure, which we marked as $\{\ldots\}$-vectors. We shall reconsider Eq. (789) and this time perform the integration by parts not only with the $\mathbf{C}^\alpha \mathbf{C}^\beta \cdot \left(f^k_{,\alpha\sigma} f^l_{,\beta} + f^k_{,\alpha} f^l_{,\beta\sigma} \right)$-terms, but also with $M^2 \cdot \left(f^k f^l_{,\sigma} + f^k_{,\sigma} f^l \right)$:

$$\delta_\gamma W = 0 = \int_V d^n x \sqrt{\gamma} \cdot \left(R^{\delta\gamma} - \frac{1}{2} R \cdot \gamma^{\delta\gamma} + \Lambda \cdot \gamma^{\delta\gamma} + \kappa \cdot T^{\delta\gamma} \right)$$

$$\times \begin{pmatrix} \delta g_{\delta\gamma} = \delta \left(\frac{\partial G^i[x_{\forall i}]}{\partial x^\delta} \cdot \frac{\partial G^j[x_{\forall i}]}{\partial x^\gamma} \gamma_{ij} \right) = \gamma_{ij} \cdot \delta \left(\frac{\partial G^i[x_{\forall i}]}{\partial x^\delta} \cdot \frac{\partial G^j[x_{\forall i}]}{\partial x^\gamma} \right) \\ = \frac{\gamma_{ij}}{4} \delta \left[\delta^i_\delta \delta^j_\gamma \begin{Bmatrix} 1 + M \cdot C_k f^k + \mathbf{C}^\alpha C_k f^k_{,\alpha} \\ 1 + M \cdot C_k f^k - \mathbf{C}^\alpha C_k f^k_{,\alpha} \\ 1 - M \cdot C_k f^k + \mathbf{C}^\alpha C_k f^k_{,\alpha} \\ 1 - M \cdot C_k f^k - \mathbf{C}^\alpha C_k f^k_{,\alpha} \end{Bmatrix} \begin{Bmatrix} 1 + M \cdot C_l f^l + \mathbf{C}^\beta C_l f^l_{,\beta} \\ 1 + M \cdot C_l f^l - \mathbf{C}^\beta C_l f^l_{,\beta} \\ 1 - M \cdot C_l f^l + \mathbf{C}^\beta C_l f^l_{,\beta} \\ 1 - M \cdot C_l f^l - \mathbf{C}^\beta C_l f^l_{,\beta} \end{Bmatrix} \right] \\ = \delta \left[C_k C_l \delta^i_\delta \delta^j_\gamma \left(1 + M^2 \cdot f^k f^l + \mathbf{C}^\alpha \mathbf{C}^\beta \cdot f^k_{,\alpha} f^l_{,\beta} \right) \right] \gamma_{ij} \\ = C_k C_l \delta^i_\delta \delta^j_\gamma \left(M^2 \cdot \left(f^k f^l_{,\sigma} + f^k_{,\sigma} f^l \right) + \mathbf{C}^\alpha \mathbf{C}^\beta \cdot \left(f^k_{,\alpha\sigma} f^l_{,\beta} + f^k_{,\alpha} f^l_{,\beta\sigma} \right) \right) \gamma_{ij} \delta^\sigma \end{pmatrix}. \tag{812}$$

$$\delta_\gamma W = 0 = \int_V d^n x \sqrt{\gamma} \cdot \left(R^{\delta\gamma} - \frac{1}{2} R \cdot \gamma^{\delta\gamma} + \Lambda \cdot \gamma^{\delta\gamma} + \kappa \cdot T^{\delta\gamma} \right)$$

$$\times \begin{pmatrix} \delta g_{\delta\gamma} = \delta \left[\gamma_{\delta\gamma} \left(1 + M^2 \cdot C_k C_l f^k f^l + \mathbf{C}^\alpha \mathbf{C}^\beta \cdot (C_k f^k)_{,\alpha} (C_l f^l)_{,\beta} \right) \right] \\ = \gamma_{\delta\gamma} \begin{pmatrix} M^2 \cdot \left(f^k C_k (C_l f^l)_{,\sigma} + (C_k f^k)_{,\sigma} C_l f^l \right) \\ + \left(\left(\mathbf{C}^\alpha (C_k f^k)_{,\alpha} \right)_{,\sigma} \mathbf{C}^\beta (C_l f^l)_{,\beta} + \mathbf{C}^\alpha (C_k f^k)_{,\alpha} \left(\mathbf{C}^\beta (C_l f^l)_{,\beta} \right)_{,\sigma} \right) \end{pmatrix} \delta^\sigma \end{pmatrix}$$

$$= \int_V d^n x \sqrt{\gamma} \cdot (\ldots)^{\delta\gamma} \times \left(\gamma_{\delta\gamma} \left(M^2 \cdot (C_k C_l f^k f^l)_{,\sigma} + \left(\mathbf{C}^\alpha \mathbf{C}^\beta \cdot (C_k f^k)_{,u} (C_l f^l)_{,\beta} \right)_{,\sigma} \right) \delta^\sigma \right)$$

$$= -\int_V d^n x \times \left(\left(M^2 \cdot \left[\sqrt{\gamma} \cdot (\ldots)^{\delta\gamma} \gamma_{\delta\gamma} \right]_{,\sigma} C_k C_l f^k f^l \right. \right.$$
$$+ \left. \left[\sqrt{\gamma} \cdot (\ldots)^{\delta\gamma} \gamma_{\delta\gamma} \right]_{,\sigma} \mathbf{C}^\alpha \mathbf{C}^\beta \left(C_k f^k \right)_{,\alpha} \left(C_l f^l \right)_{,\beta} \right) \delta^\sigma \right)$$
$$= -\int_V d^n x \times \left(\left(\left[\sqrt{\gamma} \cdot (\ldots)^{\delta\gamma} \gamma_{\delta\gamma} \right]_{,\sigma} \left(M^2 \cdot C_k C_l f^k f^l + \mathbf{C}^\alpha \mathbf{C}^\beta \left(C_k f^k \right)_{,\alpha} \left(C_l f^l \right)_{,\beta} \right) \right) \delta^\sigma \right).$$

$$(813)$$

In contrast to Eq. (789), we have considered all **C**-vectors coordinate dependent in order to stay more general. As the integrand should be zero for all cases, we obtain the equation:

$$0 = M^2 \cdot C_k C_l f^k f^l + \mathbf{C}^\alpha \mathbf{C}^\beta \left(C_k f^k \right)_{,\alpha} \left(C_l f^l \right)_{,\beta}$$
$$= M^2 \cdot C_k C_l f^k f^l + \mathbf{C}^\alpha \mathbf{C}^\beta \left(C_k f^k \right)_{,\alpha} \left(C_l f^l \right)_{,\beta}$$
$$= M^2 \cdot C_k C_l f^k f^l$$
$$+ \mathbf{C}^\alpha \mathbf{C}^\beta \left[C_{k,\alpha} f^k C_{l,\beta} f^l + C_k f^k_{,\alpha} C_l f^l_{,\beta} + C_{k,\alpha} f^k C_l f^l_{,\beta} + C_k f^k_{,\alpha} C_{l,\beta} f^l \right].$$

$$(814)$$

Now we remember that in order to obtain the Laplace operator, we had set $\mathbf{C}^\beta \mathbf{C}^\alpha = d \cdot g^{\alpha\beta}$, respectively $\mathbf{C}^\beta \mathbf{C}^\alpha = d \cdot \gamma^{\alpha\beta}$ (which is either way here). Setting this into Eq. (814) and—for simplicity—assuming the C_k, C_l not to depend on the coordinates, we have:

$$0 = C_k C_l \left(M^2 \cdot f^k f^l + d \cdot g^{\alpha\beta} f^k_{,\alpha} f^l_{,\beta} \right). \tag{815}$$

Further assuming a Minkowski metric (Eq. (795)), we can factorize with the Dirac matrices, obtaining the Dirac-like equations:

$$0 = M \cdot f^k - \sqrt{d} \cdot \gamma^\alpha f^k_{,\alpha}; \quad 0 = M \cdot f^k + \sqrt{d} \cdot \gamma^\alpha f^k_{,\alpha}. \tag{816}$$

So we think we can assume that Eq. (814) or rather:

$$0 = M^2 \cdot C_k C_l f^k f^l + d \cdot g^{\alpha\beta} \left(C_k f^k \right)_{,\alpha} \left(C_l f^l \right)_{,\beta} \tag{817}$$

is just presenting the metric form of the Dirac equation squared.

On the other hand, we note that the squared form Eq. (815) (with the constant C_k, C_l) could also be brought into the $\{\ldots\}$-vector form as follows (as we consider the Dirac analogy we here assume diagonal metrics):

$$0 = C_k C_l \left(M^2 \cdot f^k f^l + d \cdot g^{\alpha\beta} f^k_{,\alpha} f^l_{,\beta} \right) \Rightarrow 0 = \frac{1}{16} V^k \cdot V^l$$

$$V^k = \left\{ \begin{array}{l} \frac{M}{\sqrt{d}} f^k + \sqrt{g^{00}} f^k_{,0} + \sqrt{g^{11}} f^k_{,1} + \sqrt{g^{22}} f^k_{,2} + \sqrt{g^{33}} f^k_{,3} \\ \frac{M}{\sqrt{d}} f^k + \sqrt{g^{00}} f^k_{,0} + \sqrt{g^{11}} f^k_{,1} + \sqrt{g^{22}} f^k_{,2} - \sqrt{g^{33}} f^k_{,3} \\ \vdots \\ \frac{M}{\sqrt{d}} f^k - \sqrt{g^{00}} f^k_{,0} - \sqrt{g^{11}} f^k_{,1} - \sqrt{g^{22}} f^k_{,2} + \sqrt{g^{33}} f^k_{,3} \\ \frac{M}{\sqrt{d}} f^k - \sqrt{g^{00}} f^k_{,0} - \sqrt{g^{11}} f^k_{,1} - \sqrt{g^{22}} f^k_{,2} + \sqrt{g^{33}} f^k_{,3} \end{array} \right\}$$

$$V^l = \left\{ \begin{array}{l} \frac{M}{\sqrt{d}} f^l + \sqrt{g^{00}} f^l_{,0} + \sqrt{g^{11}} f^l_{,1} + \sqrt{g^{22}} f^l_{,2} + \sqrt{g^{33}} f^l_{,3} \\ \vdots \\ \frac{M}{\sqrt{d}} f^l - \sqrt{g^{00}} f^l_{,0} - \sqrt{g^{11}} f^l_{,1} - \sqrt{g^{22}} f^l_{,2} - \sqrt{g^{33}} f^l_{,3} \end{array} \right\}. \tag{818}$$

Consideration of the complete variated metric from Eq. (813) before the variation and using the constant C_k, C_l-simplification:

$$\delta g_{\delta\gamma} = \gamma_{\delta\gamma} \delta \left[\left(1 + M^2 \cdot C_k C_l f^k f^l + \mathbf{C}^\alpha \mathbf{C}^\beta \cdot \left(C_k f^k \right)_{,\alpha} \left(C_l f^l \right)_{,\beta} \right) \right]$$
$$\simeq \gamma_{\delta\gamma} \delta \left[\left(1 + C_k C_l \left(M^2 \cdot f^k f^l + \mathbf{C}^\alpha \mathbf{C}^\beta \cdot f^k_{,\alpha} f^l_{,\beta} \right) \right) \right], \tag{819}$$

shows us just "1 + (815)" before the variation. Applying the $\{\dots\}$-vectors and extending them via:

$$\delta g_{\delta\gamma} = \delta \left(\frac{1}{32} \left\{ \begin{array}{l} 1 + \sqrt{d} \cdot V^k \\ 1 - \sqrt{d} \cdot V^k \end{array} \right\} \cdot \left\{ \begin{array}{l} 1 + \sqrt{d} \cdot V^l \\ 1 - \sqrt{d} \cdot V^l \end{array} \right\} \right) \tag{820}$$

exactly gives us Eq. (819) in the $\{\dots\}$-factorized form. As we are allowed to add an arbitrary constant into the variation or as its desired total outcome automatically gives zero in the case that the $V^k V^l$-scalar products result in a constant, we would just have to demand:

$$\delta g_{\delta\gamma} = \delta \left(\frac{1}{32} \left\{ \begin{array}{l} 1 + \sqrt{d} \cdot V^k \\ 1 - \sqrt{d} \cdot V^k \end{array} \right\} \cdot \left\{ \begin{array}{l} 1 + \sqrt{d} \cdot V^l \\ 1 - \sqrt{d} \cdot V^l \end{array} \right\} \right) = 0$$
$$\Rightarrow Q^2 = M^2 \cdot C_k C_l \left(f^k f^l + d \cdot g^{\alpha\beta} f^k_{,\alpha} f^l_{,\beta} \right) = d \cdot V^k \cdot V^l, \tag{821}$$

which, apart from the difference in treating the **f**-vector, is identical with Eq. (803). Please note that the corresponding form of Eq. (821) without the use of the constant C_k, C_l-simplification would read:

$$Q^2 = M^2 \cdot C_k C_l f^k f^l + \mathbf{C}^\alpha \mathbf{C}^\beta \cdot \left(C_k f^k \right)_{,\alpha} \left(C_l f^l \right)_{,\beta} \tag{822}$$

and requires significantly different **V**- or $\{\ldots\}$-vectors, namely:

$$Q^2 = M^2 \cdot C_k C_l f^k f^l + \mathbf{C}^\alpha \mathbf{C}^\beta \cdot (C_k f^k)_{,\alpha} (C_l f^l)_{,\beta} = \frac{V^k \cdot V^l}{2}$$

$$V^k = \begin{Bmatrix} M \cdot C_k f^k + \mathbf{C}^\alpha \cdot (C_k f^k)_{,\alpha} \\ M \cdot C_k f^k - \mathbf{C}^\alpha \cdot (C_k f^k)_{,\alpha} \end{Bmatrix}; \quad V^l = \begin{Bmatrix} M \cdot C_l f^l + \mathbf{C}^\beta \cdot (C_l f^l)_{,\beta} \\ M \cdot C_l f^l - \mathbf{C}^\beta \cdot (C_l f^l)_{,\beta} \end{Bmatrix},$$

$$(823)$$

where now the base-vectors $\mathbf{C}^\alpha$ and $\mathbf{C}^\beta$ are taking care about the right factor selection during the $\{\ldots\}$-scalar product. In other words: Dirac's quaternions perform a certain factor selection just as well and thus, simply take on the job of tensor indices and base vector structures. It has to be pointed out that Eq. (823) could also be completely $\{\ldots\}$-factorized as follows:

$$0 = M^2 \cdot C_k C_l f^k f^l + \mathbf{C}^\alpha \mathbf{C}^\beta \cdot (C_k f^k)_{,\alpha} (C_l f^l)_{,\beta} - Q^2 = \frac{V_+^k \cdot V_+^l}{4}$$

$$V_+^k = \begin{Bmatrix} i \cdot Q + M \cdot C_k f^k + \mathbf{C}^\alpha \cdot (C_k f^k)_{,\alpha} \\ i \cdot Q + M \cdot C_k f^k - \mathbf{C}^\alpha \cdot (C_k f^k)_{,\alpha} \\ i \cdot Q - M \cdot C_k f^k + \mathbf{C}^\alpha \cdot (C_k f^k)_{,\alpha} \\ i \cdot Q - M \cdot C_k f^k - \mathbf{C}^\alpha \cdot (C_k f^k)_{,\alpha} \end{Bmatrix};$$

$$V_+^l = \begin{Bmatrix} i \cdot Q + M \cdot C_l f^l + \mathbf{C}^\beta \cdot (C_l f^l)_{,\beta} \\ i \cdot Q + M \cdot C_l f^l - \mathbf{C}^\beta \cdot (C_l f^l)_{,\beta} \\ i \cdot Q - M \cdot C_l f^l + \mathbf{C}^\beta \cdot (C_l f^l)_{,\beta} \\ i \cdot Q - M \cdot C_l f^l - \mathbf{C}^\beta \cdot (C_l f^l)_{,\beta} \end{Bmatrix}, \quad (824)$$

whereby it does not matter which of the addends keeps its sign while for the others the signs alternate, because we obtain the same results also for:

$$0 = M^2 \cdot C_k C_l f^k f^l + \mathbf{C}^\alpha \mathbf{C}^\beta \cdot (C_k f^k)_{,\alpha} (C_l f^l)_{,\beta} - Q^2 = \frac{V_+^k \cdot V_+^l}{4}$$

$$V_+^k = \begin{Bmatrix} i \cdot Q + M \cdot C_k f^k + \mathbf{C}^\alpha \cdot (C_k f^k)_{,\alpha} \\ i \cdot Q + M \cdot C_k f^k - \mathbf{C}^\alpha \cdot (C_k f^k)_{,\alpha} \\ -i \cdot Q + M \cdot C_k f^k + \mathbf{C}^\alpha \cdot (C_k f^k)_{,\alpha} \\ -i \cdot Q + M \cdot C_k f^k - \mathbf{C}^\alpha \cdot (C_k f^k)_{,\alpha} \end{Bmatrix};$$

$$V_+^l = \begin{Bmatrix} i \cdot Q + M \cdot C_l f^l + \mathbf{C}^\beta \cdot (C_l f^l)_{,\beta} \\ i \cdot Q + M \cdot C_l f^l - \mathbf{C}^\beta \cdot (C_l f^l)_{,\beta} \\ -i \cdot Q + M \cdot C_l f^l + \mathbf{C}^\beta \cdot (C_l f^l)_{,\beta} \\ -i \cdot Q + M \cdot C_l f^l - \mathbf{C}^\beta \cdot (C_l f^l)_{,\beta} \end{Bmatrix}, \quad (825)$$

and:

$$0 = M^2 \cdot C_k C_l \, f^k f^l + \mathbf{C}^\alpha \mathbf{C}^\beta \cdot \left(C_k f^k \right)_{,\alpha} \left(C_l f^l \right)_{,\beta} - Q^2 = \frac{V_+^{\,k} \cdot V_+^{\,l}}{4}$$

$$V_+^{\,k} = \begin{cases} i \cdot Q + M \cdot C_k f^k + \mathbf{C}^\alpha \cdot \left(C_k f^k \right)_{,\alpha} \\ -i \cdot Q + M \cdot C_k f^k + \mathbf{C}^\alpha \cdot \left(C_k f^k \right)_{,\alpha} \\ i \cdot Q - M \cdot C_k f^k + \mathbf{C}^\alpha \cdot \left(C_k f^k \right)_{,\alpha} \\ -i \cdot Q - M \cdot C_k f^k + \mathbf{C}^\alpha \cdot \left(C_k f^k \right)_{,\alpha} \end{cases} ;$$

$$V_+^{\,l} = \begin{cases} i \cdot Q + M \cdot C_l f^l + \mathbf{C}^\beta \cdot \left(C_l f^l \right)_{,\beta} \\ -i \cdot Q + M \cdot C_l f^l + \mathbf{C}^\beta \cdot \left(C_l f^l \right)_{,\beta} \\ i \cdot Q - M \cdot C_l f^l + \mathbf{C}^\beta \cdot \left(C_l f^l \right)_{,\beta} \\ -i \cdot Q - M \cdot C_l f^l + \mathbf{C}^\beta \cdot \left(C_l f^l \right)_{,\beta} \end{cases} . \tag{826}$$

The most general form, containing them all, which we only write for one of the two vectors, would be:

$$0 = M^2 \cdot C_k C_l \, f^k f^l + \mathbf{C}^\alpha \mathbf{C}^\beta \cdot \left(C_k f^k \right)_{,\alpha} \left(C_l f^l \right)_{,\beta} - Q^2 = \frac{V_+^{\,k} \cdot V_+^{\,l}}{8}$$

$$V_+^{\,k} = \begin{cases} i \cdot Q + M \cdot C_k f^k + \mathbf{C}^\alpha \cdot \left(C_k f^k \right)_{,\alpha}, \; -i \cdot Q + M \cdot C_k f^k + \mathbf{C}^\alpha \cdot \left(C_k f^k \right)_{,\alpha}, \\ i \cdot Q - M \cdot C_k f^k + \mathbf{C}^\alpha \cdot \left(C_k f^k \right)_{,\alpha}, \; -i \cdot Q - M \cdot C_k f^k + \mathbf{C}^\alpha \cdot \left(C_k f^k \right)_{,\alpha}, \\ i \cdot Q + M \cdot C_k f^k - \mathbf{C}^\alpha \cdot \left(C_k f^k \right)_{,\alpha}, \; -i \cdot Q + M \cdot C_k f^k - \mathbf{C}^\alpha \cdot \left(C_k f^k \right)_{,\alpha}, \\ i \cdot Q - M \cdot C_k f^k - \mathbf{C}^\alpha \cdot \left(C_k f^k \right)_{,\alpha}, \; -i \cdot Q - M \cdot C_k f^k - \mathbf{C}^\alpha \cdot \left(C_k f^k \right)_{,\alpha} \end{cases} . \tag{827}$$

We will not further consider the $\{\dots\}$-vectors and how to deal with them, but only needed to point out their principal structure. For more, the interested reader is referred to [133].

Thus, we can conclude that as long as we are allowed to ignore the surface integral term of the variational integral, we have the same functional outcomes* before (Eqs. (819) and (820)) and after the variation (Eq. (815)). Thereby it does not matter whether we use the simplified form with constant C_k, C_l or not. The latter only allows an easy comparison with the classical Dirac expressions. From this automatically follows that the variated metric Eq. (820) gives the same solutions as the ones we already derived from the non-variated one (e.g. Eq. (803)).

However, considering the $\{\dots\}$-vector decomposition or factorization and summing all $\{\dots\}$-components up, we realize that the result for

---

*With "functional outcome", we mean all terms containing the function vector **f** and its derivatives.

the most general form (Eq. (827)) is always zero. So, not only the scalar product, mirroring the variational task (first line of Eq. (827)) vanishes (as it is demanded within the variation procedure), but also the sum of all those vector components multiplying to the variational task. Knowing that our $\{\dots\}$-vectors are just a special sort of base vector transformations (which is to say a set of Jacobi matrices taking all sign-options into account), which can easily be seen by looking at our variational starting point (e.g. Eq. (812)), we wonder whether there is a fundamental interpretation to this—more or less—peculiar finding. In principle, of course, it is not surprising to find vectors whose components' sum gives zero, while their scalar product does not. In our case, however, things are a bit more complicated, because our vector components are all Jacobi or tensor transformation matrices. Each one of them provides a complete set of transformation rules for a tensor of rank one. Now we found that a set of Jacobi matrices summing up to a zero outcome (or a constant if we count the Kroeneckers in, e.g., Eq. (812), just being connected with the constant), apparently gives the right transformation for any metric, where, as we know, two Jacobi matrices have to be multiplied, to result in the correct outcome for a quantum gravity variation of the Einstein–Hilbert action.

If so, we are effectively calculating with zeros and still obtain "something". In fact, then the universe is calculating with zeros and still obtains all what it contains. Apparently, it, meaning the universe, has nothing at hand and still makes the most out of it, namely, everything there is (including us) asking question like "Why?" and "How?" and "What for?".

We may now try and find somewhat simpler, which is to say more practical, structures, which we will do by directly variating the metric tensor under the Einstein–Hilbert action integral.

5.3 Matrix Approach/Second-Order Approach

Even though it will not add much to the discussion about the applications and uses considered later on in this book, we here still perform a few evaluations also with respect to functions matrix approaches.

A matrix approach with respect to all three possible derivative arrangements resulting from a covariant variation (c.f. Eq. (77)) leads us to the following three variational integrals (thereby assuming that $C^\alpha C^\beta = d \cdot \gamma^{\alpha\beta}$ and—for simplicity at the moment—that $C_{jk}C_l^j$ is a matrix with constant components):

$$\delta_\gamma W = 0 = \int_V d^n x \left(R^{\delta\gamma} \left[\gamma^{\delta\gamma} \right] - \frac{1}{2} R \cdot \gamma^{\delta\gamma} + \Lambda \cdot \gamma^{\delta\gamma} + \kappa \cdot T^{\delta\gamma} \left[\gamma^{\delta\gamma} \right] \right)$$

$$\times \sqrt{\gamma} \cdot \gamma_{ij} \cdot \left(C_k^i C_l^j \cdot \left[f_\delta^k \cdot \frac{\partial f_\gamma^l}{\partial x^\sigma} + f_\gamma^l \cdot \frac{\partial f_\delta^k}{\partial x^\sigma} \right] + C^\alpha C^\beta \cdot C_k^i C_l^j \left[f_{\delta,\alpha}^k \cdot \frac{\partial f_{\gamma,\beta}^l}{\partial x^\sigma} + f_{\gamma,\beta}^l \cdot \frac{\partial f_{\delta,\alpha}^k}{\partial x^\sigma} \right] \right)$$

$$= \int_V d^n x \, (\ldots)^{\delta\gamma} \sqrt{\gamma} \cdot C_{jk} C_l^j \left(\left[f_\delta^k \cdot \frac{\partial f_\gamma^l}{\partial x^\sigma} + f_\gamma^l \cdot \frac{\partial f_\delta^k}{\partial x^\sigma} \right] + C^\alpha C^\beta \cdot \left[f_{\delta,\alpha}^k \cdot \frac{\partial f_{\gamma,\beta}^l}{\partial x^\sigma} + f_{\gamma,\beta}^l \cdot \frac{\partial f_{\delta,\alpha}^k}{\partial x^\sigma} \right] \right)$$

$$= \int_V d^n x \times C_{jk} C_l^j \left(\sqrt{\gamma} \cdot (\ldots)^{\delta\gamma} \left[f_\delta^k \cdot f_{\gamma,\sigma}^l + f_\gamma^l \cdot f_{\delta,\sigma}^k \right] - \begin{bmatrix} \left[(\ldots)^{\delta\gamma} C^\alpha C^\beta \cdot \sqrt{\gamma} \cdot f_{\delta,\alpha}^k \right]_{,\beta} \cdot f_{\gamma,\sigma}^l \\ + \left[(\ldots)^{\delta\gamma} C^\alpha C^\beta \cdot \sqrt{\gamma} \cdot f_{\gamma,\beta}^l \right]_{,\alpha} \cdot f_{\delta,\sigma}^k \end{bmatrix} \right)$$

$$= \int_V d^n x \times \sqrt{\gamma} \cdot C_{jk} C_l^j \left(\begin{matrix} \left((\ldots)^{\delta\gamma} f_\delta^k - d \cdot \left(\gamma^{\alpha\beta} \cdot f_{\delta,\alpha}^k (\ldots)^{\delta\gamma}_{,\beta} + (\ldots)^{\delta\gamma} \Delta f_\delta^k \right) \right) \cdot f_{\gamma,\sigma}^l \\ + \left((\ldots)^{\delta\gamma} f_\gamma^l - d \cdot \left(\gamma^{\alpha\beta} \cdot f_{\gamma,\beta}^l (\ldots)^{\delta\gamma}_{,\alpha} + (\ldots)^{\delta\gamma} \Delta f_\gamma^l \right) \right) \cdot f_{\delta,\sigma}^k \end{matrix} \right),$$

$$\tag{828}$$

$$\delta_\gamma W = 0$$

$$= \int_V d^n x \times \sqrt{\gamma} \cdot C_{jk} C_l^j \left(\begin{matrix} \left((\ldots)^{\delta\gamma} f_\sigma^k - d \cdot \left(\gamma^{\alpha\beta} \cdot f_{\sigma,\alpha}^k (\ldots)^{\delta\gamma}_{,\beta} + (\ldots)^{\delta\gamma} \Delta f_\sigma^k \right) \right) \cdot f_{\gamma,\delta}^l \\ + \left((\ldots)^{\delta\gamma} f_\gamma^l - d \cdot \left(\gamma^{\alpha\beta} \cdot f_{\gamma,\beta}^l (\ldots)^{\delta\gamma}_{,\alpha} + (\ldots)^{\delta\gamma} \Delta f_\gamma^l \right) \right) \cdot f_{\sigma,\delta}^k \end{matrix} \right),$$

$$\tag{829}$$

$$\delta_\gamma W = 0$$

$$= \int_V d^n x \times \sqrt{\gamma} \cdot C_{jk} C_l^j \left(\begin{matrix} \left((\ldots)^{\delta\gamma} f_\delta^k - d \cdot \left(\gamma^{\alpha\beta} \cdot f_{\delta,\alpha}^k (\ldots)^{\delta\gamma}_{,\beta} + (\ldots)^{\delta\gamma} \Delta f_\delta^k \right) \right) \cdot f_{\sigma,\gamma}^l \\ + \left((\ldots)^{\delta\gamma} f_\sigma^l - d \cdot \left(\gamma^{\alpha\beta} \cdot f_{\sigma,\beta}^l (\ldots)^{\delta\gamma}_{,\alpha} + (\ldots)^{\delta\gamma} \Delta f_\sigma^l \right) \right) \cdot f_{\delta,\gamma}^k \end{matrix} \right).$$

$$\tag{830}$$

We only briefly repeat the evaluation process for the first integral, which goes as follows:

$$\delta_\gamma W = 0 = \int_V d^n x \left(R^{\delta\gamma} \left[\gamma^{\delta\gamma} \right] - \frac{1}{2} R \cdot \gamma^{\delta\gamma} + \Lambda \cdot \gamma^{\delta\gamma} + \kappa \cdot T^{\delta\gamma} \left[\gamma^{\delta\gamma} \right] \right)$$

$$\times \sqrt{\gamma} \cdot \delta \left(C^2 \cdot \gamma_{ij} C_k^i C_l^j f_\delta^k f_\gamma^l - C^\alpha C^\beta \cdot \gamma_{ij} C_k^i C_l^j \left(f_\delta^k \right)_{,\alpha} f_{\gamma,\beta}^l \right)$$

$$= \int_V d^n x \left(R^{\delta\gamma} \left[\gamma^{\delta\gamma} \right] - \frac{1}{2} R \cdot \gamma^{\delta\gamma} + \Lambda \cdot \gamma^{\delta\gamma} + \kappa \cdot T^{\delta\gamma} \left[\gamma^{\delta\gamma} \right] \right)$$

$$\times \sqrt{\gamma} \cdot C_{jk} C_l^j \left(\left[f_\delta^k \cdot \frac{\partial f_\gamma^l}{\partial x^\sigma} + f_\gamma^l \cdot \frac{\partial f_\delta^k}{\partial x^\sigma} \right] + C^\alpha C^\beta \cdot \left[f_{\delta,\alpha}^k \cdot \frac{\partial f_{\gamma,\beta}^l}{\partial x^\sigma} + f_{\gamma,\beta}^l \cdot \frac{\partial f_{\delta,\alpha}^k}{\partial x^\sigma} \right] \right).$$

$$\tag{831}$$

Integration by parts of the second addend in the last line, performed as:

$$\int_V d^n x \, (\ldots)^{\delta\gamma} C^\alpha C^\beta \cdot \sqrt{\gamma} \cdot f^k_{\delta,\alpha} \cdot \frac{\partial f^l_{\gamma,\beta}}{\partial x^\sigma} = \int_V d^n x \underbrace{\left\{ \left[(\ldots)^{\delta\gamma} C^\alpha C^\beta \cdot \sqrt{\gamma} \cdot f^k_{\delta,\alpha}\right] \cdot \frac{\partial f^l_\gamma}{\partial x^\sigma} \right\}_{,\beta}}_{=S=0 \text{ per definition}}$$

$$- \int_V d^n x \, \left[(\ldots)^{\delta\gamma} C^\alpha C^\beta \cdot \sqrt{\gamma} \cdot f^k_{\delta,\alpha}\right]_{,\beta} \cdot \frac{\partial f^l_\gamma}{\partial x^\sigma}; \quad (\ldots)^{\delta\gamma}$$

$$= \left(R^{\delta\gamma} - \frac{1}{2} R \cdot \gamma^{\delta\gamma} + \Lambda \cdot \gamma^{\delta\gamma} + \kappa \cdot T^{\delta\gamma} \right) \tag{832}$$

yields:

$$\delta_\gamma W = 0 = \int_V d^n x \left(R^{\delta\gamma}[\gamma^{\delta\gamma}] - \frac{1}{2} R \cdot \gamma^{\delta\gamma} + \Lambda \cdot \gamma^{\delta\gamma} + \kappa \cdot T^{\delta\gamma}[\gamma^{\delta\gamma}] \right)$$

$$\times \sqrt{\gamma} \cdot C_{jk} C^j_l \left(\left[f^k_\delta \frac{\partial f^l_\gamma}{\partial x^\sigma} + f^l_\gamma \cdot \frac{\partial f^k_\delta}{\partial x^\sigma} \right] + C^\alpha C^\beta \cdot \left[f^k_{\delta,\alpha} \cdot \frac{\partial f^l_{\gamma,\beta}}{\partial x^\sigma} + f^l_{\gamma,\beta} \frac{\partial f^k_{\delta,\alpha}}{\partial x^\sigma} \right] \right)$$

$$= \int_V d^n x \times C_{jk} C^j_l \left(\sqrt{\gamma} \cdot (\ldots)^{\delta\gamma} \left[f^k_\delta \cdot f^l_{\gamma,\sigma} + f^l_\gamma \cdot f^k_{\delta,\sigma} \right] \right.$$

$$\left. - \left[\begin{array}{c} \left[(\ldots)^{\delta\gamma} C^\alpha C^\beta \cdot \sqrt{\gamma} \cdot f^k_{\delta,\alpha}\right]_{,\beta} \cdot f^l_{\gamma,\sigma} \\ + \left[(\ldots)^{\delta\gamma} C^\alpha C^\beta \cdot \sqrt{\gamma} \cdot f^l_{\gamma,\beta}\right]_{,\alpha} \cdot f^k_{\delta,\sigma} \end{array} \right] \right). \tag{833}$$

Together with $C^\alpha C^\beta = d \cdot \gamma^{\alpha\beta}$ we are able to further simplify Eq. (833):

$$\delta_\gamma W = 0 = \int_V d^n x \times$$

$$C_{jk} C_l^j \sqrt{\gamma} \cdot \left((\ldots)^{\delta\gamma} \left[f_\delta^k \cdot f_{\gamma,\sigma}^l + f_\gamma^l \cdot f_{\delta,\sigma}^k \right] - \frac{d}{\sqrt{\gamma}} \cdot \left[\begin{matrix} \left[(\ldots)^{\delta\gamma} \gamma^{\alpha\beta} \cdot \sqrt{\gamma} \cdot f_{\delta,\alpha}^k \right]_{,\beta} \cdot f_{\gamma,\sigma}^l \\ + \left[(\ldots)^{\delta\gamma} \gamma^{\alpha\beta} \cdot \sqrt{\gamma} \cdot f_{\gamma,\beta}^l \right]_{,\alpha} \cdot f_{\delta,\sigma}^k \end{matrix} \right] \right)$$

$$= \int_V d^n x \times C_{jk} C_l^j \sqrt{\gamma} \cdot \left(\begin{matrix} \left((\ldots)^{\delta\gamma} f_\delta^k - \frac{d}{\sqrt{\gamma}} \cdot \left[(\ldots)^{\delta\gamma} \gamma^{\alpha\beta} \cdot \sqrt{\gamma} \cdot f_{\delta,\alpha}^k \right]_{,\beta} \right) \cdot f_{\gamma,\sigma}^l \\ + \left((\ldots)^{\delta\gamma} f_\gamma^l - \frac{d}{\sqrt{\gamma}} \cdot \left[(\ldots)^{\delta\gamma} \gamma^{\alpha\beta} \cdot \sqrt{\gamma} \cdot f_{\gamma,\beta}^l \right]_{,\alpha} \right) \cdot f_{\delta,\sigma}^k \end{matrix} \right)$$

$$= \int_V d^n x \times C_{jk} C_l^j \sqrt{\gamma}$$

$$\cdot \left(\begin{matrix} \left((\ldots)^{\delta\gamma} f_\delta^k - \frac{d}{\sqrt{\gamma}} \cdot \left((\ldots)^{\delta\gamma}_{,\beta} \gamma^{\alpha\beta} \cdot \sqrt{\gamma} \cdot f_{\delta,\alpha}^k + (\ldots)^{\delta\gamma} \overbrace{\left[\gamma^{\alpha\beta} \cdot \sqrt{\gamma} \cdot f_{\delta,\alpha}^k \right]_{,\beta}}^{\sqrt{\gamma} \cdot \Delta f_\delta^k} \right) \right) \cdot f_{\gamma,\sigma}^l \\ + \left((\ldots)^{\delta\gamma} f_\gamma^l - \frac{d}{\sqrt{\gamma}} \cdot \left((\ldots)^{\delta\gamma}_{,\alpha} \gamma^{\alpha\beta} \cdot \sqrt{\gamma} \cdot f_{\gamma,\beta}^l + (\ldots)^{\delta\gamma} \underbrace{\left[\gamma^{\alpha\beta} \cdot \sqrt{\gamma} \cdot f_{\gamma,\beta}^l \right]_{,\alpha}}_{\sqrt{\gamma} \cdot \Delta f_\gamma^l} \right) \right) \cdot f_{\delta,\sigma}^k \end{matrix} \right)$$

$$(834)$$

and finally result in Eq. (828).

Subsequently, we obtain the sets of metric quantum equations from the variations of Eqs. (828), (829) and (830), respectively, as follows:

$$(\ldots)^{\delta\gamma} f_\delta^k - d \cdot \left(\gamma^{\alpha\beta} \cdot f_{\delta,\alpha}^k (\ldots)^{\delta\gamma}_{,\beta} + (\ldots)^{\delta\gamma} \Delta f_\delta^k \right) = 0$$
$$(\ldots)^{\delta\gamma} f_\gamma^l - d \cdot \left(\gamma^{\alpha\beta} \cdot f_{\gamma,\beta}^l (\ldots)^{\delta\gamma}_{,\alpha} + (\ldots)^{\delta\gamma} \Delta f_\gamma^l \right) = 0, \qquad (835)$$

$$(\ldots)^{\delta\gamma} f_\sigma^k - d \cdot \left(\gamma^{\alpha\beta} \cdot f_{\sigma,\alpha}^k (\ldots)^{\delta\gamma}_{,\beta} + (\ldots)^{\delta\gamma} \Delta f_\sigma^k \right) = 0$$
$$(\ldots)^{\delta\gamma} f_\gamma^l - d \cdot \left(\gamma^{\alpha\beta} \cdot f_{\gamma,\beta}^l (\ldots)^{\delta\gamma}_{,\alpha} + (\ldots)^{\delta\gamma} \Delta f_\gamma^l \right) = 0, \qquad (836)$$

$$(\ldots)^{\delta\gamma} f_\delta^k - d \cdot \left(\gamma^{\alpha\beta} \cdot f_{\delta,\alpha}^k (\ldots)^{\delta\gamma}_{,\beta} + (\ldots)^{\delta\gamma} \Delta f_\delta^k \right) = 0$$
$$(\ldots)^{\delta\gamma} f_\sigma^l - d \cdot \left(\gamma^{\alpha\beta} \cdot f_{\sigma,\beta}^l (\ldots)^{\delta\gamma}_{,\alpha} + (\ldots)^{\delta\gamma} \Delta f_\sigma^l \right) = 0. \qquad (837)$$

While for pure ordinary derivative variation we get a consistent pair of equations, we face severe inconsistency problems with respect to the variations resulting from the covariant derivative. Now we just may conclude that covariant derivative variation has to be excluded for the very inconsistency reason we found, but we should not give up this early. After all, we are looking for intrinsic solutions. Thus, even though the covariant derivative of the metric tensor (and subsequently its variation) should vanish, we could still have what we call an intelligent zero with non-vanishing wave function solutions.

Thus, now we substitute $f^l_{,\gamma} = F^l_{,\gamma}$, further assume again that we could use $(\ldots)^{\delta\gamma} \simeq D \cdot \gamma^{\delta\gamma}$ or even $(\ldots)^{\delta\gamma} = D \cdot \gamma^{\delta\gamma}$ and, after a further integration by parts obtain for Eq. (828):

$$
-\int_V d^n x \times D \cdot C_{jk} C^j_l \begin{pmatrix} \left(\sqrt{\gamma} \cdot \gamma^{\delta\gamma} F^k_{,\delta} - d \cdot \sqrt{\gamma} \cdot \left(\gamma^{\alpha\beta} \cdot F^k_{,\delta\alpha} \gamma^{\delta\gamma}_{,\beta} + \gamma^{\delta\gamma} \Delta F^k_{,\delta}\right)\right)_{,\gamma} \cdot F^l_{,\sigma} \\ + \left(\sqrt{\gamma} \cdot \gamma^{\delta\gamma} F^l_{,\gamma} - d \cdot \sqrt{\gamma} \cdot \left(\gamma^{\alpha\beta} \cdot F^l_{,\gamma\beta} \gamma^{\delta\gamma}_{,\alpha} + \gamma^{\delta\gamma} \Delta F^l_{,\gamma}\right)\right)_{,\delta} \cdot F^k_{,\sigma} \end{pmatrix}
$$

$$
= -\int_V d^n x \times D \cdot C_{jk} C^j_l \begin{pmatrix} \left(\sqrt{\gamma} \cdot \Delta F^k - d \cdot \left(\left(\sqrt{\gamma} \cdot \gamma^{\alpha\beta} \cdot F^k_{,\delta\alpha} \gamma^{\delta\gamma}_{,\beta}\right)_{,\gamma} + \sqrt{\gamma} \cdot \Delta^2 F^k\right)\right) \cdot F^l_{,\sigma} \\ + \left(\sqrt{\gamma} \cdot \Delta F^l - d \cdot \left(\left(\sqrt{\gamma} \cdot \gamma^{\alpha\beta} \cdot F^l_{,\gamma\beta} \gamma^{\delta\gamma}_{,\alpha}\right)_{,\delta} + \sqrt{\gamma} \cdot \Delta^2 F^l\right)\right) \cdot F^k_{,\sigma} \end{pmatrix}
$$

$$
= -\int_V d^n x \times D \cdot C_{jk} C^j_l \begin{pmatrix} \left(\sqrt{\gamma} \cdot \Delta F^k - d \cdot \left(\begin{pmatrix} \left(\sqrt{\gamma} \cdot \gamma^{\alpha\beta} \cdot F^k_{,\delta\alpha} \gamma^{\delta\gamma}\right)_{,\beta} \\ -\gamma^{\delta\gamma} \left(\sqrt{\gamma} \cdot \gamma^{\alpha\beta} \cdot F^k_{,\delta\alpha}\right)_{,\beta} \end{pmatrix}_{,\gamma} + \sqrt{\gamma} \cdot \Delta^2 F^k\right)\right) \cdot F^l_{,\sigma} \\ + \left(\sqrt{\gamma} \cdot \Delta F^l - d \cdot \left(\begin{pmatrix} \left(\sqrt{\gamma} \cdot \gamma^{\alpha\beta} \cdot F^l_{,\gamma\beta} \gamma^{\delta\gamma}\right)_{,\alpha} \\ -\gamma^{\delta\gamma} \left(\sqrt{\gamma} \cdot \gamma^{\alpha\beta} \cdot F^l_{,\gamma\beta}\right)_{,\alpha} \end{pmatrix}_{,\delta} + \sqrt{\gamma} \cdot \Delta^2 F^l\right)\right) \cdot F^k_{,\sigma} \end{pmatrix}.
$$

$$(838)$$

Recognizing additional Laplace operators, we can simplify a bit:

$$
0 = -\int_V d^n x \times D \cdot C_{jk} C^j_l \begin{pmatrix} \left(\sqrt{\gamma} \cdot \Delta F^k - d \cdot \left(\left(\left(\sqrt{\gamma} \cdot \gamma^{\alpha\beta} \cdot F^k_{,\delta\alpha} \gamma^{\delta\gamma}\right)_{,\beta}\right)_{,\gamma}\right)\right) \cdot F^l_{,\sigma} \\ + \left(\sqrt{\gamma} \cdot \Delta F^l - d \cdot \left(\left(\left(\sqrt{\gamma} \cdot \gamma^{\alpha\beta} \cdot F^l_{,\gamma\beta} \gamma^{\delta\gamma}\right)_{,\alpha}\right)_{,\delta}\right)\right) \cdot F^k_{,\sigma} \end{pmatrix}
$$

$$
= -\int_V d^n x \times D \cdot C_{jk} C^j_l \begin{pmatrix} \left(\sqrt{\gamma} \cdot \Delta F^k - d \cdot \left(\sqrt{\gamma} \cdot \gamma^{\alpha\beta} \cdot F^k_{,\delta\alpha} \gamma^{\delta\gamma}\right)_{,\beta\gamma}\right) \cdot F^l_{,\sigma} \\ + \left(\sqrt{\gamma} \cdot \Delta F^l - d \cdot \left(\sqrt{\gamma} \cdot \gamma^{\alpha\beta} \cdot F^l_{,\gamma\beta} \gamma^{\delta\gamma}\right)_{,\alpha\delta}\right) \cdot F^k_{,\sigma} \end{pmatrix}
$$

$$
= -\int_V d^n x \times D \cdot C_{jk} C^j_l \left(\left(\sqrt{\gamma} \cdot \Delta F^k - d \cdot \left(\sqrt{\gamma} \cdot F^{k,\beta\gamma}\right)_{,\beta\gamma}\right) \cdot F^l_{,\sigma}\right.
$$

$$
\left. + \left(\sqrt{\gamma} \cdot \Delta F^l - d \cdot \left(\sqrt{\gamma} \cdot F^{l,\alpha\delta}\right)_{,\alpha\delta}\right) \cdot F^k_{,\sigma}\right).
$$

$$(839)$$

Performing the same evaluation also for Eqs. (829) and (830) (only the terms differing from what we already did and which cannot be brought into the same forms as found in Eq. (839)) gives us:

$$
\begin{aligned}
\delta_\gamma W = 0 &= \int_V d^n x \times \sqrt{\gamma} \cdot C_{jk} C_l^j \left(\left((\ldots)^{\delta\gamma} F_{,\sigma}^k - d \cdot \left(\gamma^{\alpha\beta} \cdot F_{,\sigma\alpha}^k (\ldots)^{\delta\gamma}_{,\beta} + (\ldots)^{\delta\gamma} \Delta F_{,\sigma}^k \right) \right) \cdot F_{,\gamma\delta}^l + \cdots \right) \\[2ex]
&= -\int_V d^n x \times C_{jk} C_l^j \left(\left(\sqrt{\gamma} \cdot (\ldots)^{\delta\gamma} F_{,\sigma}^k - d \cdot \left(\gamma^{\alpha\beta} \cdot F_{,\sigma\alpha}^k (\ldots)^{\delta\gamma}_{,\beta} + (\ldots)^{\delta\gamma} \Delta F_{,\sigma}^k \right) \right)_{,\delta} \cdot F_{,\gamma}^l + \cdots \right) \\[2ex]
&= \int_V d^n x \times C_{jk} C_l^j \left(\left(\sqrt{\gamma} \cdot (\ldots)^{\delta\gamma} F_{,\sigma}^k - d \cdot \left(\gamma^{\alpha\beta} \cdot F_{,\sigma\alpha}^k (\ldots)^{\delta\gamma}_{,\beta} + (\ldots)^{\delta\gamma} \Delta F_{,\sigma}^k \right) \right)_{,\delta\gamma} \cdot F^l + \cdots \right) \\[2ex]
&= \int_V d^n x \times C_{jk} C_l^j \left(\left(\sqrt{\gamma} \cdot \gamma^{\delta\gamma} F_{,\sigma}^k - d \cdot \left(\sqrt{\gamma} \cdot \gamma^{\alpha\beta} \cdot F_{,\sigma\alpha}^k \gamma^{\delta\gamma}_{,\beta} + \sqrt{\gamma} \cdot \gamma^{\delta\gamma} \Delta F_{,\sigma}^k \right) \right)_{,\delta\gamma} \cdot F^l + \cdots \right) \\[2ex]
&= \int_V d^n x \times C_{jk} C_l^j \left(\left(\left(\left(\sqrt{\gamma} \cdot \gamma^{\delta\gamma} \right)_{,\delta} F_{,\sigma}^k \atop + \sqrt{\gamma} \cdot \gamma^{\delta\gamma} \left(F_{,\sigma}^k \right)_{,\delta} \right)_{,\gamma} \right. \right. \\[2ex]
&\quad \left. \left. -d \cdot \left(\left(\sqrt{\gamma} \cdot \gamma^{\alpha\beta} \cdot F_{,\sigma\alpha}^k \gamma^{\delta\gamma}_{,\beta} \right)_{,\delta} \atop + \left(\sqrt{\gamma} \cdot \gamma^{\delta\gamma} \right)_{,\delta} \left(\Delta F_{,\sigma}^k \right)_{,\delta} \right)_{,\gamma} \right) \cdot F^l + \cdots \right) \\[2ex]
&= \int_V d^n x \times C_{jk} C_l^j \left(\left(\left(\left(\sqrt{\gamma} \cdot \gamma^{\delta\gamma} \right)_{,\delta} F_{,\sigma}^k \atop + \sqrt{\gamma} \cdot \Delta \left(F_{,\sigma}^k \right) \right)_{,\gamma} \right. \right. \\[2ex]
&\quad \left. \left. -d \cdot \left(\left(\sqrt{\gamma} \cdot \gamma^{\alpha\beta} \cdot F_{,\sigma\alpha}^k \gamma^{\delta\gamma}_{,\beta} \right)_{,\delta\gamma} \atop + \left(\left(\sqrt{\gamma} \cdot \gamma^{\delta\gamma} \right)_{,\delta} \Delta F_{,\sigma}^k \right)_{,\gamma} + \sqrt{\gamma} \cdot \Delta^2 F_{,\sigma}^k \right) \right) \cdot F^l + \cdots \right),
\end{aligned}
$$

$$\tag{840}$$

$$
= \int_V d^n x \times C_{jk} C_l^j \left(\left(\left(\left(\sqrt{\gamma} \cdot \gamma^{\delta\gamma} \right)_{,\delta} F_{,\sigma}^k \right)_{,\gamma} + \sqrt{\gamma} \cdot \Delta \left(F_{,\sigma}^k \right) \right) - d \cdot \left(\left(\frac{\left(\sqrt{\gamma} \cdot \gamma^{\alpha\beta} \cdot F_{,\sigma\alpha}^k \gamma^{\delta\gamma} \right)_{,\beta}}{\sqrt{\gamma} \Delta F_{,\sigma}^k} - \gamma^{\delta\gamma} \left(\sqrt{\gamma} \cdot \gamma^{\alpha\beta} \cdot F_{,\sigma\alpha}^k \right)_{,\beta} \right)_{,\delta\gamma} + \left(\left(\sqrt{\gamma} \cdot \gamma^{\delta\gamma} \right)_{,\delta} \Delta F_{,\sigma}^k \right)_{,\gamma} + \sqrt{\gamma} \cdot \Delta^2 F_{,\sigma}^k \right) \right) \cdot F^l + \ldots
$$

$$
= \int_V d^n x \times C_{jk} C_l^j \left(\left(\left(\left(\sqrt{\gamma} \cdot \gamma^{\delta\gamma} \right)_{,\delta} F_{,\sigma}^k \right)_{,\gamma} + \sqrt{\gamma} \cdot \Delta \left(F_{,\sigma}^k \right) \right) - d \cdot \left(\left(\sqrt{\gamma} \cdot \gamma^{\alpha\beta} \cdot F_{,\sigma\alpha}^k \gamma^{\delta\gamma} \right)_{,\beta} - \gamma^{\delta\gamma} \left(\sqrt{\gamma} \cdot \gamma^{\alpha\beta} \cdot F_{,\sigma}^k \right)_{,\delta\gamma} + \left(\left(\sqrt{\gamma} \cdot \gamma^{\delta\gamma} \right)_{,\delta} \Delta F_{,\sigma}^k \right)_{,\gamma} + \sqrt{\gamma} \cdot \Delta^2 F_{,\sigma}^k \right) \right) \cdot F^l + \ldots
$$

$$
= \int_V d^n x \times C_{jk} C_l^j \left(\left(\left(\left(\sqrt{\gamma} \cdot \gamma^{\delta\gamma} \right)_{,\delta} F_{,\sigma}^k \right)_{,\gamma} + \sqrt{\gamma} \cdot \Delta \left(F_{,\sigma}^k \right) \right) - d \cdot \left(\sqrt{\gamma} \cdot \gamma^{\alpha\beta} \cdot F_{,\sigma\alpha}^k \gamma^{\delta\gamma} \right)_{,\beta\delta\gamma} \cdot F^l + \ldots \right.
$$

$$
\left. = \int_V d^n x \times C_{jk} C_l^j \left(\left(\left(\sqrt{\gamma} \cdot \gamma^{\delta\gamma} \right)_{,\delta} F_{,\sigma}^k \right)_{,\gamma} + \sqrt{\gamma} \cdot \Delta^2 F_{,\sigma}^k \right)_{,\gamma} - d \cdot \left(\sqrt{\gamma} \cdot \gamma^{\delta\gamma} F_{,\sigma}^{k,\beta} \right)_{,\beta\delta\gamma} \cdot F^l + \ldots \right),
$$

$$(841)$$

$$\delta_\gamma W = 0 = \int_V d^n x \times \sqrt{\gamma} \cdot C_{jk} C_l^j \left(\ldots + \left((\ldots)^{\delta\gamma} F_{,\sigma}^l - d \cdot \left(\gamma^{\alpha\beta} \cdot F_{,\sigma\beta}^l (\ldots)^{\delta\gamma}{}_{,\alpha} + (\ldots)^{\delta\gamma} \Delta F_{,\sigma}^l \right) \right) \cdot F_{,\delta\gamma}^k \right)$$

$$= -\int_V d^n x \times C_{jk} C_l^j \left(\ldots + \left(\sqrt{\gamma} \cdot \left((\ldots)^{\delta\gamma} F_{,\sigma}^l - d \cdot \left(\gamma^{\alpha\beta} \cdot F_{,\sigma\beta}^l (\ldots)^{\delta\gamma}{}_{,\alpha} + (\ldots)^{\delta\gamma} \Delta F_{,\sigma}^l \right) \right) \right)_{,\gamma} \cdot F_{,\delta}^k \right)$$

$$= \int_V d^n x \times C_{jk} C_l^j \left(\ldots + \left(\sqrt{\gamma} \cdot \left((\ldots)^{\delta\gamma} F_{,\sigma}^l - d \cdot \left(\gamma^{\alpha\beta} \cdot F_{,\sigma\beta}^l (\ldots)^{\delta\gamma}{}_{,\alpha} + (\ldots)^{\delta\gamma} \Delta F_{,\sigma}^l \right) \right) \right)_{,\gamma\delta} \cdot F^k \right)$$

$$= \int_V d^n x \times C_{jk} C_l^j \left(\left(\left(\left((\sqrt{\gamma} \cdot \gamma^{\delta\gamma})_{,\delta} F_{,\sigma}^l \right)_{,\gamma} + \sqrt{\gamma} \cdot \Delta (F_{,\sigma}^l) \right) - d \cdot \left(\sqrt{\gamma} \cdot \gamma^{\delta\gamma} F_{,\sigma}^{l,\beta} \right)_{,\beta\delta\gamma} \right) \cdot F^k + \ldots \right).$$

$$(842)$$

Finally we have conditions which can be compactified to just two equations:

$$C_{jk} C_l^j \left(\left(\sqrt{\gamma} \cdot \Delta F^k - d \cdot \left(\sqrt{\gamma} \cdot F^{k,\beta\gamma} \right)_{,\beta\gamma} \right) \cdot F_{,\sigma}^l \right.$$
$$+ \left(\sqrt{\gamma} \cdot \Delta F^l - d \cdot \left(\sqrt{\gamma} \cdot F^{l,\alpha\delta} \right)_{,\alpha\delta} \right) \cdot F_{,\sigma}^k \right) = 0$$
$$\Rightarrow \sqrt{\gamma} \cdot \Delta F^k - d \cdot \left(\sqrt{\gamma} \cdot F^{k,\beta\gamma} \right)_{,\beta\gamma} = 0$$
$$\left(\left(\sqrt{\gamma} \cdot \gamma^{\delta\gamma} \right)_{,\delta} F_{,\sigma}^k \right)_{,\gamma} + \sqrt{\gamma} \cdot \Delta \left(F_{,\sigma}^k \right) - d \cdot \left(\sqrt{\gamma} \cdot \gamma^{\delta\gamma} F_{,\sigma}^{k,\beta} \right)_{,\beta\delta\gamma} = 0.$$

$$(843)$$

It should be emphasized that these equations, in connection with the assumption of the possibility of fulfilling:

$$(\ldots)^{\delta\gamma} = \left(R^{\delta\gamma} \left[\gamma^{\delta\gamma} \right] - \frac{1}{2} R \cdot \gamma^{\delta\gamma} + \Lambda \cdot \gamma^{\delta\gamma} + \kappa \cdot T^{\delta\gamma} \left[\gamma^{\delta\gamma} \right] \right) = D \cdot \gamma^{\delta\gamma},$$

$$(844)$$

do not only give conditions for the function F, which in fact is a wave function, but also for the metric tensor

$$\sqrt{\gamma} \cdot \Delta F^k - d \cdot \left(\left(\frac{\gamma^{\lambda\kappa} \gamma_{\lambda\kappa,\gamma} \gamma^{\lambda\kappa} \gamma_{\lambda\kappa,\beta}}{4} + \frac{\left(\gamma^{\lambda\kappa} \gamma_{\lambda\kappa,\beta} \right)_{,\gamma}}{2} \right) F^{k,\gamma\beta} + F^{k,\gamma\beta}{}_{,\beta\gamma} + 2 \cdot \gamma^{\lambda\kappa} \gamma_{\lambda\kappa,\gamma} F^{k,\gamma\beta}{}_{,\beta} \right) = 0.$$

$$(845)$$

5.3.1 Generalization

Just for the reason of generality, it could be useful to avoid the simplifications (Eq. (844)) and $C_{jk} C_l^j = \text{const}_{kl}$, which renders Eq. (828) to:

$$\delta_\gamma W = 0 = \int_V d^n x \left(R^{\delta\gamma} \left[\gamma^{\delta\gamma} \right] - \frac{1}{2} R \cdot \gamma^{\delta\gamma} + \Lambda \cdot \gamma^{\delta\gamma} + \kappa \cdot T^{\delta\gamma} \left[\gamma^{\delta\gamma} \right] \right)$$

$$\times \sqrt{\gamma} \cdot \gamma_{ij} \cdot \left(C_k^i C_l^j \cdot \left[f_\delta^k \cdot \frac{\partial f_\gamma^l}{\partial x^\sigma} + f_\gamma^l \cdot \frac{\partial f_\delta^k}{\partial x^\sigma} \right] + C^\alpha C^\beta \cdot C_k^i C_l^j \left[f_{\delta,\alpha}^k \cdot \frac{\partial f_{\gamma,\beta}^j}{\partial x^\sigma} + f_{\gamma,\beta}^l \cdot \frac{\partial f_{\delta,\alpha}^k}{\partial x^\sigma} \right] \right)$$

$$= \int_V d^n x \, (\dots)^{\delta\gamma} \sqrt{\gamma} \cdot C_{jk} C_l^j \left(\left[f_\delta^k \cdot \frac{\partial f_\gamma^l}{\partial x^\sigma} + f_\gamma^l \cdot \frac{\partial f_\delta^k}{\partial x^\sigma} \right] + d \cdot \gamma^{\alpha\beta} \cdot \left[f_{\delta,\alpha}^k \cdot \frac{\partial f_\gamma^l}{\partial x^\sigma} + f_{\gamma,\beta}^l \cdot \frac{\partial f_{\delta,\alpha}^k}{\partial x^\sigma} \right] \right)$$

$$= \int_V d^n x \times \left(\sqrt{\gamma} \cdot C_{jk} C_l^j \, (\dots)^{\delta\gamma} \left[f_\delta^k \cdot f_{\gamma,\sigma}^l + f_\gamma^l \cdot f_{\delta,\sigma}^k \right] - d \cdot \left[\begin{matrix} \left[(\dots)^{\delta\gamma} C_{jk} C_l^j \gamma^{\alpha\beta} \sqrt{\gamma} \cdot f_{\delta,\alpha}^k \right]_{,\beta} \cdot f_{\gamma,\sigma}^l \\ + \left[(\dots)^{\delta\gamma} C_{jk} C_l^j \gamma^{\alpha\beta} \sqrt{\gamma} \cdot f_{\gamma,\beta}^l \right]_{,\alpha} \cdot f_{\delta,\sigma}^k \end{matrix} \right] \right)$$

$$= \int_V d^n x \times \sqrt{\gamma} \cdot \left(\begin{matrix} \left(C_{jk} C_l^j \, (\dots)^{\delta\gamma} f_\delta^k - d \cdot \left(\begin{matrix} \gamma^{\alpha\beta} \cdot f_{\delta,\alpha}^k \left((\dots)^{\delta\gamma} C_{jk} C_l^j \right)_{,\beta} \\ + C_{jk} C_l^j \, (\dots)^{\delta\gamma} \Delta f_\delta^k \end{matrix} \right) \right) \cdot f_{\gamma,\sigma}^l \\ + \left(C_{jk} C_l^j \, (\dots)^{\delta\gamma} f_\gamma^l - d \cdot \left(\begin{matrix} \gamma^{\alpha\beta} \cdot f_{\gamma,\beta}^l \left(C_{jk} C_l^j \, (\dots)^{\delta\gamma} \right)_{,\alpha} \\ + C_{jk} C_l^j \, (\dots)^{\delta\gamma} \Delta f_\gamma^l \end{matrix} \right) \right) \cdot f_{\delta,\sigma}^k \end{matrix} \right), \tag{846}$$

and leaves us with the quantum gravity equation:

$$\left(\begin{matrix} \left(C_{jk} C_l^j \, (\dots)^{\delta\gamma} f_\delta^k - d \cdot \left(\begin{matrix} \gamma^{\alpha\beta} \cdot f_{\delta,\alpha}^k \left((\dots)^{\delta\gamma} C_{jk} C_l^j \right)_{,\beta} \\ + C_{jk} C_l^j \, (\dots)^{\delta\gamma} \Delta f_\delta^k \end{matrix} \right) \right) \cdot f_{\gamma,\sigma}^l \\ + \left(C_{jk} C_l^j \, (\dots)^{\delta\gamma} f_\gamma^l - d \cdot \left(\begin{matrix} \gamma^{\alpha\beta} \cdot f_{\gamma,\beta}^l \left(C_{jk} C_l^j \, (\dots)^{\delta\gamma} \right)_{,\alpha} \\ + C_{jk} C_l^j \, (\dots)^{\delta\gamma} \Delta f_\gamma^l \end{matrix} \right) \right) \cdot f_{\delta,\sigma}^k \end{matrix} \right) = 0$$

$$\Rightarrow \left(C_{jk} C_l^j \, (\dots)^{\delta\gamma} f_\delta^k - d \cdot \left(\begin{matrix} \gamma^{\alpha\beta} \cdot f_{\delta,\alpha}^k \left((\dots)^{\delta\gamma} C_{jk} C_l^j \right)_{,\beta} \\ + C_{jk} C_l^j \, (\dots)^{\delta\gamma} \Delta f_\delta^k \end{matrix} \right) \right) \cdot f_{\gamma,\sigma}^l = 0$$

$$\Rightarrow C_{jk} C_l^j \, (\dots)^{\delta\gamma} f_\delta^k - d \cdot \left(\gamma^{\alpha\beta} \cdot f_{\delta,\alpha}^k \left((\dots)^{\delta\gamma} C_{jk} C_l^j \right)_{,\beta} + C_{jk} C_l^j \, (\dots)^{\delta\gamma} \Delta f_\delta^k \right) = 0. \tag{847}$$

Thereby it needs to be pointed out that all three equations give solutions, while the first is the most and the third is the least general one. We also realize that the first two equations are of scalar character and the third is a vector of equations. There we have conditions for all components of our function vector $\mathbf{F}^k$ if again assuming that we have $f_\gamma^l = F_{,\gamma}^l$, $f_\delta^k = F_{,\delta}^k$. Thus, like the Dirac equation, we have a system of equations for a vector wave function only that it is quantum gravity compatible. Using our recipe from above and still not assuming Eq. (844), we can still obtain a more general form than before, which reads:

$$\delta_\gamma W = 0$$

$$
= \int_V d^n x \times \left(
\begin{array}{c}
\left(\left(\sqrt{\gamma} \cdot C_{jk} C_l^j \, (\ldots)^{\delta\gamma} \, f_\delta^k - d \cdot \left(\begin{array}{c} \sqrt{\gamma} \cdot \gamma^{\alpha\beta} \cdot f_{\delta,\alpha}^k \left((\ldots)^{\delta\gamma} \, C_{jk} C_l^j \right)_{,\beta} \\ + \sqrt{\gamma} \cdot C_{jk} C_l^j \, (\ldots)^{\delta\gamma} \, \Delta f_\delta^k \end{array} \right) \right) \cdot F_{,\sigma\gamma}^l \right. \\
\left. + \left(\sqrt{\gamma} \cdot C_{jk} C_l^j \, (\ldots)^{\delta\gamma} \, f_\gamma^l - d \cdot \left(\begin{array}{c} \sqrt{\gamma} \cdot \gamma^{\alpha\beta} \cdot f_{\gamma,\beta}^l \left(C_{jk} C_l^j \, (\ldots)^{\delta\gamma} \right)_{,\alpha} \\ + \sqrt{\gamma} \cdot C_{jk} C_l^j \, (\ldots)^{\delta\gamma} \, \Delta f_\gamma^l \end{array} \right) \right) \cdot F_{,\sigma\delta}^k \right)
\end{array}
\right)
$$

$$
= \int_V d^n x \times \left(
\begin{array}{c}
\left(\left(\sqrt{\gamma} \cdot C_{jk} C_l^j \, (\ldots)^{\delta\gamma} \, F_{,\delta}^k - d \cdot \left(\begin{array}{c} \left(\sqrt{\gamma} \cdot \gamma^{\alpha\beta} \cdot F_{,\delta\alpha}^k \, (\ldots)^{\delta\gamma} \, C_{jk} C_l^j \right)_{,\beta} \\ - (\ldots)^{\delta\gamma} \, C_{jk} C_l^j \left(\sqrt{\gamma} \cdot \gamma^{\alpha\beta} \cdot F_{,\delta\alpha}^k \right)_{,\beta} \\ + \sqrt{\gamma} \cdot C_{jk} C_l^j \, (\ldots)^{\delta\gamma} \, \Delta F_{,\delta}^k \end{array} \right) \right) \cdot F_{,\sigma\gamma}^l \right. \\
\left. + \left(\sqrt{\gamma} \cdot C_{jk} C_l^j \, (\ldots)^{\delta\gamma} \, F_{,\gamma}^l - d \cdot \left(\begin{array}{c} \left(\sqrt{\gamma} \cdot \gamma^{\alpha\beta} \cdot F_{,\gamma\beta}^l C_{jk} C_l^j \, (\ldots)^{\delta\gamma} \right)_{,\alpha} \\ - C_{jk} C_l^j \, (\ldots)^{\delta\gamma} \left(\sqrt{\gamma} \cdot \gamma^{\alpha\beta} \cdot F_{,\gamma\beta}^l \right)_{,\alpha} \\ + \sqrt{\gamma} \cdot C_{jk} C_l^j \, (\ldots)^{\delta\gamma} \, \Delta F_{,\gamma}^l \end{array} \right) \right) \cdot F_{,\sigma\delta}^k \right)
\end{array}
\right)
$$

$$
= \int_V d^n x \times \left(
\begin{array}{c}
\left(\sqrt{\gamma} \cdot C_{jk} C_l^j \, (\ldots)^{\delta\gamma} \, F_{,\delta}^k - d \cdot \left(\sqrt{\gamma} \cdot \gamma^{\alpha\beta} \cdot F_{,\delta\alpha}^k \, (\ldots)^{\delta\gamma} \, C_{jk} C_l^j \right)_{,\beta} \right) \cdot F_{,\sigma\gamma}^l \\
+ \left(\sqrt{\gamma} \cdot C_{jk} C_l^j \, (\ldots)^{\delta\gamma} \, F_{,\gamma}^l - d \cdot \left(\sqrt{\gamma} \cdot \gamma^{\alpha\beta} \cdot F_{,\gamma\beta}^l C_{jk} C_l^j \, (\ldots)^{\delta\gamma} \right)_{,\alpha} \right) \cdot F_{,\sigma\delta}^k
\end{array}
\right),
$$

$$(848)$$

where now the subsequent quantum gravity equations read:

$$
\left(
\begin{array}{c}
\left(\sqrt{\gamma} \cdot C_{jk} C_l^j \, (\ldots)^{\delta\gamma} \, F_{,\delta}^k - d \cdot \left(\sqrt{\gamma} \cdot \gamma^{\alpha\beta} \cdot F_{,\delta\alpha}^k \, (\ldots)^{\delta\gamma} \, C_{jk} C_l^j \right)_{,\beta} \right) \cdot F_{,\sigma\gamma}^l \\
+ \left(\sqrt{\gamma} \cdot C_{jk} C_l^j \, (\ldots)^{\delta\gamma} \, F_{,\gamma}^l - d \cdot \left(\sqrt{\gamma} \cdot \gamma^{\alpha\beta} \cdot F_{,\gamma\beta}^l C_{jk} C_l^j \, (\ldots)^{\delta\gamma} \right)_{,\alpha} \right) \cdot F_{,\sigma\delta}^k
\end{array}
\right) = 0
$$

$$
\Rightarrow \left(\sqrt{\gamma} \cdot C_{jk} C_l^j \, (\ldots)^{\delta\gamma} \, F_{,\delta}^k - d \cdot \left(\sqrt{\gamma} \cdot \gamma^{\alpha\beta} \cdot F_{,\delta\alpha}^k \, (\ldots)^{\delta\gamma} \, C_{jk} C_l^j \right)_{,\beta} \right) \cdot F_{,\sigma\gamma}^l = 0
$$

$$
\Rightarrow \sqrt{\gamma} \cdot C_{jk} C_l^j \, (\ldots)^{\delta\gamma} \, F_{,\delta}^k - d \cdot \left(\sqrt{\gamma} \cdot \gamma^{\alpha\beta} \cdot F_{,\delta\alpha}^k \, (\ldots)^{\delta\gamma} \, C_{jk} C_l^j \right)_{,\beta} = 0. \qquad (849)
$$

Another perhaps more practical form (because it will lead us to the traditional quantum equations) could be:

$$
\sqrt{\gamma} \cdot C_{jk} C_l^j \, (\ldots)^{\delta\gamma} \, F_{,\delta}^k - d \cdot \left(\sqrt{\gamma} \cdot \gamma^{\alpha\beta} \cdot F_{,\delta\alpha}^k \, (\ldots)^{\delta\gamma} \, C_{jk} C_l^j \right)_{,\beta}
$$

$$
= \sqrt{\gamma} \cdot C_{jk} C_l^j \, (\ldots)^{\delta\gamma} \, F_{,\delta}^k - d \cdot \left((\ldots)^{\delta\gamma} \, C_{jk} C_l^j \left(\sqrt{\gamma} \cdot \gamma^{\alpha\beta} \cdot F_{,\delta\alpha}^k \right)_{,\beta} \right.
$$

$$
\left. + \sqrt{\gamma} \cdot \gamma^{\alpha\beta} \cdot F_{,\delta\alpha}^k \left((\ldots)^{\delta\gamma} \, C_{jk} C_l^j \right)_{,\beta} \right)
$$

$$= \sqrt{\gamma} \cdot C_{jk} C_l^j (\ldots)^{\delta\gamma} F_{,\delta}^k - d \cdot \left((\ldots)^{\delta\gamma} C_{jk} C_l^j \sqrt{\gamma} \cdot \Delta F_{,\delta}^k \right.$$

$$\left. + \sqrt{\gamma} \cdot \gamma^{\alpha\beta} \cdot F_{,\delta\alpha}^k \left((\ldots)^{\delta\gamma} C_{jk} C_l^j \right)_{,\beta} \right)$$

$$= \sqrt{\gamma} \cdot \left(C_{jk} C_l^j (\ldots)^{\delta\gamma} \left(F_{,\delta}^k - d \cdot \Delta F_{,\delta}^k \right) - d \cdot \gamma^{\alpha\beta} \cdot F_{,\delta\alpha}^k \left((\ldots)^{\delta\gamma} C_{jk} C_l^j \right)_{,\beta} \right)$$

$$= 0. \tag{850}$$

Now we assume the existence of the following tensor:

$$C_{jk} C_l^j (\ldots)^{\delta\gamma} \cdot \Omega_{\delta\gamma}^{kl} = \omega, \tag{851}$$

which allows us to further simplify Eq. (850) as follows:

$$\sqrt{\gamma} \cdot \left(C_{jk} C_l^j (\ldots)^{\delta\gamma} \left(F_{,\delta}^k - d \cdot \Delta F_{,\delta}^k \right) - d \cdot \gamma^{\alpha\beta} \cdot F_{,\delta\alpha}^k \left((\ldots)^{\delta\gamma} C_{jk} C_l^j \right)_{,\beta} \right) \cdot \Omega_{\delta\gamma}^{kl}$$

$$= \sqrt{\gamma} \cdot \left(\omega \cdot \left(F_{,\delta}^k - d \cdot \Delta F_{,\delta}^k \right) - d \cdot \gamma^{\alpha\beta} \cdot F_{,\delta\alpha}^k \cdot \Omega_{\delta\gamma}^{kl} \left((\ldots)^{\delta\gamma} C_{jk} C_l^j \right)_{,\beta} \right) = 0$$

$$\Rightarrow \left(F_{,\delta}^k - d \cdot \Delta F_{,\delta}^k \right) - d \cdot F_{,\delta\alpha}^k \cdot \frac{\gamma^{\alpha\beta} \cdot \Omega_{\delta\gamma}^{kl} \left((\ldots)^{\delta\gamma} C_{jk} C_l^j \right)_{,\beta}}{\omega} = 0$$

$$\Rightarrow \left(F_{,\gamma}^l - d \cdot \Delta F_{,\gamma}^l \right) - d \cdot F_{,\gamma\alpha}^l \cdot \frac{\gamma^{\alpha\beta} \cdot \Omega_{\delta\gamma}^{kl} \left((\ldots)^{\delta\gamma} C_{jk} C_l^j \right)_{,\beta}}{\omega} = 0. \tag{852}$$

We recognize the classical Klein–Gordon equation with potential, which we could write as:

$$\frac{m^2 c^2}{\hbar^2} \Psi - \Delta \Psi = V \cdot \Psi \tag{853}$$

and where we extract the following analogue:

$$\Rightarrow \frac{1}{d} = \frac{m^2 c^2}{\hbar^2}; \quad F_{,\gamma}^l \rightarrow \Psi; \quad F_{,\gamma\alpha}^l \cdot \frac{\gamma^{\alpha\beta} \cdot \Omega_{\delta\gamma}^{kl} \left((\ldots)^{\delta\gamma} C_{jk} C_l^j \right)_{,\beta}}{\omega} \rightarrow V \cdot \Psi. \tag{854}$$

We note that we can also obtain this result with the more general function matrixes f_γ^l, f_δ^k with the resulting quantum gravity equation just reading:

$$\left(f_\delta^k - d \cdot \Delta f_\delta^k \right) - d \cdot f_{\delta\alpha}^k \cdot \frac{\gamma^{\alpha\beta} \cdot \Omega_{\delta\gamma}^{kl} \left((\ldots)^{\delta\gamma} C_{jk} C_l^j \right)_{,\beta}}{\omega} = 0$$

$$\left(f_\gamma^l - d \cdot \Delta f_\gamma^l \right) - d \cdot f_{\gamma,\alpha}^l \cdot \frac{\gamma^{\alpha\beta} \cdot \Omega_{\delta\gamma}^{kl} \left((\ldots)^{\delta\gamma} C_{jk} C_l^j \right)_{,\beta}}{\omega} = 0. \tag{855}$$

If, however, considering the metric origin of f_γ^l, f_δ^k and $F_{,\gamma}^l$, $F_{,\delta}^k$, namely:

$$\delta_\gamma W = 0 = \int_V d^n x \sqrt{\gamma} \cdot \left(R^{\delta\gamma} \left[\gamma^{\delta\gamma} \right] - \frac{1}{2} R \cdot \gamma^{\delta\gamma} + \Lambda \cdot \gamma^{\delta\gamma} + \kappa \cdot T^{\delta\gamma} \left[\gamma^{\delta\gamma} \right] \right)$$

$$
\times
\left(
\begin{array}{l}
\delta g_{\delta\gamma} = \delta \left(\frac{\partial G^i [x_{\gamma i}]}{\partial x^\delta} \cdot \frac{\partial G^j [x_{\gamma i}]}{\partial x^\gamma} \gamma_{ij} \right) = \gamma_{ij} \cdot \delta \left(\frac{\partial G^i [x_{\gamma i}]}{\partial x^\delta} \cdot \frac{\partial G^j [x_{\gamma i}]}{\partial x^\gamma} \right) \\[4pt]
= \frac{\gamma_{ij}}{4} \delta \left(
\left\{
\begin{array}{l}
\delta_\delta^i + C_k^i \cdot f_\delta^k + \mathbf{C}^\alpha \cdot C_k^i \cdot f_{\delta,\alpha}^k \\
\delta_\delta^i + C_k^i \cdot f_\delta^k - \mathbf{C}^\alpha \cdot C_k^i \cdot f_{\delta,\alpha}^k \\
\delta_\delta^i - C_k^i \cdot f_\delta^k + \mathbf{C}^\alpha \cdot C_k^i \cdot f_{\delta,\alpha}^k \\
\delta_\delta^i - C_k^i \cdot f_\delta^k - \mathbf{C}^\alpha \cdot C_k^i \cdot f_{\delta,\alpha}^k
\end{array}
\right\}
\left\{
\begin{array}{l}
\delta_\gamma^j + C_l^j \cdot f_\gamma^l + \mathbf{C}^\beta \cdot C_l^j \cdot f_{\gamma,\beta}^l \\
\delta_\gamma^j + C_l^j \cdot f_\gamma^l - \mathbf{C}^\beta \cdot C_l^j \cdot f_{\gamma,\beta}^l \\
\delta_\gamma^j - C_l^j \cdot f_\gamma^l + \mathbf{C}^\beta \cdot C_l^j \cdot f_{\gamma,\beta}^l \\
\delta_\gamma^j - C_l^j \cdot f_\gamma^l - \mathbf{C}^\beta \cdot C_l^j \cdot f_{\gamma,\beta}^l
\end{array}
\right\}
\right) \\[4pt]
= \delta \left(\delta_\delta^i \delta_\gamma^j + C_k^i \cdot f_\delta^k C_l^j \cdot f_\gamma^l + \mathbf{C}^\alpha \cdot C_k^i \cdot f_{\delta,\alpha}^k \mathbf{C}^\beta \cdot C_l^j \cdot f_{\gamma,\beta}^l \right) \gamma_{ij} \\[4pt]
= \gamma_{ij} \cdot \left(C_k^i C_l^j \cdot \left[f_\delta^k \cdot \frac{\partial f_\gamma^l}{\partial x^\sigma} + f_\gamma^l \cdot \frac{\partial f_\delta^k}{\partial x^\sigma} \right] + \mathbf{C}^\alpha \mathbf{C}^\beta \cdot C_k^i C_l^j \left[f_{\delta,\alpha}^k \cdot \frac{\partial f_{\gamma,\beta}^l}{\partial x^\sigma} + f_{\gamma,\beta}^l \cdot \frac{\partial f_{\delta,\alpha}^k}{\partial x^\sigma} \right] \right) \delta^\sigma
\end{array}
\right),
$$

$$(856)$$

we realize that indeed the setting $F_{,\gamma}^l$, $F_{,\delta}^k$ would be very reasonable, but obviously not of need at all costs in order to obtain practicable quantum gravity equations.

For the reason of covariance and completeness, we also consider options with covariant derivatives within the function vector settings (Eq. (856)), where the following rather general approach shall be used:

$$\delta_\gamma W = 0 = \int_V d^n x \sqrt{\gamma} \cdot \left(R^{\delta\gamma} \left[\gamma^{\delta\gamma} \right] - \frac{1}{2} R \cdot \gamma^{\delta\gamma} + \Lambda \cdot \gamma^{\delta\gamma} + \kappa \cdot T^{\delta\gamma} \left[\gamma^{\delta\gamma} \right] \right)$$

$$
\times
\left(
\begin{array}{l}
\delta g_{\delta\gamma} = \delta \left(\frac{\partial G^i [x_{\gamma i}]}{\partial x^\delta} \cdot \frac{\partial G^j [x_{\gamma i}]}{\partial x^\gamma} \gamma_{ij} \right) = \gamma_{ij} \cdot \delta \left(\frac{\partial G^i [x_{\gamma i}]}{\partial x^\delta} \cdot \frac{\partial G^j [x_{\gamma i}]}{\partial x^\gamma} \right) \\[4pt]
= \frac{\gamma_{ij}}{4} \delta \left(
\left\{
\begin{array}{l}
\delta_\delta^i + C_k^i \cdot f_\delta^k + \mathbf{C}^\alpha \cdot C_k^i \cdot f_{\delta(:)\alpha}^k \\
\delta_\delta^i + C_k^i \cdot f_\delta^k - \mathbf{C}^\alpha \cdot C_k^i \cdot f_{\delta(:)\alpha}^k \\
\delta_\delta^i - C_k^i \cdot f_\delta^k + \mathbf{C}^\alpha \cdot C_k^i \cdot f_{\delta(:)\alpha}^k \\
\delta_\delta^i - C_k^i \cdot f_\delta^k - \mathbf{C}^\alpha \cdot C_k^i \cdot f_{\delta(:)\alpha}^k
\end{array}
\right\}
\left\{
\begin{array}{l}
\delta_\gamma^j + C_l^j \cdot f_\gamma^l + \mathbf{C}^\beta \cdot C_l^j \cdot f_{\gamma(:)\beta}^l \\
\delta_\gamma^j + C_l^j \cdot f_\gamma^l - \mathbf{C}^\beta \cdot C_l^j \cdot f_{\gamma(:)\beta}^l \\
\delta_\gamma^j - C_l^j \cdot f_\gamma^l + \mathbf{C}^\beta \cdot C_l^j \cdot f_{\gamma(:)\beta}^l \\
\delta_\gamma^j - C_l^j \cdot f_\gamma^l - \mathbf{C}^\beta \cdot C_l^j \cdot f_{\gamma(:)\beta}^l
\end{array}
\right\}
\right) \\[4pt]
= \delta \left(\delta_\delta^i \delta_\gamma^j + C_k^i \cdot f_\delta^k C_l^j \cdot f_\gamma^l + \mathbf{C}^\alpha \cdot C_k^i \cdot f_{\delta(:)\alpha}^k \mathbf{C}^\beta \cdot C_l^j \cdot f_{\gamma(:)\beta}^l \right) \gamma_{ij} \\[4pt]
= \gamma_{ij} \cdot \left(C_k^i C_l^j \cdot \left[f_\delta^k \cdot \frac{\partial f_\gamma^l}{\partial x^\sigma} + f_\gamma^l \cdot \frac{\partial f_\delta^k}{\partial x^\sigma} \right] + \mathbf{C}^\alpha \mathbf{C}^\beta \cdot C_k^i C_l^j \left[f_{\delta,\alpha}^k \cdot \frac{\partial f_{\gamma(:)\beta}^l}{\partial x^\sigma} + f_{\gamma,\beta}^l \cdot \frac{\partial f_{\delta(:)\alpha}^k}{\partial x^\sigma} \right] \right) \delta^\sigma
\end{array}
\right).
$$

$$(857)$$

Further evaluation with the setting $F_{;\gamma}^l$, $F_{;\delta}^k$, thereby assuming the existence of the following tensor:

$$(\ldots)^{\delta\gamma} \cdot \Omega_{\delta\gamma} = \omega, \qquad (858)$$

and temporarily using the alternative operator $(, ;) \equiv \begin{cases} , \\ ; \end{cases}$ for ordinary or covariant derivatives, respectively, gives us:

$$
0 = \sqrt{\gamma} \cdot C_{jk} C_l^j \, (\ldots)^{\delta\gamma} \, F_{;\delta}^k - d \cdot \left(\sqrt{\gamma} \cdot \gamma^{\alpha\beta} \cdot F_{;\delta\alpha}^k \, (\ldots)^{\delta\gamma} \, C_{jk} C_l^j \right)_{(;)\beta}
$$

$$
\xrightarrow{C_{jk} C_l^j = \gamma_{kl}} = \sqrt{\gamma} \cdot (\ldots)^{\delta\gamma} \, F_{l;\delta} - d \cdot \left(\sqrt{\gamma} \cdot \gamma^{\alpha\beta} \cdot F_{l;\delta(;)\alpha} \, (\ldots)^{\delta\gamma} \right)_{(;)\beta}
$$

$$
= \underbrace{\left\{ \sqrt{\gamma} \cdot (\ldots)^{\delta\gamma} \, F_{l;\delta} - d \cdot \left(\sqrt{\gamma} \cdot \gamma^{\alpha\beta} \cdot F_{l;\delta,\alpha} \, (\ldots)^{\delta\gamma} \right)_{,\beta} = \sqrt{\gamma} \cdot (\ldots)^{\delta\gamma} \, F_{l;\delta} - d \cdot \left(\begin{array}{c} \sqrt{\gamma} \cdot \gamma^{\alpha\beta} \cdot F_{l;\delta,\alpha} \, (\ldots)^{\delta\gamma}_{\,,\beta} \\ + (\ldots)^{\delta\gamma} \, (\sqrt{\gamma} \cdot \gamma^{\alpha\beta} \cdot F_{l;\delta,\alpha})_{,\beta} \end{array} \right) \right.}
$$

$$
= \underbrace{\sqrt{\gamma} \cdot (\ldots)^{\delta\gamma} \, F_{l;\delta} - d \cdot \left(\sqrt{\gamma} \cdot \gamma^{\alpha\beta} \cdot F_{l;\delta\alpha} \, (\ldots)^{\delta\gamma} \right)_{;\beta} = \sqrt{\gamma} \cdot (\ldots)^{\delta\gamma} \, F_{l;\delta} - d \cdot \left(\begin{array}{c} \sqrt{\gamma} \cdot \gamma^{\alpha\beta} \cdot F_{l;\delta\alpha} \, (\ldots)^{\delta\gamma}_{\;;\beta} \\ + (\ldots)^{\delta\gamma} \, (\sqrt{\gamma} \cdot \gamma^{\alpha\beta} \cdot F_{l;\delta\alpha})_{;\beta} \end{array} \right)}
$$

$$
= \underbrace{\sqrt{\gamma} \cdot (\ldots)^{\delta\gamma} \, F_{l;\delta} - d \cdot \left(\begin{array}{c} \sqrt{\gamma} \cdot \gamma^{\alpha\beta} \cdot F_{l;\delta,\alpha} \, (\ldots)^{\delta\gamma}_{\,,\beta} \\ + \sqrt{\gamma} \cdot (\ldots)^{\delta\gamma} \, \Delta F_{l;\delta} \end{array} \right) \xrightarrow{\times \Omega_{\delta\gamma}} \sqrt{\gamma} \cdot \omega \cdot F_{l;\delta} - d \cdot \left(\begin{array}{c} \sqrt{\gamma} \cdot \gamma^{\alpha\beta} \cdot F_{l;\delta,\alpha} \, (\ldots)^{\delta\gamma}_{\,,\beta} \, \Omega_{\delta\gamma} \\ + \sqrt{\gamma} \cdot \omega \cdot \Delta F_{l;\delta} \end{array} \right)}
$$

$$
= \underbrace{\sqrt{\gamma} \cdot (\ldots)^{\delta\gamma} \, F_{l;\delta} - d \cdot \left(\begin{array}{c} \sqrt{\gamma} \cdot \gamma^{\alpha\beta} \cdot F_{l;\delta\alpha} \, (\ldots)^{\delta\gamma}_{\;;\beta} \\ + (\ldots)^{\delta\gamma} \, \sqrt{\gamma} \, F_{l;\delta\alpha}^{;\alpha} \end{array} \right) \xrightarrow{\times \Omega_{\delta\gamma}} \sqrt{\gamma} \cdot \omega \cdot F_{l;\delta} - d \cdot \left(\begin{array}{c} \sqrt{\gamma} \cdot \gamma^{\alpha\beta} \cdot F_{l;\delta\alpha} \, (\ldots)^{\delta\gamma}_{\;;\beta} \, \Omega_{\delta\gamma} \\ + \sqrt{\gamma} \cdot \omega \cdot F_{l;\delta\alpha}^{;\alpha} \end{array} \right)}
$$

$$
= \left\{ \begin{array}{l} \sqrt{\gamma} \cdot \omega \cdot F_{l;\delta} - d \cdot \left(\sqrt{\gamma} \cdot \gamma^{\alpha\beta} \cdot F_{l;\delta,\alpha} \, (\ldots)^{\delta\gamma}_{\,,\beta} \, \Omega_{\delta\gamma} + \sqrt{\gamma} \cdot \omega \cdot \Delta F_{l;\delta} \right. \\ \sqrt{\gamma} \cdot \omega \cdot F_{l;\delta} - d \cdot \left(\sqrt{\gamma} \cdot \gamma^{\alpha\beta} \cdot F_{l;\delta\alpha} \, (\ldots)^{\delta\gamma}_{\;;\beta} \, \Omega_{\delta\gamma} + \sqrt{\gamma} \cdot \omega \cdot \Delta F_{l;\delta} \right) . \end{array} \right.
\tag{859}
$$

Still we have the classical Klein–Gordon equation with mass and potential.

Comparison with Eq. (853) tells us that the last line of Eq. (859) should be given as follows:

$$0 = \begin{cases} \frac{m^2 c^2}{\hbar^2} \cdot F_{l;\delta} - \gamma^{\alpha\beta} \cdot F_{l;\delta,\alpha} \left(\ldots\right)^{\delta\gamma}{}_{,\beta} \frac{\Omega_{\delta\gamma}}{\omega} + \Delta F_{l;\delta} \\ \frac{m^2 c^2}{\hbar^2} \cdot F_{l;\delta} - \gamma^{\alpha\beta} \cdot F_{l;\delta\alpha} \left(\ldots\right)^{\delta\gamma}{}_{;\beta} \frac{\Omega_{\delta\gamma}}{\omega} + \Delta F_{l;\delta} \end{cases}. \tag{860}$$

The corresponding form of Eq. (857) then reads (now with the new definitions: $\mathbf{C}^\alpha \mathbf{C}^\beta = \mathbf{e}^\alpha \mathbf{e}^\beta = \gamma^{\alpha\beta}$ and $M^2 = \frac{m^2 c^2}{\hbar^2}$):

$$\delta_\gamma W = 0 = \int_V d^n x \sqrt{\gamma} \cdot \left(R^{\delta\gamma} \left[\gamma^{\delta\gamma}\right] - \frac{1}{2} R \cdot \gamma^{\delta\gamma} + \Lambda \cdot \gamma^{\delta\gamma} + \kappa \cdot T^{\delta\gamma} \left[\gamma^{\delta\gamma}\right] \right)$$

$$\times \left(\begin{array}{l} \delta g_{\delta\gamma} = \delta \left(\frac{\partial G^l [x_{vi}]}{\partial x^\delta} \cdot \frac{\partial G^l [x_{vi}]}{\partial x^\gamma} \gamma_{ij} \right) = \gamma_{ij} \cdot \delta \left(\frac{\partial G^l [x_{vi}]}{\partial x^\delta} \cdot \frac{\partial G^l [x_{vi}]}{\partial x^\gamma} \right) \\ = \frac{\gamma_{ij}}{4} \delta \left\{ \begin{array}{l} \delta_\delta^i + M \cdot C_k^i \cdot f_\delta^k + \mathbf{C}^\alpha \cdot C_k^i \cdot f_{\delta(:)\alpha}^k \\ \delta_\delta^i + M \cdot C_k^i \cdot f_\delta^k - \mathbf{C}^\alpha \cdot C_k^i \cdot f_{\delta(:)\alpha}^k \\ \delta_\delta^i - M \cdot C_k^i \cdot f_\delta^k + \mathbf{C}^\alpha \cdot C_k^i \cdot f_{\delta(:)\alpha}^k \\ \delta_\delta^i - M \cdot C_k^i \cdot f_\delta^k - \mathbf{C}^\alpha \cdot C_k^i \cdot f_{\delta(:)\alpha}^k \end{array} \right\} \left\{ \begin{array}{l} \delta_\gamma^j + M \cdot C_l^j \cdot f_\gamma^l + \mathbf{C}^\beta \cdot C_l^j \cdot f_{\gamma(:)\beta}^l \\ \delta_\gamma^j + M \cdot C_l^j \cdot f_\gamma^l - \mathbf{C}^\beta \cdot C_l^j \cdot f_{\gamma(:)\beta}^l \\ \delta_\gamma^j - M \cdot C_l^j \cdot f_\gamma^l + \mathbf{C}^\beta \cdot C_l^j \cdot f_{\gamma(:)\beta}^l \\ \delta_\gamma^j - M \cdot C_l^j \cdot f_\gamma^l - \mathbf{C}^\beta \cdot C_l^j \cdot f_{\gamma(:)\beta}^l \end{array} \right\} \\ = \delta \left(\delta_\delta^i \delta_\gamma^j + M^2 \cdot C_k^i \cdot f_\delta^k C_l^j \cdot f_\gamma^l + \mathbf{C}^\alpha \cdot C_k^i \cdot f_{\delta(:)\alpha}^k \mathbf{C}^\beta \cdot C_l^j \cdot f_{\gamma(:)\beta}^l \right) \gamma_{ij} \\ = \gamma_{ij} \cdot \left(M^2 \cdot C_k^i C_l^j \cdot \left[f_\delta^k \cdot \frac{\partial f_\gamma^l}{\partial x^\sigma} + f_\gamma^l \cdot \frac{\partial f_\delta^k}{\partial x^\sigma} \right] + \mathbf{C}^\alpha \mathbf{C}^\beta \cdot C_k^i C_l^j \left[f_{\delta,\alpha}^k \cdot \frac{\partial f_{\gamma(:)\beta}^l}{\partial x^\sigma} + f_{\gamma,\beta}^l \cdot \frac{\partial f_{\delta(:)\alpha}^k}{\partial x^\sigma} \right] \right) \delta^\sigma \end{array} \right).$$

$$\tag{861}$$

We may see the splitting up of the l, k, δ, γ in connection with the introduction of our matrix object $\Omega_{\delta\gamma}^{kl}$ in Eq. (852) as problematic and so we shall proceed with our investigation of other transformations and options for simplifications of (850). Thus, "activating" the simplification Eq. (844) and going back to Eq. (850) gives:

$$\left(\begin{array}{l} \left(\sqrt{\gamma} \cdot C_{jk} C_l^j \gamma^{\delta\gamma} F_{,\delta}^k - d \cdot \left(\sqrt{\gamma} \cdot \gamma^{\alpha\beta} \cdot F_{,\delta\alpha}^k \gamma^{\delta\gamma} C_{jk} C_l^j \right)_{,\beta} \right) \cdot F_{,\sigma\gamma}^l \\ + \left(\sqrt{\gamma} \cdot C_{jk} C_l^j \gamma^{\delta\gamma} F_{,\gamma}^l - d \cdot \left(\sqrt{\gamma} \cdot \gamma^{\alpha\beta} \cdot F_{,\gamma\beta}^l C_{jk} C_l^j \gamma^{\delta\gamma} \right)_{,\alpha} \right) \cdot F_{,\sigma\delta}^k \end{array} \right) = 0$$

$$\Rightarrow \left(\sqrt{\gamma} \cdot C_{jk} C_l^j \gamma^{\delta\gamma} F_{,\delta}^k - d \cdot \left(\sqrt{\gamma} \cdot \gamma^{\alpha\beta} \cdot F_{,\delta\alpha}^k \gamma^{\delta\gamma} C_{jk} C_l^j \right)_{,\beta} \right) \cdot F_{,\sigma\gamma}^l = 0$$

$$\Rightarrow \sqrt{\gamma} \cdot C_{jk} C_l^j \gamma^{\delta\gamma} F_{,\delta}^k - d \cdot \left(\sqrt{\gamma} \cdot \gamma^{\alpha\beta} \cdot F_{,\delta\alpha}^k \gamma^{\delta\gamma} C_{jk} C_l^j \right)_{,\beta} = 0. \tag{862}$$

As the result is to be seen under the integral Eq. (848), we could perform another integration by parts and obtain:

$$
\int_V d^n x \times \left(\left(\sqrt{\gamma} \cdot C_{jk} C_l^j \gamma^{\delta\gamma} F_{,\delta}^k - d \cdot \left(\sqrt{\gamma} \cdot \gamma^{\alpha\beta} \cdot F_{,\delta\alpha}^k \gamma^{\delta\gamma} C_{jk} C_l^j \right)_{,\beta} \right) \cdot F_{,\sigma\gamma}^l \right.
$$
$$
\left. + \left(\sqrt{\gamma} \cdot C_{jk} C_l^j \gamma^{\delta\gamma} F_{,\gamma}^l - d \cdot \left(\sqrt{\gamma} \cdot \gamma^{\alpha\beta} \cdot F_{,\gamma\beta}^l C_{jk} C_l^j \gamma^{\delta\gamma} \right)_{,\alpha} \right) \cdot F_{,\sigma\delta}^k \right) \tag{863}
$$

$$
= -\int_V d^n x \times \left(\left(\sqrt{\gamma} \cdot C_{jk} C_l^j \gamma^{\delta\gamma} F_{,\delta}^k - d \cdot \left(\sqrt{\gamma} \cdot \gamma^{\alpha\beta} \cdot F_{,\delta\alpha}^k \gamma^{\delta\gamma} C_{jk} C_l^j \right)_{,\beta} \right)_{,\gamma} \cdot F_{,\sigma}^l \right.
$$
$$
\left. + \left(\sqrt{\gamma} \cdot C_{jk} C_l^j \gamma^{\delta\gamma} F_{,\gamma}^l - d \cdot \left(\sqrt{\gamma} \cdot \gamma^{\alpha\beta} \cdot F_{,\gamma\beta}^l C_{jk} C_l^j \gamma^{\delta\gamma} \right)_{,\alpha} \right)_{,\delta} \cdot F_{,\sigma}^k \right)
$$

$$
= -\int_V d^n x \times \left((\Delta F_l - d \cdot \Delta^2 F_l) \cdot F_{,\sigma}^l + (\Delta F_l - d \cdot \Delta^2 F_l) \cdot F_{,\sigma}^k \right) = 0
$$

$$
\Rightarrow (\Delta F_l - d \cdot \Delta^2 F_l) \cdot F_{,\sigma}^l + (\Delta F_l - d \cdot \Delta^2 F_l) \cdot F_{,\sigma}^k = 0
$$

$$
\Rightarrow \Delta F_l - d \cdot \Delta^2 F_l = 0.
$$

As the matrix $C_{jk} C_l^j$ is not fixed yet, we might like to set it as:

$$
C_{jk} C_l^j = \gamma_{kl} \tag{864}
$$

and then obtain the equations:

$$\left(\begin{array}{l} \left(\sqrt{\gamma} \cdot \gamma_{kl} \gamma^{\delta\gamma} F^k_{,\delta} - d \cdot \left(\sqrt{\gamma} \cdot \gamma^{\alpha\beta} \cdot F^k_{,\delta\alpha} \gamma^{\delta\gamma} \gamma_{kl} \right)_{,\beta} \right) \cdot F^l_{,\sigma\gamma} \\ + \left(\sqrt{\gamma} \cdot \gamma_{kl} \gamma^{\delta\gamma} F^l_{,\gamma} - d \cdot \left(\sqrt{\gamma} \cdot \gamma^{\alpha\beta} \cdot F^l_{,\gamma\beta} \gamma_{kl} \gamma^{\delta\gamma} \right)_{,\alpha} \right) \cdot F^k_{,\sigma\delta} \end{array} \right) = 0$$

$$\Rightarrow \left(\sqrt{\gamma} \cdot \gamma_{kl} \gamma^{\delta\gamma} F^k_{,\delta} - d \cdot \left(\sqrt{\gamma} \cdot \gamma^{\alpha\beta} \cdot F^k_{,\delta\alpha} \gamma^{\delta\gamma} \gamma_{kl} \right)_{,\beta} \right) \cdot F^l_{,\sigma\gamma} = 0$$

$$\Rightarrow \sqrt{\gamma} \cdot \gamma_{kl} \gamma^{\delta\gamma} F^k_{,\delta} - d \cdot \left(\sqrt{\gamma} \cdot \gamma^{\alpha\beta} \cdot F^k_{,\delta\alpha} \gamma^{\delta\gamma} \gamma_{kl} \right)_{,\beta}$$

$$= \sqrt{\gamma} \cdot \gamma^{\delta\gamma} \left(\left(\gamma_{kl} F^k \right)_{,\delta} - \gamma_{kl,\delta} F^k \right)$$

$$- d \cdot \left(\sqrt{\gamma} \cdot \gamma^{\alpha\beta} \cdot \left(\begin{array}{l} \left(\gamma_{kl} F^k \right)_{,\delta\alpha} - \gamma_{kl,\delta\alpha} \\ - \gamma_{kl,\delta} F^k_{,\alpha} - \gamma_{kl,\alpha} F^k_{,\delta} \end{array} \right) \gamma^{\delta\gamma} \right)_{,\beta} = 0. \tag{865}$$

This does not look like a great improvement, but in cases of metrics with constant components (Minkowski), we have an extreme simplification, namely:

$$\sqrt{\gamma} \cdot \gamma^{\delta\gamma} F_{l,\delta} - d \cdot \gamma^{\delta\gamma} \left(\sqrt{\gamma} \cdot \gamma^{\alpha\beta} \cdot F_{l,\delta\alpha} \right)_{,\beta}$$

$$= \sqrt{\gamma} \cdot \gamma^{\delta\gamma} F_{l,\delta} - d \cdot \sqrt{\gamma} \cdot \gamma^{\delta\gamma} \Delta F_{l,\delta} = 0. \tag{866}$$

As this would still be under the integral Eq. (848), we could integrate by parts and obtain:

$$\int_V d^n x \times \left(\left(\sqrt{\gamma} \cdot \gamma^{\delta\gamma} F_{l,\delta} - d \cdot \sqrt{\gamma} \cdot \gamma^{\delta\gamma} \Delta F_{l,\delta} \right) \cdot F^l_{,\sigma\gamma} \right.$$

$$\left. + \left(\sqrt{\gamma} \cdot \gamma^{\delta\gamma} F_{k,\gamma} - d \cdot \sqrt{\gamma} \cdot \gamma^{\delta\gamma} \Delta F_{k,\gamma} \right) \cdot F^k_{,\sigma\delta} \right)$$

$$= - \int_V d^n x \times \left(\left(\sqrt{\gamma} \cdot \gamma^{\delta\gamma} F_{l,\delta} - d \cdot \sqrt{\gamma} \cdot \gamma^{\delta\gamma} \Delta F_{l,\delta} \right)_{,\gamma} \cdot F^l_{,\sigma} \right.$$

$$\left. + \left(\sqrt{\gamma} \cdot \gamma^{\delta\gamma} F_{k,\gamma} - d \cdot \sqrt{\gamma} \cdot \gamma^{\delta\gamma} \Delta F_{k,\gamma} \right)_{,\delta} \cdot F^k_{,\sigma} \right)$$

$$= - \int_V d^n x \times \left(\left(\Delta F_l - d \cdot \Delta^2 F_l \right) \cdot F^l_{,\sigma} + \left(\Delta F_l - d \cdot \Delta^2 F_l \right) \cdot F^k_{,\sigma} \right) = 0$$

$$\Rightarrow \left(\Delta F_l - d \cdot \Delta^2 F_l \right) \cdot F^l_{,\sigma} + \left(\Delta F_l - d \cdot \Delta^2 F_l \right) \cdot F^k_{,\sigma} = 0$$

$$\Rightarrow \Delta F_l - d \cdot \Delta^2 F_l = 0. \tag{867}$$

5.3.2 Appendix to Matrix Approaches and Second-Order Approaches

In this appendix, being more a storage, we collect a few equations and transformations we needed with respect to the functional matrix approaches considered in section 5.2. Thereby the starting point was:

$$\delta_\gamma W = 0 = \int_V d^n x \left(R^{\delta\gamma} \left[\gamma^{\delta\gamma} \right] - \frac{1}{2} R \cdot \gamma^{\delta\gamma} + \Lambda \cdot \gamma^{\delta\gamma} + \kappa \cdot T^{\delta\gamma} \left[\gamma^{\delta\gamma} \right] \right)$$

$$\times \sqrt{\gamma} \cdot \left(C^2 \cdot \left(\left[f_{j\delta} \cdot \frac{\partial f_\gamma^j}{\partial x^\sigma} + f_\gamma^j \cdot \frac{\partial f_{j\delta}}{\partial x^\sigma} \right] + C^\alpha C^\beta \cdot \gamma_{ij} \cdot \left[f_{\delta,\alpha}^i \cdot \frac{\partial f_{\gamma,\beta}^j}{\partial x^\sigma} + f_{\gamma,\beta}^j \cdot \frac{\partial f_{\delta,\alpha}^i}{\partial x^\sigma} \right] \right.\right.$$

$$= \int_V d^n x \times \sqrt{\gamma} \cdot C^2 \cdot (\ldots)^{\delta\gamma} \left[f_{j\delta} \cdot \frac{\partial f_\gamma^j}{\partial x^\sigma} + f_\gamma^j \cdot \frac{\partial f_{j\delta}}{\partial x^\sigma} \right] - \left[\begin{matrix} (\ldots)^{\delta\gamma} C^\alpha C^\beta \cdot \sqrt{\gamma} \cdot f_{j\delta,\alpha} \end{matrix} \right]_{,\beta} \cdot \frac{\partial f_\gamma^j}{\partial x^\sigma} \\ + \left[(\ldots)^{\delta\gamma} C^\alpha C^\beta \cdot \sqrt{\gamma} \cdot f_{\gamma,\beta}^j \right]_{,\alpha} \cdot \frac{\partial f_{j\delta}}{\partial x^\sigma} \right]$$

$$= \int_V d^n x \times \sqrt{\gamma} \cdot C^2 \cdot (\ldots)^{\delta\gamma} \left[f_{j\delta} \cdot f_{\gamma,\sigma}^j + f_\gamma^j \cdot f_{j\delta,\sigma} \right] - \left[\begin{matrix} (\ldots)^{\delta\gamma} C^\alpha C^\beta \cdot \sqrt{\gamma} \cdot f_{j\delta,\alpha} \end{matrix}_{,\beta} \cdot f_{\gamma,\sigma}^j \\ + \left[(\ldots)^{\delta\gamma} C^\alpha C^\beta \cdot \sqrt{\gamma} \cdot f_{\gamma,\beta}^j \right]_{,\alpha} \cdot f_{j\delta,\sigma} \right]$$

$$= \int_V d^n x \times \sqrt{\gamma} \cdot \left(\begin{matrix} \left(C^2 \cdot (\ldots)^{\delta\gamma} f_{j\delta} - d \cdot (\gamma^{\alpha\beta} \cdot f_{j\delta,\alpha} (\ldots)^{\delta\gamma}_{,\beta} + (\ldots)^{\delta\gamma} \Delta f_{j\delta}) \right) \cdot f_{\gamma,\sigma}^j \\ + \left(C^2 \cdot (\ldots)^{\delta\gamma} f_\gamma^j - d \cdot (\gamma^{\alpha\beta} \cdot f_{\gamma,\beta}^j (\ldots)^{\delta\gamma}_{,\alpha} + (\ldots)^{\delta\gamma} \Delta f_\gamma^j) \right) \cdot f_{j\delta,\sigma} \end{matrix} \right) ,$$

$$\tag{868}$$

$$\delta_\gamma W = 0 = \int_V d^n x \times \sqrt{\gamma} \cdot \left(\begin{matrix} \left(C^2 \cdot (\ldots)^{\delta\gamma} f_{j\sigma} - d \cdot (\gamma^{\alpha\beta} \cdot f_{j\sigma,\alpha} (\ldots)^{\delta\gamma}_{,\beta} + (\ldots)^{\delta\gamma} \Delta f_{j\sigma}) \right) \cdot f_{\gamma,\delta}^j \\ + \left(C^2 \cdot (\ldots)^{\delta\gamma} f_\gamma^j - d \cdot (\gamma^{\alpha\beta} \cdot f_{\gamma,\beta}^j (\ldots)^{\delta\gamma}_{,\alpha} + (\ldots)^{\delta\gamma} \Delta f_\gamma^j) \right) \cdot f_{j\sigma,\delta} \end{matrix} \right) ,$$

$$\tag{869}$$

$$\delta_\gamma W = 0 = \int_V d^n x \times \sqrt{\gamma} \cdot \left(\begin{matrix} \left(C^2 \cdot (\ldots)^{\delta\gamma} f_{j\delta} - d \cdot (\gamma^{\alpha\beta} \cdot f_{j\delta,\alpha} (\ldots)^{\delta\gamma}_{,\beta} + (\ldots)^{\delta\gamma} \Delta f_{j\delta}) \right) \cdot f_{\sigma,\gamma}^j \\ + \left(C^2 \cdot (\ldots)^{\delta\gamma} f_\sigma^j - d \cdot (\gamma^{\alpha\beta} \cdot f_{\sigma,\beta}^j (\ldots)^{\delta\gamma}_{,\alpha} + (\ldots)^{\delta\gamma} \Delta f_\sigma^j) \right) \cdot f_{j\delta,\gamma} \end{matrix} \right) .$$

$$\tag{870}$$

We only briefly repeat the evaluation process for the first integral, which goes as follows:

$$\delta_\gamma W = 0 = \int_V d^n x \left(R^{\delta\gamma} \left[\gamma^{\delta\gamma} \right] - \frac{1}{2} R \cdot \gamma^{\delta\gamma} + \Lambda \cdot \gamma^{\delta\gamma} + \kappa \cdot T^{\delta\gamma} \left[\gamma^{\delta\gamma} \right] \right)$$

$$\times \sqrt{\gamma} \cdot \delta \left(C^2 \cdot \gamma_{ij} f^i_\delta f^j_\gamma - C^\alpha C^\beta \cdot \left(\gamma_{ij} f^i_\delta \right)_{,\alpha} f^j_{\gamma,\beta} \right)$$

$$= \int_V d^n x \left(R^{\delta\gamma} \left[\gamma^{\delta\gamma} \right] - \frac{1}{2} R \cdot \gamma^{\delta\gamma} + \Lambda \cdot \gamma^{\delta\gamma} + \kappa \cdot T^{\delta\gamma} \left[\gamma^{\delta\gamma} \right] \right)$$

$$\times \sqrt{\gamma} \cdot \left(C^2 \cdot \left[f_{j\delta} \cdot \frac{\partial f^j_\gamma}{\partial x^\sigma} + f^j_\gamma \cdot \frac{\partial f_{j\delta}}{\partial x^\sigma} \right] + C^\alpha C^\beta \right.$$

$$\left. \cdot \left[f_{j\delta,\alpha} \cdot \frac{\partial f^j_{\gamma,\beta}}{\partial x^\sigma} + f^j_{\gamma,\beta} \cdot \frac{\partial f_{j\delta,\alpha}}{\partial x^\sigma} \right] \right). \tag{871}$$

Integration by parts of the second addend in the last line, performed as:

$$\int_V d^n x (\ldots)^{\delta\gamma} C^\alpha C^\beta \cdot \sqrt{\gamma} \cdot \left(\gamma_{ij} \cdot f^i_\delta \right)_{,\alpha} \cdot \frac{\partial f^j_{\gamma,\beta}}{\partial x^\sigma}$$

$$= \underbrace{\int_V d^n x \left\{ \left[(\ldots)^{\delta\gamma} C^\alpha C^\beta \cdot \sqrt{\gamma} \cdot f_{j\delta,\alpha} \right] \cdot \frac{\partial f^j_\gamma}{\partial x^\sigma} \right\}_{,\beta}}_{=S=0 \text{ per definition}}$$

$$- \int_V d^n x \left[(\ldots)^{\delta\gamma} C^\alpha C^\beta \cdot \sqrt{\gamma} \cdot f_{j\delta,\alpha} \right]_{,\beta} \cdot \frac{\partial f^j_\gamma}{\partial x^\sigma}; \quad (\ldots)^{\delta\gamma}$$

$$= \left(R^{\delta\gamma} - \frac{1}{2} R \cdot \gamma^{\delta\gamma} + \Lambda \cdot \gamma^{\delta\gamma} + \kappa \cdot T^{\delta\gamma} \right) \tag{872}$$

yields:

$$\delta_\gamma W = 0 = \int_V d^n x \left(R^{\delta\gamma} \left[\gamma^{\delta\gamma} \right] - \frac{1}{2} R \cdot \gamma^{\delta\gamma} + \Lambda \cdot \gamma^{\delta\gamma} + \kappa \cdot T^{\delta\gamma} \left[\gamma^{\delta\gamma} \right] \right)$$

$$\times \sqrt{\gamma} \cdot \left(C^2 \cdot \left[f_{j\delta} \cdot \frac{\partial f^j_\gamma}{\partial x^\sigma} + f^j_\gamma \cdot \frac{\partial f_{j\delta}}{\partial x^\sigma} \right] \right.$$

$$\left. + C^\alpha C^\beta \cdot \gamma_{ij} \cdot \left[f^i_{\delta,\alpha} \cdot \frac{\partial f^j_{\gamma,\beta}}{\partial x^\sigma} + f^j_{\gamma,\beta} \cdot \frac{\partial f^i_{\delta,\alpha}}{\partial x^\sigma} \right] \right)$$

$$= \int_V d^n x \times \left(\sqrt{\gamma} \cdot C^2 \cdot (\ldots)^{\delta\gamma} \left[f_{j\delta} \cdot \frac{\partial f^j_\gamma}{\partial x^\sigma} + f^j_\gamma \cdot \frac{\partial f_{j\delta}}{\partial x^\sigma} \right] \right.$$

$$\left. - \left[\begin{array}{l} \left[(\ldots)^{\delta\gamma} C^\alpha C^\beta \cdot \sqrt{\gamma} \cdot f_{j\delta,\alpha} \right]_{,\beta} \cdot \frac{\partial f^j_\gamma}{\partial x^\sigma} \\ + \left[(\ldots)^{\delta\gamma} C^\alpha C^\beta \cdot \sqrt{\gamma} \cdot f^j_{\gamma,\beta} \right]_{,\alpha} \cdot \frac{\partial f_{j\delta}}{\partial x^\sigma} \end{array} \right] \right). \tag{873}$$

Together with $C^\alpha C^\beta = d \cdot \gamma^{\alpha\beta}$, we are able to further simplify Eq. (833):

$$\delta_\gamma W = 0 = \int_V d^n x$$

$$\times \sqrt{\gamma} \cdot \left(C^2 \cdot (\ldots)^{\delta\gamma} \left[f_{j\delta} \cdot \frac{\partial f_\gamma^j}{\partial x^\sigma} + f_\gamma^j \cdot \frac{\partial f_{j\delta}}{\partial x^\sigma} \right] - \frac{d}{\sqrt{\gamma}} \cdot \left[\begin{array}{l} \left[(\ldots)^{\delta\gamma} \gamma^{\alpha\beta} \cdot \sqrt{\gamma} \cdot f_{j\delta,\alpha}\right]_{,\beta} \cdot \frac{\partial f_\gamma^j}{\partial x^\sigma} \\ + \left[(\ldots)^{\delta\gamma} \gamma^{\alpha\beta} \cdot \sqrt{\gamma} \cdot f_{\gamma,\beta}^j\right]_{,\alpha} \cdot \frac{\partial f_{j\delta}}{\partial x^\sigma} \end{array} \right] \right)$$

$$= \int_V d^n x \sqrt{\gamma} \cdot \left(\begin{array}{l} \left(C^2 \cdot (\ldots)^{\delta\gamma} f_{j\delta} - \frac{d}{\sqrt{\gamma}} \cdot \left[(\ldots)^{\delta\gamma} \gamma^{\alpha\beta} \cdot \sqrt{\gamma} \cdot f_{j\delta,\alpha}\right]_{,\beta} \right) \cdot \frac{\partial f_\gamma^j}{\partial x^\sigma} \\ + \left(C^2 \cdot (\ldots)^{\delta\gamma} f_\gamma^j - \frac{d}{\sqrt{\gamma}} \cdot \left[(\ldots)^{\delta\gamma} \gamma^{\alpha\beta} \cdot \sqrt{\gamma} \cdot f_{\gamma,\beta}^j\right]_{,\alpha} \right) \cdot \frac{\partial f_{j\delta}}{\partial x^\sigma} \end{array} \right)$$

$$= \int_V d^n x \times \sqrt{\gamma} \cdot \left\{ \underbrace{C^2 \cdot (\ldots)^{\delta\gamma} f_{j\delta} - \frac{d}{\sqrt{\gamma}} \cdot \left(\gamma^{\alpha\beta} \cdot \sqrt{\gamma} \cdot f_{j\delta,\alpha} (\ldots)^{\delta\gamma}_{,\beta} + (\ldots)^{\delta\gamma} \left[\gamma^{\alpha\beta} \cdot \sqrt{\gamma} \cdot f_{j\delta,\alpha}\right]_{,\beta} \right)}_{\sqrt{\gamma} \cdot \Delta f_{j\delta}} \cdot \frac{\partial f_\gamma^j}{\partial x^\sigma} \right.$$

$$\left. + \underbrace{C^2 \cdot (\ldots)^{\delta\gamma} f_\gamma^j - \frac{d}{\sqrt{\gamma}} \cdot \left(\gamma^{\alpha\beta} \cdot \sqrt{\gamma} \cdot f_{\gamma,\beta}^j (\ldots)^{\delta\gamma}_{,\alpha} + (\ldots)^{\delta\gamma} \left[\gamma^{\alpha\beta} \cdot \sqrt{\gamma} \cdot f_{\gamma,\beta}^j\right]_{,\alpha} \right)}_{\sqrt{\gamma} \cdot \Delta f_{j\delta}} \cdot \frac{\partial f_{j\delta}}{\partial x^\sigma} \right\} \tag{874}$$

and finally result in Eq. (828).

Subsequently, we obtain the sets of quantum equations of the kind:

$$C^2 \cdot (\ldots)^{\delta\gamma} F_{j\delta} - D \cdot \left(\gamma^{\alpha\beta} \cdot F_{j\delta,\alpha} (\ldots)^{\delta\gamma}_{,\beta} + (\ldots)^{\delta\gamma} \Delta F_{j\delta} \right) = 0$$

$$C^2 \cdot (\ldots)^{\delta\gamma} F_\gamma^j - D \cdot \left(\gamma^{\alpha\beta} \cdot F_{\gamma,\beta}^j (\ldots)^{\delta\gamma}_{,\alpha} + (\ldots)^{\delta\gamma} \Delta F_\gamma^j \right) = 0. \tag{875}$$

In addition, we could variate as follows:

$$\delta_\gamma W = 0 = \int_V d^n x \times \sqrt{\gamma} \cdot \left(\begin{array}{l} \left(C^2 \cdot (\ldots)^{\delta\gamma} F_{,j\delta} - d \cdot \left(\gamma^{\alpha\beta} \cdot F_{,j\delta\alpha} (\ldots)^{\delta\gamma}_{,\beta} + (\ldots)^{\delta\gamma} \Delta F_{,j\delta} \right) \right) \cdot F^{,j}_{,\gamma\sigma} \\ + \left(C^2 \cdot (\ldots)^{\delta\gamma} F^{,j}_{,\gamma} - d \cdot \left(\gamma^{\alpha\beta} \cdot F^{,j}_{,\gamma\beta} (\ldots)^{\delta\gamma}_{,\alpha} + (\ldots)^{\delta\gamma} \Delta F^{,j}_{,\gamma} \right) \right) \cdot F_{,j\delta\sigma} \end{array} \right)$$

$$= -\int_V d^n x \times \left(\begin{array}{l} \left((C^2 \cdot \sqrt{\gamma} \cdot (\ldots)^{\delta\gamma} F_{,j\delta} - d \cdot \sqrt{\gamma} \cdot \left(\gamma^{\alpha\beta} \cdot F_{,j\delta\alpha} (\ldots)^{\delta\gamma}_{,\beta} + (\ldots)^{\delta\gamma} \Delta F_{,j\delta} \right) \right)_{,\gamma} \cdot F^{,j}_{,\sigma} \\ + \left(C^2 \cdot \sqrt{\gamma} \cdot (\ldots)^{\delta\gamma} F^{,j}_{,\gamma} - d \cdot \sqrt{\gamma} \cdot \left(\gamma^{\alpha\beta} \cdot F^{,j}_{,\gamma\beta} (\ldots)^{\delta\gamma}_{,\alpha} + (\ldots)^{\delta\gamma} \Delta F^{,j}_{,\gamma} \right) \right)_{,\delta} \cdot F_{,j\sigma} \end{array} \right), \quad (876)$$

Now we assume again that we could use $(\ldots)^{\delta\gamma} \cong D \cdot \gamma^{\delta\gamma}$ or even $(\ldots)^{\delta\gamma} = D \cdot \gamma^{\delta\gamma}$ and obtain:

$$0 = -\int_V d^n x \times D \cdot \left(\begin{array}{l} \left(\sqrt{\gamma} \cdot C^2 \cdot \gamma^{\delta\gamma} F_{,j\delta} - d \cdot \sqrt{\gamma} \cdot \left(\gamma^{\alpha\beta} \cdot F_{,j\delta\alpha} \gamma^{\delta\gamma}_{,\beta} + \gamma^{\delta\gamma} \Delta F_{,j\delta} \right) \right)_{,\gamma} \cdot F^{,j}_{,\sigma} \\ + \left(\sqrt{\gamma} \cdot C^2 \cdot \gamma^{\delta\gamma} F^{,j}_{,\gamma} - d \cdot \sqrt{\gamma} \cdot \left(\gamma^{\alpha\beta} \cdot F^{,j}_{,\gamma\beta} \gamma^{\delta\gamma}_{,\alpha} + \gamma^{\delta\gamma} \Delta F^{,j}_{,\gamma} \right) \right)_{,\delta} \cdot F_{,j\sigma} \end{array} \right)$$

$$= -\int_V d^n x \times D \cdot \left(\begin{array}{l} \left(C^2 \cdot \sqrt{\gamma} \cdot \Delta F_{,j} - d \cdot \left(\sqrt{\gamma} \cdot \gamma^{\alpha\beta} \cdot F_{,j\delta\alpha} \gamma^{\delta\gamma}_{,\beta} \right)_{,\gamma} + d \cdot \sqrt{\gamma} \cdot \Delta^2 F_{,j} \right) \cdot F^{,j}_{,\sigma} \\ + \left(C^2 \cdot \sqrt{\gamma} \cdot \Delta F^{,j} - d \cdot \left(\sqrt{\gamma} \cdot \gamma^{\alpha\beta} \cdot F^{,j}_{,\gamma\beta} \gamma^{\delta\gamma}_{,\alpha} \right)_{,\delta} + d \cdot \sqrt{\gamma} \cdot \Delta^2 F^{,j} \right) \cdot F_{,j\sigma} \end{array} \right)$$

$$= -\int_V d^n x \times D \cdot \left(\begin{array}{l} C^2 \cdot \sqrt{\gamma} \cdot \Delta F_{,j} - d \cdot \left(\begin{array}{l} \left(\sqrt{\gamma} \cdot \gamma^{\alpha\beta} \cdot F_{,j\delta\alpha} \gamma^{\delta\gamma} \right)_{,\beta} \\ -\gamma^{\delta\gamma} \left(\sqrt{\gamma} \cdot \gamma^{\alpha\beta} \cdot F_{,j\delta\alpha} \right)_{,\beta} \end{array} \right)_{,\gamma} + d \cdot \sqrt{\gamma} \cdot \Delta^2 F_{,j} \right) \cdot F^{,j}_{,\sigma} \\ + \left(C^2 \cdot \sqrt{\gamma} \cdot \Delta F^{,j} - d \cdot \left(\begin{array}{l} \left(\sqrt{\gamma} \cdot \gamma^{\alpha\beta} \cdot F^{,j}_{,\gamma\beta} \right)_{,\alpha} \\ -\gamma^{\delta\gamma} \left(\sqrt{\gamma} \cdot \gamma^{\alpha\beta} \cdot F^{,j}_{,\gamma\beta} \right)_{,\alpha} \end{array} \right)_{,\delta} + d \cdot \sqrt{\gamma} \cdot \Delta^2 F^{,j} \right) \cdot F_{,j\sigma} \end{array} \right). \quad (877)$$

Recognizing additional Laplace operators, we can again simplify:

$$
0 = -\int_V d^n x \times D \cdot \left(
\begin{aligned}
&\left(C^2 \cdot \sqrt{\gamma} \cdot \Delta F_{,j} - d \cdot \left(\begin{matrix}\left(\sqrt{\gamma} \cdot \gamma^{\alpha\beta} \cdot F_{,j\delta a}\gamma^{\delta\gamma}\right)_{,\beta} \\ -\sqrt{\gamma} \cdot \gamma^{\delta\gamma}\Delta F_{,j\delta}\end{matrix}\right) + d \cdot \sqrt{\gamma} \cdot \Delta^2 F_{,j}\right)_{,\gamma} \cdot F^{,j}_{,\sigma} \\
&+ \left(C^2 \cdot \sqrt{\gamma} \cdot \Delta F^{,j} - d \cdot \left(\begin{matrix}\left(\gamma^{\alpha\beta} \cdot F^{,j}_{,\gamma\beta}\gamma^{\delta\gamma}\right)_{,\alpha} \\ -\sqrt{\gamma} \cdot \gamma^{\delta\gamma}\Delta F^{,j}_{,\gamma}\end{matrix}\right) + d \cdot \sqrt{\gamma} \cdot \Delta^2 F^{,j}\right)_{,\delta} \cdot F_{,j\sigma}
\end{aligned}
\right)
$$

$$
= -\int_V d^n x \times D \cdot \left(
\begin{aligned}
&\left(C^2 \cdot \sqrt{\gamma} \cdot \Delta F_{,j} - d \cdot \left(\sqrt{\gamma} \cdot \gamma^{\alpha\beta} \cdot F_{,j\delta a}\gamma^{\delta\gamma}\right)_{,\beta\gamma}\right) \cdot F^{,j}_{,\sigma} \\
&+ \left(C^2 \cdot \sqrt{\gamma} \cdot \Delta F^{,j} - d \cdot \left(\sqrt{\gamma} \cdot \gamma^{\alpha\beta} \cdot F^{,j}_{,\gamma\beta}\gamma^{\delta\gamma}\right)_{,\alpha\delta}\right) \cdot F_{,j\sigma}
\end{aligned}
\right)
$$

$$
= \int_V d^n x \times D \cdot \left(
\left(d \cdot \left(\sqrt{\gamma} \cdot \Delta F^{,\gamma}_{,j}\right)_{,\gamma} - C^2 \cdot \sqrt{\gamma} \cdot \Delta F_{,j}\right) \cdot F^{,j}_{,\sigma} + \left(d \cdot \left(\sqrt{\gamma} \cdot \Delta F^{,j\delta}\right)_{,\delta} - C^2 \cdot \sqrt{\gamma} \cdot \Delta F^{,j}\right) \cdot F_{,j\sigma}\right).
$$

$$(878)$$

Performing the same evaluation also for Eqs. (829) and (830) (only the terms differing from what we already did and which cannot be brought into the same forms as found in Eq. (839)) gives us:

$$
\delta_\gamma W = 0 = \int_V d^n x \times \sqrt{\gamma} \cdot \left(\left(C^2 \cdot (\ldots)^{\delta\gamma} F_{,j\sigma} - d \cdot \left(\gamma^{\alpha\beta} \cdot F_{,j\sigma a}(\ldots)^{\delta\gamma}_{,\beta} + (\ldots)^{\delta\gamma}\,\Delta F_{,j\sigma}\right) \right) \cdot F^{,j}_{,\gamma\delta} + \ldots \right)
$$

$$
= -\int_V d^n x \times D \cdot \left(\left(C^2 \cdot \sqrt{\gamma} \cdot \gamma^{\delta\gamma} F_{,j\sigma} - d \cdot \sqrt{\gamma} \cdot \left(\gamma^{\alpha\beta} \cdot F_{,j\sigma a}\gamma^{\delta\gamma}_{,\beta} + \gamma^{\delta\gamma}\,\Delta F_{,j\sigma}\right)\right)_{,\delta} \cdot F^{,j}_{,\gamma} + \ldots \right)
$$

$$
= -\int_V d^n x \times D \cdot \left(\left(C^2 \cdot \sqrt{\gamma} \cdot \gamma^{\delta\gamma} F_{,\sigma j} - d \cdot \sqrt{\gamma} \cdot \left(\gamma^{\alpha\beta} \cdot F_{,\sigma j a}\gamma^{\delta\gamma}_{,\beta} + \gamma^{\delta\gamma}\,\Delta F_{,\sigma j}\right)\right)_{,\delta} \cdot F^{,\sigma}_{,\gamma} + \ldots \right), \quad (879)
$$

$$\delta_\gamma W = 0 = \int_V d^n x \times \sqrt{\gamma} \cdot \left(\ldots + \left(C^2 \cdot (\ldots)^{\delta\gamma} F_{,\sigma}^{,j} - d \cdot \left(\gamma^{\alpha\beta} \cdot F_{,\sigma\beta}^{,j} (\ldots)^{\delta\gamma} \right)_{,\alpha} + (\ldots)^{\delta\gamma} \Delta F_{,\sigma}^{,j} \right) \right) \cdot F_{,j\delta\gamma}$$

$$= -\int_V d^n x \times D \cdot \left(\ldots + \left(C^2 \cdot \sqrt{\gamma} \cdot \gamma^{\delta\gamma} F_{,j}^{,\sigma} - d \cdot \sqrt{\gamma} \cdot \left(\gamma^{\alpha\beta} \cdot F_{,j\beta}^{,\sigma} \gamma^{\delta\gamma} + \gamma^{\delta\gamma} \Delta F_{,j}^{,\sigma} \right) \right)_{,\gamma} \cdot F_{,\sigma\delta} \right). \quad (880)$$

We boldly interchanged σ and j in Eqs. (840) and (842) in the third and the second line, respectively. Now we fix the index j, which so far was not determined, to one of the indices δ or γ. In essence, without the previous swap, this move could also just be seen as a condition for the variation with respect to σ. In result, we have:

$$\delta_\gamma W = 0 = -\int_V d^n x \times D \cdot \left(\left(C^2 \cdot \sqrt{\gamma} \cdot \gamma^{\delta\gamma} F_{,\sigma j} - d \cdot \sqrt{\gamma} \cdot \left(\gamma^{\alpha\beta} \cdot F_{,\sigma j\alpha} \gamma^{\delta\gamma}_{,\beta} + \gamma^{\delta\gamma} \Delta F_{,\sigma j} \right) \right)_{,\delta} \cdot F_{,\gamma}^{,\sigma} + \ldots \right)$$

$$= -\int_V d^n x \times D \cdot \left(\left(C^2 \cdot \sqrt{\gamma} \cdot \gamma^{\delta\gamma} F_{,\sigma\gamma} - d \cdot \sqrt{\gamma} \cdot \left(\gamma^{\alpha\beta} \cdot F_{,\sigma\gamma\alpha} \gamma^{\delta\gamma}_{,\beta} + \gamma^{\delta\gamma} \Delta F_{,\sigma\gamma} \right) \right)_{,\delta} \cdot F_{,\gamma}^{,\sigma} + \ldots \right)$$

$$= -\int_V d^n x \times D \cdot \left(\left(C^2 \cdot \sqrt{\gamma} \cdot \Delta F_{,\sigma} - d \cdot \left(\sqrt{\gamma} \cdot \gamma^{\alpha\beta} \cdot F_{,\sigma\gamma\alpha} \gamma^{\delta\gamma}_{,\beta} - d \cdot \sqrt{\gamma} \cdot \Delta^2 F_{,\sigma} \right) \cdot F_{,\gamma}^{,\sigma} + \ldots \right) \right)$$

$$= -\int_A d^n x \times D \cdot \left(\left(C^2 \cdot \sqrt{\gamma} \cdot \Delta F_{,\sigma} - d \cdot \left(\begin{array}{l} \left(\sqrt{\gamma} \cdot \gamma^{\alpha\beta} \cdot F_{,\sigma\gamma} \gamma^{\delta\gamma} \right)_{,\beta} \\ -\left(\sqrt{\gamma} \cdot \gamma^{\alpha\beta} \cdot F_{,\sigma\alpha} \right)_{,\beta} \gamma^{\delta\gamma} \end{array} \right)_{,\delta} - d \cdot \Delta^2 F_{,\sigma} \right) \cdot F_{,\gamma}^{,\sigma} + \ldots \right)$$

$$= -\int_A d^n x \times \sqrt{\gamma} \cdot D \cdot \left(\left(C^2 \cdot \sqrt{\gamma} \cdot \Delta F_{,\sigma} - d \cdot \left(\sqrt{\gamma} \cdot \gamma^{\alpha\beta} \cdot F_{,\sigma\gamma\alpha} \gamma^{\delta\gamma} \right)_{,\beta\delta} \right) \cdot F_{,\gamma}^{,\sigma} + \ldots \right)$$

$$= -\int_A d^n x \times \sqrt{\gamma} \cdot D \cdot \left(\left(C^2 \cdot \sqrt{\gamma} \cdot \Delta F_{,\sigma} - d \cdot \left(\sqrt{\gamma} \cdot \Delta F_{,\sigma}^{,\beta} \right)_{,\beta} \right) \cdot F_{,\gamma}^{,\sigma} + \ldots \right), \quad (881)$$

$$\delta_\gamma W = 0 = -\int_V d^n x \times \sqrt{\gamma} \cdot d \cdot \left(\dots + \left(C^2 \cdot \gamma^{\delta\gamma} F^{\,\sigma}_{,\delta} - D \cdot \left(\gamma^{\alpha\beta} \cdot F^{\,\sigma}_{,\delta\beta} \gamma^{\delta\gamma}_{,\alpha} + \gamma^{\delta\gamma} \Delta F^{\,\sigma}_{,\delta} \right) \right)_{,\gamma} \cdot F_{,\sigma\delta} \right)$$

$$= -\int_V d^n x \times \sqrt{\gamma} \cdot d \cdot \left(\dots + \left(C^2 \cdot \gamma^{\delta\gamma} F^{\,\sigma}_{,\delta} - D \cdot \left(\gamma^{\alpha\beta} \cdot F^{\,\sigma}_{,\delta\beta} \gamma^{\delta\gamma}_{,\alpha} + \gamma^{\delta\gamma} \Delta F^{\,\sigma}_{,\delta} \right) \right)_{,\gamma} \cdot F_{,\sigma\delta} \right)$$

$$= -\int_V d^n x \times \sqrt{\gamma} \cdot d \cdot \left(\left(C^2 \cdot \Delta F^{\,\sigma\beta} - D \cdot \left(\Delta F^{\,\sigma\beta} \right)_{,\beta} \right) \cdot F_{,\sigma\delta} + \dots \right). \tag{882}$$

Finally we have conditions which can be compactified to just two equations:

$$\left(D \cdot \left(\sqrt{\gamma} \cdot \Delta F^{\,\gamma}_{,j} \right)_{,\gamma} - C^2 \cdot \sqrt{\gamma} \cdot \Delta F_{,j} \right) \cdot F^{\,j}_{,\sigma}$$

$$+ \left(D \cdot \left(\sqrt{\gamma} \cdot \Delta F^{\,j\delta} \right)_{,\delta} - C^2 \cdot \sqrt{\gamma} \cdot \Delta F^{\,j} \right) \cdot F_{,j\sigma} = 0$$

$$\left(D \cdot \left(\sqrt{\gamma} \cdot \Delta F_{,\sigma}^{\,\beta} \right)_{,\beta} - C^2 \cdot \sqrt{\gamma} \cdot \Delta F_{,\sigma} \right) \cdot F^{\,\sigma}_{,\gamma} = 0;$$

$$\left(D \cdot \left(\sqrt{\gamma} \cdot \Delta F^{\,\sigma\beta} \right)_{,\beta} - C^2 \cdot \sqrt{\gamma} \cdot \Delta F^{\,\sigma\beta} \right) \cdot F_{,\sigma\delta} = 0$$

$$\Rightarrow \left(D \cdot \left(\sqrt{\gamma} \cdot \Delta F^{\,\gamma}_{,j} \right)_{,\gamma} - C^2 \cdot \sqrt{\gamma} \cdot \Delta F_{,j} \right) \cdot F^{\,j}_{,\sigma} = 0;$$

$$\left(D \cdot \left(\sqrt{\gamma} \cdot \Delta F^{\,j\delta} \right)_{,\delta} - C^2 \cdot \sqrt{\gamma} \cdot \Delta F^{\,j} \right) \cdot F_{,j\sigma} = 0$$

$$\Rightarrow 0 = D \cdot \left(\sqrt{\gamma} \cdot \Delta F^{\,\gamma}_{,j} \right)_{,\gamma} - C^2 \cdot \sqrt{\gamma} \cdot \Delta F_{,j};$$

$$0 = D \cdot \left(\sqrt{\gamma} \cdot \Delta F^{\,j\delta} \right)_{,\delta} - C^2 \cdot \sqrt{\gamma} \cdot \Delta F^{\,j}. \tag{883}$$

It should be emphasized that these equations (last line above) in connection with the assumption of the possibility of fulfilling:

$$(\dots)^{\delta\gamma} = \left(R^{\delta\gamma} \left[\gamma^{\delta\gamma} \right] - \frac{1}{2} R \cdot \gamma^{\delta\gamma} + \Lambda \cdot \gamma^{\delta\gamma} + \kappa \cdot T^{\delta\gamma} \left[\gamma^{\delta\gamma} \right] \right) = D \cdot \gamma^{\delta\gamma}, \tag{884}$$

give conditions not only for the function F, which in fact is a wave function, but also for the metric tensor.

We only briefly repeat the evaluation process for the first integral, which goes as follows:

$$\delta_\gamma W = 0 = \int_V d^n x \left(R^{\delta\gamma} \left[\gamma^{\delta\gamma} \right] - \frac{1}{2} R \cdot \gamma^{\delta\gamma} + \Lambda \cdot \gamma^{\delta\gamma} + \kappa \cdot T^{\delta\gamma} \left[\gamma^{\delta\gamma} \right] \right)$$

$$\times \sqrt{\gamma} \cdot \delta \left(C^2 \cdot \gamma_{ij} \left(F_{,k} \gamma^{ki} \right)_{,\delta} \left(F_{,l} \gamma^{lj} \right)_{,\gamma} - C^\alpha C^\beta \cdot \left(\gamma_{ij} \left(F_{,k} \gamma^{ki} \right)_{,\delta} \right)_{,\alpha} \left(F_{,l} \gamma^{lj} \right)_{,\gamma\beta} \right)$$

$$= \int_V d^n x \left(R^{\delta\gamma} \left[\gamma^{\delta\gamma} \right] - \frac{1}{2} R \cdot \gamma^{\delta\gamma} + \Lambda \cdot \gamma^{\delta\gamma} + \kappa \cdot T^{\delta\gamma} \left[\gamma^{\delta\gamma} \right] \right)$$

$$\times \sqrt{\gamma} \cdot \left(\begin{array}{l} C^2 \cdot \left[\gamma_{ij} \left(F_{,k} \gamma^{ki} \right)_{,\delta} \cdot \dfrac{\partial (F_{,l} \gamma^{lj})_{,\gamma}}{\partial x^\sigma} + \left(F_{,l} \gamma^{lj} \right)_{,\gamma} \cdot \dfrac{\partial \left(\gamma_{ij} (F_{,k} \gamma^{ki})_{,\delta} \right)}{\partial x^\sigma} \right] \\[2ex] + C^\alpha C^\beta \cdot \left[\left(\gamma_{ij} \left(F_{,k} \gamma^{ki} \right)_{,\delta} \right)_{,\alpha} \cdot \dfrac{\partial (F_{,l} \gamma^{lj})_{,\gamma\beta}}{\partial x^\sigma} + \left(F_{,l} \gamma^{lj} \right)_{,\gamma\beta} \cdot \dfrac{\partial \left(\gamma_{ij} (F_{,k} \gamma^{ki})_{,\delta} \right)_{,\alpha}}{\partial x^\sigma} \right] \end{array} \right). \tag{885}$$

Integration by parts of the second addend in the last line, performed as:

$$\int_V d^n x \,(\ldots)^{\delta\gamma} C^\alpha C^\beta \cdot \sqrt{\gamma} \cdot (\gamma_{ij} \cdot f_\delta^i)_{,\alpha} \cdot \frac{\partial f_{\gamma,\beta}^j}{\partial x^\sigma} = \int_V d^n x \left\{ \underbrace{[(\ldots)^{\delta\gamma} C^\alpha C^\beta \cdot \sqrt{\gamma} \cdot f_{j\delta,\alpha}]}_{=S=0 \text{ per definition}} \cdot \frac{\partial f_\gamma^j}{\partial x^\sigma} \right\}_{,\beta}$$

$$- \int_V d^n x \,[(\ldots)^{\delta\gamma} C^\alpha C^\beta \cdot \sqrt{\gamma} \cdot f_{j\delta,\alpha}]_{,\beta} \cdot \frac{\partial f_\gamma^j}{\partial x^\sigma};$$

$$(\ldots)^{\delta\gamma} = \left(R^{\delta\gamma} - \frac{1}{2} R \cdot \gamma^{\delta\gamma} + \Lambda \cdot \gamma^{\delta\gamma} + \kappa \cdot T^{\delta\gamma} \right), \qquad (886)$$

yields:

$$\delta_\gamma W = 0 = \int_V d^n x \left(R^{\delta\gamma} [\gamma^{\delta\gamma}] - \frac{1}{2} R \cdot \gamma^{\delta\gamma} + \Lambda \cdot \gamma^{\delta\gamma} + \kappa \cdot T^{\delta\gamma} [\gamma^{\delta\gamma}] \right)$$

$$\times \sqrt{\gamma} \cdot \left(C^2 \cdot \left[\gamma_{ij} \left(F_{,k}\gamma^{ki} \right)_{,\delta} \cdot \frac{\partial \left(F_{,l}\gamma^{lj} \right)_{,\gamma}}{\partial x^\sigma} + \left(F_{,l}\gamma^{lj} \right)_{,\gamma} \cdot \frac{\partial \left(\gamma_{ij} \left(F_{,k}\gamma^{ki} \right)_{,\delta} \right)}{\partial x^\sigma} \right] \right.$$

$$\left. + C^\alpha C^\beta \cdot \left[\left(\gamma_{ij} \left(F_{,k}\gamma^{ki} \right)_{,\delta} \right) \cdot \frac{\partial \left(F_{,l}\gamma^{lj} \right)_{,\gamma\beta}}{\partial x^\sigma} + \left(F_{,l}\gamma^{lj} \right)_{,\gamma\beta} \cdot \frac{\partial \left(\gamma_{ij} \left(F_{,k}\gamma^{ki} \right)_{,\delta} \right)_{,\alpha}}{\partial x^\sigma} \right] \right)$$

$$= \int_V d^n x \times \left(\sqrt{\gamma} \cdot C^2 \cdot (\ldots)^{\delta\gamma} \left[\gamma_{ij} \left(F_{,k}\gamma^{ki} \right)_{,\delta} \cdot \left(\gamma_{ij} \left(F_{,k}\gamma^{ki} \right)_{,\delta} \right) \cdot \left(F_{,l}\gamma^{lj} \right)_{,\gamma} \right] \right.$$

$$\left. - \left[\left[(\ldots)^{\delta\gamma} C^\alpha C^\beta \cdot \sqrt{\gamma} \cdot \left(\gamma_{ij} \left(F_{,k}\gamma^{ki} \right)_{,\delta} \right) \cdot \frac{\partial \left(F_{,l}\gamma^{lj} \right)_{,\gamma}}{\partial x^\sigma} \right]_{,\alpha}^{,\beta} \right. \right.$$

$$\left. \left. + \left[(\ldots)^{\delta\gamma} C^\alpha C^\beta \cdot \sqrt{\gamma} \cdot \left(F_{,l}\gamma^{lj} \right)_{,\gamma\beta} \cdot \frac{\partial \left(\gamma_{ij} \left(F_{,k}\gamma^{ki} \right)_{,\delta} \right)}{\partial x^\sigma} \right]_{,\alpha} \right] \right). \qquad (887)$$

Together with $C^\alpha C^\beta = d \cdot \gamma^{\alpha\beta}$, we are able to further simplify Eq. (833):

$$
\delta_\gamma W = 0 = \int_V d^n x \times \left(\sqrt{\gamma} \cdot C^2 \cdot (\ldots)^{\delta\gamma} \left[\gamma_{ij} \left(F_{,k}\gamma^{ki} \right)_{,\delta} \cdot \frac{\partial \left(F_{,l}\gamma^{lj} \right)_{,\gamma}}{\partial x^\sigma} + \left(F_{,l}\gamma^{lj} \right)_{,\gamma} \cdot \frac{\partial \left(\gamma_{ij} \left(F_{,k}\gamma^{ki} \right)_{,\delta} \right)}{\partial x^\sigma} \right] \right.
$$

$$
\left. -d \cdot \left[\left[(\ldots)^{\delta\gamma} \gamma^{\alpha\beta} \cdot \sqrt{\gamma} \cdot \left(\gamma_{ij} \left(F_{,k}\gamma^{ki} \right)_{,\delta} \right)_{,\alpha} \right]_{,\beta} \cdot \frac{\partial \left(F_{,l}\gamma^{lj} \right)_{,\gamma}}{\partial x^\sigma} \right. \right.
$$

$$
\left. \left. + \left[(\ldots)^{\delta\gamma} \gamma^{\alpha\beta} \cdot \sqrt{\gamma} \cdot \left(F_{,l}\gamma^{lj} \right)_{,\gamma\beta} \right]_{,\alpha} \cdot \frac{\partial \left(\gamma_{ij} \left(F_{,k}\gamma^{ki} \right)_{,\delta} \right)}{\partial x^\sigma} \right] \right) \tag{888}
$$

$$
= \int_V d^n x \times \left(\left(\sqrt{\gamma} \cdot C^2 \cdot (\ldots)^{\delta\gamma} \gamma_{ij} \left(F_{,k}\gamma^{ki} \right)_{,\delta} - d \left[(\ldots)^{\delta\gamma} \gamma^{\alpha\beta} \sqrt{\gamma} \cdot \left(\gamma_{ij} \left(F_{,k}\gamma^{ki} \right)_{,\delta} \right)_{,\alpha} \right]_{,\beta} \right) \cdot \frac{\partial \left(F_{,l}\gamma^{lj} \right)_{,\gamma}}{\partial x^\sigma} \right.
$$

$$
\left. + \left(\sqrt{\gamma} \cdot C^2 \cdot (\ldots)^{\delta\gamma} \left(F_{,l}\gamma^{lj} \right)_{,\gamma} - d \left[(\ldots)^{\delta\gamma} \gamma^{\alpha\beta} \sqrt{\gamma} \cdot \left(F_{,l}\gamma^{lj} \right)_{,\gamma\beta} \right]_{,\alpha} \right) \cdot \frac{\partial \left(\gamma_{ij} \left(F_{,k}\gamma^{ki} \right)_{,\delta} \right)}{\partial x^\sigma} \right) \tag{}
$$

$$
= \int_V d^n x \times \left(\left(\sqrt{\gamma} \cdot C^2 \cdot (\ldots)^{\delta\gamma} \gamma_{ij} \left(F_{,k}\gamma^{ki} \right)_{,\delta} - d \underbrace{\left[(\ldots)^{\delta\gamma}_{,\beta} \gamma^{\alpha\beta} \cdot \sqrt{\gamma} \cdot \left(\gamma_{ij} \left(F_{,k}\gamma^{ki} \right)_{,\delta} \right)_{,\alpha} + (\ldots)^{\delta\gamma} \left[\gamma^{\alpha\beta} \sqrt{\gamma} \cdot \left(\gamma_{ij} \left(F_{,k}\gamma^{ki} \right)_{,\delta} \right)_{,\alpha} \right]_{,\beta} \right]}_{\sqrt{\gamma} \cdot \Delta \left(\gamma_{ij} \left(F_{,k}\gamma^{ki} \right)_{,\delta} \right)} \right) \cdot \frac{\partial \left(F_{,l}\gamma^{lj} \right)_{,\gamma}}{\partial x^\sigma} \right.
$$

$$
\left. + \left(\sqrt{\gamma} \cdot C^2 \cdot (\ldots)^{\delta\gamma} \left(F_{,l}\gamma^{lj} \right)_{,\gamma} - d \underbrace{\left[(\ldots)^{\delta\gamma}_{,\alpha} \gamma^{\alpha\beta} \cdot \sqrt{\gamma} \cdot \left(F_{,l}\gamma^{lj} \right)_{,\gamma\beta} + (\ldots)^{\delta\gamma} \left[\gamma^{\alpha\beta} \sqrt{\gamma} \cdot \left(F_{,l}\gamma^{lj} \right)_{,\gamma\beta} \right]_{,\alpha} \right]}_{\sqrt{\gamma} \cdot \Delta \left(F_{,l}\gamma^{lj} \right)_{,\gamma}} \right) \cdot \frac{\partial \left(\gamma_{ij} \left(F_{,k}\gamma^{ki} \right)_{,\delta} \right)}{\partial x^\sigma} \right). \tag{689}
$$

And with Eq. (844), finally we obtain:

$$
= D \int_V d^n x \left\{ \sqrt{\gamma} \cdot C^2 \cdot \gamma^{\delta\gamma} \gamma_{ij} \left(F_{,k}\gamma^{ki} \right)_{,\delta} - d \left(\begin{array}{l} \gamma^{\delta\gamma}_{,\beta}\gamma^{\alpha\beta} \cdot \sqrt{\gamma} \cdot \left(\gamma_{ij} \left(F_{,k}\gamma^{ki} \right)_{,\delta} \right)_{,\alpha} \\ + \gamma^{\delta\gamma}_{,\alpha}\gamma^{\alpha\beta} \cdot \sqrt{\gamma} \cdot \Delta \left(\gamma_{ij} \left(F_{,k}\gamma^{ki} \right)_{,\delta} \right) \\ + \gamma^{\delta\gamma} \sqrt{\gamma} \cdot \Delta \left(F_{,l}\gamma^{lj} \right)_{,\gamma} \end{array} \right) \cdot \frac{\partial \left(F_{,l}\gamma^{lj} \right)_{,\gamma}}{\partial x^\alpha} \right.
$$

$$
\left. + \left(\gamma^{\delta\gamma}_{,\alpha}\gamma^{\alpha\beta} \cdot \sqrt{\gamma} \cdot \left(F_{,l}\gamma^{lj} \right)_{,\gamma\beta} + \gamma^{\delta\gamma} \sqrt{\gamma} \cdot \Delta \left(F_{,l}\gamma^{lj} \right)_{,\gamma} \right) \cdot \frac{\partial \left(\gamma_{ij} \left(F_{,k}\gamma^{ki} \right)_{,\delta} \right)}{\partial x^\sigma} \right\}
$$

$$
= D \int_V d^n x \left\{ \sqrt{\gamma} \cdot C^2 \cdot \gamma^{\delta\gamma} \gamma_{ij} \left(F_{,k}\gamma^{ki} \right)_{,\delta} - d \left(\begin{array}{l} \gamma^{\delta\gamma} \gamma^{\alpha\beta} \cdot \sqrt{\gamma} \cdot \left(\gamma_{ij} \left(F_{,k}\gamma^{ki} \right)_{,\delta} \right)_{,\alpha\beta} \\ - \gamma^{\delta\gamma} \left(\gamma^{\alpha\beta} \cdot \sqrt{\gamma} \cdot \left(\gamma_{ij} \left(F_{,k}\gamma^{ki} \right)_{,\delta} \right)_{,\alpha} \right)_{,\beta} \\ + \gamma^{\delta\gamma} \gamma^{\alpha\beta} \cdot \sqrt{\gamma} \cdot \Delta \left(F_{,l}\gamma^{lj} \right)_{,\gamma\beta} \end{array} \right) \cdot \frac{\partial \left(F_{,l}\gamma^{lj} \right)_{,\gamma}}{\partial x^\alpha} \right.
$$

$$
\left. + \left(\gamma^{\delta\gamma} \gamma^{\alpha\beta} \left(\gamma^{\alpha\beta} \cdot \sqrt{\gamma} \cdot \left(F_{,l}\gamma^{lj} \right)_{,\gamma\beta} \right)_{,\alpha} - \gamma^{\delta\gamma} \sqrt{\gamma} \cdot \Delta \left(F_{,l}\gamma^{lj} \right)_{,\gamma} \right) \cdot \frac{\partial \left(\gamma_{ij} \left(F_{,k}\gamma^{ki} \right)_{,\delta} \right)}{\partial x^\sigma} \right\} .
$$

$$(890)$$

$$
\begin{aligned}
= D \int_V d^n x \Bigg(& \left[\sqrt{\gamma} \cdot C^2 \cdot \gamma^{\delta\gamma}\gamma_{ij}\left(F_{,k}\gamma^{ki}\right)_{,\delta} - d \begin{pmatrix} \gamma^{\delta\gamma}\gamma^{\alpha\beta} \cdot \sqrt{\gamma} \cdot \left(\gamma_{ij}\left(F_{,k}\gamma^{ki}\right)_{,\delta}\right)_{,\alpha} \\ -\gamma^{\delta\gamma}\sqrt{\gamma}\cdot\Delta\left(\gamma_{ij}\left(F_{,k}\gamma^{ki}\right)_{,\delta}\right) \\ +\gamma^{\delta\gamma}\sqrt{\gamma}\sqrt{\gamma}\cdot\Delta\left(\gamma_{ij}\left(F_{,k}\gamma^{ki}\right)_{,\delta}\right) \end{pmatrix}_{,\beta} \right] \cdot \frac{\partial\left(F_{,l}\gamma^{li}\right)_{,\gamma}}{\partial x^\sigma} \\
&+ \begin{pmatrix} \gamma^{\delta\gamma}\gamma^{\alpha\beta}\cdot\sqrt{\gamma}\cdot\left(F_{,l}\gamma^{lj}\right)_{,\gamma\beta},\alpha \\ -\gamma^{\delta\gamma}\sqrt{\gamma}\sqrt{\gamma}\cdot\Delta\left(F_{,l}\gamma^{lj}\right)_{,\gamma} \\ +\gamma^{\delta\gamma}\sqrt{\gamma}\sqrt{\gamma}\cdot\Delta\left(F_{,l}\gamma^{lj}\right)_{,\gamma} \end{pmatrix} \cdot \frac{\partial\left(\gamma_{ij}\left(F_{,k}\gamma^{ki}\right)_{,\delta}\right)}{\partial x^\sigma}
\end{aligned}
$$

$$
\begin{aligned}
= D \int_V d^n x \Bigg(& \left(\sqrt{\gamma}\cdot C^2 \cdot \gamma^{\delta\gamma}\gamma_{ij}\left(F_{,k}\gamma^{ki}\right)_{,\delta} - d\left(\gamma^{\delta\gamma}\gamma^{\alpha\beta}\cdot\sqrt{\gamma}\cdot\left(\gamma_{ij}\left(F_{,k}\gamma^{ki}\right)_{,\delta}\right)_{,\alpha}\right)_{,\beta} \right) \cdot \frac{\partial\left(F_{,l}\gamma^{li}\right)_{,\gamma}}{\partial x^\sigma} \\
&+ \left(\sqrt{\gamma}\cdot C^2 \cdot \gamma^{\delta\gamma}\gamma_{ij}\left(F_{,k}\gamma^{ki}\right)_{,\gamma} - d\left(\gamma^{\delta\gamma}\gamma^{\alpha\beta}\cdot\sqrt{\gamma}\cdot\left(F_{,l}\gamma^{lj}\right)_{,\gamma\beta},\alpha\right) \right) \cdot \frac{\partial\left(\gamma_{ij}\left(F_{,k}\gamma^{ki}\right)_{,\delta}\right)}{\partial x^\sigma}
\end{aligned}
$$

$$\tag{891}$$

Integration by parts yields:

$$= D \int_V d^n x \left(d \left(\left(\gamma^{\delta\gamma} \gamma^{\alpha\beta} \cdot \sqrt{\gamma} \cdot \left(\gamma_{ij} \left(F_{,k} \gamma^{ki} \right)_{,\delta} \right)_{,\alpha} \right)_{,\beta} \right) - \sqrt{\gamma} \cdot C^2 \cdot \gamma^{\delta\gamma} \gamma_{ij} \left(F_{,k} \gamma^{ki} \right)_{,\delta} \right)_{,\gamma} \cdot \frac{\partial \left(F_{,l} \gamma^{lj} \right)}{\partial x^\sigma}$$

$$+ d \left(\left(\gamma^{\delta\gamma} \gamma^{\alpha\beta} \cdot \sqrt{\gamma} \cdot \left(F_{,l} \gamma^{lj} \right)_{,\gamma\beta} \right)_{,\alpha} \right) - \sqrt{\gamma} \cdot C^2 \cdot \gamma^{\delta\gamma} \left(F_{,l} \gamma^{lj} \right)_{,\gamma} \right)_{,\delta} \cdot \int d^n x^\delta \frac{\partial \left(\gamma_{ij} \left(F_{,k} \gamma^{ki} \right)_{,\delta} \right)}{\partial x^\sigma}$$

$$= D \int_V d^n x \left(d \left(\gamma^{\delta\gamma} \gamma^{\alpha\beta} \cdot \sqrt{\gamma} \cdot \left(\gamma_{ij} \left(F_{,k} \gamma^{ki} \right)_{,\delta} \right)_{,\alpha} \right)_{,\beta\gamma} - \sqrt{\gamma} \cdot C^2 \cdot \Delta \left(\gamma_{ij} \left(F_{,k} \gamma^{ki} \right) \right) \cdot \frac{\partial \left(F_{,l} \gamma^{lj} \right)}{\partial x^\sigma}$$

$$+ d \left(\gamma^{\delta\gamma} \gamma^{\alpha\beta} \cdot \sqrt{\gamma} \cdot \left(F_{,l} \gamma^{lj} \right)_{,\gamma\beta} \right)_{,\alpha\delta} - \sqrt{\gamma} \cdot C^2 \cdot \Delta \left(F_{,l} \gamma^{lj} \right) \cdot \int d^n x^\delta \frac{\partial \left(\gamma_{ij} \left(F_{,k} \gamma^{ki} \right)_{,\delta} \right)}{\partial x^\sigma}$$

$$= D \int_V d^n x \left(d \left(\gamma^{\delta\gamma} \gamma^{\alpha\beta} \cdot \sqrt{\gamma} \cdot \left(\gamma_{ij} \left(F^{,i} \right)_{,\delta} \right)_{,\alpha} \right)_{,\beta\gamma} - \sqrt{\gamma} \cdot C^2 \cdot \Delta F_{,j} \cdot \frac{\partial \left(F_{,l} \gamma^{lj} \right)}{\partial x^\sigma} \right.$$

$$+ d \left(\gamma^{\alpha\beta} \cdot \sqrt{\gamma} \cdot \left(F^{,j} \right)_{,\delta} \underbrace{ \right)_{,\beta} }_{\sqrt{\gamma} \cdot \Delta \left(\left(F^{,j} \right)^\delta \right)} {}_{,\alpha\delta} - \sqrt{\gamma} \cdot C^2 \cdot \Delta \left(F^{,j} \right) \cdot \int d^n x^\delta \frac{\partial \left(\gamma_{ij} \left(F_{,k} \gamma^{ki} \right)_{,\delta} \right)}{\partial x^\sigma} \right) .$$

$$\tag{892}$$

In the end, we have the following equation to fulfill:

$$d\left(\gamma^{\delta\gamma}\gamma^{\alpha\beta}\cdot\sqrt{\gamma}\cdot\left(\gamma_{ij}\left(F^{\cdot i}\right)_{,\delta}\right)_{,\alpha}\right)_{,\beta\gamma}-\sqrt{\gamma}\cdot C^2\cdot\Delta F_{,j}=0$$

$$d\sqrt{\gamma}\cdot\Delta\left(\left(F^{\cdot j}\right)^{,\delta}\right)_{,\delta}-\sqrt{\gamma}\cdot C^2\cdot\Delta\left(F^{\cdot j}\right)=0. \tag{893}$$

We realize that we should not drag the metric γ_{ij} into the variation. Instead, we perform it as follows:

We only briefly repeat the evaluation process for the first integral, which goes as follows:

$$\delta_\gamma W = 0 = \int_V d^n x \left(R^{\delta\gamma}\left[\gamma^{\delta\gamma}\right]-\frac{1}{2}R\cdot\gamma^{\delta\gamma}+\Lambda\cdot\gamma^{\delta\gamma}+\kappa\cdot T^{\delta\gamma}\left[\gamma^{\delta\gamma}\right]\right)$$

$$\times\sqrt{\gamma}\cdot\delta\left(C^2\cdot\gamma_{ij}\left(F_{,k}\gamma^{ki}\right)_{,\delta}\left(F_{,l}\gamma^{lj}\right)_{,\gamma}-C^\alpha C^\beta\cdot\gamma_{ij}\left(F_{,k}\gamma^{ki}\right)_{,\delta\alpha}\left(F_{,l}\gamma^{lj}\right)_{,\gamma\beta}\right)$$

$$=\int_V d^n x \left(R^{\delta\gamma}\left[\gamma^{\delta\gamma}\right]-\frac{1}{2}R\cdot\gamma^{\delta\gamma}+\Lambda\cdot\gamma^{\delta\gamma}+\kappa\cdot T^{\delta\gamma}\left[\gamma^{\delta\gamma}\right]\right)$$

$$\times\sqrt{\gamma}\cdot\left(\begin{array}{c}C^2\cdot\gamma_{ij}\left[\left(F_{,k}\gamma^{ki}\right)_{,\delta}\cdot\dfrac{\partial\left(F_{,l}\gamma^{lj}\right)_{,\gamma}}{\partial x^\sigma}+\left(F_{,l}\gamma^{lj}\right)_{,\gamma}\cdot\dfrac{\partial\left(F_{,k}\gamma^{ki}\right)_{,\delta}}{\partial x^\sigma}\right]\\[2mm]+C^\alpha C^\beta\cdot\gamma_{ij}\left[\left(F_{,k}\gamma^{ki}\right)_{,\delta\alpha}\cdot\dfrac{\partial\left(F_{,l}\gamma^{lj}\right)_{,\gamma\beta}}{\partial x^\sigma}+\left(F_{,l}\gamma^{lj}\right)_{,\gamma\beta}\cdot\dfrac{\partial\left(F_{,k}\gamma^{ki}\right)_{,\delta\alpha}}{\partial x^\sigma}\right]\end{array}\right).$$

$$\tag{894}$$

Integration by parts of the second addend in the last line, performed as:

$$\int_V d^n x \left(\ldots\right)^{\delta\gamma}C^\alpha C^\beta\cdot\sqrt{\gamma}\cdot\left(\gamma_{ij}\cdot f_\delta^i\right)_{,\alpha}\cdot\frac{\partial f_{\gamma,\beta}^j}{\partial x^\sigma}$$

$$=\underbrace{\int_V d^n x\left\{\left[\left(\ldots\right)^{\delta\gamma}C^\alpha C^\beta\cdot\sqrt{\gamma}\cdot f_{j\delta,\alpha}\right]\cdot\frac{\partial f_\gamma^j}{\partial x^\sigma}\right\}_{,\beta}}_{=S=0\text{ per definition}}$$

$$-\int_V d^n x\left[\left(\ldots\right)^{\delta\gamma}C^\alpha C^\beta\cdot\sqrt{\gamma}\cdot f_{j\delta,\alpha}\right]_{,\beta}\cdot\frac{\partial f_\gamma^j}{\partial x^\sigma};\quad\left(\ldots\right)^{\delta\gamma}$$

$$=\left(R^{\delta\gamma}-\frac{1}{2}R\cdot\gamma^{\delta\gamma}+\Lambda\cdot\gamma^{\delta\gamma}+\kappa\cdot T^{\delta\gamma}\right) \tag{895}$$

yields:

$$\delta_\gamma W = 0 = \int_V d^n x \left(R^{\delta\gamma} \left[\gamma^{\delta\gamma}\right] - \frac{1}{2} R \cdot \gamma^{\delta\gamma} + \Lambda \cdot \gamma^{\delta\gamma} + \kappa \cdot T^{\delta\gamma} \left[\gamma^{\delta\gamma}\right] \right)$$

$$\times \sqrt{\gamma} \cdot \left(\begin{array}{c} C^2 \cdot \gamma_{ij} \left[\left(F_{,k}\gamma^{ki}\right)_{,\delta} \cdot \frac{\partial\left(F_{,l}\gamma^{lj}\right)_{,\gamma}}{\partial x^\sigma} + \left(F_{,l}\gamma^{lj}\right)_{,\gamma} \cdot \frac{\partial\left(F_{,k}\gamma^{ki}\right)_{,\delta}}{\partial x^\sigma} \right] \\ +C^\alpha C^\beta \cdot \gamma_{ij} \left[\left(F_{,k}\gamma^{ki}\right)_{,\delta\alpha} \cdot \frac{\partial\left(F_{,l}\gamma^{lj}\right)_{,\gamma\beta}}{\partial x^\sigma} + \left(F_{,l}\gamma^{lj}\right)_{,\gamma\beta} \cdot \frac{\partial\left(F_{,k}\gamma^{ki}\right)_{,\delta\alpha}}{\partial x^\sigma} \right] \end{array} \right)$$

$$= \int_V d^n x \times \left(\begin{array}{c} \sqrt{\gamma} \cdot C^2 \cdot (...)^{\delta\gamma} \gamma_{ij} \left[\left(F_{,k}\gamma^{ki}\right)_{,\delta} \cdot \frac{\partial\left(F_{,l}\gamma^{lj}\right)_{,\gamma}}{\partial x^\sigma} + \left(F_{,l}\gamma^{lj}\right)_{,\gamma} \cdot \frac{\partial\left(F_{,k}\gamma^{ki}\right)_{,\delta}}{\partial x^\sigma} \right] \\ \left[\begin{array}{c} \left[(...)^{\delta\gamma} C^\alpha C^\beta \cdot \sqrt{\gamma} \cdot \gamma_{ij} \left(F_{,k}\gamma^{ki}\right)_{,\delta\alpha} \right]_{,\beta} \cdot \frac{\partial\left(F_{,l}\gamma^{lj}\right)_{,\gamma}}{\partial x^\sigma} \\ + \left[(...)^{\delta\gamma} C^\alpha C^\beta \cdot \sqrt{\gamma} \cdot \gamma_{ij} \left(F_{,l}\gamma^{lj}\right)_{,\gamma\beta} \right]_{,\alpha} \cdot \frac{\partial\left(F_{,k}\gamma^{ki}\right)_{,\delta}}{\partial x^\sigma} \end{array} \right] \end{array} \right).$$

$$(896)$$

Together with $C^\alpha C^\beta = d \cdot \gamma^{\alpha\beta}$, we are able to further simplify Eq. (833):

$$\delta_\gamma W = 0$$

$$= \int_V d^n x \times \left(\begin{array}{c} \sqrt{\gamma} \cdot C^2 \cdot (...)^{\delta\gamma} \gamma_{ij} \left[\left(F_{,k}\gamma^{ki}\right)_{,\delta} \cdot \frac{\partial\left(F_{,l}\gamma^{lj}\right)_{,\gamma}}{\partial x^\sigma} + \left(F_{,l}\gamma^{lj}\right)_{,\gamma} \cdot \frac{\partial\left(F_{,k}\gamma^{ki}\right)_{,\delta}}{\partial x^\sigma} \right] \\ -d \cdot \left[\begin{array}{c} \left[(...)^{\delta\gamma} \gamma^{\alpha\beta} \cdot \sqrt{\gamma} \cdot \gamma_{ij} \left(F_{,k}\gamma^{ki}\right)_{,\delta\alpha} \right]_{,\beta} \cdot \frac{\partial\left(F_{,l}\gamma^{lj}\right)_{,\gamma}}{\partial x^\sigma} \\ + \left[(...)^{\delta\gamma} \gamma^{\alpha\beta} \cdot \sqrt{\gamma} \cdot \gamma_{ij} \left(F_{,l}\gamma^{lj}\right)_{,\gamma\beta} \right]_{,\alpha} \cdot \frac{\partial\left(F_{,k}\gamma^{ki}\right)_{,\delta}}{\partial x^\sigma} \end{array} \right] \end{array} \right)$$

$$= \int_V d^n x \times$$

$$\left(\begin{array}{c} \left(\sqrt{\gamma} \cdot C^2 \cdot (...)^{\delta\gamma} \gamma_{ij} \left(F_{,k}\gamma^{ki}\right)_{,\delta} - d \left[(...)^{\delta\gamma} \gamma^{\alpha\beta} \cdot \sqrt{\gamma} \cdot \gamma_{ij} \left(F_{,k}\gamma^{ki}\right)_{,\delta\alpha} \right]_{,\beta} \right) \cdot \frac{\partial\left(F_{,l}\gamma^{lj}\right)_{,\gamma}}{\partial x^\sigma} \\ + \left(\sqrt{\gamma} \cdot C^2 \cdot (...)^{\delta\gamma} \gamma_{ij} \left(F_{,l}\gamma^{lj}\right)_{,\gamma} - d \left[(...)^{\delta\gamma} \gamma^{\alpha\beta} \cdot \sqrt{\gamma} \cdot \gamma_{ij} \left(F_{,l}\gamma^{lj}\right)_{,\gamma\beta} \right]_{,\alpha} \right) \cdot \frac{\partial\left(F_{,k}\gamma^{ki}\right)_{,\delta}}{\partial x^\sigma} \end{array} \right).$$

$$(897)$$

$$= \int_V d^n x \times$$

$$\left(\begin{array}{c} \left(\sqrt{\gamma} \cdot C^2 \cdot (...)^{\delta\gamma} \gamma_{ij} \left(F_{,k}\gamma^{ki}\right)_{,\delta} - d \left(\begin{array}{c} (...)^{\delta\gamma} \gamma_{ij} \cdot \overbrace{\left[\gamma^{\alpha\beta} \cdot \sqrt{\gamma} \cdot \left(F_{,k}\gamma^{ki}\right)_{,\delta\alpha}\right]_{,\beta}}^{\sqrt{\gamma}\cdot\Delta\left(F_{,k}\gamma^{ki}\right)_{,\delta}} \\ + \left[(...)^{\delta\gamma} \gamma_{ij}\right]_{,\beta} \cdot \gamma^{\alpha\beta} \cdot \sqrt{\gamma} \cdot \left(F_{,k}\gamma^{ki}\right)_{,\delta\alpha} \end{array} \right) \right) \cdot \frac{\partial\left(F_{,l}\gamma^{lj}\right)_{,\gamma}}{\partial x^\sigma} \\ + \left(\sqrt{\gamma} \cdot C^2 \cdot (...)^{\delta\gamma} \gamma_{ij} \left(F_{,l}\gamma^{lj}\right)_{,\gamma} - d \left(\begin{array}{c} (...)^{\delta\gamma} \gamma_{ij} \cdot \overbrace{\left[\gamma^{\alpha\beta} \cdot \sqrt{\gamma} \cdot \left(F_{,l}\gamma^{lj}\right)_{,\gamma\beta}\right]_{,\alpha}}^{\sqrt{\gamma}\cdot\Delta\left(F_{,l}\gamma^{lj}\right)_{,\gamma}} \\ + \left[(...)^{\delta\gamma} \gamma_{ij}\right]_{,\alpha} \cdot \gamma^{\alpha\beta} \cdot \sqrt{\gamma} \cdot \left(F_{,l}\gamma^{lj}\right)_{,\gamma\beta} \end{array} \right) \right) \cdot \frac{\partial\left(F_{,k}\gamma^{ki}\right)_{,\delta}}{\partial x^\sigma} \end{array} \right).$$

$$(898)$$

And with Eq. (844), we obtain:

$$
= D \int_V d^n x \left[\left(\sqrt{\gamma} \cdot C^2 \cdot \gamma^{\delta\gamma} \gamma_{ij} \left(F_{,k}\gamma^{ki}\right)_{,\delta} - d \left(\begin{array}{l} \gamma^{\delta\gamma}_{,\beta}\gamma^{\alpha\beta} \cdot \sqrt{\gamma} \cdot \left(\gamma_{ij}\left(F_{,k}\gamma^{ki}\right)_{,\delta}\right)_{,\alpha} \\[4pt] +\gamma^{\delta\gamma}\sqrt{\gamma} \cdot \Delta\left(\gamma_{ij}\left(F_{,k}\gamma^{ki}\right)_{,\delta}\right) \end{array} \right) \right) \cdot \frac{\partial\left(F_{,l}\gamma^{lj}\right)_{,\gamma}}{\partial x^a} \right.
$$

$$
\left. + \left(\sqrt{\gamma} \cdot C^2 \cdot \gamma^{\delta\gamma} \left(F_{,l}\gamma^{lj}\right)_{,\gamma} - d \left(\begin{array}{l} \gamma^{\delta\gamma}_{,\alpha}\gamma^{\alpha\beta} \cdot \sqrt{\gamma} \cdot \left(F_{,l}\gamma^{lj}\right)_{,\gamma\beta} \\[4pt] +\gamma^{\delta\gamma}\sqrt{\gamma} \cdot \Delta\left(F_{,l}\gamma^{lj}\right) \end{array} \right) \right) \cdot \frac{\partial\left(\gamma_{ij}\left(F_{,k}\gamma^{ki}\right)_{,\delta}\right)}{\partial x^a} \right]
$$

$$
= D \int_V d^n x \left[\left(\sqrt{\gamma} \cdot C^2 \cdot \gamma^{\delta\gamma} \gamma_{ij} \left(F_{,k}\gamma^{ki}\right)_{,\delta} - d \left(\begin{array}{l} \gamma^{\delta\gamma}\gamma^{\alpha\beta} \cdot \sqrt{\gamma} \cdot \left(\gamma_{ij}\left(F_{,k}\gamma^{ki}\right)_{,\delta}\right)_{,\alpha}\Big)_{,\beta} \\[4pt] -\gamma^{\delta\gamma}\left(\gamma^{\alpha\beta}\sqrt{\gamma} \cdot \left(\gamma_{ij}\left(F_{,k}\gamma^{ki}\right)_{,\delta}\right)_{,\alpha}\right)_{,\beta} \\[4pt] +\gamma^{\delta\gamma}\sqrt{\gamma} \cdot \Delta\left(\gamma_{ij}\left(F_{,k}\gamma^{ki}\right)_{,\delta}\right) \end{array} \right) \right) \cdot \frac{\partial\left(F_{,l}\gamma^{lj}\right)_{,\gamma}}{\partial x^a} \right.
$$

$$
\left. + \left(\sqrt{\gamma} \cdot C^2 \cdot \gamma^{\delta\gamma} \left(F_{,l}\gamma^{lj}\right)_{,\gamma} - d \left(\begin{array}{l} \gamma^{\delta\gamma}\gamma^{\alpha\beta} \cdot \sqrt{\gamma} \cdot \left(F_{,l}\gamma^{lj}\right)_{,\gamma\beta}\Big)_{,\alpha} \\[4pt] -\gamma^{\delta\gamma}\left(\gamma^{\alpha\beta}\sqrt{\gamma} \cdot \left(F_{,l}\gamma^{lj}\right)_{,\gamma\beta}\right)_{,\alpha} \\[4pt] +\gamma^{\delta\gamma}\sqrt{\gamma} \cdot \Delta\left(F_{,l}\gamma^{lj}\right) \end{array} \right) \right) \cdot \frac{\partial\left(\gamma_{ij}\left(F_{,k}\gamma^{ki}\right)_{,\delta}\right)}{\partial x^a} \right] . \quad (899)
$$

$$= D \int_V d^n x \left[\underbrace{ \sqrt{\gamma}\cdot C^2\cdot\gamma^{\delta\gamma}\gamma_{ij}\left(F_{,k}\gamma^{ki}\right)_{,\delta} - d \left(\begin{array}{l} \gamma^{\delta\gamma}\gamma^{\alpha\beta}\cdot\sqrt{\gamma}\cdot\left(\gamma_{ij}\left(F_{,k}\gamma^{ki}\right)_{,\delta}\right)_{,\alpha} \\[4pt] -\gamma^{\delta\gamma}\sqrt{\gamma}\sqrt{\gamma}\cdot\Delta\left(\gamma_{ij}\left(F_{,k}\gamma^{ki}\right)_{,\delta}\right) \\[4pt] +\gamma^{\delta\gamma}\sqrt{\gamma}\sqrt{\gamma}\cdot\Delta\left(\gamma_{ij}\left(F_{,k}\gamma^{ki}\right)_{,\delta}\right) \end{array} \right)_{,\beta} \cdot \frac{\partial\left(F_{,l}\gamma^{lj}\right)_{,\gamma}}{\partial x^\sigma} } \right.$$

$$\left. + \left(\sqrt{\gamma}\cdot C^2\cdot\gamma^{\delta\gamma}\left(F_{,l}\gamma^{lj}\right)_{,\gamma} - d \left(\begin{array}{l} \gamma^{\delta\gamma}\gamma^{\alpha\beta}\cdot\sqrt{\gamma}\cdot\left(F_{,l}\gamma^{lj}\right)_{,\gamma\beta} \\[4pt] -\gamma^{\delta\gamma}\sqrt{\gamma}\sqrt{\gamma}\cdot\Delta\left(F_{,l}\gamma^{lj}\right)_{,\gamma} \\[4pt] +\gamma^{\delta\gamma}\sqrt{\gamma}\sqrt{\gamma}\cdot\Delta\left(F_{,l}\gamma^{lj}\right)_{,\gamma} \end{array} \right)_{,\alpha} \cdot \frac{\partial\left(\gamma_{ij}\left(F_{,k}\gamma^{ki}\right)_{,\delta}\right)}{\partial x^\sigma} \right) \right]$$

$$= D \int_V d^n x \left[\left(\sqrt{\gamma}\cdot C^2\cdot\gamma^{\delta\gamma}\gamma_{ij}\left(F_{,k}\gamma^{ki}\right)_{,\delta} - d \left(\gamma^{\delta\gamma}\gamma^{\alpha\beta}\cdot\sqrt{\gamma}\cdot\left(\gamma_{ij}\left(F_{,k}\gamma^{ki}\right)_{,\delta}\right)_{,\alpha} \right)_{,\beta} \cdot \frac{\partial\left(F_{,l}\gamma^{lj}\right)_{,\gamma}}{\partial x^\sigma} \right) \right.$$

$$\left. + \left(\sqrt{\gamma}\cdot C^2\cdot\gamma^{\delta\gamma}\left(F_{,l}\gamma^{lj}\right)_{,\gamma} - d \left(\left(\gamma^{\delta\gamma}\gamma^{\alpha\beta}\cdot\sqrt{\gamma}\cdot\left(F_{,l}\gamma^{lj}\right)_{,\gamma\beta} \right)_{,\alpha} \right) \cdot \frac{\partial\left(\gamma_{ij}\left(F_{,k}\gamma^{ki}\right)_{,\delta}\right)}{\partial x^\sigma} \right) \right] . \quad (900)$$

Integration by parts yields:

$$
= D \int_V d^n x \left(d\left(\left(\gamma^{\delta\gamma}\gamma^{\alpha\beta} \cdot \sqrt{\gamma} \cdot \left(\gamma_{ij}\left(F_{,k}\gamma^{ki}\right)_{,\delta}\right)_{,\alpha}\right)_{,\beta} - \sqrt{\gamma}\cdot C^2 \cdot \gamma^{\delta\gamma}\gamma_{ij}\left(F_{,k}\gamma^{ki}\right)_{,\delta}\right)_{,\gamma} \cdot \frac{\partial\left(F_{,l}\gamma^{lj}\right)}{\partial x^\sigma}\right.
$$
$$
\left. + d\left(\left(\gamma^{\delta\gamma}\gamma^{\alpha\beta} \cdot \sqrt{\gamma} \cdot \left(F_{,l}\gamma^{lj}\right)_{,\gamma\beta}\right)_{,\alpha} - \sqrt{\gamma}\cdot C^2 \cdot \gamma^{\delta\gamma}\left(F_{,l}\gamma^{lj}\right)_{,\gamma}\right)_{,\delta} \cdot \int d^n x^\delta \frac{\partial\left(\gamma_{ij}\left(F_{,k}\gamma^{ki}\right)_{,\delta}\right)}{\partial x^\sigma} \cdot \frac{\partial\left(F_{,l}\gamma^{lj}\right)}{\partial x^\sigma}\right)
$$

$$
= D \int_V d^n x \left(d\left(\gamma^{\delta\gamma}\gamma^{\alpha\beta} \cdot \sqrt{\gamma} \cdot \left(\gamma_{ij}\left(F_{,k}\gamma^{ki}\right)_{,\delta}\right)_{,\alpha}\right)_{,\beta\gamma} - \sqrt{\gamma}\cdot C^2 \cdot \Delta\left(\gamma_{ij}\left(F_{,k}\gamma^{ki}\right)\right) \cdot \frac{\partial\left(F_{,l}\gamma^{lj}\right)}{\partial x^\sigma}\right.
$$
$$
\left. + d\left(\gamma^{\delta\gamma}\gamma^{\alpha\beta} \cdot \sqrt{\gamma} \cdot \left(F_{,l}\gamma^{lj}\right)_{,\gamma\beta}\right)_{,\alpha\delta} - \sqrt{\gamma}\cdot C^2 \cdot \Delta\left(F_{,l}\gamma^{lj}\right) \cdot \int d^n x^\delta \frac{\partial\left(\gamma_{ij}\left(F_{,k}\gamma^{ki}\right)_{,\delta}\right)}{\partial x^\sigma}\right)
$$

$$
= D \int_V d^n x \left(\underbrace{d\left(\gamma^{\delta\gamma}\gamma^{\alpha\beta} \cdot \sqrt{\gamma} \cdot \left(\gamma_{ij}\left(F^{,i}\right)_{,\delta}\right)_{,\alpha}\right)_{,\beta\gamma}}_{\sqrt{\gamma}\cdot\Delta\left(\left(F^{,j}\right)^{\delta}\right)} - \sqrt{\gamma}\cdot C^2 \cdot \Delta F_{,j} \cdot \frac{\partial\left(F_{,l}\gamma^{lj}\right)}{\partial x^\sigma}\right.
$$
$$
\left. + d\left(\gamma^{\alpha\beta} \cdot \sqrt{\gamma} \cdot \left(F^{,j}\right)_{,\beta}^{\,\delta}\right)_{,\alpha\delta} - \sqrt{\gamma}\cdot C^2 \cdot \Delta\left(F^{,j}\right) \cdot \int d^n x^\delta \frac{\partial\left(\gamma_{ij}\left(F_{,k}\gamma^{ki}\right)_{,\delta}\right)}{\partial x^\sigma}\right) .
$$

$$(901)$$

In the end, we have the following equation:

$$d \left(\gamma^{\delta\gamma} \gamma^{\alpha\beta} \cdot \sqrt{\gamma} \cdot \left(\gamma_{ij} \left(F^{\cdot i} \right)_{,\delta} \right)_{,\alpha} \right)_{,\beta\gamma} - \sqrt{\gamma} \cdot C^2 \cdot \Delta F_{,j} = 0$$

$$d \sqrt{\gamma} \cdot \Delta \left(\left(F^{\cdot j} \right)^{,\delta} \right)_{,\delta} - \sqrt{\gamma} \cdot C^2 \cdot \Delta \left(F^{\cdot j} \right) = 0. \tag{902}$$

5.4 Direct Metric Variation

5.4.1 Direct Metric Variation: Simplest Approach

Our next goal is to derive metric Klein–Gordon equations with scalar functions f which should not require an f-derivative to give the metric equivalent of the potential V.

Avoiding the base vectors, we could also just start our metric variation in a much more direct manner, namely:

$$\delta_\gamma W = 0 = \int_V d^n x \sqrt{g} \cdot \left(R^{\delta\gamma} - \frac{1}{2} R \cdot g^{\delta\gamma} + \Lambda \cdot g^{\delta\gamma} + \kappa \cdot T^{\delta\gamma} \right)$$

$$\times \left(\delta g_{\delta\gamma} = g_{\delta\gamma} \delta_\sigma \left(1 + h^\sigma \cdot f + \overbrace{C^{\sigma\alpha}}^{=d \cdot \sqrt{g} \cdot g^{\sigma\alpha}} \cdot f_{,\alpha} \right) = g_{\delta\gamma} \left(h^\sigma \cdot f_{,\sigma} + d \cdot \sqrt{g} \cdot \Delta f \right) \right).$$

$$\tag{903}$$

For the reason of simplicity, we assumed $\mathbf{h}^\sigma$ to be a constant vector in the metric in question. After performing the integration by parts for the integrand with the terms $h^\sigma \cdot f_{,\sigma}$ as follows:

$$\int_V d^n x \, (\dots)^{\delta\gamma} \sqrt{g} \cdot g_{\delta\gamma} f_{,\sigma}$$

$$= \underbrace{\int_V d^n x \left\{ \left[(\dots)^{\delta\gamma} \sqrt{g} \cdot g_{\delta\gamma} \right] \cdot f \right\}_{,\sigma}}_{=S=0 \text{ per definition}}$$

$$- \int_V d^n x \left[(\dots)^{\delta\gamma} \sqrt{g} \cdot g_{\delta\gamma} \right]_{,\sigma} \cdot f; \quad (\dots)^{\delta\gamma}$$

$$= \left(R^{\delta\gamma} - \frac{1}{2} R \cdot g^{\delta\gamma} + \Lambda \cdot g^{\delta\gamma} + \kappa \cdot T^{\delta\gamma} \right),$$

$$\tag{904}$$

we can reshape Eq. (903) to:

$$0 = \int_V d^n x \left(\sqrt{g} \cdot (\ldots)^{\delta\gamma} g_{\delta\gamma} d \cdot \sqrt{g} \cdot \Delta f - h^\sigma \cdot \left[\sqrt{g} \cdot (\ldots)^{\delta\gamma} g_{\delta\gamma} \right]_{,\sigma} \cdot f \right)$$

$$0 = \int_V d^n x \left(g \cdot (\ldots)^{\delta\gamma} g_{\delta\gamma} d \cdot \Delta f - h^\sigma \cdot \left[\sqrt{g} \cdot (\ldots)^{\delta\gamma} g_{\delta\gamma} \right]_{,\sigma} \cdot f \right). \quad (905)$$

From the last line, we extract the integrand and set it to zero, and factorizing out the scalar term $g \cdot (\ldots)^{\delta\gamma} g_{\delta\gamma}$ finally gives us:

$$d \cdot \Delta f - h^\sigma \cdot \frac{\left[\sqrt{g} \cdot (\ldots)^{\delta\gamma} g_{\delta\gamma} \right]_{,\sigma}}{g \cdot (\ldots)^{\delta\gamma} g_{\delta\gamma}} \cdot f = 0$$

$$\Rightarrow \Delta f - \frac{h^\sigma}{d \cdot \sqrt{g}} \cdot \left(\frac{g^{\delta\gamma} g_{\delta\gamma,\sigma}}{2} + \frac{\left[(\ldots)^{\delta\gamma} g_{\delta\gamma} \right]_{,\sigma}}{(\ldots)^{\delta\gamma} g_{\delta\gamma}} \right) \cdot f = 0$$

$$\xrightarrow{n=4 \quad \& \quad 4 \cdot \Lambda + \kappa \cdot T^{\delta\gamma} g_{\delta\gamma} = 0} \Delta f - \frac{h^\sigma}{d \cdot \sqrt{g}} \cdot \left(\frac{g^{\delta\gamma} g_{\delta\gamma,\sigma}}{2} + \frac{R_{,\sigma}}{R} \right) \cdot f = 0. \quad (906)$$

As desired, we now have the perfectly mirrored Klein–Gordon equation, and comparison with Eq. (723) leaves us with the correlation(s):

$$\frac{h^\sigma}{d \cdot \sqrt{g}} \cdot \left(\frac{g^{\delta\gamma} g_{\delta\gamma,\sigma}}{2} + \frac{R_{,\sigma}}{R} \right) = \frac{m^2 \cdot c^2}{\hbar^2} - V. \quad (907)$$

In the case of non-constant h^σ, the evaluation goes as follows:

$$\delta_\gamma W = 0 = \int_V d^n x \sqrt{g} \cdot \left(R^{\delta\gamma} - \frac{1}{2} R \cdot g^{\delta\gamma} + \Lambda \cdot g^{\delta\gamma} + \kappa \cdot T^{\delta\gamma} \right)$$

$$\times \left(\delta g_{\delta\gamma} = g_{\delta\gamma} \delta_\sigma \left(1 + h^\sigma \cdot f + \overbrace{C^{\sigma\alpha}}^{=d \cdot \sqrt{g} \cdot g^{\sigma\alpha}} \cdot f_{,\alpha} \right) = g_{\delta\gamma} \left((h^\sigma \cdot f)_{,\sigma} + d \cdot \sqrt{g} \cdot \Delta f \right) \right).$$

$$(908)$$

After performing the integration by parts for the integrand now with the terms $(h^\sigma \cdot f)_{,\sigma}$ as follows:

$$\int_V d^n x (\ldots)^{\delta\gamma} \sqrt{g} \cdot g_{\delta\gamma} (h^\sigma \cdot f)_{,\sigma} = \underbrace{\int_V d^n x \left\{ \left[(\ldots)^{\delta\gamma} \sqrt{g} \cdot g_{\delta\gamma} \right] \cdot (h^\sigma \cdot f) \right\}_{,\sigma}}_{=S=0 \text{ per definition}}$$

$$- \int_V d^n x \left[(\ldots)^{\delta\gamma} \sqrt{g} \cdot g_{\delta\gamma} \right]_{,\sigma} \cdot (h^\sigma \cdot f); \quad (\ldots)^{\delta\gamma}$$

$$= \left(R^{\delta\gamma} - \frac{1}{2} R \cdot g^{\delta\gamma} + \Lambda \cdot g^{\delta\gamma} + \kappa \cdot T^{\delta\gamma} \right), \quad (909)$$

we can reshape Eq. (908) and end up with the same result as before, namely:

$$0 = \int_V d^n x \left(\sqrt{g} \cdot (\ldots)^{\delta\gamma} g_{\delta\gamma} d \cdot \sqrt{g} \cdot \Delta f - \left[\sqrt{g} \cdot (\ldots)^{\delta\gamma} g_{\delta\gamma} \right]_{,\sigma} \cdot (h^\sigma \cdot f) \right)$$

$$0 = \int_V d^n x \left(g \cdot (\ldots)^{\delta\gamma} g_{\delta\gamma} d \cdot \Delta f - h^\sigma \cdot \left[\sqrt{g} \cdot (\ldots)^{\delta\gamma} g_{\delta\gamma} \right]_{,\sigma} \cdot f \right). \qquad (910)$$

5.4.1.1 Towards "Dirac"

Factorization under the integral Eq. (910) and a bit of reshaping leads to the following differential equation of first order (Dirac-like).

$$0 = \int_V d^n x \left(\sqrt{g} \cdot (\ldots)^{\delta\gamma} g_{\delta\gamma} d \cdot \sqrt{g} \cdot \Delta f - h^\sigma \cdot \left[\sqrt{g} \cdot (\ldots)^{\delta\gamma} g_{\delta\gamma} \right]_{,\sigma} \cdot f \right)$$

$$0 = \int_V d^n x \left(\sqrt{g} \cdot (\ldots)^{\delta\gamma} g_{\delta\gamma} d \cdot \left(\sqrt{g} \cdot g^{\sigma\alpha} f_{,\alpha} \right)_{,\sigma} - h^\sigma \cdot \left[\sqrt{g} \cdot (\ldots)^{\delta\gamma} g_{\delta\gamma} \right]_{,\sigma} \cdot f \right)$$

$$= - \int_V d^n x \left(\left[\sqrt{g} \cdot (\ldots)^{\delta\gamma} g_{\delta\gamma} \right]_{,\sigma} d \cdot \sqrt{g} \cdot g^{\sigma\alpha} f_{,\alpha} + h^\sigma \cdot \left[\sqrt{g} \cdot (\ldots)^{\delta\gamma} g_{\delta\gamma} \right]_{,\sigma} \cdot f \right)$$

$$= - \int_V d^n x \left(\left[\sqrt{g} \cdot (\ldots)^{\delta\gamma} g_{\delta\gamma} \right]_{,\sigma} \left(d \cdot \sqrt{g} \cdot g^{\sigma\alpha} f_{,\alpha} + h^\sigma \cdot f \right) \right)$$

$$\Rightarrow d \cdot \sqrt{g} \cdot g^{\sigma\alpha} f_{,\alpha} + h^\sigma \cdot f = \left(d \cdot \sqrt{g} \cdot g^{\sigma\alpha} \partial_\alpha + h^\sigma \right) f = 0. \qquad (911)$$

We will not discuss this equation, but go on and investigate vector forms for the function f.

5.4.2 Direct Metric Variation: Vector Approach

Our next goal is to derive metric Klein–Gordon equations with a function-vector $\mathbf{f}^\sigma$ which should not require an f-derivative to give the metric equivalent of the potential V.

Avoiding the base vectors, we could also just start our metric variation in a much more direct manner, namely:

Starting brief and simple . . .

$$\delta_\gamma W = 0 = \int_V d^n x \sqrt{g} \cdot \left(R^{\delta\gamma} - \frac{1}{2} R \cdot g^{\delta\gamma} + \Lambda \cdot g^{\delta\gamma} + \kappa \cdot T^{\delta\gamma} \right)$$

$$\times \left(\delta g_{\delta\gamma} = g_{\delta\gamma} \delta_\sigma \left(1 + h^\sigma h_\beta f^\beta + C^{\sigma\alpha} \cdot \left(h_\beta f^\beta \right)_{,\alpha} \right) \right.$$

$$= g_{\delta\gamma} \left(h^\sigma h_\beta f^\beta_{,\sigma} + d \cdot \sqrt{g} \cdot \Delta \left(h_\beta f^\beta \right) \right) \right). \tag{912}$$

After performing the integration by parts for the integrand with the terms $h^\sigma h_\beta f^\beta_{,\sigma}$ as follows:

$$\int_V d^n x (\ldots)^{\delta\gamma} \sqrt{g} \cdot g_{\delta\gamma} h^\sigma h_\beta f^\beta_{,\sigma} = \underbrace{\int_V d^n x \left\{ \left[(\ldots)^{\delta\gamma} \sqrt{g} \cdot g_{\delta\gamma} \right] \cdot h^\sigma h_\beta f^\beta \right\}_{,\sigma}}_{=S=0 \text{ per definition}}$$

$$- \int_V d^n x \left[(\ldots)^{\delta\gamma} \sqrt{g} \cdot g_{\delta\gamma} \right]_{,\sigma} \cdot h^\sigma h_\beta f^\beta; \quad (\ldots)^{\delta\gamma}$$

$$= \left(R^{\delta\gamma} - \frac{1}{2} R \cdot g^{\delta\gamma} + \Lambda \cdot g^{\delta\gamma} + \kappa \cdot T^{\delta\gamma} \right), \tag{913}$$

we can reshape Eq. (912) to:

$$0 = \int_V d^n x \left(g \cdot (\ldots)^{\delta\gamma} g_{\delta\gamma} d \cdot \Delta \left(h_\beta f^\beta \right) - h^\sigma \left[\sqrt{g} \cdot (\ldots)^{\delta\gamma} g_{\delta\gamma} \right]_{,\sigma} \cdot h_\beta f^\beta \right). \tag{914}$$

By the way, we note that Eq. (914) could also have a somewhat more general origin, thereby using all possible and vector-forms:

$$\delta_\gamma W = 0 = \int_V d^n x \sqrt{g} \cdot \left(R^{\delta\gamma} - \frac{1}{2} R \cdot g^{\delta\gamma} + \Lambda \cdot g^{\delta\gamma} + \kappa \cdot T^{\delta\gamma} \right)$$

$$\times \left(\begin{array}{c} \delta g_{\delta\gamma} = g_{\delta\gamma} \delta_\sigma \left(1 + f^\sigma + h^\sigma h_\beta f^\beta + H^{\sigma\beta} f_\beta + C^{\sigma\alpha} \cdot \left(h_\beta f^\beta \right)_{,\alpha} \right. \\ \left. + C^{\beta\alpha} \cdot \left(h_\beta f^\sigma \right)_{,\alpha} + A^\alpha \cdot f^\sigma_{,\alpha} + B^{\alpha\beta\sigma} \cdot f_{\beta,\alpha} \right) \\ = g_{\delta\gamma} \delta_\sigma \left(\begin{array}{c} 1 + f^\sigma + h^\sigma h_\beta f^\beta + H^{\sigma\beta} f_\beta + C^{\sigma\alpha} \cdot \left(h_\beta f^\beta \right)_{,\alpha} + C^{\beta\alpha} \cdot \left(h_\beta f^\sigma \right)_{,\alpha} + A^\alpha \cdot f^\sigma_{,\alpha} \\ + C^{\sigma\beta} \cdot \left(h^\alpha f_\beta \right)_{,\alpha} + C^{\sigma\alpha} \cdot \left(h^\beta f_\beta \right)_{,\alpha} + C^{\alpha\beta} \cdot \left(h^\sigma f_\beta \right)_{,\alpha} + C^{\beta\alpha} \cdot \left(h^\sigma f_\beta \right)_{,\alpha} \end{array} \right) \end{array} \right). \tag{915}$$

From the last line, we extract the integrand and set it to zero, and factorizing out the scalar term $g \cdot (\ldots)^{\delta \gamma} g_{\delta \gamma}$ finally gives us:

$$d \cdot \Delta f - h^\sigma \cdot \frac{\left[\sqrt{g} \cdot (\ldots)^{\delta \gamma} g_{\delta \gamma} \right]_{,\sigma}}{g \cdot (\ldots)^{\delta \gamma} g_{\delta \gamma}} \cdot f = 0$$

$$\Rightarrow \Delta f - \frac{h^\sigma}{d \cdot \sqrt{g}} \cdot \left(\frac{g^{\delta \gamma} g_{\delta \gamma, \sigma}}{2} + \frac{\left[(\ldots)^{\delta \gamma} g_{\delta \gamma} \right]_{,\sigma}}{(\ldots)^{\delta \gamma} g_{\delta \gamma}} \right) \cdot f = 0$$

$$\underset{\xrightarrow{\quad n=4 \quad \& \quad 4 \cdot \Lambda + \kappa \cdot T^{\delta \gamma} g_{\delta \gamma} = 0 \quad}}{} \Delta f - \frac{h^\sigma}{d \cdot \sqrt{g}} \cdot \left(\frac{g^{\delta \gamma} g_{\delta \gamma, \sigma}}{2} + \frac{R_{,\sigma}}{R} \right) \cdot f = 0. \quad (916)$$

As desired, we now have the perfectly mirrored Klein–Gordon equation, and comparison with Eq. (723) leaves us with the correlation(s):

$$\frac{h^\sigma}{d \cdot \sqrt{g}} \cdot \left(\frac{g^{\delta \gamma} g_{\delta \gamma, \sigma}}{2} + \frac{R_{,\sigma}}{R} \right) = \frac{m^2 \cdot c^2}{\hbar^2} - V. \quad (917)$$

5.4.2.1 Towards "Dirac"

This time, the "factorization" and second-time application of the integration by parts results in the following first-order differential equations:

$$0 = \int_V d^n x \left(\sqrt{g} \cdot (\ldots)^{\delta \gamma} g_{\delta \gamma} \cdot \left(C^{\sigma \alpha} \left(h_\beta f^\beta \right)_{,\alpha} \right)_{,\sigma} - \left[\sqrt{g} \cdot (\ldots)^{\delta \gamma} g_{\delta \gamma} \right]_{,\sigma} \cdot h^\sigma h_\beta f^\beta \right)$$

$$= \int_V d^n x \left(- \left[\sqrt{g} \cdot (\ldots)^{\delta \gamma} g_{\delta \gamma} \right]_{,\sigma} \cdot \left(C^{\sigma \alpha} \left(h_\beta f^\beta \right)_{,\alpha} \right) - \left[\sqrt{g} \cdot (\ldots)^{\delta \gamma} g_{\delta \gamma} \right]_{,\sigma} h^\sigma h_\beta f^\beta \right)$$

$$= - \int_V d^n x \left[\sqrt{g} \cdot (\ldots)^{\delta \gamma} g_{\delta \gamma} \right]_{,\sigma} \left(C^{\sigma \alpha} \left(h_\beta f^\beta \right)_{,\alpha} + h^\sigma h_\beta f^\beta \right)$$

$$\Rightarrow C^{\sigma \alpha} \left(h_\beta f^\beta \right)_{,\alpha} + h^\sigma h_\beta f^\beta = C^{\sigma \alpha} h_{\beta, \alpha} f^\beta + C^{\sigma \alpha} h_\beta f^\beta_{,\alpha} + h^\sigma h_\beta f^\beta = 0$$

$$\underset{\xrightarrow{\quad h^\sigma h_\beta \equiv M \cdot \delta^\sigma_\beta \, ; \, C^{\sigma \alpha} = d \sqrt{g} \cdot g^{\sigma \alpha} \quad}}{} = d \sqrt{g} \cdot g^{\sigma \alpha} \left(h_{\beta, \alpha} f^\beta + h_\beta f^\beta_{,\alpha} \right) + M \cdot f^\sigma = 0. \quad (918)$$

5.4.2.2 Alternative elastic outcome

Just adding one suitable part of the more general vector approach (Eq. (915)) results in equations mirroring the properties of elastic spaces:

$$\delta_\gamma W = 0 = \int_V d^n x \sqrt{g} \cdot \left(R^{\delta\gamma} - \frac{1}{2} R \cdot g^{\delta\gamma} + \Lambda \cdot g^{\delta\gamma} + \kappa \cdot T^{\delta\gamma} \right)$$

$$\times \left(\begin{array}{l} \delta g_{\delta\gamma} = g_{\delta\gamma} \delta_\sigma \left(1 + M \cdot f^\sigma + h_\beta \left(\overbrace{C^{\sigma\alpha}}^{=d\cdot\sqrt{g}\cdot g^{\sigma\alpha}} f^\beta_{,\alpha} + \overbrace{C^{\beta\alpha}}^{=d\cdot\sqrt{g}\cdot g^{\beta\alpha}} f^\sigma_{,\alpha} \right) \right) \\ = g_{\delta\gamma} \left(M \cdot f^\sigma_{,\sigma} + d \cdot \left[h_\beta \left(\sqrt{g} \cdot g^{\sigma\alpha} f^\beta_{,\alpha} + \sqrt{g} \cdot g^{\beta\alpha} f^\sigma_{,\alpha} \right) \right]_{,\sigma} \right) \\ = g_{\delta\gamma} \left(M \cdot f^\sigma_{,\sigma} + d \cdot \left[\left(\sqrt{g} \cdot g^{\sigma\alpha} \left(h_\beta f^\beta \right) \right)_{,\alpha} - \sqrt{g} \cdot g^{\sigma\alpha} h_{\beta,\alpha} f^\beta + \sqrt{g} \cdot g^{\beta\alpha} h_\beta f^\sigma_{,\alpha} \right]_{,\sigma} \right) \\ = g_{\delta\gamma} \left(M \cdot f^\sigma_{,\sigma} + d \cdot \left(\sqrt{g} \cdot \Delta \left(h_\beta f^\beta \right) - \sqrt{g} \cdot g^{\sigma\alpha} h_{\beta,\alpha} f^\beta + \left(\sqrt{g} \cdot g^{\beta\alpha} h_\beta f^\sigma_{,\alpha} \right)_{,\sigma} \right) \right) \end{array} \right),$$

$$(919)$$

where the subsequent first-order differential equation reads:

$$0 = \int_V d^n x \left(\sqrt{g} \cdot (\ldots)^{\delta\gamma} g_{\delta\gamma} \cdot \left(C^{\sigma\alpha} \left(f^\beta \right)_{,\alpha} + C^{\beta\alpha} \left(f^\sigma \right)_{,\alpha} \right)_{,\sigma} - \left[\sqrt{g} \cdot (\ldots)^{\delta\gamma} g_{\delta\gamma} \right]_{,\sigma} \cdot f^\sigma \right)$$

$$= \int_V d^n x \left(- \left[\sqrt{g} \cdot (\ldots)^{\delta\gamma} g_{\delta\gamma} \right]_{,\sigma} \cdot \left(C^{\sigma\alpha} \left(h_\beta f^\beta \right)_{,\alpha} \right) - \left[\sqrt{g} \cdot (\ldots)^{\delta\gamma} g_{\delta\gamma} \right]_{,\sigma} h^\sigma h_\beta f^\beta \right)$$

$$= - \int_V d^n x \left[\sqrt{g} \cdot (\ldots)^{\delta\gamma} g_{\delta\gamma} \right]_{,\sigma} \left(C^{\sigma\alpha} \left(h_\beta f^\beta \right)_{,\alpha} + h^\sigma h_\beta f^\beta \right)$$

$$\Rightarrow C^{\sigma\alpha} \left(h_\beta f^\beta \right)_{,\alpha} + h^\sigma h_\beta f^\beta = 0. \tag{920}$$

5.4.2.3 Trying to fix h_β

Now we try to find a suitable candidate for the vector $\mathbf{h}_\beta$ and, assuming that base vectors may do a good enough job, substitute as follows:

$$\delta_\gamma W = 0 = \int_V d^n x \sqrt{g} \cdot \left(R^{\delta\gamma} - \frac{1}{2} R \cdot g^{\delta\gamma} + \Lambda \cdot g^{\delta\gamma} + \kappa \cdot T^{\delta\gamma} \right)$$

$$\times \left(\begin{array}{l} \delta g_{\delta\gamma} = g_{\delta\gamma} \delta_\sigma \left(1 + \mathbf{M} \cdot \mathbf{f}^\sigma + \mathbf{h}_\beta \left(\overbrace{C^{\sigma\alpha}}^{=d\cdot\sqrt{g}\cdot g^{\sigma\alpha}} \mathbf{f}^\beta_{,\alpha} + \overbrace{C^{\beta\alpha}}^{=d\cdot\sqrt{g}\cdot g^{\beta\alpha}} \mathbf{f}^\sigma_{,\alpha} \right) \right) \\ = g_{\delta\gamma} \left(\mathbf{M} \cdot \mathbf{f}^\sigma_{,\sigma} + d \cdot \left[\mathbf{h}_\beta \left(\sqrt{g} \cdot g^{\sigma\alpha} \mathbf{f}^\beta_{,\alpha} + \sqrt{g} \cdot g^{\beta\alpha} \mathbf{f}^\sigma_{,\alpha} \right) \right]_{,\sigma} \right) \\ = g_{\delta\gamma} \left(\mathbf{M} \cdot \mathbf{f}^\sigma_{,\sigma} + d \cdot \left[\left(\sqrt{g} \cdot g^{\sigma\alpha} \left(\mathbf{h}_\beta \mathbf{f}^\beta \right) \right)_{,\alpha} - \sqrt{g} \cdot g^{\sigma\alpha} \mathbf{h}_{\beta,\alpha} \mathbf{f}^\beta + \sqrt{g} \cdot g^{\beta\alpha} \mathbf{h}_\beta \mathbf{f}^\sigma_{,\alpha} \right]_{,\sigma} \right) \\ = g_{\delta\gamma} \left(\mathbf{M} \cdot \mathbf{f}^\sigma_{,\sigma} + d \cdot \left(\sqrt{g} \cdot \Delta \left(\mathbf{h}_\beta \mathbf{f}^\beta \right) - \sqrt{g} \cdot g^{\sigma\alpha} \mathbf{h}_{\beta,\alpha} \mathbf{f}^\beta + \left(\sqrt{g} \cdot g^{\beta\alpha} \mathbf{h}_\beta \mathbf{f}^\sigma_{,\alpha} \right)_{,\sigma} \right) \right) \\ \xrightarrow{\mathbf{h}_\beta = \mathbf{g}_\beta} = g_{\delta\gamma} \left(\mathbf{M} \cdot \mathbf{f}^\sigma_{,\sigma} + d \cdot \left(\sqrt{g} \cdot \Delta \left(\mathbf{g}_\beta \mathbf{f}^\beta \right) - \sqrt{g} \cdot g^{\sigma\alpha} \mathbf{g}_{\beta,\alpha} \mathbf{f}^\beta + \left(\sqrt{g} \cdot g^{\beta\alpha} \mathbf{g}_\beta \mathbf{f}^\sigma_{,\alpha} \right)_{,\sigma} \right) \right) \end{array} \right).$$

$$(921)$$

Now the subsequent first-order differential equations would read:

$$0 = \int_V d^n x \left(\sqrt{g} \cdot (\ldots)^{\delta \gamma} g_{\delta \gamma} \cdot \left(\mathbf{M} \cdot \mathbf{f}^{\sigma}{}_{,\sigma} + d \cdot \left[\mathbf{g}_{\beta} \left(\sqrt{g} \cdot g^{\sigma \alpha} \mathbf{f}^{\beta}{}_{,\alpha} + \sqrt{g} \cdot g^{\beta \alpha} \mathbf{f}^{\sigma}{}_{,\alpha} \right) \right]_{,\sigma} \right) \right)$$

$$0 = \int_V d^n x \left(-\left[\sqrt{g} \cdot (\ldots)^{\delta \gamma} g_{\delta \gamma} \right]_{,\sigma} \cdot \left(\mathbf{M} \cdot \mathbf{f}^{\sigma} + d \cdot \left[\mathbf{g}_{\beta} \left(\sqrt{g} \cdot g^{\sigma \alpha} \mathbf{f}^{\beta}{}_{,\alpha} + \sqrt{g} \cdot g^{\beta \alpha} \mathbf{f}^{\sigma}{}_{,\alpha} \right) \right] \right) \right)$$

$$\Rightarrow \mathbf{M} \cdot \mathbf{f}^{\sigma} + d \cdot \left[\mathbf{g}_{\beta} \left(\sqrt{g} \cdot g^{\sigma \alpha} \mathbf{f}^{\beta}{}_{,\alpha} + \sqrt{g} \cdot g^{\beta \alpha} \mathbf{f}^{\sigma}{}_{,\alpha} \right) \right] = 0. \tag{922}$$

Without loss of generality, we may assume that $\mathbf{f}^{\beta} = f \cdot \mathbf{g}^{\beta}$, which does give us:

$$\begin{aligned}
0 &= \mathbf{M} \cdot \mathbf{g}^{\sigma} f + d \cdot \left[\mathbf{g}_{\beta} \left(\sqrt{g} \cdot g^{\sigma \alpha} \left(\mathbf{g}^{\beta} f \right)_{,\alpha} + \sqrt{g} \cdot g^{\beta \alpha} \left(\mathbf{g}^{\sigma} f \right)_{,\alpha} \right) \right] \\
&= \mathbf{M} \cdot \mathbf{g}^{\sigma} f + d \cdot \left[\mathbf{g}_{\beta} \left(\sqrt{g} \cdot g^{\sigma \alpha} \left(\mathbf{g}^{\beta}{}_{,\alpha} f + \mathbf{g}^{\beta} f_{,\alpha} \right) + \sqrt{g} \cdot g^{\beta \alpha} \left(\mathbf{g}^{\sigma}{}_{,\alpha} f + \mathbf{g}^{\sigma} f_{,\alpha} \right) \right) \right] \\
&= \mathbf{M} \cdot \mathbf{g}^{\sigma} f + d \cdot \left[\left(\sqrt{g} \cdot g^{\sigma \alpha} \left(\mathbf{g}_{\beta} \mathbf{g}^{\beta}{}_{,\alpha} f + n \cdot f_{,\alpha} \right) + \sqrt{g} \cdot g^{\beta \alpha} \left(\mathbf{g}_{\beta} \mathbf{g}^{\sigma}{}_{,\alpha} f + \delta^{\sigma}_{\beta} f_{,\alpha} \right) \right) \right] \\
&= \mathbf{M} \cdot \mathbf{g}^{\sigma} f + d \cdot \left[\left(\sqrt{g} \cdot g^{\sigma \alpha} \left(\mathbf{g}_{\beta} \mathbf{g}^{\beta}{}_{,\alpha} f + (n+1) \cdot f_{,\alpha} \right) + \sqrt{g} \cdot g^{\beta \alpha} \mathbf{g}_{\beta} \mathbf{g}^{\sigma}{}_{,\alpha} f \right) \right] \\
&= \mathbf{M} \cdot \mathbf{g}^{\sigma} f + d \cdot \sqrt{g} \cdot \left[\left(g^{\sigma \alpha} \left(\mathbf{g}_{\beta} \mathbf{g}^{\beta}{}_{,\alpha} f + (n+1) \cdot f_{,\alpha} \right) + \mathbf{g}^{\alpha} \mathbf{g}^{\sigma}{}_{,\alpha} f \right) \right]. \tag{923}
\end{aligned}$$

As the simplest example, we may consider here the Minkowski metric for a t,x,y,z space-time and obtain the periodic solutions:

$$f = e^{5 \cdot d \cdot c \cdot (\pm x \pm y \pm z \pm i \cdot c \cdot t)}. \tag{924}$$

Thereby we have used the fact that in Eq. (921) we could have defined arbitrary sign constellations with respect to our setting $C^{\sigma \alpha} = d \cdot \sqrt{g} \cdot g^{\sigma \alpha}; C^{\beta \alpha} = d \cdot \sqrt{g} \cdot g^{\beta \alpha}$. The degrees of freedom residing within this substitution could also be used to introduce flexible real factors $c_{(\sigma \alpha)}$ in connection with each metric component, which would in the Minkowski-case lead to a more general solution, namely:

$$f = e^{5 \cdot d \cdot c \cdot \left(\pm c_{(11)} \cdot x \pm c_{(22)} \cdot y \perp c_{(33)} \cdot z \pm c_{(00)} \cdot i \cdot c \cdot t \right)}. \tag{925}$$

5.4.3 Direct Metric Variation: Matrix Approach

Avoiding the base vectors, we could also just start our metric variation in a much more direct manner, namely:

$$\delta_\gamma W = 0 = \int_V d^n x \sqrt{g} \cdot \left(R^{\delta\gamma} - \frac{1}{2} R \cdot g^{\delta\gamma} + \Lambda \cdot g^{\delta\gamma} + \kappa \cdot T^{\delta\gamma} \right)$$

$$\times \left(\delta g_{\delta\gamma} = g_{\delta\gamma} \delta_\sigma \left(1 + h_{\alpha\beta} f^{\alpha\beta} H^\sigma + h_{\alpha\beta} \left(f^{\sigma\beta} H^\alpha + f^{\alpha\sigma} H^\beta \right) \right. \right.$$

$$\left. \left. + C^{\sigma\alpha} \cdot \left(h_{\alpha\beta} f^{\alpha\beta} \right)_{,\alpha} + \cdots \right) \right). \tag{926}$$

However, as we do not intend to further investigate this approach in here, we leave it to the interested reader to perform and discuss the necessary evaluations.

Use

Chapter 6

A Problematic Matter or What Is the Matter with Matter

With this, we should have enough material to sum up our derivation of the world formula and start on some applications. However, before we can do that, we have to discuss the matter term $T^{\delta\gamma}$, or, as it is also called, the energy momentum tensor from the Einstein–Hilbert action integrand $R^{\delta\gamma} - \frac{1}{2}R \cdot \gamma^{\delta\gamma} + \Lambda \cdot \gamma^{\delta\gamma} + \kappa \cdot T^{\delta\gamma}$. Hilbert [137] and Einstein [138] had to postulate this term and introduced it into the General Theory of Relativity. They were not able to give a fundamental origin other than the suggestion that "matter simply had to be there".

From the previous chapter, we know that the extension of the variation of the Einstein–Hilbert action [137] surprisingly automatically brings in mass and potential forms (e.g., Eqs. (860) and (861)) which we can interpret as matter. On the other hand, we also saw matter appear via our functional wrapper approach (e.g., Eq. (85)) as yet another form of metric variation [128, 129].

For convenience, we briefly repeat both derivations here. We start with a brief repetition of the derivation of the Klein–Gordon equation out of the functional wrapper approach.

The World Formula: A Late Recognition of David Hilbert's Stroke of Genius
Norbert Schwarzer
Copyright © 2022 Jenny Stanford Publishing Pte. Ltd.
ISBN 978-981-4877-20-6 (Hardcover), 978-1-003-14644-5 (eBook)
www.jennystanford.com

6.1 The Other Metric Origin of the Klein–Gordon Equation

The metric transformation (84) directly results in a Ricci scalar (Eq. (106)) resembling the Klein–Gordon equation (note: C_{N2} is a constant which depends on the dimension):

$$G_{\alpha\beta} = F\left[f\left[t, x, y, z\right]\right] \cdot \delta_\alpha^i \delta_\beta^j g_{ij} \quad \Leftrightarrow \quad \left[\Delta_g + \frac{\overbrace{R^* \cdot F\left[f\right]^2}^{= \frac{M^2 c^2}{\hbar^2} + V}}{f \cdot C_{N2} \cdot \frac{\partial F[f]}{\partial f}}\right] f = 0$$

$$0 = \frac{\partial\left(\sqrt{-g}\left[R - 2\kappa L_M - 2\Lambda\right]\right)}{\partial n}$$

$$= \left[R - 2\kappa L_M - 2\Lambda\right] \cdot \frac{\partial\sqrt{-g}}{\partial n} + \sqrt{-g} \cdot \frac{\partial\left[R - 2\kappa L_M - 2\Lambda\right]}{\partial n}. \quad (927)$$

Thereby, as shown in the book part "Trials", the "driving force" for the extremalization of the Ricci scalar results from the classical Einstein–Hilbert action via an additional* variation with respect to the dimension of space, the centers of gravity, which is to say their number and individual positions, the coordinates and their possible transformations.

The non-linear term $\frac{R^* \cdot F[f]^2}{f \cdot C_{N2} \cdot \frac{\partial F[f]}{\partial f}}$ in the first line in Eq. (927) might be seen as controversial, but it was shown in [129–132] that a setting $R^* = 0$ (flat space) together with the introduction of suitably chosen additional (potentially entangled) dimensions would solve the problem and perfectly reproduce the classical and completely linear Klein–Gordon equation. For completeness, we are going to repeat the derivation of the mass term in the following two subsections.

6.2 The Metric Origin of Scalar Fields like the Higgs Field and Symmetry Breaking

In [129], we were able to show that a Minkowski metric:

---

*Variation in addition to the metric tensor g_{ij}.

$$g_{\alpha\beta} = \begin{pmatrix} -c^2 & 0 & 0 & 0 \\ 0 & 1 & 0 & 0 \\ 0 & 0 & 1 & 0 \\ 0 & 0 & 0 & 1 \end{pmatrix}, \tag{928}$$

which we subject to the following transformation:

$$G_{\alpha\beta} = F\left[f\left[t, x, y, z\right]\right]_{\alpha\beta}^{ij} g_{ij} \to G_{\alpha\beta} = F\left[f\left[t, x, y, z\right]\right] \cdot \delta_{\alpha}^i \delta_{\beta}^j g_{ij}, \tag{929}$$

results in the following Ricci scalar:

$$R^* = \frac{1}{F\left[f\right]^3} \cdot \left(-C_{N2} \cdot F\left[f\right] \cdot \frac{\partial F\left[f\right]}{\partial f} \cdot \Delta_g f^2\right) \xrightarrow{4D} R^* \cdot \frac{\left(2 \cdot C_2 + C_1 \cdot f^2\right)^3}{8 \cdot C_2 \cdot C_1}$$

$$= -3 \cdot \Delta_g f^2. \tag{930}$$

Please note that the functional wrapper $F\left[f\right]$ reading:

$$F\left[f\right] = C_1 \cdot f^2 + \frac{C_1^2 \cdot f^4}{4 \cdot C_2} + C_2, \tag{931}$$

with arbitrary constants C_i, assures the appearance of linear Laplace operator terms for the resulting Ricci scalar of the transformed metric in (929) (c.f. condition (100) and solve the equation for $F\left[f\right]$ in 4 dimensions).

Now we assume f to be a constant, which automatically leads to the simple equation:

$$R^* \cdot \frac{\left(2 \cdot C_2 + C_1 \cdot f^2\right)^3}{8 \cdot C_2 \cdot C_1} = 0. \tag{932}$$

Also assuming that the Ricci curvature R^* shall be proportional to f^n, with an arbitrary exponent $n \geq 0$, we obtain the familiar trivial solution of $f_0 = 0$. However, from Eq. (14) we also obtain the non-trivial ground states:

$$f^2 = -\frac{2 \cdot C_2}{C_1}. \tag{933}$$

As the constants C_i are arbitrary, we could simply set them as follows:

$$2 \cdot C_2 = -\mu^2; \quad C_1 = 2 \cdot \lambda. \tag{934}$$

This gives us the additional ground state solutions directly in the classical Higgs field style, namely:

$$\left(f_{1,2}\right)^2 = \frac{\mu^2}{2 \cdot \lambda} \quad \Rightarrow \quad f_{1,2} = \pm\frac{\mu}{\sqrt{2 \cdot \lambda}}. \tag{935}$$

The measured value for the $f_{1,2}$ is known to be (en.wikipedia.org/wiki/Higgs_boson):

$$|f_{1,2}| = \frac{\mu}{\sqrt{2 \cdot \lambda}} = \frac{246 GeV}{\sqrt{2} \cdot c^2}. \tag{936}$$

Rewriting Eq. (13) with the use of Eq. (16) as:

$$R^* \cdot \frac{\left(-\mu^2 + 2 \cdot \lambda \cdot f^2\right)^3}{8 \cdot \mu^2 \cdot 2 \cdot \lambda} = f^n \cdot \frac{\left(2 \cdot \lambda \cdot f^2 - \mu^2\right)^3}{16 \cdot \mu^2 \cdot \lambda} = 3 \cdot \Delta_g f^2 \tag{937}$$

gives us the total Higgs–Ricci curvature connection. It also, automatically, gives us a curvature value R^* at the ground state, which is:

$$R^* \sim |f_{1,2}|^n = \left(\frac{\mu}{\sqrt{2 \cdot \lambda}}\right)^n = \left(\frac{246 GeV}{\sqrt{2} \cdot c^2}\right)^n. \tag{938}$$

Now, with the Higgs mass m_H known to be 125 GeV/c² and the fact that we have [4]:

$$m_H = \sqrt{2 \cdot \mu^2} = \sqrt{4 \cdot \lambda \cdot (f_{1,2})^2}, \tag{939}$$

we can obtain:

$$\lambda \simeq 0.13; \quad \mu \simeq 88.8 \, GeV/c^2. \tag{940}$$

A dimensional analysis in Eq. (13) shows that an assumption of $R^* \sim f^n$ ($n \geq 0$) requires the introduction of a constant q, with the dimension $1/(\text{length}^2 * \text{mass}^{2+n})$, resulting in:

$$q \cdot f^n \cdot \frac{\left(2 \cdot \lambda \cdot f^2 - \mu^2\right)^3}{48 \cdot \mu^2 \cdot \lambda} = \Delta_g f^2. \tag{941}$$

From Eq. (13) and the results given in [96, 110, 124], we can directly deduce that there are Higgs field-options also in other dimensions and that symmetry breaking can occur in arbitrary N-dimensional space-times except for $N = 1, 2$.

Knowing that the dimension of the cosmological constant Λ is length^{-2}, we might rewrite Eq. (941) as follows:

$$\frac{\Lambda}{m_q^{2+n}} \cdot f^n \cdot \frac{\left(2 \cdot \lambda \cdot f^2 - \mu^2\right)^3}{48 \cdot \mu^2 \cdot \lambda} = \Delta_g f^2. \tag{942}$$

Here m_q shall denote a constant mass. For $n = 0$, the Ricci curvature R^* becomes independent on f and the option for a symmetric ground state with constant f vanishes. In this case, the occurrence of a Higgs field with broken symmetry is not an option, but a must.

6.3 The Metric Origin of Mass

It was shown in [129] and [130] that mass automatically results from certain entanglements of dimensions. As an example, we choose a 6-dimensional space-time in t, x, y, z, u, v and entangle the coordinates u and v as shown in the appendix to this subsection.

We result in the following Ricci scalar:

$$R^* = \frac{1}{F[f]^3} \cdot \left(-C_{N2} \cdot F[f] \cdot \frac{\partial F[f]}{\partial f} \cdot \Delta_g f \right);$$

$$F[f] = C_f + f[t, r, \vartheta, \varphi]$$

$$\overset{6D}{\longrightarrow} R^* = \frac{5}{f^2} \cdot \left(\frac{C_{v0}^2}{5} \cdot f + \frac{\partial^2 f}{c^2 \cdot \partial t^2} - \Delta_{3D-\text{sphere}} f \right);$$

$$f = f[t, r, \vartheta, \varphi]; \quad C_f = 0. \tag{943}$$

Here $\Delta_{3D-\text{sphere}} f$ denotes the Laplace operator in spherical coordinates.

As for $R^* = 0$ and $\frac{C_{v0}^2}{5} = \frac{m^2 \cdot c^2}{\hbar^2}$ (m = mass, c = speed of light in vacuum, $\hbar$ = reduced Planck constant), we clearly have obtained the quantum Klein–Gordon equation and we can easily conclude that mass can obviously be constructed by a suitable set of additional, entangled coordinates. As the whole in fact is a distributed field, we also—along the way—have obtained the Higgs field [129]. Because the function $g[v]$ just gives a bell-shaped localized distribution, it is also assured that the additional coordinates u and v are compactified, because the arbitrary constants C_{v1} and C_{v2} can always be chosen in such a way that $g[v]$ takes on the character of a Dirac delta distribution. Thus, the coordinates would be omnipresent in the space-time t, r, ϑ, φ, but do not appear in elongated form (on bigger scale), which means that u and v are invisible but only act via the production of mass in a 3+1 space-time.

6.3.1 Anisotropy of Inertia

It should be pointed out that with $f[\ldots]$ also depending on u and / or v we would automatically end up in certain mass- or inertia-anisotropies.

6.3.2 Appendix to the Metric Origin of Mass via Entanglement

In [129–132] we introduced special de Sitter vacuum metrics, which gave mass terms in the corresponding Ricci scalar based Klein–Gordon equation (99) in various dimensions. For completeness, we repeat the main results here.

Introducing a metric for the coordinates $t, r, \vartheta, \varphi, u, v$ of the form:

$$
g^6_{\alpha\beta} = \begin{pmatrix}
-c^2 & 0 & 0 & 0 & 0 & 0 \\
0 & 1 & 0 & 0 & 0 & 0 \\
0 & 0 & r^2 & 0 & 0 & 0 \\
0 & 0 & 0 & r^2 \cdot \sin[\vartheta] & 0 & 0 \\
0 & 0 & 0 & 0 & g[v] & 0 \\
0 & 0 & 0 & 0 & 0 & g[v]
\end{pmatrix} \cdot f[t, r, \vartheta, \varphi],
\tag{944}
$$

with the entanglement solution for u and v as follows:

$$
g[v] = \frac{C^2_{v1}}{C^2_{v0} \cdot (1 + \cosh[C_{v1} \cdot (v + C_{v2})])},
\tag{945}
$$

gives us the following Ricci scalar from Eq. (99) with condition (100):

$$
R^* = \frac{1}{F[f]^3} \cdot \left(-C_{N2} \cdot F[f] \cdot \frac{\partial F[f]}{\partial f} \cdot \Delta_g f \right); \quad F[f] = C_f + f[t, r, \vartheta, \varphi]
$$

$$
\xrightarrow{6D} R^* = \frac{5}{f^2} \cdot \left(\frac{C^2_{v0}}{5} \cdot f + \frac{\partial^2 f}{c^2 \cdot \partial t^2} - \Delta_{3D-\text{sphere}} f \right); \quad f = f[t, r, \vartheta, \varphi]; \quad C_f = 0.
\tag{946}
$$

Here $\Delta_{3D-\text{sphere}} f$ denotes the Laplace operator in spherical coordinates.

As for $R^* = 0$ and $\frac{C^2_{v0}}{5} = \frac{m^2 \cdot c^2}{\hbar^2}$ (m = mass, c = speed of light in vacuum, $\hbar$ = reduced Planck constant), we have obtained the classical quantum Klein–Gordon equation with mass and therefore conclude that mass can obviously be constructed by a suitable set of additional, entangled coordinates.

In connection with a non-vanishing cosmological constant, we have also shown in [130] that with:

$$
g^6_{\alpha\beta} = \begin{pmatrix}
-c^2 \cdot h[r] & 0 & 0 & 0 & 0 & 0 \\
0 & \frac{1}{h[r]} & 0 & 0 & 0 & 0 \\
0 & 0 & r^2 & 0 & 0 & 0 \\
0 & 0 & 0 & r^2 \cdot \sin[\vartheta] & 0 & 0 \\
0 & 0 & 0 & 0 & g[v] & 0 \\
0 & 0 & 0 & 0 & 0 & g[v]
\end{pmatrix}
\tag{947}
$$

with : $\quad h[r] = 1 + \frac{C^2_v \cdot r^2}{6} - r^2 \cdot \frac{\Lambda}{3} - \frac{C_r}{r}; \quad C^2_v = \Lambda,$

we could have a complete vacuum solution, with:

$$G^{\alpha\beta} + \Lambda g^{\alpha\beta} = R^{\alpha\beta} - \frac{1}{2} R g^{\alpha\beta} + \Lambda g^{\alpha\beta} = 0. \tag{948}$$

In this case, the condition $C_v^2 = \Lambda$ provides the necessary balance between matter and cosmological constant. As the constant C_r would be arbitrary, it could be set to zero in order to avoid any singularities.

We see that the metric in Eq. (947) would be a vacuum solution of the Einstein field equations but still the structure (universe) it describes contains matter. Assuming that the condition (948) should be a boundary condition for the object described by Eq. (947), we result in a balance rule for the mass parameter C_v and the cosmological constant. The object would have a resulting non-zero Ricci curvature of:

$$R = 3 \cdot \Lambda. \tag{949}$$

6.4 The Metric Origin of Spin and Spin Fields

Again, the recipe to evaluate the intrinsic parameter of spin requires the use of entangled dimensions [131]. However, as the evaluation is perfectly similar to the one we performed in connection with the mass and the Higgs field above, we just refer to the previous sub-section above and its appendix.

Similar to the metric approach (Eq. (944)) or—already more general— Eq. (947), we could construct a metric as follows:

$$g_{\alpha\beta}^6 = \begin{pmatrix} -c^2 \cdot h\,[r] & 0 & 0 & 0 & 0 & 0 \\ 0 & \frac{1}{h[r]} & 0 & 0 & 0 & 0 \\ 0 & 0 & r^2 & 0 & 0 & 0 \\ 0 & 0 & 0 & r^2 \cdot \sin[\vartheta] & 0 & 0 \\ 0 & 0 & 0 & 0 & a^2 \cdot g\,[v] & 0 \\ 0 & 0 & 0 & 0 & 0 & a^2 \end{pmatrix}$$

$$\text{with}: \quad h\,[r] = 1 + \frac{S^2 \cdot r^2}{3 \cdot a^2} - r^2 \cdot \frac{\Lambda}{3} - \frac{C_r}{r}; \quad S^2 = \frac{a^2 \cdot \Lambda}{2}. \tag{950}$$

Now we shall consider S a spin-like matter option in addition to our masses as given above.

This time we have:

$$g_{1,2}\,[v] = \begin{cases} C_{S1} \sin\,[S\,(v - C_{S2})]^2 \\ C_{S1} \cos\,[S\,(v - C_{S2})]^2 \end{cases}. \tag{951}$$

For more information and discussion (especially with respect to the extension to arbitrary coordinates and the flatness problem of our universe), the reader is referred to the original papers [129–132] and the next sections of this book.

Chapter 7

Solving the Flatness Problem

7.1 Introduction

It has long been a mystery why our universe seems to be so extremely close to the flat case, meaning close to an almost vanishing curvature [182, 183]. The standard cosmological model, mainly based on the Friedmann universe [186, 187], predicts that there should have been either a perfectly flat universe or an extreme increase or a decrease of the parameter of the density ρ since the beginning of time (assumed to be the Big Bang). The equation containing the matter density and comparing it with its critical value guaranteeing a perfectly flat universe ρ_c in classical cosmology reads:

$$\left(\frac{\rho_c}{\rho} - 1\right) \cdot \rho \cdot a\,[t]^2 = \text{const.} \tag{952}$$

Thereby $a[t]$ gives the scale parameter of the universe, which—as the universe expands—is time-dependent. For simplicity, we cite from the corresponding Wikipedia page [183]:

"The right hand side of...(952) contains constants only and therefore the left hand side must remain constant throughout the evolution of the universe.

As the universe expands, the scale factor $a[t]$ increases, but the density ρ decreases as matter (or energy) becomes spread out. For the standard model of the universe which contains mainly matter and radiation for most of its history, ρ decreases more quickly than $a[t]^2$ increases, and so the factor

The World Formula: A Late Recognition of David Hilbert's Stroke of Genius
Norbert Schwarzer
Copyright © 2022 Jenny Stanford Publishing Pte. Ltd.
ISBN 978-981-4877-20-6 (Hardcover), 978-1-003-14644-5 (eBook)
www.jennystanford.com

$\rho * a[t]^2$ will decrease. Since the time of the Planck era, shortly after the Big Bang, this term has decreased by a factor of around 10^{60} (according to ref. *Peter Coles; Francesco Lucchin (1997). Cosmology. Chichester: Wiley. ISBN 978-0-471-95473-6* given in [186]) and so $\left(\frac{\rho_c}{\rho} - 1\right)*$ must have increased by a similar amount to retain the constant value of their product."

Most interestingly and in contrast to this theoretical prediction, however, the density was found to still be very near the critical values of a flat universe.

The most prominent attempt to cure the situation is been seen in the so-called inflation theory [186–188], but there are also other attempts using anisotropic gravity approaches or non-constant speed of light assumptions (e.g., [188, 190, 191]).

Here now we want to investigate the possibility of additional dimensions leading to suitable mass-, spin- and Higgs fields, providing the necessary conditions for a balanced universe with nearly flat characteristics for principally infinite time scales.

7.2 A Cosmological Balance between Spin and Cosmological Constant

As we saw in the appendix of section 5.3 (there performed for the mass field), it is also possible to construct a spin-like field as follows:

$$g_{\alpha\beta}^6 = \begin{pmatrix} -c^2 \cdot h[r] & 0 & 0 & 0 & 0 & 0 \\ 0 & \dfrac{1}{h[r]} & 0 & 0 & 0 & 0 \\ 0 & 0 & r^2 & 0 & 0 & 0 \\ 0 & 0 & 0 & r^2 \cdot \sin[\vartheta] & 0 & 0 \\ 0 & 0 & 0 & 0 & a^2 \cdot g[v] & 0 \\ 0 & 0 & 0 & 0 & 0 & a^2 \end{pmatrix}$$

$$\text{with:} \quad h[r] = 1 + \frac{S^2 \cdot r^2}{3 \cdot a^2} - r^2 \cdot \frac{\Lambda}{3} - \frac{C_r}{r}; \quad S^2 = \frac{a^2 \cdot \Lambda}{2}. \tag{953}$$

We could consider this a spin-like matter option in addition to our masses as given above.

This time we have:

$$g_{1,2}[v] = \left\{ \begin{array}{l} C_{S1} \sin\left[S\left(v - C_{S2}\right)\right]^2 \\ C_{S1} \cos\left[S\left(v - C_{S2}\right)\right]^2 \end{array} \right\}, \tag{954}$$

which gives us the following Ricci scalar:

$$R^* = \frac{1}{F[f]^3} \cdot \left(-C_{N2} \cdot F[f] \cdot \frac{\partial F[f]}{\partial f} \cdot \Delta_g f \right);$$

$$F[f] = C_f + f[t, r, \vartheta, \varphi] \xrightarrow{6D} R^* = \frac{5}{f^2} \cdot \left(\frac{2}{5} \cdot \frac{S^2}{a^2} \cdot f + \frac{\partial^2 f}{c^2 \cdot \partial t^2} - \Delta_{3D-\text{sphere}} f \right);$$

$$f = f[t, r, \vartheta, \varphi]; \quad C_f = 0. \tag{955}$$

With the parameter S being connected with the angular function of the coordinate v, we consider it some kind of spin. The field of the entangled coordinates u and v we therefore name a spin-field.

With Eq. (950), we could obtain a complete vacuum solution again. As before with mass, the condition $S^2 = \frac{a^2 \cdot \Lambda}{2}$ provides the necessary balance between spin-originated matter and cosmological constant. As before, we have the constant Ricci scalar as given in Eq. (949), which happens to be the usual Ricci value for a de Sitter space in 6 dimensions.

7.3 Extension to Higher Dimensions

In the following—for better association—we will use the denomination $C_v^2 = M^2$.

By adding more pairwise entangled dimensions in the form of Eq. (945) or Eq. (951), we always end up with a Ricci scalar equal to the correspondingly equivalent de Sitter space situation in n dimensions, which would be:

$$R = \frac{2 \cdot n \cdot \Lambda}{n - 2} \Rightarrow \lim_{n \to \infty} R = 2 \cdot \Lambda. \tag{956}$$

In order to fulfill the vacuum Einstein field equations, we have to have the following functions $h[r]$ then:

$$h[r] = 1 - r^2 \cdot \frac{2 \cdot \Lambda}{3 \cdot (n - 2)} - \frac{C_r}{r}. \tag{957}$$

In addition, the masses and "spin" parameters M_i and S_i have to be suitable multiples of Λ. This, however, always couples these parameters to the cosmological constant and leads to a balance between masses, spins and potentially other matter objects one might code with forms of entangled dimensions to the "dark energy" being associated with Λ. Thus, in order to give an example, in the case of $3 + 1$ ordinary space-time dimensions plus

q pairs of entangled dimensions of the M-type (Eq. (945)), we obtain the following dependencies for all M_i:

$$M_i^2 = \frac{2}{q+1} \cdot \Lambda. \tag{958}$$

The S-type entanglement (Eq. (951)) would require:

$$S_i^2 = \frac{\Lambda}{q+1}. \tag{959}$$

The same result is obtained for any mixture of the M- and S-type. As the connection between total dimension n and the number for entangled pairs q would be $n = 4 + 2*q$, we have a direct connection between dimension n and the dependency of the matter parameters M and S to the cosmological constant. It appears only logical to assume that similar dependencies are resulting from other forms of entanglement with potentially higher numbers of dimensions being involved. With the dependencies (958) and (959), we find the following sums:

$$\sum_{i=1}^{q} M_i^2 = q \cdot \frac{2 \cdot \Lambda}{q+1} = \left(\frac{n}{2} - 2\right) \cdot \frac{2 \cdot \Lambda}{\frac{n}{2} - 1} = \frac{n-4}{n-2} \cdot 2 \cdot \Lambda \Rightarrow \lim_{n \to \infty} \sum_{i=1}^{q} M_i^2 = 2 \cdot \Lambda$$

$$\sum_{i=1}^{q} S_i^2 = q \cdot \frac{\Lambda}{q+1} = \left(\frac{n}{2} - 2\right) \cdot \frac{\Lambda}{\frac{n}{2} - 1} = \frac{n-4}{n-2} \cdot \Lambda \Rightarrow \lim_{n \to \infty} \sum_{i=1}^{q} S_i^2 = \Lambda.$$

$$\tag{960}$$

7.4 Cosmological Balance and the Flatness Problem

For simplicity, we concentrate on the M-fields only, and by combination of the first line of Eq. (960) with Eq. (956), we see that there is a clear correlation between the total mass of our special (mass-containing) de Sitter universe with its Ricci scalar:

$$\sum_{i=1}^{q} M_i^2 = \frac{n-4}{n-2} \cdot 2 \cdot \Lambda = \frac{n-4}{n} \cdot R \Rightarrow \lim_{n \to \infty} \sum_{i=1}^{q} M_i^2 = 2 \cdot \Lambda = R. \tag{961}$$

Assuming a small curvature R (as we obviously find it in our universe) for this de Sitter space being a constant, we always have a corresponding sum of M_i just proportional to this very Ricci scalar. And, of course, with R being small, our M-sum would appear close to the critical sum for the flat space-time. This, in contrast to the standard models mentioned in the introduction to this section, would easily explain the so-called flatness problem of our universe, because matter and cosmological constant would balance each other.

7.5 Conclusions

It was found that the so-called flatness problem could be solved by the assumption of a constant Ricci scalar for a mass containing de Sitter space. Thereby the mass (or other forms of matter) has been produced by the pairwise entanglement of additional dimensions.

Chapter 8

Anti-Gravity

One surprising finding of the extended variation of the Einstein–Hilbert action is the discovery of repulsion terms in gravitational interactions [185]. Here we only briefly repeat the evaluation and thereby apply the simple scalar variation approach from section 5.1.3 in the form (733).

As the term $\frac{R_{,\alpha} f^{,\alpha}}{R}$ represents the metric equivalent for the classical potential V, we should—in principle—be able to apply a (any) classical quantum solution for f and obtain a reasonable outcome for the new potential in comparison with the classical one. In fact, choosing the Schrödinger hydrogen problem (e.g. [170]) and applying the ground state solution with $f \sim e^{-r/a}$, we indeed obtain very complicated expressions, but the graphics show that in fact we have a very good agreement with the classical $V[r] = 1/r$ Coulomb potential (see Fig. 8.1, $a = 1$).

So, we just reproduced the classical Coulomb potential for the hydrogen atom almost perfectly within the metric picture. Thereby we have assumed that the quantum effects are small (just as they truly are, because otherwise they would bend the space-time like bigger massive objects or even black holes do). But what if considering black holes and their hydrogen-like metric interaction fields, thereby assuming that we have a similar or even the same f-ground state solution as we classically know from the Schrödinger hydrogen?

Well, this has been shown in Fig. 8.2.

The World Formula: A Late Recognition of David Hilbert's Stroke of Genius
Norbert Schwarzer
Copyright © 2022 Jenny Stanford Publishing Pte. Ltd.
ISBN 978-981-4877-20-6 (Hardcover), 978-1-003-14644-5 (eBook)
www.jennystanford.com

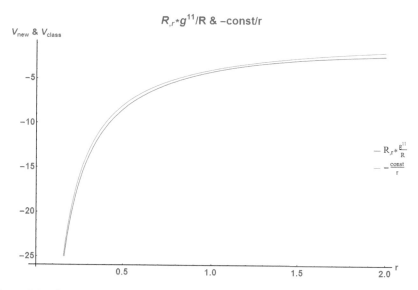

Figure 8.1 Comparison between classical Coulomb potential $1/r$ and the new metric "potential", resulting from $\frac{R_{,\alpha}f^{,\alpha}}{R}$ for the Schrödinger hydrogen ground state solution. Thereby we explicitly chose the constant const a bit imperfect in order to make the two lines in the graphic distinguishable.

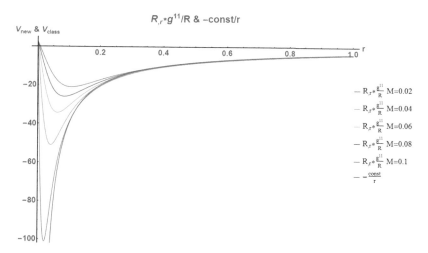

Figure 8.2 Comparison between classical Coulomb potential $1/r$ and the new metric "potential" for a variety of mass parameters M, resulting from $\frac{R_{,\alpha}f^{,\alpha}}{R}$ for the Schrödinger hydrogen ground state solution. Just for orientation, we also presented the classical Coulomb potential $V = -$const$/r$ (c.f. legend at the right hand side of the diagram).

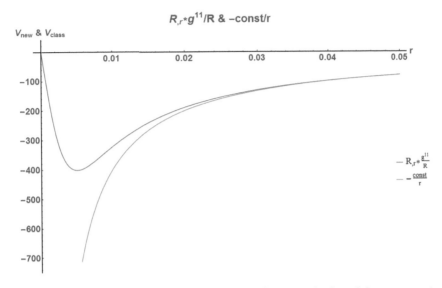

Figure 8.3 Comparison between classical Coulomb potential $1/r$ and the new metric "potential", resulting from $\frac{R_{,\alpha}f^{,\alpha}}{R}$ for the Schrödinger hydrogen ground state solution near the center $r = 0$. The metric parameter M was chosen to be $M = 0.005$.

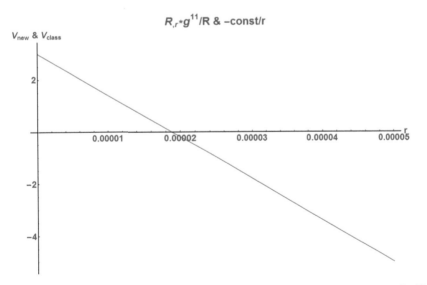

Figure 8.4 Zoomed center area for the new metric "potential", resulting from $\frac{R_{,\alpha}f^{,\alpha}}{R}$ for the Schrödinger hydrogen ground state solution. The metric parameter M was chosen to be $M = 0.005$.

Now we suddenly see that the interaction goes from attraction to repulsion, if moving very close to the center at $r = 0$.

Ok, here one might say that this is just an effect in stronger (gravity) fields, but, NO! It is a general thing. We only have it that in weaker interactions like the usual Coulomb fields the repulsion kicks in extremely late . . . so late in fact, that it can neither be seen nor—probably—measured. Even in fairly strong fields with parameters for M as big as 0.005 (half of the M as chosen in Fig. 8.1) we detect the kicking in of the repulsion only very near the center (Figs. 8.3 and 8.4).

Chapter 9

The Expansion of the Universe

Now we wonder what would happen in cases where the parameter M exceeds a certain limit. To our surprise (see Fig. 9.1), we find that the energetic minimum that we saw in Figs. 8.2 and 8.3 first moves further outward towards bigger r for increasing M and for $M > 1$ disappears and the potential shows only repulsion. We even make out regions where the resulting potential would lead to an accelerated expansion.

We think that we here have found a possible explanation for the observations of our universe's accelerated expansion [184]. From Eq. (724), we take the analogy that:

$$\frac{M^2}{d} = \frac{m^2 \cdot c^2}{\hbar^2}. \tag{962}$$

Setting $d = 1$ and considering m more a mass density rather than a mass, we may simply assume that a certain mass density leading to $M > 1$ leads to an expansion and automatically mass density change in such a way that the universe could find a steady state. Thereby we probably have some kind of oscillation among expansion and contraction.

Above we investigated a variety of states with different M parameters with $M < 1$ and found that there is always a repulsion kicking in the closer one moves towards the center $r = 0$. Interestingly, the bigger the parameter M, the further the potential minimum moves to bigger r and gets ever shallower (Fig. 9.1).

The World Formula: A Late Recognition of David Hilbert's Stroke of Genius
Norbert Schwarzer
Copyright © 2022 Jenny Stanford Publishing Pte. Ltd.
ISBN 978-981-4877-20-6 (Hardcover), 978-1-003-14644-5 (eBook)
www.jennystanford.com

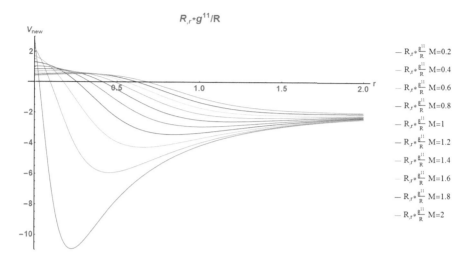

Figure 9.1 Comparison between a variety of new metric "potentials" for a central field at ground state level with different mass parameters M as resulting from $\frac{R_{,\alpha} f^{,\alpha}}{R}$ for the Schrödinger hydrogen ground state solution. Surprisingly, for certain $M > 1$, the local minimum seems to disappear. But on closer investigation (Fig. 9.2), we see that there is always such a minimum at a certain (further and further out) r position and we obtain a center positioned additional minimum.

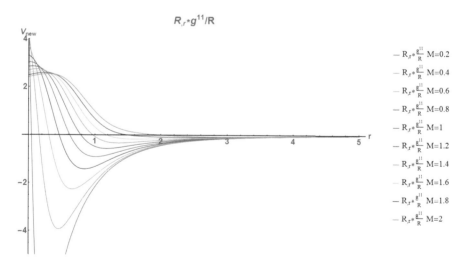

Figure 9.2 Potentials from Fig. 9.1 with better resolution in y direction and some absolute shift to a zero behavior for $r \to \infty$. Now we see that there are still minima for $M > 1$, but that there are also center minima from certain M onwards.

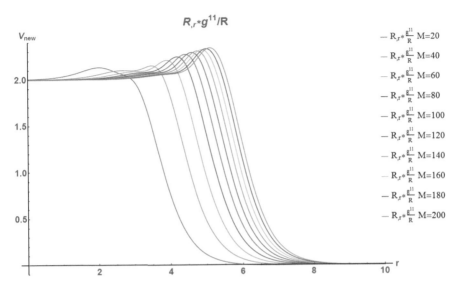

Figure 9.3 Comparison between a variety of new metric "potentials" for a central field at ground state level with different mass parameters M as resulting from $\frac{R_{\alpha}f^{,\alpha}}{R}$ for the Schrödinger hydrogen ground state solution. In contrast to Figs. 9.1 and 9.2, here we choose fairly big M values and obtain rather interesting center behavior.

Now we expanded our investigation to even bigger M and found that for $M > 1$ there is a peculiar change in behavior at $r = 0$, where in dependence on the size of M we also obtain a center minimum (Figs. 9.1 to 9.4).

9.1 A Few Thoughts at This Point

As according to [184] our universe is still being found in a state of accelerated expansion, either it must be extremely huge or its state differs significantly from the ground state we have assumed here. At any rate, our extended variation of the Einstein–Hilbert action has led us to solutions which would not only explain the expansion behavior of our universe but also provide interesting potential structures with respect to spontaneous symmetry breaking and the Higgs field [140].

In this context, another explanation appears meaningful, too, namely the one based on a grainy substructure of our space-time. Just as suggested in the public stories by Bodan [1–3], the space-time could be considered to

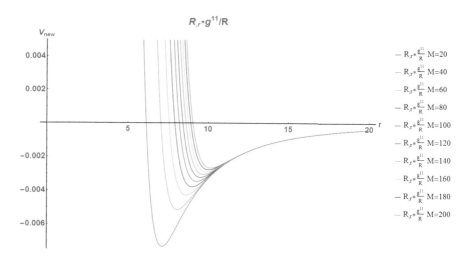

Figure 9.4 Potentials from Fig. 9.3 with better resolution in the y direction around the x-axis.

consist of many spherical entities with mass M. Depending on the parameter of M, these entities would tend to reach a certain optimum size of radius R. Being currently on their way to getting expanded, the whole universe, which consists of these "little beauties", would expand, too, and the observer within that very universe (himself consisting of the spherical entities and therefore, unable to see them directly with his means at hand) would just register this expansion, assuming the universe is growing, while in fact "only" its cells are. Now the question would come up, of course, why these cells are acting so seemingly well-coordinated. The answer to this could already have been given in [79–84] by our assumption of the creation of all these cells in approximately the same moment at the beginning of our universe in a state of great pressure (perhaps within the center of a huge black hole [84]). It was shown that one single cell could split up in many if sufficiently— uniformly—compressed. With all the resulting new cells just having the right M parameter and being very similar (because they were created in the same instant and all under—almost—the same conditions), we would easily end up with exactly the right structure to describe our "peculiar" universe.

As a nice side effect, we would have a background of spherical universal cells of just the right field behavior for the Higgs field (Figs. 9.1 and 9.2 for $M > 1$ and Fig. 9.3).

Considering all cells (or the majority of them) to be in the ground state as investigated in this book (Figs. 8.1 to 9.4), we would also be able to predict the future of our universe. Assuming namely (due to the existence of the Higgs field with its Mexican hat structure) an omnipresent parameter $M >$ 1 for "most of the cells", we automatically have extremely shallow minima at certain $r = R$. Thus, the cells currently exploding outwards will continue to do so for a very long time, simply because the slope of the potential outside the minimum is so low. Then the expansion slows down and finally comes to a hold, and then the cells will eventually start to shrink again. Now it will depend on the amount of inner friction how far this shrinking will go, whether a kind of pulsating leads to an endless oscillation of the cells around the $r = R$ size (and in consequence, also the universe) or whether everything just comes to a standstill at the $r = R$-minima positions for all cells (perhaps after a few ever more feeble pulses). Bringing in some kind of "secondary quantization" there may even be another ground state of endless lowest energy vibration of some kind.

Chapter 10

An Elastic World Formula

Tree formed by wind.

The World Formula: A Late Recognition of David Hilbert's Stroke of Genius
Norbert Schwarzer
Copyright © 2022 Jenny Stanford Publishing Pte. Ltd.
ISBN 978-981-4877-20-6 (Hardcover), 978-1-003-14644-5 (eBook)
www.jennystanford.com

Even though we already saw that we could derive a simple quantum gravity equation sporting the character and structure of the fundamental equations of linear elasticity from the starting point of the Einstein–Hilbert action, we want to repeat our evaluation in a somewhat more rigorous and general manner. In its classical form, the Einstein–Hilbert action [137, 138] reads:

$$\delta_g W = 0 = \delta_g \int_V d^n x \left(\sqrt{g}\, [R - 2\kappa L_M - 2\Lambda] \right). \tag{963}$$

Its traditional evaluation, as first performed by Hilbert in [137], results in the Einstein field equations, which we give here explicitly under the integral of the variation task (Eq. (1)):

$$\delta_\gamma W = 0 = \int_V d^n x \sqrt{g} \cdot \left(R^{\delta\gamma} - \frac{1}{2} R \cdot g^{\delta\gamma} + \Lambda \cdot g^{\delta\gamma} + \kappa \cdot T^{\delta\gamma} \right) \delta g_{\delta\gamma}. \tag{964}$$

For simplicity only we have ignored the sign under the square root of the metric determinant and will keep doing so throughout the chapter and the rest of the book. While in many previous examples we only considered scalar approaches, here we seek to perform the variation under the integral in vector functional form (applying a vector function $\mathbf{f}^k$ and its derivatives) and align it similar to our functional wrapper approach, closely mirroring the usual tensor transformations:

$$\delta_\gamma W = 0$$

$$= \int_V d^n x \sqrt{g} \cdot \left(R^{\delta\gamma} - \frac{1}{2} R \cdot g^{\delta\gamma} + \Lambda \cdot g^{\delta\gamma} + \kappa \cdot T^{\delta\gamma} \right) \times \left(\delta g_{\delta\gamma} = \gamma_{ij} \delta F^{ij}_{\delta\gamma} \right)$$

$$\tag{965}$$

as follows:

$$\delta_\gamma W = 0 = \int_V d^n x \sqrt{g} \cdot \left(R^{\delta\gamma} - \frac{1}{2} R \cdot g^{\delta\gamma} + \Lambda \cdot g^{\delta\gamma} + \kappa \cdot T^{\delta\gamma} \right)$$

$$\times \left(\begin{array}{l}
\delta g_{\delta\gamma} = g_{\delta\gamma} \delta_\sigma \left(1 + M \cdot h^\sigma h_\beta f^\beta + h_\beta \left(\overbrace{C^{\sigma a}}^{=d_1 \cdot \sqrt{g} \cdot g^{\sigma a}} f^\beta_{,\alpha} + \overbrace{C^{\beta a}}^{=d_2 \cdot \sqrt{g} \cdot g^{\beta a}} f^\sigma_{,\alpha} \right) \right) \\[2ex]
- g_{\delta\gamma} \left(M \cdot \left(h^\sigma h_\beta f^\beta \right)_{,\sigma} + \left[h_\beta \left(d_1 \sqrt{g} \cdot g^{\sigma a} f^\beta_{,\alpha} + d_2 \sqrt{g} \cdot g^{\beta a} f^\sigma_{,\alpha} \right) \right]_{,\sigma} \right) \\[2ex]
= g_{\delta\gamma} \left(M \cdot \left(h^\sigma h_\beta f^\beta \right)_{,\sigma} + \left(\begin{array}{l} \left[(d_1 \sqrt{g} \cdot g^{\sigma a} f^\beta_{,\alpha} + d_2 \sqrt{g} \cdot g^{\beta a} f^\sigma_{,\alpha}) \right] h_{\beta,\sigma} \\ + h_\beta \left(d_1 \sqrt{g} \cdot g^{\sigma a} f^\beta_{,\alpha} + d_2 \sqrt{g} \cdot g^{\beta a} f^\sigma_{,\alpha} \right)_{,\sigma} \end{array} \right) \right) \\[2ex]
= g_{\delta\gamma} \left(M \cdot \left(h^\sigma f^\beta h_{\beta,\sigma} + h_\beta \left(h^\sigma f^\beta \right)_{,\sigma} \right) + \left(\begin{array}{l} \left(d_1 \sqrt{g} \cdot g^{\sigma a} f^\beta_{,\alpha} + d_2 \sqrt{g} \cdot g^{\beta a} f^\sigma_{,\alpha} \right) h_{\beta,\sigma} \\ + h_\beta \left(d_1 \sqrt{g} \cdot \Delta f^\beta + d_2 \left(\sqrt{g} \cdot g^{\beta a} f^\sigma_{,\alpha} \right)_{,\sigma} \right) \end{array} \right) \right)
\end{array} \right).$$

$$\tag{966}$$

Please note that our handling of the metric $g_{\delta\gamma}$ and its variation $\delta g_{\delta\gamma}$ requires some elaboration. We demand the metric tensor $g_{\delta\gamma}$ to be constructed as follows:

$$
\begin{aligned}
g_{\delta\gamma} &= \gamma_{ij}\delta_\delta^i\delta_\gamma^j \left(1 + M \cdot h^\sigma h_\beta f^\beta + h_\beta \left(d_1 \cdot \sqrt{g} \cdot g^{\sigma\alpha} f^\beta{}_{,\alpha} + d_2 \cdot \sqrt{g} \cdot g^{\beta\alpha} f^\sigma{}_{,\alpha}\right)\right) \\
&= \gamma_{\delta\gamma} \left(1 + M \cdot h^\sigma h_\beta f^\beta + h_\beta \left(d_1 \cdot \sqrt{g} \cdot g^{\sigma\alpha} f^\beta{}_{,\alpha} + d_2 \cdot \sqrt{g} \cdot g^{\beta\alpha} f^\sigma{}_{,\alpha}\right)\right),
\end{aligned}
$$
$$(967)$$

which reminds us of the usual covariant tensor coordinate transformation. We assume the variation only to act on the function vector $\mathbf{f}^k$ and its derivatives, which gives us:

$$
\begin{aligned}
\delta g_{\delta\gamma} &= \gamma_{\delta\gamma}\delta\left(1 + M \cdot h^\sigma h_\beta f^\beta + h_\beta \left(d_1 \cdot \sqrt{g} \cdot g^{\sigma\alpha} f^\beta{}_{,\alpha} + d_2 \cdot \sqrt{g} \cdot g^{\beta\alpha} f^\sigma{}_{,\alpha}\right)\right) \\
&= \frac{g_{\delta\gamma}\delta\left(M \cdot h^\sigma h_\beta f^\beta + h_\beta \left(Cd_1 \cdot \sqrt{g} \cdot g^{\sigma\alpha} f^\beta{}_{,\alpha} + d_2 \cdot \sqrt{g} \cdot g^{\beta\alpha} f^\sigma{}_{,\alpha}\right)\right)}{\left(1 + M \cdot h^\sigma h_\beta f^\beta + h_\beta \left(d_1 \cdot \sqrt{g} \cdot g^{\sigma\alpha} f^\beta{}_{,\alpha} + d_2 \cdot \sqrt{g} \cdot g^{\beta\alpha} f^\sigma{}_{,\alpha}\right)\right)} \\
&\xrightarrow{1 \gg M \cdot h^\sigma h_\beta f^\beta + h_\beta \left(d_1 \cdot \sqrt{g} \cdot g^{\sigma\alpha} f^\beta{}_{,\alpha} + d_2 \cdot \sqrt{g} \cdot g^{\beta\alpha} f^\sigma{}_{,\alpha}\right)} \\
&\simeq g_{\delta\gamma}\delta\left(M \cdot h^\sigma h_\beta f^\beta + h_\beta \left(d_1 \cdot \sqrt{g} \cdot g^{\sigma\alpha} f^\beta{}_{,\alpha} + d_2 \cdot \sqrt{g} \cdot g^{\beta\alpha} f^\sigma{}_{,\alpha}\right)\right).
\end{aligned}
$$
$$(968)$$

Due to the approximation in the last line, we need not bother about the potentially problematic division by a vector at the moment, but in order to reuse Eq. (968) also in the non-approximated form, we may just demand that the properties of h_σ are such that we could write:

$$
\begin{aligned}
\delta g_{\delta\gamma} &= \gamma_{\delta\gamma}\delta\left(1 + M \cdot h^\sigma h_\beta f^\beta + h_\beta \left(d_1 \cdot \sqrt{g} \cdot g^{\sigma\alpha} f^\beta{}_{,\alpha} + d_2 \cdot \sqrt{g} \cdot g^{\beta\alpha} f^\sigma{}_{,\alpha}\right)\right) \\
&= \frac{g_{\delta\gamma}\delta\left(M \cdot h^\sigma h_\beta f^\beta + h_\beta \left(d_1 \cdot \sqrt{g} \cdot g^{\sigma\alpha} f^\beta{}_{,\alpha} + d_2 \cdot \sqrt{g} \cdot g^{\beta\alpha} f^\sigma{}_{,\alpha}\right)\right)}{h_\sigma \left(1 + M \cdot h^\sigma h_\beta f^\beta + h_\beta \left(d_1 \cdot \sqrt{g} \cdot g^{\sigma\alpha} f^\beta{}_{,\alpha} + d_2 \cdot \sqrt{g} \cdot g^{\beta\alpha} f^\sigma{}_{,\alpha}\right)\right)}.
\end{aligned}
$$
$$(969)$$

We also give the corresponding evaluation of the general form (Eq. (965)), where we have:

$$
g_{\delta\gamma} = \gamma_{ij} F_{\delta\gamma}^{ij} \;\Rightarrow\; \gamma_{ij} = g_{\delta\gamma}\left[F^{-1}\right]_{ij}^{\delta\gamma} \;\Rightarrow\; \delta g_{\delta\gamma} = \gamma_{ij}\delta F_{\delta\gamma}^{ij} = g_{\delta\gamma}\left[F^{-1}\right]_{ij}^{\alpha\beta}\delta F_{\alpha\beta}^{ij}.
$$
$$(970)$$

Now we reorder the last line in Eq. (966) and obtain:

$$\delta_\gamma W = 0 = \int_V d^n x \sqrt{g} \cdot \left(R^{\delta\gamma} - \frac{1}{2} R \cdot g^{\delta\gamma} + \Lambda \cdot g^{\delta\gamma} + \kappa \cdot T^{\delta\gamma} \right)$$
$$\times \left(g_{\delta\gamma} \left(\begin{array}{c} \left(M \cdot h^\sigma f^\beta h_{\beta,\sigma} + d_1 \sqrt{g} \cdot g^{\sigma\alpha} f^\beta{}_{,\alpha} + d_2 \sqrt{g} \cdot g^{\beta\alpha} f^\sigma{}_{,\alpha} \right) h_{\beta,\sigma} \\ + h_\beta \left(M \cdot \left(h^\sigma f^\beta \right)_{,\sigma} + d_1 \sqrt{g} \cdot \Delta f^\beta + d_2 \left(\sqrt{g} \cdot g^{\beta\alpha} f^\sigma{}_{,\alpha} \right)_{,\sigma} \right) \end{array} \right) \right).$$

(971)

Demanding the variated metric to give zero, we would have:

$$\delta_\gamma W = 0 = \left(\begin{array}{c} \left(M \cdot h^\sigma f^\beta h_{\beta,\sigma} + d_1 \sqrt{g} \cdot g^{\sigma\alpha} f^\beta{}_{,\alpha} + d_2 \sqrt{g} \cdot g^{\beta\alpha} f^\sigma{}_{,\alpha} \right) h_{\beta,\sigma} \\ + h_\beta \left(M \cdot \left(h^\sigma f^\beta \right)_{,\sigma} + d_1 \sqrt{g} \cdot \Delta f^\beta + d_2 \left(\sqrt{g} \cdot g^{\beta\alpha} f^\sigma{}_{,\alpha} \right)_{,\sigma} \right) \end{array} \right),$$

(972)

where we could multiply with the contravariant h^β yielding:

$$0 = \left(\begin{array}{c} \left(M \cdot h^\sigma f^\beta h_{\beta,\sigma} + d_1 \sqrt{g} \cdot g^{\sigma\alpha} f^\beta{}_{,\alpha} + d_2 \sqrt{g} \cdot g^{\beta\alpha} f^\sigma{}_{,\alpha} \right) \overbrace{h_{\beta,\sigma} h^\beta}^{=C} \\ + h^\beta h_\beta \cdot \left(M \cdot \left(h^\sigma f^\beta \right)_{,\sigma} + d_1 \sqrt{g} \cdot \Delta f^\beta + d_2 \left(\sqrt{g} \cdot g^{\beta\alpha} f^\sigma{}_{,\alpha} \right)_{,\sigma} \right) \end{array} \right).$$

(973)

Just for the moment assuming h_β to be a vector of constants, we obtain:

$$0 = M \cdot \left(h^\sigma f^\beta \right)_{,\sigma} + d_1 \sqrt{g} \cdot \Delta f^\beta + d_2 \left(\sqrt{g} \cdot g^{\beta\alpha} f^\sigma{}_{,\alpha} \right)_{,\sigma}$$
$$= \frac{M \cdot \left(h^\sigma f^\beta \right)_{,\sigma}}{d_1 \sqrt{g}} + \Delta f^\beta + \frac{d_2 \left(\sqrt{g} \cdot g^{\beta\alpha} f^\sigma{}_{,\alpha} \right)_{,\sigma}}{d_1 \sqrt{g}}.$$

(974)

Now writing the whole in Cartesian coordinates:

$$0 = \frac{M \cdot \left(h^\sigma f^\beta \right)_{,\sigma}}{d_1} + \Delta f^\beta + \frac{d_2 \left(g^{\beta\alpha} f^\sigma{}_{,\alpha} \right)_{,\sigma}}{d_1}$$

(975)

and we recognize the fundamental equation of elasticity [152–154] in the isotropic case with volume and external forces if we would just set $\frac{d_2}{d_1} = \frac{1}{1-2\cdot v}$ (with v giving Poisson's ratio). Of course, we could already have discovered the elastic terms in the full equation (Eq. (973)), but due to its complexity it is a bit more difficult to make it out in there.

We may not like to have the first derivative of the $\mathbf{f}^\beta$-vector in Eq. (974) (or its Cartesian version, Eq. (975)) and thus, reshape the equation by performing an integration by parts on the corresponding term $\frac{M \cdot \left(h^\sigma f^\beta \right)_{,\sigma}}{d_1 \sqrt{g}}$. This gives us:

$$\delta W = 0 = \int_V d^n x \sqrt{g} \cdot \overbrace{\left(R^{\delta\gamma} - \frac{1}{2} R \cdot g^{\delta\gamma} + \Lambda \cdot g^{\delta\gamma} + \kappa \cdot T^{\delta\gamma} \right)}^{(\dots)^{\delta\gamma}}$$

$$\times \left(g_{\delta\gamma} h_\beta \left(M \cdot \left(h^\sigma f^\beta \right)_{,\sigma} + d_1 \sqrt{g} \cdot \Delta f^\beta + d_2 \left(\sqrt{g} \cdot g^{\beta\alpha} f^\sigma{}_{,\alpha} \right)_{,\sigma} \right) \right)$$

$$= \int_V d^n x \times \left(\sqrt{g} \cdot (\dots)^{\delta\gamma} g_{\delta\gamma} h_\beta \left(M \cdot \left(h^\sigma f^\beta \right)_{,\sigma} \right.\right.$$

$$\left.\left. + d_1 \sqrt{g} \cdot \Delta f^\beta + d_2 \left(\sqrt{g} \cdot g^{\beta\alpha} f^\sigma{}_{,\alpha} \right)_{,\sigma} \right) \right)$$

$$= \int_V d^n x \times \left(\begin{array}{c} \sqrt{g} \cdot (\dots)^{\delta\gamma} g_{\delta\gamma} h_\beta \left(d_1 \sqrt{g} \cdot \Delta f^\beta + d_2 \left(\sqrt{g} \cdot g^{\beta\alpha} f^\sigma{}_{,\alpha} \right)_{,\sigma} \right) \\ -M \cdot \left[\sqrt{g} \cdot (\dots)^{\delta\gamma} g_{\delta\gamma} h_\beta \right]_{,\sigma} h^\sigma f^\beta \end{array} \right).$$

$$(976)$$

Without the approximation used in Eq. (968), we would have:

$$\delta W = 0 = \int_V d^n x \sqrt{g} \cdot \overbrace{\left(R^{\delta\gamma} - \frac{1}{2} R \cdot g^{\delta\gamma} + \Lambda \cdot g^{\delta\gamma} + \kappa \cdot T^{\delta\gamma} \right)}^{(\dots)^{\delta\gamma}}$$

$$\times \left(\underbrace{\frac{g_{\delta\gamma} h_\beta \left(M \cdot \left(h^\sigma f^\beta \right)_{,\sigma} + d_1 \sqrt{g} \cdot \Delta f^\beta + d_2 \left(\sqrt{g} \cdot g^{\beta\alpha} f^\sigma{}_{,\alpha} \right)_{,\sigma} \right)}{h_\sigma \left(1 + M \cdot h^\sigma h_\beta f^\beta + h_\beta \left(d_1 \cdot \sqrt{g} \cdot g^{\sigma\alpha} f^\beta{}_{,\alpha} + d_2 \cdot \sqrt{g} \cdot g^{\beta\alpha} f^\sigma{}_{,\alpha} \right) \right)}}_{\begin{bmatrix} \vdots \end{bmatrix}} \right)$$

$$= \int_V d^n x \times \left(\sqrt{g} \cdot \frac{(\dots)^{\delta\gamma}}{\begin{bmatrix} \vdots \end{bmatrix}} g_{\delta\gamma} h_\beta \left(M \cdot \left(h^\sigma f^\beta \right)_{,\sigma} + d_1 \sqrt{g} \cdot \Delta f^\beta + d_2 \left(\sqrt{g} \cdot g^{\beta\alpha} f^\sigma{}_{,\alpha} \right)_{,\sigma} \right) \right)$$

$$= \int_V d^n x \times \left(\begin{array}{c} \sqrt{g} \cdot \frac{(\dots)^{\delta\gamma}}{\begin{bmatrix} \vdots \end{bmatrix}} g_{\delta\gamma} h_\beta \left(d_1 \sqrt{g} \cdot \Delta f^\beta + d_2 \left(\sqrt{g} \cdot g^{\beta\alpha} f^\sigma{}_{,\alpha} \right)_{,\sigma} \right) \\ -M \cdot \left[\sqrt{g} \cdot \frac{(\dots)^{\delta\gamma}}{\begin{bmatrix} \vdots \end{bmatrix}} g_{\delta\gamma} h_\beta \right]_{,\sigma} h^\sigma f^\beta \end{array} \right). \quad (977)$$

The resulting equations would now read:

$$\sqrt{g} \cdot \frac{(\ldots)^{\delta\gamma}}{\begin{bmatrix} \vdots \end{bmatrix}} g_{\delta\gamma} h_\beta \left(d_1 \sqrt{g} \cdot \Delta f^\beta + d_2 \left(\sqrt{g} \cdot g^{\beta\alpha} f^\sigma_{,\alpha} \right)_{,\sigma} \right)$$

$$- M \cdot \left[\sqrt{g} \cdot \frac{(\ldots)^{\delta\gamma}}{\begin{bmatrix} \vdots \end{bmatrix}} g_{\delta\gamma} h_\beta \quad h^\sigma f^\beta \right]_{,\sigma} = 0. \tag{978}$$

As we know that in 4 dimensions we could simplify:

$$(\ldots)^{\delta\gamma} g_{\delta\gamma} \xrightarrow{\Lambda=0;\ T^{\delta\gamma}=0} -R, \tag{979}$$

we would end up with the following interesting elastic quantum gravity equation for our space-time:

$$\sqrt{g} \cdot \frac{R}{\begin{bmatrix} \vdots \end{bmatrix}} h_\beta \left(d_1 \sqrt{g} \cdot \Delta f^\beta + d_2 \left(\sqrt{g} \cdot g^{\beta\alpha} f^\sigma_{,\alpha} \right)_{,\sigma} \right) - M \cdot \left[\sqrt{g} \cdot \frac{R}{\begin{bmatrix} \vdots \end{bmatrix}} h_\beta \quad h^\sigma f^\beta \right]_{,\sigma} = 0. \tag{980}$$

The vector h_β could still be considered to be a vector of constants, because this would not compromise the generality of our equations, as any functional issue taken away from h_β by our const-demand would simply be taken on by f (and/or the metric). This gives us the following system of vector equations:

$$0 = \sqrt{g} \cdot \frac{R}{\begin{bmatrix} \vdots \end{bmatrix}} \left(d_1 \sqrt{g} \cdot \Delta f^\beta + d_2 \left(\sqrt{g} \cdot g^{\beta\alpha} f^\sigma_{,\alpha} \right)_{,\sigma} \right) - M \cdot \left[\sqrt{g} \cdot \frac{R}{\begin{bmatrix} \vdots \end{bmatrix}} \quad h^\sigma f^\beta \right]_{,\sigma}$$

$$\Delta f^\beta + \frac{d_2}{d_1 \sqrt{g}} \left(\sqrt{g} \cdot g^{\beta\alpha} f^\sigma_{,\alpha} \right)_{,\sigma} - \frac{M}{d_1 \cdot g \cdot \frac{R}{\begin{bmatrix} \vdots \end{bmatrix}}} \cdot \left[\sqrt{g} \cdot \frac{R}{\begin{bmatrix} \vdots \end{bmatrix}} \quad h^\sigma f^\beta \right]_{,\sigma} = 0. \tag{981}$$

The appearance of the fundamental equation of elasticity may already be taken as a good sign with respect to the procedure applied here in order to obtain quantum equations out of the Einstein–Hilbert action.

It was shown elsewhere that from equations of the type Eq. (981), applying the so-called n-function approach [160–163], we are able to evaluate very particle-like (leptons and quarks) solutions.

However, it should be seen as a drawback that our approach (Eq. (966)) looks rather arbitrary. We can change this when starting very close to the usual tensor transformations as follows:

$$\delta_\gamma W = 0 = \int_V d^n x \sqrt{g} \cdot \left(R^{\delta\gamma} - \frac{1}{2} R \cdot g^{\delta\gamma} + \Lambda \cdot g^{\delta\gamma} + \kappa \cdot T^{\delta\gamma} \right)$$

$$\times \begin{pmatrix} \delta g_{\delta\gamma} = \delta \left(\frac{\partial G^i [x_{\forall i}]}{\partial x^\delta} \cdot \frac{\partial G^j [x_{\forall i}]}{\partial x^\gamma} \gamma_{ij} \right) = g_{ij} \cdot \delta \left(\frac{\partial G^i [x_{\forall i}]}{\partial x^\delta} \cdot \frac{\partial G^j [x_{\forall i}]}{\partial x^\gamma} \right) \\ = \frac{g_{\delta\gamma}}{8} \delta_{k|l} \left(\{\ldots\}^k \{\ldots\}^l \right) \\ = g_{\delta\gamma} \delta_{k|l} \left[\left(C^k C^l + M^2 \cdot f^k f^l + \mathbf{C}^\alpha \mathbf{C}^\beta \cdot f^k_{,\alpha} f^l_{,\beta} + C^k C^l f^\alpha_{,\alpha} f^\beta_{,\beta} \right) \right] \\ = g_{\delta\gamma} \begin{pmatrix} M^2 \cdot \left(f^k \left(f^l_{,l} \right) + f^k_{,k} f^l \right) \\ +\mathbf{C}^\alpha \mathbf{C}^\beta \cdot \left(f^k_{,\alpha k} f^l_{,\beta} + f^k_{,\alpha} f^l_{,\beta l} \right) + C^k C^l \left(f^\alpha_{,\alpha k} f^\beta_{,\beta} + f^\alpha_{,\alpha} f^\beta_{,\beta l} \right) \end{pmatrix} \delta_{k|l} \end{pmatrix} \cdot$$

$$\tag{982}$$

Thereby the $\{\ldots\}$-vectors read:

$$\{\ldots\}^j = \begin{Bmatrix} C^j + M \cdot f^j + \mathbf{C}^\alpha f^j_{,\alpha} + C^j f^\alpha_{,\alpha}, \ C^j + M \cdot f^j + \mathbf{C}^\alpha f^j_{,\alpha} - C^j f^\alpha_{,\alpha} \\ C^j + M \cdot f^j - \mathbf{C}^\alpha f^j_{,\alpha} + C^j f^\alpha_{,\alpha}, \ C^j + M \cdot f^j - \mathbf{C}^\alpha f^j_{,\alpha} - C^j f^\alpha_{,\alpha} \\ C^j - M \cdot f^j + \mathbf{C}^\alpha f^j_{,\alpha} + C^j f^\alpha_{,\alpha}, \ C^j - M \cdot f^j + \mathbf{C}^\alpha f^j_{,\alpha} - C^j f^\alpha_{,\alpha} \\ C^j - M \cdot f^j - \mathbf{C}^\alpha f^j_{,\alpha} + C^j f^\alpha_{,\alpha}, \ C^j - M \cdot f^j - \mathbf{C}^\alpha f^j_{,\alpha} - C^j f^\alpha_{,\alpha} \end{Bmatrix} \cdot$$

$$\tag{983}$$

After performing integration by parts as follows:

$$\delta_\gamma W = 0 = \int_V d^n x$$

$$\times \begin{pmatrix} \sqrt{g} \cdot (\ldots)^{\delta\gamma} g_{\delta\gamma} M^2 \cdot \left(f^k \left(f^l_{,l} \right) + f^k_{,k} f^l \right) - \left[\sqrt{g} \cdot (\ldots)^{\delta\gamma} g_{\delta\gamma} \mathbf{C}^\alpha \mathbf{C}^\beta f^l_{,\beta} \right]_{,\alpha} f^k_{,k} \\ - \left[\sqrt{g} \cdot (\ldots)^{\delta\gamma} g_{\delta\gamma} \mathbf{C}^\alpha \mathbf{C}^\beta f^k_{,\alpha} \right]_{,\beta} f^l_{,l} + \sqrt{g} \cdot (\ldots)^{\delta\gamma} g_{\delta\gamma} C^k C^l \left(f^\alpha_{,\alpha k} f^\beta_{,\beta} + f^\alpha_{,\alpha} f^\beta_{,\beta l} \right) \end{pmatrix} \delta_{k|l}$$

$$\xrightarrow{f^\beta_{,\beta} = f^l_{,l} = f^\alpha_{,\alpha} = f^k_{,k}}$$

$$= \int_V d^n x \begin{pmatrix} \left(\sqrt{g} \cdot (\ldots)^{\delta\gamma} g_{\delta\gamma} \left(M^2 \cdot f^k + C^k C^l f^\beta_{,\beta l} \right) - \left[\sqrt{g} \cdot (\ldots)^{\delta\gamma} g_{\delta\gamma} \mathbf{C}^\alpha \mathbf{C}^\beta f^k_{,\alpha} \right]_{,\beta} \right) f^l_{,l} \\ + \left(\sqrt{g} \cdot (\ldots)^{\delta\gamma} g_{\delta\gamma} \left(M^2 \cdot f^l + C^k C^l f^\alpha_{,\alpha k} \right) - \left[\sqrt{g} \cdot (\ldots)^{\delta\gamma} g_{\delta\gamma} \mathbf{C}^\alpha \mathbf{C}^\beta f^l_{,\beta} \right]_{,\alpha} \right) f^k_{,k} \end{pmatrix}$$

$$\xrightarrow{\mathbf{C}^\alpha \mathbf{C}^\beta = g^{\alpha\beta}}$$

$$= \int_V d^n x \begin{pmatrix} \left(\begin{matrix} \sqrt{g} \cdot (\ldots)^{\delta\gamma} g_{\delta\gamma} \left(M^2 \cdot f^k + C^k C^l f^\beta_{,\beta l} \right) \\ - [(\ldots)^{\delta\gamma} g_{\delta\gamma}]_{,\beta} \sqrt{g} \cdot g^{\alpha\beta} f^k_{,\alpha} - (\ldots)^{\delta\gamma} g_{\delta\gamma} \sqrt{g} \cdot \Delta f^k \end{matrix} \right) f^l_{,l} \\ + \left(\begin{matrix} \sqrt{g} \cdot (\ldots)^{\delta\gamma} g_{\delta\gamma} \left(M^2 \cdot f^l + C^k C^l f^\alpha_{,\alpha k} \right) \\ - [(\ldots)^{\delta\gamma} g_{\delta\gamma}]_{,\alpha} \sqrt{g} \cdot g^{\alpha\beta} f^l_{,\beta} - (\ldots)^{\delta\gamma} g_{\delta\gamma} \sqrt{g} \cdot \Delta f^l \end{matrix} \right) f^k_{,k} \end{pmatrix},$$

$$\tag{984}$$

we have once again the typical elasticity terms $C^k C^l f^\alpha{}_{,\alpha k}$ and Δf^k, which, with the grouping:

$$
0 = \int_V d^n x \sqrt{g} \cdot \left(
\begin{array}{l}
\left((\dots)^{\delta\gamma} g_{\delta\gamma} \left(M^2 \cdot f^k + C^k C^l f^\beta{}_{,\beta l} - \Delta f^k \right) - \left[(\dots)^{\delta\gamma} g_{\delta\gamma} \right]_{,\beta} g^{\alpha\beta} f^k{}_{,\alpha} \right) f^l{}_{,l} \\
+ \left((\dots)^{\delta\gamma} g_{\delta\gamma} \left(M^2 \cdot f^l + C^k C^l f^\alpha{}_{,\alpha k} - \Delta f^l \right) - \left[(\dots)^{\delta\gamma} g_{\delta\gamma} \right]_{,\alpha} g^{\alpha\beta} f^l{}_{,\beta} \right) f^k{}_{,k}
\end{array}
\right),
$$

$$(985)$$

allow us to extract the following system of quantum gravity equations:

$$
0 = (\dots)^{\delta\gamma} g_{\delta\gamma} \left(M^2 \cdot f^k + C^k C^l f^\beta{}_{,\beta l} - \Delta f^k \right) - \left[(\dots)^{\delta\gamma} g_{\delta\gamma} \right]_{,\beta} g^{\alpha\beta} f^k{}_{,\alpha}.
$$

$$(986)$$

Now, in principle, we could start and try to interpret the results but there is still the problematic step in the second line of Eq. (982), which requires some elaboration. Performed 100% correctly, the transformation should read:

$$
\delta g_{\delta\gamma} = \delta \left(\frac{\partial G^i [x_{\forall i}]}{\partial x^\delta} \cdot \frac{\partial G^j [x_{\forall i}]}{\partial x^\gamma} \gamma_{ij} \right) = \gamma_{ij} \cdot \delta \left(\frac{\partial G^i [x_{\forall i}]}{\partial x^\delta} \cdot \frac{\partial G^j [x_{\forall i}]}{\partial x^\gamma} \right),
$$

$$(987)$$

where we insist that the variation only acts on the Jacobi matrices and not on the original metric tensor γ_{ij}. However, we know that we can always write this tensor as:

$$
g_{\delta\gamma} \left(\frac{\partial G^i [x_{\forall i}]}{\partial x^\delta} \cdot \frac{\partial G^j [x_{\forall i}]}{\partial x^\gamma} \right)^{-1} = \gamma_{ij}
$$

$$
= \frac{g_{\delta\gamma} \delta_i^\delta \delta_j^\gamma}{C_k C_l \left(C^k C^l + M^2 \cdot f^k f^l + C^\alpha C^\beta \cdot f^k{}_{,\alpha} f^l{}_{,\beta} + C^k C^l f^\alpha{}_{,\alpha} f^\beta{}_{,\beta} \right)}
$$

$$
\overbrace{C_k C_l C^k C^l}^{=1} \gg C_k C_l \left(M^2 \cdot f^k f^l + C^\alpha C^\beta \cdot f^k{}_{,\alpha} f^l{}_{,\beta} + C^k C^l f^\alpha{}_{,\alpha} f^\beta{}_{,\beta} \right) \atop \longrightarrow \simeq g_{ij} \qquad (988)
$$

and so obtain the approximation in the second line of Eq. (982). Without this approximation, the evaluation would just become a little bit more complex, but it is still possible, because instead of Eq. (982) we have to write:

$$\delta_\gamma W = 0 = \int_V d^n x \sqrt{g} \left(R^{\delta\gamma} - \frac{1}{2} R \cdot g^{\delta\gamma} + \Lambda \cdot g^{\delta\gamma} + \kappa \cdot T^{\delta\gamma} \right)$$

$$\times$$

$$\delta g_{\delta\gamma} = \delta \left(\frac{\partial G^i [x_{vi}]}{\partial x^\delta} \cdot \frac{\partial G^i [x_{vi}]}{\partial x^\gamma} \gamma_{ij} \right) = \gamma_{ij} \cdot \delta \left(\frac{\partial G^i [x_{vi}]}{\partial x^\delta} \cdot \frac{\partial G^i [x_{vi}]}{\partial x^\gamma} \right)$$

$$= \underbrace{\overline{C_k C_l \left(C^k C^l + M^2 \cdot f^k f^l + \mathbf{C}^\alpha \mathbf{C}^\beta \cdot f^k_{,\alpha} f^l_{,\beta} + C^k C^l f^\alpha_{,\alpha} f^\beta_{,\beta} \right)}}_{\boxed{\cdots}} \delta \left(\frac{\partial G^i [x_{vi}]}{\partial x^\delta} \cdot \frac{\partial G^i [x_{vi}]}{\partial x^\gamma} \right)^{g_{\delta\gamma} \delta^\gamma_i \delta^\gamma_j}$$

$$= \frac{g_{\delta\gamma}}{8 \cdot \boxed{\cdots}} \delta_{k|l} \left(\{\cdots\}^k \{\cdots\}^l \right)$$

$$= \frac{g_{\delta\gamma}}{\boxed{\cdots}} \delta_{k|l} \left[\left(C^k C^l + M^2 \cdot f^k f^l + \mathbf{C}^\alpha \mathbf{C}^\beta \cdot f^k_{,\alpha} f^l_{,\beta} + C^k C^l f^\alpha_{,\alpha} f^\beta_{,\beta} \right) \right]$$

$$= \frac{g_{\delta\gamma}}{\boxed{\cdots}} \left(\begin{array}{l} M^2 \cdot \left(f^k \left(f^l_{,l} \right) + f^k_{,k} f^l \right) \\ + \mathbf{C}^\alpha \mathbf{C}^\beta \cdot \left(f^k_{,\alpha k} f^l_{,\beta} + f^k_{,\alpha} f^l_{,\beta l} \right) + C^k C^l \left(f^\alpha_{,\alpha k} f^\beta_{,\beta} + f^\alpha_{,\alpha} f^\beta_{,\beta l} \right) \end{array} \right)^{\delta^{k|l}}.$$

$$\tag{989}$$

This only slightly changes our results from Eqs. (985) and (986) as follows:

$$
0 = \int_V d^n x \sqrt{g} \cdot \left(\left[(\ldots)^{\delta\gamma}\, \frac{g_{\delta\gamma}}{\boxed{\cdots}} \left(M^2 \cdot f^k + C^k C^l f^\beta{}_{,\beta l} - \Delta f^k \right) - \left[(\ldots)^{\delta\gamma}\, \frac{g_{\delta\gamma}}{\boxed{\cdots}}\, g^{\alpha\beta} f^k{}_{,\alpha} \right]_{,\beta} \right] f^l{}_{,l} \right.
$$
$$
\left. + \left[(\ldots)^{\delta\gamma}\, \frac{g_{\delta\gamma}}{\boxed{\cdots}} \left(M^2 \cdot f^l + C^k C^l f^\alpha{}_{,\alpha k} - \Delta f^l \right) - \left[(\ldots)^{\delta\gamma}\, \frac{g_{\delta\gamma}}{\boxed{\cdots}}\, g^{\alpha\beta} f^l{}_{,\beta} \right]_{,\alpha} \right] f^k{}_{,k} \right), \tag{990}
$$

giving us the following "elastic" quantum gravity equation:

$$
0 = (\ldots)^{\delta\gamma}\, \frac{g_{\delta\gamma}}{\boxed{\cdots}} \left(M^2 \cdot f^k + C^k C^l f^\beta{}_{,\beta l} - \Delta f^k \right) - \left[(\ldots)^{\delta\gamma}\, \frac{g_{\delta\gamma}}{\boxed{\cdots}}\, g^{\alpha\beta} f^k{}_{,\alpha} \right]_{,\beta} \tag{991}
$$

$$\frac{\left(R^{\delta\gamma} - \frac{1}{2} R \cdot g^{\delta\gamma} + \Lambda \cdot g^{\delta\gamma} + \kappa \cdot T^{\delta\gamma}\right) g_{\delta\gamma}}{1 + C_k C_l \left(M^2 \cdot f^k f^l + C^\alpha C^\beta \cdot f^k_{,\alpha} f^l_{,\beta} + C^k C^l f^\alpha_{,\alpha} f^\beta_{,\beta}\right)} \left(M^2 \cdot f^k + C^k C^l f^\beta_{,\beta l} - \Delta f^k\right)$$

$$- \left[\frac{\left(R^{\delta\gamma} - \frac{1}{2} R \cdot g^{\delta\gamma} + \Lambda \cdot g^{\delta\gamma} + \kappa \cdot T^{\delta\gamma}\right) g_{\delta\gamma}}{1 + C_k C_l \left(M^2 \cdot f^k f^l + C^\alpha C^\beta \cdot f^k_{,\alpha} f^l_{,\beta} + C^k C^l f^\alpha_{,\alpha} f^\beta_{,\beta}\right)}\right]_{,\beta} g^{\alpha\beta} f^k_{,\alpha} = 0.$$

$$(992)$$

10.1 A Fundamental Top-Down Approach and Two Problems

It is quite clear that a most fundamental origin of the "elastic" equations could be used as some kind of top-down approach in any kind of materials science problem suitable for a continuous approach.

Considering the result in Eq. (986) or its least approximated partner in Eq. (992)—apart from the linear Ricci scalar approach, which can be generalized according to Eq. (372)—as being most general, here we want to briefly discuss its potential. We easily see that the classical fundamental equations of elasticity are obtained with $M = 0$ and:

$$\left[\frac{\left(R^{\delta\gamma} - \frac{1}{2} R \cdot g^{\delta\gamma} + \Lambda \cdot g^{\delta\gamma} + \kappa \cdot T^{\delta\gamma}\right) g_{\delta\gamma}}{1 + C_k C_l \left(M^2 \cdot f^k f^l + C^\alpha C^\beta \cdot f^k_{,\alpha} f^l_{,\beta} + C^k C^l f^\alpha_{,\alpha} f^\beta_{,\beta}\right)}\right]_{,\beta} = 0, \quad (993)$$

just leaving us with:

$$\frac{\left(R^{\delta\gamma} - \frac{1}{2} R \cdot g^{\delta\gamma} + \Lambda \cdot g^{\delta\gamma} + \kappa \cdot T^{\delta\gamma}\right) g_{\delta\gamma}}{1 + C_k C_l \left(M^2 \cdot f^k f^l + C^\alpha C^\beta \cdot f^k_{,\alpha} f^l_{,\beta} + C^k C^l f^\alpha_{,\alpha} f^\beta_{,\beta}\right)} \left(C^k C^l f^\beta_{,\beta l} - \Delta f^k\right) = 0.$$

$$(994)$$

Now we have the interesting situation that we could have either the Einstein field equations satisfied, which is to say:

$$R^{\delta\gamma} - \frac{1}{2} R \cdot g^{\delta\gamma} + \Lambda \cdot g^{\delta\gamma} + \kappa \cdot T^{\delta\gamma} = 0 \qquad (995)$$

or the elastic equations:

$$C^k C^l f^\beta_{,\beta l} - \Delta f^k = 0, \qquad (996)$$

where for a setting $C^k C^l = -\frac{c^k c^l}{1-2\cdot\nu}$ we can immediately recognize the typical elasticity terms, because we obtain:

$$\frac{c^k c^l}{1 - 2 \cdot \nu} f^\beta_{,\beta l} + \Delta f^k = 0. \qquad (997)$$

As solutions to these equations were already presented elsewhere [160–162], thereby using an extension of Neuber's famous three-function approach [182, 183], we refrain from repeating them here, but only discuss the extensions in cases of non-vanishing M and the term in Eq. (993). We are also interested in the consequences of de facto having two options for the satisfaction of Eq. (994). We shall start with the latter problem.

10.1.1 The Two Options Problem or Einstein's (and Hilbert's) True "Blunder"

If coming from the classical continuum mechanical field of elasticity, it appears most peculiar that there suddenly should be two possibilities to solve the same problem. If considering a simple contact or bending problem, it should still hold that all forces have to be balanced in order to achieve an equilibrium. It simply cannot be that the presence of the Einstein field equation term makes solving the force-equilibrium problem obsolete.

Something has to be wrong here!

Now we remember, however, that Einstein and Hilbert had artificially introduced the energy momentum tensor, respectively the Lagrange density of matter into their derivations in order to always be able to balance everything just within their term, namely Eq. (995).

Could it be possible that this actually was a mistake?

Could it be that this, in contrast to Einstein's introduction of the cosmological constant, was even the true and only blunder?

After all our more comprehensive equations, we would not need such a term. We could easily avoid the whole matter business completely and just live with a total (elastic) gravity equation of the kind:

$$
\frac{\left(R^{\delta\gamma} - \frac{1}{2} R \cdot g^{\delta\gamma} + \Lambda \cdot g^{\delta\gamma} \right) g_{\delta\gamma}}{1 + C_k C_l \left(M^2 \cdot f^k f^l + \mathbf{C}^\alpha \mathbf{C}^\beta \cdot f^k_{,\alpha} f^l_{,\beta} + C^k C^l f^\alpha_{,\alpha} f^\beta_{,\beta} \right)} \left(M^2 \cdot f^k + C^k C^l f^\beta_{,\beta l} - \Delta f^k \right)
$$
$$
- \left[\frac{\left(R^{\delta\gamma} - \frac{1}{2} R \cdot g^{\delta\gamma} + \Lambda \cdot g^{\delta\gamma} \right) g_{\delta\gamma}}{1 + C_k C_l \left(M^2 \cdot f^k f^l + \mathbf{C}^\alpha \mathbf{C}^\beta \cdot f^k_{,\alpha} f^l_{,\beta} + C^k C^l f^\alpha_{,\alpha} f^\beta_{,\beta} \right)} \right]_{,\beta} g^{\alpha\beta} f^k_{,\alpha} = 0.
$$
$$(998)$$

Whenever now the vacuum condition is not fulfilled and thus,

$$
R^{\delta\gamma} - \frac{1}{2} R \cdot g^{\delta\gamma} + \Lambda \cdot g^{\delta\gamma} \neq 0, \tag{999}
$$

the "rest" of Eq. (998), which in our vector functional approach f^k would interestingly be a quasi-elastic equation, takes over and still solves the whole ensemble. Even more interestingly, the additional terms sport all features of the classical quantum equations.

This, however, is usually seen as a quantum gravity equation, a "Theory of Everything" or "world formula"*.

10.1.2 Gravitational Fields

It is well-known from the classical theory of elasticity (e.g., [192]) that apart from the elastic field terms $\frac{c^k c^l}{1-2\cdot v} f^\beta_{,\beta l} + \Delta f^k$, we can also have external forces as well as gravitational or density fields. Thus, it is rather obvious that the other terms in Eq. (998), meaning:

$$\frac{\left(R^{\delta\gamma} - \frac{1}{2}R \cdot g^{\delta\gamma} + \Lambda \cdot g^{\delta\gamma}\right) g_{\delta\gamma}}{1 + C_k C_l \left(M^2 \cdot f^k f^l + \mathbf{C}^\alpha \mathbf{C}^\beta \cdot f^k_{,\alpha} f^l_{,\beta} + C^k C^l f^\alpha_{,\alpha} f^\beta_{,\beta}\right)} M^2 \cdot f^k$$

and

$$\left[\frac{\left(R^{\delta\gamma} - \frac{1}{2}R \cdot g^{\delta\gamma} + \Lambda \cdot g^{\delta\gamma}\right) g_{\delta\gamma}}{1 + C_k C_l \left(M^2 \cdot f^k f^l + \mathbf{C}^\alpha \mathbf{C}^\beta \cdot f^k_{,\alpha} f^l_{,\beta} + C^k C^l f^\alpha_{,\alpha} f^\beta_{,\beta}\right)}\right]_{,\beta} g^{\alpha\beta} f^k_{,\alpha},$$

$$(1000)$$

are standing for such external or gravitational fields.

The significant difference to the classical theory, however, is that now everything belongs together, which is to say, there is no "external" or "gravitational", there is only one metric and its variational vector function $\mathbf{f}^k$, apparently making it all.

In other words: The elastic fields themselves produce gravitation or gravity fields and this then couples back into the elastic deformation.

10.2 The "Elastic" Dirac Equation: Fixing the Sign Convention

Taking our hypothesis and considerations from the previous sections, we know that the variation of a constant should also give a zero outcome. Thus, taking Eq. (989), we should also be able to fulfill the total variational task by

---

*This author does not like those grandiloquent expressions, especially as they do not even remotely cover the situation. But as they are so widely used in connection with combinations of Einstein field and quantum equations, we dare to mention them here, too, but insist that this is only be done for completeness and recognition.

demanding:

$$B^{kl} = C^k C^l + M^2 \cdot f^k f^l + \mathbf{C}^\alpha \mathbf{C}^\beta \cdot f^k_{,\alpha} f^l_{,\beta} + C^k C^l f^\alpha_{,\alpha} f^\beta_{,\beta}$$
$$= C^k C^l + M^2 \cdot f^k f^l + g^{\alpha\beta} \cdot f^k_{,\alpha} f^l_{,\beta} + C^k C^l f^\alpha_{,\alpha} f^\beta_{,\beta}, \quad (1001)$$

with B^{kl} being a matrix of constants.

As only first-order derivatives play a role in Eq. (1001), we may consider this a Dirac-type quantum gravity equation.

Solutions to this equation for the "particle at rest" situation were already discussed in section "5.2.2.1 Towards "Dirac" again and a metric Pauli exclusion principle". Another interesting solution would be the Schwarzschild metric with an assumed purely r-dependent **f**-vector, where—only for simplicity—we set all components to zero and just demand $f^1 = f[r]$. In addition we assume $C^k = 0$ and set $B^{11} = Q^2$.

$$f[r] = \frac{\pm Q \cdot \tan\left[M \cdot \left(\sqrt{r \cdot (r - r_s)} - \left(r_s \cdot \text{artanh}\left[\sqrt{\frac{r}{r - r_s}}\right] \pm C_r\right)\right)\right]}{M \cdot \sqrt{\sec\left[M \cdot \left(\sqrt{r \cdot (r - r_s)} - \left(r_s \cdot \text{artanh}\left[\sqrt{\frac{r}{r - r_s}}\right] \pm C_r\right)\right)\right]^2}}$$

$$= \pm \frac{Q}{M} \cdot \tan\left[M \cdot \left(\sqrt{r \cdot (r - r_s)} - \left(r_s \cdot \text{artanh}\left[\sqrt{\frac{r}{r - r_s}}\right] \pm C_r\right)\right)\right]$$

$$\times \sqrt{\cos\left[M \cdot \left(\sqrt{r \cdot (r - r_s)} - \left(r_s \cdot \text{artanh}\left[\sqrt{\frac{r}{r - r_s}}\right] \pm C_r\right)\right)\right]^2}.$$
$$(1002)$$

In order to guarantee that the function does not grow infinitely with r, this solution requires a positive metric r-component from Eq. (1001) resulting in the equation:

$$B^{kl} = Q^2 = M^2 \cdot f^k f^l + \frac{f^k_{,\alpha} f^l_{,\beta}}{1 - \frac{r_s}{r}}. \quad (1003)$$

This automatically fixes the sign convention to {-,+,+,+} for the dimensions {time t, space x, space y, space z}.

Comparison with Eq. (804) shows the similarity with the particle at rest, but also reveals a principal difference. So, apart from the fact that we do not have a time-dependency, we now also find a structural transition position for $r = r_s$. Figure 10.1 shows the distribution along r for the cases $M = r_s = 1$, $Q = \pm1$ and $C_r = \pm\pi/4$. We recognize the structural transition at $r = r_s$ and zoom it out in Fig. 10.2.

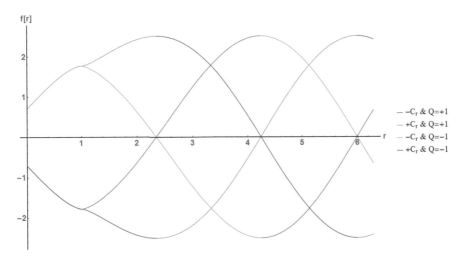

Figure 10.1 $f[r]$ for a Schwarzschild metric according to our solution (Eq. (1002)). Only the real parts of $f[r]$ are shown.

Just as we saw it with the particle at rest solution with the time argument, we find again that we can obtain the coding of matter and anti-matter character just completely via the Q, obviously being the elementary charge. In order to still obtain four non-degenerate* solutions, we apply the parameter C_r in $\pm$ as shown in Fig. 10.4.

In order to investigate the effect of Schwarzschild radius r_s and mass M, we vary these parameters simultaneously in Fig. 10.3. We observe higher frequencies and lower amplitudes of the oscillations of $f[r]$ with increasing masses $M = r_s$. We also see that with increasing masses the absolute distortion (and nothing else is $f[r]$) increases.

Now we want to investigate whether we can also obtain the right behavior for spin-combinations, by assuming (similar to our investigations from section "5.2.2.1 Towards "Dirac" again and a metric Pauli exclusion principle") that the parameter C_r has something to do with spin.

---

*Please note that in contrast to the classical quantum theory, here we use the expression "degenerate" not for undistinguishable energy states but for undistinguishable f-solutions.

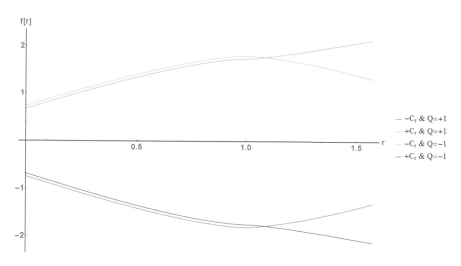

Figure 10.2 Zoomed region $r < r_s$ of $f[r]$ for a Schwarzschild metric according to our solution (Eq. (1002)). For illustration, we have separated the otherwise degenerated solutions 1/4 and 2/3 by a tiny artificial difference. Only the real parts of $f[r]$ are shown.

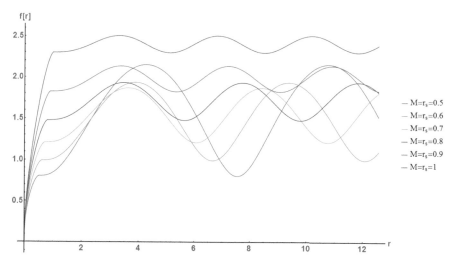

Figure 10.3 $f[r]$ for a Schwarzschild metric according to our solution (Eq. (1002)) for a variety of "masses" $M = r_s$ parameter choices. In contrast to the two previous figures, we have shown the absolute values of $f[r]$.

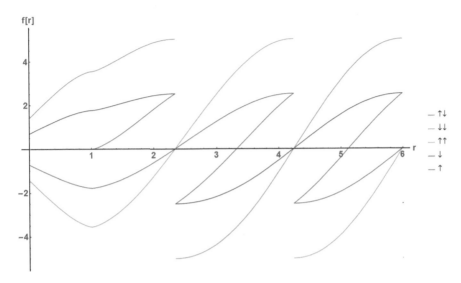

Figure 10.4 "Ripped photon" solutions to Schwarzschild particles obtained from Eq. (1002) only for $Q = +1$, but $C_r = \pm\pi/4$ in various spin-pairings. For comparison, we also show the single up and down states again.

Just as we found it for the particle at rest, it is clear from Figs. 10.1 and 10.2 that matter and anti-matter solutions cancel each other out. Thus, as it is the more interesting situation, we are now investigating combinations of matter solutions with different C_r-states. This is shown in Fig. 10.4, where we clearly see that anti-parallel mixtures (blue) are leading to significantly lower overall (integrated) oscillations than the parallel pairings (yellow and green) and even the singular states (red and lilac in Fig. 10.4). Associating, as one classically does, the integrated oscillations with energy, the anti-parallel pairings are significantly favored against parallel ones. This clearly is spin behavior of elementary particles, following the Pauli exclusion principle [175] and we might conclude to have found C_r as an elementary spin-parameter, but this time it is for a Schwarzschild object and not for the classical Dirac particle at rest.

SCALES AND ZERO-SUMS

Chapter 11

The Origin of Time

Some of the aspects and results we are going to present in this section have already been shown previously in this book. But as the topic merits a compact presentation, we are risking a bit of redundancy and repeat them here. This will also help to get a holistic picture and supports the understanding.

11.1 An Introduction in a Somewhat Informal and Illustrative Form

We take it that the reader is familiar with the problem of quantum reality, measurement and the so-called collapse of the wave function.

One could say that an electron is not there until someone measures it.

Bohr and Einstein had quite some "goes" about this and Bohr always won. Einstein, however, never liked the whole thing (he was even cited with: "the moon isn't only there when somebody looks at it") and when he finally came out with his Einstein–Podolsky–Rosen, or EPR, paradox [176, 177], everybody thought he had a point. But then John Bell came with his famous inequality equation [178]; measurements were performed on the EPR problem and the EPR paradox was proven to be a reality (bitter to some, though) [179–181].

The World Formula: A Late Recognition of David Hilbert's Stroke of Genius
Norbert Schwarzer
Copyright © 2022 Jenny Stanford Publishing Pte. Ltd.
ISBN 978-981-4877-20-6 (Hardcover), 978-1-003-14644-5 (eBook)
www.jennystanford.com

This author often wondered about the "conflict", because it has happened so often in science that in the end someone found out that they all had missed something, meaning, a simple change of paradigm would perhaps solve the problem. To answer this question, one should first ask, what could have been the cause for the problem?

Obviously, it was/is our understanding of time and space!

Let us here concentrate solely on the problem of what is time. This author was of the opinion it all had to do with our notion (understanding) of time. In both QT and GTR, namely, time has been planted into the theories like a geranium into a flowerpot. None of the two theories properly answers the question where it comes from. On the other hand, the Einstein field equations get perfectly along without any time. Flat space does not need time and the fact that we have introduced the Minkowski metric simply was to assure the right Lorentz transformations. In order to do so, one has to introduce the so-called sign convention, where the time component of the metric has a different sign than the spatial components. And now putting the problem of the appearance of time very simply one could ask:

Where does the minus (in fact, it is $-c^2$, with c denoting the speed of light in vacuum) for the metric time component come from?

In order to answer that question, we started with a simple flat space, where all metric components are equal (just like an $n-D$ Cartesian space) [123]. Any quantization with functions of the type:

$$f[x, y, z, \ldots] = f[cx\,{}^*x + xy\,{}^*y + cz\,{}^*z + \ldots]$$

leads to conditions $cx^2 + cy^2 + cz^2 + \ldots = 0$, which means that at least one coordinate must become time-like.

Fine! So quantization brings time, but with time being so omnipresent, perhaps there even is a deeper layer.

Let's go back to the Cartesian flat space and simply assume that two coordinates are forced to do something together. Let's "entangle" them in a most general manner. Our reasoning for this conjecture: If something happens, then it has to mean a change of one dimension with respect to the other. Thus, entangling them would assure this happening, right? Doing this mathematically and looking for the most general solution for the Einstein field equations forces two things to happen:

(a) The entanglement becomes a wave (moving forward and backward).
(b) One of the entangled dimensions must become time-like.

Note that this dimension becomes time-like due to the entanglement; it did not had to be so before.

What has this to do with the measurement, the wave function collapse and all this?

Well, here comes in the change of paradigm. The states of entanglement are the states with time. These are also the states of the wave functions. Then the process of measurement destroys the entanglement, thus collapses the wave function assuring the entanglement and provides a certain metric as (temporarily static) result. With time itself doing the coupling (and only becoming time due to it), all the "ghostly or spooky interaction at a distance" problems are solved. There is also no need for Einstein's "hidden parameters" as the communication is a direct metric one. The components of the one dimension taking on the position of time and the one staying spatial are forming ONE system or entity (or "effective dimension") as long as the entanglement is not broken up.

Simply changing the paradigm of time from being something which is always there to something which comes and goes together with processes and action, meaning, not the action goes with time, but it actually makes the time, brings in quite some understanding of originally very fishy things.

Interestingly, the moment one "activates" the mechanics of "the process triggers time and not the other way round", one automatically gets a very fundamental solution within the apparatus of this concept. The protagonists of the "holographic principle" will probably like to hear it: This fundamental solution is based on a 2D, which is to say pairwise entanglement. Investigating the properties of this fundamental and most omnipresent solution gives:

(a) the Heisenberg principle
(b) an upper speed limit, which is the speed of light, of course and
(c) a connection between smallest resolvable structures and the kurtosis of the solution**

**Meaning, with this entanglement solution being so omnipresent, there is principally no way to access structures underneath a certain scale.

In other words:

There is a limit in size to things we can touch or, as some more religious people might like to put it, "God is hiding something from our curious gaze"!

11.2 A Variety of Timely Dimensions

As it was shown in the most recent papers [127, 128, 129] of the series of "Science Riddles" that matter fields can just be obtained via suitably entangled dimensions and as in this context also the flatness problem of the universe [182, 183] could be solved [130], we here want to investigate the option of entangled dimensions also with respect to the character of time.

It was demonstrated in [123] that we always need at least one coordinate being of "different" (imaginary) character in order to fulfill fundamental quantum equations resulting from an extended variation of the Einstein–Hilbert action with respect to the dimension of space [116, 145].

A similar result can be obtained directly as solution for the Einstein field equations for any metric of diagonal shape subjected to a transformation of the kind:

$$G_{\alpha\beta} = F\left[f\left[t, x, y, z, \ldots, \xi_k, \ldots, \xi_n\right]\right]^{ij}_{\alpha\beta} g_{ij}$$
$$\rightarrow G_{\alpha\beta} = F\left[f\left[t, x, y, z, \ldots, \xi_k, \ldots, \xi_n\right]\right] \cdot \delta^i_\alpha \delta^j_\beta g_{ij}. \qquad (1004)$$

Here the functions F have to be set to $F\left[f[\ldots]\right] = f[c_{t=0} {}^* t + c_{x=1} {}^* x + \ldots + c_k {}^* \xi_k]$ with the condition:

$$\sum_{i=0}^{n} c_i^2 = 0. \qquad (1005)$$

Please note that the indices i, j do not necessarily have to run from 0 to n, with n denoting the dimension of the space-time we are considering. In fact, also the functions $f[\ldots]$ might depend only on those coordinates we want to appear entangled. This gives us $F\left[f[\ldots]\right] = f[c_p {}^* \xi_p + \ldots + c_i {}^* \xi_i + \ldots + c_k {}^* \xi_k]$ with

$$G_{\alpha\beta} = \sum_{i,j=p}^{k} F\left[f\left[\xi_p, \ldots, \xi_i, \ldots, \xi_k\right]\right] \cdot \delta^i_\alpha \delta^j_\beta g_{ij} \qquad (1006)$$

and the condition:

$$\sum_{i=p}^{k} c_i^2 = 0. \qquad (1007)$$

According to the quantum approaches in [123], we need at least one coordinate to be of imaginary character and thus, sticking out this way from the others. The character of the function for $g[\ldots]$ depends on the dimension of the metric it is applied on. Here is an example.

Introducing a metric for the coordinates $t, \tau, T, r, \vartheta, \varphi$ of the form:

$$g^6_{\alpha\beta} = \begin{pmatrix} g[t, \tau, T] \equiv g & 0 & 0 & 0 & 0 & 0 \\ 0 & g & 0 & 0 & 0 & 0 \\ 0 & 0 & g & 0 & 0 & 0 \\ 0 & 0 & 0 & 1 & 0 & 0 \\ 0 & 0 & 0 & 0 & r^2 & 0 \\ 0 & 0 & 0 & 0 & 0 & r^2 \cdot \sin[\vartheta] \end{pmatrix} \cdot f[r, \vartheta, \varphi], \quad (1008)$$

with the following function for $g[\ldots]$:

$$g[t, \tau, T] = \frac{C_1}{(c_0 \cdot t + c_1 \cdot \tau + c_2 \cdot T + C_2)^2}; \quad c_0^2 + c_1^2 + c_2^2 = 0; \quad f[\ldots] = 1,$$
$$(1009)$$

directly solves the vacuum Einstein field equations. We see that at least one of the c_i has to be imaginary and thus would make the corresponding coordinate t, τ or T to appear time-like. Choosing a general $f[\ldots]$ as introduced in Eq. (1008) and demanding $c_0^2 + c_1^2 + c_2^2 = 0$ as set in Eq. (945) in order to fulfill the Einstein field equations gives us the following Ricci scalar from [14]:

$$R^* = \frac{1}{F[f]^3} \cdot \left(-C_{N2} \cdot F[f] \cdot \frac{\partial F[f]}{\partial f} \cdot \Delta_g f \right); \quad F[f] = C_f + f[r, \vartheta, \varphi]$$

$$\xrightarrow{6D} R^* = -\frac{1}{f^2} \cdot \left(6 \cdot \frac{(c_0^2 + c_1^2 + c_2^2)}{C_1} \cdot f + 5 \cdot \Delta_{3D-sphere} f \right); \quad f = f[r, \vartheta, \varphi];$$

$$C_f = 0. \quad (1010)$$

Here, $\Delta_{3D-sphere} f$ denotes the Laplace operator in spherical coordinates.

As for $R^* = 0$ and $6 \cdot \frac{(c_0^2 + c_1^2 + c_2^2)}{C_1} = \frac{C_{v0}^2}{5} = \frac{m^2 \cdot c^2}{\hbar^2}$ (m = mass, c = speed of light in vacuum, $\hbar$ = reduced Planck constant), we clearly have obtained the quantum Klein–Gordon equation and we can easily conclude that mass can obviously be constructed by a suitable set of additional, entangled coordinates. Please note that in contrast to the entanglement of 2 dimensions resulting in mass-terms (e.g., [127] and section 5.1 here), where no time-like coordinate was needed, this time we obtain a bell-shaped localized distribution for $g[\ldots]$ only at certain positions of the time-like coordinate. At these positions, let us choose here $t = t_0$; however, it can be easily assured that the additional coordinates T and τ are compactified, because the arbitrary constants C_1 and C_2 can always be chosen in such a way that $g[t = t_0, \ldots]$ takes on the character of delta-like distributions. Thus, the coordinates T and τ would be omnipresent in the space r, ϑ, φ, but do not appear in elongated form (on bigger scale), which means that T and τ are invisible but only act via the production of mass in a $3 + 1$ space-time.

It can be shown that one can always find vacuum solutions to any kind of entanglement approach (Eq. (1006)) in an arbitrary dimensional space-time with arbitrary many dimensions being entangled. We always find solutions of the type $g[...]$ as given in Eq. (945) or Eq. (1009). So, for instance, if entangling 4 dimensions in a 6-dimensional space with Cartesian coordinates t, τ, T, x, y, z, we would have:

$$g^6_{\alpha\beta} = \begin{pmatrix} g\,[t,\tau,T,x] \equiv g & 0 & 0 & 0 & 0 & 0 \\ 0 & g & 0 & 0 & 0 & 0 \\ 0 & 0 & g & 0 & 0 & 0 \\ 0 & 0 & 0 & g & 0 & 0 \\ 0 & 0 & 0 & 0 & 1 & 0 \\ 0 & 0 & 0 & 0 & 0 & 1 \end{pmatrix}, \tag{1011}$$

with the following solution function for $g[...]$:

$$g\,[t,\tau,T,x] = \frac{C_1}{c_0 \cdot t + c_1 \cdot \tau + c_2 \cdot T + c_3 \cdot x + C_2}; \quad c_0^2 + c_1^2 + c_2^2 + c_3^2 = 0. \tag{1012}$$

The entanglement of 6 dimensions as chosen here:

$$g^6_{\alpha\beta} = \begin{pmatrix} g\,[t,\tau,T,x,y,z] \equiv g & 0 & 0 & 0 & 0 & 0 \\ 0 & g & 0 & 0 & 0 & 0 \\ 0 & 0 & g & 0 & 0 & 0 \\ 0 & 0 & 0 & g & 0 & 0 \\ 0 & 0 & 0 & 0 & g & 0 \\ 0 & 0 & 0 & 0 & 0 & g \end{pmatrix} \tag{1013}$$

requires:

$$g\,[t,\tau,T,x,y,z] = \frac{C_1}{(c_0 \cdot t + c_1 \cdot \tau + c_2 \cdot T + c_3 \cdot x + c_4 \cdot y + c_5 \cdot z + C_2)^2}$$
$$c_0^2 + c_1^2 + c_2^2 + c_3^2 + c_4^2 + c_5^2 = 0 \tag{1014}$$

if one still wants to solve the vacuum Einstein field equations.

In order to find the simplest of such solutions, where obviously only 2 dimensions are entangled, we assume a Cartesian system in n dimensions t,

$x_1, x_2,\ldots$ and combine 2 dimensions as follows:

$$g_{\alpha\beta}^n = \begin{pmatrix} F\,[f\,[t,\,x_1]] & 0 & 0 & 0 & \cdots & 0 \\ 0 & F\,[f\,[t,\,x_1]] & 0 & 0 & \cdots & 0 \\ 0 & 0 & 1 & 0 & \cdots & 0 \\ 0 & 0 & 0 & 1 & \cdots & 0 \\ & \cdots & \cdots & \cdots & \ddots & 0 \\ 0 & 0 & 0 & 0 & 0 & 1 \end{pmatrix}. \qquad (1015)$$

A very general solution can be found via:

$$F\,[f] = C_1 \cdot e^{C_2 \cdot f}; \quad f\,[t,\,x_1] = C_+\,[x_1 + i \cdot t] + C_-\,[x_1 - i \cdot t]. \qquad (1016)$$

Thereby C_1 and C_2 denote arbitrary constants and the $C_\pm[\ldots]$ stand for arbitrary functions. We recognize the characteristic wave functions. As before, we have 1 dimension (here we chose t), which had to become time-like in order to fulfill the Einstein field equations. In general, we can have many such 2-dimensional entanglements within one and the same metric, via:

$$g_{\alpha\beta}^n = \begin{pmatrix} F\,[f_0\,[t,\,x_1]] & 0 & 0 & 0 & \cdots & 0 \\ 0 & F\,[f_0\,[t,\,x_1]] & 0 & 0 & \cdots & 0 \\ 0 & 0 & F\,[f_1\,[x_1,\,x_2]] & 0 & \cdots & 0 \\ 0 & 0 & 0 & F\,[f_1\,[x_1,\,x_2]] & \cdots & 0 \\ & \cdots & \cdots & \cdots & \ddots & 0 \\ 0 & 0 & 0 & 0 & 0 & 1 \end{pmatrix} \qquad (1017)$$

or

$$g_{\alpha\beta}^n = \begin{pmatrix} F\,[f_0\,[t,\,x_1]] & 0 & 0 & 0 & \cdots & 0 \\ 0 & F\,[f_2\,[x_1,\,x_2]] & 0 & 0 & \cdots & 0 \\ 0 & 0 & F\,[f_2\,[x_1,\,x_2]] & 0 & \cdots & 0 \\ 0 & 0 & 0 & F\,[f_0\,[x_1,\,x_2]] & \cdots & 0 \\ & \cdots & \cdots & \cdots & \ddots & 0 \\ 0 & 0 & 0 & 0 & 0 & 1 \end{pmatrix} \qquad (1018)$$

and so on, with all $f_i[\ldots]$ being of the wave-kind as given in Eq. (1016).

It should be noted that the solutions of the kind Eqs. (945), (1013), (1017) and (1018) are taking on very complicated forms in curvilinear coordinates. The usual transformation rules, as given here for the covariant metric tensor

$$g_{\delta\gamma} = \frac{\partial G^\alpha\,[f\,[x_k]]}{\partial x^\delta}\frac{\partial G^\beta\,[f\,[x_k]]}{\partial x^\gamma} g_{\alpha\beta}, \qquad (1019)$$

lead to fulfilled complex structures even in simple coordinate systems like the spherical one. The author was therefore not able to find a suitable entanglement solution, e.g., for the Schwarzschild metric. Perhaps the skilled and interested reader might see to find such a solution. Nevertheless, we intend to discuss spherical objects with respect to their time dimension(s) further below in this chapter.

11.3 Interpretation of the Result

Seeing any process or action as the change of a property/dimension with respect to another property/dimension automatically leads to an entanglement of the two. As this can also be seen as being the case for more properties/dimensions changing with respect to others, we can—in principle—describe all processes and actions as entanglement. As shown above, in this case at least one of the entangled properties/dimensions has to become time-like. It is important to note that such a time-like appearance thereby comes with the entanglement (and thus the process or action). It has not necessarily been there before. Destruction of the entanglement—e.g., caused by a measurement process—breaks up the connection between the two (or more) connected properties/dimensions and thereby also freezes the formerly time-like property/dimension at a certain (non-entangled) metric state.

With time itself being part of the entanglement, it also becomes clear why the EPR paradox is no paradox [177] at all. Two quantum objects being created in one process and being separated over a long distance are connected as long as they are entangled. The entanglement even leads to the time the process "has". We easily see this when observing the solution in Eq. (1016). There we have two waves which would run apart, but with the process time t itself providing the connection, "communication" among the two objects is assured as long as the wave function is not destroyed (e.g., due to a measurement process).

11.4 Principal Consequences

Assuming the simplest and fundamental 2D-entanglement solutions (Eq. (1015)) to be omnipresent within this universe (at least as omnipresent as there seems to be time), we now want to evaluate what consequence such solutions would have on the properties of our universe.

We demand the functions $f[...]$ of the 2D-entanglements (Eq. (1016)) to be such that integration over the whole space is possible and that an expectation value and a variance do exist. The simplest form for such a function would then be of the kind:

$$f[t, x_1] = -c_0 \cdot ((x_1 - x_0) + c \cdot (t - t_0))^2 - c_0 \cdot ((x_1 - x_0) - c \cdot (t - t_0))^2$$
$$= -2 \cdot c_0 \cdot ((x_1 - x_0)^2 + c^2 (t - t_0)^2), \tag{1020}$$

where we find the requested integrability for real $c_0 > 0$. Please note that for simplicity and recognition, we have applied the general solution (1016), but substituted $t \rightarrow i^* c^* t$, leading to the solutions (1020) and the familiar metric form with the following sign structure:

$$g_{\alpha\beta}^n = \begin{pmatrix} -c^2 \cdot F[f[t, x_1]] & 0 & 0 & 0 & \cdots & 0 \\ 0 & F[f[t, x_1]] & 0 & 0 & \cdots & 0 \\ 0 & 0 & 1 & 0 & \cdots & 0 \\ 0 & 0 & 0 & 1 & \cdots & 0 \\ \cdots & \cdots & \cdots & \cdots & \ddots & 0 \\ 0 & 0 & 0 & 0 & 0 & 1 \end{pmatrix}. \tag{1021}$$

For simplicity, we shall denote $x_1 = x$. Setting C_2 in Eq. (1016) $C_2 = 1$, we have the total function $F[...]$ as:

$$F[f] = C_1 \cdot e^{-2 \cdot c_0 \cdot ((x-x_0)^2 + c^2(t-t_0)^2)} = C_1 \cdot e^{-2 \cdot c_0 \cdot ((x-x_0)^2 + c^2(t-t_0)^2)}. \tag{1022}$$

The expectation values of $F[...]$ for x and t can be given via the integrals:

$$E[x] = \int_{-\infty}^{+\infty} \int_{-\infty}^{+\infty} x \cdot F[f[t, x]] \, dx \cdot dt$$

$$E[t] = \int_{-\infty}^{+\infty} \int_{-\infty}^{+\infty} t \cdot F[f[t, x]] \, dx \cdot dt, \tag{1023}$$

which we demand to be x_0 and t_0, respectively. It should be noted that the boundaries of the integration could be a reason for discussion, because—as said—time only comes in existence in the moment of action, which is to say, in the moment the entanglement between t and x is activated. It also ends with the entanglement and thus, should have a finite starting and ending point. Here, however, for simplicity, we explicitly intend to avoid such a discussion and assume the time of the entanglement to be active significantly longer than the standard deviation from the expectation value, being given as the square root of $E[t^2] = \Delta t^2$ (see below in Eq. (1025)). In other words: As long as Δt is much smaller than the total duration of the entanglement, we can apply Eq. (1023). This fixes C_1 to:

$$C_1 = \frac{2 \cdot c_0 \cdot c}{\pi} \tag{1024}$$

and results in the following variances:

$$\Delta x^2 = \int\limits_{-\infty}^{+\infty} \int\limits_{-\infty}^{+\infty} x^2 \cdot F\left[f\left[t, x\right]\right] dx \cdot dt = x_0^2 + \frac{1}{4 \cdot c_0}$$

$$\Delta t^2 = \int\limits_{-\infty}^{+\infty} \int\limits_{-\infty}^{+\infty} t^2 \cdot F\left[f\left[t, x\right]\right] dx \cdot dt = t_0^2 + \frac{1}{4 \cdot c^2 \cdot c_0}$$

$$\Delta x \cdot \Delta t = \int\limits_{-\infty}^{+\infty} \int\limits_{-\infty}^{+\infty} t \cdot x \cdot F\left[f\left[t, x\right]\right] dx \cdot dt = x_0 \cdot t_0. \tag{1025}$$

In addition, we know from [36] that the distribution $F[f]$, investigated in connection with the Cauchy–Schwarz or Hölder inequality, which is given as:

$$|E\left[x \cdot t\right]|^2 \leq E\left[x^2\right] \cdot E\left[t^2\right], \tag{1026}$$

results in a Heisenberg uncertainty principle. Using Eq. (1025) and setting it into Eq. (349), we have:

$$|x_0 \cdot t_0|^2 \leq \left(x_0^2 + \frac{1}{4 \cdot c_0}\right) \cdot \left(t_0^2 + \frac{1}{4 \cdot c^2 \cdot c_0}\right), \tag{1027}$$

where we easily see that this is fulfilled for $c > 0$ and $c_0 > 0$. Now we want to go on from there and intend to obtain other principal limits. First, we form the square roots in Eq. (1027):

$$|x_0 \cdot t_0| \leq \sqrt{\left(x_0^2 + \frac{1}{4 \cdot c_0}\right) \cdot \left(t_0^2 + \frac{1}{4 \cdot c^2 \cdot c_0}\right)}. \tag{1028}$$

Division of both sides in Eq. (1028) by Δt^2 leads to:

$$\left(\frac{\Delta x}{\Delta t}\right)_{\text{min}} = \frac{|x_0|}{\sqrt{\left(t_0^2 + \frac{1}{4 \cdot c^2 \cdot c_0}\right)}} \leq \sqrt{\frac{\left(x_0^2 + \frac{1}{4 \cdot c_0}\right)}{\left(t_0^2 + \frac{1}{4 \cdot c^2 \cdot c_0}\right)}} = \left(\frac{\Delta x}{\Delta t}\right)_{\text{max}} \equiv c. \tag{1029}$$

Thereby we have obtained an upper limit for a velocity and assume that it shall be the maximum velocity in our universe, which is the speed of light in vacuum c. We easily see that the condition $x_0 = c^* t_0$ gives the correct result on the right hand side of our inequality (1029) with:

$$\sqrt{\frac{\left(x_0^2 + \frac{1}{4 \cdot c_0}\right)}{\left(t_0^2 + \frac{1}{4 \cdot c^2 \cdot c_0}\right)}} = \sqrt{\frac{\left(c^2 \cdot t_0^2 + \frac{1}{4 \cdot c_0}\right)}{\left(t_0^2 + \frac{1}{4 \cdot c^2 \cdot c_0}\right)}} = \sqrt{\frac{c^2 \cdot \left(t_0^2 + \frac{1}{4 \cdot c^2 \cdot c_0}\right)}{\left(t_0^2 + \frac{1}{4 \cdot c^2 \cdot c_0}\right)}} = c. \quad (1030)$$

Knowing that the dimension of c_0 is meters^{-2} and having demanded above that it should be a huge value, we might simply assume that $c_0 = 1/x_0^2 = 1/\ell_P^2$. For completeness, however, we should also look for other possibilities. So, with the maximum speed allocated and still one constant left to be determined, we might wonder what the smallest possible velocity could be. A suitable assumption could be the quotient of the smallest length scale in our universe (Planck length $\ell_P = \sqrt{\frac{\hbar \cdot G}{c^3}} \approx 1.616229 \times 10^{-35}$ meters) and the age of the universe, being $A = 13.81 \times 10^9$ years. From Eq. (1029), we get the equation:

$$\left(\frac{\Delta x}{\Delta t}\right)_{min} \equiv \frac{\ell_P}{A} = \frac{|x_0|}{\sqrt{\left(t_0^2 + \frac{1}{4 \cdot c^2 \cdot c_0}\right)}} = \frac{c \cdot |x_0|}{\sqrt{\left(x_0^2 + \frac{1}{4 \cdot c_0}\right)}} = \frac{c \cdot \ell_P}{\sqrt{\left(\ell_P^2 + \frac{1}{4 \cdot c_0}\right)}}$$

$$(1031)$$

and obtain:

$$\left(\frac{\Delta x}{\Delta t}\right)_{min} \equiv \frac{\ell_P}{A} = 3.711 \times 10^{-53} \frac{\text{meters}}{\text{second}} = \frac{c \cdot \ell_P}{\sqrt{\left(l_P^2 + \frac{1}{4 \cdot c_0}\right)}}$$

$$\Rightarrow c_0 = 1.46655 \times 10^{-53} \text{meters}^{-2}. \quad (1032)$$

We recognize not only the right dimension of the cosmological constant Λ (which is no surprise, of course), but also the suitable value in close proximity to the known Λ ($\Lambda = 1.1056 \times 10^{-52}$ meters^{-2}, en.wikipedia.org/wiki/Cosmological_constant). This value, however, would be in total contrast to our assumption of the small standard deviation of the expectation value as discussed above in connection with the integration boundaries.

Thus, so our temporary conclusion, our currently best suggestion for the constant c_0 would be:

$$c_0 = \frac{1}{x_0^2} \xrightarrow{x_0 = \ell_P} = \frac{1}{\ell_P^2}. \quad (1033)$$

This, however, is not very satisfying as—apart from the dimensional aspect—it does not give any structural or metrical hint about the origin or limits of c_0. Thus, in order to find reasonable metric conditions for the determination of c_0, we also consider the third (skewness) and the fourth (kurtosis) moment:

$$\int_{-\infty}^{+\infty}\int_{-\infty}^{+\infty} x^3 \cdot F\,[f\,[t, x]]\,dx \cdot dt = x_0^3 + \frac{3 \cdot x_0}{4 \cdot c_0}$$

$$\int_{-\infty}^{+\infty}\int_{-\infty}^{+\infty} t^3 \cdot F\,[f\,[t, x]]\,dx \cdot dt = t_0^3 + \frac{3 \cdot t_0}{4 \cdot c \cdot c_0}, \qquad (1034)$$

$$\int_{-\infty}^{+\infty}\int_{-\infty}^{+\infty} x^4 \cdot F\,[f\,[t, x]]\,dx \cdot dt = x_0^4 + \frac{3 \cdot x_0^2}{2 \cdot c_0} + \frac{3}{16 \cdot c_0^2}$$

$$\int_{-\infty}^{+\infty}\int_{-\infty}^{+\infty} t^4 \cdot F\,[f\,[t, x]]\,dx \cdot dt = t_0^4 + \frac{3 \cdot t_0^2}{2 \cdot c^2 \cdot c_0} + \frac{3}{16 \cdot c^4 \cdot c_0^2}. \qquad (1035)$$

We see that the skewness would have the right dimension for the construction of the gravitational or Newton's constant $G = 6.672 \times 10^{-11}$ $\mathrm{m^3 kg^{-1} s^{-2}}$ except that we currently miss a proper way to describe mass. One might, however, consider the kurtosis for x as evaluated in Eq. (1035) to provide the variance of the surface area, which we assume to be ℓ_P^2. As a result, we obtain an unreasonable negative value for c_0:

$$c_0 = -\frac{1}{8 \cdot x_0^2} \xrightarrow{x_0 = \ell_P} = -\frac{1}{8 \cdot \ell_P^2}. \qquad (1036)$$

When assuming the kurtosis to be:

$$c_k \cdot x_0^4 = \int_{-\infty}^{+\infty}\int_{-\infty}^{+\infty} x^4 \cdot F\,[f\,[t, x]]\,dx \cdot dt = x_0^4 + \frac{3 \cdot x_0^2}{2 \cdot c_0} + \frac{3}{16 \cdot c_0^2}, \qquad (1037)$$

with an yet to fix constant c_k, we obtain:

$$c_0 = \frac{3}{4 \cdot x_0^2 \cdot \left(\sqrt{3 \cdot (2 + c_k)} - 3\right)} \xrightarrow{x_0 = l_P} = \frac{3}{4 \cdot l_P^2 \cdot \left(\sqrt{3 \cdot (2 + c_k)} - 3\right)}. \qquad (1038)$$

This gives us reasonable values c_0 for all $c_k > 1$ and means that, in order to obtain reasonable values for the constant c_0, the kurtosis of our distributions must be positive and bigger than:

$$x_0^4 < \int_{-\infty}^{+\infty}\int_{-\infty}^{+\infty} x^4 \cdot F\,[f\,[t, x]]\,dx \cdot dt = x_0^4 + \frac{3 \cdot x_0^2}{2 \cdot c_0} + \frac{3}{16 \cdot c_0^2}. \qquad (1039)$$

Geometrically this means that our distribution (1022) has to be more dominated by the outer regions than a corresponding normal distribution.

As it is reasonable to consider the constant c_0 being independent on the expectation value x_0, we should simply demand that:

$$\ell_P^4 < \int\limits_{-\infty}^{+\infty} \int\limits_{-\infty}^{+\infty} x^4 \cdot F\left[f\left[t, x\right]\right] dx \cdot dt = \frac{3}{16 \cdot c_0^2}. \qquad (1040)$$

11.5 What Happens with Time in a Black Hole?

It was shown in [116, 145, 132] and section 3.8 in this book that by extending the variation of the Einstein–Hilbert action with respect to the number of the dimension—in principle—we also seek a maximum volume or surface for a given dimension. For instance, taking the radius of a Schwarzschild object, we would look for the corresponding dimension making the volume* or surface integral an extremum. As the determinant g of the Schwarzschild metric is just equal to the one of an n-sphere with the additional time-dimension to be integrated, we can easily use the surface integral of n-spheres, which reads:

$$S_n = 2 \cdot \frac{\pi^{\frac{n}{2}}}{\Gamma\left[\frac{n}{2}\right]} \cdot r^{n-1}. \qquad (1041)$$

11.5.1 The Bekenstein Bit-Problem [146]

In [145], where we wanted to investigate the problem of finding the most fundamental Turing machine, we also had to take care of the question how the universe stores information. One of the most famous problems in this context has been seen in the Bekenstein Bit-problem, where it was found that black holes can store information, but it has been a mystery how these objects actually do this. In [116, 145] and section 3.8 here, we have shown that bit-like information is stored as dimensions and that each bit becomes 1 dimension. For convenience, we will repeat parts of the evaluation here.

In the early 1970s, J. Bekenstein [147] investigated the connection between black hole surface area and information. Thereby he simply

---

*While using the volume integral in [116, 132, 145], we here intend to show that similar results are also possible with the surface. The reason for this follows from the well-known recurrence relations between surface and volume for n-spheres (e.g. en.wikipedia.org/wiki/N-sphere).

considered the surfaces change of a black hole which would be hit by a photon just of the same size as the black hole. His idea was that with such a geometric constellation the outcome of the information would just consist of the information whether the photon fell into the black hole or whether it did not. Thus, it would be a 1-bit information. His calculations led him to the funny proportionality of area and information. He found that the number of bits, coded by a certain black hole, is proportional to the surface area of this very black hole if measured in Planck area ℓ_P^2. In fact, the dependency how one bit of information changes the area of the black hole (ΔA) reads:

$$\Delta A = 32 \cdot \pi^2 \cdot \ell_P^2 + 64 \cdot \pi^3 \cdot \frac{\ell_P^4}{r_s^2}. \qquad (1042)$$

Ignoring the extremely small second term, one could just assume our black hole to be constructed out of many such bit surface pieces. Thus, we could write:

$$q \cdot \Delta A = q \cdot 32 \cdot \pi^2 \cdot \ell_P^2 = 4 \cdot \pi \cdot r_s^2 \quad \Rightarrow \quad r_s^2 = q \cdot 8 \cdot \pi \cdot \ell_P^2, \qquad (1043)$$

where r_s gives the radius of the black hole. We see that our black hole radius is proportional to the square root of the bits q thrown into it.

Now we want to compare the dependency $r_s[q]$ with the radii $r_{max}[n]$ resulting in maximum surface of n-spheres for a certain number of space-time dimensions n.

Note: the correct solution for the evaluation of the Schwarzschild radius r_s as function of the bits thrown into a black hole object (if using Eq. (1042)) would be:

$$r_s = 2 \cdot \ell_P \cdot \sqrt{\pi \cdot \left(q + \sqrt{q \cdot (1+q)}\right)}! \qquad (1044)$$

We find a perfect fit to the n-spheres with maximized surface to a given radius (dots in Fig. 11.1) with a Planck length of $\ell_P = 0.07878133250775303^*$ (see the line graph in Fig. 11.1).

Our finding not only connects the intrinsic dimension of a black hole with its mass respectively its surface but also gives an explanation to the hitherto unsolved problem of "what are the micro states of a black hole giving it temperature and allowing it to store information". According to the evaluation in this section, these microstates are just various states of dimensions realized within the black hole in dependence on the number

---

*Please note that this value came out slightly different when using the volume [116, 132, 145], where we obtained $\ell_P = 0.07881256452824544$.

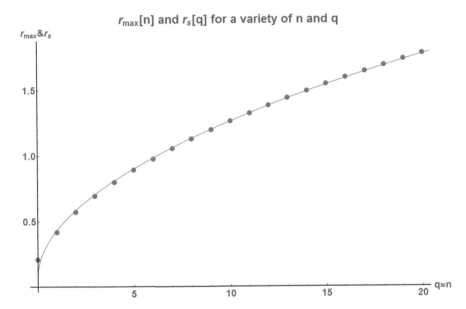

Figure 11.1 Radius r_{max} for which at a certain number of dimensions the n-sphere has maximum surface in dependence on n compared with the increase of the Schwarzschild radius r_s of a black hole in dependence on the number of bits q thrown into it by using Eq. (1044). We find that $q = n$.

of bits it contains (and thus, its mass). The bigger the number of bits, the higher the intrinsic dimensions the black hole has. In fact, the connection even is a direct one and only seems to deviate from the simple direct proportionality for very low numbers of masses*, respectively Schwarzschild radii r_s, respectively numbers of bits q the black hole has swallowed.

This finding also gives us a direct connection between a principal mathematical law (the maximum volume as function of the dimension for a given radius of an n-sphere) to the number of bits a black hole contains, to the mass or Schwarzschild radius of this very black hole and the number and character of microstates the black hole actually uses to internally code the bits.

---

*Besides, this deviation is also suggested by the Bekenstein finding summed up in Eq. (1042), where we could assume the second term to become of importance at lower numbers of r_s.

It has to be pointed out that the expression "intrinsic dimension" truly stands for the part of space for $r < r_s$. For more information about this aspect and its handling, the reader is referred to [116, 145].

11.5.2 Consequences with Respect to the Dimension of Time

Even though we currently do not have a suitable metric solution for the black hole with pairwise entangled dimensions, we will still try to discuss the problem of time for these objects.

We start with the metric of an n-dimensional spherical object (potentially a black hole):

$$g_{\alpha\beta}^n = \begin{pmatrix} -c^2 \cdot g\,[r] & 0 & 0 & 0 & \cdots & 0 \\ 0 & \frac{1}{g[r]} & 0 & 0 & \cdots & 0 \\ 0 & 0 & r^2 & 0 & \cdots & 0 \\ 0 & 0 & 0 & r^2 \cdot \sin^2 \varphi_1 & \cdots & 0 \\ \cdots & \cdots & \cdots & \cdots & \ddots & 0 \\ 0 & 0 & 0 & 0 & 0 & g_{nn} \end{pmatrix}$$

$$g_{44} = r^2 \cdot \sin^2 \varphi_1 \cdot \sin^2 \varphi_2; \quad g_{nn} = r^2 \cdot \prod_{j=1}^{n-2} \sin^2 \varphi_j; \quad g\,[r] = 1 - \frac{r_s^{n-3}}{r^{n-3}}.$$

$$(1045)$$

In 4 dimensions, the metric would be [142]:

$$g_{\alpha\beta}^{\text{Schwarzschild}} = \begin{pmatrix} -c^2 \left(1 - \frac{2 \cdot G \cdot M}{c^2 \cdot r}\right) & 0 & 0 & 0 \\ 0 & \left(1 - \frac{2 \cdot G \cdot M}{c^2 \cdot r}\right)^{-1} & 0 & 0 \\ 0 & 0 & r^2 & 0 \\ 0 & 0 & 0 & r^2 \cdot \sin^2 \vartheta \end{pmatrix}.$$

$$(1046)$$

Here r gives the spherical radius, M stands for the gravitational mass of the Schwarzschild object and the constants c and G arc the speed of light in vacuum and the gravitational or Newton constant, respectively. One easily sees that the solution above has two singularities, namely one at $r = r_s = \frac{2 \cdot M \cdot G}{c^2}$, which is a result of our choice of coordinates and could be transformed away, and at $r = 0$. The parameter r_s is called the Schwarzschild radius.

We see that the classical time component $g_{00} = -c^2 \cdot g\,[r]$ vanishes at $r = r_s$ and becomes space-like for $r < r_s$, which is inside the black hole. There, however, the r-component g_{11} would become time-like and thus, we would still have a "classical" time dimension behind the event horizon.

But what about our entangled time? Could it still be obtained inside a black hole?

In order to investigate this question, we introduce a p-dimensional black hole ($p \geq 4$) within a space of $n > 6$ dimensions as follows:

$$
g_{\alpha\beta}^n =
\begin{pmatrix}
-c^2 \cdot g\,[r] & 0 & 0 & 0 & \cdots & 0 \\
0 & \frac{1}{g[r]} & 0 & 0 & \cdots & 0 \\
0 & 0 & r^2 & 0 & \cdots & 0 \\
0 & 0 & 0 & r^2 \cdot \sin^2 \varphi_1 & \cdots & 0 \\
\cdots & \cdots & \cdots & \cdots & \ddots & 0 \\
0 & 0 & 0 & 0 & 0 & g_{nn}
\end{pmatrix}
$$

$$
g_{44} = r^2 \cdot \sin^2 \varphi_1 \cdot \sin^2 \varphi_2; \quad g\,[r] = 1 - \frac{r_s^{p-3}}{r^{p-3}}
$$

$$
g_{aa} = g_{nn} = F\,[f\,[x_a, x_n]] \quad a = n - 1. \tag{1047}
$$

As before, we have the general solution:

$$
F\,[f] = C_1 \cdot e^{C_2 \cdot f}; \quad f\,[x_a, x_n] = C_+\,[x_n + i \cdot x_a] + C_-\,[x_n - i \cdot x_a]. \tag{1048}
$$

Thus, as long as our black hole has additional degrees of freedom, independent on its spherical geometry, we can easily have entangled time coordinates as shown above.

But how could it be understood that a black hole of a certain mass or radius should contain a corresponding number of dimensions? Obviously the 4-dimensional black hole would not be a very massive and therefore not a very stable object (c.f. [116] and see Schwarzschild radius for $n = q = 4$ in Fig. 11.1).

The way to overcome that particular problem would be to say goodbye to the idea of black holes of a certain radius existing independent on the number of dimensions they contain. Thereby especially the number of dimensions making out the Schwarzschild n-sphere are of importance. The simplest way to construct objects being compatible with the Bekenstein-observation and our finding of the $r_s\,[q = n]$-dependency would be a radius-dependent change of dimensions. When moving towards the black hole, the dimension near the event horizon increases until it reaches the necessary n at the corresponding $r_s\,[q = n]$. While it appears to be a kind of logic to assume an $n[r]$-dependency being simply inverse to Eq. (1044) (note that we have $q = n$) for $r < r_s$, we can only guess* what a suitable approach for the

---

*Respectively experimentally determine.

outside region $r > r_s$ might be. For simplicity, here we assume an exponential behavior. Let the inverse function to Eq. (1044) be denoted as $n[r_s]$; we approach the total and $n[r]$-variable metric via the Schwarzschild function:

$$g[r] = 1 - \frac{r_s^{N[r]-3}}{r^{N[r]-3}}; \quad n[r] = \begin{cases} n[r] \text{ for } r < r_s \\ C \cdot \left(n[r_s] \cdot e^{-\left(\frac{r}{r_s}\right)^p} + 4 \right) \text{ for } r \geq r_s \end{cases}$$

$$\text{with:} \quad \left[C \cdot \left(n[r_s] \cdot e^{-\left(\frac{r}{r_s}\right)^p} + 4 \right) \right]_{r=r_s} = n[r_s]; \quad p > 1. \quad (1049)$$

This way we would end up with the usual outside behavior for spherical massive objects, but avoid the singularity for $r = 0$. In addition, we obtain a solution being able to store the Bekenstein-bits q by the means of dimensions n via the simple relation $n = q$.

11.6 A Timely Conclusion

We conclude that time is not simply there, but that it comes with processes as entanglement. Without entanglement, there is no time.

11.7 Questions to the Skilled and Interested Reader and One Answer

In [131], we asked the following:

How could we find a pairwise entangled solution for the Schwarzschild metric in order to explain the time coordinate there? That is, how can we "derive at a sign convention" for massive objects instead of postulating it?

The attentive reader may have realized that at least the second part of the question was partially answered in section 10.2.

11.8 Centers of Gravity as Origin of Time

Having derived holomorphic functions with pairwise entangled dimensions as possible origin of time, we shall ask why—on earth—dimensions should at all want to entangle. The answer "simply because it is possible" does not suffice.

However, in recalling the possibility that there can always be centers of gravity (c.f. sections 4.5) with base vector transformation matrices like:

$$\frac{\partial G^j \, [x_k]}{\partial x^\gamma} \mathbf{e}_j = \frac{\partial G^j \, [x_k - \xi_{kj}]}{\partial x^\gamma} \mathbf{e}_j, \tag{1050}$$

(with the ξ_{kj} denoting the various centers of gravity), we could either treat the corresponding multiple center problem as one of the number of spatial dimensions n times the number of gravity centers N being permanently variated, or as an ensemble of gravity centers moving along certain paths in time. The second presentation would coincide with our classical "notion", while the first one only leads to the pretence of time via the permanent variation of the centers of gravity ξ_{kj}. Now we assume the centers of gravity to be "invisible", respectively unresolvable with the means at hand in our scalar part of the universe (c.f. what was said about the pieces of space in chapter 9). Then we would not be able to distinguish the two forms of presentation and may just interpret the permanent variation of the gravity centers' positions as time within our space.

If so, however, we should also be able to formulate this picture in such a way that it would mirror or come close to our entanglement approach here. Finding the connection is simple if we assume to have a time dimension parallel to each spatial dimension and that to us these various times simply cannot be resolved just as we cannot resolve the variation of the centers of gravity (and the centers themselves). Then we would have a great number of holomorphic functions residing in our 6-dimensional metric as follows:

$$g^6_{\alpha\beta} = \begin{pmatrix} c_1^2 \cdot \sum_{\forall j} f_{1j} & 0 & 0 & 0 & 0 & 0 \\ 0 & \sum_{\forall j} f_{1j} & 0 & 0 & 0 & 0 \\ 0 & 0 & c_2^2 \cdot \sum_{\forall k} f_{2k} & 0 & 0 & 0 \\ 0 & 0 & 0 & \sum_{\forall k} f_{2k} & 0 & 0 \\ 0 & 0 & 0 & 0 & c_3^2 \cdot \sum_{\forall l} f_{3l} & 0 \\ 0 & 0 & 0 & 0 & 0 & \sum_{\forall l} f_{3l} \end{pmatrix} ;$$

$$f_{1j} = F_j \, [c_1 \cdot t_1 + i \cdot x_1] ; \quad f_{2k} = F_k \, [c_2 \cdot t_2 + i \cdot x_2] ; \quad f_{3l} = F_l \, [c_3 \cdot t_3 + i \cdot x_3] . \tag{1051}$$

Now the space consisting of permanently variated pieces or centers of gravity would be described by a set of holomorphic functions. We may even see those functions as waves if we substitute the imaginary number by a suitable sign convention.

It might have to do with the simplest form for the wrapper approach (c.f. Eq. (120)) in connection with the Ricci scalar that at least on bigger spaces this sums up to just 6 dimensions.

Insisting on only 1 time dimension leads to more complicated mixed solutions like the ones derived in section 11.2 for the combination of three coordinates, but in essence it is quite the same story, namely:

What in reality is the variated positioning or movement of centers of gravity, being underneath any scale we can detect, appears to us as time within a space of uncertainties just being formed by the permanent jitter "underneath". Thereby these uncertainties are determined by or limited to (which means that they cannot be smaller than) the smallest scale accessible to us. In consequence, instead of using the Fourier-like space-time (Eq. (1051)), we should also be able to just describe everything via a more simple space with enough dimensions and centers of gravity as hinted via the base vector transformation matrix in Eq. (1050). Both descriptions are sufficiently correct from our limited perspective and applicable within our scalar part of the universe, but only one of the two truly describes the reality everywhere and for all scales.

Chapter 12

A Time before Time or What Was before the Big Bang

The question to be asked how—for God's sake—if there really was a Big Bang, all this, meaning everything there is and what we call the universe, could have fitted into a mathematical point. Usually the association is of some kind of huge force and pressure, which—naturally—has to grow to an infinite size if it really needs to hold an entire universe like ours in a piece of "volume" with no size at all. So, obviously, there is a problem.

The simple answer to this question can directly be extracted from the Einstein–Hilbert action, which we here give again to save us the referencing within this book:

$$\delta_g W = 0 = \delta_g \int_V d^n x \left(\sqrt{g} \left[R - 2\kappa L_M - 2\Lambda \right] \right). \tag{1052}$$

The reader immediately recognizes the simplest solution to our volume problem, namely the determinant g of the covariant metric tensor $g_{\delta\gamma}$. Simply by demanding the primordial (which means here directly at the moment or "slightly before" the Big Bang—which is a bit of a joke, because there was no time as we understand it) space with the condition $g = 0$, the Einstein–Hilbert action would automatically be fulfilled. And as there is no condition known to this author which could possibly prohibit the $g = 0$-condition, we might see this as one possibility (a simple one) to answer the question of

The World Formula: A Late Recognition of David Hilbert's Stroke of Genius
Norbert Schwarzer
Copyright © 2022 Jenny Stanford Publishing Pte. Ltd.
ISBN 978-981-4877-20-6 (Hardcover), 978-1-003-14644-5 (eBook)
www.jennystanford.com

how the universe might have managed the "big squeeze" to a mathematical point at, respectively right before its beginning. A bit more complicated could have been the achievement of totally metric-free spaces, and the ignorant (politicians usually) may consider such spaces empty or even "nothingness", but we will see that this is not quite right.

12.1 A Variety of Timely Dimensions

It was shown in the previous chapters (mainly in chapters 6 to 10), but also in the more recent papers of the series "Science Riddles" [126–128] that matter fields can just be obtained via suitably entangled dimensions. In this context, too, the flatness problem of the universe (e.g., [183, 184]) could be solved (see the corresponding chapter "7 Solving the Flatness Problem" above and [130]). Then, in [131] we investigated the option of entangled dimensions also with respect to the character of time.

It was already demonstrated in [125] that we need at least one coordinate being of "different" (imaginary) character in order to fulfill fundamental quantum equations resulting from an extended variation of the Einstein–Hilbert action with respect to the dimension of space [116, 145]. We achieved this by entangling dimensions (mainly chapter 6 here).

In order to find the simplest of such solutions, where obviously only 2 dimensions are entangled, we assume a Cartesian system in n dimensions t, $x_1, x_2, \ldots$ and combine 2 dimensions as follows:

$$g^n_{\alpha\beta} = \begin{pmatrix} F\,[f\,[t, x_1]] & 0 & 0 & 0 & \cdots & 0 \\ 0 & F\,[f\,[t, x_1]] & 0 & 0 & \cdots & 0 \\ 0 & 0 & 1 & 0 & \cdots & 0 \\ 0 & 0 & 0 & 1 & \cdots & 0 \\ \cdots & \cdots & \cdots & \cdots & \ddots & 0 \\ 0 & 0 & 0 & 0 & 0 & 1 \end{pmatrix}. \tag{1053}$$

A very general solution can be found via:

$$F\,[f] = C_1 \cdot e^{C_2 \cdot f}; \quad f\,[t, x_1] = \begin{pmatrix} C_+\,[x_1 + i \cdot t] + C_-\,[x_1 - i \cdot t] \\ C_+\,[t + i \cdot x_1] + C_-\,[t - i \cdot x_1] \end{pmatrix}. \tag{1054}$$

Thereby C_1 and C_2 denote arbitrary constants and $_\pm[\ldots]$ stand for arbitrary functions. We recognize the characteristic wave functions. As before, we have

1 dimension which had to become time-like in order to fulfill the Einstein field equations. A generalization of the solution to many entangled pairs was given in [131].

It should be noted that the solutions of the kind (Eqs. (1015) and (1016)) are taking on very complicated forms in curvilinear coordinates. The usual transformation rules, as here given for the covariant metric tensor

$$g_{\delta\gamma} = \frac{\partial G^\alpha \left[f \left[x_k \right] \right]}{\partial x^\delta} \frac{\partial G^\beta \left[f \left[x_k \right] \right]}{\partial x^\gamma} g_{\alpha\beta}, \tag{1055}$$

lead to fulfilled complex structures even in simple coordinate systems like the spherical one. The author was therefore not able to find a suitable entanglement solution, e.g., for the Schwarzschild metric. Nevertheless, we discussed spherical objects with respect to their time dimension(s) in [131].

Here, now we want to discuss a different, more general or (if one wants to say so) deeper way of finding general solutions.

12.2 A Different Allocation and Interpretation of the Entangled Solutions

Interpreting the solutions not as components of the metric tensor but as the fundamental solutions of the base vectors might change our picture of quantum reality and space completely. Such an approach is motivated by our recent derivations of a quaternion-free Dirac equation [133] and the factorization of the Einstein field equations [134], where the base vectors play an important role.

Applying Eq. (1055), we know that the base vectors $\mathbf{g}_\delta$ to a certain metric are given as:

$$\mathbf{g}_\delta = \frac{\partial G^\alpha \left[f \left[x_k \right] \right]}{\partial x^\delta} \mathbf{e}_\alpha. \tag{1056}$$

Here the vectors $\mathbf{e}_\alpha$ shall denote the base vectors of a Cartesian system of the right (in principle arbitrary) dimension. We also know that the base vectors form the metric via the following scalar product:

$$g_{\delta\gamma} = \frac{\partial G^\alpha \left[f \left[x_k \right] \right]}{\partial x^\delta} \frac{\partial G^\beta \left[f \left[x_k \right] \right]}{\partial x^\gamma} g_{\alpha\beta} = \mathbf{g}_\delta \cdot \mathbf{g}_\gamma. \tag{1057}$$

Now, for simplicity, we assume to have a 2-dimensional space in t and x and base vectors with functions of the type (1016) as follows:

$$G^\alpha \left[f \left[x_k \right] \right] = G^\alpha \left[x, t \right] = \left(\frac{C_- \left[x - i \cdot t \right]}{C_+ \left[x + i \cdot t \right]} \right)^\alpha = \left(\frac{F \left[x + i \cdot t \right]}{F \left[i \cdot x - t \right]} \right)^\alpha$$

$$\Rightarrow \mathbf{g}_\delta = \frac{\partial G^0 \left[x, t \right]}{\partial x^\delta} \mathbf{e}_0 + \frac{\partial G^1 \left[x, t \right]}{\partial x^\delta} \mathbf{e}_1$$

$$\Rightarrow \left(\begin{array}{l} \mathbf{g}_0 = \frac{\partial F \left[x + i \cdot t \right]}{\partial x} \mathbf{e}_0 + \frac{\partial F \left[i \cdot x - t \right]}{\partial x} \mathbf{e}_1 = F' \left(\mathbf{e}_0 + i \cdot \mathbf{e}_1 \right) \\ \mathbf{g}_1 = \frac{\partial F \left[x + i \cdot t \right]}{\partial t} \mathbf{e}_0 + \frac{\partial F \left[i \cdot x - t \right]}{\partial t} \mathbf{e}_1 = F' \left(i \cdot \mathbf{e}_0 - \mathbf{e}_1 \right) \end{array} \right).$$

$$(1058)$$

Here F stands for an arbitrary function of one argument and F' denotes the derivative with respect to its argument. We recognize the holomorphic character of the argument of F. Evaluating the metric (1057), we obtain:

$$g_{\delta \gamma} = \mathbf{g}_\delta \cdot \mathbf{g}_\gamma = \left(\begin{array}{cc} 0 & 0 \\ 0 & 0 \end{array} \right). \qquad (1059)$$

We conclude that base vectors constructed from entangled or Dirac* solutions (Eq. (1016)) can be arranged in such a way that the resulting metric would be identical zero. Space would have no distance with respect to the dimensions entangled in such a way. Please note that other possible combinations for the same zero-metric result in 2 dimensions would be:

$$\begin{cases} \mathbf{g}_0 = \frac{\partial F \left[x - i \cdot t \right]}{\partial x} \mathbf{e}_0 + \frac{\partial F \left[-i \cdot x - t \right]}{\partial x} \mathbf{e}_1 = F' \left(\mathbf{e}_0 - i \cdot \mathbf{e}_1 \right) \\ \mathbf{g}_1 = \frac{\partial F \left[x - i \cdot t \right]}{\partial t} \mathbf{e}_0 + \frac{\partial F \left[-i \cdot x - t \right]}{\partial t} \mathbf{e}_1 = F' \left(-i \cdot \mathbf{e}_0 - \mathbf{e}_1 \right) \\ \mathbf{g}_0 = \frac{\partial F \left[x + i \cdot t \right]}{\partial x} \mathbf{e}_0 + \frac{\partial F \left[-i \cdot x + t \right]}{\partial x} \mathbf{e}_1 = F' \left(\mathbf{e}_0 - i \cdot \mathbf{e}_1 \right) \\ \mathbf{g}_1 = \frac{\partial F \left[x + i \cdot t \right]}{\partial t} \mathbf{e}_0 + \frac{\partial F \left[-i \cdot x + t \right]}{\partial t} \mathbf{e}_1 = F' \left(i \cdot \mathbf{e}_0 + \mathbf{e}_1 \right) \\ \mathbf{g}_0 = \frac{\partial F \left[x - i \cdot t \right]}{\partial x} \mathbf{e}_0 + \frac{\partial F \left[i \cdot x + t \right]}{\partial x} \mathbf{e}_1 = F' \left(\mathbf{e}_0 + i \cdot \mathbf{e}_1 \right) \\ \mathbf{g}_1 = \frac{\partial F \left[x - i \cdot t \right]}{\partial t} \mathbf{e}_0 + \frac{\partial F \left[i \cdot x + t \right]}{\partial t} \mathbf{e}_1 = F' \left(-i \cdot \mathbf{e}_0 + \mathbf{e}_1 \right) \end{cases} \qquad (1060)$$

12.3 Metric-Free Spaces in Higher Dimensions

While in 2 dimensions, the request for a metric-free space binds three of four coefficients of our base vectors (two coefficients each), we have more free

---

*It was shown in [133, 134] that we obtain the same wave-type solutions for the Dirac and the generalized Dirac equation.

parameters in higher dimensions. So, for instance, in 3 dimensions, we would only have the conditions:

$$\mathbf{g}_\delta = \begin{cases} \mathbf{g}_0 = e_0^0 \cdot \mathbf{e}_0 + e_0^1 \cdot \mathbf{e}_1 + e_0^2 \cdot \mathbf{e}_2 \\ \mathbf{g}_1 = e_1^0 \cdot \mathbf{e}_0 + e_1^1 \cdot \mathbf{e}_1 + e_1^2 \cdot \mathbf{e}_2 \ ; \quad \text{with}: \\ \mathbf{g}_2 = e_2^0 \cdot \mathbf{e}_0 + e_2^1 \cdot \mathbf{e}_1 + e_2^2 \cdot \mathbf{e}_2 \end{cases} \quad (1061)$$

$$\text{I} \Rightarrow \begin{cases} e_1^0 = -\dfrac{i \cdot e_0^0 \cdot e_0^1}{\sqrt{\left(e_0^0\right)^2 + \left(e_0^2\right)^2}}, \ e_0^1 = -i \cdot \sqrt{\left(e_0^0\right)^2 + \left(e_0^2\right)^2}, \ e_2^0 = \dfrac{e_0^0 \cdot e_0^2}{e_0^2} \\ e_2^1 = \dfrac{i \cdot \left(e_0^0\right)^2 e_2^2}{e_0^2 \cdot \sqrt{\left(e_0^0\right)^2 + \left(e_0^2\right)^2}} + \dfrac{i \cdot e_0^2 \cdot e_2^2}{\sqrt{\left(e_0^0\right)^2 + \left(e_0^2\right)^2}}, \ e_1^2 = -\dfrac{i \cdot e_0^2 \cdot e_0^1}{\sqrt{\left(e_0^0\right)^2 + \left(e_0^2\right)^2}} \end{cases}, \quad (1062)$$

$$\text{II} \Rightarrow \begin{cases} e_1^0 = \dfrac{i \cdot e_0^0 \cdot e_0^1}{\sqrt{\left(e_0^0\right)^2 + \left(e_0^2\right)^2}}, \ e_0^1 = i \cdot \sqrt{\left(e_0^0\right)^2 + \left(e_0^2\right)^2}, \ e_2^0 = \dfrac{e_0^0 \cdot e_0^2}{e_0^2} \\ e_2^1 = -\dfrac{i \cdot \left(e_0^0\right)^2 e_2^2}{e_0^2 \cdot \sqrt{\left(e_0^0\right)^2 + \left(e_0^2\right)^2}} - \dfrac{i \cdot e_0^2 \cdot e_2^2}{\sqrt{\left(e_0^0\right)^2 + \left(e_0^2\right)^2}}, \ e_1^2 = \dfrac{i \cdot e_0^2 \cdot e_0^1}{\sqrt{\left(e_0^0\right)^2 + \left(e_0^2\right)^2}} \end{cases}, \quad (1063)$$

with subsequently always 4 coefficients to choose freely.

As we saw in [133, 134] that we can easily obtain solutions of the type Eq. (1016) also for higher-order entanglement (meaning that more coordinates are involved), we conclude that with higher orders of entanglement we have more and more options to achieve metric-free solutions.

12.4 Interpretation of the Result

Seeing any process or action as the change of a property/dimension with respect to another property/dimension automatically leads to an entanglement of the two. As this can also be seen as being the case for more properties/dimensions changing with respect to others, we can—in principle—describe all processes and actions as entanglements. As shown in [131], in this case at least one of the entangled properties/dimensions has to become time-like. It is important to note that such a time-like appearance thereby comes with the entanglement (and thus the process or action). It has not necessarily been there before. Destruction of the entanglement—e.g., caused by a measurement process—breaks up the connection between the two (or more) connected properties/dimensions and thereby also freezes the formerly time-like property/dimension at a certain (non-entangled) metric state.

Now it was shown here that the solutions producing the entanglement could be used for the construction of base vectors leading to zero-metric spaces or space-times. As a vanishing metric automatically fulfills the Einstein field equations and as our base vector solutions are resulting from the Dirac equation, one would have obtained a quantum compatible gravity solution. With the solutions having no metric, they would also lack of a distance, which automatically renders the Einstein–Podolsky–Rosen (EPR) paradox [176, 177] into no paradox at all. It holds what was said in chapter "10 Two quantum objects being created in one process and being separated over an apparently long distance for an outside observer are connected without being distant in any measurable way as long as they are entangled." We easily see this when observing the solution in Eq. (1058) with the corresponding metric in Eq. (1059). There, probably within or as a part of "ordinary" space, we have two waveforms which interlock in such a way that space cannot form any measurable distance. We have to assume that a measurement process, destroying the entanglement, and thus the waveforms, would also bring back the ordinary (measurable or metric) space.

We further assume that such solutions are omnipresent, and as a result we obtain:

(a) the creation of time (see discussion in chapter 10 here and [131]),
(b) the Heisenberg uncertainty [131] and chapter 6,
(c) the speed of light maximum [131], and now as shown here,
(d) a rather foam-like space.

Thus, with solutions of intrinsic manifolds being entangled and therefore having no measurable distance, quantum effects do not seem to make the space just grainy, but rather porous.

Furthermore, the structure of the solutions in 2D (Eqs. (1058) and (1060)) and 3D (Eq. (1061)) leads to the suspicion that elementary particles might be constructed this way.

Last but not least, one might conclude that—perhaps—the whole universe once erupted from a state of pure metric-free entanglement and thus, as shown here, had no size, as there was no measurable distance. With more and more dimensions falling out of the entanglement during a cooling period, metric appeared and so did distance and our universe started to swell in a measurable manner.

12.5 Conclusions with Respect to the Primordial Universe

We conclude that there is no space with any measurable distance within certain systems of entangled dimensions. In fact, with this type of entanglement there is no distance "within" manifolds of dimensions of such an entanglement. We also conclude that the primordial universe probably existed in such a "non-metric" state.

Chapter 13

Why Is Gravity So Weak?

Foto: P. Heuer-Schwarzer
Location: Prasonisi (Rhodos)
Rider: Dr. Schwarzer

Using the weakness of gravity.

The World Formula: A Late Recognition of David Hilbert's Stroke of Genius
Norbert Schwarzer
Copyright © 2022 Jenny Stanford Publishing Pte. Ltd.
ISBN 978-981-4877-20-6 (Hardcover), 978-1-003-14644-5 (eBook)
www.jennystanford.com

In this chapter, we will see that our almost miniscule extension of the Einstein–Hilbert action [137], as introduced earlier in this book (c.f. especially chapter 10), also helps us to understand the huge differences between the force of gravity and the other fundamental interactions, which are the strong, the electromagnetic and the weak interaction.

While the strong interaction is about 100 times more powerful than the electromagnetic force, we have the weak interaction to be about 1000 times weaker than the electromagnetic one. These differences, however, are nothing in comparison to gravity, which is 10^{35} times weaker than the weakest weak interaction.

Where does this huge difference come from?

Why does gravity stick out so much?

These questions will be answered in this chapter.

13.1 An Elastic World Formula with Scale

Here we will not just repeat the evaluation from chapter 10, but also point out an important extension, which may help us to understand the differences among the fundamental forms of interactions.

We derive a simple quantum gravity equation sporting the character and structure of the fundamental equations of linear elasticity. Our starting point shall be the Einstein–Hilbert action in its classical form [137, 138]:

$$\delta_g W = 0 = \delta_g \int_V d^n x \left(\sqrt{-g} \left[R - 2\kappa L_M - 2\Lambda \right] \right). \tag{1064}$$

Its traditional evaluation, as first performed by Hilbert in [137], results in the Einstein field equations, which we here explicitly give under the integral of the variation task (Eq. (1064)):

$$\delta_\gamma W = 0 = \int_V d^n x \sqrt{g} \cdot \left(R^{\delta\gamma} - \frac{1}{2} R \cdot g^{\delta\gamma} + \Lambda \cdot g^{\delta\gamma} + \kappa \cdot T^{\delta\gamma} \right) \delta g_{\delta\gamma}. \tag{1065}$$

For simplicity only, we have ignored the sign under the square root of the metric determinant and will keep doing so throughout the chapter. As before, we here seek to perform the variation under the integral in vector functional form (applying a vector function $\mathbf{f}^k$ and its derivatives) and align it similar to our functional wrapper approach as follows:

$$\delta_\gamma W = 0 = \int_V d^n x \sqrt{g} \cdot \left(R^{\delta\gamma} - \frac{1}{2} R \cdot g^{\delta\gamma} + \Lambda \cdot g^{\delta\gamma} + \kappa \cdot T^{\delta\gamma} \right)$$

$$\times$$

$$\delta g_{\delta\gamma} = g_{\delta\gamma}\delta_\sigma \left(1 + M \cdot h^\sigma h_\beta f^\beta + h_\beta \left(\underbrace{C^{\sigma\alpha}}_{=d_1\sqrt{g}\cdot g^{\sigma\alpha}} f^\beta{}_{,\alpha} + \underbrace{C^{\beta\alpha}}_{=d_2\sqrt{g}\cdot g^{\beta\alpha}} f^\sigma{}_{,\alpha} \right) \right)$$

$$= g_{\delta\gamma} \left(M \cdot \left(h^\sigma h_\beta f^\beta \right)_{,\sigma} + \left[h_\beta \left(d_1\sqrt{g} \cdot g^{\sigma\alpha} f^\beta{}_{,\alpha} + d_2\sqrt{g} \cdot g^{\beta\alpha} f^\sigma{}_{,\alpha} \right) \right]_{,\sigma} \right)$$

$$= g_{\delta\gamma} \left(M \cdot \left(h^\sigma h_\beta f^\beta \right)_{,\sigma} + \left(\begin{array}{c} \left[\left(d_1\sqrt{g} \cdot g^{\sigma\alpha} f^\beta{}_{,\alpha} + d_2\sqrt{g} \cdot g^{\beta\alpha} f^\sigma{}_{,\alpha} \right) \right] h_{\beta,\sigma} \\ + h_\beta \left(d_1\sqrt{g} \cdot g^{\sigma\alpha} f^\beta{}_{,\alpha} + d_2\sqrt{g} \cdot g^{\beta\alpha} f^\sigma{}_{,\alpha} \right)_{,\sigma} \end{array} \right) \right)$$

$$= g_{\delta\gamma} \left(M \cdot \left(h^\sigma f^\beta h_{\beta,\sigma} + h_\beta \left(h^\sigma f^\beta \right)_{,\sigma} \right) + \left(\begin{array}{c} \left(d_1\sqrt{g} \cdot g^{\sigma\alpha} f^\beta{}_{,\alpha} + d_2\sqrt{g} \cdot g^{\beta\alpha} f^\sigma{}_{,\alpha} \right) h_{\beta,\sigma} \\ + h_\beta \left(d_1\sqrt{g} \cdot \Delta f^\beta + d_2 \left(\sqrt{g} \cdot g^{\beta\alpha} f^\sigma{}_{,\alpha} \right)_{,\sigma} \right) \end{array} \right) \right). \tag{1066}$$

Please note that our handling of the metric $g_{\delta\gamma}$ and its variation $\delta g_{\delta\gamma}$ requires some elaboration. We demand the metric tensor $g_{\delta\gamma}$ to be constructed as follows:

$$
\begin{aligned}
g_{\delta\gamma} &= \gamma_{ij}\delta^i_\delta\delta^j_\gamma \left(1 + M \cdot h^\sigma h_\beta f^\beta + h_\beta \left(d_1 \cdot \sqrt{g} \cdot g^{\sigma\alpha} f^\beta{}_{,\alpha} + d_2 \cdot \sqrt{g} \cdot g^{\beta\alpha} f^\sigma{}_{,\alpha}\right)\right)\\
&= \gamma_{\delta\gamma} \left(1 + M \cdot h^\sigma h_\beta f^\beta + h_\beta \left(d_1 \cdot \sqrt{g} \cdot g^{\sigma\alpha} f^\beta{}_{,\alpha} + d_2 \cdot \sqrt{g} \cdot g^{\beta\alpha} f^\sigma{}_{,\alpha}\right)\right),
\end{aligned}
$$
(1067)

which reminds us of the usual covariant tensor coordinate transformation. We assume the variation only to act on the function vector $\mathbf{f}^k$ and its derivatives, which gives us:

$$
\begin{aligned}
\delta g_{\delta\gamma} &= \gamma_{\delta\gamma}\delta \left(1 + M \cdot h^\sigma h_\beta f^\beta + h_\beta \left(d_1 \cdot \sqrt{g} \cdot g^{\sigma\alpha} f^\beta{}_{,\alpha} + d_2 \cdot \sqrt{g} \cdot g^{\beta\alpha} f^\sigma{}_{,\alpha}\right)\right)\\
&= \frac{g_{\delta\gamma}\delta \left(M \cdot h^\sigma h_\beta f^\beta + h_\beta \left(Cd_1 \cdot \sqrt{g} \cdot g^{\sigma\alpha} f^\beta{}_{,\alpha} + d_2 \cdot \sqrt{g} \cdot g^{\beta\alpha} f^\sigma{}_{,\alpha}\right)\right)}{\left(1 + M \cdot h^\sigma h_\beta f^\beta + h_\beta \left(d_1 \cdot \sqrt{g} \cdot g^{\sigma\alpha} f^\beta{}_{,\alpha} + d_2 \cdot \sqrt{g} \cdot g^{\beta\alpha} f^\sigma{}_{,\alpha}\right)\right)}\\
&\quad \xrightarrow{1 \gg M \cdot h^\sigma h_\beta f^\beta + h_\beta \left(d_1 \cdot \sqrt{g}\cdot g^{\sigma\alpha} f^\beta{}_{,\alpha} + d_2 \cdot \sqrt{g}\cdot g^{\beta\alpha} f^\sigma{}_{,\alpha}\right)}\\
&\simeq g_{\delta\gamma}\delta \left(M \cdot h^\sigma h_\beta f^\beta + h_\beta \left(d_1 \cdot \sqrt{g} \cdot g^{\sigma\alpha} f^\beta{}_{,\alpha} + d_2 \cdot \sqrt{g} \cdot g^{\beta\alpha} f^\sigma{}_{,\alpha}\right)\right).
\end{aligned}
$$
(1068)

Due to the approximation in the last line, we need not bother about the potentially problematic division by a vector at the moment, but in order to reuse Eq. (1066) also in the non-approximated form, we may just demand that the properties of h_σ are such that we could write:

$$
\begin{aligned}
\delta g_{\delta\gamma} &= \gamma_{\delta\gamma}\delta \left(1 + M \cdot h^\sigma h_\beta f^\beta + h_\beta \left(d_1 \cdot \sqrt{g} \cdot g^{\sigma\alpha} f^\beta{}_{,\alpha} + d_2 \cdot \sqrt{g} \cdot g^{\beta\alpha} f^\sigma{}_{,\alpha}\right)\right)\\
&= \frac{g_{\delta\gamma}\delta \left(M \cdot h^\sigma h_\beta f^\beta + h_\beta \left(d_1 \cdot \sqrt{g} \cdot g^{\sigma\alpha} f^\beta{}_{,\alpha} + d_2 \cdot \sqrt{g} \cdot g^{\beta\alpha} f^\sigma{}_{,\alpha}\right)\right)}{h_\sigma \left(1 + M \cdot h^\sigma h_\beta f^\beta + h_\beta \left(d_1 \cdot \sqrt{g} \cdot g^{\sigma\alpha} f^\beta{}_{,\alpha} + d_2 \cdot \sqrt{g} \cdot g^{\beta\alpha} f^\sigma{}_{,\alpha}\right)\right)}.
\end{aligned}
$$
(1069)

Now we rearrange the last line in Eq. (1066) and obtain:

$$
\begin{aligned}
\delta_\gamma W = 0 = &\int_V d^n x \sqrt{g} \cdot \left(R^{\delta\gamma} - \frac{1}{2} R \cdot g^{\delta\gamma} + \Lambda \cdot g^{\delta\gamma} + \kappa \cdot T^{\delta\gamma}\right)\\
&\times \left(g_{\delta\gamma} \left(\begin{array}{l} \left(M \cdot h^\sigma f^\beta h_{\beta,\sigma} + d_1\sqrt{g} \cdot g^{\sigma\alpha} f^\beta{}_{,\alpha} + d_2\sqrt{g} \cdot g^{\beta\alpha} f^\sigma{}_{,\alpha}\right) h_{\beta,\sigma}\\ +h_\beta \left(M \cdot \left(h^\sigma f^\beta\right)_{,\sigma} + d_1\sqrt{g} \cdot \Delta f^\beta + d_2 \left(\sqrt{g} \cdot g^{\beta\alpha} f^\sigma{}_{,\alpha}\right)_{,\sigma}\right)\end{array}\right)\right).
\end{aligned}
$$
(1070)

Demanding the variated metric to give zero, we would have:

$$\delta_\gamma W = 0 = \left(\begin{array}{c} \left(M \cdot h^\sigma \, f^\beta h_{\beta,\sigma} + d_1 \sqrt{g} \cdot g^{\sigma\alpha} \, f^\beta{}_{,\alpha} + d_2 \sqrt{g} \cdot g^{\beta\alpha} \, f^\sigma{}_{,\alpha} \right) h_{\beta,\sigma} \\ + h_\beta \left(M \cdot \left(h^\sigma \, f^\beta \right)_{,\sigma} + d_1 \sqrt{g} \cdot \Delta f^\beta + d_2 \left(\sqrt{g} \cdot g^{\beta\alpha} \, f^\sigma{}_{,\alpha} \right)_{,\sigma} \right) \end{array} \right), \tag{1071}$$

where we could multiply with the contravariant $\mathbf{h}^\beta$ yielding:

$$0 = \left(\begin{array}{c} \overbrace{\left(M \cdot h^\sigma \, f^\beta h_{\beta,\sigma} + d_1 \sqrt{g} \cdot g^{\sigma\alpha} \, f^\beta{}_{,\alpha} + d_2 \sqrt{g} \cdot g^{\beta\alpha} \, f^\sigma{}_{,\alpha} \right) h_{\beta,\sigma} h^\beta}^{=C} \\ + h^\beta h_\beta \cdot \left(M \cdot \left(h^\sigma \, f^\beta \right)_{,\sigma} + d_1 \sqrt{g} \cdot \Delta f^\beta + d_2 \left(\sqrt{g} \cdot g^{\beta\alpha} \, f^\sigma{}_{,\alpha} \right)_{,\sigma} \right) \end{array} \right). \tag{1072}$$

Just for the moment assuming $\mathbf{h}_\beta$ to be a vector of constants, we obtain:

$$0 = M \cdot \left(h^\sigma \, f^\beta \right)_{,\sigma} + d_1 \sqrt{g} \cdot \Delta f^\beta + d_2 \left(\sqrt{g} \cdot g^{\beta\alpha} \, f^\sigma{}_{,\alpha} \right)_{,\sigma}$$

$$= \frac{M \cdot \left(h^\sigma \, f^\beta \right)_{,\sigma}}{d_1 \sqrt{g}} + \Delta f^\beta + \frac{d_2 \left(\sqrt{g} \cdot g^{\beta\alpha} \, f^\sigma{}_{,\alpha} \right)_{,\sigma}}{d_1 \sqrt{g}}. \tag{1073}$$

Now we write the whole in Cartesian coordinates:

$$0 = \frac{M \cdot \left(h^\sigma \, f^\beta \right)_{,\sigma}}{d_1} + \Delta f^\beta + \frac{d_2 \left(g^{\beta\alpha} \, f^\sigma{}_{,\alpha} \right)_{,\sigma}}{d_1} \tag{1074}$$

and we recognize the fundamental equation of elasticity [152, 153] in the isotropic case with volume and external forces if we would just set $\frac{d_2}{d_1} = \frac{1}{1-2\cdot\nu}$ (with ν giving the Poisson's ratio). Of course, we could already have discovered the elastic terms in the full Eq. (1072), but due to its complexity, it is a bit more difficult to make it out in there.

We may not like to have the first derivative of the $\mathbf{f}^\beta$-vector in Eq. (1073) (or its Cartesian version Eq. (1074)) and thus, reshape the equation by performing an integration by parts on the corresponding term $\frac{M \cdot \left(h^\sigma \, f^\beta \right)_{,\sigma}}{d_1 \sqrt{g}}$.

Without the approximation introduced in Eq. (1068), we then have:

$$
\delta W = 0 = \int_V d^n x \sqrt{g} \cdot \overbrace{\left(R^{\delta\gamma} - \frac{1}{2} R \cdot g^{\delta\gamma} + \Lambda \cdot g^{\delta\gamma} + \kappa \cdot T^{\delta\gamma} \right)}^{(\ldots)^{\delta\gamma}}
$$

$$
\times \left(\frac{g_{\delta\gamma} h_\beta \left(M \cdot \left(h^\sigma f^\beta \right)_{,\sigma} + d_1 \sqrt{g} \cdot \Delta f^\beta + d_2 \left(\sqrt{g} \cdot g^{\beta\alpha} f^\sigma{}_{,\alpha} \right)_{,\sigma} \right)}{h_\sigma \left(1 + M \cdot h^\sigma h_\beta f^\beta + h_\beta \left(d_1 \cdot \sqrt{g} \cdot g^{\sigma\alpha} f^\beta{}_{,\alpha} + d_2 \cdot \sqrt{g} \cdot g^{\beta\alpha} f^\sigma{}_{,\alpha} \right) \right)} \right)
\atop \begin{bmatrix} \vdots \end{bmatrix}
$$

$$
= \int_V d^n x \times \left(\sqrt{g} \cdot \frac{(\ldots)^{\delta\gamma}}{\begin{bmatrix} \vdots \end{bmatrix}} g_{\delta\gamma} h_\beta \left(M \cdot \left(h^\sigma f^\beta \right)_{,\sigma} + d_1 \sqrt{g} \cdot \Delta f^\beta + d_2 \left(\sqrt{g} \cdot g^{\beta\alpha} f^\sigma{}_{,\alpha} \right)_{,\sigma} \right) \right)
$$

$$
= \int_V d^n x \times \left(\begin{array}{l} \left(\sqrt{g} \cdot \dfrac{(\ldots)^{\delta\gamma}}{\begin{bmatrix} \vdots \end{bmatrix}} g_{\delta\gamma} h_\beta \left(d_1 \sqrt{g} \cdot \Delta f^\beta + d_2 \left(\sqrt{g} \cdot g^{\beta\alpha} f^\sigma{}_{,\alpha} \right)_{,\sigma} \right) \right) \\[2em] -M \cdot \left[\sqrt{g} \cdot \dfrac{(\ldots)^{\delta\gamma}}{\begin{bmatrix} \vdots \end{bmatrix}} g_{\delta\gamma} h_\beta \quad h^\sigma f^\beta \right]_{,\sigma} \end{array} \right) .
\tag{1075}
$$

The resulting equations would now read:

$$
\sqrt{g} \cdot \frac{(\ldots)^{\delta\gamma}}{\begin{bmatrix} \vdots \end{bmatrix}} g_{\delta\gamma} h_\beta \left(d_1 \sqrt{g} \cdot \Delta f^\beta + d_2 \left(\sqrt{g} \cdot g^{\beta\alpha} f^\sigma{}_{,\alpha} \right)_{,\sigma} \right) - M \cdot \left[\sqrt{g} \cdot \frac{(\ldots)^{\delta\gamma}}{\begin{bmatrix} \vdots \end{bmatrix}} g_{\delta\gamma} h_\beta \quad h^\sigma f^\beta \right]_{,\sigma} = 0.
$$

$$
\tag{1076}
$$

As we know from our considerations in the previous chapters (e.g., chapter 10) that in 4 dimensions we could simplify:

$$
(\ldots)^{\delta\gamma} g_{\delta\gamma} \xrightarrow{\Lambda=0;\ T^{\delta\gamma}=0} -R,
\tag{1077}
$$

we would end up with the following interesting elastic quantum gravity equation for our space-time:

$$\sqrt{g} \cdot \frac{R}{\begin{bmatrix} \vdots \end{bmatrix}} h_\beta \left(d_1 \sqrt{g} \cdot \Delta f^\beta + d_2 \left(\sqrt{g} \cdot g^{\beta\alpha} f^\sigma{}_{,\alpha} \right)_{,\sigma} \right) - M \cdot \begin{bmatrix} \sqrt{g} \cdot \frac{R}{\begin{bmatrix} \vdots \end{bmatrix}} h_\beta \end{bmatrix}_{,\sigma} h^\sigma f^\beta = 0.$$

$$(1078)$$

The vector $\mathbf{h}_\beta$ could still be considered to be a vector of constants, because this would not compromise the generality of our equations, as any functional issue taken away from $\mathbf{h}_\beta$ by our const-demand would simply be taken on by f (and/or the metric). This gives us the following system of vector equations:

$$0 = \sqrt{g} \cdot \frac{R}{\begin{bmatrix} \vdots \end{bmatrix}} \left(d_1 \sqrt{g} \cdot \Delta f^\beta + d_2 \left(\sqrt{g} \cdot g^{\beta\alpha} f^\sigma{}_{,\alpha} \right)_{,\sigma} \right) - M \cdot \begin{bmatrix} \sqrt{g} \cdot \frac{R}{\begin{bmatrix} \vdots \end{bmatrix}} \end{bmatrix}_{,\sigma} h^\sigma f^\beta$$

$$\text{or} \quad \Rightarrow \quad \Delta f^\beta + \frac{d_2}{d_1 \sqrt{g}} \left(\sqrt{g} \cdot g^{\beta\alpha} f^\sigma{}_{,\alpha} \right)_{,\sigma} - \frac{M}{d_1 \cdot g \cdot \frac{R}{\begin{bmatrix} \vdots \end{bmatrix}}} \cdot \begin{bmatrix} \sqrt{g} \cdot \frac{R}{\begin{bmatrix} \vdots \end{bmatrix}} \end{bmatrix}_{,\sigma} h^\sigma f^\beta = 0.$$

$$(1079)$$

It was shown elsewhere that from equations of the type (1079), applying the so-called n-function approach [160–162], we are able to evaluate very particle like (leptons and quarks) solutions.

For generalization, we now assume the product of Ricci scalar and square root of the metric determinant to consist of summands of various scales leading to the Einstein–Hilbert action as follows:

$$\delta_g W = 0 = \delta_g \int_V d^n x \left(\sqrt{g_1} \cdot (R_1 - 2\Lambda_1) + \sqrt{g_2} \cdot (R_2 - 2\Lambda_2) + \sqrt{g_3} \cdot (R_3 - 2\Lambda_3) + \ldots \right).$$

$$(1080)$$

We explicitly ignored the matter term because we think that the extended action (Eq. (1080)) already contains it due to one of the addends. The various addends $\sqrt{g_k} \cdot (R_k - 2\Lambda_k)$ may be understood as terms of a Fourier series with different wavelengths or—as it is more appropriate here—scales.

Now we simply assume that the total action could be made to zero by demanding the various addends to deliver zero simultaneously. As an example, we just consider only two addends $\sqrt{g_1} \cdot (R_1 - 2\Lambda_1)$, $\sqrt{g_2} \cdot (R_2 - 2\Lambda_2)$ and demand the cosmological constants to vanish. Performing the derivation as given above, we end up with two equations (Eq. (1079)), namely one for $\sqrt{g_1} \cdot R_1$ and one for $\sqrt{g_2} \cdot R_2$, reading:

$$\delta_\gamma W = 0 = \int_V d^n x \sqrt{g_1} \cdot \left(R_1{}^{\delta\gamma} - \frac{1}{2} R_1 \cdot g^{\delta\gamma} \right) \delta g_{1\delta\gamma}, \qquad (1081)$$

$$\delta_\gamma W = 0 = \int_V d^n x \sqrt{g_2} \cdot \left(R_2{}^{\delta\gamma} - \frac{1}{2} R_2 \cdot g^{\delta\gamma} \right) \delta g_{2\delta\gamma}. \qquad (1082)$$

While for Eq. (1081), we assume to satisfy the equation via the Einstein field equations, which is to say:

$$0 = R_1{}^{\delta\gamma} - \frac{1}{2} R_1 \cdot g^{\delta\gamma}, \qquad (1083)$$

we could have the quantum Eq. (1079) for the case given in Eq. (1082), leading to (in 4 dimensions):

$$\Delta f^\beta + \frac{d_2}{d_1 \sqrt{g_2}} \left(\sqrt{g_2} \cdot g_2^{\beta\alpha} f^\sigma{}_{,\alpha} \right)_{,\sigma} - \frac{M}{d_1 \cdot g_2 \cdot \frac{R_2}{\begin{bmatrix} \vdots \end{bmatrix}_2}} \cdot \left[\sqrt{g_2} \cdot \frac{R_2}{\begin{bmatrix} \vdots \end{bmatrix}_2} \right]_{,\sigma} h^\sigma f^\beta = 0$$

$$\text{with}: \quad \begin{bmatrix} \vdots \end{bmatrix}_2 = h_\sigma \left(1 + M \cdot h^\sigma h_\beta f^\beta + h_\beta \left(d_1 \cdot \sqrt{g_2} \cdot g_2^{\sigma\alpha} f^\beta{}_{,\alpha} + d_2 \cdot \sqrt{g_2} \cdot g_2^{\beta\alpha} f^\sigma{}_{,\alpha} \right) \right).$$

$$(1084)$$

Obviously, the last equation is a quantum or—rather—a quantum gravity equation, which in a flat metric goes over to an almost ordinary Klein–Gordon and, thus, quantum equation. By assuming a very different metric curvature (much smaller) for the case in Eq. (1083) on larger scales in comparison to Eq. (1084) on smaller ones, we might have found a fairly simple explanation for the fact that gravity, determined by Eq. (1083), could be so much weaker than the other fundamental forces.

As it has been demonstrated in the previous chapters of this book that there is a great variety of equations one can extract from the variation of the metric tensor and Eq. (1084) should be seen as just one of these many options, we assume to be able to describe all stronger interactions (strong, electromagnetic and weak) via forms of $\delta g_{\delta\gamma}$. This way, it would be of little wonder that the three forces are so close together, while gravity, being governed by the completely different equation (Eq. (1083)), could easily sport an extremely different strength (or in this case weakness) than the other fundamental forces.

It is interesting to note that only a split-up in scale leading to the simultaneous conditions in Eqs. (1083) and (1084) would make up the fundamental properties of our universe and it might be worthwhile to think about the possibility that the so called "symmetry break", being assumed to be an essential part of the very early universe, may just have been such a scalar split-up. As a by-product, one would also obtain a nice explanation for the big difference of gravity to the other interactions. On the other hand, we may assume that very strong gravitational fields reverse the Ricci scalar split-up, thereby avoiding the singularity of classical black holes.

13.2 Consequences

We were able to show that quantum gravity equations, sporting the character of equations of elasticity, can directly be obtained from the classical Einstein–Hilbert action by just a small extension of the variation under the action integral. By assuming a sum of different Ricci scalars, one can thereby construct situations of simultaneous appearance of quantum, quantum gravity and pure Einstein field equations simply acting on different scales of the spatial curvature. This way, it is not difficult to explain the huge difference between the force of gravity and the other fundamental forces, which are strong, weak and electromagnetic.

TEACHING

Chapter 14

The Other Applications

From a true world formula, it can be expected to give answers to just everything. However, as we saw in the simplest cases how quickly the math of the world formula-approach can become very complex, we have to admit that in some cases its application is not only cumbersome, but perhaps even foolish or just impossible. One question, however, which definitively requires the most holistic viewpoint there can be probably is the question of "What is good?"

14.1 What Is Good?

The true and ultimate "good" can always only be a holistic good. Thus, only a holistic consideration of the system and the situation we are interested in will give us the correct answer to our question of "What is good?"

Unfortunately, as we have seen in the previous chapters, the answer to that question is as complicated as the answer to everything—in fact, it IS the answer to everything and it does not seem to be 42*. And as "everything" really means everything, it also—naturally—means GOOD. It was shown in almost cruel examples that the consequences of this can be very disturbing,

*C.f. Douglas Adams' famous book *The Hitchhiker's Guide to the Galaxy*.

The World Formula: A Late Recognition of David Hilbert's Stroke of Genius
Norbert Schwarzer
Copyright © 2022 Jenny Stanford Publishing Pte. Ltd.
ISBN 978-981-4877-20-6 (Hardcover), 978-1-003-14644-5 (eBook)
www.jennystanford.com

sometimes even unbearable (e.g., c.f. [1–8]). Let alone the uncertainties residing in every system in this universe clearly reduces every do-gooder attitude about "knowing what is right and good" to absurdity.

It requires the most holistic model to answer the question and even after having it done as best as one could, one needs to be aware that there are:

(a) multitudes of solutions with similar good minima,
(b) principal uncertainties to all solutions, which increase the further one needs to extrapolate into the future
(c) principally incomplete sets of input parameters,
(d) no satisfying good definitions of "good", and
(e) intrinsic degrees of freedom with respect to the variational functions and the choice of coordinates (c.f. chapter 5).

And the peradventure about what is good does not stop at the principle uncertainty of everything. It is also determined by the factor of sustainability. For many things and decisions, even in the simplest fields, sustainability cannot be guaranteed for long time spans. Things are getting much more difficult in complex scenarios and environments like ecosystems or human societies.

Nevertheless, "THE GOOD", which is to say the optimum solution, can theoretically be found by a comprehensive consideration of the system in question, taking into account all its degrees of freedom (properties) and treating them as dimensions in a generalized space or space-time. Theoretically "letting them loose" leads to the governing question where to find the minimum, and, as shown at the beginning of this book, results in the generalized Einstein field equations. Their solutions are all also possible solutions to the problem under consideration and now it only requires the setting of boundary conditions or the incorporation of individual properties to the dimensions in order to separate impractical from realistic solutions. The technique was given in [86], section "Generalization of Einstein-field-equation-solutions".

Thus, theoretically, we have a "Theory of Everything" at hand, which principally allows us to find the universal good answer to any question. Practically, however, this might prove to be a rather complicated and tedious process—mathematically. In this context, the process of mathematically answering the question "What is good?" is not much different from answering it in any other way, respectively with any other method. The difference would be, however, that one would have something of substance and proof instead of the usual politician's and do-gooder's waffle.

Bearing this in mind, it is almost shocking to register the impertinence with which do-gooders, politicians and media-liars are trying to tell us what is good. Especially those who have not the slightest inkling of how to mirror and holistically reflect reality, but who are brimful with dim dogma and stupid ideology, often govern the discussion about "good". The question why such types govern such discussions is easily answered: because they cannot do anything else to earn a living than to do to politics, spiritual work or any other usually public and thus, completely parasitic job.

Let us take for instance the fact that they want to force us to believe that welcoming as many poor migrants as possible (usually with no education at all and from completely uncivilized origin or even criminals) among our midst actually is a VERY GOOD ACT. However, when evaluating things through and taking everything into account, when performing a few estimations about costs, possible future developments, one soon realizes that the supposedly "good deed" in fact is a devastatingly destructive atrocity to mankind in general.

In fact, in some cases the bad consequences are obviously visible and so close that the do-gooders' usually arrogant attitude and their self-declared superiority over the rest of the society often is the much more unbearable thing than the bad situation those do-gooders intend to cure. A wonderful explanation to this situation in connection with bad migration was given here:

www.youtube.com/watch?v=LPjzfGChGlE

The author is of the opinion that shortsighted do-gooder-ship including the misleading and brainwashing of people should be punished by law. Stupidity on purpose is bad enough already, but calling this stupidity "political correctness", "equality rights", "social justice" or just "being good" simply is a crime.

It will be shown in the next section that "good" cannot be equality, because the perfect equality is death. "Good" cannot be the maximum inequality either, because this state is just chaos or "heat death" (and so the other extreme of equality). No, the universal "good" of a system is the state in which this system can unfold its full evolutionary or innovative potential in most holistic and multiple manners. It is eminent for that fact that no society can be called "good" if it hinders or punishes even the simplest forms of progress, innovative thought and act and development. Extreme religiousness always falls under that category and as most monotheistic religions and left ideologies easily develop tendencies to such totalitarianism, they cannot be "good". Islamic societies are most prominent among this and therefore have to be refused and sincerely fought.

While the mathematical formulation for perfect equality is simple (c.f. [86]), the author struggles in finding such a thing for the perfect state of chance, motivation and positive development. Of course, the maximum need for change arises inside a singularity, where it can easily be understood that the infinite gradient has a tendency to relax and smooth, but that is just relaxation and not evolution. In other words, the amount of information would not increase. The true "good" of a system has to be an invariant, because it should not change when the coordinate system is changed. The invariants we seek are those of the metric tensor g_{ij}, because this is the one governing and determining it all. It defines the structure of the system. However, as we see from the form of the Einstein–Hilbert action (Eq. (1064)), being determined by the Ricci scalar, we might also consider certain functions of the invariants of g_{ij} to form the invariants defining "good". In the end, however, all this can be brought back to the invariants of the governing metric tensor. The metric tensor g_{ij} of an N-dimensional space has N eigenvalues λ_i (with $i, j = 1, 2, \ldots, N$). The eigenvalues are the N solutions of the characteristic equation for the metric tensor g_{ij}, which reads:

$$0 = p\left(g_{ij}\right) = \det\left[g_{ij} - \lambda \cdot \delta_{ij}\right]; \quad \delta_{ij} = \begin{pmatrix} 1 & \cdots & 0 \\ \vdots & \ddots & \vdots \\ 0 & \cdots & 1 \end{pmatrix}. \tag{1085}$$

From this equation, the invariants can be obtained as the coefficients for the various powers of λ. So we usually define the coefficient to λ^{N-1} as the first invariant I_1, the second invariant I_2 is the coefficient to λ^{N-2} and so on. With the help of the eigenvalues the invariants of g_{ij} can be obtained via:

$$I_1 = \lambda_1 + \lambda_2 + \lambda_3 + \ldots = \sum_{i=1}^{N} \lambda_i$$

$$I_2 = \lambda_1\lambda_2 + \lambda_1\lambda_3 + \lambda_2\lambda_3 + \ldots = \frac{1}{2}\left(\sum_{i,j=1}^{N} \lambda_j \cdot \lambda_i - \sum_{i=1}^{N} \lambda_i^2\right)$$

$$= \frac{1}{2}\left(\left(\sum_{i=1}^{N} \lambda_i\right)^2 - \sum_{i=1}^{N} \lambda_i^2\right)$$

$$I_3 = \lambda_1\lambda_2\lambda_3 + \lambda_1\lambda_3\lambda_4 + \lambda_2\lambda_3\lambda_4 + \dots$$

$$= \frac{1}{6}\left(\sum_{i=1}^{N}\lambda_i\right)^3 - \frac{1}{2}\left(\sum_{i=1}^{N}\lambda_i\right)\left(\sum_{i=1}^{N}\lambda_i^2\right) + \frac{1}{3}\left(\sum_{i=1}^{N}\lambda_i^3\right)$$

$$\vdots$$

$$I_k = \frac{(-1)^k}{k!} \cdot B_k\left[s_1, -1!s_2, 2!s_3, \dots, (-1)^{k-1}(k-1)!s_k\right]; \quad s_k = \sum_{i=1}^{N}\lambda_i^k$$

$$\vdots$$

$$I_N = \lambda_1\lambda_2\lambda_3\lambda_4\dots = \prod_{i=1}^{N}\lambda_i, \tag{1086}$$

where the B_k give the Bell polynomials.

It should be noted that with the Cayley–Hamilton theorem, one could immediately extract the following system of equations related to the characteristic Eq. (1085):

$$0 = p\left(g_{ij}\right) = \left(g_{ij}\right)^N + I_1 \cdot \left(g_{ij}\right)^{N-1} + I_2 \cdot \left(g_{ij}\right)^{N-2} + \dots$$
$$+ I_{N-1} \cdot g_{ij} + (-1)^N \cdot \det\left[g_{ij}\right] \cdot \delta_{ij}. \tag{1087}$$

For illustration and convenience, we give the invariants of a 3-dimensional strain tensor γ_{ij} according to our complex base-vectors given above. We might see it as the spatial part of a 4-dimensional space-time-metric in General Theory of Relativity (GTR). We assume a presentation of γ_{ij} via its base vectors $\boldsymbol{\in}_i$ as follows:

$$\gamma_{ij} = \frac{1}{2}\left(\boldsymbol{\in}_i \cdot \boldsymbol{\in}_j\right); \quad |\gamma_{ij}| = \frac{1}{2}\left(\boldsymbol{\in}_i \cdot \boldsymbol{\in}_j^*\right); \quad i, j = 1, 2, 3. \tag{1088}$$

Only concentrating on the spatial metric of a 4-dimensional space-time, the three invariants of this spatial part can now be evaluated in the usual way from the eigensystem of the strain tensor as given in Eq. (1088). The three invariants read:

$$I_\gamma = \gamma_{ii} = \gamma_{11} + \gamma_{22} + \gamma_{33} = \lambda_1 + \lambda_2 + \lambda_3; \quad |I_\gamma| = |\gamma_{ii}|; \quad i, j = 1, 2, 3$$

$$II_\gamma = \frac{1}{2}\left((\gamma_{ii})^2 - \gamma_{ji}\gamma_{ij}\right) = \lambda_1\lambda_2 + \lambda_1\lambda_3 + \lambda_2\lambda_3;$$

$$|II_\gamma| = \frac{1}{2}\left((|\gamma_{ii}|)^2 - |\gamma_{ji}||\gamma_{ij}|\right)$$

$$III_\gamma = \det\gamma_{ij} = \lambda_1\lambda_2\lambda_3; \quad |III_\gamma| = \det|\gamma_{ij}|, \tag{1089}$$

where we have used the three eigenvalues λ_i. Please note that in cases of purely diagonal metrics, which are quite often objects of interest in GTR problems, these eigenvalues are just the corresponding diagonal metric components, which is to say:

$$\{\lambda_1, \lambda_2, \lambda_3\} = \frac{1}{2}\{€_1 \cdot €_1, €_2 \cdot €_2, €_3 \cdot €_3\};$$

$$\{|\lambda_1|, |\lambda_2|, |\lambda_3|\} = \frac{1}{2}\left\{€_1 \cdot €_1^*, €_2 \cdot €_2^*, €_3 \cdot €_3^*\right\}. \tag{1090}$$

However, here, for the reason of completeness, we keep the general form.

We postulate that "good" is a function $G[I_1, I_2, \ldots, I_N]$ of the metric tensor invariants and only with this metric known (completely known!) and also knowing the function $G[\ldots]$, we are able to tell what the true "good" of a given system might be.

To sum it up: NOBODY, no priest, no imam, no politician, no do-gooder, not even the pope himself does truly know what good is. We can only "guess", evaluate as far as possible and try our best to do the right thing and act in such a way that the results reach an optimum of "goodness". Of course, this is an impossibly difficult task, but it would be foolish not to try. What we should not do, however, is to try and force our next to believe in the same good as we do. This simply is, because he might know more than we do and he might have his reasons to doubt that our "good" actually is a lasting one and a holistic good one, too.

But the likelihood that somebody is closer to the answer about the holistic good, increases dramatically if this somebody has done some REAL calculations following a Theory-of-Everything approach* and extracted his conclusions from these calculations.

This author does not trust people and their suggestions if they start these suggestions with a "friendly face" argument or something like "according to my belief". He, meanwhile, downright rejects such suggestions if they end with "Yes we can" or "Wir schaffen das".

The true seeker should not make up his mind, but rather evaluate an opinion.

This approach, where we see "good" as the holistic extremum of a system, stands in contrast to most classical definitions of "good" being more or

---

*With this, the author does not necessarily mean the quantum gravity approach as presented in this book, but any sufficiently holistic ansatz.

less a "declaration" [193] or being based on such declarations. Following the Hamilton principle, the holistic "good" is the **minimum of suffering** by considering everything which does matter or which does potentially matter to the system including the timely development. And one should not argue that "suffering" is nothing one could not clearly place, because it can easily be defined, e.g., by the amount of negative stress leading to flow of certain hormones in almost all living entities or by the accumulation of defects, failure and destruction. Of course, this only holds for a holistic metric analysis including time. Thus, temporary failure does not automatically exclude a certain solution, because it still could be a "good" solution in total. It only requires to find the extremum just as the Einstein field equations automatically provide. There is nothing to agree upon and there is nothing to define or to declare, the true "good" IS DEFINED via a function $G[...]$ of the invariants of a given system and its environment. Of course, the fact that this simple definition will ruin all the false shepherds'* foundations of existence, destroy their parasitic way of life and clearly show their total uselessness will not help this concept to get widely known.

However, for all those critics, mainly from the fields of the so-called "snivel-sciences" we give three hints they should think about before starting to rant:

(A) Is there a more general form of describing any system as the one we have introduced, namely the arbitrary set of arbitrary properties?

(B) What other law than a minimum-principle (perhaps in a generalized $f(R)$ form instead of the linear R as used in Eq. (86)) do you intend to introduce?

(C) Perhaps having introduced another law, how do you explain the discrepancy to the real world this other law produces?

Of course, many snivel-scientists will not bother about these questions, but simply criticize the whole concept, because it challenges their way of life, which is to say their parasitic existence, living on the costs of the hardworking part of the society.

Thus, criticism to our approach is for sure and this especially holds as this concept itself helps to expose the very kind of "humanism" which is, in fact, no humanism at all, but only a certain form of parasitism. Almost all

---

*Among them, we have most of the imams, priests, shamans, politicians, spiritual leaders, ideologists, genderixe, social justice worriers, do-gooders, Marxists, etc.

protagonists of the so-called "political correctness", "friendly face" actions and "we must help" attitudes do belong to this kind of necrotic tissue of our modern society. Most such "humanistic" categories only exist as an excuse or a justification for various higher forms of exploitation of groups of hosts by a certain "elite of cadgers". These political parasites and their symbiotic partners form the various help-sectors (social industry, asylum industry, churches, etc.), the brainwashing media and the false shepherd-churches would stand no chance in cheating their prey any longer if everything would be put under a truly holistic process of analyzing its global content of "real goodness".

Two excellent examples would be the current "ethic" discussion about economic migration plus the ideology of Marxism and similar "equality ideologies". For these two examples and related topics, we refer to the original literature [44, 86, 100, 101, 105].

14.2 Quantum Gravity Computer or Is There an Ultimate Turing Machine?

The answer to the question in the headline is "Yes, there is!" [70, 71, 145].

14.3 Can the World Formula-Approach Be Used for Optimum Decision Making?

Yes, see [100]!

14.4 Is There a Way to Bring Ethic Problems into Mathematical Form?

The author thinks yes, there is [101, 105].

14.5 Theoretical Biology and Evolutionary Stable Strategies

See [67].

Chapter 15

How to Derive a World Formula

In the first part of this chapter, we will just sum up (in a very compact form) what is of need for the derivation of a world formula "to a certain problem". Thereby the author is aware of the irony residing in the choice of the words in "…" signs, which should be understood as a certain restriction or rather a reasonable confinement of the problem. Obviously one does not need to consider all scales and dimensions there can be in general if only being interested in a miniscule aspect of this universe. Which is to say: if only being interested in the analysis of the mechanical contact problem of a pen's tip on a sheet of paper, it is probably of low importance what the wife of the holder of the pen would be thinking about his writing. This holds even more so if she had been already dead for a couple of years. But well, her influence should not be neglected if the holder of the pen thinks about her, thereby emotionally being carried away and pressing the pen hard onto the sheet of paper. So hard perhaps, that it leads to small fractures. Then, of course, the woman is not negligible.

The recipe shall be summed up as follows:

Find all properties and degrees of freedom of the system you intend to consider and treat them as dimensions of the space in which you are going to deal with the matter in question.

The World Formula: A Late Recognition of David Hilbert's Stroke of Genius
Norbert Schwarzer
Copyright © 2022 Jenny Stanford Publishing Pte. Ltd.
ISBN 978-981-4877-20-6 (Hardcover), 978-1-003-14644-5 (eBook)
www.jennystanford.com

(a) Write down the corresponding Einstein–Hilbert action in the most general manner, which is:

$$\delta_g W = 0 = \delta_g \int_V d^n x \left(\sqrt{g} \left[f [R] - 2 \cdot \kappa \cdot L_M - 2 \cdot \Lambda \right] \right).$$

Thereby note that the matter term L_M is not fundamental and only of artificial character. So whenever possible try to avoid it. Matter will come in automatically with the further (complete) variation of the metric tensor. In fact, the holistic evaluation of the action does reveal it as it will be demonstrated later within this chapter (section 15.5 and following).

(b) Check whether you can simplify to linearized Ricci scalars, potentially on various levels of scales, leading to (here already given without the Lagrange matter density L_M):

$$\delta_g W = 0 = \delta_g \int_V d^n x \left(\sqrt{g_1} \cdot (R_1 - 2\Lambda_1) + \sqrt{g_2} \cdot (R_2 - 2\Lambda_2) \right.$$
$$\left. + \sqrt{g_3} \cdot (R_3 - 2\Lambda_3) + \cdots \right).$$

(c) Now derive the corresponding Einstein field equations, which gives:

i. in the $f[R]$-case:

$$\delta_g W = 0 = \delta_g \int_V d^n x \sqrt{g} \, (\ldots)^{\delta \gamma} \, \delta g_{\delta \gamma}; \quad (\ldots)^{\delta \gamma} = (\ldots)_{\mu \nu} \, g^{\delta \mu} g^{\gamma \nu}$$

$$\text{with:} \quad (\ldots)_{\mu \nu} = F'[R] \cdot R_{\mu \nu} - \frac{1}{2} F[R] \cdot g_{\mu \nu}$$
$$+ F''[R] \left[R_{;\mu \nu} - \Delta_g R \cdot g_{\mu \nu} \right] + F'''[R] \left[R_{;\mu} R_{;\nu} - R^{;\sigma} R_{;\sigma} \cdot g_{\mu \nu} \right]$$

ii. in the linear R-case:

$$\delta_g W = 0 = \delta_g \int_V d^n x \sqrt{g} \, (\ldots)^{\delta \gamma} \, \delta g_{\delta \gamma}; \quad (\ldots)^{\delta \gamma} = R^{\delta \gamma} - \frac{1}{2} R \cdot g^{\delta \gamma} + \Lambda \cdot g^{\delta \gamma}$$

iii. in the case of multiple scales:

$$\delta_g W = 0$$
$$= \delta_g \int_V d^n x \left(\sqrt{g_1} \, (\ldots)_1^{\delta \gamma} \, \delta g_{1 \delta \gamma} + \sqrt{g_2} \, (\ldots)_2^{\delta \gamma} \, \delta g_{2 \delta \gamma} + \sqrt{g_3} \, (\ldots)_3^{\delta \gamma} \, \delta g_{3 \delta \gamma} + \ldots \right)$$

$$(\ldots)_k^{\delta \gamma} = R_k^{\delta \gamma} - \frac{1}{2} R_k \cdot g_k^{\delta \gamma} + \Lambda \cdot g_k^{\delta \gamma}$$

iv. one should not forget that also the constant under the integral may be seen as a possibility, which might just be written as:

$$\delta_g W = 0 = \delta_g \int_V d^n x \left(\sqrt{g} \left[f \left[R \right] - 2 \cdot \kappa \cdot L_M - 2 \cdot \Lambda \right] \right)$$

$$= \delta_g \int_V d^n x \ \text{const}$$

(d) Check for the possibility to variate with respect to:

  i. The number of main dimensions
  ii. The number of centers of gravity
  iii. The positions of the centers of gravity
  iv. The intrinsic degrees of freedom residing within the metric. Thereby you need to be aware that the most general form of metric variation should be given as:

$$\delta_\gamma W = 0 = \int_V d^n x \sqrt{g} \cdot (\ldots)^{\delta \gamma} \times \left(\delta g_{\delta \gamma} = g_{ij} \delta F^{ij}_{\delta \gamma} \right)$$

(e) Solve the resulting addends under the integral either with respect to the metric (see the various types of Einstein field equations under point c), or achieve a zero-outcome for the total variational task via the options given under point d). Please note that there are often multiple options. Make sure you consider them all, because these options are making up your—most holistic—model of the problem, which you intend to consider via a world formula approach.

15.1 The Simplest Example

The main headline for this part of the book was chosen to be "Teaching". Now we all know that the best way of teaching is via examples and therefore we want to finish our book with the simplest example of a world formula this author could think of.

We assume that we are able to describe our problem, which in this case should be the whole universe (after all we are very humble), via a set of properties and seek them to form an ensemble of states all being defined by the extremal principle given by the Einstein–Hilbert action in its simplest form:

$$\delta W = 0 = \delta \int_V d^n x \left(\sqrt{-G} \cdot R \right).$$

(1091)

Thereby G shall denote the determinant of the metric tensor $G_{\alpha\beta}$. Further assuming that the action W, being a constant, would automatically variate to zero, we may demand:

$$\sqrt{-G} \cdot R = \text{const} = K.$$

(1092)

Now we use the simplest form of metric adaptation we could think of, namely:

$$G_{\alpha\beta} = F\left[f\left[t, x, y, z, \ldots, \xi_k, \ldots, \xi_n\right]\right]_{\alpha\beta}^{ij} g_{ij}$$
$$\rightarrow G_{\alpha\beta} = F\left[f\left[t, x, y, z, \ldots, \xi_k, \ldots, \xi_n\right]\right] \cdot \delta_\alpha^i \delta_\beta^j g_{ij}$$

(1093)

and evaluate the resulting Ricci scalar first in n and then simplify to 4 dimensions:

$$R = \overset{\overset{\text{diagonal}}{g_{\alpha\beta}}}{\longrightarrow} = \frac{1}{F[f]^3} \cdot \left(\begin{pmatrix} C_{N1} \cdot \left(\frac{\partial F[f]}{\partial f} \right)^2 - C_{N2} \cdot F[f] \cdot \frac{\partial^2 F[f]}{\partial f^2} \end{pmatrix} \cdot (\tilde{\nabla}_g f)^2 \\ -C_{N2} \cdot F[f] \cdot \frac{\partial F[f]}{\partial f} \cdot \Delta_g f \right)$$

$$\overset{n=4}{\longrightarrow} = \frac{1}{F[f]^3} \cdot \left(\begin{pmatrix} \frac{3}{2} \cdot \left(\frac{\partial F[f]}{\partial f} \right)^2 - 3 \cdot F[f] \cdot \frac{\partial^2 F[f]}{\partial f^2} \end{pmatrix} \cdot (\tilde{\nabla}_g f)^2 - 3 \cdot F[f] \cdot \frac{\partial F[f]}{\partial f} \cdot \Delta_g f \right)$$

$$\overset{\frac{3}{2} \cdot \left(\frac{\partial F[f]}{\partial f} \right)^2 - 3 \cdot F[f] \cdot \frac{\partial^2 F[f]}{\partial f^2} = 0}{\longrightarrow} = -\frac{3}{F[f]^2} \cdot \left(\frac{\partial F[f]}{\partial f} \cdot \Delta_g f \right).$$

(1094)

Thereby we have used the following function $F[f]$:

$$F[f] = C_1 \cdot f + \frac{C_1^2 \cdot f^2}{4 \cdot C_2} + C_2$$

(1095)

in order to satisfy the condition in the third line (on the arrow), reading:

$$C_{N1} \cdot \left(\frac{\partial F[f]}{\partial f} \right)^2 - C_{N2} \cdot F[f] \cdot \frac{\partial^2 F[f]}{\partial f^2} = 0.$$

(1096)

Please note that the Laplace operator in the third line of Eq. (1094) is to be performed with respect to the untampered metric $g_{\alpha\beta}$ and not $G_{\alpha\beta}$! This we tried to make clear by the index g.

Now we can set R from Eq. (1094) into our condition (Eq. 1092) and end up with the equation:

$$\frac{3}{F\,[f]^2}\cdot\left(\frac{\partial F\,[f]}{\partial f}\cdot\Delta_g f\right)=\frac{3\cdot\left(C_1+\frac{C_1^2\cdot f}{2\cdot C_2}\right)}{\left(C_1\cdot f+\frac{C_1^2\cdot f^2}{4\cdot C_2}+C_2\right)^2}\cdot\Delta_g f=-\frac{K}{\sqrt{-G}}=\frac{K_0}{\sqrt{G}}$$

$$\Rightarrow\frac{3\cdot\left(C_1+\frac{C_1^2\cdot f}{2\cdot C_2}\right)}{\left(C_1\cdot f+\frac{C_1^2\cdot f^2}{4\cdot C_2}+C_2\right)^2}\cdot\Delta_g f=\frac{K_0}{\left(C_1\cdot f+\frac{C_1^2\cdot f^2}{4\cdot C_2}+C_2\right)^2\cdot\sqrt{g}}$$

$$\Rightarrow 0=\frac{K_0}{\sqrt{g}}\cdot\left(C_1+\frac{C_1^2\cdot f}{2\cdot C_2}\right)^{-1}-3\cdot\Delta_g f. \qquad (1097)$$

In order to see that this equation already contains the classical Klein–Gordon equation, we may want to rush there and assume that the constant C_2 is huge compared to C_1 and thus, we have:

$$0=\frac{K_0}{\sqrt{g}}\cdot\left(C_1+\frac{C_1^2\cdot f}{2\cdot C_2}\right)^{-1}-3\cdot\Delta_g f$$

$$\xrightarrow{\frac{C_1^2\cdot f}{2\cdot C_2}\ll C_1}\simeq\frac{K_0}{\sqrt{g}}\cdot\left(\frac{1}{C_1}-\frac{f}{2\cdot C_2}+\dots\right)-3\cdot\Delta_g f$$

$$\xrightarrow{h\equiv\frac{1}{C_1}-\frac{f}{2\cdot C_2}}=\frac{K_0}{\sqrt{g}}\cdot h+6\cdot C_2\cdot\Delta_g h. \qquad (1098)$$

Applying the definition of the Laplace operator in curvilinear coordinates, we can reshape the last equation as follows:

$$0=\frac{K_0}{\sqrt{g}}\cdot h+6\cdot C_2\cdot\Delta_g h=\frac{K_0}{\sqrt{g}}\cdot h+\frac{6\cdot C_2}{\sqrt{g}}\cdot\left(\sqrt{g}\cdot g^{\alpha\beta}h_{,\beta}\right)_{,\alpha}$$

$$\Rightarrow 0=K_0\cdot h+6\cdot C_2\cdot\left(\sqrt{g}\cdot g^{\alpha\beta}h_{,\beta}\right)_{,\alpha}. \qquad (1099)$$

We can conclude that Eq. (1097) already is (a relatively simple) quantum gravity equation, but, most interestingly, it is not truly becoming non-linear before the function f becomes big enough to bring the term $\frac{C_1^2\cdot f}{2\cdot C_2}$ into the scale range of C_1. This is exactly the current notion of quantum theory versus General Theory of Relativity. Only with the quantum effects becoming huge, there would be a need for quantum gravity, which would then be accessible via Eq. (1097). Finding general solutions to Eq. (1097), however, seems to be relatively difficult, as we obtain cumbersome differential equations already in the second step of our Taylor expansion of the term $\left(C_1+\frac{C_1^2\cdot f}{2\cdot C_2}\right)^{-1}$. Here is the example for the second-order approximation:

$$0 = \frac{K_0}{\sqrt{g}} \cdot \left(C_1 + \frac{C_1^2 \cdot f}{2 \cdot C_2} \right)^{-1} - 3 \cdot \Delta_g f$$

$$\xrightarrow{\frac{C_1^2 \cdot f}{2 \cdot C_2} \ll C_1} \simeq \frac{K_0}{\sqrt{g}} \cdot \left(\frac{1}{C_1} - \frac{f}{2 \cdot C_2} + \frac{C_1 \cdot f^2}{4 \cdot C_2^2} + \dots \right) - 3 \cdot \Delta_g f. \quad (1100)$$

For completeness, we now want to incorporate mass via entangled dimensions as derived in section 6.3 and thereby need to move to 6 dimensions. The corresponding evaluation would be:

$$R \xrightarrow{g_{\alpha\beta}^{\text{diagonal}}} = \frac{1}{F[f]^3} \cdot \left(\left(C_{N1} \cdot \left(\frac{\partial F[f]}{\partial f} \right)^2 - C_{N2} \cdot F[f] \cdot \frac{\partial^2 F[f]}{\partial f^2} \right) \cdot \left(\tilde{\nabla}_g f \right)^2 \right.$$
$$\left. - C_{N2} \cdot F[f] \cdot \frac{\partial F[f]}{\partial f} \cdot \Delta_g f \right)$$

$$\xrightarrow{n=6} = -\frac{5}{F[f]^2} \cdot \left(\frac{\partial^2 F[f]}{\partial f^2} \cdot \left(\tilde{\nabla}_g f \right)^2 + \frac{\partial F[f]}{\partial f} \cdot \Delta_g f \right)$$

$$\xrightarrow{\frac{\partial^2 F[f]}{\partial f^2} = 0} = -\frac{5}{F[f]^2} \cdot \left(\frac{\partial F[f]}{\partial f} \cdot \Delta_g f \right). \quad (1101)$$

Setting $F[f] = C_1 + f$ and again demanding $\sqrt{-G} \cdot R = \text{const} = K$ leads to the same principal structure as we had in a 4-dimensional space-time:

$$-\frac{5}{F[f]^2} \cdot \left(\frac{\partial F[f]}{\partial f} \cdot \Delta_g f \right) = \frac{5}{(C_1 + f)^2} \cdot \Delta_g f = -\frac{K}{\sqrt{-G}} = \frac{K_0}{\sqrt{G}}$$

$$\Rightarrow \frac{5}{(C_1 + f)^2} \cdot \Delta_g f = \frac{K_0}{(C_1 + f)^3 \cdot \sqrt{g}}$$

$$\Rightarrow 0 = \frac{K_0}{\sqrt{g}} \cdot (C_1 + f)^{-1} - 5 \cdot \Delta_g f. \quad (1102)$$

We note that for the case of $K = 0$ we have two solutions, namely:

$$n = 4: \quad 0 = \left(C_1 + \frac{C_1^2 \cdot f}{2 \cdot C_2} \right) \cdot 3 \cdot \Delta_g f \Rightarrow \begin{cases} 0 = 1 + \frac{C_1 \cdot f}{2 \cdot C_2} \Rightarrow f = -2 \cdot \frac{C_2}{C_1} \\ 0 = \Delta_g f \end{cases}$$

$$n = 6: \quad 0 = (C_1 + f) \cdot 5 \cdot \Delta_g f \Rightarrow \begin{cases} 0 = C_1 + f \Rightarrow f = -C_1 \\ 0 = \Delta_g f \end{cases}. \quad (1103)$$

We see that not only the classical Klein–Gordon equation emerges without any approximation, but also that there is an additional Higgs-like solution the moment f would be a constant (c.f. section 6.2). Of course, with f being a constant, also $0 = \Delta_g f$ would automatically be fulfilled, but the important point is the fact that such a scalar or Higgs-like solution exists as well.

We see that already our simplest world formula approach brought two essential things classical quantum theory had to postulate:

(A) The Klein–Gordon quantum equation
(B) A scalar field (possibly the Higgs field)

Only for simplicity, we concentrate on the $n = 4$-case with the following discussion:

With the backward transformation $h \to f$ from our substitution used in Eq. (1099) and the subsequent full solution for the metric $G_{\alpha\beta}$:

$$h \equiv \frac{1}{C_1} - \frac{f}{2 \cdot C_2} \Rightarrow f = 2 \cdot C_2 \left(\frac{1}{C_1} - h \right)$$

$$\Rightarrow F[f] \to F[h] = C_2 \cdot (C_1 \cdot h - 2)^2$$

$$\Rightarrow G_{\alpha\beta} = C_2 \cdot (C_1 \cdot h - 2)^2 \cdot g_{\alpha\beta} = C_2 \cdot (C_1 \cdot h[x_{k=0,1,2,3}] - 2)^2 \cdot g_{\alpha\beta},$$

$$(1104)$$

we automatically also have a complete solution to the variation of the Einstein–Hilbert action. Any function h, which satisfies the condition in the last line of Eq. (1099) (definitively being a quantum equation) and for which also holds that $C_1 \cdot h \ll 2$, built into a metric tensor according to the second line in Eq. (1104), also fulfills the extremal condition for the Einstein–Hilbert action and therefore has to be considered to be a solution of the General Theory of Relativity, too. This indeed makes our result (Eq. (1099)) a world formula or Theory of Everything equation.

For completeness, we now also derive the corresponding equations for the case in 6 dimensions. Thereby we have from Eq. (1102):

$$0 = \frac{K_0}{\sqrt{g}} \cdot (C_1 + f)^{-1} - 5 \cdot \Delta_g f = \frac{K_0}{C_1 \cdot \sqrt{g}} \cdot \left(1 - \frac{f}{C_1} + \frac{f^2}{C_1^2} - \ldots \right) - 5 \cdot \Delta_g f$$

$$\xrightarrow{C_1 \gg f} \approx \frac{K_0}{C_1 \cdot \sqrt{g}} \cdot \left(1 - \frac{f}{C_1} \right) - 5 \cdot \Delta_g f$$

$$\Rightarrow 0 = \frac{K_0}{C_1 \cdot \sqrt{g}} \cdot \left(1 - \frac{f}{C_1} \right) - 5 \cdot \Delta_g f \xrightarrow{h=1-\frac{f}{C_1}} 0 = \frac{K_0}{C_1 \cdot \sqrt{g}} \cdot h + 5 \cdot C_1 \cdot \Delta_g h$$

$$\Rightarrow 0 = K_0 \cdot h + 5 \cdot C_1^2 \cdot \left(\sqrt{g} \cdot g^{\alpha\beta} h_{,\beta} \right)_{,\alpha}.$$

$$(1105)$$

The resulting metric is similar to the 4-dimensional case, namely:

$$h \equiv h = 1 - \frac{f}{C_1} \Rightarrow f = C_1 (1 - h) \Rightarrow F[f] \to F[h] = C_1 \cdot (2 - h)$$

$$\Rightarrow G_{\alpha\beta} = C_1 \cdot (2 - h) \cdot g_{\alpha\beta} = C_1 \cdot (2 - h[x_{k=0,1,2,3}]) \cdot g_{\alpha\beta}.$$

$$(1106)$$

As it is widely assumed that the universe should have some kind of flat ground state, we obviously have a quite reasonable structure. While the metric $g_{\alpha\beta}$ may be seen as a pure gravity motivated solution (perhaps to the classical Einstein field equations (EFE)), the "perturbated" metric $G_{\alpha\beta}$ would effectively consist of the classical EFE solution plus the perturbation, with the latter definitively being of quantum origin. In other words: In the approximated case where we have $1 \gg \frac{C_1 \cdot f}{2 \cdot C_2}$ (in 4 dimensions) or $f \ll C_1$ (in 6 dimensions), the quantum fields add upon the gravitational metric distortions. In the flat case, the quantum field even is the only source of distortion. Only for very strong quantum fields, where the conditions $1 \gg \frac{C_1 \cdot f}{2 \cdot C_2}$, $f \ll C_1$ and so on do not hold, we need to worry about the original Eqs. (1097) and (1102) (both last line).

The same principal result is also obtained for all other numbers of dimensions except $n = 2$. However, we shall leave it to the interested reader to perform the evaluations as presented for $n = 4$ and $n = 6$ also for other numbers of dimensions, thereby applying the results presented in section 3.3.4.

15.2 Extraction of the Dirac Equation (Trials)

With the derivation of the Klein–Gordon equations in various dimension from our metric approach as shown in section 15.1 above, we might already be satisfied and simply assume that, just as Dirac did [143], his famous first-order differential equation would evolve as a square root out of the Klein–Gordon equation. Having derived the latter, the task could be considered completed. However, as said earlier also in this book (c.f. section 3.11.7), we merely consider the success of the Dirac equation a "peculiar accident" and think that there is a deeper metric meaning behind it. Not being satisfied with the complicated second-order differential equations resulting from our quantum gravity approach above (c.f. Eqs. (1097) and (1102) (both last line)) anyway, we here intend to derive the Dirac equation in a different way than done by Dirac in [143]. In order to easily compare our result with the original work of Dirac, however, we restrict the consideration here to the case of 4 dimensions with respect to the finish and also stick to Cartesian coordinates in order to be able to apply the original Dirac matrices. In section 15.2.1, we are then going to generalize the recipe and in order to keep the task there simple and also in order to provide a good connection to the derivations here, we will already keep the general metric terms and expressions wherever possible.

Later in section 15.7, we will show how a truly metric Dirac equation can be derived in an arbitrary number of dimensions and arbitrary coordinates.

For the simplest possible extraction of the Dirac equation, we go back to our Einstein–Hilbert action (Eq. (1091)) and perform the variation with respect to the functional $F[f]$ as follows:

$$\delta W = 0 = \delta \int_V d^n x \left(\sqrt{-G} \cdot R \right) = \delta \left(\delta_{F[f] \to f} \int_V d^n x \left(\sqrt{-G} \cdot R \right) \right).$$

(1107)

Again using the adapted metric (Eq. (1093)), the "inner" variation $\delta_{F[f] \to f}$ automatically makes all first derivatives of $F[f]$ with respect to f to zero (in other words: we are looking for minima of the functional F with respect to its core function f), leaving Eq. (1094) as follows (number of dimensions at first arbitrary, then fixed to $n = 4$ and $n = 6$):

$$R = \xrightarrow{g_{\alpha\beta}^{\text{diagonal}}} = -\frac{C_{N2}}{F[f]^2} \cdot \frac{\partial^2 F[f]}{\partial f^2} \cdot \left(\tilde{\nabla}_g f \right)^2 \begin{cases} \xrightarrow{n=4} = -\frac{3}{F[f]^2} \cdot \frac{\partial^2 F[f]}{\partial f^2} \cdot \left(\tilde{\nabla}_g f \right)^2 \\ \xrightarrow{n=6} = -\frac{5}{F[f]^2} \cdot \frac{\partial^2 F[f]}{\partial f^2} \cdot \left(\tilde{\nabla}_g f \right)^2 \end{cases}.$$

(1108)

Now we simply take it that independent on the function F the total action would give zero if the operator $\left(\tilde{\nabla}_g f \right)^2$ results in zero. Now we assume $f = f[t, x, y, z]$ with the four coordinates t, x, y, z to be arranged in the Minkowski metric and find that the operator $\left(\tilde{\nabla}_g f \right)^2$ can be given as:

$$\left(\tilde{\nabla}_g f \right)^2 = f_{,\alpha} g^{\alpha\beta} f_{,\beta} = -\frac{(\partial_t f)^2}{c^2} + (\partial_x f)^2 + (\partial_y f)^2 + (\partial_z f)^2.$$

(1109)

In the case of 6 dimensions, we assume the same dependency for f, but have two additional dimensions u and v which are taken to be entangled, thereby producing mass as evaluated in section 6.3. This time, the operator result for the Ricci scalar (Eq. (1094)) reads:

$$R = -\frac{5}{F[f]^2} \cdot \frac{\partial^2 F[f]}{\partial f^2} \cdot \left(\tilde{\nabla}_g f \right)^2 = \frac{C}{F[f]^2} \cdot f_{,\alpha} g^{\alpha\beta} f_{,\beta}$$

$$= \frac{C}{F[f]^2} \cdot \left(-\frac{(\partial_t f)^2}{c^2} + (\partial_x f)^2 + (\partial_y f)^2 + (\partial_z f)^2 - M^2 (C_1 + f)^2 \right) = 0.$$

(1110)

We realize that factorization is possible as a straight quaternion root for Eq. (1109) and, by the means of the third binomial equation combined with that root (so just as Dirac did in [143]), also for Eq. (1110). This leads to a factorized Eq. (1110) as follows:

$$(\gamma^\sigma \partial_\sigma f - i \cdot (C_1 + f) \cdot M)(\gamma^\sigma \partial_\sigma f + i \cdot (C_1 + f) \cdot M) = 0. \qquad (1111)$$

Thereby we have used the Dirac matrices:

$$\gamma^0 = \begin{pmatrix} 1 & & & \\ & 1 & & \\ & & -1 & \\ & & & -1 \end{pmatrix}; \quad \gamma^1 = \begin{pmatrix} & & & 1 \\ & & 1 & \\ & -1 & & \\ -1 & & & \end{pmatrix}$$

$$\gamma^2 = \begin{pmatrix} & & & -i \\ & & i & \\ & i & & \\ -i & & & \end{pmatrix}; \quad \gamma^3 = \begin{pmatrix} & & 1 & \\ & & & -1 \\ -1 & & & \\ & 1 & & \end{pmatrix}; \quad I = \begin{pmatrix} 1 & & & \\ & 1 & & \\ & & 1 & \\ & & & 1 \end{pmatrix}. $$

$$(1112)$$

It suffices that just one of the two factors in Eq. (1111) gives zero in order to fully satisfy the Einstein–Hilbert action condition (Eq. (1091)), which is to say, we only need either:

$$\gamma^\sigma \partial_\sigma f - i \cdot (C_1 + f) \cdot M = 0 \qquad (1113)$$

or:

$$\gamma^\sigma \partial_\sigma f + i \cdot (C_1 + f) \cdot M = 0. \qquad (1114)$$

These are the classical Dirac equations from [143] only that this time they are true roots, resulting from the inner variation of the Einstein–Hilbert action and not postulated Klein–Gordon operator decompositions as in the Dirac theory. The parameter M is given as $M = \frac{m \cdot c}{\hbar}$ with c = speed of light in vacuum, m = rest mass, $\hbar$ = reduced Planck constant.

"Sadly only" that we did not automatically obtain the function f as vector. Because, just as Dirac realized, we see that the 2-rank structure of the Dirac matrices forces the functions f to be vectors or some 4-component forms. But we will see in the following that our new origin for the Dirac equation might not need such an (after all, postulated) extension.

15.2.1 Towards a Curvilinear Generalization

Observing the structure of the $\left(\tilde{\nabla}_g f\right)^2$ operator in Eq. (1109), we see that a general factorization should be possible in curvilinear coordinates by the means of the base vector decomposition of the metric in the following form:

$$\left(\tilde{\nabla}_g f\right)^2 = f_{,\alpha}\mathbf{e}^\alpha \cdot \mathbf{e}^\beta f_{,\beta} \xrightarrow{\quad g_{\alpha\beta}^{\text{diagonal}} \quad} = f_{,\alpha}\mathbf{e}^\alpha \cdot \mathbf{e}^\alpha f_{,\alpha}. \qquad (1115)$$

Now we can separate the two factors and the correct (Dirac-compatible) product simply results from the scalar product of the corresponding base vectors $\mathbf{e}^\alpha$, $\mathbf{e}^\beta$. Thus, a generalized form of the Dirac equations (1113) and (1114) should just be:

$$f_{,\alpha} \mathbf{e}^\alpha - i \cdot (C_1 + f) \cdot \mathbf{M} = 0 \tag{1116}$$

and:

$$\mathbf{e}^\beta f_{,\beta} + i \cdot (C_1 + f) \cdot \mathbf{M} = 0, \tag{1117}$$

respectively (with: $\mathbf{M} \cdot \mathbf{M} = M^2$). Please note that only one of the two equations (1116) or (1117) needs to be fulfilled. We have to realize that in the metric picture, apparently, not the function f becomes a vector, but the parameter M, which is associated with mass, spin (c.f. section 6.4) and possibly also other "matter things". This has quite some consequences regarding our notion of inertia. One might even call it a paradigm shift, but we shall leave it to the reader to move forward at this point, while we intend to go on with our investigation of other options in connection with possible Dirac extractions out of the Einstein–Hilbert action.

Without the classical "spinor" appearance for the f-solutions above, one might wonder where now the spin could come from, but here we remind of the appearance of $l = 1/2*k$-solutions ($k = 1, 2, 3, \ldots$) for the main momentum of the classical Schrödinger hysitron solution in section 3.4 and the entanglement of dimensions leading to spin-like parameters in section 6.4.

It should be explicitly mentioned, however, that, just as shown in sections 5.2 and 10.2 there is also the option of actually extending the scalar f-approach (Eq. (1093)) into a vector approach. But as we do not see the need right now, because this here should be "the simplest" example, we refrain from investigating this possibility here and leave it for later (c.f. section 15.7.1) instead.

15.2.2 A Few Thoughts about Further Variations

Please note that in principle, our resulting equation from Eq. (1110) is still under the variational integral, thereby reading (no variation with respect to the metric as this has been considered already done):

$$\delta W = 0 = \delta \int_V d^n x \left(\sqrt{-G} \cdot R \right)$$

$$= -\int_V d^n x \left(\sqrt{-F\,[f]^n} \cdot g \cdot \frac{C_{N2}}{F\,[f]^2} \cdot \frac{\partial^2 F\,[f]}{\partial f^2} \cdot \delta \left(f_{,\alpha} g^{\alpha\beta} f_{,\beta} \right) \right)$$

$$= -\int_V d^n x \left(\sqrt{-F\,[f]^n} \cdot g \cdot \frac{C_{N2}}{F\,[f]^2} \cdot \frac{\partial^2 F\,[f]}{\partial f^2} \cdot g^{\alpha\beta} \delta \left(f_{,\alpha} f_{,\beta} \right) \right).$$

$$(1118)$$

From there, we could move on as follows:

$$\delta W = 0$$

$$= -\int_V d^n x \left(\overbrace{\sqrt{F\,[f]^n} \cdot \frac{C_{N2}}{F\,[f]^2} \cdot \frac{\partial^2 F\,[f]}{\partial f^2}}^{(\ldots)} \cdot \left(\sqrt{-g} \cdot g^{\alpha\beta} f_{,\alpha} \cdot \delta f_{,\beta} + \sqrt{-g} \cdot g^{\alpha\beta} f_{,\beta} \cdot \delta f_{,\alpha} \right) \right)$$

$$= \int_V d^n x \left(\left[(\ldots) \cdot \sqrt{-g} \cdot g^{\alpha\beta} f_{,\alpha} \right]_{,\beta} + \left[(\ldots) \cdot \sqrt{-g} \cdot g^{\alpha\beta} f_{,\beta} \right]_{,\alpha} \right) \cdot \delta f$$

$$= \int_V d^n x \sqrt{-g} \cdot \left((\ldots)_{,\beta} \cdot g^{\alpha\beta} f_{,\alpha} + (\ldots)_{,\alpha} \cdot g^{\alpha\beta} f_{,\beta} + 2 \cdot \Delta f \right) \cdot \delta f.$$

$$(1119)$$

This holds in any number of dimensions, whereas the constants C_{N2} are different for each n and are given in the table in section 15.4 (in fact, we simply have: $C_{N2} = n - 1$, with n giving the number of dimensions). The reader may have realized that we have used integration by parts in the second line and that it is possible to further compactify the last line as follows:

$$\delta W = 0 = \int_V d^n x \sqrt{-g} \cdot \left((\ldots)_{,\beta} \cdot g^{\alpha\beta} f_{,\alpha} + (\ldots)_{,\alpha} \cdot g^{\alpha\beta} f_{,\beta} + 2 \cdot \Delta f \right) \cdot \delta f$$

$$= \int_V d^n x \sqrt{-g} \cdot 2 \cdot \left((\ldots)^{,\alpha} f_{,\alpha} + \Delta f \right) \cdot \delta f$$

$$\Rightarrow 0 = (\ldots)^{,\alpha} f_{,\alpha} + \Delta f = \left(\sqrt{F\,[f]^n} \cdot \frac{C_{N2}}{F\,[f]^2} \cdot \frac{\partial^2 F\,[f]}{\partial f^2} \right)^{,\alpha} f_{,\alpha} + \Delta f.$$

$$(1120)$$

Similarly, the integration by parts in Eq. (1119) may also be performed as follows:

$$\delta W = 0 = -\int_V d^n x \left(\overbrace{\sqrt{F\,[f]}^n \cdot \frac{C_{N2}}{F\,[f]^2} \cdot \frac{\partial^2 F\,[f]}{\partial f^2}} \cdot \left(\sqrt{-g} \cdot g^{\alpha\beta} f_{,\alpha} \cdot \delta f_{,\beta} + \sqrt{-g} \cdot g^{\alpha\beta} f_{,\beta} \cdot \delta f_{,\alpha} \right) \right)$$

$$= \int_V d^n x \left(\left[(\ldots) \cdot \sqrt{-g} \cdot g^{\alpha\beta} f_{,\alpha} + (\ldots) \cdot \sqrt{-g} \cdot g^{\alpha\beta} f_{,\beta} \right]_{,\beta} + \left[(\ldots) \cdot \sqrt{-g} \cdot g^{\alpha\beta} f_{,\beta} \right]_{,\alpha} \right) \cdot \delta f$$

$$= \int_V d^n x \sqrt{-g} \cdot \left((\ldots)_{,\beta} \cdot g^{\alpha\beta} f_{,\alpha} + (\ldots)_{,\alpha} \cdot g^{\alpha\beta} f_{,\beta} + 2 \cdot \Delta f \right) \cdot \delta f. \tag{1121}$$

By assuming that the metric $g_{\alpha\beta}$ as previously obtained from a variational task of the very same Einstein–Hilbert action, which itself was only varied with respect to the metric not including the functional $F\,[f]$, namely:

$$\delta_g W = 0 = \delta_g \int_V d^n x \left(\sqrt{-g} \cdot R \right), \tag{1122}$$

we shall consider the complete task consisting of Eq. (1122), leading to the Einstein field equations, and Eq. (1120) (last line) as a quantum gravity set of equations. The functions F, f and the metric $g_{\alpha\beta}$ are thereby the set's solutions.

It is kind of interesting to notice that exactly in 4 dimensions the last line of Eq. (1120) simplifies to:

$$\xrightarrow{n=4} 0 = (\ldots)^{,\alpha} f_{,\alpha} + \Delta f = i \cdot \left(C_{N2} \cdot \frac{\partial^2 F\,[f]}{\partial f^2} \right)^{,\alpha} f_{,\alpha} + \Delta f. \tag{1123}$$

Properly observing the variation Eq. (1118), we can conclude that also a constant solution to the kernel should fulfill the extremal task. Then we would look for a solution to the following equation:

$$\delta W = 0 = \delta \int_V d^n x \left(\sqrt{-G} \cdot R \right) \Rightarrow \delta f_{,\alpha} g^{\alpha\beta} f_{,\beta} = 0 \Rightarrow f_{,\alpha} g^{\alpha\beta} f_{,\beta} = \text{const} = C_D^2.$$

$$\tag{1124}$$

Interestingly we already considered such solutions for the particle at rest case in a flat space situation (section 5.2.2) and for the $f\,[r]$-case for a Schwarzschild metric (section 10.2) In both cases, we obtained sets of solutions showing suitable behavior with respect to the Pauli principle [175].

So it looks like we have a choice of either solving the extremal principle via Eq. (1124) and the subsequent Dirac equations, or via the Klein–Gordon like form (Eq. (1120), last line).

15.2.3 Schwarzschild Metric for Entertainment

For illustration, we want to consider the stationary Schwarzschild problem with radius dependency only, which is to say $f = f[r]$. For simplicity, we bring Eq. (1124) into the Dirac form:

$$f_{,\alpha} \mathbf{e}^{\alpha} - i \cdot (C_1 + f) \cdot \mathbf{M} = \mathbf{C}_D^-, \tag{1125}$$

$$\mathbf{e}^{\beta} f_{,\beta} + i \cdot (C_1 + f) \cdot \mathbf{M} = \mathbf{C}_D^+ \tag{1126}$$

and demand $-i \cdot C_1 \cdot \mathbf{M} = \mathbf{C}_D^-$, $i \cdot C_1 \cdot \mathbf{M} = \mathbf{C}_D^+$. The resulting equations give the following solutions:

$$f[r] = C_r \cdot e^{\pm i \cdot M \cdot \left(r \cdot \sqrt{1 - \frac{r_s}{r}} + \frac{1}{2} \cdot r_s \cdot \log\left[2 \cdot r \cdot \left(1 + \sqrt{1 - \frac{r_s}{r}} \right) - r_s \right] \right)}. \tag{1127}$$

Thereby we have assumed the following metric with "entangled mass" M (c.f. section 6.3):

$$g_{\alpha\beta}^6 = \begin{pmatrix} -c^2 \cdot h[r] & 0 & 0 & 0 & 0 & 0 \\ 0 & \frac{1}{h[r]} & 0 & 0 & 0 & 0 \\ 0 & 0 & r^2 & 0 & 0 & 0 \\ 0 & 0 & 0 & r^2 \cdot \sin[\vartheta] & 0 & 0 \\ 0 & 0 & 0 & 0 & g[v] & 0 \\ 0 & 0 & 0 & 0 & 0 & g[v] \end{pmatrix}$$

$$\text{with:} \quad h[r] = 1 - \frac{r_s}{r}; \quad g[v] = \frac{C_{v1}^2}{C_{v0}^2 \cdot (1 + \cosh[C_{v1} \cdot (v + C_{v2})])};$$

$$\frac{C_{v0}^2}{5} = M^2. \tag{1128}$$

Figures 15.1 to 15.6 give some insight into the metric distortion caused by the function $f[r]$ for a variety of parameter settings. For simplicity, we fixed $C_r = 1$.

For better orientation and in order to make the absolute curves distinguishable from the real curves, a small constant was added to the $|f[r]|$ graphs in all of the figures.

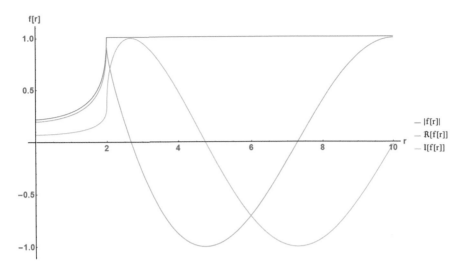

Figure 15.1 Absolute (blue), real (yellow) and imaginary (green) part of $f[r]$ according to the Dirac Schwarzschild solution (Eq. (1127)) for $M = 0.5$ and $r_s = 2$.

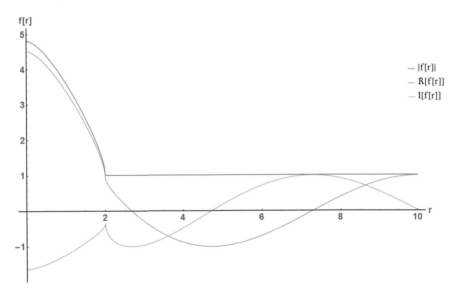

Figure 15.2 Absolute (blue), real (yellow) and imaginary (green) part of $f[r]$ according to the Dirac Schwarzschild solution (Eq. (1127)) for $M = -0.5$ and $r_s = 2$.

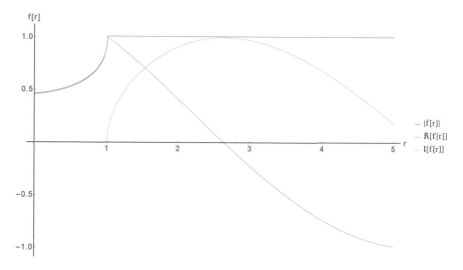

Figure 15.3 Absolute (blue), real (yellow) and imaginary (green) part of $f[r]$ according to the Dirac Schwarzschild solution (Eq. (1127)) for $M = 0.5$ and $r_s = 1$.

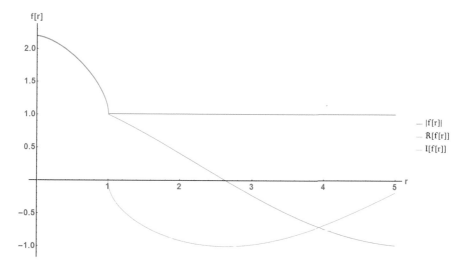

Figure 15.4 Absolute (blue), real (yellow) and imaginary (green) part of $f[r]$ according to the Dirac Schwarzschild solution (Eq. (1127)) for $M = -0.5$ and $r_s = 1$.

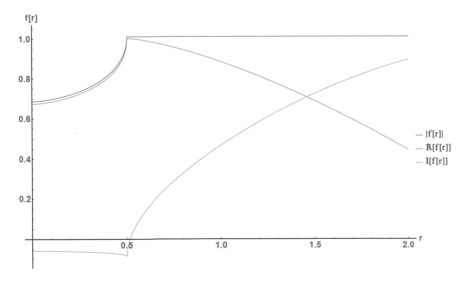

Figure 15.5 Absolute (blue), real (yellow) and imaginary (green) part of $f[r]$ according to the Dirac Schwarzschild solution (Eq. (1127)) for $M = 0.5$ and $r_s = 0.5$.

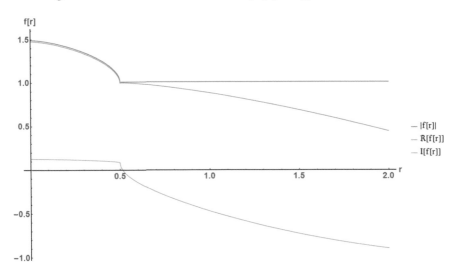

Figure 15.6 Absolute (blue), real (yellow) and imaginary (green) part of $f[r]$ according to the Dirac Schwarzschild solution (Eq. (1127)) for $M = -0.5$ and $r_s = 0.5$.

15.3 Extraction of More Dirac-Like Equations

This time our starting point shall be the metric result in Eq. (1094). Instead of finding a wrapper function $F[f]$ satisfying condition (Eq. (1096)), we now seek for an $F = (C1+f)^q$ in such a way that Eq. (1094) becomes factorizable. We start with the attempt to have something like $\left(\tilde{\nabla}_g f\right)^2 + \frac{(C_1+f)^2 \cdot const}{\sqrt{g}} = 0$:

$$
R \overset{g_{\alpha\beta}^{\text{diagonal}}}{=\!=\!=} \frac{1}{F[f]^3} \cdot \left(\begin{pmatrix} C_{N1} \cdot \left(\frac{\partial F[f]}{\partial f}\right)^2 - C_{N2} \cdot F[f] \cdot \frac{\partial^2 F[f]}{\partial f^2} \end{pmatrix} \cdot \left(\tilde{\nabla}_g f\right)^2 \\ - C_{N2} \cdot F[f] \cdot \frac{\partial F[f]}{\partial f} \cdot \Delta_g f \right)
$$

$$
= \left(C_{N3} \cdot (C_1 + f)^{-q-2} \cdot \left(\tilde{\nabla}_g f\right)^2 - \frac{C_{N2}}{F[f]^2} \cdot \frac{\partial F[f]}{\partial f} \cdot \Delta_g f \right)
$$

$$
= (C_1 + f)^{-q-2} \cdot \left(C_{N3} \cdot \left(\tilde{\nabla}_g f\right)^2 - C_{N4} \cdot (C_1 + f) \cdot \Delta_g f \right)
$$

$$
\overset{C_{N4} \cdot \Delta_g f = \frac{C_{N5}}{\sqrt{g}} \cdot (C_1+f)}{=\!=\!=} C_{N3} \cdot (C_1 + f)^{-q-2} \cdot \left(\tilde{\nabla}_g f\right)^2 - \frac{C_{N5}}{\sqrt{g}} \cdot (C_1 + f)^{-q} \equiv \frac{K}{\sqrt{(C_1 + f)^{q \cdot n} \cdot g}}
$$

$$
\overset{q=q;\, n=2}{\underset{K_0^2 = K + C_{N5}}{=\!=\!=}} C_{N3} \cdot \left(\tilde{\nabla}_g f\right)^2 \equiv \frac{(C_1 + f)^2 \cdot K_0^2}{\sqrt{g}} \quad \Rightarrow \quad 0 = C_{N3} \cdot \left(\tilde{\nabla}_g f\right)^2 - \frac{(C_1 + f)^2 \cdot K_0^2}{\sqrt{g}}.
$$

$$(1129)$$

Please note that this time we choose the constant K to be a square of a constant leading to K_0^2. This we only did for convenience to incorporate the eigenvalue of the Klein–Gordon (part) equation $\overset{C_{N5} \cdot \Delta_g f = \frac{C_{N5}}{\sqrt{g}} \cdot (C_1+f)}{\longrightarrow}$ in Eq. (1129), which is just there as a condition, and with respect to the upcoming evaluations. Unfortunately, this way is only possible in 2 dimensions (but leaves q arbitrary). It should also be noted that due to the necessary fixing of $n = 2$ (no other solutions are possible to obtain $\left(\tilde{\nabla}_g f\right)^2 + \frac{(C_1+f)^2 \cdot const}{\sqrt{g}} = 0$, because as one can easily prove they would all result in $q = 0$), we automatically only have $C_{3N} = C_{4N} = 1$ (c.f. table below).

Nevertheless, we want to factorize the last equation in order to get something like:

$$
0 = \left(\sqrt{\sqrt{g} \cdot C_{N3}} \cdot \left[\gamma^g \tilde{\nabla}_g\right] f - (C_1 + f) \cdot K_0 \right)
$$
$$
\times \left(\sqrt{\sqrt{g} \cdot C_{N3}} \cdot \left[\gamma^g \tilde{\nabla}_g\right] f + (C_1 + f) \cdot K_0 \right), \quad (1130)
$$

where the structure of the terms $\left[\gamma^g \tilde{\nabla}_g\right] f$ is yet unknown. However, one can easily prove that in Cartesian coordinates the Pauli matrices σ^j would do the job. We would obtain the following:

$$0 = \left(\sqrt{\sqrt{g} \cdot C_{N3}} \cdot \sigma^j \partial_j f + (C_1 + f) \cdot K_0 \right)$$

$$\times \left(\sqrt{\sqrt{g} \cdot C_{N3}} \cdot \sigma^j \partial_j f + (C_1 + f) \cdot K_0 \right). \tag{1131}$$

Each of the two factors in Eq. (1131) would then be a solution to the last line in Eq. (1129) via:

$$0 = \left(\sqrt{\sqrt{g} \cdot C_{N3}} \cdot \sigma^j \partial_j f - (C_1 + f) \cdot K_0 \right)$$

$$0 = \left(\sqrt{\sqrt{g} \cdot C_{N3}} \cdot \sigma^j \partial_j f + (C_1 + f) \cdot K_0 \right). \tag{1132}$$

From the many solutions one can obtain from Eq. (1132), only those are valid which also satisfy the eigenvalue equation in the fourth line of Eq. (1129), meaning:

$$0 = \Delta_g f - \frac{C_{N5}}{C_{N4} \cdot \sqrt{g}} \cdot (C_1 + f). \tag{1133}$$

Not being satisfied with the situation that our approach only works in 2 dimensions, we adapt Eq. (1129) as follows:

$$R \xrightarrow{\overset{\text{diagonal}}{g_{\alpha\beta}}} = (C_1 + f)^{-q-2} \cdot \left(C_{N3} \cdot \left(\tilde{\nabla}_g f \right)^2 - C_{N4} \cdot (C_1 + f) \cdot \Delta_g f \right)$$

$$\xrightarrow{C_{N4} \cdot \Delta_g f = C_{N4} \cdot C_{N3} \cdot A^2 \cdot (C_1 + f)} = C_{N3} \cdot (C_1 + f)^{-q-2} \cdot \left(\tilde{\nabla}_g f \right)^2 - C_{N4} \cdot C_{N3} \cdot A^2 \cdot (C_1 + f)^{-q}$$

$$\equiv \frac{K}{\sqrt{(C_1 + f)^{q \cdot n} \cdot g}}$$

$$\underset{n=\text{arbitrary}}{\overset{q = \frac{4}{n-2}}{\Longrightarrow}} C_{N3} \cdot \left(\tilde{\nabla}_g f \right)^2 - (C_1 + f)^2 \cdot C_{N4} \cdot C_{N3} \cdot A^2 = \frac{K}{\sqrt{g}} \quad / : C_{N3}$$

$$\xrightarrow{K/C_{N3}=K_0^2} \left(\tilde{\nabla}_g f \right)^2 - (C_1 + f)^2 \cdot C_{N4} \cdot A^2 = \frac{K_0^2}{\sqrt{g}}. \tag{1134}$$

This time factorization is possible by the means of the third binomial equation (so just as Dirac did in [143]) and leads to:

$$\left([\gamma^g \tilde{\nabla}_g] f - (C_1 + f) \cdot \sqrt{C_{N4}} \cdot A \right) \left([\gamma^g \tilde{\nabla}_g] f + (C_1 + f) \cdot \sqrt{C_{N4}} \cdot A \right) = \frac{K_0^2}{\sqrt{g}}. \tag{1135}$$

Going over to Cartesian, respectively Minkowski coordinates where our Nabla operator expression from Eq. (1134) becomes:

$$\left(\tilde{\nabla}_g f \right)^2 = f_{,\alpha} g^{\alpha\beta} f_{,\beta} = -\frac{(\partial_t f)^2}{c^2} + (\partial_x f)^2 + (\partial_y f)^2 + (\partial_z f)^2, \tag{1136}$$

one can easily prove that the Dirac matrices, here now as:

$$\gamma_{Dirac}^{\beta} \simeq i \cdot \gamma^{\beta} \quad \text{with}: \quad \gamma^0 = \begin{pmatrix} 1 & & & \\ & 1 & & \\ & & -1 & \\ & & & -1 \end{pmatrix}; \quad \gamma^1 = \begin{pmatrix} & & & 1 \\ & & 1 & \\ & -1 & & \\ -1 & & & \end{pmatrix};$$

$$\gamma^2 = \begin{pmatrix} & & & -i \\ & & i & \\ & i & & \\ -i & & & \end{pmatrix}; \quad \gamma^3 = \begin{pmatrix} & & 1 & \\ & & & -1 \\ -1 & & & \\ & 1 & & \end{pmatrix}; \quad I = \begin{pmatrix} 1 & & & \\ & 1 & & \\ & & 1 & \\ & & & 1 \end{pmatrix},$$

$$(1137)$$

would just do the job and we could write Eq. (1130) as follows:

$$\left(\gamma^\sigma \partial_\sigma f - (C_1 + f) \cdot \sqrt{C_{N4}} \cdot A \right) \left(\gamma^\sigma \partial_\sigma f + (C_1 + f) \cdot \sqrt{C_{N4}} \cdot A \right) = \frac{K_0^2}{\sqrt{g}}.$$

$$(1138)$$

Now we may also want to factorize the right hand side of Eq. (1138) and write:

$$\frac{K_0^2}{\sqrt{g}} = \frac{\mathbf{K}_0^-}{\sqrt{\sqrt{g}}} \cdot \frac{\mathbf{K}_0^+}{\sqrt{\sqrt{g}}}. \tag{1139}$$

We will mention further below that the product in Eq. (1138) could also be a scalar product (on both sides). We pointed this out by using bold K symbols in Eq. (1139). This should—of course—also render f and C_1 vectors, but we will leave the corresponding discussion for later and refrain from allocating corresponding markings to the symbols (here the reader is referred to section 15.7).

It suffices to allocate the two factors in Eq. (1138) in such a way that the subsequent product gives back the correct square of the last line in Eq. (1134), which we achieve via:

$$\gamma^\sigma \partial_\sigma f - (C_1 + f) \cdot \sqrt{C_{N4}} \cdot A = \frac{\mathbf{K}_0^-}{\sqrt{\sqrt{g}}}$$

$$\gamma^\sigma \partial_\sigma f + (C_1 + f) \cdot \sqrt{C_{N4}} \cdot A = \frac{\mathbf{K}_0^+}{\sqrt{\sqrt{g}}}. \tag{1140}$$

Please note that this would also solve our quantum gravity condition (Eq. (1092)) with the additional boundary of f also fulfilling the Klein–Gordon-like equation:

$$C_{N4} \cdot \Delta_g f = C_{N4} \cdot C_{N3} \cdot A^2 \cdot (C_1 + f)$$

$$\Rightarrow 0 = C_{N3} \cdot A^2 \cdot (C_1 + f) - \frac{\left(\sqrt{g} \cdot g^{\alpha\beta} f_{,\beta} \right)_{,\alpha}}{\sqrt{g}}. \tag{1141}$$

Thus, as already said above with respect to the case $n = 2$, from the set of possible solutions to Eq. (1140), we have to select those which are also eigenvalue solutions to Eq. (1141). In the following, we will now try to find a suitable combination of the two conditions.

As the constant A is "arbitrary"*, we can try to choose it such that there is a simultaneous solution to both equations, namely the Dirac-like Eq. (1138), respectively Eq. (1140) and the Klein–Gordon Eq. (1141). In order to properly prove this, we substitute $h = C_1 + f$, making Eq. (1140) to:

$$0 = \left(\gamma^\sigma \partial_\sigma - \sqrt{C_{N4}} \cdot A\right) h. \tag{1142}$$

Now, as a first trial, we apply the co- or complementary Dirac operator $\left(\gamma^\sigma \partial_\sigma + \sqrt{C_{N4}} \cdot A\right) \cdot k_0$ on both sides of Eq. (1142) and obtain:

$$0 = k_0 \cdot \left(\gamma^\sigma \partial_\sigma + \sqrt{C_{N4}} \cdot A\right)\left(\gamma^\sigma \partial_\sigma - \sqrt{C_{N4}} \cdot A\right) h = k_0 \cdot \left(\Delta_g - C_{N4} \cdot A^2\right) h$$

$$\Rightarrow 0 = k_0 \cdot \left(\Delta_g - C_{N4} \cdot A^2\right)(C_1 + f) \xrightarrow{k_0 = -1} = C_{N4} \cdot A^2 (C_1 + f) - \Delta_g f. \tag{1143}$$

Comparison with Eq. (1141) tells us that unless we do not have $C_{N3} = C_{N4}$, there is a discrepancy between our result from Eq. (1143) and the Klein–Gordon condition we needed to introduce in Eq. (1134). For convenience, here we repeat (see, e.g., [97]) the corresponding table of constants for the Ricci scalar evaluation according to the first line in Eq. (1129) (or, correspondingly and already driven a bit forward in the evaluation, the first line in Eq. (1134)):

N = number of dimensions	C_{N1}	C_{N2}	C_{N3}	C_{N4}
2	1	1	1**	1**
3	3/2	2	0	8
4	3/2	3	0	6
5	1	4	0	16/3
6	0	5	0	5
7	−3/2	6	0	24/5
8	−7/2	7	0	14/3
9	−6	8	0	32/7
10	−9	9	0	9/2

**As in the case of $n = 2$ the condition $q = \frac{4}{n-2}$ from Eq. (1134) cannot be applied, we used condition $q =$ arbitrary.

---

*Thereby we have to be aware that the Klein–Gordon equation (Eq. (1141)) is an eigenvalue equation and the eigenvalues $C_{N3} \cdot A^2$ are usually discrete. That is why we have used the "..." signs for the word "arbitrary".

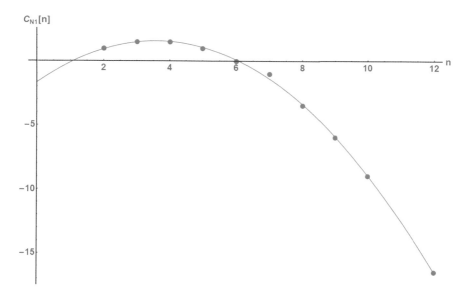

Figure 15.7 Coefficient C_{N1} from the transformed Ricci scalar as given in Eq. (1129) (first line) as function of the number of dimensions n. While the dots give the exact values (c.f. table above), the line shows a parabolic fit with an approach: $C_{N1} = c_0 + c_1*n + c_2*n^2$ (with c_0, c_1, c_2 as fitting constants).

As a side note, we point out that $C_{N2} = n - 1$, while there is a nearly parabolic behavior found for C_{N1} (see Fig. 15.7). Most interestingly, we find the maxima for C_{N1} at $n = 3$ and $n = 4$. We also find that the sum of the coefficients $C_{N1} + C_{N2}$ has its maximum at $n = 5$, 6 and 7 with $C_{N1} + C_{N2} = 5$ and give $C_{N1} + C_{N2} = 0$ for $n = 10$, which peculiarly is the number of dimensions most string- and brane-theoreticians consider as the right number of dimensions for our universe (see Fig. 15.8).

As we see that the combination $C_{N3} = C_{N4}$, does not exist (at least not for the cases we are interested in here, being $2 < n \leq 10$), we must conclude that the classical Dirac way of factorizing the Klein–Gordon equation is unsuitable for our purposes here. What is more, as we only find $C_{N3} = 0$—for all integer n—we are back with the "rather ugly" Laplace operator equation from section 15.1.

So, what now?

Well, obviously we require a better way to factorize the Klein–Gordon equation leaving us with a greater structural flexibility. As shown in [8] and quite some occasions in this book already (e.g., section 5.2 with the

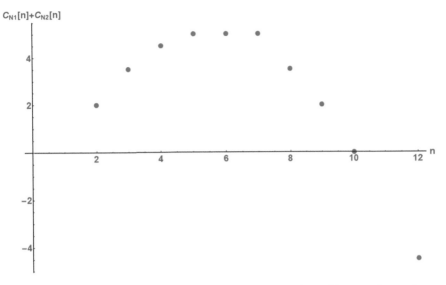

Figure 15.8 Sum of coefficients $C_{N1} + C_{N2}$ from the transformed Ricci scalar as given in Eq. (1129) (first line) as function of the number of dimensions n.

subsections "Towards Dirac"), we apply a vectorial type of root extraction to Eq. (1138), thereby avoiding the Dirac path of using the third binomial formula. The reader may easily prove that the following is correct:

$$\frac{1}{2} \left\{ \begin{array}{l} \gamma^\sigma \partial_\sigma f - (C_1 + f) \cdot \sqrt{C_{N4}} \cdot A \\ \gamma^\sigma \partial_\sigma f + (C_1 + f) \cdot \sqrt{C_{N4}} \cdot A \end{array} \right\} \left\{ \begin{array}{l} \gamma^\sigma \partial_\sigma f - (C_1 + f) \cdot \sqrt{C_{N4}} \cdot A \\ \gamma^\sigma \partial_\sigma f + (C_1 + f) \cdot \sqrt{C_{N4}} \cdot A \end{array} \right\}$$

$$= \left(\gamma^\sigma \partial_\sigma f - (C_1 + f) \cdot \sqrt{C_{N4}} \cdot A \right) \left(\gamma^\sigma \partial_\sigma f + (C_1 + f) \cdot \sqrt{C_{N4}} \cdot A \right)$$

$$= \frac{K_0^2}{\sqrt{g}} = \frac{\mathbf{K}_0^-}{\sqrt{\sqrt{g}}} \cdot \frac{\mathbf{K}_0^+}{\sqrt{\sqrt{g}}}. \tag{1144}$$

We could even have something rather asymmetric like:

$$\frac{1}{2} \left\{ \begin{array}{l} a \cdot \gamma^\sigma \partial_\sigma f - b \cdot (C_1 + f) \\ a \cdot \gamma^\sigma \partial_\sigma f + b \cdot (C_1 + f) \end{array} \right\} \left\{ \begin{array}{l} c \cdot \gamma^\sigma \partial_\sigma f - d \cdot (C_1 + f) \\ c \cdot \gamma^\sigma \partial_\sigma f + d \cdot (C_1 + f) \end{array} \right\} \xrightarrow[\ b \cdot d = -C_{N4} \cdot A^2\ ]{a \cdot c = 1}$$

$$= \left(\gamma^\sigma \partial_\sigma f - (C_1 + f) \cdot \sqrt{C_{N4}} \cdot A \right) \left(\gamma^\sigma \partial_\sigma f + (C_1 + f) \cdot \sqrt{C_{N4}} \cdot A \right)$$

$$= \left(\tilde{\nabla}_g f \right)^2 - (C_1 + f)^2 \cdot C_{N4} \cdot A^2 = \frac{K_0^2}{\sqrt{g}} = \frac{\mathbf{K}_0^-}{\sqrt{\sqrt{g}}} \cdot \frac{\mathbf{K}_0^+}{\sqrt{\sqrt{g}}}. \tag{1145}$$

Thus, we can construct a factorization of the last line of Eq. (1134) in the form of Eq. (1145) and correspondingly decompose the Klein–Gordon equation

(Eq. (1141)) just as follows:

$$\frac{1}{2} \left\{ \begin{array}{c} a_0 \cdot \gamma^\sigma \partial_\sigma - b_0 \\ a_0 \cdot \gamma^\sigma \partial_\sigma + b_0 \end{array} \right\} \left\{ \begin{array}{c} c \cdot \gamma^\sigma \partial_\sigma f - d\,(C_1 + f) \\ c \cdot \gamma^\sigma \partial_\sigma f + d\,(C_1 + f) \end{array} \right\} \xrightarrow[b_0 \cdot d = C_{N3} \cdot A^2]{a_0 \cdot c = -1}$$

$$= C_{N3} \cdot A^2 \,(C_1 + f) - \Delta_g f. \tag{1146}$$

This time the second factor on the left hand side of our decomposed Klein–Gordon equation agrees perfectly with the second factor of the left hand side of the vectorially decomposed last line of Eq. (1134) as given in the first line of Eq. (1145). As the system is underdetermined, we could just set $a = c$ and $b = -d$ and consequently obtain the other parameters as follows:

$$c = 1; \quad d = \sqrt{C_{N4}} \cdot A; \quad a_0 = -1; \quad b_0 = C_{N3} \cdot \sqrt{\frac{1}{C_{N4}}} \cdot A. \tag{1147}$$

Now we only need a suitable allocation of the vectors on the left with the corresponding ones on the right in Eq. (1145), leading us to:

$$\left\{ \begin{array}{c} c \cdot \gamma^\sigma \partial_\sigma f - d \cdot (C_1 + f) \\ c \cdot \gamma^\sigma \partial_\sigma f + d \cdot (C_1 + f) \end{array} \right\} = \sqrt{\frac{2}{\sqrt{g}}} \left\{ \begin{array}{c} \mathbf{K}_0^- \\ \mathbf{K}_0^+ \end{array} \right\} \quad \text{with}: K_0^2 = \mathbf{K}_0^- \cdot \mathbf{K}_0^- + \mathbf{K}_0^+ \cdot \mathbf{K}_0^+.$$

$$\tag{1148}$$

These clearly are Dirac equations.

We may see it as a drawback that, so far, we only find solutions (equations to solve) for the cases $n = 2$ (Eq. (1132)). So, we go back to Eq. (1129), but this time we apply the following q-parameters and approaches for the Laplace operator term:

$$R = \xrightarrow[]{g_{\alpha\beta}^{\text{diagonal}}} = (C_1 + f)^{-q-2} \cdot \left(C_{N3} \cdot \left(\tilde{\nabla}_g f \right)^2 - C_{N4} \cdot (C_1 + f) \cdot \Delta_g f \right)$$

$$\xrightarrow[]{C_{N4} \cdot \Delta_g f = C_{N3} \cdot \frac{A^2}{\sqrt{g}}} = C_{N3} \cdot (C_1 + f)^{-q-2} \cdot \left(\tilde{\nabla}_g f \right)^2 - \frac{C_{N3}}{\sqrt{g}} \cdot A^2 \cdot (C_1 + f)^{-q-1}$$

$$= -\frac{K}{\sqrt{(C_1 + f)^{q \cdot n} \cdot g}}$$

$$\xRightarrow[n = \text{arbitrary}]{q = \frac{2}{n-2}} C_{N3} \cdot \left(\tilde{\nabla}_g f \right)^2 - (C_1 + f) \cdot \frac{C_{N3}}{\sqrt{g}} \cdot A^2 = (C_1 + f) \cdot \frac{K}{\sqrt{g}} \quad \Big/ : C_{N3}$$

$$\xrightarrow[]{K/C_{N3} = K_0^2} \left(\tilde{\nabla}_g f \right)^2 - \frac{(C_1 + f)}{\sqrt{g}} \cdot \left(K_0^2 + A^2 \right) = 0. \tag{1149}$$

In 4 dimensions, using the Dirac matrices, we now could just write:

$$\frac{1}{2}\left\{\begin{matrix}a\cdot\gamma^\sigma\partial_\sigma f-b\cdot\sqrt{C_1+f}\\a\cdot\gamma^\sigma\partial_\sigma f+b\cdot\sqrt{C_1+f}\end{matrix}\right\}\left\{\begin{matrix}c\cdot\gamma^\sigma\partial_\sigma f-d\cdot\sqrt{C_1+f}\\c\cdot\gamma^\sigma\partial_\sigma f+d\cdot\sqrt{C_1+f}\end{matrix}\right\}\xrightarrow[b\cdot d=-\frac{\left(K_0^2+A^2\right)}{\sqrt{g}}]{a\cdot c=1}$$

$$=\left(\gamma^\sigma\partial_\sigma f-\sqrt{(C_1+f)\cdot\frac{\left(K_0^2+A^2\right)}{\sqrt{g}}}\right)\left(\gamma^\sigma\partial_\sigma f+\sqrt{(C_1+f)\cdot\frac{\left(K_0^2+A^2\right)}{\sqrt{g}}}\right)$$

$$=\left(\tilde{\nabla}_g f\right)^2-(C_1+f)\cdot\frac{\left(K_0^2+A^2\right)}{\sqrt{g}}=0,\tag{1150}$$

where we recognize the Dirac equations again.

For completeness, we give the corresponding coefficients in the following table.

$N=$ **number of dimensions**	C_{N1}	C_{N2}	C_{N3}	C_{N4}
2	1	1	1**	1**
3	3/2	2	2	4
4	3/2	3	3/2	3
5	1	4	4/3	8/3
6	0	5	5/4	5/2
7	-3/2	6	6/5	12/5
8	-7/2	7	7/6	7/3
9	-6	8	8/7	16/7
10	-9	9	9/8	9/4

**As in the case of $n=2$ the condition $q=\frac{2}{n-2}$ from (1149) cannot be applied, we used condition $q=$ arbitrary.

We realize the following structures of the series:

$$C_{N3}=\left(\begin{matrix}1^{**}&\text{for}&n=2\\\frac{n-1}{n-2}&\text{for}&n>2\end{matrix}\right);\quad C_{N4}=\left(\begin{matrix}1^{**}&\text{for}&n=2\\2\cdot\frac{n-1}{n-2}&\text{for}&n>2\end{matrix}\right).\tag{1151}$$

Insisting on having the Laplace term developed to an eigenequation, we may evolve Eq. (1129) into the following:

$$R\xrightarrow[]{\overset{\text{diagonal}}{g_{\alpha\beta}}}=(C_1+f)^{-q-2}\cdot\left(C_{N3}\cdot\left(\tilde{\nabla}_g f\right)^2-C_{N4}\cdot(C_1+f)\cdot\Delta_g f\right)$$

$$\xrightarrow[]{C_{N4}\cdot\Delta_g f=-C_{N3}\cdot A^2\cdot(C_1+f)}=C_{N3}\cdot(C_1+f)^{-q-2}\cdot\left(\tilde{\nabla}_g f\right)^2+C_{N3}\cdot A^2\cdot(C_1+f)^{-q}$$

$$\equiv\frac{K}{\sqrt{(C_1+f)^{q\cdot n}\cdot g}}$$

$$\xrightarrow[n=\text{arbitrary}]{q=\frac{2}{n-2}} C_{N3} \cdot \left(\tilde{\nabla}_g f\right)^2 + (C_1 + f)^2 \cdot C_{N3} \cdot A^2$$

$$= (C_1 + f) \cdot \frac{K}{\sqrt{g}} \quad / : C_{N3}$$

$$\xrightarrow{K/C_{N3}=K_0^2} \left(\tilde{\nabla}_g f\right)^2 + (C_1 + f)^2 \cdot A^2 - (C_1 + f) \cdot \frac{K_0^2}{\sqrt{g}} = 0. \tag{1152}$$

This time, it is obviously much more difficult to find a suitable vectorial root. But the reader may prove that the following is correct:

$$\frac{1}{2} \left\{ \begin{array}{l} \gamma^\sigma \partial_\sigma f + A \cdot (C_1 + f) \\ \gamma^\sigma \partial_\sigma f - A \cdot (C_1 + f) \end{array} \right\} \left\{ \begin{array}{l} \gamma^\sigma \partial_\sigma f + A \cdot (C_1 + f) \\ \gamma^\sigma \partial_\sigma f - A \cdot (C_1 + f) \end{array} \right\}$$

$$= \frac{\mathbf{K_0}}{\sqrt{\sqrt{g}}} \sqrt{C_1 + f} \cdot \frac{\mathbf{K_0}}{\sqrt{\sqrt{g}}} \sqrt{C_1 + f}$$

$$\Leftrightarrow \quad \left(\tilde{\nabla}_g f\right)^2 + (C_1 + f)^2 \cdot A^2 = (C_1 + f) \cdot \frac{K_0^2 = \mathbf{K_0} \cdot \mathbf{K_0}}{\sqrt{g}}; \quad \mathbf{K_0} = \left\{ \begin{array}{l} K_0^+ \\ K_0^- \end{array} \right\}$$

$$\Rightarrow \quad \left\{ \begin{array}{l} \gamma^\sigma \partial_\sigma f + A \cdot (C_1 + f) \\ \gamma^\sigma \partial_\sigma f - A \cdot (C_1 + f) \end{array} \right\} = \left\{ \begin{array}{l} K_0^+ \\ K_0^- \end{array} \right\} \sqrt{2 \cdot \frac{C_1 + f}{\sqrt{g}}}. \tag{1153}$$

Similarly, we could also expand our approach to a 4-vector with now demanding a vanishing result for the scalar product:

$$\frac{1}{4} \left\{ \begin{array}{l} \gamma^\sigma \partial_\sigma f + A \cdot (C_1 + f) + i \cdot \frac{K_0}{\sqrt{\sqrt{g}}} \sqrt{C_1 + f} \\ \gamma^\sigma \partial_\sigma f + A \cdot (C_1 + f) - i \cdot \frac{K_0}{\sqrt{\sqrt{g}}} \sqrt{C_1 + f} \\ \gamma^\sigma \partial_\sigma f - A \cdot (C_1 + f) + i \cdot \frac{K_0}{\sqrt{\sqrt{g}}} \sqrt{C_1 + f} \\ \gamma^\sigma \partial_\sigma f - A \cdot (C_1 + f) - i \cdot \frac{K_0}{\sqrt{\sqrt{g}}} \sqrt{C_1 + f} \end{array} \right.$$

$$\left. \begin{array}{l} \gamma^\sigma \partial_\sigma f + A \cdot (C_1 + f) + i \cdot \frac{K_0}{\sqrt{\sqrt{g}}} \sqrt{C_1 + f} \\ \gamma^\sigma \partial_\sigma f + A \cdot (C_1 + f) - i \cdot \frac{K_0}{\sqrt{\sqrt{g}}} \sqrt{C_1 + f} \\ \gamma^\sigma \partial_\sigma f - A \cdot (C_1 + f) + i \cdot \frac{K_0}{\sqrt{\sqrt{g}}} \sqrt{C_1 + f} \\ \gamma^\sigma \partial_\sigma f - A \cdot (C_1 + f) - i \cdot \frac{K_0}{\sqrt{\sqrt{g}}} \sqrt{C_1 + f} \end{array} \right\}$$

$$= \left(\tilde{\nabla}_g f\right)^2 + (C_1 + f)^2 \cdot A^2 - (C_1 + f) \cdot \frac{K_0^2}{\sqrt{g}} = 0. \tag{1154}$$

15.4 Avoiding the Quaternions and Going to Curvilinear Coordinates (but Diagonal Metrics)

Just as Dirac had realized, we conclude from the 2-rank structure of the Dirac matrices that f has to be a vector of functions rather than a scalar, but as this book is already quite long, we leave this more or less formal discussion for later (or the interested reader, whoever comes first). We also refrain from investigating the vector issue here, because it was shown several times in this book already that there is also an alternative way to factorize Eq. (1138), thereby not necessarily needing Dirac's quaternions (Eq. (1137)). In some cases, the corresponding square equations (last line in Eqs. (1129) and (1134)) had even been solved and then given the typical vectors or spinors as solutions with the correct Pauli behavior (e.g., section 5.2 with the subsections "Towards Dirac" or section 10.2). In section 15.2.1, we also saw that there was no need for an **f**-vector structure because the correct factorization could be performed via a suitable metric tensor decomposition. Nevertheless, it is still possible to reasonably introduce the function-vector when seeking for suitable internal structures of the scalar f. This will be considered in the sections 15.5 and 15.7.

Here now we once again want to investigate the vectorial root extraction (e.g., [8]) as alternative way for the factorization of operators like Eq. (1110) without quaternions or base vectors. The reader may prove that the following is correct:

$$
V = \begin{cases}
i \cdot M\left(C_1 + f\right) + i \cdot \frac{\partial_t f}{c} + \partial_x f + \partial_y f + \partial_z f \\
i \cdot M\left(C_1 + f\right) + i \cdot \frac{\partial_t f}{c} + \partial_x f + \partial_y f - \partial_z f \\
i \cdot M\left(C_1 + f\right) + i \cdot \frac{\partial_t f}{c} + \partial_x f - \partial_y f + \partial_z f \\
i \cdot M\left(C_1 + f\right) + i \cdot \frac{\partial_t f}{c} + \partial_x f - \partial_y f - \partial_z f \\
i \cdot M\left(C_1 + f\right) + i \cdot \frac{\partial_t f}{c} - \partial_x f + \partial_y f + \partial_z f \\
i \cdot M\left(C_1 + f\right) + i \cdot \frac{\partial_t f}{c} - \partial_x f + \partial_y f - \partial_z f \\
i \cdot M\left(C_1 + f\right) + i \cdot \frac{\partial_t f}{c} - \partial_x f - \partial_y f + \partial_z f \\
\vdots \\
i \cdot M\left(C_1 + f\right) - i \cdot \frac{\partial_t f}{c} - \partial_x f - \partial_y f - \partial_z f
\end{cases}
$$

$$
\Rightarrow \frac{1}{16} V \cdot V = \left(-\frac{\left(\partial_t f\right)^2}{c^2} + \left(\partial_x f\right)^2 + \left(\partial_y f\right)^2 + \left(\partial_z f\right)^2 - M^2 \left(C_1 + f\right)^2 \right) = 0.
$$

$$(1155)$$

This author sees it as an interesting fact that the correct scalar product appears out of a vector where all derivative components would sum up to zero (a zero-sum). Even a total zero-sum would be possible with just a 32-component $\{\dots\}$-vector of the structure as given in Eq. (1155). We already saw this with our approaches in section 5.2. One may conclude at this point that the zero-sum, with its zero-summing components being properly separated in vector components, always sums-up in such a way that the zero is not being realized at once, seems to be a fundamental concept within this universe.

In other words: The various dimensions form vector components, which, if being summed-up straight away, would give zero. But with the universe keeping these vector components properly separated into different dimensions, the zero has not been realized. It needs to be pointed out that only this separation—obviously—makes it possible for something to be. Without it, which is to say without the separation, nothing could exist. This is essential and it totally contradicts all equality ideologies [44].

15.5 Going Back to the Elastic Space-Time and Forming Particles

15.5.1 A Few Auxiliary Calculations

Just as in the sections before, we apply the extended Einstein–Hilbert action (Eq. (1091)) and by incorporating Eq. (1093), the first line of Eq. (1094) and the first identity of Eq. (1136), we obtain:

$$\delta W = 0 = \delta \int_V d^n x \left(\sqrt{-G} \cdot R \right)$$

$$= \delta_{g,f} \int_V d^n x \, \frac{\sqrt{-F\,[f]}^{\,n} \cdot g}{F\,[f]^3} \cdot \left(\begin{array}{l} \left(C_{N1} \cdot \left(\frac{\partial F[f]}{\partial f} \right)^2 - C_{N2} \cdot F\,[f] \cdot \frac{\partial^2 F[f]}{\partial f^2} \right) \cdot f_{,\alpha} g^{\alpha\beta} f_{,\beta} \\ -C_{N2} \cdot F\,[f] \cdot \frac{\partial F[f]}{\partial f} \cdot \Delta_g f \end{array} \right).$$

$$\tag{1156}$$

Even without knowing $F\,[\dots]$ and actually performing the variation with respect to f, we immediately see that we have an elastic equation as part of the total variational result. This we simply deduce from our derivations in section 15.2.2, namely Eqs. (1118) to (1120). The total variation can be written as:

$$\delta W = 0 = \overbrace{\delta_g \int_V d^n x \frac{\sqrt{-F[f]^n \cdot g}}{F[f]^3} \cdot \left(\begin{array}{c} \left(C_{N1} \cdot \left(\frac{\partial F[f]}{\partial f} \right)^2 - C_{N2} \cdot F[f] \cdot \frac{\partial^2 F[f]}{\partial f^2} \right) \cdot f_{,\alpha} g^{\alpha\beta} f_{,\beta} \\ -C_{N2} \cdot F[f] \cdot \frac{\partial F[f]}{\partial f} \cdot \Delta_g f \end{array} \right)}^{\delta_g W}$$

$$+ \underbrace{\delta_f \int_V d^n x \frac{\sqrt{-F[f]^n \cdot g}}{F[f]^3} \cdot \left(\begin{array}{c} \left(C_{N1} \cdot \left(\frac{\partial F[f]}{\partial f} \right)^2 - C_{N2} \cdot F[f] \cdot \frac{\partial^2 F[f]}{\partial f^2} \right) \cdot f_{,\alpha} g^{\alpha\beta} f_{,\beta} \\ -C_{N2} \cdot F[f] \cdot \frac{\partial F[f]}{\partial f} \cdot \Delta_g f \end{array} \right)}_{\delta_f W}$$

$$\Rightarrow \delta_f W = \delta_{f_{,\alpha}|f_{,\beta}|\Delta_g f} W = \int_V d^n x \frac{\sqrt{F[f]^n}}{F[f]^3} \cdot \left(\begin{array}{c} \left(C_{N1} \cdot \left(\frac{\partial F[f]}{\partial f} \right)^2 - C_{N2} \cdot F[f] \cdot \frac{\partial^2 F[f]}{\partial f^2} \right) \times \\ (\sqrt{-g} \cdot g^{\alpha\beta} f_{,\alpha} \cdot \delta f_{,\beta} + \sqrt{-g} \cdot g^{\alpha\beta} f_{,\beta} \cdot \delta f_{,\alpha}) \cdot \\ -C_{N2} \cdot F[f] \cdot \frac{\partial F[f]}{\partial f} \cdot \sqrt{-g} \cdot \delta \Delta_g f \end{array} \right)$$

$$(1157)$$

Without loss of generality, we may also consider f an n-form f of some kind. However, in order not to compromise the simplicity of Eq. (1157), we leave the symbol f for the moment.

Now we define:

$$\frac{\sqrt{F[f]^n}}{F[f]^3} \cdot \left(C_{N1} \cdot \left(\frac{\partial F[f]}{\partial f} \right)^2 - C_{N2} \cdot F[f] \cdot \frac{\partial^2 F[f]}{\partial f^2} \right) \equiv -(\dots) \quad (1158)$$

and obtain:

$$\Rightarrow \delta_f W = \delta_{f_{,\alpha}|f_{,\beta}|\Delta_g f} W = \int_V d^n x \left(\begin{array}{c} \overbrace{-(\dots) \times (\sqrt{-g} \cdot g^{\alpha\beta} f_{,\alpha} \cdot \delta f_{,\beta} + \sqrt{-g} \cdot g^{\alpha\beta} f_{,\beta} \cdot \delta f_{,\alpha})}^{\text{here integration by parts}} \\ -C_{N2} \cdot \frac{\sqrt{F[f]^n}}{F[f]^2} \cdot \frac{\partial F[f]}{\partial f} \cdot \sqrt{-g} \cdot \delta \Delta_g f \end{array} \right)$$

$$= \int_V d^n x \sqrt{-g} \cdot \left(\begin{array}{c} ((\dots)_{,\beta} \cdot g^{\alpha\beta} f_{,\alpha} + (\dots)_{,\alpha} \cdot g^{\alpha\beta} f_{,\beta} + 2 \cdot (\dots) \cdot \Delta_g f) \cdot \delta f \\ -C_{N2} \cdot \frac{\sqrt{F[f]^n}}{F[f]^2} \cdot \frac{\partial F[f]}{\partial f} \cdot \delta \Delta_g f \end{array} \right)$$

$$= \int_V d^n x \sqrt{-g} \cdot \left(2 \cdot ((\dots)_{,\beta} \cdot g^{\alpha\beta} f_{,\alpha} + (\dots) \cdot \Delta_g f) \cdot \delta f - C_{N2} \cdot \frac{\sqrt{F[f]^n}}{F[f]^2} \cdot \frac{\partial F[f]}{\partial f} \cdot \delta \Delta_g f \right).$$

$$(1159)$$

Now we come back to what was said a few lines above and introduce an n-form $f \rightarrow \mathbf{f}$ as follows:

$$\delta_f W = \int_V d^n x \sqrt{-g} \cdot \left(\left(\overbrace{(\dots f \dots)}^{\equiv(\dots)}_{,\beta} \cdot g^{\alpha\beta} f_{,\alpha} + (\dots f \dots)_{,\alpha} \cdot g^{\alpha\beta} f_{,\beta} + 2 \cdot (\dots) \Delta_g f \right) \cdot \delta f \right.$$
$$\left. - C_{N2} \cdot \frac{\sqrt{F[f]^n}}{F[f]^2} \cdot \frac{\partial F[f]}{\partial f} \cdot \delta \Delta_g f \right)$$
$$= \int_V d^n x \sqrt{-g} \cdot \left(2 \cdot ((\dots)_{,\beta} \cdot g^{\alpha\beta} f_{,\alpha} + (\dots) \Delta_g f) \cdot \delta f - C_{N2} \cdot \frac{\sqrt{F[f]^n}}{F[f]^2} \cdot \frac{\partial F[f]}{\partial f} \cdot \delta \Delta_g f \right).$$

$$(1160)$$

Thereby we could easily choose a form as follows:

$$G^\gamma_{,\sigma} = \mathbf{f} \Rightarrow \delta_f W = \int_V d^n x \sqrt{-g} \cdot \left(\begin{array}{c} 2 \cdot ((\dots)_{,\beta} \cdot g^{\alpha\beta} G^\gamma_{,\sigma\alpha} + (\dots) \Delta_g G^\gamma_{,\sigma}) \cdot (\delta G^\gamma_{,\sigma})^\sigma_\gamma \\ - C_{N2} \cdot \frac{\sqrt{F[G^\gamma_{,\sigma}]^n}}{F[G^\gamma_{,\sigma}]^2} \cdot \frac{\partial F[G^\gamma_{,\sigma}]}{\partial G^\gamma_{,\sigma}} \cdot \delta_\gamma \Delta_g G^\gamma_{,\sigma} \end{array} \right).$$

$$(1161)$$

The new indices could also be α and β. This, however, distinguishes the addends in the first line of Eq. (1160) and we should change Eq. (1161) to:

$$\overbrace{\frac{1}{2} \cdot (G^\alpha_{,\sigma} + G^\beta_{,\sigma})}^{\ddots} = \mathbf{f}$$

$$\Rightarrow \delta_{\mathbf{f}} W = \int_V d^n x \sqrt{-g} \cdot \left(\begin{array}{c} \frac{1}{2} \cdot \left(\begin{array}{c} (\dots)_{,\beta} \cdot g^{\alpha\beta} G^\alpha_{,\sigma\alpha} + (\dots)_{,\beta} \cdot g^{\alpha\beta} G^\alpha_{,\sigma\beta} \\ + (\dots)_{,\beta} \cdot g^{\alpha\beta} G^\beta_{,\sigma\alpha} + (\dots)_{,\beta} \cdot g^{\alpha\beta} G^\beta_{,\sigma\beta} \\ + 2 (\dots) \Delta_g \left(G^\alpha_{,\sigma} + G^\beta_{,\sigma} \right) \end{array} \right) \cdot (\delta G^{\alpha|\beta}_{,\sigma})^\sigma_{\alpha|\beta} \\ - C_{N2} \cdot \frac{\sqrt{F\left[\frac{1}{2} \cdot (G^\alpha_{,\sigma} + G^\beta_{,\sigma})\right]^n}}{F\left[\begin{smallmatrix} \ddots \end{smallmatrix}\right]^2} \cdot \frac{\partial F\left[\begin{smallmatrix} \ddots \end{smallmatrix}\right]}{\partial \begin{smallmatrix} \ddots \end{smallmatrix}} \cdot \delta_\gamma \Delta_g \frac{1}{2} \cdot \left(G^\alpha_{,\sigma} + G^\beta_{,\sigma} \right) \end{array} \right).$$

$$(1162)$$

As this looks quite awkward, we reduce our approach for G to a scalar and reevaluate Eq. (1161) with the simplest possible form, which is a divergence, giving us:

$$G^\sigma_{,\sigma} = f$$

$$\Rightarrow \delta_f W = \int_V d^n x \sqrt{-g} \cdot \left(\begin{array}{c} 2 \cdot \overbrace{((\dots)_{,\beta} \cdot g^{\alpha\beta} G^\sigma_{,\sigma\alpha} + (\dots) \Delta_g G^\sigma_{,\sigma})}^{\text{elasticity}} \cdot (\delta G^\sigma_{,\sigma}) \\ - C_{N2} \cdot \frac{\sqrt{F[G^\sigma_{,\sigma}]^n}}{F[G^\sigma_{,\sigma}]^2} \cdot \frac{\partial F[G^\sigma_{,\sigma}]}{\partial G^\sigma_{,\sigma}} \cdot \delta \Delta_g G^\sigma_{,\sigma} \end{array} \right).$$

$$(1163)$$

Here now we clearly recognize the structure of the fundamental equation of elasticity in the first line of the integrand.

However, being interested in a potentially more general derivation, we apply the results from chapter 4 and variate in a more general manner with respect to the coordinates x^σ. This does change the last line in Eq. (1157) as follows:

$$\Rightarrow \delta_f W = \delta_\sigma W$$

$$= \int_V d^n x \left(\overbrace{\left(-(\dots)_{,\sigma} \cdot \sqrt{-g} \cdot g^{\alpha\beta} f_{,\alpha} f_{,\beta} - (\dots) \cdot \left(\begin{array}{c} \sqrt{-g} \cdot g^{\alpha\beta} f_{,\alpha} \cdot f_{,\beta\sigma} \\ +\sqrt{-g} \cdot g^{\alpha\beta} f_{,\beta} \cdot f_{,\alpha\sigma} \end{array} \right) \right)}^{\text{here integration by parts}} \delta x^\sigma \\ -C_{N2} \cdot \sqrt{-g} \cdot \delta \left[\frac{\sqrt{F[f]}^n}{F[f]^2} \cdot \frac{\partial F[f]}{\partial f} \cdot \Delta_g f \right] \right)$$

$$= \int_V d^n x \sqrt{-g} \cdot \left(\begin{array}{c} \left((\dots)_{,\sigma\beta} \cdot g^{\alpha\beta} f_{,\alpha} + 2 \cdot (\dots)_{,\sigma} \cdot \Delta_g f + (\dots)_{,\sigma\alpha} \cdot g^{\alpha\beta} f_{,\beta} \\ + (\dots)_{,\alpha} \cdot g^{\alpha\beta} f_{,\sigma\beta} + (\dots)_{,\beta} \cdot g^{\alpha\beta} f_{,\sigma\alpha} + 2 \cdot (\dots) \cdot \Delta_g f_{,\sigma} \end{array} \right) \cdot f \cdot \delta x^\sigma \\ -C_{N2} \cdot \delta \left[\frac{\sqrt{F[f]}^n}{F[f]^2} \cdot \frac{\partial F[f]}{\partial f} \cdot \Delta_g f \right] \right).$$

$$(1164)$$

Thereby we have assumed that the variation with respect to the metric has already been taken care of in the term for $\delta_g W$ and that therefore only F and f have to be considered within $\delta_f W$. We have also ignored any potential terms resulting from the integration by parts with respect to δx^σ. Please note that with respect to the integration by parts we have quite some options. Here is one alternative to Eq. (1164):

$$\Rightarrow \delta_f W = \delta_\sigma W$$

$$= \int_V d^n x \left(\left(-(\dots)_{,\sigma} \cdot \sqrt{-g} \cdot g^{\alpha\beta} f_{,\alpha} f_{,\beta} - (\dots) \cdot \overbrace{\left(\begin{array}{c} \sqrt{-g} \cdot g^{\alpha\beta} f_{,\alpha} \cdot f_{,\beta\sigma} \\ +\sqrt{-g} \cdot g^{\alpha\beta} f_{,\beta} \cdot f_{,\alpha\sigma} \end{array} \right)}^{\text{here integration by parts}} \right) \delta x^\sigma \\ -C_{N2} \cdot \sqrt{-g} \cdot \delta \left[\frac{\sqrt{F[f]}^n}{F[f]^2} \cdot \frac{\partial F[f]}{\partial f} \cdot \Delta_g f \right] \right)$$

$$= \int_V d^n x \left(\begin{array}{c} \left((\sqrt{-g} \cdot (\dots) \cdot g^{\alpha\beta} f_{,\alpha})_{,\sigma} \cdot f_{,\beta} + (\sqrt{-g} \cdot (\dots) \cdot g^{\alpha\beta} f_{,\beta})_{,\sigma} \cdot f_{,\alpha} \right) \cdot \delta x^\sigma \\ -(\dots)_{,\sigma} \cdot \sqrt{-g} \cdot g^{\alpha\beta} f_{,\alpha} f_{,\beta} \\ -C_{N2} \cdot \sqrt{-g} \cdot \delta \left[\frac{\sqrt{F[f]}^n}{F[f]^2} \cdot \frac{\partial F[f]}{\partial f} \cdot \Delta_g f \right] \end{array} \right)$$

$$= \int_V d^n x \left(\begin{array}{c} \left(2 \cdot (\sqrt{-g} \cdot (\dots) \cdot g^{\alpha\beta} f_{,\alpha})_{,\sigma} - (\dots)_{,\sigma} \cdot \sqrt{-g} \cdot g^{\alpha\beta} f_{,\alpha} \right) \cdot f_{,\beta} \cdot \frac{\delta x^\sigma}{2} \\ \left(2 \cdot (\sqrt{-g} \cdot (\dots) \cdot g^{\alpha\beta} f_{,\beta})_{,\sigma} - (\dots)_{,\sigma} \cdot \sqrt{-g} \cdot g^{\alpha\beta} f_{,\beta} \right) \cdot f_{,\alpha} \cdot \frac{\delta x^\sigma}{2} \\ -C_{N2} \cdot \sqrt{-g} \cdot \delta \left[\frac{\sqrt{F[f]}^n}{F[f]^2} \cdot \frac{\partial F[f]}{\partial f} \cdot \Delta_g f \right] \end{array} \right).$$

$$(1165)$$

Another suitable development might be:

$$\Rightarrow \delta_f W = \delta_\sigma W$$

$$= \int_V d^n x \left(\left(\overbrace{- (\ldots)_{,\sigma} \cdot \sqrt{-g} \cdot g^{\alpha\beta} f_{,\alpha} f_{,\beta}}^{\text{here integration by parts}} - (\ldots) - (\ldots) \cdot \left(\begin{matrix} \sqrt{-g} \cdot g^{\alpha\beta} f_{,\alpha} \cdot f_{,\beta\sigma} \\ + \sqrt{-g} \cdot g^{\alpha\beta} f_{,\beta} \cdot f_{,\alpha\sigma} \end{matrix} \right) \right) \delta x^\sigma \\ - C_{N2} \cdot \sqrt{-g} \cdot \delta_\sigma \left[\frac{\sqrt{F[f]}^n}{F[f]^2} \cdot \frac{\partial F[f]}{\partial f} \cdot \Delta_g f \right] \right)$$

$$= \int_V d^n x \left(\left((\ldots) \cdot \left(\sqrt{-g} \cdot g^{\alpha\beta} g^{\sigma\chi} f_{,\alpha} f_{,\beta} \right)_{,\sigma} - (\ldots) \cdot \left(\begin{matrix} \sqrt{-g} \cdot g^{\alpha\beta} f_{,\alpha} \cdot f_{,\beta\sigma} \\ + \sqrt{-g} \cdot g^{\alpha\beta} f_{,\beta} \cdot f_{,\alpha\sigma} \end{matrix} \right) \right) \delta x_\chi \\ - C_{N2} \cdot \sqrt{-g} \cdot \delta_\sigma \left[\frac{\sqrt{F[f]}^n}{F[f]^2} \cdot \frac{\partial F[f]}{\partial f} \cdot \Delta_g f \right] \right).$$

$$(1166)$$

If we want to variate the f- and the g-part all together under one integral, the whole should read:

$$\Rightarrow \delta_{f|g} W = \delta_\sigma W$$

$$= \int_V d^n x \left(\left(- ((\ldots) \cdot \sqrt{-g} \cdot g^{\alpha\beta})_{,\sigma} \cdot f_{,\alpha} f_{,\beta} - (\ldots) \cdot \left(\begin{matrix} \sqrt{-g} \cdot g^{\alpha\beta} f_{,\alpha} \cdot f_{,\beta\sigma} \\ + \sqrt{-g} \cdot g^{\alpha\beta} f_{,\beta} \cdot f_{,\alpha\sigma} \end{matrix} \right) \right) \delta x^\sigma \\ - C_{N2} \cdot \delta_\sigma \left[\frac{\sqrt{F[f]}^n}{F[f]^2} \cdot \frac{\partial F[f]}{\partial f} \cdot \sqrt{-g} \cdot \Delta_g f \right] \right)$$

$$= \int_V d^n x \left(\left(- ((\ldots) \cdot \sqrt{-g} \cdot g^{\alpha\beta})_{,\sigma} \cdot f_{,\alpha} f_{,\beta} - (\ldots) \cdot \left(\begin{matrix} \sqrt{-g} \cdot g^{\alpha\beta} f_{,\alpha} \cdot f_{,\beta\sigma} \\ + \sqrt{-g} \cdot g^{\alpha\beta} f_{,\beta} \cdot f_{,\alpha\sigma} \end{matrix} \right) \right) g^{\sigma\chi} \delta x_\chi \\ - C_{N2} \cdot \delta_\sigma \left[\frac{\sqrt{F[f]}^n}{F[f]^2} \cdot \frac{\partial F[f]}{\partial f} \cdot \sqrt{-g} \cdot \Delta_g f \right] \right)$$

$$= \int_V d^n x \left(\left(\overbrace{-g^{\sigma\chi} ((\ldots) \cdot \sqrt{-g} \cdot g^{\alpha\beta})_{,\sigma} \cdot f_{,\alpha} f_{,\beta}}^{\text{here integration by parts}} - (\ldots) \cdot \left(\begin{matrix} \sqrt{-g} \cdot g^{\alpha\beta} f_{,\alpha} \cdot f_{,\beta\sigma} \\ + \sqrt{-g} \cdot g^{\alpha\beta} f_{,\beta} \cdot f_{,\alpha\sigma} \end{matrix} \right) g^{\sigma\chi} \right) \delta x_\chi \\ - C_{N2} \cdot \delta_\sigma \left[\frac{\sqrt{F[f]}^n}{F[f]^2} \cdot \frac{\partial F[f]}{\partial f} \cdot \sqrt{-g} \cdot \Delta_g f \right] \right)$$

$$= \int_V d^n x \left(\left(\begin{matrix} (\ldots) \cdot \sqrt{-g} \cdot g^{\alpha\beta} \cdot (f_{,\alpha} f_{,\beta} \cdot g^{\sigma\chi}_{,\sigma} + g^{\sigma\chi} (f_{,\alpha} f_{,\beta\sigma} + f_{,\alpha\sigma} f_{,\beta})) \\ - (\ldots) \cdot \left(\begin{matrix} \sqrt{-g} \cdot g^{\alpha\beta} f_{,\alpha} \cdot f_{,\beta\sigma} \\ + \sqrt{-g} \cdot g^{\alpha\beta} f_{,\beta} \cdot f_{,\alpha\sigma} \end{matrix} \right) \end{matrix} \right) \delta x_\chi \\ - C_{N2} \cdot \delta_\sigma \left[\frac{\sqrt{F[f]}^n}{F[f]^2} \cdot \frac{\partial F[f]}{\partial f} \cdot \sqrt{-g} \cdot \Delta_g f \right] \right)$$

$$= \int_V d^n x \left((\ldots) \cdot \sqrt{-g} \cdot g^{\alpha\beta} \cdot \left(f_{,\alpha} f_{,\beta} \cdot g^{\sigma\chi}_{,\sigma} + \overbrace{(f_{,\alpha} f_{,\beta\sigma} + f_{,\alpha\sigma} f_{,\beta}) - (f_{,\alpha} f_{,\beta\sigma} + f_{,\beta} f_{,\alpha\sigma})}^{=0} \right) \delta x_\chi \\ - C_{N2} \cdot \delta_\sigma \left[\frac{\sqrt{F[f]}^n}{F[f]^2} \cdot \frac{\partial F[f]}{\partial f} \cdot \sqrt{-g} \cdot \Delta_g f \right] \right)$$

$$= \int_V d^n x \left((\ldots) \cdot \sqrt{-g} \cdot g^{\alpha\beta} \cdot (f_{,\alpha} f_{,\beta} \cdot g^{\sigma\chi}_{,\sigma}) \delta x_\chi - C_{N2} \cdot \delta_\sigma \left[\frac{\sqrt{F[f]}^n}{F[f]^2} \cdot \frac{\partial F[f]}{\partial f} \cdot \sqrt{-g} \cdot \Delta_g f \right] \right).$$

$$(1167)$$

Similarly, one could expand the $-C_{N2} \cdot \delta_\sigma \left[\frac{\sqrt{F[f]}^n}{F[f]^2} \cdot \frac{\partial F[f]}{\partial f} \cdot \sqrt{-g} \cdot \Delta_g f \right]$-term, where we abbreviate as follows:

$$\left[C_{N2} \cdot \frac{\sqrt{F[f]}^n}{F[f]^2} \cdot \frac{\partial F[f]}{\partial f} \cdot \sqrt{-g} \right] \cdot \Delta_g f = [\cdots] \cdot \Delta_g f \qquad (1168)$$

and now evaluate:

$$\delta_\sigma W = \int_V d^n x \, ((\ldots) \cdot \sqrt{-g} \cdot g^{\alpha\beta} \cdot \left(f_{,\alpha} f_{,\beta} \cdot g^{\sigma\chi}{}_{,\sigma} \right) \delta x_\chi$$

$$-C_{N2} \cdot \delta_\sigma \left[\frac{\sqrt{F[f]}^n}{F[f]^2} \cdot \frac{\partial F[f]}{\partial f} \cdot \sqrt{-g} \cdot \Delta_g f \right])$$

$$= \int_V d^n x \, ((\ldots) \cdot \sqrt{-g} \cdot g^{\alpha\beta} \cdot f_{,\alpha} f_{,\beta} \cdot g^{\sigma\chi}{}_{,\sigma}$$

$$- \left([\cdots] \right)_{,\sigma} \cdot g^{\sigma\chi} \Delta_g f - \left([\cdots] \right) \cdot g^{\sigma\chi} \left(\Delta_g f \right)_{,\sigma}) \delta x_\chi$$

$$= \int_V d^n x \, ((\ldots) \cdot \sqrt{-g} \cdot g^{\alpha\beta} \cdot f_{,\alpha} f_{,\beta} \cdot g^{\sigma\chi}{}_{,\sigma}$$

$$+ [\cdots] \cdot \left(g^{\sigma\chi} \Delta_g f \right)_{,\sigma} - \left([\cdots] \right) \cdot g^{\sigma\chi} \left(\Delta_g f \right)_{,\sigma}) \delta x_\chi$$

$$= \int_V d^n x \, ((\ldots) \cdot \sqrt{-g} \cdot g^{\alpha\beta} \cdot f_{,\alpha} f_{,\beta} \cdot g^{\sigma\chi}{}_{,\sigma} + [\cdots] \cdot g^{\sigma\chi}{}_{,\sigma} \cdot \Delta_g f) \delta x_\chi$$

$$= \int_V d^n x \, ((\ldots) \cdot \sqrt{-g} \cdot g^{\alpha\beta} \cdot f_{,\alpha} f_{,\beta} + [\cdots] \cdot \Delta_g f) \cdot g^{\sigma\chi}{}_{,\sigma} \delta x_\chi$$

$$= \int_V d^n x \sqrt{-g} \cdot \left((\ldots) \cdot g^{\alpha\beta} \cdot f_{,\alpha} f_{,\beta} + C_{N2} \cdot \frac{\sqrt{F[f]}^n}{F[f]^2} \cdot \frac{\partial F[f]}{\partial f} \cdot \Delta_g f \right) \cdot g^{\sigma\chi}{}_{,\sigma} \delta x_\chi.$$

$$(1169)$$

Also incorporating Eq. (1158), we obtain:

$$\delta_\sigma W = \int_V d^n x \sqrt{-g} \cdot \frac{\sqrt{F[f]}^n}{F[f]^2} \cdot \left(\begin{array}{l} \left(C_{N2} \cdot \frac{\partial^2 F[f]}{\partial f^2} - \frac{C_{N1}}{F[f]} \cdot \left(\frac{\partial F[f]}{\partial f} \right)^2 \right) \cdot g^{\alpha\beta} \cdot f_{,\alpha} f_{,\beta} \\ + C_{N2} \cdot \frac{\partial F[f]}{\partial f} \cdot \Delta_g f \end{array} \right) \cdot g^{\sigma\chi}{}_{,\sigma} \delta x_\chi.$$

$$(1170)$$

In the simplest case $F[f] = \text{const} + f$ we could write:

$$\delta_\sigma W = \int_V d^n x \sqrt{-g} \cdot \frac{\sqrt{f}^n}{f^2} \cdot \left(C_{N2} \cdot \Delta_g f - \frac{C_{N1}}{f} \cdot g^{\alpha\beta} \cdot f_{,\alpha} f_{,\beta} \right) \cdot g^{\sigma\chi}{}_{,\sigma} \delta x_\chi = 0$$

$$0 = C_{N2} \cdot \Delta_g f - \frac{C_{N1}}{f} \cdot g^{\alpha\beta} \cdot f_{,\alpha} f_{,\beta} \Rightarrow 0 = C_{N2} \cdot f \cdot \Delta_g f - C_{N1} \cdot g^{\alpha\beta} \cdot f_{,\alpha} f_{,\beta}.$$

$$(1171)$$

Expansion of the Laplace operator leads to:

$$0 = C_{N2} \cdot f \cdot \Delta_g f - C_{N1} \cdot g^{\alpha\beta} \cdot f_{,\alpha} f_{,\beta}$$

$$\Rightarrow 0 = C_{N2} \cdot f \cdot \frac{1}{\sqrt{-g}} \cdot \left(\sqrt{-g} \cdot g^{\alpha\beta} \cdot f_{,\alpha}\right)_{,\beta} - C_{N1} \cdot g^{\alpha\beta} \cdot f_{,\alpha} f_{,\beta}$$

$$= C_{N2} \cdot \left(\left(f \cdot g^{\alpha\beta} \cdot f_{,\alpha}\right)_{,\beta} - \sqrt{-g} \cdot g^{\alpha\beta} \cdot f_{,\alpha} \cdot \left(f \cdot \frac{1}{\sqrt{-g}}\right)_{,\beta} \right) - C_{N1} \cdot g^{\alpha\beta} \cdot f_{,\alpha} f_{,\beta}$$

$$= C_{N2} \cdot \left(\begin{array}{c} f \cdot g^{\alpha\beta} \cdot f_{,\alpha\beta} + f_{,\beta} \cdot g^{\alpha\beta} \cdot f_{,\alpha} + f \cdot g^{\alpha\beta}{}_{,\beta} \cdot f_{,\alpha} \\ -\sqrt{-g} \cdot g^{\alpha\beta} \cdot f_{,\alpha} \cdot \left(f \cdot \frac{1}{\sqrt{-g}}\right)_{,\beta} \end{array} \right) - C_{N1} \cdot g^{\alpha\beta} \cdot f_{,\alpha} f_{,\beta}$$

$$= C_{N2} \cdot \left(\begin{array}{c} f \cdot g^{\alpha\beta} \cdot f_{,\alpha\beta} + f \cdot g^{\alpha\beta}{}_{,\beta} \cdot f_{,\alpha} + \overbrace{f_{,\beta} \cdot g^{\alpha\beta} \cdot f_{,\alpha} - g^{\alpha\beta} \cdot f_{,\alpha} \cdot f_{,\beta}}^{=0} \\ -\sqrt{-g} \cdot g^{\alpha\beta} \cdot f_{,\alpha} \cdot f \cdot \left(\frac{1}{\sqrt{-g}}\right)_{,\beta} \end{array} \right)$$
$$-C_{N1} \cdot g^{\alpha\beta} \cdot f_{,\alpha} f_{,\beta}$$

$$= C_{N2} \cdot \left(f \cdot g^{\alpha\beta} \cdot f_{,\alpha\beta} + f \cdot g^{\alpha\beta}{}_{,\beta} \cdot f_{,\alpha} - \sqrt{-g} \cdot g^{\alpha\beta} \cdot f_{,\alpha} \cdot f \cdot \frac{g \cdot g^{\mu\nu} g_{\mu\nu,\beta}}{2 \cdot \left(\sqrt{-g}\right)^3} \right)$$
$$-C_{N1} \cdot g^{\alpha\beta} \cdot f_{,\alpha} f_{,\beta}$$

$$= C_{N2} \cdot \left(f \cdot g^{\alpha\beta} \cdot f_{,\alpha\beta} + f \cdot g^{\alpha\beta}{}_{,\beta} \cdot f_{,\alpha} - g^{\alpha\beta} \cdot f_{,\alpha} \cdot f \cdot \frac{g^{\mu\nu} g_{\mu\nu,\beta}}{2} \right) - C_{N1} \cdot g^{\alpha\beta} \cdot f_{,\alpha} f_{,\beta}.$$

$$(1172)$$

Assuming a flat space or an almost flat space with Cartesian like coordinates yields:

$$0 = C_{N2} \cdot \left(f \cdot g^{\alpha\beta} \cdot f_{,\alpha\beta} + f \cdot g^{\alpha\beta}{}_{,\beta} \cdot f_{,\alpha} - g^{\alpha\beta} \cdot f_{,\alpha} \cdot f \cdot \frac{g^{\mu\nu} g_{\mu\nu,\beta}}{2} \right)$$
$$-C_{N1} \cdot g^{\alpha\beta} \cdot f_{,\alpha} f_{,\beta}$$
$$= C_{N2} \cdot f \cdot g^{\alpha\beta} \cdot f_{,\alpha\beta} - C_{N1} \cdot g^{\alpha\beta} \cdot f_{,\alpha} f_{,\beta}. \qquad (1173)$$

Remembering that this in principle is under the integral (1170), we can once again apply integration by parts and try to further simplify the equation above. In order to do so, however, we want to go back to the non-approximated case. Thus, starting from Eq. (1156), setting $F[f] = f$ and once again going for a coordinate variation with respect to δx_χ, but keeping everything general otherwise, yields:

$$\delta_\sigma W = 0 = \int_V d^n x \sqrt{-g} \cdot \frac{\sqrt{f^n}}{f^2} \cdot \left(C_{N2} \cdot \Delta_g f - \frac{C_{N1}}{f} \cdot g^{\alpha\beta} \cdot f_{,\alpha} f_{,\beta} \right) \cdot g^{\sigma\chi}{}_{,\sigma} \delta x_\chi$$

$$\xrightarrow{n=4} 0 = 2 \cdot f \cdot \Delta_g f - 3 \cdot f_{,\beta} g^{\alpha\beta} f_{,\alpha}$$

$$\xrightarrow{n=6} 0 = \Delta_g f. \qquad (1174)$$

Using integration by parts on Eq. (1171) gives us:

$$\delta_\sigma W = 0 = \int_V d^n x \sqrt{-g} \cdot \frac{\sqrt{f^n}}{f^2} \cdot \left(C_{N2} \cdot \Delta_g f - \frac{C_{N1}}{f} \cdot g^{\alpha\beta} \cdot f_{,\alpha} f_{,\beta} \right) \cdot g^{\sigma\chi}{}_{,\sigma} \delta x_\chi$$

$$= \int_V d^n x \left(C_{N2} \cdot \sqrt{-g} \cdot \frac{\sqrt{f^n}}{f^2} \cdot \Delta_g f + C_{N1} \cdot \left(\sqrt{-g} \cdot \frac{\sqrt{f^n}}{f^3} \cdot g^{\sigma\chi}{}_{,\sigma} g^{\alpha\beta} \cdot f_{,\alpha} \right)_{,\beta} f \right) \cdot \delta x_\chi$$

$$= \int_V d^n x \left(C_{N2} \cdot \sqrt{-g} \cdot \frac{\sqrt{f^n}}{f^3} \cdot \Delta_g f + C_{N1} \cdot \left(\left(\frac{\sqrt{f^n}}{f^3} \cdot g^{\sigma\chi}{}_{,\sigma} \right)_{,\beta} \cdot g^{\sigma\chi}{}_{,\sigma} \cdot \Delta_g f \right. \right.$$
$$\left. \left. + \sqrt{-g} \cdot \frac{\sqrt{f^n}}{f^3} \cdot g^{\sigma\chi}{}_{,\sigma} \cdot \Delta_g f \right) \right) \cdot f \cdot \delta x_\chi$$

$$= \int_V d^n x \sqrt{-g} \cdot \frac{\sqrt{f^n}}{f^3} \cdot \left((C_{N2} + C_{N1} \cdot g^{\sigma\chi}{}_{,\sigma}) \cdot \Delta_g f + C_{N1} \cdot \left(\left(\frac{\sqrt{f^n}}{f^3} \cdot g^{\sigma\chi}{}_{,\sigma} \right)_{,\beta} \frac{f^3}{\sqrt{f^n}} g^{\alpha\beta} \cdot f_{,\alpha} \right) \right) \cdot f \cdot \delta x_\chi$$

$$= \int_V d^n x \sqrt{-g} \cdot \frac{\sqrt{f^n}}{f^3} \cdot \left((C_{N2} + C_{N1} \cdot g^{\sigma\chi}{}_{,\sigma}) \cdot \Delta_g f + C_{N1} \cdot \left(\frac{n-6}{2 \cdot f} \cdot g^{\sigma\chi}{}_{,\sigma} + g^{\sigma\chi}{}_{,\sigma\beta} \right) g^{\alpha\beta} \cdot f_{,\alpha} \right) \cdot f \cdot \delta x_\chi$$

$$\Rightarrow 0 = (C_{N2} + C_{N1} \cdot g^{\sigma\chi}{}_{,\sigma}) \cdot \Delta_g f + C_{N1} \cdot \left(\frac{n-6}{2 \cdot f} \cdot g^{\sigma\chi}{}_{,\sigma} + g^{\sigma\chi}{}_{,\sigma\beta} \right) g^{\alpha\beta} \cdot f_{,\alpha}. \tag{1175}$$

As this does not seem to be very helpful (except in the case $n = 6$ where we already have the simple Eq. (1174)), we try the general form (Eq. (1170)), which leads us to:

$$\delta_\sigma W = 0$$

$$= \int_V d^n x \sqrt{-g} \cdot \left(\overbrace{ \begin{array}{l} \dfrac{\sqrt{F[f]^n}}{F[f]^2} \cdot \left(C_{N2} \cdot \dfrac{\partial^2 F[f]}{\partial f^2} - \dfrac{C_{N1}}{F[f]} \cdot \left(\dfrac{\partial F[f]}{\partial f} \right)^2 \right) \cdot g^{\alpha\beta} \cdot f_{,\alpha} f_{,\beta} \\ + C_{N2} \cdot \dfrac{\sqrt{F[f]^n}}{F[f]^2} \cdot \dfrac{\partial F[f]}{\partial f} \cdot \Delta_g f \end{array} }^{\{\ldots\}} \right) \cdot g^{\sigma\chi}{}_{,\sigma} \delta x_\chi$$

$$\underbrace{}_{[\ldots]}$$

$$= \int_V d^n x \left(\sqrt{-g} \cdot \{\ldots\} \cdot g^{\sigma\chi}{}_{,\sigma} \cdot g^{\alpha\beta} \cdot f_{,\alpha} f_{,\beta} + \sqrt{-g} \cdot [\ldots] \cdot g^{\sigma\chi}{}_{,\sigma} \cdot \Delta_g f \right) \delta x_\chi$$

$$= \int_V d^n x \left(-\left(\sqrt{-g} \cdot \{\ldots\} \cdot g^{\sigma\chi}{}_{,\sigma} \cdot g^{\alpha\beta} \cdot f_{,\alpha} \right)_{,\beta} + \sqrt{-g} \cdot \frac{[\ldots]}{f} \cdot g^{\sigma\chi}{}_{,\sigma} \cdot \Delta_g f \right) \cdot f \cdot \delta x_\chi$$

$$= \int_V d^n x \sqrt{-g} \cdot \left(g^{\sigma\chi}{}_{,\sigma} \cdot \left(\frac{[\ldots]}{f} - \{\ldots\} \right) \cdot \Delta_g f - (\{\ldots\} \cdot g^{\sigma\chi}{}_{,\sigma})_{,\beta} \cdot g^{\alpha\beta} \cdot f_{,\alpha} \right) \cdot f \cdot \delta x_\chi.$$

$$(1176)$$

Here now we realize a great flexibility to derive quantum gravity equations of sufficiently convenient forms, simply by adapting the wrapping functions $F[f]$ in accordance with the number of dimensions and our needs, respectively intentions.

For those who miss the classical Hilbert form, we recall the general starting point (Eq. (1156)) and apply the chain rule for differentiation, yielding:

$$\delta W = 0$$

$$= \delta_\sigma \int_V d^n x \left(\sqrt{-G} \cdot R \right) = \int_V d^n x \left(\sqrt{-G} \cdot \left[R^{\alpha\beta} - \frac{R}{2} G^{\alpha\beta} \right] \right) \delta_\sigma G_{\alpha\beta}$$

$$= \int_V d^n x \left(\sqrt{-G} \cdot \left[R^{\alpha\beta} - \frac{R}{2} G^{\alpha\beta} \right] \right) \left(\frac{\partial F[f]}{\partial f} \cdot g_{\alpha\beta} f_{,\sigma} + F[f] \cdot g_{\alpha\beta,\sigma} \right) g^{\sigma\chi} \delta x_\chi.$$

$$(1177)$$

It is very important to point out that the Ricci scalar and the Ricci tensor in the equation above are the corresponding Ricci-curvatures with respect to the transformed metric $G_{\alpha\beta}$ and not the undisturbed metric $g_{\alpha\beta}$ (c.f. section 3.3 and chapter 16).

Integration by parts in Eq. (1177) leads us to either:

$$
\delta W = 0 = \int_V d^n x \left(\sqrt{-g} \cdot \overbrace{\sqrt{F[f]}^n \left[R^{\alpha\beta} - \frac{R}{2} G^{\alpha\beta} \right]}^{[\cdots]^{\alpha\beta}} \right) \left(\frac{\partial F[f]}{\partial f} \cdot g_{\alpha\beta} f_{,\sigma} + F[f] \cdot g_{\alpha\beta,\sigma} \right) g^{\sigma\chi} \delta x_\chi
$$

$$
= \int_V d^n x \sqrt{-g} \left([\cdots]^{\alpha\beta} \cdot \frac{\partial F[f]}{\partial f} \cdot g^{\sigma\chi} g_{\alpha\beta} f_{,\sigma} + [\cdots]^{\alpha\beta} \cdot F[f] \cdot g^{\sigma\chi} g_{\alpha\beta,\sigma} \right) \frac{\partial}{\partial x^\chi} \delta x
$$

$$
= \overbrace{\underset{\text{Surface}}{\int} d^n x \sqrt{-g} \left([\cdots]^{\alpha\beta} \cdot \frac{\partial F[f]}{\partial f} \cdot g^{\sigma\chi} g_{\alpha\beta} f_{,\sigma} + [\cdots]^{\alpha\beta} \cdot F[f] \cdot g^{\sigma\chi} g_{\alpha\beta,\sigma} \right) \delta x}^{=0}
$$

$$
- \int_V d^n x \left(\sqrt{-g} \left([\cdots]^{\alpha\beta} \cdot \frac{\partial F[f]}{\partial f} \cdot g^{\sigma\chi} g_{\alpha\beta} f_{,\sigma} + [\cdots]^{\alpha\beta} \cdot F[f] \cdot g^{\sigma\chi} g_{\alpha\beta,\sigma} \right) \right)_{,\chi} \delta x
$$

$$
= - \int_V d^n x \sqrt{-g} \left(\begin{array}{c} ([\cdots]^{\alpha\beta} \cdot g_{\alpha\beta})_{,\chi} \, g^{\sigma\chi} \frac{\partial F[f]}{\partial f} f_{,\sigma} + [\cdots]^{\alpha\beta} \cdot g_{\alpha\beta} \cdot \Delta_g F[f] \\ + ([\cdots]^{\alpha\beta} \cdot F[f])_{,\chi} \cdot g^{\sigma\chi} g_{\alpha\beta,\sigma} + [\cdots]^{\alpha\beta} \cdot F[f] \cdot \Delta_g g_{\alpha\beta} \end{array} \right)_{,\chi} \delta x,
$$

$$
\tag{1178}
$$

or:

$$
\delta W = 0 = \int_V d^n x \left(\sqrt{-g} \cdot \overbrace{\sqrt{F[f]}^n \left[R^{\alpha\beta} - \frac{R}{2} G^{\alpha\beta} \right]}^{[\cdots]^{\alpha\beta}} \right) \left(\frac{\partial F[f]}{\partial f} \cdot g_{\alpha\beta} f_{,\sigma} + F[f] \cdot g_{\alpha\beta,\sigma} \right) g^{\sigma\chi} \delta x_\chi
$$

$$
= \int_V d^n x \left(\sqrt{-g} \cdot [\cdots]^{\alpha\beta} \cdot \frac{\partial F[f]}{\partial f} \cdot g^{\sigma\chi} g_{\alpha\beta} f_{,\sigma} - \left(\sqrt{-g} \cdot [\cdots]^{\alpha\beta} \cdot F[f] \cdot g^{\sigma\chi} \right)_{,\sigma} g_{\alpha\beta} \right) \delta x_\chi
$$

$$
= \int_V d^n x \left(\sqrt{-g} \cdot [\cdots]^{\alpha\beta} \cdot \frac{\partial F[f]}{\partial f} \cdot g^{\sigma\chi} g_{\alpha\beta} f_{,\sigma} - \left(\begin{array}{c} \sqrt{-g} \cdot [\cdots]^{\alpha\beta} g^{\sigma\chi} \cdot \frac{\partial F[f]}{\partial f} \cdot f_{,\sigma} \\ + (\sqrt{-g} \cdot [\cdots]^{\alpha\beta} g^{\sigma\chi})_{,\sigma} \cdot F[f] \end{array} \right) g_{\alpha\beta} \right) \delta x_\chi
$$

$$
= - \int_V d^n x \left((\sqrt{-g} \cdot [\cdots]^{\alpha\beta} g^{\sigma\chi})_{,\sigma} \cdot F[f] \cdot g_{\alpha\beta} \right) \delta x_\chi
$$

$$
\Rightarrow \left(\sqrt{-g} \cdot \sqrt{F[f]}^n \left[R^{\alpha\beta} - \frac{R}{2} G^{\alpha\beta} \right] g^{\sigma\chi} \right)_{,\sigma} \cdot g_{\alpha\beta}
$$

$$
= \left(\sqrt{-g} \cdot \sqrt{F[f]}^n \left[R^{\alpha\beta} - \frac{R}{2} \cdot \frac{g^{\alpha\beta}}{F[f]} \right] g^{\sigma\chi} \right)_{,\sigma} \cdot g_{\alpha\beta} = 0,
\tag{1179}
$$

or:

$$\delta W = 0 = \int_V d^n x \underbrace{\left[\sqrt{-g} \cdot \sqrt{F[f]}^n \left[R^{\alpha\beta} - \frac{R}{2} G^{\alpha\beta} \right]}_{[\cdots]^{\alpha\beta}} \left(\frac{\partial F[f]}{\partial f} \cdot g_{\alpha\beta} f_{,\sigma} + F[f] \cdot g_{\alpha\beta,\sigma} \right) g^{\sigma\chi} \delta x_\chi \right]}$$

$$= \int_V d^n x \left(-\sqrt{-g} \cdot [\cdots]^{\alpha\beta} \cdot \frac{\partial F[f]}{\partial f} \cdot g^{\sigma\chi} g_{\alpha\beta} f_{,\sigma} + \sqrt{-g} \cdot [\cdots]^{\alpha\beta} \cdot F[f] \cdot g^{\sigma\chi} g_{\alpha\beta,\sigma} \right) \delta x_\chi$$

$$= \int_V d^n x \left(-\left(\sqrt{-g} \cdot [\cdots]^{\alpha\beta} \cdot g^{\sigma\chi} g_{\alpha\beta} \right)_{,\sigma} F[f] + \sqrt{-g} \cdot [\cdots]^{\alpha\beta} \cdot F[f] \cdot g^{\sigma\chi} g_{\alpha\beta,\sigma} \right) \delta x_\chi$$

$$= \int_A d^n x \left(-\left(\frac{\left(\sqrt{-g} \cdot [\cdots]^{\alpha\beta} \cdot g^{\sigma\chi} \right)_{,\sigma} g_{\alpha\beta}}{+ \left(\sqrt{-g} \cdot [\cdots]^{\alpha\beta} \cdot g^{\sigma\chi} \right) g_{\alpha\beta,\sigma}} \right) F[f] + \sqrt{-g} \cdot [\cdots]^{\alpha\beta} \cdot F[f] \cdot g^{\sigma\chi} g_{\alpha\beta,\sigma} \right) \delta x_\chi$$

$$= -\int_V d^n x \left(\left(\sqrt{-g} \cdot [\cdots]^{\alpha\beta} g^{\sigma\chi} \right)_{,\sigma} \cdot F[f] \cdot g_{\alpha\beta} \right) \delta x_\chi$$

$$\Rightarrow \left(\sqrt{-g} \cdot \sqrt{F[f]}^n \left[R^{\alpha\beta} - \frac{R}{2} G^{\alpha\beta} \right] g^{\sigma\chi} \right)_{,\sigma} \cdot g_{\alpha\beta}$$

$$= \left(\sqrt{-g} \cdot \sqrt{F[f]}^n \left[R^{\alpha\beta} - \frac{R}{2} \cdot \frac{g^{\alpha\beta}}{F[f]} \right] g^{\sigma\chi} \right)_{,\sigma} \cdot g_{\alpha\beta} = 0.$$

$$(1180)$$

15.5.2 Deriving a Simple Equation of Elastic Space-Time

Now we have enough material for the derivation of quantum gravity equations sporting the typical structure of the equation of elasticity. In order to have an even more general derivation, we use Eq. (1170) and apply the procedure from the end of the previous subsection:

$$\delta_\sigma W = 0$$

$$= \int_V d^n x \sqrt{-g} \cdot \underbrace{\left(\frac{\sqrt{F[f]^n}}{F[f]^2} \cdot \left(C_{N2} \cdot \frac{\partial^2 F[f]}{\partial f^2} - \frac{C_{N1}}{F[f]} \cdot \left(\frac{\partial F[f]}{\partial f} \right)^2 \right) \cdot g^{\alpha\beta} \cdot f_{,\alpha} f_{,\beta} + C_{N2} \cdot \frac{\sqrt{F[f]^n}}{F[f]^2} \cdot \frac{\partial F[f]}{\partial f} \cdot \Delta_g f \right)}_{[\dots]} \cdot g^{\sigma\chi}{}_{,\sigma} \, \delta x_\chi$$

$$= \int_V d^n x \sqrt{-g} \cdot \left(\{\dots\} \cdot g^{\alpha\beta} \cdot f_{,\alpha} f_{,\beta} + [\dots] \cdot \Delta_g f \right) \cdot g^{\sigma\chi}{}_{,\sigma} \, \delta x_\chi. \tag{1181}$$

Now we assume the scalar f to be a divergence $G^\sigma{}_{,\sigma}$ and incorporate this into Eq. (1181) plus the simpler form (because of $F[f] = f$) (Eq. (1174)):

$$0 = \{\dots\} \cdot g^{\alpha\beta} \cdot (G^\sigma{}_{,\sigma})_{,\alpha} (G^\sigma{}_{,\sigma})_{,\beta} + [\dots] \cdot \Delta_g (G^\sigma{}_{,\sigma}), \tag{1182}$$

$$0 = C_{N2} \cdot \Delta_g (G^\sigma{}_{,\sigma}) - \frac{C_{N1}}{(G^\sigma{}_{,\sigma})} \cdot G^\sigma{}_{,\sigma\beta} \, g^{\alpha\beta} \, G^\sigma{}_{,\sigma\alpha}$$

$$\xrightarrow{n=4} 0 = 2 \cdot (G^\sigma{}_{,\sigma}) \cdot \Delta_g (G^\sigma{}_{,\sigma}) - 3 \cdot G^\sigma{}_{,\sigma\beta} \, g^{\alpha\beta} \, G^\sigma{}_{,\sigma\alpha}$$

$$\xrightarrow{n=6} 0 = \Delta_g (G^\sigma{}_{,\sigma}). \tag{1183}$$

Applying integration by parts to the product term with the first derivative would just also give us a Laplace operator. As this is not exactly what we were aiming for, we simply consider another scalar term, formed from **G**-vectors, namely $G^\lambda G_\lambda + \alpha \cdot h \cdot G^\lambda{}_{,\lambda}$. Setting this into Eq. (1177) gives us:

$$\delta W = 0 = \int_V d^n x \left(\sqrt{-G} \cdot \left[R^{\alpha\beta} - \frac{R}{2} G^{\alpha\beta} \right] \right)$$

$$\times \left(\frac{\partial F[f]}{\partial f} \cdot g_{\alpha\beta} \left(G^\lambda G_\lambda + \alpha \cdot h \cdot G^\lambda{}_{,\lambda} \right)_{,\sigma} + F[f] \cdot g_{\alpha\beta,\sigma} \right) g^{\sigma\chi} \delta x_\chi$$

$$= \int_V d^n x \left(\sqrt{-G} \cdot \left[R^{\alpha\beta} - \frac{R}{2} G^{\alpha\beta} \right] \right)$$

$$\times \left(\frac{\partial F[f]}{\partial f} \cdot g_{\alpha\beta} \left(G^\lambda{}_{,\sigma} G_\lambda + G^\lambda G_{\lambda,\sigma} + \alpha \cdot h_{,\sigma} \cdot G^\lambda{}_{,\lambda} + \alpha \cdot h \cdot G^\lambda{}_{,\lambda\sigma} \right) + F[f] \cdot g_{\alpha\beta,\sigma} \right) g^{\sigma\chi} \delta x_\chi$$

$$\xrightarrow{G^\lambda = h^{,\lambda}, G_\lambda = h_{,\lambda}} = \int_V d^n x \left(\sqrt{-G} \cdot \left[R^{\alpha\beta} - \frac{R}{2} G^{\alpha\beta} \right] \right)$$

$$\times \left(\frac{\partial F[f]}{\partial f} \cdot g_{\alpha\beta} \left(h^{,\lambda}{}_{,\sigma} h_{,\lambda} + h^{,\lambda} h_{,\lambda\sigma} + \alpha \cdot h_{,\sigma} \cdot h^{,\lambda}{}_{,\lambda} + \alpha \cdot h \cdot h^{,\lambda}{}_{,\lambda\sigma} \right) + F[f] \cdot g_{\alpha\beta,\sigma} \right) g^{\sigma\chi} \delta x_\chi.$$

(1184)

Looking for some suitable integration by parts may lead us to:

$$\frac{\partial F[f]}{\partial f} g_{\alpha\beta} g^{\sigma\chi} \cdot \sqrt{F[f]^n} \cdot \left[R^{\alpha\beta} - \tfrac{R}{2} G^{\alpha\beta}\right] = \{\ldots\}^{\sigma\chi}$$

$$= \int_V d^n x \left(\{\ldots\}^{\sigma\chi} \left(h^{,\lambda}_{,\sigma} h_{,\lambda} + h^{,\lambda} h_{,\lambda\sigma} + \alpha \cdot h_{,\sigma} \cdot h^{,\lambda}_{,\lambda} + \alpha \cdot h \cdot h^{,\lambda}_{,\lambda\sigma} \right) \right.$$
$$\left. + \sqrt{-g} \cdot \sqrt{F[f]^n} \cdot \left[R^{\alpha\beta} - \tfrac{R}{2} G^{\alpha\beta}\right] F[f] \cdot g^{\sigma\chi} g_{\alpha\beta,\sigma} \right) \delta x_\chi$$

$$= \int_V d^n x \left(\left(\sqrt{-g} \cdot \{\ldots\}^{\sigma\chi} \left(g^{\lambda\gamma} h_{,\gamma}\right)_{,\sigma} \right)_{,\lambda} + \left(\sqrt{-g} \cdot \{\ldots\}^{\sigma\chi} g^{\lambda\gamma} h_{,\lambda\sigma} \right)_{,\gamma} \right.$$
$$-\alpha \cdot \left(\sqrt{-g} \cdot \{\ldots\}^{\sigma\chi} \left(g^{\lambda\gamma} h_{,\gamma}\right)_{,\lambda} \right)_{,\sigma}$$
$$\left. + \sqrt{-g} \cdot \{\ldots\}^{\sigma\chi} \alpha \cdot h \cdot h^{,\lambda}_{,\lambda\sigma} + \sqrt{-g} \cdot \sqrt{F[f]^n} \cdot \left[R^{\alpha\beta} - \tfrac{R}{2} G^{\alpha\beta}\right] F[f] \cdot g^{\sigma\chi} g_{\alpha\beta,\sigma} \right) \cdot h \; \delta x_\chi$$

$$= \int_V d^n x \left(\sqrt{-g} \cdot \{\ldots\}^{\sigma\chi}_{,\lambda} g^{\lambda\gamma} h_{,\gamma\sigma} + \sqrt{-g} \cdot \{\ldots\}^{\sigma\chi}_{,\gamma} g^{\lambda\gamma} h_{,\lambda\sigma} + \cdots \right.$$
$$-\sqrt{-g} \cdot \{\ldots\}^{\sigma\chi} \Delta_g h_{,\sigma} + \sqrt{-g} \cdot \{\ldots\}^{\sigma\chi}_{,\gamma} g^{\lambda\gamma} h_{,\lambda\sigma} + \cdots$$
$$+ \sqrt{-g} \cdot \{\ldots\}^{\sigma\chi} \alpha \cdot h \cdot g^{\lambda\gamma} h_{,\gamma\lambda\sigma} + \alpha \cdot \left(\sqrt{-g} \cdot \{\ldots\}^{\sigma\chi} g^{\lambda\gamma} \right)_{,\sigma} h_{,\gamma\lambda} + \alpha \cdot \sqrt{-g} \cdot \{\ldots\}^{\sigma\chi} g^{\lambda\gamma} h_{,\gamma\lambda\sigma}$$
$$\left. + \sqrt{-g} \cdot \sqrt{F[f]^n} \cdot \left[R^{\alpha\beta} - \tfrac{R}{2} G^{\alpha\beta}\right] F[f] \cdot g^{\sigma\chi} g_{\alpha\beta,\sigma} \right) \cdot h \; \delta x_\chi.$$

$$(1185)$$

Here, too, however, the result is not truly mirroring the fundamental equation of elasticity as all third-order derivatives not resulting in Laplace operators vanish:

$$= \int_V d^n x \left(-\left(\sqrt{-g} \cdot \{\ldots\}^{\sigma\chi}_{,\lambda} g^{\lambda\gamma} h_{,\gamma\sigma} + \sqrt{-g} \cdot \{\ldots\}^{\sigma\chi} \Delta_g h_{,\sigma} + \sqrt{-g} \cdot \{\ldots\}^{\sigma\chi}_{,\gamma} g^{\lambda\gamma} h_{,\lambda\sigma} + \cdots \right. \right.$$
$$+ \sqrt{-g} \cdot \{\ldots\}^{\sigma\chi} \Delta_g h_{,\sigma} + \alpha \cdot \left(\sqrt{-g} \cdot \{\ldots\}^{\sigma\chi} g^{\lambda\gamma} \right)_{,\sigma} h_{,\gamma\lambda}$$
$$\left. \left. + \sqrt{-g} \cdot \sqrt{F[f]^n} \cdot \left[R^{\alpha\beta} - \tfrac{R}{2} G^{\alpha\beta}\right] F[f] \cdot g^{\sigma\chi} g_{\alpha\beta,\sigma} \right) \cdot h \right) \delta x_\chi. \quad (1186)$$

Something obviously is missing. Either there is no elastic space-time option (not even in approximated form) or our approach is incomplete.

Remembering that the variation should not exclude certain options, we extend our approach in such a way that only the condition of W remaining a scalar will be fulfilled. Still staying as simple as possible, though, we start with Eq. (1177) and apply a special variation for the function f:

$$\delta W = 0 = \int_V d^n x \left(\sqrt{-G} \cdot \left[R^{\alpha\beta} - \frac{R}{2} G^{\alpha\beta} \right] \right)$$

$$\times \left(\frac{\frac{\partial F[f]}{\partial f} \cdot g_{\alpha\beta}}{1+\alpha} \left(\left(\sqrt{-g} \cdot g^{\lambda\gamma} h_{,\lambda\gamma}\right)_{,\sigma} + \frac{\alpha}{2 \cdot n} \cdot \left(\begin{array}{c} \left(\sqrt{-g} \cdot g^{\lambda\gamma} h_{,\lambda\sigma}\right)_{,\gamma} \\ + \left(\sqrt{-g} \cdot g^{\lambda\gamma} h_{,\gamma\sigma}\right)_{,\lambda} \end{array} \right) \right) + F[f] \cdot g_{\alpha\beta,\sigma} \right) g^{\sigma\chi} \delta x_\chi$$

$$= \int_V d^n x \left(\sqrt{-G} \cdot \left[R^{\alpha\beta} - \frac{R}{2} G^{\alpha\beta} \right] \right)$$

$$\times \left(\frac{\frac{\partial F[f]}{\partial f} \cdot g_{\alpha\beta}}{1+\alpha} \left(\left(\sqrt{-g} \cdot g^{\lambda\gamma} h_{,\lambda\gamma}\right)_{,\sigma} + \frac{\alpha}{n} \cdot \sqrt{-g} \cdot \Delta_g\left(h_{,\sigma}\right) \right) + F[f] \cdot g_{\alpha\beta,\sigma} \right) g^{\sigma\chi} \delta x_\chi$$

$$= \int_V d^n x \left(\sqrt{-G} \cdot \left[R^{\alpha\beta} - \frac{R}{2} G^{\alpha\beta} \right] \right)$$

$$\times \left(\frac{\frac{\partial F[f]}{\partial f} \cdot g_{\alpha\beta}}{1+\alpha} \left(\left(\sqrt{-g} \cdot g^{\lambda\gamma}\right)_{,\sigma} h_{,\lambda\gamma} + \sqrt{-g} \cdot g^{\lambda\gamma} h_{,\lambda\gamma\sigma} + \frac{\alpha}{n} \cdot \sqrt{-g} \cdot \Delta_g\left(h_{,\sigma}\right) \right) + F[f] \cdot g_{\alpha\beta,\sigma} \right) g^{\sigma\chi} \delta x_\chi.$$

$$(1187)$$

In this context, f would have to be considered as an internally structured scalar, which is subjected to a more general variation as follows:

$$f = \sqrt{-g} \cdot g^{\lambda\gamma} h_{,\lambda\gamma}; \quad \delta \to \frac{\delta_\sigma}{1+\alpha} \cdot \left(1 + \frac{\alpha}{2 \cdot n} \cdot (\delta_\gamma^\sigma \delta_\sigma^\gamma + \delta_\lambda^\sigma \delta_\sigma^\lambda)\right)$$

$$\Rightarrow \delta f = \frac{\delta_\sigma}{1+\alpha} \cdot \left(1 + \frac{\alpha}{2 \cdot n} \cdot (\delta_\gamma^\sigma \delta_\sigma^\gamma + \delta_\lambda^\sigma \delta_\sigma^\lambda)\right) \sqrt{-g} \cdot g^{\lambda\gamma} h_{,\lambda\gamma}$$

$$= \frac{\delta_\sigma}{1+\alpha} \cdot \left(\sqrt{-g} \cdot g^{\lambda\gamma} h_{,\lambda\gamma} + \frac{\alpha}{2 \cdot n} \cdot (\delta_\gamma^\sigma \sqrt{-g} \cdot g^{\lambda\gamma} h_{,\lambda\gamma} \delta_\sigma^\gamma + \delta_\lambda^\sigma \sqrt{-g} \cdot g^{\lambda\gamma} h_{,\lambda\gamma} \delta_\sigma^\lambda)\right)$$

$$= \frac{1}{1+\alpha} \cdot \left(\delta_\sigma \left(\sqrt{-g} \cdot g^{\lambda\gamma} h_{,\lambda\gamma}\right) + \frac{\alpha}{2 \cdot n} \cdot (\delta_\gamma^\sigma \delta_\sigma \left(\sqrt{-g} \cdot g^{\lambda\gamma} h_{,\lambda\gamma} \delta_\sigma^\gamma\right) + \delta_\lambda^\sigma \delta_\sigma \left(\sqrt{-g} \cdot g^{\lambda\gamma} h_{,\lambda\gamma} \delta_\sigma^\lambda\right))\right)$$

$$= \frac{1}{1+\alpha} \cdot \left(\delta_\sigma \left(\sqrt{-g} \cdot g^{\lambda\gamma} h_{,\lambda\gamma}\right) + \frac{\alpha}{2 \cdot n} \cdot (\delta_\gamma \left(\sqrt{-g} \cdot g^{\lambda\gamma} h_{,\lambda\gamma} \delta_\sigma^\gamma\right) + \delta_\lambda \left(\sqrt{-g} \cdot g^{\lambda\gamma} h_{,\lambda\gamma} \delta_\sigma^\lambda\right))\right)$$

$$= \frac{1}{1+\alpha} \cdot \left(\delta_\sigma \left(\sqrt{-g} \cdot g^{\lambda\gamma} h_{,\lambda\gamma}\right) + \frac{\alpha}{2 \cdot n} \cdot (\delta_\gamma \left(\sqrt{-g} \cdot g^{\lambda\gamma} h_{,\lambda\sigma}\right) + \delta_\lambda \left(\sqrt{-g} \cdot g^{\lambda\gamma} h_{,\sigma\gamma}\right))\right)$$

$$= \frac{1}{1+\alpha} \cdot \left(\left(\sqrt{-g} \cdot g^{\lambda\gamma} h_{,\lambda\gamma}\right)_{,\sigma} + \frac{\alpha}{2 \cdot n} \cdot \left(\left(\sqrt{-g} \cdot g^{\lambda\gamma} h_{,\lambda\sigma}\right)_{,\gamma} + \left(\sqrt{-g} \cdot g^{\lambda\gamma} h_{,\gamma\sigma}\right)_{,\lambda}\right)\right) \delta x^\sigma.$$

$$(1188)$$

Similarly, one may also keep the simple variation and consider the intrinsic degrees of f instead:

$$f = \frac{1}{1+\alpha} \cdot \left(\sqrt{-g} \cdot g^{\lambda\gamma} h_{,\lambda\gamma} + \frac{\alpha}{2 \cdot n} \cdot (\sqrt{-g} \cdot g^{\lambda\gamma} h_{,\lambda\gamma} \delta_\sigma^\gamma \delta_\gamma^\sigma + \sqrt{-g} \cdot g^{\lambda\gamma} h_{,\lambda\gamma} \delta_\sigma^\lambda \delta_\lambda^\sigma)\right); \quad \delta \to \delta_\sigma$$

$$\Rightarrow \delta f = \frac{\delta_\sigma}{1+\alpha} \cdot \left(\sqrt{-g} \cdot g^{\lambda\gamma} h_{,\lambda\gamma} + \frac{\alpha}{2 \cdot n} \cdot (\sqrt{-g} \cdot g^{\lambda\gamma} h_{,\lambda\gamma} \delta_\sigma^\gamma \delta_\gamma^\sigma + \sqrt{-g} \cdot g^{\lambda\gamma} h_{,\lambda\gamma} \delta_\sigma^\lambda \delta_\lambda^\sigma)\right)$$

$$= \frac{\delta_\sigma}{1+\alpha} \cdot \left(\sqrt{-g} \cdot g^{\lambda\gamma} h_{,\lambda\gamma} + \frac{\alpha}{2 \cdot n} \cdot (\delta_\gamma^\sigma \sqrt{-g} \cdot g^{\lambda\gamma} h_{,\lambda\gamma} \delta_\sigma^\gamma + \delta_\lambda^\sigma \sqrt{-g} \cdot g^{\lambda\gamma} h_{,\lambda\gamma} \delta_\sigma^\lambda)\right)$$

$$= \frac{1}{1+\alpha} \cdot \left(\delta_\sigma \left(\sqrt{-g} \cdot g^{\lambda\gamma} h_{,\lambda\gamma}\right) + \frac{\alpha}{2 \cdot n} \cdot (\delta_\gamma^\sigma \delta_\sigma \left(\sqrt{-g} \cdot g^{\lambda\gamma} h_{,\lambda\gamma} \delta_\sigma^\gamma\right) + \delta_\lambda^\sigma \delta_\sigma \left(\sqrt{-g} \cdot g^{\lambda\gamma} h_{,\lambda\gamma} \delta_\sigma^\lambda\right))\right)$$

$$= \frac{1}{1+\alpha} \cdot \left(\delta_\sigma \left(\sqrt{-g} \cdot g^{\lambda\gamma} h_{,\lambda\gamma}\right) + \frac{\alpha}{2 \cdot n} \cdot (\delta_\gamma \left(\sqrt{-g} \cdot g^{\lambda\gamma} h_{,\lambda\gamma} \delta_\sigma^\gamma\right) + \delta_\lambda \left(\sqrt{-g} \cdot g^{\lambda\gamma} h_{,\lambda\gamma} \delta_\sigma^\lambda\right))\right)$$

$$= \frac{1}{1+\alpha} \cdot \left(\delta_\sigma \left(\sqrt{-g} \cdot g^{\lambda\gamma} h_{,\lambda\gamma}\right) + \frac{\alpha}{2 \cdot n} \cdot (\delta_\gamma \left(\sqrt{-g} \cdot g^{\lambda\gamma} h_{,\lambda\sigma}\right) + \delta_\lambda \left(\sqrt{-g} \cdot g^{\lambda\gamma} h_{,\sigma\gamma}\right))\right)$$

$$= \frac{1}{1+\alpha} \cdot \left(\left(\sqrt{-g} \cdot g^{\lambda\gamma} h_{,\lambda\gamma}\right)_{,\sigma} + \frac{\alpha}{2 \cdot n} \cdot \left(\left(\sqrt{-g} \cdot g^{\lambda\gamma} h_{,\lambda\sigma}\right)_{,\gamma} + \left(\sqrt{-g} \cdot g^{\lambda\gamma} h_{,\gamma\sigma}\right)_{,\lambda}\right)\right) \delta x^\sigma.$$

$$(1189)$$

The reader may prove easily that without the variation f just reduces to the following scalar:

$$f = \sqrt{-g} \cdot g^{\lambda\gamma} h_{,\lambda\gamma}. \quad (1190)$$

Thus, we have introduced just an intelligent 1 as factor, namely in:

$$1 = \frac{1}{1+\alpha} \cdot \left(1 + \frac{\alpha}{2 \cdot n} \cdot (n+n)\right); \quad 1 = \frac{n}{n} = \frac{\delta_\sigma^\gamma \delta_\gamma^\sigma}{n} = \frac{\delta_\sigma^\lambda \delta_\lambda^\sigma}{n}. \quad (1191)$$

With the identity $G^\lambda = h^{,\lambda}; G_\lambda = h_{,\lambda}$ we recognize the essentials of the governing equation of elasticity in the last line of Eq. (1187). Thereby the vectors $G^\lambda = h^{,\lambda}; G_\lambda = h_{,\lambda}$ represent simple displacement vectors. The similarity becomes obvious (c.f. [148–158]) when assuming an almost flat space and very nearly Cartesian coordinates, which simplifies Eq. (1187) as follows:

$$\delta W = 0 \simeq \int_V d^n x \left(\sqrt{-G} \cdot \left[R^{\alpha\beta} - \frac{R}{2} G^{\alpha\beta} \right] \right) \sqrt{-g}$$

$$\cdot \left(\frac{\frac{\partial F[f]}{\partial f} \cdot g_{\alpha\beta}}{1+\alpha} \left(g^{\lambda\gamma} h_{,\lambda\gamma\sigma} + \frac{\alpha}{n} \cdot \Delta_g \left(h_{,\sigma} \right) \right) \right) g^{\sigma\chi} \delta x_\chi$$

$$\Rightarrow 0 = g^{\lambda\gamma} h_{,\sigma\lambda\gamma} + \frac{\alpha}{n} \cdot \Delta_g \left(h_{,\sigma} \right), \tag{1192}$$

where for $\frac{\alpha}{n} = 1 - 2 \cdot \nu$ (ν giving Poisson's ratio, see, e.g., [152]) we recognize the typical linear elastic equation for isotropic bodies in the absence of gravitational forces (thereby we only refer to the gravitational forces in the classical 3-dimensional linear theory of elasticity, as being meant in classical textbooks like [192] and not the quantum-gravimetric fields considered and derived here). Just as with the classical elasticity we find a vanishing Laplace operator for the case of incompressibility $\nu = 0.5$, where only shear would be allowed. We may assume that elementary particles with vanishing of very low masses would be described by such shear fields (see section 15.5.3).

Repeating the evaluation with an approach for f as follows:

$$f = \frac{\sqrt{-g}}{1+\beta} \cdot \left(h_{,\lambda} g^{\lambda\gamma} h_{,\gamma} + \frac{\beta}{2 \cdot n} \cdot \left(\begin{array}{c} h_{,\lambda} g^{\lambda\gamma} h_{,\gamma} \delta^\gamma_\sigma \delta^\sigma_\gamma \\ + h_{,\lambda} g^{\lambda\gamma} h_{,\gamma} \delta^\lambda_\sigma \delta^\sigma_\lambda \end{array} \right) \right); \quad \delta \to \delta_\sigma$$

$$\Rightarrow \delta f = \frac{1}{1+\beta} \cdot \left(\begin{array}{c} \left(\sqrt{-g} \cdot h_{,\lambda} g^{\lambda\gamma} h_{,\gamma} \right)_{,\sigma} \\ + \frac{\beta}{2 \cdot n} \cdot \left(\left(\sqrt{-g} \cdot h_{,\lambda} g^{\lambda\gamma} h_{,\sigma} \right)_{,\gamma} + \left(\sqrt{-g} \cdot h_{,\sigma} g^{\lambda\gamma} h_{,\gamma} \right)_{,\lambda} \right) \end{array} \right),$$

$$\tag{1193}$$

leads us to:

$$\delta W = 0 = \int_V d^n x \left(\sqrt{-G} \cdot \left[R^{\alpha\beta} - \frac{R}{2} G^{\alpha\beta} \right] \right)$$

$$\times \left(\begin{array}{c} \frac{\frac{\partial F[f]}{\partial f} \cdot g_{\alpha\beta}}{1+\beta} \left(\left(\sqrt{-g} \cdot h_{,\lambda} g^{\lambda\gamma} h_{,\gamma} \right)_{,\sigma} + \frac{\beta}{2 \cdot n} \cdot \left(\begin{array}{c} \left(\sqrt{-g} \cdot h_{,\lambda} g^{\lambda\gamma} h_{,\sigma} \right)_{,\gamma} \\ + \left(\sqrt{-g} \cdot h_{,\sigma} g^{\lambda\gamma} h_{,\gamma} \right)_{,\lambda} \end{array} \right) \right) \\ + F[f] \cdot g_{\alpha\beta,\sigma} \end{array} \right) g^{\sigma\chi} \delta x_\chi$$

$$= \int_V d^n x \left(\sqrt{-G} \cdot \left[R^{\alpha\beta} - \frac{R}{2} G^{\alpha\beta} \right] \right)$$

$$\times \left(\frac{\frac{\partial F[f]}{\partial f} \cdot g_{\alpha\beta}}{1+\beta} \left(\begin{array}{l} \left(\sqrt{-g} \cdot g^{\lambda\gamma}\right)_{,\sigma} h_{,\lambda} h_{,\gamma} + \sqrt{-g} \cdot g^{\lambda\gamma} \left(h_{,\lambda\sigma} h_{,\gamma} + h_{,\lambda} h_{,\gamma\sigma}\right) \\ + \frac{\beta}{n} \cdot h_{,\sigma} \sqrt{-g} \cdot \Delta_g h + \frac{\beta}{2 \cdot n} \cdot g^{\lambda\gamma} \sqrt{-g} \cdot \left(h_{,\lambda} h_{,\sigma\gamma} + h_{,\sigma\lambda} h_{,\gamma}\right) \end{array} \right) \right) g^{\sigma\chi} \delta x_\chi$$
$$+F[f] \cdot g_{\alpha\beta,\sigma}$$

$$= \int_V d^n x \left(\sqrt{-G} \cdot \left[R^{\alpha\beta} - \frac{R}{2} G^{\alpha\beta} \right] \right)$$

$$\times \left(\frac{\frac{\partial F[f]}{\partial f} \cdot g_{\alpha\beta}}{1+\beta} \left(\begin{array}{l} \left(\sqrt{-g} \cdot g^{\lambda\gamma}\right)_{,\sigma} h_{,\lambda} h_{,\gamma} + \frac{\beta}{n} \cdot h_{,\sigma} \sqrt{-g} \cdot \Delta_g h \\ + \sqrt{-g} \cdot g^{\lambda\gamma} \left(1 + \frac{\beta}{2 \cdot n}\right) \left(h_{,\lambda\sigma} h_{,\gamma} + h_{,\lambda} h_{,\gamma\sigma}\right) \end{array} \right) + F[f] \cdot g_{\alpha\beta,\sigma} \right) g^{\sigma\chi} \delta x_\chi .$$

$$(1194)$$

Integration by parts helps us to linearize parts of the integrand:

$$\sqrt{-G} \cdot \left[R^{\alpha\beta} - \frac{R}{2} G^{\alpha\beta} \right] g^{\sigma\chi} = \left(\ddots \right)^{\alpha\beta\sigma\chi} \quad ; \quad \frac{\frac{\partial F[f]}{\partial f} \cdot g_{\alpha\beta}}{1+\beta} = A_{\alpha\beta}$$

$$\longrightarrow \delta W = 0 = \int_V d^n x \left(\ddots \right)^{\alpha\beta\sigma\chi}$$

$$\times \left(A_{\alpha\beta} \left(\begin{array}{l} \left(\sqrt{-g} \cdot g^{\lambda\gamma}\right)_{,\sigma} h_{,\lambda} h_{,\gamma} + \frac{\beta}{n} \cdot h_{,\sigma} \sqrt{-g} \cdot \Delta_g h \\ + \sqrt{-g} \cdot g^{\lambda\gamma} \left(1 + \frac{\beta}{2 \cdot n}\right) \left(h_{,\lambda\sigma} h_{,\gamma} + h_{,\lambda} h_{,\gamma\sigma}\right) \end{array} \right) + F[f] \cdot g_{\alpha\beta,\sigma} \right) \delta x_\chi$$

$$= \int_V d^n x \left(\begin{array}{l} \left(\ddots \right)^{\alpha\beta\sigma\chi} F[f] \cdot g_{\alpha\beta,\sigma} \\ -h \cdot \left(\begin{array}{l} \left(\left(\ddots \right)^{\alpha\beta\sigma\chi} A_{\alpha\beta} \left(\sqrt{-g} \cdot g^{\lambda\gamma}\right)_{,\sigma} h_{,\lambda} \right)_{,\gamma} \\ + \frac{\beta}{n} \cdot \left(\left(\ddots \right)^{\alpha\beta\sigma\chi} A_{\alpha\beta} \sqrt{-g} \cdot \Delta_g h \right)_{,\sigma} \\ + \left(1 + \frac{\beta}{2 \cdot n}\right) \left(\begin{array}{l} \left(\left(\ddots \right)^{\alpha\beta\sigma\chi} A_{\alpha\beta} \sqrt{-g} \cdot g^{\lambda\gamma} h_{,\lambda\sigma} \right)_{,\gamma} \\ + \left(\left(\ddots \right)^{\alpha\beta\sigma\chi} A_{\alpha\beta} \sqrt{-g} \cdot g^{\lambda\gamma} h_{,\gamma\sigma} \right)_{,\lambda} \end{array} \right) \end{array} \right) \end{array} \right) \delta x_\chi .$$

$$(1195)$$

Once again, considering the case of an almost flat space and very nearly Cartesian coordinates, the integrand simplifies to:

$$\approx \int_V d^n x \left(-h \cdot \left(\begin{array}{l} \frac{\beta}{n} \cdot \left(\left(\ddots \right)^{\alpha\beta\sigma\chi} A_{\alpha\beta} \sqrt{-g} \cdot \Delta_g h \right)_{,\sigma} \\ + \left(1 + \frac{\beta}{2 \cdot n}\right) \left(\ddots \right)^{\alpha\beta\sigma\chi} \sqrt{-g} \cdot g^{\lambda\gamma} \left(2 \cdot A_{\alpha\beta} h_{,\lambda\gamma\sigma} + \left(\begin{array}{l} A_{\alpha\beta,\gamma} h_{,\lambda\sigma} \\ + A_{\alpha\beta,\lambda} h_{,\gamma\sigma} \end{array} \right) \right) \end{array} \right) \right) \delta x_\chi$$

$$= \int_V d^n x \left(-h \cdot \left(\ddots \right)^{\alpha\beta\sigma\chi} \sqrt{-g} \cdot \left(\begin{array}{l} \frac{\beta}{n} \cdot \left(A_{\alpha\beta,\sigma} \cdot \Delta_g h + A_{\alpha\beta} \cdot \left(\Delta_g h\right)_{,\sigma}\right) \\ + \left(1 + \frac{\beta}{2 \cdot n}\right) g^{\lambda\gamma} \left(\begin{array}{l} 2 \cdot A_{\alpha\beta} h_{,\lambda\gamma\sigma} \\ + \left(A_{\alpha\beta,\gamma} h_{,\lambda\sigma} + A_{\alpha\beta,\lambda} h_{,\gamma\sigma} \right) \end{array} \right) \end{array} \right) \right) \delta x_\chi$$

$$\xrightarrow{F[f]=f} = \int\limits_V d^n x \left(-h \cdot \left(\stackrel{\cdot\cdot}{\cdot}\right)^{\alpha\beta\sigma\chi} \sqrt{-g} \cdot A_{\alpha\beta} \cdot \left(\frac{\frac{\beta}{n}\cdot(\Delta_g h)_{,\sigma}}{+2\cdot\left(1+\frac{\beta}{2\cdot n}\right)g^{\lambda\gamma}h_{,\lambda\gamma\sigma}}\right)\right) \delta x_\chi$$

$$\Rightarrow 0 = \frac{\beta}{n}\cdot(\Delta_g h)_{,\sigma} +2\cdot\left(1+\frac{\beta}{2\cdot n}\right)g^{\lambda\gamma}h_{,\lambda\gamma\sigma} \simeq \frac{\beta}{n}\cdot\Delta_g(h_{,\sigma}) +2\cdot\left(1+\frac{\beta}{2\cdot n}\right)g^{\lambda\gamma}h_{,\lambda\gamma\sigma}.$$

$$(1196)$$

As before in Eq. (1192), we recognize the typical features of the fundamental equation of linear elasticity, only that this time we have the connection to Poisson's ratio via:

$$\frac{\beta}{(2\cdot n+\beta)} = 1-2\cdot\nu \quad\Rightarrow\quad \beta = n\cdot\left(\frac{1}{\nu}-2\right); \quad \nu = \frac{n}{2\cdot n+\beta}. \quad (1197)$$

We note that for $\beta = 0$ we would have incompressibility. Setting the result for β into our approach for f in Eq. (1193):

$$f = \frac{\sqrt{-g}}{1+n\cdot\left(\frac{1}{\nu}-2\right)} \cdot \left(h_{,\lambda}g^{\lambda\gamma}h_{,\gamma} + \frac{\left(\frac{1}{\nu}-2\right)}{2}\cdot\left(h_{,\lambda}g^{\lambda\gamma}h_{,\gamma}\delta^\sigma_\sigma\delta^\sigma_\gamma + h_{,\lambda}g^{\lambda\gamma}h_{,\gamma}\delta^\lambda_\sigma\delta^\sigma_\lambda\right)\right); \quad \delta\to\delta_\sigma$$

$$\Rightarrow f = \frac{\sqrt{-g}\cdot\nu}{\nu+n\cdot(1-2\cdot\nu)} \cdot \left(h_{,\lambda}g^{\lambda\gamma}h_{,\gamma} + \left(\frac{1}{2\cdot\nu}-1\right)\cdot\left(h_{,\lambda}g^{\lambda\gamma}h_{,\gamma}\delta^\sigma_\sigma\delta^\sigma_\gamma + h_{,\lambda}g^{\lambda\gamma}h_{,\gamma}\delta^\lambda_\sigma\delta^\sigma_\lambda\right)\right)$$

$$= \frac{\sqrt{-g}}{\nu+n\cdot(1-2\cdot\nu)} \cdot \left(\nu\cdot h_{,\lambda}g^{\lambda\gamma}h_{,\gamma} + \left(\frac{1}{2}-\nu\right)\cdot\left(h_{,\lambda}g^{\lambda\gamma}h_{,\gamma}\delta^\sigma_\sigma\delta^\sigma_\gamma + h_{,\lambda}g^{\lambda\gamma}h_{,\gamma}\delta^\lambda_\sigma\delta^\sigma_\lambda\right)\right)$$

$$\Rightarrow \delta f = \frac{1}{\nu+n\cdot(1-2\cdot\nu)} \left(\frac{(1-2\cdot\nu)\cdot\left(\left(\sqrt{-g}\cdot h_{,\lambda}g^{\lambda\gamma}h_{,\sigma}\right)_{,\gamma}+\left(\sqrt{-g}\cdot h_{,\sigma}g^{\lambda\gamma}h_{,\gamma}\right)_{,\lambda}\right)}{2} +\nu\cdot\left(\sqrt{-g}\cdot h_{,\lambda}g^{\lambda\gamma}h_{,\gamma}\right)_{,\sigma}\right)$$

$$(1198)$$

results in the surprising constellation that, with a vanishing Poisson's ratio, the variation of f is dominated by the scalar f's intrinsic degree of freedom, namely the part:

$$\lim_{\nu\to\infty}\delta f = \frac{\left(\sqrt{-g}\cdot h_{,\lambda}g^{\lambda\gamma}h_{,\sigma}\right)_{,\gamma} + \left(\sqrt{-g}\cdot h_{,\sigma}g^{\lambda\gamma}h_{,\gamma}\right)_{,\lambda}}{2\cdot n}. \quad (1199)$$

15.5.2.1 Conclusions with respect to our elastic quantum gravity equations

We note, that in contrast to the f-situation, our intrinsic vector-approaches lead to some surprising equations not just sporting the fundamental structures of the classical equations of elasticity, but also revealing the inner degrees of freedom for a metric variation $F[f]$, whereby f shows itself as structurally complex. This is the same situation as Dirac found when moving from the Klein–Gordon equation to his very own equation, which—of course—was the Dirac equation [143]. There, too, the formerly scalar function f became a vector function.

But can our elastic equations truly be the metric equivalent to the Dirac equation?

If so, we should expect to be able to derive some particle solutions, which we will attempt within the next subsection.

15.5.3 From the Elastic Equation to Particles

15.5.3.1 Potential elementary particle solutions?

In the previous section, we have shown that a metric wrapper of the kind:

$$G_{\delta\gamma} = F\,[f] \cdot g_{\delta\gamma} \qquad (1200)$$

subjected to the usual Hilbert variation and applying the chain rule during the variation process results in quantum gravity equations as follows:

$$\delta W = 0 = \delta_\sigma \int_V d^n x \left(\sqrt{-G} \cdot R\right) = \int_V d^n x \left(\sqrt{-G} \cdot \left[R^{\alpha\beta} - \frac{R}{2} G^{\alpha\beta}\right]\right) \delta_\sigma G_{\alpha\beta}$$

$$= \int_V d^n x \left(\sqrt{-G} \cdot \left[R^{\alpha\beta} - \frac{R}{2} G^{\alpha\beta}\right]\right) \left(\frac{\partial F\,[f]}{\partial f} \cdot g_{\alpha\beta}\, f_{,\sigma} + F\,[f] \cdot g_{\alpha\beta,\sigma}\right) g^{\sigma\chi} \delta x_\chi.$$

$$(1201)$$

Allowing the scalar function f to be internally structured (e.g., being a divergence of a vector), brought us to equations with the typical structure known from the fundamental equations of elasticity (e.g., see Eq. (1192)), where the internal vectors G_α forming the scalar f revealed themselves as displacement vectors. We suspect these vectors and their fields to describe elementary particles. However, as our resulting elastic equations from the previous section and their derivations are quite complicated and as we are only interested in the nearly flat space solutions anyway, we want to use simpler paths to obtain them.

In chapters 2 to 6, we considered quite a few options for the variation of the metric tensor. Here we repeat one which we consider most important in connection with elementary particle physics (c.f. section 2.2).

Our starting point shall be the usual tensor transformation rule for the covariant metric tensor:

$$g_{\delta\gamma} = \mathbf{g}_\delta \cdot \mathbf{g}_\gamma = \frac{\partial G^\alpha\,[x_k]}{\partial x^\delta} \frac{\partial G^\beta\,[x_k]}{\partial x^\gamma} g_{\alpha\beta}. \qquad (1202)$$

The base vectors $\mathbf{g}_\delta$ to a certain metric are given as:

$$\mathbf{g}_\delta = \frac{\partial G^\alpha\,[x_k]}{\partial x^\delta}\mathbf{e}_\alpha, \qquad (1203)$$

where the functions $G^\alpha[\ldots]$ denote arbitrary functions of the coordinates x_k. Here the vectors $\mathbf{e}_\alpha$ shall denote the base vectors of a fundamental coordinate system of the right (in principle arbitrary) number of dimension. Thus, we have the variation for $\delta g_{\delta\gamma}$ in Eqs. (1202) and (1203) as follows:

$$\delta g_{\delta\gamma} = \delta\left(\mathbf{g}_\delta \cdot \mathbf{g}_\gamma\right) = \mathbf{g}_\delta \cdot \delta\mathbf{g}_\gamma + \delta\mathbf{g}_\delta \cdot \mathbf{g}_\gamma$$

$$= \frac{\partial G^\alpha\,[x_k]}{\partial x^\delta}\mathbf{e}_\alpha \cdot \delta\left(\frac{\partial G^\beta\,[x_k]}{\partial x^\gamma}\mathbf{e}_\beta\right) + \delta\left(\frac{\partial G^\alpha\,[x_k]}{\partial x^\delta}\mathbf{e}_\alpha\right) \cdot \frac{\partial G^\beta\,[x_k]}{\partial x^\gamma}\mathbf{e}_\beta.$$

$$(1204)$$

Now we introduce two additional degrees of freedom, namely:

(a) that neither the number of dimensions in which the base vectors exist and form a complete transformation (Eq. (1204)) needs to be the same as the metric space they define,

(b) nor that the variation δ is defined or fixed in any way. It shall simply mean that we do some kind of infinitesimal "shaking" or "rocking" of the metric system in question. Thereby we note that this is just the lax verbal formulation of our mathematically correct procedures from the previous subsection, where we went for the inner degrees of freedom of the scalar f.

In order to properly account for point (a), we shall rewrite Eqs. (1203) and (1204) as:

$$\delta g_{\delta\gamma} = \delta\left(\mathbf{g}_\delta \cdot \mathbf{g}_\gamma\right) = \mathbf{g}_\delta \cdot \delta\mathbf{g}_\gamma + \delta\mathbf{g}_\delta \cdot \mathbf{g}_\gamma$$

$$= \frac{\partial G^i\,[x_k]}{\partial x^\delta}\mathbf{e}_i \cdot \delta\left(\frac{\partial G^j\,[x_k]}{\partial x^\gamma}\mathbf{e}_j\right) + \delta\left(\frac{\partial G^i\,[x_k]}{\partial x^\delta}\mathbf{e}_i\right) \cdot \frac{\partial G^j\,[x_k]}{\partial x^\gamma}\mathbf{e}_j$$

$$\mathbf{g}_\delta = \frac{\partial G^j\,[x_k]}{\partial x^\delta}\mathbf{e}_j. \qquad (1205)$$

Thereby the Latin indices are running to a different (potentially higher) number of dimensions N than the Greek indices, which shall be defined for a space or space-time of n dimensions.

A very general variation of Eq. (1205) could be as follows:

$$\delta g_{\delta\gamma} = \delta\left(\mathbf{g}_\delta \cdot \mathbf{g}_\gamma\right) = \mathbf{g}_\delta \cdot \delta\mathbf{g}_\gamma + \delta\mathbf{g}_\delta \cdot \mathbf{g}_\gamma$$

$$= \frac{\partial G^i\,[x_k]}{\partial x^\delta}\mathbf{e}_i \cdot \left(\frac{\partial^2 G^j\,[x_k]}{\partial x^\alpha \partial x^\gamma}\mathbf{e}_j\right)\delta x^\alpha + \left(\frac{\partial^2 G^i\,[x_k]}{\partial x^\alpha \partial x^\delta}\mathbf{e}_i\right)\delta x^\alpha \cdot \frac{\partial G^j\,[x_k]}{\partial x^\gamma}\mathbf{e}_j.$$

$$(1206)$$

We note that setting Eq. (1206) equal to zero automatically fulfills the Einstein field equations, which is to say Eq. (86) (or the classical form in Eq. (87)). In section 2.2, we demonstrated that the general variation in Eq. (1206) can be treated with techniques known from the theory of elasticity [152] to solve the equations. These techniques, known there as the 3-function-ansatz and the Galerkin approach, were introduced in the thirties of the last century by Heinz Neuber [153]. It appears tempting to suggest an extension of this ansatz to an n-function approach with n giving the number of dimensions of the space-time under consideration.

For simplicity, we only give the ansatz in Cartesian coordinates:

$$G^j\,[x_k] = g^{jl}\partial_l G\,[x_k]; \quad \Delta G\,[x_k] = 0$$

$$G^j\,[x_k] = x_\xi \cdot g^{jl}\partial_l G\,[x_k] + \alpha \cdot G\,[x_k]; \quad \xi \text{ any of } 0, 1, \ldots, n-1$$

$$G^j\,[x_k] = \left\{ \ldots, \overbrace{\frac{\partial G\,[x_k]}{\partial x_\zeta}}^{\text{pos } \xi}, \ldots, -\overbrace{\frac{\partial G\,[x_k]}{\partial x_\xi}}^{\text{pos } \zeta}, \ldots \right\}; \quad \forall (\ldots, \ldots) = 0$$

$$G^j\,[x_k] = \left\{ \ldots, -\overbrace{\frac{\partial G\,[x_k]}{\partial x_\zeta}}^{\text{pos } \xi}, \ldots, \overbrace{\frac{\partial G\,[x_k]}{\partial x_\xi}}^{\text{pos } \zeta}, \ldots \right\}; \quad \forall (\ldots, \ldots) = 0.$$

$$(1207)$$

Please note that we kept the notation for the function $G[x_k]$ from the previous sections. It should not be mixed up with the metric determinant G.

Any combination of the basic solutions (Eq. (1207)) is also a solution to an elasticity equation in n dimensions. Please note that, depending on the number of dimensions n, there can be many such shear-type solutions (especially spin-like solutions as given in section 2.2 (see Eqs. (69) to (72)). We see that we can construct non-trivial solutions to Eq. (1206) out of combinations of harmonic functions and their derivatives. In order to satisfy Eq. (1206) in its currently given form via:

$$\sum_{\alpha=0}^{n-1} \frac{\partial^2 G^j\,[x_k]}{\partial x^\alpha \partial x^\gamma}\mathbf{e}_j = 0 = \sum_{\alpha=0}^{n-1} \frac{\partial^2 G^i\,[x_k]}{\partial x^\alpha \partial x^\delta}\mathbf{e}_i, \qquad (1208)$$

we have to set $\alpha = -1$.

Thereby we have to perform the variation as follows:

$$\delta g_{\delta\gamma} = \delta \left(\mathbf{g}_\delta \cdot \mathbf{g}_\gamma \right) = \mathbf{g}_\delta \cdot \delta \mathbf{g}_\gamma + \delta \mathbf{g}_\delta \cdot \mathbf{g}_\gamma$$

$$= \frac{\partial G^i\,[x_k]}{\partial x^\delta}\mathbf{e}_i \cdot \left(\frac{\partial^2 G^j\,[x_k]}{\partial x^\beta \partial x^\gamma}\mathbf{e}_j \right) \delta x^\beta + \left(\frac{\partial^2 G^i\,[x_k]}{\partial x^\chi \partial x^\delta}\mathbf{e}_i \right) \delta x^\chi \cdot \frac{\partial G^j\,[x_k]}{\partial x^\gamma}\mathbf{e}_j$$

$$= \frac{\partial G^i\,[x_k]}{\partial x^\gamma}\mathbf{e}_i C_\delta^\gamma \cdot \left(\frac{\partial^2 G^j\,[x_k]}{\partial x^\beta \partial x^\gamma}\mathbf{e}_j \right) C_\chi^\beta \delta x^\chi + \left(\frac{\partial^2 G^i\,[x_k]}{\partial x^\chi \partial x^\delta}\mathbf{e}_i \right) \delta x^\chi \cdot \frac{\partial G^j\,[x_k]}{\partial x^\gamma}\mathbf{e}_j.$$

$$(1209)$$

A simple exchange of the dummy indices in Eq. (1209) leads us to:

$$= \frac{\partial G^j\,[x_k]}{\partial x^\gamma}\mathbf{e}_j C_\delta^\gamma \cdot \left(\frac{\partial^2 G^i\,[x_k]}{\partial x^\beta \partial x^\gamma}\mathbf{e}_i \right) C_\chi^\beta \delta x^\chi + \left(\frac{\partial^2 G^i\,[x_k]}{\partial x^\chi \partial x^\delta}\mathbf{e}_i \right) \delta x^\chi \cdot \frac{\partial G^j\,[x_k]}{\partial x^\gamma}\mathbf{e}_j$$

$$= \left(C_\delta^\gamma C_\chi^\beta \cdot \left(\frac{\partial^2 G^i\,[x_k]}{\partial x^\beta \partial x^\gamma}\mathbf{e}_i \right) + \left(\frac{\partial^2 G^i\,[x_k]}{\partial x^\chi \partial x^\delta}\mathbf{e}_i \right) \right) \delta x^\chi \cdot \frac{\partial G^j\,[x_k]}{\partial x^\gamma}\mathbf{e}_j, \qquad (1210)$$

and assuming $C_\delta^\gamma C_\chi^\beta = b^2 \cdot \delta_\delta^\gamma \delta_\chi^\beta$ yields the equation:

$$= \left(b^2 \left(\frac{\partial^2 G^i\,[x_k]}{\partial x^\chi \partial x^\delta}\mathbf{e}_i \right) + \left(\frac{\partial^2 G^i\,[x_k]}{\partial x^\chi \partial x^\delta}\mathbf{e}_i \right) \right) \delta x^\chi \cdot \frac{\partial G^j\,[x_k]}{\partial x^\gamma}\mathbf{e}_j$$

$$= (b^2 + 1) \left(\frac{\partial^2 G^i\,[x_k]}{\partial x^\chi \partial x^\delta}\mathbf{e}_i \right) \delta x^\chi \cdot \frac{\partial G^j\,[x_k]}{\partial x^\gamma}\mathbf{e}_j$$

$$\Rightarrow \sum_{\chi=0}^{n-1} \frac{\partial^2 G^i\,[x_k]}{\partial x^\chi \partial x^\delta}\mathbf{e}_i = 0. \qquad (1211)$$

We need to point out that the variation with respect to the base vectors is not the only variation one could apply to the Einstein–Hilbert action. There is also the possibility to variate with respect to the number of dimensions and the position of centers of gravity. It was demonstrated in sections 3.8 and 4.5 that this gives us a connection to the classical thermodynamics. Here, however, this extension is not of need.

15.5.3.2 Spin due to shear → neutrino (?)

In previous papers of our series "Einstein had it..." and "Science Riddles", we already have suspected spin as some kind of shear [57, 65, 81, 90, 103, 129]. Here now with the solutions of the type in line 3 and 4 in Eq. (1207) and comparing this with corresponding situations in mechanical contact problems (e.g., [150, 151]), we think that the best way to start and look for spin structures is via these types of solutions.

As all of the solutions in the *n*-function-ansatz (Eq. (1207)) satisfy Eq. (1209) (and more—c.f. [139]) as long as only the condition $\Delta G[x_k] = 0$ is fulfilled, we can feed any harmonics into the approach. In order to keep things as simple as possible, we here apply a Minkowski metric in Cartesian coordinates:

$$g_{\alpha\beta} = \begin{pmatrix} -c^2 & 0 & 0 & 0 \\ 0 & 1 & 0 & 0 \\ 0 & 0 & 1 & 0 \\ 0 & 0 & 0 & 1 \end{pmatrix} \tag{1212}$$

and solve our Laplace problem for $\Delta G[x_k] = 0$ with a separation approach leading us to:

$$G[x_k] = e^{i \cdot h \cdot t} \cdot \sin[a \cdot x] \cdot \sin[b \cdot y] \cdot \sin[d \cdot z]; \quad h = c \cdot \sqrt{a^2 + b^2 + d^2}. \tag{1213}$$

In order to localize the resulting field or object in question, we apply Fourier transforms. So, one suitable form might be the following choice with the Dawson integral $F[a]$:

$$e^{-\frac{x^2}{\delta}} = \int_{-\infty}^{+\infty} da \ \sqrt{\frac{2 \cdot \delta}{\pi}} \cdot F\left[\frac{\sqrt{\delta}}{2} \cdot a\right] \cdot \sin[a \cdot x]; \quad F[a] = e^{-a^2} \int_0^a d\xi \ e^{\xi^2}. \tag{1214}$$

This results in an approach for G as follows:

$$G[x_k] = \int_{-\infty}^{+\infty} da \ \sqrt{\frac{2 \cdot \delta}{\pi}} \cdot F\left[\frac{\sqrt{\delta}}{2} \cdot a\right] \cdot \sin[a \cdot x]$$

$$\cdot \int_{-\infty}^{+\infty} db \ \sqrt{\frac{2 \cdot \delta}{\pi}} \cdot F\left[\frac{\sqrt{\delta}}{2} \cdot b\right] \cdot \sin[b \cdot y]$$

$$\times \int_{-\infty}^{+\infty} dd \ \sqrt{\frac{2 \cdot \delta}{\pi}} \cdot F\left[\frac{\sqrt{\delta}}{2} \cdot d\right] \cdot \sin[d \cdot z] \cdot e^{i \cdot c \cdot \sqrt{a^2 + b^2 + d^2} \cdot t}. \tag{1215}$$

Of course, it might be necessary to adjust the F-functions due to the $e^{i \cdot c \cdot \sqrt{a^2 + b^2 + d^2} \cdot t}$-term, but this does not matter here as we only intended to achieve some kind of localization of the field in a most general manner. As the functional choice for F is not restricted, we can in principle also have the general function $G[\ldots]$ as follows:

$$G[x_k] = \int\limits_{-\infty}^{+\infty} dd \int\limits_{-\infty}^{+\infty} db$$

$$\cdot \int\limits_{-\infty}^{+\infty} da \; F[a, b, d] \cdot \sin[a \cdot x] \cdot \sin[b \cdot y] \cdot \sin[d \cdot z] \cdot e^{i \cdot c \cdot \sqrt{a^2+b^2+d^2} \cdot t}.$$

$$(1216)$$

Having established a way to obtain fairly arbitrary spatial field distributions via Eq. (1216), we now want to investigate options of obtaining the property of spin according to Eq. (1207) via:

$$G^j[x_k] = \pm \left\{ 0, \overbrace{\frac{\partial G}{\partial x_\zeta}}^{\text{pos } \xi = x}, \overbrace{-\frac{\partial G}{\partial x_\xi}}^{\text{pos } \zeta = y}, 0 \right\} = \pm \left\{ 0, \frac{\partial G}{\partial y}, -\frac{\partial G}{\partial x}, 0 \right\}; \quad G \equiv G[x_k],$$

$$(1217)$$

where we assume the z-axis as spin direction.

In order to have the simplest start, we will not rush to complicated distributions with Eq. (1216), but simply apply the classical point force potential with:

$$G[x_k] = \frac{1}{-c^2 \cdot t^2 + x^2 + y^2 + z^2}. \qquad (1218)$$

As it does not make much sense to dig into discussions about the character of time compared to space, and as we want to see a way to incorporate the toolbox of potential theory into our approach as quickly as possible, we set $c = 1$ and further assume t to be purely imaginary, which is to say $t \to i*t$. This allows us to move on with the point potential of:

$$G[x_k] = \frac{1}{t^2 + x^2 + y^2 + z^2}. \qquad (1219)$$

Now we have to find a way to construct a function $G[\ldots]$ which for $t \to \infty$ does not vanish. Integrating Eq. (1219) with respect to t would give us the requested:

$$H = \int G[x_k] dt = \frac{\arctan\left[\frac{t}{\sqrt{x^2+y^2+z^2}}\right]}{\sqrt{x^2+y^2+z^2}}; \quad \lim_{t \to \infty} H = \frac{\pi}{2}\sqrt{\frac{1}{x^2+y^2+z^2}}.$$

$$(1220)$$

We recognize the potential of a point charge as limit for $t \to \infty$. The resulting total quantum gravity solution (Eq. (1217)) in the case of $t \to \infty$ would then read:

$$\lim_{t \to \infty} G^j[x_k] = \mp C_s \cdot \frac{\pi}{2} \cdot \left\{ 0, \frac{y}{(x^2 + y^2 + z^2)^{3/2}}, -\frac{x}{(x^2 + y^2 + z^2)^{3/2}}, 0 \right\}$$

$$= \mp C_s \cdot \frac{\pi}{2} \cdot \frac{1}{(x^2 + y^2 + z^2)^{3/2}} \cdot \{0, y, -x, 0\}. \tag{1221}$$

Thereby C_s stands for a suitable constant. The complete solution can be given as:

$$G^j[x_k] = \pm C_s \cdot \left(\frac{\frac{t}{(x^2+y^2+z^2)^2\left(1+\frac{t^2}{x^2+y^2+z^2}\right)}}{+ \frac{\arctan\left[\frac{t}{\sqrt{x^2+y^2+z^2}}\right]}{(x^2+y^2+z^2)^{3/2}}} \right) \cdot \{0, y, -x, 0\}. \tag{1222}$$

15.5.3.3 Electric charge due to contact solutions → electron and positron (?)

Now we apply the same harmonic function H as used in the subsection above, set it as the G-harmonic $G = H$:

$$G = H = \int \frac{dt}{t^2 + x^2 + y^2 + z^2} = \frac{\arctan\left[\frac{t}{\sqrt{x^2+y^2+z^2}}\right]}{\sqrt{x^2 + y^2 + z^2}};$$

$$\lim_{t \to \infty} G = \frac{1}{2}\pi \sqrt{\frac{1}{x^2 + y^2 + z^2}} \tag{1223}$$

and consider the first solution in Eq. (1207) in 4 dimensions t, x, y, z in the following form:

$$G^j = \partial^j G; \quad \partial^j = \delta^{jl}\partial_l = \partial_j. \tag{1224}$$

With the function $f[\ldots]$ given as:

$$f = f[t, x, y, z]$$

$$= \frac{t}{(x^2 + y^2 + z^2)^2 \left(1 + \frac{t^2}{x^2+y^2+z^2}\right)} + \frac{\arctan\left[\frac{t}{\sqrt{x^2+y^2+z^2}}\right]}{(x^2 + y^2 + z^2)^{3/2}}, \tag{1225}$$

the vector field $\mathbf{G}^j$ does read:

$$G^j[x_k] = C_f \cdot f \cdot \left\{ \frac{1}{f \cdot (t^2 + x^2 + y^2 + z^2)}, x, y, z \right\}. \tag{1226}$$

The attentive reader recognizes the potential of a point charge for the limiting case of $t \to \infty$, because we have the following limit:

$$f_\infty = \lim_{t\to\infty} f\,[t, x, y, z] = \frac{\pi}{2} \cdot \frac{1}{\left(x^2 + y^2 + z^2\right)^{3/2}}, \tag{1227}$$

$$\lim_{t\to\infty} G^j\,[x_k] = C_f \cdot f_\infty \cdot \{0, x, y, z\}$$

$$= \frac{\pi}{2} \cdot \frac{C_f}{\left(x^2 + y^2 + z^2\right)^{3/2}} \cdot \{0, x, y, z\}. \tag{1228}$$

We realize that with the constant:

$$C_f = \frac{2}{\pi} \cdot \frac{Q}{4 \cdot \pi \cdot \varepsilon_0}, \tag{1229}$$

we have in Eq. (1228) the electric field of a point charge with charge Q. The constant ε_0 is the electric field constant.

The resulting vector field for a combined solution (Eqs. (1228) and (1221)) has been illustrated in Fig. 15.9.

It is not illogical to assume that Eq. (1226) together with Eq. (1222) describes or at least has something to do with the metric form of the electron or the positron, leading to the total solution:

$$G^j\,[x_k] = C_f \cdot f \cdot \left\{ \frac{1}{f \cdot (t^2 + x^2 + y^2 + z^2)}, x \pm \frac{y}{\chi}, y \mp \frac{x}{\chi}, z \right\}$$

$$\xrightarrow{\lim_{t\to\infty}} C_f \cdot f_\infty \cdot \left\{ 0, x \pm \frac{y}{\chi}, y \mp \frac{x}{\chi}, z \right\}$$

$$G^j\,[x_k] = C_f \cdot f \cdot \left\{ \frac{1}{f \cdot (t^2 + x^2 + y^2 + z^2)}, x \pm \frac{z}{\chi}, y, z \mp \frac{x}{\chi} \right\}$$

$$\xrightarrow{\lim_{t\to\infty}} C_f \cdot f_\infty \cdot \left\{ 0, x \pm \frac{x}{\chi}, y, z \mp \frac{x}{\chi} \right\}$$

$$G^j\,[x_k] = C_f \cdot f \cdot \left\{ \frac{1}{f \cdot (t^2 + x^2 + y^2 + z^2)}, x, y \mp \frac{z}{\chi}, z \pm \frac{y}{\chi} \right\}$$

$$\xrightarrow{\lim_{t\to\infty}} C_f \cdot f_\infty \cdot \left\{ 0, x, y \mp \frac{z}{\chi}, z \pm \frac{y}{\chi} \right\}; \quad \text{with:} \quad \frac{1}{\chi} = \frac{C_s}{i \cdot C_f}. \tag{1230}$$

For simplicity and brevity, we assume in the following a unit system where we can set:

$$\frac{1}{\chi} = \frac{C_s}{C_f} = \frac{1}{i}. \tag{1231}$$

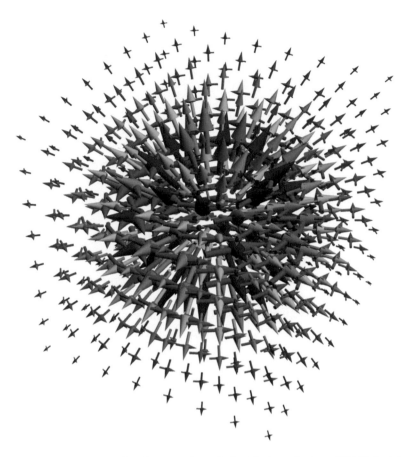

Figure 15.9 Vector field for the particle coded with the solutions (1228) (rainbow colored arrows) and (1221) (grey arrows).

From the structure of f we directly extract that $+t$- and $-t$-solutions are anti-symmetric except for their t-component, because for pairings of $+t$- and $-t$-solutions with equal spins we would obtain:

$$G^j\,[+t,\ldots] + G^j\,[-t,\ldots] = C_f \cdot \left\{\frac{2}{(t^2 + x^2 + y^2 + z^2)},\, 0,\, 0,\, 0\right\}$$

$$\xrightarrow{\underset{t\to\infty}{\lim}} C_f \cdot \{0,\, 0,\, 0,\, 0\}. \tag{1232}$$

In the case of anti-parallel spin pairings, our result would be as follows (we only present the evaluation for the case of spin in z):

$$G^j[+t,\ldots]\uparrow +G^j[-t,\ldots]\downarrow = C_f \cdot f \cdot \left\{ \frac{1}{f \cdot (t^2 + x^2 + y^2 + z^2)}, x + \frac{y}{i}, y - \frac{x}{i}, z \right\}$$

$$-C_f \cdot f \cdot \left\{ \frac{-1}{f \cdot (t^2 + x^2 + y^2 + z^2)}, x - \frac{y}{i}, y + \frac{x}{i}, z \right\}$$

$$= 2 \cdot C_f \cdot f \cdot \left\{ \frac{1}{f \cdot (t^2 + x^2 + y^2 + z^2)}, \frac{y}{i}, -\frac{x}{i}, 0 \right\} \xrightarrow{\lim_{t \to \infty}} 2 \cdot C_f \cdot f_\infty \cdot \left\{ 0, \frac{y}{i}, -\frac{x}{i}, 0 \right\}.$$

$$(1233)$$

Considering the result in Eq. (1233) of photonic character, we might conclude that in order to account for the law of conserved energy, only the annihilation process (Eq. (1233)) would be allowed, while Eq. (1232) could only appear in connection with virtual particles.

15.5.3.4 Postulation

We postulate that in an *n*-dimensional space with time there is always a process which integrates fundamental harmonic solutions with respect to the time dimension. We further postulate that there is a certain maximum of successive integrations of this kind, simply because the higher the number of integrations, the more and more unstable this renders the resulting harmonics, respectively the particles they describe, which is to say that these particles show divergent behavior with time. Each of these integrations shall be considered a state of excitement of the fundamental solution.

At the moment, we give no further justification for these postulations but simply consider them as trials and apply them to the fundamental point solution (Eq. (1219)). Starting with three "exciting integrations" gives us the following three states of excitement H_1, H_2, H_3 and the ground state H_0:

$$H_1 = \int G[x_k]dt = \frac{\arctan\left[\frac{t}{r}\right]}{r}; \quad \lim_{t \to \pm\infty} H_1 = \pm\frac{\pi}{2 \cdot r}; \quad r^2 = x^2 + y^2 + z^2,$$

$$(1234)$$

$$H_2 = \int H_1[x_k]dt = \frac{t \cdot \arctan\left[\frac{t}{r}\right]}{r} - \frac{1}{2}\log\left[t^2 + r^2\right]; \quad \lim_{t \to \pm\infty} H_2 = \frac{\infty}{r},$$

$$(1235)$$

$$H_3 = \int H_2[x_k]dt = \frac{(t^2 - r^2) \cdot \arctan\left[\frac{t}{r}\right] + t \cdot r\left(1 - \log\left[t^2 + r^2\right]\right)}{2 \cdot r};$$

$$\lim_{t \to \pm\infty} H_3 = \pm\frac{\infty}{r}.$$

$$(1236)$$

$$H_0 = G\left[x_k\right] = \frac{1}{t^2 + x^2 + y^2 + z^2}. \tag{1237}$$

15.5.3.5 The three generations of particles

We find that only the H_1 state gives stable solutions, while H_0 vanishes in time and H_2, H_3 do grow infinitely, which, so our **third postulation**, refers to consequently unstable objects (particles).

As the limiting procedure for all excited states shows, we always obtain the spatial distribution of an electrical charged point particle, only that the states H_2, H_3 are unstable, because the solutions grow with time. Thus, together with the spin solution (Eq. (1222)), we suspect to have the electron or positron with Eq. (1230) and the muon, anti-muon, tauon and anti-tauon with $(C_f = 1)$:

$$G^j\left[x_k\right]_2 = f_2 \cdot \left\{\frac{H_1}{f_2}, x \pm \frac{y}{i}, y \mp \frac{x}{i}, z\right\} \xrightarrow[t \to \pm\infty]{\lim} f_{2\infty} \cdot \left\{\frac{\pi}{2 \cdot r \cdot f_{2\infty}}, x \pm \frac{y}{i}, y \mp \frac{x}{i}, z\right\}$$

$$G^j\left[x_k\right]_2 = f_2 \cdot \left\{\frac{H_1}{f_2}, x \pm \frac{z}{i}, y, z \mp \frac{x}{i}\right\} \xrightarrow[t \to \pm\infty]{\lim} f_{2\infty} \cdot \left\{\frac{\pi}{2 \cdot r \cdot f_{2\infty}}, x \pm \frac{z}{i}, y, z \mp \frac{x}{i}\right\}$$

$$G^j\left[x_k\right]_2 = f_2 \cdot \left\{\frac{H_1}{f_2}, x, y \mp \frac{z}{i}, z \pm \frac{y}{i}\right\} \xrightarrow[t \to \pm\infty]{\lim} f_{2\infty} \cdot \left\{\frac{\pi}{2 \cdot r \cdot f_{2\infty}}, x, y \mp \frac{z}{i}, z \pm \frac{y}{i}\right\}, \tag{1238}$$

$$G^j\left[x_k\right]_3 = f_3 \cdot \left\{\frac{H_2}{f_3}, x \pm \frac{y}{i}, y \mp \frac{x}{i}, z\right\} \xrightarrow[t \to \pm\infty]{\lim} \pm f_{3\infty} \cdot \left\{\frac{\infty}{r \cdot f_{3\infty}}, x \pm \frac{y}{i}, y \mp \frac{x}{i}, z\right\}$$

$$G^j\left[x_k\right]_3 = f_3 \cdot \left\{\frac{H_2}{f_3}, x \pm \frac{z}{i}, y, z \mp \frac{x}{i}\right\} \xrightarrow[t \to \pm\infty]{\lim} \pm f_{3\infty} \cdot \left\{\frac{\infty}{r \cdot f_{3\infty}}, x \pm \frac{z}{i}, y, z \mp \frac{x}{i}\right\}$$

$$G^j\left[x_k\right]_3 = f_3 \cdot \left\{\frac{H_2}{f_3}, x, y \mp \frac{z}{i}, z \pm \frac{y}{i}\right\} \xrightarrow[t \to \pm\infty]{\lim} \pm f_{3\infty} \cdot \left\{\frac{\infty}{r \cdot f_{3\infty}}, x, y \mp \frac{z}{i}, z \pm \frac{y}{i}\right\}, \tag{1239}$$

respectively (please remember the unit-simplification (1231) and $C_f = 1$). Thereby we used the following abbreviations:

$$f_2 = -\frac{\left(r + t \cdot \arctan\left[\frac{t}{r}\right]\right)}{r^3}; \quad f_{2\infty} = -\frac{\infty}{r^3}, \tag{1240}$$

$$f_3 = -\frac{t \cdot r + \left(t^2 + r^2\right) \cdot \arctan\left[\frac{t}{r}\right]}{2 \cdot r^3}; \quad f_{3\infty} = -\frac{\infty}{r^3}. \tag{1241}$$

15.5.3.6 Back to the neutrino: About its oscillations

With the thesis of the three particle generations having something to do with excited states as introduced in the last subsection, we face a problem with respect to the potential neutrino solution (Eq. (1222)). Taking the recipe from above, namely, Eq. (1222) would code the electron neutrino, while:

$$G^j [x_k]_2 = C_s \cdot f_2 \cdot \{0, \pm y, \mp x, 0\} \xrightarrow[t \to \pm \infty]{\lim} C_s \cdot f_{2\infty} \cdot \{0, \pm y, \mp x, 0\}$$

$$G^j [x_k]_2 = C_s \cdot f_2 \cdot \{0, \pm z, 0, \mp x\} \xrightarrow[t \to \pm \infty]{\lim} C_s \cdot f_{2\infty} \cdot \{0, \pm z, 0, \mp x\}$$

$$G^j [x_k]_2 = C_s \cdot f_2 \cdot \{0, 0, \mp z, \pm y\} \xrightarrow[t \to \pm \infty]{\lim} C_s \cdot f_{2\infty} \cdot \{0, 0, \mp z, \pm y\}, \qquad (1242)$$

$$G^j [x_k]_3 = C_s \cdot f_3 \cdot \{0, \pm y, \mp x, 0\} \xrightarrow[t \to \pm \infty]{\lim} \pm C_s \cdot f_{3\infty} \cdot \{0, \pm y, \mp x, 0\}$$

$$G^j [x_k]_3 = C_s \cdot f_3 \cdot \{0, \pm z, 0, \mp x\} \xrightarrow[t \to \pm \infty]{\lim} \pm C_s \cdot f_{3\infty} \cdot \{0, \pm z, 0, \mp x\}$$

$$G^j [x_k]_3 = C_s \cdot f_3 \cdot \{0, 0, \mp z, \pm y\} \xrightarrow[t \to \pm \infty]{\lim} \pm C_s \cdot f_{3\infty} \cdot \{0, 0, \mp z, \pm y\} \qquad (1243)$$

would stand for the muon and the tauon neutrino, respectively. Thereby we still have the unstable functions f_2 and f_3 forcing the muon and the tauon neutrino to decay after a certain amount of time. However, it is known that the neutrinos are oscillating, which means that it was measured that they have the ability to change into each other [194, 195]. Now, with the hypothesis of the excited states for the μ and the τ neutrino, we have to take the oscillations not as a possibility, but as a must. Within the process, the neutrinos probably interact with the also permanently jittering environment and therefore no fragments (decay-products) are of need. Any excess energy can simply be absorbed by the background of a non-zero energy universe. From this, we shall conclude that neutrino oscillations require a certain minimum of background energy to allow for such interchanges. Thereby the background also is responsible for the fact that effectively the neutrinos appear as massless in the current energetic state of our universe. In a universe of too low energy, however, the oscillations would be forbidden and the excited states would simply decay into the lowest and stable state neutrinos, which are the electron neutrinos.

Another conclusion from this hypothesis would immediately be that at higher universal temperatures, respectively background energies, the interaction with this background should also allow electrons, muons and tauons to oscillate. The same probably also holds—of course —for the quarks, only that there the background energy needs to be even higher than for the massive leptons.

15.5.3.7 An asymmetry

We see that while the first and third states of excitement are anti-symmetric with respect to $+t$ and $-t$, we have symmetry with the second state of excitement. Assuming that within a very young universe there were equal amounts of $+t$- and $-t$-solutions and at a certain stage a recombination/annihilation of $(+t)+(-t)$-solutions would have taken place, we would have found that particles being in the second state of excitement would behave differently than Eq. (1233), namely:

$$G^j\,[+t,\ldots]_2\,\uparrow\,+G^j\,[-t,\ldots]_2\,\downarrow = C_f \cdot f_2 \cdot \left(\left\{ \frac{H_1}{f_2}, x + \frac{y}{\chi}, y - \frac{x}{\chi}, z \right\} + \left\{ -\frac{H_1}{f_2}, x - \frac{y}{\chi}, y + \frac{x}{\chi}, z \right\} \right)$$

$$= 2 \cdot C_f \cdot f_2 \cdot \{0, x, y, z\} \xrightarrow{\underset{t\to\infty}{\lim}} 2 \cdot C_f \cdot f_{2\infty} \cdot \{0, x, y, z\}$$

$$G^j\,[+t,\ldots]_2\,\uparrow\,+G^j\,[-t,\ldots]_2\,\uparrow = C_f \cdot f_2 \cdot \left(\left\{ \frac{H_1}{f_2}, x + \frac{y}{\chi}, y - \frac{x}{\chi}, z \right\} + \left\{ -\frac{H_1}{f_2}, x + \frac{y}{\chi}, y - \frac{x}{\chi}, z \right\} \right)$$

$$= 2 \cdot C_f \cdot f_2 \cdot \left\{ 0, x + \frac{y}{\chi}, y - \frac{x}{\chi}, z \right\} \xrightarrow{\underset{t\to\infty}{\lim}} 2 \cdot C_f \cdot f_{2\infty} \cdot \left\{ 0, x + \frac{y}{\chi}, y - \frac{x}{\chi}, z \right\}.$$

$$(1244)$$

In other words, the annihilation might have been incomplete and later disintegration of the remaining products as given in Eq. (1244) might have favored matter particles due to a dominating time-vector in $+t$ within the universe as a whole.

There we might have a starting point for the explanation of the matter anti-matter asymmetry we observe in our universe. The splitting-up should have occurred in an era of energy density allowing for the heavier particles like non-neutrino leptons and quarks to oscillate just as we nowadays observe neutrino oscillations. Here we might have one point to test the whole theory via experiment and/or observation.

15.5.3.8 Towards metric quark solutions

Now we apply the same harmonic function H as used in the subsection above, set it as the G-harmonic $G = H$

$$G = H = \int \frac{dt}{t^2 + x^2 + y^2 + z^2} = \frac{\arctan\left[\frac{t}{\sqrt{x^2+y^2+z^2}}\right]}{\sqrt{x^2 + y^2 + z^2}};$$

$$\lim_{t \to \infty} G = \frac{1}{2}\pi \sqrt{\frac{1}{x^2 + y^2 + z^2}} \qquad (1245)$$

and consider the second solution in Eq. (1207) in 4 dimensions t, x, y, z in the following form:

$$G^j = \frac{1}{3}\left((x + y + z) \cdot \partial^j G - 3 \cdot G\right) = \frac{(x + y + z)}{3} \cdot \partial^j G - G; \quad \partial^j = \delta^{jl}\partial_l = \partial_j.$$

$$(1246)$$

Please note that we have already incorporated the condition $\alpha = -1$. With the function $f[\ldots]$ given as:

$$f = f[t, x, y, z] = \frac{t}{\left(x^2 + y^2 + z^2\right)^2 \left(1 + \frac{t^2}{x^2+y^2+z^2}\right)} + \frac{\arctan\left[\frac{t}{\sqrt{x^2+y^2+z^2}}\right]}{\left(x^2 + y^2 + z^2\right)^{3/2}},$$

$$(1247)$$

the vector field G^j does read:

$$G^j [x_k] = C_f \cdot \left(f \cdot \frac{(x + y + z)}{3} \cdot \left\{\frac{1}{f \cdot (t^2 + x^2 + y^2 + z^2)}, x, y, z\right\} - \{G, G, G, G\}\right).$$

$$(1248)$$

The attentive reader recognizes the potential of a point charge for the limiting case of $t \to \infty$ in the second vector term with the function G_∞:

$$f_\infty = \lim_{t \to \infty} f[t, x, y, z] = \frac{\pi}{2} \cdot \frac{1}{(x^2 + y^2 + z^2)^{3/2}}; \quad G_\infty = \lim_{t \to \infty} G = \frac{\pi}{2} \frac{1}{\sqrt{x^2 + y^2 + z^2}},$$

$$(1249)$$

$$\lim_{t \to \infty} G^j [x_k] = C_f \cdot \left(f_\infty \cdot \frac{(x + y + z)}{3} \cdot \{0, x, y, z\} - \{G_\infty, G_\infty, G_\infty, G_\infty\}\right).$$

$$(1250)$$

Together with the first term, however, it is clear that this cannot be the coding for a simple electric point charge. Here one might be tempted to assume to have a structural recipe for the quarks, but with the harmonic (1245) we are unable to see why there should be a form of interaction forcing the particles to behave like agapornis [80], which is to say inseparables.

However, if assuming that the lowest state of existence for agapornis (confined) particles would be a state with H_2, we would de facto have a reason to firmly bind these particles together. In order to also have three generations with these particles, we require a further state of "excitement integration", which would be:

$$H_4 = \int H_3 dt = \frac{2t \left(t^2 - 3 \cdot r^2\right) \cdot \arctan\left[\frac{t}{r}\right] + r \left(5t^2 + \left(r^2 - 3t^2\right) \log\left[t^2 + r^2\right]\right)}{12 \cdot r};$$

$$\lim_{t \to \pm\infty} H_4 = \frac{\infty}{r}.$$

(1251)

A discussion of how the confinement could mathematically be realized was given in [161] by applying a harmonic solution of the kind given in Eq. (1213).

15.5.3.9 Testing the theory

Apart from the option to test the theory outlined here by the means of the matter anti-matter asymmetry in our universe as hinted in the subsection "An Asymmetry" above, we could also try and investigate the time dependencies of the particles. Either directly after creation or briefly before annihilation, the supposed-to-be-stable particles (e.g., electrons and positrons) following the t-dependencies as given in states (1234) to (1236) and (1251) should appear significantly different than during their $t \to \infty$-period of existence. One requires to observe the evolution of these particles as a suitable assumption for the time scales, which then automatically leads to the lifetime of the heavier entities. Thus, experimentalists would face the problem of resolving electron properties within the range $<<10^{-6}$ seconds after their creation.

In other words, one would have to make the actual creation process of positron-electron pairs visible in order to see whether or whether not the theory outlined here provides anything of use.

A third testing option arises from the fact that in the current hypothesis to explain excited particle states as timely integrations, there is no condition to be seen which would restrict the number of such "exciting integrations". It is therefore to be concluded that in principle also heavier leptons than the tauon should exist. However, as their lifetime will most likely be extremely short and their probability to even come into existence will surely be overshadowed by many lighter particles, it is of little wonder that so far no observations of such particles have been reported.

Last but not least, we expect a permanent creation and annihilation of the quark-particles within mesons and hadrons. As the only explanation for the confinement of the quarks has been seen in the manifestation of excited states (1234) to (1236) and (1251), there, so must be concluded, has to be an infinite ongoing of birth and decease of these particles in a circle of permanent interchange. Observations of this cyclic interior within the mesons or hadrons should also help us to reveal the intrinsic nature of the particles in connection with our corresponding metric solutions.

15.6 Summing Up the Simplest Example

Simply by demanding that the integrand of the Einstein–Hilbert action gives a constant, we were able to derive quantum equations in Klein–Gordon and Dirac style for any arbitrary number of dimensions $n \geq 2$ (n integer) and any arbitrary metric. We therefore conclude that not only the Einstein field equations, respectively the General Theory of Relativity resides within the Einstein–Hilbert action, but also quantum theory. The problem just was to get it out from there.

As a byproduct, it was found that the metric equivalent of the Dirac equation, where we obtain function vectors instead of scalar functions, is just some kind of elastic equation of space-time, with then those vector functions being equivalent to displacement vectors in the theory of elasticity. The introduction of quaternions is not needed anymore.

15.7 Using a Base Vector Approach → Leading Us to a Metric Dirac Equation

Even though we already obtained the Dirac equation by simply considering the Ricci scalar of the transformed (wrapped or scaled) metric tensor $G_{\delta\gamma} = F[f] \cdot g_{\delta\gamma}$ in section 15.2 in connection with a Minkowski space-time, we are still interested to obtain first-order metric equations in Dirac style (but potentially without the need to resort to quaternions) in the most general manner possible.

Thus, for completeness, we also want to consider the variation given in Eq. (1177) with an f-setting of the kind:

$$f = \mathbf{e}^\xi \cdot \mathbf{G}_\xi = \frac{\mathbf{e}^\xi \cdot \mathbf{G}_\xi + \alpha \cdot \mathbf{e}^\xi \cdot \mathbf{G}_\xi + \beta \cdot g^{\xi\rho}\mathbf{e}_\rho \cdot \mathbf{G}_\xi}{1+\alpha+\beta}$$

$$= \frac{\mathbf{e}^\xi \cdot \mathbf{G}_\xi + \frac{\alpha}{n}\cdot\delta^\sigma_\xi\mathbf{e}^\xi \cdot \mathbf{G}_\xi\delta^\xi_\sigma + \frac{\beta}{n}\cdot\delta^\rho_\sigma\delta^{\xi}_\rho g^{\xi\rho}\mathbf{e}_\rho \cdot \mathbf{G}_\xi}{1+\alpha+\beta}$$

$$\Rightarrow \delta_\sigma f = \frac{\delta_\sigma}{1+\alpha+\beta}\cdot\left(\mathbf{e}^\xi \cdot \mathbf{G}_\xi + \frac{\alpha}{n}\cdot\delta^\sigma_\xi\mathbf{e}^\xi \cdot \mathbf{G}_\xi\delta^\xi_\sigma + \frac{\beta}{n}\cdot\delta^\rho_\sigma\delta^{\xi}_\rho g^{\xi\rho}\mathbf{e}_\rho \cdot \mathbf{G}_\xi\right)$$

$$= \frac{1}{1+\alpha+\beta}\cdot\left(\delta_\sigma\left(\mathbf{e}^\xi \cdot \mathbf{G}_\xi\right) + \frac{\alpha}{n}\cdot\delta^\sigma_\xi\delta_\sigma\left(\mathbf{e}^\xi \cdot \mathbf{G}_\xi\delta^\xi_\sigma\right) + \frac{\beta}{n}\cdot\delta^\sigma_\rho\delta_\sigma\left(\delta^\rho_\sigma g^{\xi\rho}\mathbf{e}_\rho \cdot \mathbf{G}_\xi\right)\right)$$

$$= \frac{1}{1+\alpha+\beta}\cdot\left(\delta_\sigma\left(\mathbf{e}^\xi \cdot \mathbf{G}_\xi\right) + \frac{\alpha}{n}\cdot\delta_\sigma\left(\mathbf{e}^\xi \cdot \mathbf{G}_\sigma\right) + \frac{\beta}{n}\cdot\delta_\rho\left(g^{\xi\rho}\mathbf{e}_\sigma \cdot \mathbf{G}_\xi\right)\right)$$

$$= \frac{1}{1+\alpha+\beta}\cdot\left(\left(\mathbf{e}^\xi \cdot \mathbf{G}_\xi\right)_{,\sigma} + \frac{\alpha}{n}\cdot\left(\mathbf{e}^\xi \cdot \mathbf{G}_\sigma\right)_{,\xi} + \frac{\beta}{n}\cdot\left(g^{\xi\rho}\mathbf{e}_\sigma \cdot \mathbf{G}_\xi\right)_{,\rho}\right)\delta x^\sigma,$$

$$\tag{1252}$$

where $\mathbf{e}^\xi$ denotes the components of a contravariant base vector and $\mathbf{G}_\xi$ stands for the components of contravariant base vector-like form.

Subjecting this to the Hilbert variation and applying the chain rule during the variation process, results in:

$$\delta W = 0 = \delta_\sigma \int_V d^n x \left(\sqrt{-G} \cdot R \right) = \int_V d^n x \left(\sqrt{-G} \cdot \left[R^{\alpha\beta} - \frac{R}{2} G^{\alpha\beta} \right] \right) \delta_\sigma G_{\alpha\beta}$$

$$= \int_V d^n x \left(\sqrt{-G} \cdot \left[R^{\alpha\beta} - \frac{R}{2} G^{\alpha\beta} \right] \right)$$

$$\times \left(\frac{\partial F[f]}{\partial f} \cdot g_{\alpha\beta} \frac{\left(\mathbf{e}^\xi \cdot \mathbf{G}_\xi + \frac{\alpha}{n} \cdot \delta_\xi^\sigma \mathbf{e}^\xi \cdot \mathbf{G}_\xi \delta_\sigma^\xi + \frac{\beta}{n} \cdot \delta_\sigma^\rho \delta_\rho^\sigma g^{\xi\rho} \mathbf{e}_\rho \cdot \mathbf{G}_\xi \right)}{1 + \alpha + \beta} + F[f] \cdot g_{\alpha\beta,\sigma} \right)_{,\sigma} g^{\sigma\chi} \delta x_\chi. \quad (1253)$$

Using the results from Eq. (1252), we obtain:

$$= \int_V d^n x \left(\sqrt{-G} \cdot \left[R^{\alpha\beta} - \frac{R}{2} G^{\alpha\beta} \right] \right)$$

$$\times \left(\frac{\partial F[f]}{\partial f} \cdot g_{\alpha\beta} \frac{\partial_\sigma \left(\mathbf{e}^\xi \cdot \mathbf{G}_\xi \right) + \frac{\alpha}{n} \cdot \delta_\xi^\sigma \partial_\sigma \left(\mathbf{e}^\xi \cdot \mathbf{G}_\xi \delta_\sigma^\xi \right) + \frac{\beta}{n} \cdot \delta_\rho^\sigma \partial_\sigma \left(\delta_\sigma^\rho g^{\xi\rho} \mathbf{e}_\rho \cdot \mathbf{G}_\xi \right)}{1 + \alpha + \beta} + F[f] \cdot g_{\alpha\beta,\sigma} \right) g^{\sigma\chi} \delta x_\chi$$

$$= \int_V d^n x \left(\sqrt{-G} \cdot \left[R^{\alpha\beta} - \frac{R}{2} G^{\alpha\beta} \right] \right)$$

$$\times \left(\frac{\partial F[f]}{\partial f} \cdot g_{\alpha\beta} \frac{\left(\mathbf{e}^\xi \cdot \mathbf{G}_\xi \right)_{,\sigma} + \frac{\alpha}{n} \cdot \left(\mathbf{e}^\xi \cdot \mathbf{G}_\sigma \right)_{,\xi} + \frac{\beta}{n} \cdot \left(g^{\xi\rho} \mathbf{e}_\sigma \cdot \mathbf{G}_\xi \right)_{,\rho}}{1 + \alpha + \beta} + F[f] \cdot g_{\alpha\beta,\sigma} \right) g^{\sigma\chi} \delta x_\chi$$

$$= \int_V d^n x \left(\sqrt{-G} \cdot \left[R^{\alpha\beta} - \frac{R}{2} G^{\alpha\beta} \right] \right)$$

$$\times \left(\frac{\frac{\partial F[f]}{\partial f} \cdot g_{\alpha\beta}}{1 + \alpha + \beta} \left(\mathbf{e}^\xi_{,\sigma} \cdot \mathbf{G}_\xi + \mathbf{e}^\xi \cdot \mathbf{G}_{\xi,\sigma} + \frac{\alpha}{n} \cdot \left(\mathbf{e}^\xi_{,\xi} \cdot \mathbf{G}_\sigma + \mathbf{e}^\xi \cdot \mathbf{G}_{\sigma,\xi} \right) + \frac{\beta}{n} \left(g^{\xi\rho}_{,\rho} \mathbf{e}_\sigma \cdot \mathbf{G}_\xi + g^{\xi\rho} \mathbf{e}_{\sigma,\rho} \cdot \mathbf{G}_\xi + g^{\xi\rho} \mathbf{e}_\sigma \cdot \mathbf{G}_{\xi,\rho} \right) \right) + F[f] \cdot g_{\alpha\beta,\sigma} \right) g^{\sigma\chi} \delta x_\chi. \quad (1254)$$

Assuming an almost flat space-time in Cartesian-like coordinates again, where the metric derivatives may be zero, but not necessarily the derivatives of the base vectors, leads us to a set of first-order differential equations, where we cannot help noticing that there is quite some matrix similarity to the classical Dirac equation [143]:

$$
= \int_V d^n x \left(\sqrt{-G} \cdot \left[R^{\alpha\beta} - \frac{R}{2} G^{\alpha\beta} \right] \right) \times \left(\frac{\frac{\partial F[f]}{\partial f} \cdot g_{\alpha\beta}}{1 + \alpha + \beta} \left(\begin{array}{c} \mathbf{e}^{\xi}_{,\sigma} \cdot \mathbf{G}_{\xi} + \mathbf{e}^{\xi} \cdot \mathbf{G}_{\xi,\sigma} \\ + \frac{\alpha}{n} \cdot \left(\mathbf{e}^{\xi}_{,\xi} \cdot \mathbf{G}_{\sigma} + \mathbf{e}^{\xi} \cdot \mathbf{G}_{\sigma,\xi} \right) \\ + \frac{\beta}{n} \cdot \left(g^{\xi\rho} \mathbf{e}_{\sigma,\rho} \cdot \mathbf{G}_{\xi} + g^{\xi\rho} \mathbf{e}_{\sigma} \cdot \mathbf{G}_{\xi,\rho} \right) \end{array} \right) \right) g^{\sigma\chi} \delta x_\chi
$$

$$\tag{1255}$$

$$
\Rightarrow 0 = \mathbf{e}^{\xi}_{,\sigma} \cdot \mathbf{G}_{\xi} + \mathbf{e}^{\xi} \cdot \mathbf{G}_{\xi,\sigma} + \frac{\alpha}{n} \cdot \left(\mathbf{e}^{\xi}_{,\xi} \cdot \mathbf{G}_{\sigma} + \mathbf{e}^{\xi} \cdot \mathbf{G}_{\sigma,\xi} \right) + \frac{\beta}{n} \cdot \left(g^{\xi\rho} \mathbf{e}_{\sigma,\rho} \cdot \mathbf{G}_{\xi} + g^{\xi\rho} \mathbf{e}_{\sigma} \cdot \mathbf{G}_{\xi,\rho} \right).
$$

$$\tag{1256}$$

This suspicion becomes obvious when imagining all derivatives of base vectors as masses and/or potentials. In order to better understand the result, we apply the following approach for $\mathbf{G}_{\xi}$:

$$
\mathbf{G}_{\xi} = \mathbf{e}_{\xi} \cdot h
$$

and obtain:

$$= \int_V d^n x \left(\sqrt{-G} \cdot \left[R^{\alpha\beta} - \frac{R}{2} G^{\alpha\beta} \right] \right.$$

$$\times \left(\frac{\frac{\partial F[f]}{\partial f} \cdot g_{\alpha\beta}}{1+\alpha+\beta} \left(\mathbf{e}_{\xi,\sigma} \cdot \mathbf{e}_\xi \cdot h + \mathbf{e}_\xi \cdot (\mathbf{e}_\xi \cdot h)_{,\sigma} + \frac{\alpha}{n} \left(\mathbf{e}_{\xi,\xi} \cdot \mathbf{e}_\sigma \cdot h + \mathbf{e}_\xi \cdot (\mathbf{e}_\xi \cdot h)_{,\rho} \right) \right. \right.$$

$$\left. \left. + \frac{\beta}{n} \left(g^{\xi\rho}_{,\rho} \mathbf{e}_\sigma \cdot \mathbf{e}_\xi \cdot h + g^{\xi\rho} \mathbf{e}_{\sigma,\rho} \cdot \mathbf{e}_\xi \cdot h + g^{\xi\rho} \mathbf{e}_\sigma \cdot (\mathbf{e}_\xi \cdot h)_{,\rho} \right) \right) + F[f] \cdot g_{\alpha\beta,\sigma} \right) g^{\sigma\chi} \delta x_\chi$$

$$= \int_V d^n x \left(\sqrt{-G} \cdot \left[R^{\alpha\beta} - \frac{R}{2} G^{\alpha\beta} \right] \right.$$

$$\times \left(\frac{\frac{\partial F[f]}{\partial f} \cdot g_{\alpha\beta}}{1+\alpha+\beta} \left(\mathbf{e}^\xi_{,\sigma} \cdot \mathbf{e}_\xi \cdot h + \mathbf{e}^\xi \cdot (\mathbf{e}_{\xi,\sigma} \cdot h + \mathbf{e}_\xi \cdot h_{,\sigma}) \right. \right.$$

$$+ \frac{\alpha}{n} \left(\mathbf{e}^\xi_{,\xi} \cdot \mathbf{e}_\sigma \cdot h + \mathbf{e}^\xi \cdot (\mathbf{e}_{\sigma,\xi} \cdot h + \mathbf{e}_\sigma \cdot h_{,\xi}) \right)$$

$$\left. \left. + \frac{\beta}{n} \left(g^{\xi\rho}_{,\rho} \mathbf{e}_\sigma \cdot \mathbf{e}_\xi \cdot h + g^{\xi\rho} \mathbf{e}_{\sigma,\rho} \cdot \mathbf{e}_\xi \cdot h + g^{\xi\rho} \mathbf{e}_\sigma \cdot (\mathbf{e}_\xi \cdot h)_{,\rho} \right) \right) + F[f] \cdot g_{\alpha\beta,\sigma} \right) g^{\sigma\chi} \delta x_\chi$$

$$= \int_V d^n x \left(\sqrt{-G} \cdot \left[R^{\alpha\beta} - \frac{R}{2} G^{\alpha\beta} \right] \right.$$

$$\times \left(\frac{\frac{\partial F[f]}{\partial f} \cdot g_{\alpha\beta}}{1+\alpha+\beta} \left((\mathbf{e}^\xi_{,\sigma} \cdot \mathbf{e}_\xi + \mathbf{e}^\xi \cdot \mathbf{e}_{\xi,\sigma}) \cdot h + n \cdot h_{,\sigma} \right. \right.$$

$$+ \frac{\alpha}{n} \left((\mathbf{e}^\xi_{,\xi} \cdot \mathbf{e}_\sigma + \mathbf{e}^\xi \cdot \mathbf{e}_{\sigma,\xi}) \cdot h + \delta^\xi_\sigma \cdot h_{,\xi} \right)$$

$$\left. \left. + \frac{\beta}{n} \left((g_{\sigma\xi} g^{\xi\rho}_{,\rho} + \mathbf{e}_{\sigma,\rho} \cdot \mathbf{e}^\rho + g^{\xi\rho} \mathbf{e}_\sigma \cdot \mathbf{e}_{\xi,\rho}) \cdot h + \delta^\rho_\sigma \cdot h_{,\rho} \right) \right) + F[f] \cdot g_{\alpha\beta,\sigma} \right) g^{\sigma\chi} \delta x_\chi$$

$$= \int_V d^n x \left(\sqrt{-G} \cdot \left[R^{\alpha\beta} - \frac{R}{2} G^{\alpha\beta} \right] \right.$$

$$\times \left(\frac{\frac{\partial F[f]}{\partial f} \cdot g_{\alpha\beta}}{1+\alpha+\beta} \left((\mathbf{e}^\xi_{,\sigma} \cdot \mathbf{e}_\xi + \mathbf{e}^\xi \cdot \mathbf{e}_{\xi,\sigma}) \cdot h + \left(n + \frac{\alpha}{n} + \frac{\beta}{n}\right) \cdot h_{,\sigma} \right. \right.$$

$$+ \frac{\alpha}{n} \left(\mathbf{e}^\xi_{,\xi} \cdot \mathbf{e}_\sigma + \mathbf{e}^\xi \cdot \mathbf{e}_{\sigma,\xi} \right) \cdot h$$

$$\left. \left. + \frac{\beta}{n} \left(g_{\sigma\xi} g^{\xi\rho}_{,\rho} \cdot \mathbf{e}^\rho + \mathbf{e}_{\sigma,\rho} \cdot \mathbf{e}^\rho + g^{\xi\rho} \mathbf{e}_\sigma \cdot \mathbf{e}_{\xi,\rho} \right) \cdot h \right) + F[f] \cdot g_{\alpha\beta,\sigma} \right) g^{\sigma\chi} \delta x_\chi.$$

$$(1257)$$

The reader notes that once more we have introduced an intelligent constant n, because we have made use of the following identity:

$$f = \mathbf{e}^\xi \cdot \mathbf{G}_\xi = n \cdot h = \mathbf{e}^\xi \cdot \mathbf{e}_\xi \cdot h = \dots, \tag{1258}$$

with the rest of the development of intelligent constants in f just following Eq. (1252). In order to move forward, we set $\alpha = \beta = 0$ and simplify Eq. (1257) as follows:

$$= \int_V d^n x \left(\sqrt{-G} \cdot \left[R^{\alpha\beta} - \frac{R}{2} G^{\alpha\beta} \right] \right) \times \left(\frac{\partial F\,[f]}{\partial f} \cdot g_{\alpha\beta} \partial_\sigma \left(\mathbf{e}^\xi \cdot \mathbf{e}_\xi \cdot h \right) + F\,[f] \cdot g_{\alpha\beta,\sigma} \right) g^{\sigma\chi} \delta x_\chi$$

$$= \int_V d^n x \left(\sqrt{-G} \cdot \left[R^{\alpha\beta} - \frac{R}{2} G^{\alpha\beta} \right] \right)$$

$$\times \left(\frac{\partial F\,[f]}{\partial f} \cdot g_{\alpha\beta} \left(\mathbf{e}^\xi_{,\sigma} \cdot \mathbf{e}_\xi \cdot h + \mathbf{e}^\xi \cdot \mathbf{e}_{\xi,\sigma} \cdot h + n \cdot h_{,\sigma} \right) + F\,[f] \cdot g_{\alpha\beta,\sigma} \right) g^{\sigma\chi} \delta x_\chi. \tag{1259}$$

Setting $F\,[f] = f$, we can further simplify and obtain a seemingly meaningless result:

$$= \int_V d^n x \left(\sqrt{-G} \cdot \left[R^{\alpha\beta} - \frac{R}{2} G^{\alpha\beta} \right] \right)$$

$$\times \left(g_{\alpha\beta} \left(\overbrace{\mathbf{e}^\xi_{,\sigma} \cdot \mathbf{e}_\xi \cdot h + \mathbf{e}^\xi \cdot \mathbf{e}_{\xi,\sigma} \cdot h}^{=\left(\mathbf{e}^\xi \cdot \mathbf{e}_\xi\right)_{,\sigma} \cdot h = 0} + n \cdot h_{,\sigma} \right) + n \cdot h \cdot g_{\alpha\beta,\sigma} \right) g^{\sigma\chi} \delta x_\chi$$

$$= \int_V d^n x \left(\sqrt{-G} \cdot \left[R^{\alpha\beta} - \frac{R}{2} G^{\alpha\beta} \right] \right) \times n \cdot \left(g_{\alpha\beta} h_{,\sigma} + h \cdot g_{\alpha\beta,\sigma} \right) g^{\sigma\chi} \delta x_\chi. \tag{1260}$$

However, things clear up a little bit, when assuming an after all perhaps not so unusual constellation for the curvature terms, being proportional to the metric tensor, namely:

$$\xrightarrow{\left[R^{\alpha\beta} - \frac{R}{2} G^{\alpha\beta} \right] \simeq \frac{\omega}{n^2} \cdot g^{\alpha\beta}} = \int_V d^n x \left(\sqrt{-G} \right) \frac{\omega}{n^2} \cdot g^{\alpha\beta} n \cdot \left(g_{\alpha\beta} h_{,\sigma} + h \cdot g_{\alpha\beta,\sigma} \right) g^{\sigma\chi} \delta x_\chi$$

$$= \int_V d^n x \left(\sqrt{-G} \right) \cdot \left(g^{\sigma\chi} h_{,\sigma} + \frac{\omega}{n} \cdot h \cdot g^{\alpha\beta} g_{\alpha\beta,\sigma} g^{\sigma\chi} \right) \delta x_\chi$$

$$\Rightarrow 0 = g^{\sigma\chi} h_{,\sigma} + \frac{\omega}{n} \cdot h \cdot g^{\alpha\beta} g_{\alpha\beta,\sigma} g^{\sigma\chi}. \tag{1261}$$

Now we recognize quite some similarity to the Dirac equation, only that h is not a vector as it should be, but only a scalar. In order to move forward from here, we keep the $\mathbf{G}_\xi$: but maintain all other simplifications like the simple metric-proportional curvature assumption, $F[f] = f$ and $\alpha = \beta = 0$. This gives us:

$$
\underrightarrow{\left[R^{\alpha\beta} - \frac{R}{2}G^{\alpha\beta}\right] \simeq \frac{\omega}{n}\cdot g^{\alpha\beta}} = \int_V d^n x \sqrt{-G} \cdot \frac{\omega}{n} g^{\alpha\beta} \times \left(g_{\alpha\beta}\left(\mathbf{e}^\xi_{,\sigma}\cdot\mathbf{G}_\xi + \mathbf{e}^\xi\cdot\mathbf{G}_{\xi,\sigma}\right) + \mathbf{e}^\xi\cdot\mathbf{G}_\xi\cdot g_{\alpha\beta,\sigma}\right)g^{\sigma\chi}\delta x_\chi
$$

$$
= \int_V d^n x \sqrt{-G} \times \omega\left(\left(\mathbf{e}^\xi_{,\sigma}\cdot\mathbf{G}_\xi + \mathbf{e}^\xi\cdot\mathbf{G}_{\xi,\sigma}\right) + \mathbf{e}^\xi\cdot\mathbf{G}_\xi\cdot g^{\alpha\beta}g_{\alpha\beta,\sigma}\right)g^{\sigma\chi}\delta x_\chi
$$

$$
\Rightarrow 0 = \mathbf{e}^\xi\cdot\mathbf{G}_{\xi,\sigma}g^{\sigma\chi} + \mathbf{G}_\xi\cdot\left(\mathbf{e}^\xi\cdot g^{\alpha\beta}g_{\alpha\beta,\sigma} + \mathbf{e}^\xi_{,\sigma}\right)g^{\sigma\chi} \quad / \quad \cdot\mathbf{e}_\varsigma
$$

$$
0 = \delta^\xi_\varsigma \mathbf{G}_{\xi,\sigma}g^{\sigma\chi} + \mathbf{G}_\xi\cdot\left(\delta^\xi_\varsigma g^{\alpha\beta}g_{\alpha\beta,\sigma} + \mathbf{e}_\varsigma\cdot\mathbf{e}^\xi_{,\sigma}\right)g^{\sigma\chi}
$$

$$
= \mathbf{G}_{\varsigma,\sigma}g^{\sigma\chi} + \mathbf{G}_\xi\cdot\left(\delta^\xi_\varsigma g^{\alpha\beta}g_{\alpha\beta,\sigma} + \mathbf{e}_\varsigma\cdot\mathbf{e}^\xi_{,\sigma}\right)g^{\sigma\chi}. \tag{1262}
$$

Here we can see that an introduction of an arbitrary constant vector $\mathbf{E}$ and an approach of the kind:

$$
\mathbf{G}_\xi = \mathbf{E}\cdot G_\xi \tag{1263}
$$

should bring us forward, because now we obtain the following from the variation (simply insert Eq. (1263) into Eq. (1262)):

$$
\underrightarrow{\left[R^{\alpha\beta} - \frac{R}{2}G^{\alpha\beta}\right] \simeq \frac{\omega}{n}\cdot g^{\alpha\beta}} = \int_V d^n x \sqrt{-G} \cdot \frac{\omega}{n} g^{\alpha\beta}
$$

$$
\times \left(g_{\alpha\beta}\left(\mathbf{e}^\xi_{,\sigma}\cdot\mathbf{E}\cdot G_\xi + \mathbf{e}^\xi\cdot\mathbf{E}\cdot G_{\xi,\sigma}\right) + \mathbf{e}^\xi\cdot\mathbf{E}\cdot G_\xi\cdot g_{\alpha\beta,\sigma}\right)g^{\sigma\chi}\delta x_\chi
$$

$$
\Rightarrow 0 = \mathbf{e}^\xi\cdot\mathbf{E}\cdot G_{\xi,\sigma}g^{\sigma\chi} + \mathbf{E}\cdot G_\xi\cdot\left(\mathbf{e}^\xi\cdot g^{\alpha\beta}g_{\alpha\beta,\sigma} + \mathbf{e}^\xi_{,\sigma}\right)g^{\sigma\chi}/\cdot\mathbf{e}_\varsigma
$$

$$
= \mathbf{E}\cdot G_{\varsigma,\sigma}g^{\sigma\chi} + \mathbf{E}\cdot G_\xi\cdot\left(\delta^\xi_\varsigma g^{\alpha\beta}g_{\alpha\beta,\sigma} + \mathbf{e}_\varsigma\cdot\mathbf{e}^\xi_{,\sigma}\right)g^{\sigma\chi}
$$

$$
= G_{\varsigma,\sigma}g^{\sigma\chi} + G_\xi\cdot\left(\delta^\xi_\varsigma g^{\alpha\beta}g_{\alpha\beta,\sigma} + \mathbf{e}_\varsigma\cdot\mathbf{e}^\xi_{,\sigma}\right)g^{\sigma\chi}. \tag{1264}
$$

This clearly seems to resemble a metric Dirac equation.

Keen on performing the evaluation without the approximation $\left[R^{\alpha\beta} - \frac{R}{2}G^{\alpha\beta}\right] \simeq \frac{\omega}{n}\cdot g^{\alpha\beta}$, we need to remind ourselves that the Ricci curvature tensor $R^{\alpha\beta}$ and scalar R as used in this section 15.7 so far are based on the transformed metric $G_{\alpha\beta}$. Their appearance, if being given in dependence on the untransformed metric $g_{\alpha\beta}$, was already discussed in section 3.3 (c.f. also chapter 16). From there, we extract the following:

$$\delta W = 0 = \delta_\sigma \int_V d^n x \left(\sqrt{-G} \cdot R^* \right) = \delta_\sigma \int_V d^n x \left(\sqrt{-G} \cdot \left(R^*_* + R^{**}_* \right) \right)$$

$$\xrightarrow{F^{\xi\zeta}_{\alpha\beta} \to F} = \delta_\sigma \int_V d^n x \left(\frac{\sqrt{-G}}{F[f]} \cdot \left(R + R^{**} \right) \right) = \delta_\sigma \int_V d^n x \left(\frac{\sqrt{-g} \cdot F[f]^n}{F[f]} \cdot \left(R + \overbrace{R^{**}}^{=-2\kappa L_M} \right) \right)$$

$$= \int_V d^n x \, \delta_g \left(\sqrt{-g} \cdot (R - 2\kappa L_M) \right) \times \delta_\sigma \left(\frac{\sqrt{F[f]^n}}{F[f]} g_{\alpha\beta} \right)$$

$$= \int_V d^n x \left(\sqrt{-g} \cdot \left[R^{\alpha\beta} - \frac{R}{2} g^{\alpha\beta} + \kappa T^{\alpha\beta} \right] \right) \times \delta_\sigma \left(\frac{\sqrt{F[f]^n}}{F[f]} g_{\alpha\beta} \right)$$

$$= \int_V d^n x \left(\sqrt{-g} \cdot \left[R^{\alpha\beta} - \frac{R}{2} g^{\alpha\beta} + \kappa T^{\alpha\beta} \right] \right)$$

$$\cdot F[f]^{\frac{n-4}{2}} \cdot \left(\frac{\partial F[f]}{\partial f} \cdot g_{\alpha\beta} f_{,\sigma} + F[f] \cdot g_{\alpha\beta,\sigma} \right) g^{\sigma\chi} \delta x_\chi. \tag{1265}$$

Please see chapter 16 and section 3.3 with respect to the definition and derivation of the various Ricci scalars R, R^*, R^{**}, R^*_*, R^{**}_*. Thereby the energy momentum tensor has to be evaluated from:

$$\int_V d^n x \, \delta_g \left(\sqrt{-g} \cdot (-2L_M) \right) = \int_V d^n x \left(\sqrt{-g} \cdot T^{\alpha\beta} \right) \cdot \delta_\sigma \tag{1266}$$

with the matter Lagrange density given via:

$$L_M \equiv -\frac{R^{**}}{2\kappa} = -\frac{F[f] \cdot R^{**}_*}{2\kappa} = \frac{g^{\alpha\beta}}{2\kappa} \left(\begin{array}{c} \Gamma^\mu_{\sigma\alpha} \Gamma^{**\sigma}_{\beta\mu} - \Gamma^\sigma_{\alpha\beta} \Gamma^{**\mu}_{\sigma\mu} + \Gamma^{**\mu}_{\sigma\alpha} \Gamma^\sigma_{\beta\mu} - \Gamma^{**\sigma}_{\alpha\beta} \Gamma^\mu_{\sigma\mu} \\ -\Gamma^{**\sigma}_{\alpha\beta,\sigma} + \Gamma^{**\sigma}_{\beta\sigma,\alpha} + \Gamma^{**\mu}_{\sigma\alpha} \Gamma^{**\sigma}_{\beta\mu} - \Gamma^{**\sigma}_{\alpha\beta} \Gamma^{**\mu}_{\sigma\mu} \end{array} \right) \tag{1267}$$

and the "$F[f]$-transformed" affine connection:

$$\Gamma^{**\gamma}_{\alpha\beta} \equiv \frac{g^{\gamma\sigma}}{2 \cdot F[f]} \left(F[f]_{,\beta} \cdot g_{\sigma\alpha} + F[f]_{,\alpha} \cdot g_{\sigma\beta} - F[f]_{,\sigma} \cdot g_{\alpha\beta} \right). \tag{1268}$$

Inserting Eq. (1263) into the last line of Eq. (1265) now leads us to the non-approximated result:

$$= \int_V d^n x \left(\sqrt{-g} \cdot \left[R^{\alpha\beta} - \frac{R}{2} g^{\alpha\beta} + \kappa T^{\alpha\beta} \right] \right) \cdot F[f]^{\frac{n-4}{2}}$$

$$\times \left(\left(g_{\alpha\beta} \left(\mathbf{e}^\xi_{,\sigma} \cdot \mathbf{E} \cdot G_\xi + \mathbf{e}^\xi \cdot \mathbf{E} \cdot G_{\xi,\sigma} \right) + \mathbf{e}^\xi \cdot \mathbf{E} \cdot G_\xi \cdot g_{\alpha\beta,\sigma} \right) \right) g^{\sigma\chi} \delta x_\chi$$

$$= \int_V d^n x \sqrt{-g} \cdot F[f]^{\frac{n-4}{2}} \left(\left(\overbrace{\left[R^{\alpha\beta} - \frac{R}{2} g^{\alpha\beta} + \kappa T^{\alpha\beta} \right]}^{[\cdots]^{\alpha\beta}} g_{\alpha\beta} \left(\mathbf{e}^\xi_{,\sigma} \cdot \mathbf{E} \cdot G_\xi + \mathbf{e}^\xi \cdot \mathbf{E} \cdot G_{\xi,\sigma} \right) \right. \right.$$
$$\left. \left. + \mathbf{e}^\xi \cdot \mathbf{E} \cdot G_\xi \cdot \left[R^{\alpha\beta} - \frac{R}{2} g^{\alpha\beta} + \kappa T^{\alpha\beta} \right] g_{\alpha\beta,\sigma} \right) \right) g^{\sigma\chi} \delta x_\chi$$

$$= \int_V d^n x \sqrt{-g} \cdot F[f]^{\frac{n-4}{2}} \left(\left(\overbrace{\left[R \left(1 - \frac{n}{2} \right) + \kappa T^{\alpha\beta} g_{\alpha\beta} \right]}^{\equiv K} \left(\mathbf{e}^\xi_{,\sigma} \cdot \mathbf{E} \cdot G_\xi + \mathbf{e}^\xi \cdot \mathbf{E} \cdot G_{\xi,\sigma} \right) \right. \right.$$
$$\left. \left. + \mathbf{e}^\xi \cdot \mathbf{E} \cdot G_\xi \cdot [\cdots]^{\alpha\beta} g_{\alpha\beta,\sigma} \right) \right) g^{\sigma\chi} \delta x_\chi$$

$$\Rightarrow 0 = \mathbf{e}^\xi \cdot \mathbf{E} \cdot K \cdot G_{\xi,\sigma} g^{\sigma\chi} + \mathbf{E} \cdot G_\xi \cdot \left(\mathbf{e}^\xi \cdot [\cdots]^{\alpha\beta} g_{\alpha\beta,\sigma} + K \cdot \mathbf{e}^\xi_{,\sigma} \right) g^{\sigma\chi} \quad / \quad \cdot \mathbf{e}_\varsigma$$

$$= \mathbf{E} \cdot G_{\varsigma,\sigma} g^{\sigma\chi} \cdot K + \mathbf{E} \cdot G_\xi \cdot \left(\delta^\xi_\varsigma [\cdots]^{\alpha\beta} g_{\alpha\beta,\sigma} + K \cdot \mathbf{e}_\varsigma \cdot \mathbf{e}^\xi_{,\sigma} \right) g^{\sigma\chi}$$

$$= G_{\varsigma,\sigma} g^{\sigma\chi} \cdot K + G_\xi \cdot \left(\delta^\xi_\varsigma [\cdots]^{\alpha\beta} g_{\alpha\beta,\sigma} + K \cdot \mathbf{e}_\varsigma \cdot \mathbf{e}^\xi_{,\sigma} \right) g^{\sigma\chi}. \tag{1269}$$

Here we now have the full metric first-order differential quantum gravity equation in an arbitrary number of dimensions n as follows (or a "Dirac quantum gravity equation"):

$$0 = G_{\varsigma,\sigma} g^{\sigma\chi} \cdot \left[R \left(1 - \frac{n}{2} \right) + \kappa T^{\alpha\beta} g_{\alpha\beta} \right]$$
$$+ G_\xi \cdot \left(\begin{array}{c} \delta^\xi_\varsigma \left[R^{\alpha\beta} - \frac{R}{2} g^{\alpha\beta} + \kappa T^{\alpha\beta} \right] g_{\alpha\beta,\sigma} \\ + \left[R \left(1 - \frac{n}{2} \right) + \kappa T^{\alpha\beta} g_{\alpha\beta} \right] \cdot \mathbf{e}_\varsigma \cdot \mathbf{e}^\xi_{,\sigma} \end{array} \right) g^{\sigma\chi}. \tag{1270}$$

As also the energy momentum tensor is now given in completely metric manner (see Eqs. (1266) to (1268)), we may consider our task of deriving the metric Dirac equation as completed. The next step could be the consideration of a variety of examples, perhaps in connection with potential particle solutions.

However, as we already have some suitable particle equations by the means of our "elastic" derivations in section 15.5, we leave the discussion of Eqs. (1254) to (1264) to the interested reader.

15.7.1 Transition to the Classical Dirac Equation

The true and most classical Dirac equation as derived by Dirac [143] and already metrically derived here in section 15.2 should—if our assumptions are correct—also (somehow) reside inside Eq. (1270). Using the abbreviations introduced in Eq. (1269), we can develop our metric equation (Eq. (1270)) as follows:

$$0 = G_{\varsigma,\sigma} g^{\sigma\chi} \cdot K + G_\xi \cdot \left(\delta_\varsigma^\xi \left[R^{\alpha\beta} - \frac{R}{2} g^{\alpha\beta} + \kappa T^{\alpha\beta} \right] g_{\alpha\beta,\sigma} + K \cdot \mathbf{e}_\varsigma \cdot \mathbf{e}^\xi_{,\sigma} \right) g^{\sigma\chi} \Big/ : K$$

$$= G_{\varsigma,\sigma} g^{\sigma\chi} + G_\xi \cdot \left(\delta_\varsigma^\xi \left[R^{\alpha\beta} - \frac{R}{2} g^{\alpha\beta} + \kappa T^{\alpha\beta} \right] \frac{g_{\alpha\beta,\sigma}}{K} + \mathbf{e}_\varsigma \cdot \mathbf{e}^\xi_{,\sigma} \right) g^{\sigma\chi}$$

$$= \left(G_{\varsigma,\sigma} + G_\xi \cdot \left(\delta_\varsigma^\xi \left[R^{\alpha\beta} - \frac{R}{2} g^{\alpha\beta} + \kappa T^{\alpha\beta} \right] \frac{g_{\alpha\beta,\sigma}}{K} + \mathbf{e}_\varsigma \cdot \mathbf{e}^\xi_{,\sigma} \right) \right) \mathbf{e}^\sigma \cdot \mathbf{e}^\chi$$

$$= \left(G_{\varsigma,\sigma} + G_\xi \cdot \left(\delta_\varsigma^\xi \left[R^{\alpha\beta} - \frac{R}{2} g^{\alpha\beta} + \kappa T^{\alpha\beta} \right] \frac{g_{\alpha\beta,\sigma}}{K} + \overbrace{\mathbf{e}_\varsigma \cdot \mathbf{e}^\xi_{,\sigma}}^{\equiv \delta_\varsigma^\xi P_\sigma} \right) \right) \mathbf{e}^\sigma$$

$$= G_{\varsigma,\sigma} \mathbf{e}^\sigma + G_\xi \cdot \delta_\varsigma^\xi \left(\left[R^{\alpha\beta} - \frac{R}{2} g^{\alpha\beta} + \kappa T^{\alpha\beta} \right] \frac{g_{\alpha\beta,\sigma}}{K} + P_\sigma \right) \mathbf{e}^\sigma$$

$$= G_{\varsigma,\sigma} \mathbf{e}^\sigma + G_\varsigma \cdot \left(\left[R^{\alpha\beta} - \frac{R}{2} g^{\alpha\beta} + \kappa T^{\alpha\beta} \right] \frac{g_{\alpha\beta,\sigma}}{K} + P_\sigma \right) \mathbf{e}^\sigma. \tag{1271}$$

We note that the second addend in the last line is just a vector of scalars. Let us name this scalar vector **M**. Perhaps it could stand for mess (yes mess, with an "*e*"—for justification of this peculiar naming c.f. section 15.2.1). This gives us:

$$0 = G_{\varsigma,\sigma} \mathbf{e}^\sigma + G_\varsigma \cdot \mathbf{M}. \tag{1272}$$

Now we take it for granted that the direction of **M** is of no importance and that therefore also another equation exists which does read:

$$0 = G_{\varsigma,\sigma} \mathbf{e}^\sigma - G_\varsigma \cdot \mathbf{M}. \tag{1273}$$

We realize that both Eqs. (1272) and (1273) can be brought into the typical quantum mechanical operator form, namely:

$$0 = G_{\varsigma,\sigma} \mathbf{e}^\sigma + G_\varsigma \cdot \mathbf{M} \quad \Rightarrow \quad 0 = \overbrace{(\mathbf{e}^\sigma \cdot \partial_\sigma + \mathbf{M})}^{0^+_{(\sigma)}} G_\varsigma$$

$$0 = G_{\varsigma,\sigma} \mathbf{e}^\sigma - G_\varsigma \cdot \mathbf{M} \quad \Rightarrow \quad 0 = \underbrace{(\mathbf{e}^\sigma \cdot \partial_\sigma - \mathbf{M})}_{0^-_{(\sigma)}} G_\varsigma, \tag{1274}$$

with the operators $0^+_{(\sigma)}$, $0^-_{(\sigma)}$. We note that the index σ, as a dummy index, is now arbitrary. By applying one of the two operators on the corresponding other one, thereby choosing different dummies, one obtains:

$$0 = O^+_{(\sigma)} O^-_{(\rho)} G_\varsigma = (\mathbf{e}^\sigma \cdot \partial_\sigma + \mathbf{M})(\mathbf{e}^\rho \cdot \partial_\rho - \mathbf{M}) G_\varsigma = \left(\mathbf{e}^\sigma \partial_\sigma \cdot \mathbf{e}^\rho \partial_\rho - \overbrace{\mathbf{M} \cdot \mathbf{M}}^{=M^2} \right) G_\varsigma$$

$$0 = O^-_{(\rho)} O^+_{(\sigma)} G_\varsigma = (\mathbf{e}^\rho \cdot \partial_\rho - \mathbf{M})(\mathbf{e}^\sigma \cdot \partial_\sigma + \mathbf{M}) G_\varsigma = (\mathbf{e}^\rho \partial_\rho \cdot \mathbf{e}^\sigma \partial_\sigma - \mathbf{M} \cdot \mathbf{M}) G_\varsigma.$$

$$(1275)$$

Assuming that the derivatives do not act on the base vectors (e.g., because they are Minkowski like just as it was originally assumed by Dirac in [143]), we can further simplify to:

$$0 = \left(\mathbf{e}^\sigma \partial_\sigma \cdot \mathbf{e}^\rho \partial_\rho - M^2 \right) G_\varsigma = \left(\mathbf{e}^\sigma \cdot \mathbf{e}^\rho \partial_\sigma \partial_\rho - M^2 \right) G_\varsigma = \left(g^{\sigma\rho} \partial_\sigma \partial_\rho - M^2 \right) G_\varsigma$$

$$0 = \left(\mathbf{e}^\rho \partial_\rho \cdot \mathbf{e}^\sigma \partial_\sigma - M^2 \right) G_\varsigma = \left(\mathbf{e}^\rho \cdot \mathbf{e}^\sigma \partial_\rho \partial_\sigma - M^2 \right) G_\varsigma = \left(g^{\rho\sigma} \partial_\rho \partial_\sigma - M^2 \right) G_\varsigma.$$

$$(1276)$$

In our derivation (Eqs. (1272) to (1276)), we recognize exactly the way the classical Dirac operators act, too.

Again, we leave the generalization and further discussion to the interested reader.

Chapter 16

Generalization and Interpretation

16.1 Generalization

16.1.1 The Mixed Form

So far, in order to obtain metric (quantum gravity) equivalents to the main quantum equations like Klein–Gordon and Dirac, it totally suffices to just apply the following fairly simple extension to the metric tensor:

$$G_{\delta\gamma} = F\left[f\right] \cdot g_{\delta\gamma}. \tag{1277}$$

Subjecting this to the Hilbert variation and applying the chain rule during the variation process results in:

$$\delta W = 0 = \delta_\sigma \int_V d^n x \left(\sqrt{-G} \cdot R\right) = \int_V d^n x \left(\sqrt{-G} \cdot \left[R^{\alpha\beta} - \frac{R}{2}G^{\alpha\beta}\right]\right) \delta_\sigma G_{\alpha\beta}$$

$$= \int_V d^n x \left(\sqrt{-G} \cdot \left[R^{\alpha\beta} - \frac{R}{2}G^{\alpha\beta}\right]\right) \left(\frac{\partial F\left[f\right]}{\partial f} \cdot g_{\alpha\beta} f_{,\sigma} + F\left[f\right] \cdot g_{\alpha\beta,\sigma}\right) g^{\sigma\chi} \delta x_\chi$$

$$\tag{1278}$$

and, depending on the approach for the scalar f and functional wrapper $F\left[f\right]$ can give us a great variety of quantum gravity equations (see especially chapter 15).

A logic and most tensor-like extension of Eq. (1277) would be:

$$G_{\delta\gamma} = F\left[f\right]_{\delta\gamma}^{\alpha\beta} \cdot g_{\alpha\beta}. \tag{1279}$$

The World Formula: A Late Recognition of David Hilbert's Stroke of Genius
Norbert Schwarzer
Copyright © 2022 Jenny Stanford Publishing Pte. Ltd.
ISBN 978-981-4877-20-6 (Hardcover), 978-1-003-14644-5 (eBook)
www.jennystanford.com

The variation process would then give us:

$$\delta W = 0 = \delta_\sigma \int_V d^n x \left(\sqrt{-G} \cdot R \right) = \int_V d^n x \left(\sqrt{-G} \cdot \left[R^{\delta\gamma} - \frac{R}{2} G^{\delta\gamma} \right] \right) \delta_\sigma G_{\delta\gamma}$$

$$= \int_V d^n x \left(\sqrt{-G} \cdot \left[R^{\delta\gamma} - \frac{R}{2} G^{\delta\gamma} \right] \right) \left(\frac{\partial F \, [f]_{\delta\gamma}^{\alpha\beta}}{\partial f} \cdot g_{\alpha\beta} \, f_{,\sigma} + F \, [f]_{\delta\gamma}^{\alpha\beta} \cdot g_{\alpha\beta,\sigma} \right) g^{\sigma\chi} \delta x_\chi.$$

$$(1280)$$

We see that this changes neither the principal structure of Eq. (1278) nor our fundamental results from the previous section, which is to say "the simplest world formula". As a logical consequence, we have to conclude that with Eq. (1280) we have a very fundamental and general quantum gravity equation for an arbitrary number of dimensions.

From Eq. (1280), a scalar equation could be extracted as follows:

$$\delta W = 0 = \int_V d^n x \left(\sqrt{-G} \cdot \left[R^{\delta\gamma} - \frac{R}{2} G^{\delta\gamma} \right] \right) \left(\frac{\partial F \, [f]_{\delta\gamma}^{\alpha\beta}}{\partial f} \cdot g_{\alpha\beta} \, f_{,\sigma} + F \, [f]_{\delta\gamma}^{\alpha\beta} \cdot g_{\alpha\beta,\sigma} \right) g^{\sigma\chi} \delta x_\chi$$

$$= \int_V d^n x \left(\sqrt{-G} \cdot \left[R^{\delta\gamma} - \frac{R}{2} G^{\delta\gamma} \right] \right) \left(\frac{\partial F \, [f]_{\delta\gamma}^{\alpha\beta}}{\partial f} \cdot g_{\alpha\beta} \, f_{,\sigma} + F \, [f]_{\delta\gamma}^{\alpha\beta} \cdot g_{\alpha\beta,\sigma} \right) \mathbf{e}^\sigma \cdot \mathbf{e}^\chi \delta x_\chi$$

$$\Rightarrow 0 = \left(\sqrt{-G} \cdot \left[R^{\delta\gamma} - \frac{R}{2} G^{\delta\gamma} \right] \right) \left(\frac{\partial F \, [f]_{\delta\gamma}^{\alpha\beta}}{\partial f} \cdot g_{\alpha\beta} \, f_{,\sigma} + F \, [f]_{\delta\gamma}^{\alpha\beta} \cdot g_{\alpha\beta,\sigma} \right) \mathbf{e}^\sigma \cdot \mathbf{E}.$$

$$(1281)$$

Thereby **E** shall just be a suitable constant vector.

It should be noted here that we always have the option to add a cosmological constant to the variation making the first factor under the integrand to:

$$\left[R^{\delta\gamma} - \frac{R}{2} G^{\delta\gamma} \right] \rightarrow \left[R^{\delta\gamma} - \frac{R}{2} G^{\delta\gamma} + \Lambda \cdot G^{\delta\gamma} \right] \qquad (1282)$$

and thus changing the whole variation as follows:

$$\delta W = 0 = \delta_\sigma \int_V d^n x \left(\sqrt{-G} \cdot (R - 2 \cdot \Lambda) \right)$$

$$= \int_V d^n x \left(\sqrt{-G} \cdot \left[R^{\delta\gamma} - \frac{R}{2} G^{\delta\gamma} + \Lambda \cdot G^{\delta\gamma} \right] \right) \delta_\sigma G_{\delta\gamma}$$

$$= \int_V d^n x \left(\sqrt{-G} \cdot \left[R^{\delta\gamma} - \frac{R}{2} G^{\delta\gamma} + \Lambda \cdot G^{\delta\gamma} \right] \right)$$

$$\times \left(\frac{\partial F \, [f]_{\delta\gamma}^{\alpha\beta}}{\partial f} \cdot g_{\alpha\beta} \, f_{,\sigma} + F \, [f]_{\delta\gamma}^{\alpha\beta} \cdot g_{\alpha\beta,\sigma} \right) g^{\sigma\chi} \delta x_\chi. \qquad (1283)$$

16.1.2 The Matter Form

To some scientists, it may seem more convenient to obtain the variation in the classical form where the f-disturbances appear as matter term within the factor of the Einstein field equations. As the evaluation was already performed for the simple case in Eq. (1277), here we repeat it only for the generalized transformation (1279). Thereby, we assume R to be the Ricci scalar to the unperturbed (non-transformed) metric $g_{\alpha\beta}$. On the other hand, we know the Ricci scalar R to be defined by the metric tensor via:

$$R = R_{\alpha\beta}g^{\alpha\beta} = \left(\Gamma^\sigma_{\alpha\beta,\sigma} - \Gamma^\sigma_{\beta\sigma,\alpha} - \Gamma^\mu_{\sigma\alpha}\Gamma^\sigma_{\beta\mu} + \Gamma^\sigma_{\alpha\beta}\Gamma^\mu_{\sigma\mu}\right)g^{\alpha\beta}, \tag{1284}$$

with:

$$\Gamma^\gamma_{\alpha\beta} = \frac{g^{\gamma\sigma}}{2}\left(g_{\sigma\alpha,\beta} + g_{\sigma\beta,\alpha} - g_{\alpha\beta,\sigma}\right). \tag{1285}$$

Similarly, we have for R^*:

$$R^* = R^*{}_{\alpha\beta}G^{\alpha\beta} = \left(\Gamma^{*\sigma}_{\alpha\beta,\sigma} - \Gamma^{*\sigma}_{\beta\sigma,\alpha} - \Gamma^{*\mu}_{\sigma\alpha}\Gamma^{*\sigma}_{\beta\mu} + \Gamma^{*\sigma}_{\alpha\beta}\Gamma^{*\mu}_{\sigma\mu}\right)G^{\alpha\beta}$$
$$= \left(\Gamma^{*\sigma}_{\alpha\beta,\sigma} - \Gamma^{*\sigma}_{\beta\sigma,\alpha} - \Gamma^{*\mu}_{\sigma\alpha}\Gamma^{*\sigma}_{\beta\mu} + \Gamma^{*\sigma}_{\alpha\beta}\Gamma^{*\mu}_{\sigma\mu}\right)g_{\xi\zeta}F\left[f\right]^{\xi\zeta\alpha\beta}, \tag{1286}$$

with

$$\Gamma^{*\gamma}_{\alpha\beta} = \frac{g_{\xi\zeta}F[f]^{\xi\zeta\gamma\sigma}}{2}\left(\left[F[f]^{\rho\tau}_{\sigma\alpha}\cdot g_{\rho\tau}\right]_{,\beta} + \left[F[f]^{\rho\tau}_{\sigma\beta}\cdot g_{\rho\tau}\right]_{,\alpha} - \left[F[f]^{\rho\tau}_{\alpha\beta}\cdot g_{\rho\tau}\right]_{,\sigma}\right)$$
$$= \frac{g_{\xi\zeta}}{2}\left(F[f]^{\xi\zeta\gamma\sigma}F[f]^{\rho\tau}_{\sigma\alpha}\cdot g_{\rho\tau,\beta} + F[f]^{\xi\zeta\gamma\sigma}F[f]^{\rho\tau}_{\sigma\beta}\cdot g_{\rho\tau,\alpha} - F[f]^{\xi\zeta\gamma\sigma}F[f]^{\rho\tau}_{\alpha\beta}\cdot g_{\rho\tau,\sigma}\right)$$
$$+ \frac{g_{\xi\zeta}F[f]^{\xi\zeta\gamma\sigma}}{2}\left(F[f]^{\rho\tau}_{\sigma\alpha,\beta}\cdot g_{\rho\tau} + F[f]^{\rho\tau}_{\sigma\beta,\alpha}\cdot g_{\rho\tau} - F[f]^{\rho\tau}_{\alpha\beta}{}_{,\sigma}\cdot g_{\rho\tau}\right)$$
$$\equiv \Gamma^{*\gamma}_{*\alpha\beta} + \frac{g_{\xi\zeta}F[f]^{\xi\zeta\gamma\sigma}}{2}\left(F[f]^{\rho\tau}_{\sigma\alpha,\beta}\cdot g_{\rho\tau} + F[f]^{\rho\tau}_{\sigma\beta,\alpha}\cdot g_{\rho\tau} - F[f]^{\rho\tau}_{\alpha\beta}{}_{,\sigma}\cdot g_{\rho\tau}\right)$$
$$\equiv \Gamma^{*\gamma}_{*\alpha\beta} + \Gamma^{**\gamma}_{\alpha\beta}. \tag{1287}$$

Setting this into the second line in Eq. (1286) yields:

$$R^* = R^*{}_{\alpha\beta}G^{\alpha\beta} = \left(\Gamma^{*\sigma}_{\alpha\beta,\sigma} - \Gamma^{*\sigma}_{\beta\sigma,\alpha} - \Gamma^{*\mu}_{\sigma\alpha}\Gamma^{*\sigma}_{\beta\mu} + \Gamma^{*\sigma}_{\alpha\beta}\Gamma^{*\mu}_{\sigma\mu}\right)G^{\alpha\beta}$$

$$= \left(\begin{array}{c} \Gamma^{*\sigma}_{*\alpha\beta,\sigma} - \Gamma^{*\sigma}_{*\beta\sigma,\alpha} - \Gamma^{*\mu}_{\sigma\alpha}\Gamma^{*\sigma}_{*\beta\mu} + \Gamma^{*\sigma}_{*\alpha\beta}\Gamma^{*\mu}_{*\sigma\mu} \\ -\Gamma^{*\mu}_{*\sigma\alpha}\Gamma^{**\sigma}_{\beta\mu} + \Gamma^{*\sigma}_{*\alpha\beta}\Gamma^{**\mu}_{\sigma\mu} - \Gamma^{**\mu}_{\sigma\alpha}\Gamma^{*\sigma}_{*\beta\mu} + \Gamma^{**\sigma}_{\alpha\beta}\Gamma^{*\mu}_{*\sigma\mu} \\ +\Gamma^{**\sigma}_{\alpha\beta,\sigma} - \Gamma^{**\sigma}_{\beta\sigma,\alpha} - \Gamma^{**\mu}_{\sigma\alpha}\Gamma^{**\sigma}_{\beta\mu} + \Gamma^{**\sigma}_{\alpha\beta}\Gamma^{**\mu}_{\sigma\mu} \end{array}\right) g_{\xi\zeta}F[f]^{\xi\zeta\alpha\beta}$$

$$\underbrace{}_{\equiv R^*_*}$$

$$= \left(\Gamma^{*\sigma}_{*\alpha\beta,\sigma} - \Gamma^{*\sigma}_{*\beta\sigma,\alpha} - \Gamma^{*\mu}_{*\sigma\alpha}\Gamma^{*\sigma}_{*\beta\mu} + \Gamma^{*\sigma}_{*\alpha\beta}\Gamma^{*\mu}_{*\sigma\mu}\right)g_{\xi\zeta}F[f]^{\xi\zeta\alpha\beta}$$

$$+ \left(\begin{array}{c} -\Gamma^{*\mu}_{*\sigma\alpha}\Gamma^{**\sigma}_{\beta\mu} + \Gamma^{*\sigma}_{*\alpha\beta}\Gamma^{**\mu}_{\sigma\mu} - \Gamma^{**\mu}_{\sigma\alpha}\Gamma^{*\sigma}_{*\beta\mu} + \Gamma^{**\sigma}_{\alpha\beta}\Gamma^{*\mu}_{*\sigma\mu} \\ +\Gamma^{**\sigma}_{\alpha\beta,\sigma} - \Gamma^{**\sigma}_{\beta\sigma,\alpha} - \Gamma^{**\mu}_{\sigma\alpha}\Gamma^{**\sigma}_{\beta\mu} + \Gamma^{**\sigma}_{\alpha\beta}\Gamma^{**\mu}_{\sigma\mu} \end{array}\right) g_{\xi\zeta}F[f]^{\xi\zeta\alpha\beta}$$

$$\underbrace{}_{\equiv R^{**}_*}$$

$$\equiv R^*_* + R^{**}_*. \tag{1288}$$

This makes the classical Hilbert variation problem to:

$$\delta W = 0 = \delta_\sigma \int_V d^n x \left(\sqrt{-G} \cdot R^* \right) = \delta_\sigma \int_V d^n x \left(\sqrt{-G} \cdot \left(R_*^* + R_*^{**} \right) \right),$$

(1289)

from where we directly obtain the usual (classical) structure for the Ricci scalar plus the matter Lagrange density term with $R_*^{**} = -2\kappa L_M$ under the integral via:

$$\delta W = 0 = \delta_\sigma \int_V d^n x \left(\sqrt{-G} \cdot R^* \right) = \delta_\sigma \int_V d^n x \left(\sqrt{-G} \cdot \left(R_*^* + R_*^{**} \right) \right)$$

$$= \int_V d^n x \, \delta_g \left(\sqrt{-G} \left[R_*^* - 2\kappa L_M \right] \right) \times \delta_\sigma g_{\alpha\beta}.$$

(1290)

Together with the cosmological constant option, the latter result would read:

$$\delta W = 0 = \delta_\sigma \int_V d^n x \left(\sqrt{-G} \cdot \left(R^* - 2 \cdot \Lambda \right) \right)$$

$$= \delta_\sigma \int_V d^n x \left(\sqrt{-G} \cdot \left(R_*^* + R_*^{**} - 2 \cdot \Lambda \right) \right)$$

$$= \int_V d^n x \, \delta_g \left(\sqrt{-G} \left[R_*^* - 2\kappa L_M - 2 \cdot \Lambda \right] \right) \times \delta_\sigma g_{\alpha\beta}. \quad (1291)$$

We realize that in principle this is also a "mixed form" with two factors, just as derived in the subsection above, only that this time we have the familiar matter Lagrangian density and/or the energy momentum tensor back in our equation.

We find that we have obtained the starting point for the derivation of the Einstein field equations with matter in classical form just out of the generalized metric set-up (Eq. (1279)), now—in contrast to the vacuum metric $g_{\alpha\beta}$—with respect to the matter metric $G_{\alpha\beta}$. This will not change the structure of the matter Einstein field equations as long as the variation has not been extended beyond the usual metric variation, which is to say as long as the internal structure of $G_{\alpha\beta}$ (c.f. Eq. (1279)) has not been revealed by an explicit search for it.

We also immediately realize that with a complexly transformed metric $G_{\alpha\beta}$ as obtained with Eq. (1279), the classical Einstein, respectively Hilbert approach can only be an approximation as it reads:

$$\delta W = 0 = \int_V d^n x \, \delta_g \left(\sqrt{-g} \left[R - 2\kappa L_M - 2 \cdot \Lambda \right] \right) \times \delta_\sigma g_{\alpha\beta}, \quad (1292)$$

while the correct form would be given with the last line of Eq. (1291). This means that the matter equations of Einstein and Hilbert are results from the following approximation of the integrand under the Einstein–Hilbert action:

$$\sqrt{-G} \cdot \left[R^*_* - 2 \cdot \kappa \cdot L_M - 2 \cdot \Lambda \right] \xrightarrow[\text{approximation}]{\text{Einstein \& Hilbert}} \sqrt{-g} \cdot \left[R - 2 \cdot \kappa \cdot L_M - 2 \cdot \Lambda \right].$$

(1293)

It was shown in section 3.3 that in the simple case of Eq. (1277), we just have $R^*_* = \frac{R}{F[f]}$ and:

$$R^{**}_* = \left(\begin{array}{c} -\Gamma^\mu_{\sigma\alpha} \Gamma^{**\sigma}_{\beta\mu} + \Gamma^\sigma_{\alpha\beta} \Gamma^{**\mu}_{\sigma\mu} - \Gamma^{**\mu}_{\sigma\alpha} \Gamma^\sigma_{\beta\mu} + \Gamma^{**\sigma}_{\alpha\beta} \Gamma^\mu_{\sigma\mu} \\ +\Gamma^{**\sigma}_{\alpha\beta,\sigma} - \Gamma^{**\sigma}_{\beta\sigma,\alpha} - \Gamma^{**\mu}_{\sigma\alpha} \Gamma^{**\sigma}_{\beta\mu} + \Gamma^{**\sigma}_{\alpha\beta} \Gamma^{**\mu}_{\sigma\mu} \end{array} \right) \frac{g^{\alpha\beta}}{F[f]} \equiv \frac{R^{**}}{F[f]}.$$

(1294)

In this case, a dramatic simplification is possible, because the integrand of the Hilbert variation (Eq. (1278)) reads:

$$\delta W = 0 = \delta_\sigma \int_V d^n x \left(\sqrt{-G} \cdot R^* \right) = \delta_\sigma \int_V d^n x \left(\sqrt{-G} \cdot \left(R^*_* + R^{**}_* \right) \right)$$

$$\xrightarrow{F^{\xi\zeta}_{\alpha\beta} \to F} = \delta_\sigma \int_V d^n x \left(\frac{\sqrt{-G}}{F[f]} \cdot (R + R^{**}) \right)$$

$$= \delta_\sigma \int_V d^n x \left(\frac{\sqrt{-g} \cdot F[f]^n}{F[f]} \cdot \left(R + \overbrace{R^{**}}^{=-2\kappa L_M} \right) \right)$$

$$= \int_V d^n x\, \delta_g \left(\sqrt{-g} \cdot (R - 2\kappa L_M) \right) \times \delta_\sigma \left(\frac{\sqrt{F[f]^n}}{F[f]} g_{\alpha\beta} \right)$$

$$= \int_V d^n x \left(\sqrt{-g} \cdot \left[R^{\alpha\beta} - \frac{R}{2} g^{\alpha\beta} + \kappa T^{\alpha\beta} \right] \right) \times \delta_\sigma \left(\frac{\sqrt{F[f]^n}}{F[f]} g_{\alpha\beta} \right)$$

$$= \int_V d^n x \left(\sqrt{-g} \cdot \left[R^{\alpha\beta} - \frac{R}{2} g^{\alpha\beta} + \kappa T^{\alpha\beta} \right] \right) \cdot F[f]^{\frac{n-4}{2}}$$

$$\cdot \left(\frac{\partial F[f]}{\partial f} \cdot g_{\alpha\beta} f_{,\sigma} + F[f] \cdot g_{\alpha\beta,\sigma} \right) g^{\sigma\chi} \delta x_\chi.$$

(1295)

This time, there is no approximation to the classical Einstein–Hilbert form (Eq. (1293)) as we have had it for the general metric in Eq. (1279). In other words: In cases of simple metric variations as given in Eq. (1277) and the matter Lagrange density given via:

$$L_M \equiv -\frac{R^{**}}{2\kappa} = -\frac{F\,[f] \cdot R^{**}_*}{2\kappa}$$

$$= \frac{g^{\alpha\beta}}{2\kappa} \begin{pmatrix} \Gamma^{\mu}_{\sigma\alpha}\Gamma^{**\sigma}_{\beta\mu} - \Gamma^{\sigma}_{\alpha\beta}\Gamma^{**\mu}_{\sigma\mu} + \Gamma^{**\mu}_{\sigma\alpha}\Gamma^{\sigma}_{\beta\mu} - \Gamma^{**\sigma}_{\alpha\beta}\Gamma^{\mu}_{\sigma\mu} \\ -\Gamma^{**\sigma}_{\alpha\beta\,,\sigma} + \Gamma^{**\sigma}_{\beta\sigma\,,\alpha} + \Gamma^{**\mu}_{\sigma\alpha}\Gamma^{**\sigma}_{\beta\mu} - \Gamma^{**\sigma}_{\alpha\beta}\Gamma^{**\mu}_{\sigma\mu} \end{pmatrix}, \quad (1296)$$

the classical Einstein field equations for matter are exact quantum gravity equations with respect to the gravity part (meaning, the gravity factor under the variational integral). Solving these equations also solves the corresponding quantum gravity problem with the metric in Eq. (1277).

16.2 Interpretation

16.2.1 The Two Factors for Gravity and Quantum Are—Almost—Independent

Comparing our resulting Eq. (1280) with the classical Hilbert result (Eq. (1280), first line under the integrand) or just the Einstein field equations, we realize that we have two factors (thereby not counting $\sqrt{-G} \cdot g^{\sigma\chi}\delta x_\chi$) under the integral. As already one suffices to make the whole variation problem to zero, and as the first one presents the Einstein field equations, while the second one (as shown in chapter 15) gives the quantum equations, we have to conclude that there is no need for an actual quantum gravity set of special solutions. Yes indeed, the metric terms in $\left[R^{\delta\gamma} - \frac{R}{2}G^{\delta\gamma}\right]$ contain the function f and the quantum-type equation part $\left(\frac{\partial F[f]^{\alpha\beta}_{\delta\gamma}}{\partial f} \cdot g_{\alpha\beta}\,f_{,\sigma} + F\,[f]^{\alpha\beta}_{\delta\gamma} \cdot g_{\alpha\beta,\sigma}\right)$—of course—contains the metric, meaning the original unperturbed metric $g_{\alpha\beta}$, but without loss of generality one can fix one inside the factor of the other and solve it with respect to either the gravity metric or the quantum function f. The author suspects that exactly this fact allows for any development (evolution) within this universe. Without the factorial structure—so this author thinks—the whole system would just find one minimum and get stuck there for eternity. With the factorial structure, on the other hand, the system can permanently choose between minima states being achieved via either $\left[R^{\delta\gamma} - \frac{R}{2}G^{\delta\gamma}\right] = 0$ (gravity-based solutions/minima) or $\left(\frac{\partial F[f]^{\alpha\beta}_{\delta\gamma}}{\partial f} \cdot g_{\alpha\beta}\,f_{,\sigma} + F\,[f]^{\alpha\beta}_{\delta\gamma} \cdot g_{\alpha\beta,\sigma}\right) = 0$ (quantum-based solutions/minima).

It also holds, however, that the combined form $\left[R^{\delta\gamma} - \frac{R}{2}G^{\delta\gamma}\right]\left(\frac{\partial F[f]_{\delta\gamma}^{\alpha\beta}}{\partial f} \cdot g_{\alpha\beta} f_{,\sigma} + F[f]_{\delta\gamma}^{\alpha\beta} \cdot g_{\alpha\beta,\sigma}\right) = 0$ gives only n-conditions, while the separated terms result in n^2 and more equations. We could even construct a complete scalar via Eq. (1281).

Thus, gravity and quantum together have much more degrees of freedom and subsequently more options to evolve than each of the two being kept apart. One may see here a very fundamental form of marriage with the potential offspring being of by far greater diversity than without the concept of a blessed connection. This may be a hard blow to the gender-gagas, but it is simple very first-principle physics.

16.2.2 The Meaning of ...

Already with the first discoveries and developments within the field of theoretical quantum theory, the question about the interpretation of the results became eminent. The structure of the general product variation in Eq. (1279) tells us that any non-covariant transformation of a tensor results in quantum effects. With its simple form in Eq. (1277), we recognize just a general metric scale. With the resulting quantum equations being wave equations, we can also deduce that quantum effects are obviously just ripples along the various dimensions. These ripples are leading to an omnipresent jitter of space-time, being subjected to the metric system ...no matter what the metric actually describes. Of course, as we can see from the structure of both Eqs. (1277) and (1279), F could always just be a constant, reducing the quantum factor to just $F_{\delta\gamma}^{\alpha\beta} \cdot g_{\alpha\beta,\sigma}$, with all components of $F_{\delta\gamma}^{\alpha\beta}$ being constants. In this simple case, no quantum effects can be observed and we have to have $\left[R^{\delta\gamma} - \frac{R}{2}G^{\delta\gamma}\right] = 0$ to be fulfilled, which is to say gravity and the vacuum Einstein field equations govern the situation. With $F_{\delta\gamma}^{\alpha\beta}$ not being a matrix of constants, however, gravity can be ignored as long as the equation $\left(\frac{\partial F[f]_{\delta\gamma}^{\alpha\beta}}{\partial f} \cdot g_{\alpha\beta} f_{,\sigma} + F[f]_{\delta\gamma}^{\alpha\beta} \cdot g_{\alpha\beta,\sigma}\right) = 0$ for the quantum term within the Einstein–Hilbert action has been fulfilled.

Thus, we conclude:

(A) quantum gravity is just considering the solutions of the Einstein–Hilbert action in a somewhat more general manner, using more degrees of freedom, than Einstein and Hilbert did.

(B) Thereby gravity can be seen as the solution of the metric to the equation $\left[R^{\delta\gamma} - \frac{R}{2}G^{\delta\gamma}\right] = 0$ (this equation also includes the Einstein field matter equations as shown in sections 3.3 and 16.1.2).

(C) Quantum effects are solutions of the function f to the equation $\left(\frac{\partial F[f]^{\alpha\beta}_{\delta\gamma}}{\partial f} \cdot g_{\alpha\beta} f_{,\sigma} + F[f]^{\alpha\beta}_{\delta\gamma} \cdot g_{\alpha\beta,\sigma}\right) = 0$, leading to a jitter of the dimensions described by the metric $g_{\alpha\beta}$.

(D) As the two terms $\left[R^{\delta\gamma} - \frac{R}{2}G^{\delta\gamma}\right]$ (gravity) and $\left(\frac{\partial F[f]^{\alpha\beta}_{\delta\gamma}}{\partial f} \cdot g_{\alpha\beta} f_{,\sigma} + F[f]^{\alpha\beta}_{\delta\gamma} \cdot g_{\alpha\beta,\sigma}\right)$ (quantum) are factors within the same product, it suffices that one of the two vanishes in order to satisfy the whole generalized Einstein–Hilbert action, which is to say, in order to give complete quantum gravity solutions.

(E) The dramatic decrease of boundary conditions for the product case via either $\left[R^{\delta\gamma} - \frac{R}{2}G^{\delta\gamma}\right]\left(\frac{\partial F[f]^{\alpha\beta}_{\delta\gamma}}{\partial f} \cdot g_{\alpha\beta} f_{,\sigma} + F[f]^{\alpha\beta}_{\delta\gamma} \cdot g_{\alpha\beta,\sigma}\right) = 0$ or even the scalar condition in Eq. (1281) (last line) automatically leads to the conclusion that the married couple of gravity and quantum can be much more productive, because it has many more degrees of freedom, than the solitary alone. Thereby it requires the true couple consisting of gravity and quantum. In other words: The properly married ones are simply merrier and of greater evolutionary value.

Chapter 17

Outlook: A Small Selection of Project Ideas Using the World Formula Approach

In this final chapter of the book, we want to present a selection of project ideas, which should be based on world formula approaches. Thereby our intention is lesser in the direction of completeness but rather with respect to demonstrate the holistic character of the theory. It is therefore for a good reason that we start with the seemingly remote field of psychology.

As the following project ideas are from a variety of authors, we present them in either abstract or short-paper form, with the corresponding references directly allocated to the abstract or short paper.

17.1 The New Space-Time of Psychology

P. Heuer-Schwarzer and N. Schwarzer

Brief Description/Abstract

(A) It obviously is no accident that our notion of psychological aspects is expressible in terms which we usually relate to materials science. Obviously, it is not just an association or a metaphorical description

The World Formula: A Late Recognition of David Hilbert's Stroke of Genius
Norbert Schwarzer
Copyright © 2022 Jenny Stanford Publishing Pte. Ltd.
ISBN 978-981-4877-20-6 (Hardcover), 978-1-003-14644-5 (eBook)
www.jennystanford.com

when we speak about mental stress and strain, psychological defects, mismatches, dislocations or—rather general—elastic and inelastic deformations. Taking the many forms of cognitive dissonances alone and considering their deforming effects regarding various time and spatial levels and scales, we cannot ignore the striking similarity to the same or mirroring situations in the world of energy and matter.

So we ask:

Is it possible that there is a much deeper connection than this similarity of words and expressions?

Before falling under the suspicion of esoteric waffle, the authors want to point out that—after all—the origin of all our psychological activities is based on the interaction of electromagnetism. Thus, there may well be such a deeper connection, because the evolution of bigger and bigger brains [1] has had its starting point within the same fundamental laws, which are creating, forming and permanently reshaping our matter surrounding. It is therefore most likely that the matter-based interactions, which bring about our psychological self (our feelings, self-awareness, consciousness and our whole cognitive being), are following the same rules. Then, however, it should also be possible to apply these rules to the field of psychology.

(B) Assuming the existence of a true "Theory of Everything" [2, 3], one should, of course, be able to extract from this theory all physical interactions, which are effectively creating the complex pattern of a psychological entity. Additionally incorporating what was said under point A, it has to be possible to formulate a psychology in consistent mathematical form (similarly to [4] in classical medicine or [5] in philosophy) and taking its fundamental origin, one might name it a quantum gravity psychology.

(C) The new approach should be seen as an essential part of the holistic "Virtual Patient" concept as proposed by Leuenberger [6].

1. N. Schwarzer, *The Relativistic Quantum Bible: Genesis and Revelation*, www.amazon.com, ASIN: B01M1CJH1B.
2. N. Schwarzer, *Worldformula*, www.amazon.com, ISBN: 9781673032567.
3. N. Schwarzer, *The Theory of Everything: Quantum and Relativity is Everywhere – A Fermat Universe*, Jenny Stanford Publishing, 2020, ISBN-10: 9814774472.

4. N. Schwarzer, *Einstein Had It, but He Did Not See It, Part LXIX: The Hippocratic Oath in Mathematical Form and Why – So Often – It Will Be of No Use*, www.amazon.com, ASIN: B07KDSMNSK.
5. N. Schwarzer, *Philosophical Engineering, Part 1: The Honest Non-Parasitic Philosopher and the Universal "GOOD" Derived from a Theory of Everything*, www.amazon.com, ASIN: B07KNWRDYW.
6. H. Leuenberger, What is life?: a new human model of life, disease and death – a challenge for artificial intelligence and bioelectric medicine specialists, *SWISS PHARMA*, 2019, **41**, Nr.1, see www.ifiip.ch.

17.2 Towards a Deeper Understanding of Socio-Economy

N. Schwarzer

Stresses and strains in societies are not principally different from their counterparts in materials science [1]. The reason for this lies in the fact that both materials and socio-economic spaces consist of substructures (let us just name them atoms or molecules), with their interaction forming the actual material space or socio-economy. Displacing atoms against each other automatically leads to strains and—depending on the strength and character of the interaction—subsequently also to stresses. Of course, while in materials science the displacements are truly seen as actual movements or changes of positions of the atoms, they mean any kind of dimensional misfit with a generalized model.

We saw in chapter 15 of this book that the quantum gravity equivalent of the Dirac equation sports displacements as just the essential elements for all quantum effects. Quantum gravity, on the other hand, is a "theory of everything", and as such it should also be possible to formulate a quantum metric (quantum gravity) discretion for societies.

Putting it all together, it becomes evident that the way forward to obtain most holistic and truthful, which is to say ideology-free, models for socio-economic simulations is to be seen in a Dirac-like displacement vector model as derived in section 15.5 of this book. Thereby the number of dimensions is determined by not only the number of properties the socio-economy one intends to develop shall contain, but also the number of gravity centers/atoms/entities/individuals it will possess (c.f. chapter 4 of this book).

1. N. Schwarzer, *Philosophical Engineering, Part 1: The Honest Non-Parasitic Philosopher and the Universal "GOOD" Derived from a Theory of Everything*, www.amazon.com, ASIN: B07KNWRDYW.

17.3 Why Ideology-Affine Societies Are Per Se Unethical?

N. Schwarzer

More than 2370 years ago, Aristotle made it clear in his great book about "Nicomachean Ethics" [1] that ethic in general is just a certain way to live. In fact, according to Aristotle [1], a perfect ethic life is just the optimum good life. Considering systems of individuals within their societies in just THE most general manner, namely via a world formula approach (e.g., c.f. section 17.8), it can clearly be derived that the optimum GOOD inside a universe like ours, which is based on extremal principles, can always only be the minimum of suffering. This minimum, on the other hand, can only be found if it has been sought for in the most holistic manner, which is to say: The search process can only be complete if it is not restricted (e.g., by some kind of political correctness, ideology, religion, esoteric or otherwise self-binding intellectual imprisonment).

In other words: Only with all effects, facts, interactions, etc., being taken into account, there is a chance to find the optimum good at all. Any attempt to find the system's good (best) state without considering it ALL, thereby taking everything most truthfully into account is in vain. Or even clearer: Any do-gooderish act, flooding the—mainly extremely short-sighted—do-gooder's bloodstream with endorphins due to her or his do-gooder-dim-doing actions is usually the very opposite of the holistic good and, thus, nothing else but emotional parasitism, as it generates the "good feeling" of the do-gooder-parasite on the costs of others.

Thesis: Ideology-Affine Societies Are per se Unethical

Proof:

(A) True Ethic behavior of a society requires the most holistic optimization of that very society, thereby seeking the minimum state of suffering within a global variation process. That is just the definition of Aristotle's ethic in a mathematical form [2].

(B) The explicit or purposeful exclusion of certain topics, aspects, truths, facts, etc., from the variation or optimization process hinders or even completely contradicts the ethical quest. This is self-evident, because the exclusion of real-existing degrees of freedom from an optimization process automatically renders the process inholistic and thus, depending on the amount of "thought-control" and variation restrictions, from inconsistency via pretence or total nonsense to disastrous failure.

(C) It is the nature of ideologies always to exclude certain real-existing truths, simply because these truths would reveal the purely ideological (non-scientifically or just rubbish) character of the latter.

(D) From points A to C now directly follows that societies being affine to certain ideologies or even being dominated by such are per se unethical.

Now we just give examples for three ideologies being partially or even completely based on lies, and the informed reader will easily be able to see the reality-based circumstantial evidence for our general and comprehensive proof as outlined above.

(1) The Gender ideology: There are so many publications about the OBVIOUS fact that this ideology is so self-evidently stupid that we here only refer to a very mathematical proof about the piled-up nonsense content of this rubbish doctrine [3, 4].

(2) The Climate ideology: e.g., see [4, 5].

(3) Marxism: see [6] or simply study the history of the last century.

1. *Nicomachean Ethics*, by Aristotle, 350 BC, translated by W. D. Ross, www.virtuescience.com/nicomachean-ethics.html.

2. N. Schwarzer, *Philosophical Engineering, Part 1: The Honest Non-Parasitic Philosopher and the Universal "GOOD" Derived from a Theory of Everything*, www.amazon.com, ASIN: B07KNWRDYW.

3. N. Schwarzer, *The ANTI-GENDER Proof: Mathematical Proof against One of the Most Stupid and Dangerous Ideologies, Disguising Itself as "Science"*, www.amazon.com, ASIN: B07KPD9BD4.

4. N. Schwarzer, T. Bodan, *Sherlock, Watson, Stalin, Part 2: The Hell of Gender, Merkel, Communism and the Dictatorship of Parasites*, www.amazon.com, ASIN: B07BJ9PZWM.

5. N. Schwarzer, Climate religion, https://principia-scientific.org/climate-religion-why-is-it-so-simple-to-cheat-the-mass.

6. N. Schwarzer, *Einstein Had It, but He Did Not See It, Part XIII: Why Equality Is Always Hostile to Life*, www.amazon.com, ASIN: B07B6QZLMH.

17.4 Water More Important than CO_2

A. D. Wieck

Lehrstuhl für Angewandte Festkörperphysik, Ruhr-Universität Bochum, Universitätsstraße 150, D-44780 Bochum, Germany

The current climate change discussion is based on the assumption that the CO_2 molecule has a strong optical impact in the infrared spectrum emitted by the earth. However, this assumption has never been discussed scientifically or in the public; everybody takes this as a fact and nearly nobody compares the CO_2-optical modes with the ones of other molecules in the atmosphere. I would like to discuss exactly this important detail a bit more here:

In the CO_2 molecule, the two bonds between the carbon atom and the two oxygen atoms are aligned; it is a linear molecule. The oxygen atoms attract negative electron charge from the carbon atom in a symmetric way. Therefore, the charge distribution within the molecule is of the form $- + +-$. The two dipoles are thus directing in opposite directions and the overall dipole momentum of the CO_2 molecule is zero. This means that the electric field of a passing electromagnetic wave cannot couple linearly to this molecule; any coupling is of higher order, i.e., to the quadrupole moment or higher. It is well known that under these circumstances, the interaction between the molecule and the electromagnetic wave is greatly hindered, if not to say quasi-forbidden or extremely weak. As a result, CO_2 in the atmosphere has a minimal impact on the transmission of infrared radiation and this is even more pronounced since its relative content is only roughly 0.04% in the air.

On the contrary, the water molecule, H_2O, is a non-linear, but angled, bent object: It looks like Mickey-Mouse's head as the oxygen atom, the two ears being the hydrogen atoms, which are positively charged $+ \_ +$. Consequently, Mickey's chin is negatively charged and there is in this way a huge static, vertical dipole moment, which strongly couples to any electromagnetic wave, especially in the near and mid-infrared. At 20°C and 60% humidity, there is 10.4 g/m^3 H$_2$O in the air; with an air density of 1.3 kg/m^3, this is 0.8%, 20 times more than CO$_2$. In clouds, we have 100% air humidity, and this yields 1.3% water in the air, which is 33 times more than CO$_2$.

In other words: Clouds have by their water density 33 times more impact than CO$_2$ AND water is as a dipolar molecule by far more infrared (IR) active than the quadrupole-time CO$_2$. The product yields orders of magnitude more IR-impact by water than by CO$_2$. If humanity wants to reduce the IR-impact of the atmosphere, it would be more reasonable to ban clouds rather than reduce CO$_2$ by 0.01%! Even school kids know that a clear, cloud-free night ends up in very low temperatures, while a cloudy sky guarantees moderate temperatures in the early morning.

17.4.1 Comment by N. Schwarzer

In addition to this, one needs to point out the fact that the possible influence of CO$_2$ as a greenhouse gas on the climate of the earth (even theoretically) is only 1/114 of the uncertainty of the cloud coverage [1, 2] (Mark: it is the uncertainty and not the absolute value). So, rather than following lies about CO$_2$, climate activists should watch man-made defects in the water cycle as these have by far greater effects. But of course, perturbations of the water cycle would immediately cast a critical view on things like e-mobility (Li mining), solar cell production, wind power plants and the TRUE ecological footprint and quite many other greenish projects, being sold to the tax-payer as "climate rescue" maneuvers, while in fact they are nothing else but lie- and ideology-based skullduggeries. In some cases, they also serve as bamboozle strategies to distract the mass from far more important or even life-threatening problems such as migration being forced on the population of many western countries. In this context, it is important to also try to understand why people are so easily made to believe often even the greatest nonsense. We will therefore do this in section 17.5.

1. www.quora.com/What-does-Michael-Mann-s-court-battle-loss-mean-to-the-notion-of-climate-change.
2. P. Frank, Propagation of error and reliability of global air temperature projection, *Front. Earth Sci.*, 2019, **7**, 223, https://doi.org/10.3389/

feart.2019.00223 or www. frontiersin.org/articles/10.3389/feart. 2019.00223/full.

17.5 Why Is It So Simple to Cheat the Mass?

N. Schwarzer

The author's question: "Is this just a fractal . . . or the typical outcome of a "climate simulation"?"

17.5.1 Abstract

There is never much of need to create a religion. What in the antique times just needed to be the golden calf, somebody called holy, nowadays just requires a made-up (hockey stick) curve somebody declared to be of scientific origin. As long as there are parasites who can feast on the outcome of the movement potentially erupting from such fairy tales or downright lies, there will also always be enough believers willingly and dumbly following the story, no matter how illogical or inconsistent, if not to say stupid or even dangerous, it may be.

> ... and immediately we have a new religion, including the pious followers and—naturally—the parasitic "leaders", who use the chance to press every penny out of everybody dim enough to trust them.

But why is it so simple to make people believe often even the greatest rubbish?

The question why man believes respectively where his spirituality comes from is an important one. It is so important, not because of the many cruel wars fought over belief and the millions of books written about the infinite number of forms of spirituality (as many as there are thinking entities in this universe), but because it is just one fundamental aspect of the thing which brought us about: Evolution. In fact, evolution forced us to develop an inner spiritual core and thus, belief, which is to say spiritual belief, is not an option, but a must.

The question why man MUST believe was already answered in [1], but as many people do not like or partially are even afraid of math, we reproduce the essentials here.

1. N. Schwarzer, *The Relativistic Quantum Bible: Genesis and Revelation*, www.amazon.com, ASIN: B01M1CJH1B.

17.5.2 Why Does Man Believe?

Are people with a certain belief worth more than others just because of that belief?

Our dogmatic answer to this question is NO.

Are priests, shamans, imams, rabbis or any other leader of a group of such believers above other men just because of their leading role for this group of believers? I mean above other men if the aspect of measuring is not connected with this leading role?

Again, our dogmatic answer to this question is NO.

Here is just one example to make our point absolutely clear:

The pope is an ordinary man (Ask him, and he will tell you so himself!), being elected by other ordinary men to do a certain job. In his case, the job is to lead the Catholic Church. The Catholic Church, on the other hand, is nothing special. It is an organization, a kind of permanent gathering or community of people who all share a certain belief. In this case, it is "the" Christian belief. Thereby we emphasized "the" in order to point out that "the" Christian belief does not exist. It was, is and always will be a complex composition of thoughts, ideas and ideals, being splattered over a rather non-conform community. Even though some of these people will tell you and everybody else that they are special, simply because they have that belief and because they are part of this community, named "Church", you should ALWAYS treat their "assertion about being special" with great care. The same just holds for all "Churches".

Why?

Because Moslems, Jews, other Christians (other than the Catholic ones), Hindus and so on will all say the same about themselves. They will all claim to be special above others. But with everybody being special above the other, because of such different beliefs, NO ONE is.

Things would be much easier here if we were talking about different political parties or supporters of certain sports teams, where it is not about various religious beliefs but about different political opinions or sportive groups.

But why is this so?

Why does everything which has to do with religion seem so much more difficult to discuss?

The answer to this requires us to investigate our own evolution. We will not be able to understand our own partially irrational, if not to say funny, behavior regarding religion if we do not understand the origin of spirituality.

As this is a very complicated topic, full of mines to tread on, the author will leave the elaboration to a fictive entity. It might help to imagine this purely fictive entity as something remotely similar to a child. Let us assume a very young "nothingness", just realizing what the funny thing named evolution had come up with. Ok, some of you might perhaps find it easier to imagine this nothingness as a kind of a child God.

17.5.3 A Very Young Nothingness and Her/His Questions Explain to Us Spirituality

There was a moment when evolution had the completely insane idea to develop consciousness. Evolution realized, which is to say the process of mutation and selection "discovered" that life forms equipped with consciousness actually were not the fittest. Thereby it needs to be pointed out that consciousness was not the thing evolution had aimed for. It only wanted to have cleverness. With cleverness, evolution was able to form the species that were able to use tools, perhaps even create some themselves. With cleverness, evolution could bring about entities that would be able to help evolution with the increasingly complicated inventions necessary to really make the difference. Consciousness, on the other hand, merely came as a by-product nobody really had wished for. One may say that consciousness was the true original sin. A sin, by the way, nobody should be blamed for, but evolution. After all, if we all had stayed dumb like earthworms, the word "sin" would not have even existed.

At first sight, it was good to have some kind of awareness or self-awareness, but within the cruel reality, it often became a disadvantage to "feel" respectively to consciously recognize what was going on and what life or being consciously alive really meant. So, evolution had to learn it the hard way, that it could create sufficiently fit life forms, which also have consciousness only when adding yet another piece to that recipe, and this was spirituality. Only spirituality allowed these self-aware and feeling life forms to put all those difficult or impossible to understand, often cruel or even unbearable experiences to a place in those bigger and bigger growing brains where it (meaning all the mess, cruelty and madness around) would hurt a bit less.

Of course, evolution never intended to make this storage for spiritual thought bigger than absolutely necessary. After all, evolution only wanted to create fitter life forms and had no interest whatsoever in bringing dim and passive believers into being. Thus, the organ of spirituality only ever was thought as a temporary storage room for everything which could not be explained straight away. Unfortunately, however, it turned out that explanations for some things took longer than the typical life span of such life forms. In fact, there were problems where even the typical life span of a whole species would not suffice. Thus, there was always something in these storages. These storages were never empty, and the bigger brains, keen to get these unexplainable things explained, were extremely susceptible to all sorts of prophets who promised to have solutions, answers, hope. And so the epoch of the shamans, priests, imams, Gretas and CO_2-smellers, esoterics, astrologists, politicians, politic scientists, soul-experts, climate-apostles, philosophers and coffee shop owners had begun.

Did you all understand that?

No!

Ok then, let us try it a bit differently.

Spirituality is a complex organ that allows us to put things we do not quite comprehend into a state of:

"Ok, I am not getting the gist right now, but I simply blame a fictive entity for this, I name this entity God and so I need not worry about the unexplainable thing anymore!"

Now you understood, right?

Yes!

Very good!

The good thing about spirituality for all those, shamans, politicians and so on now is that we can easily manipulate people because of this organ. You simply need to assure that the broad mass thinks something like:

"Ok, I am not getting the gist right now, but I simply believe that this or that politician does somehow know what he or she is doing . . . May God help them!

Or that this or that priest, who always claims to have a direct connection to God, knows much better what is good for me . . . and so I need not worry about myself and my own life anymore!"

To make this actually work, one has to train that organ, of course. It is a bit like a muscle. You also have to train it to make it strong. There are many ways to train the spiritual organ, but in essence it is just a clever mix of act, carrot, stick and lies.

As a shaman, the moment you lose control over that organ of your people, you also lose them. This organ namely is the only way to truly have power over them all.

Why on earth do you think mediocre politicians are flooding Germany and the EU right now with such dim but super-believing "refugees"?

Why on earth do you think the pope and his colleagues from the other Churches want you to welcome more and more and more of those equally godly people from the south?

You only ever need enough of such uncritical believers becoming dominant among the other people and then . . .

Well, then for the leaders in Church and politics, there is no need anymore to actually do anything—whatever it is—right. Oh no, it totally suffices to cry something like:

"Amen, Amen, I tell you: **Yes we can!**" or in a more recent case:

"We can make it!"

But is it fair to abuse this spiritual organ in such a way?

Our dogmatic answer to this question is NO.

Does the mere existence of the organ of spirituality prove the existence of the divine?

Again, our dogmatic answer to this question is NO.

To cut it short, we state: Spirituality is a product of evolution. Without it, consciousness could not have evolved, at least not in the way it did with a certain species on earth named *Homo sapiens*. Of course, the foundation for the spiritual organ was laid earlier, long before men appeared on earth and probably right at the beginning when evolution started to experiment with bigger and somewhat more sophisticated brains . . . perhaps just right after the earthworms, I guess. Thus, the "spiritual organ" is so deep down in our thought and brain structure that we do find it extremely difficult to stay rational if it comes to any topic having to do with spirituality.

Some people will say that the author is godless, but in fact, I am the opposite. Those who call me godless are only afraid of me, because I am telling you to think for yourself instead of blindly following modern shamans, priests, imams and all those who claim to have messages from God for you.

Yes, in fact, this "I have a message from God for you" does even sound mad, does it not?

Does it not just sound as mad as "I can smell CO_2!"?

Or let us have the typical: "I have a message from God for you . . . it might even be two messages, I cannot tell, because sometimes God is only whispering . . ."

"How wonderful", I want to answer, "and did you make sure it is not coming from somebody else? From the inside of your own head, for instance, or God may help us, from the devil himself?"

They are God's message machines. What an utter pile of rubbish.

No, they have no messages from God and especially they do have none for you.

If there was a God and he had a message for you, he would give it straight to you.

Why?

Because this is the shortest way and trust me, God does know that!!!

And why would he know that?

Because he is God, you idi. . . whatever . . . and because this whole universe always goes for the shortest ways, finds the extrema, the optima [2].

This is what we have just seen in the "New Genesis" in [1]. God, if he exists and if it was he who created the universe, created this as an extremal ensemble of properties, thereby making it the most effective and innovative life- and thought-production machine there can be [1, 3, 4]. That his creation, with its inbuilt minimal principle, also provides the fairest conditions possible was elaborated in [5].

Why would just God, of all entities, miss that important point in choosing the very ineffective way of only using those self-proclaimed "chosen ones" to send messages from him to you?

In fact, nobody ever had a message from God, not more than you had such messages yourself anyway.

It was evolution and a human being which developed the idea of God, it were human beings who carried that idea on to find solace in difficult times and it were human beings who abused the idea of God.

Shame on them!

They abused it to gain control over their fellow next, over their sisters and brothers for their own advantage—Shame on them! And of course, they must fight people like the present author who actually want to take that control from them in order give it back to the person to whom it belongs, namely you.

We state: Everybody who tells you that he or she has a message from God for you is most certainly lying and only wants to gain control over you. Do not give him or her that control!!!

So, now we are ready for the next question:

Does this all mean that there is no God, respectively no divine?

Again, our dogmatic answer to this question is NO. Because if something is in your head, it does not mean that it could not also be real.

Thus, the existence of spirituality only proves one thing: Evolution has made a lot of interesting inventions, but the mere idea of the divine neither proves nor negates its existence.

However, as hinted above, the existence of that spirituality in our brain structures makes it difficult to rationally discuss the topic of the divine. This holds the more as there are other dependencies on the existence of the different forms of the divine. In order to understand that, you simply have to look at the economic dependencies of the Churches and their leaders. Simply imagine the financial disasters all those Churches and their preaching "holy men" would be in, the moment nobody believes in them anymore. The moment nobody believes anymore that these people have messages coming from the divine, they would simply have to stop their comfortable lives and start to work and scrape a living as ordinary men do. But they cannot do this as most of them are completely unfit for such an ordinary life. Consequently, they all have to fight nail and tooth to keep their position as just the top-liars among us. They are parasites, and as the only way to keep this position, they have to suppress the one thing they fear most: You starting to think for yourself.

So, how neutral can these people be if it comes to honestly answering the question about the existence of God?

And even worse, how neutral can these people be, whose complete life depends on their special connection to the divine (a most specific divine, of course, mostly), whose physical and spiritual existence totally and absolutely depends upon their uniqueness, their disparity from ordinary men, if it comes to talking to, hearing from or obtaining messages from God?

Do you think you should give the words of an ordinary man, like the pope surely is, any more attention or credit than anybody else (perhaps yourself or your best friend)? Do you think you should stop thinking for yourself, freeze in making up your own mind, only because a certain other person, a spiritual leader, has done some thinking for you . . . or pretends to have done so?

Assuming you have voted "No" in answering all these questions above, it is time for a bit more thinking.

Whenever in its long history has the Church, any Church, not put its own economic interests VERY high up, if not to say above anything else, including fate and wealth of whole civilizations?

Thus, and especially when also taking the history into account, can a Church, any Church, be considered a trustworthy and neutral judge on the matter of the divine?

And having gotten this far, can a spiritual leader, any spiritual leader, who is totally dependent on your belief in the divine and his special connection to it, ever be considered a trustworthy and neutral judge on any matter in society?

Yes, the Church leaders will not like to read all this, but in order to show also them that I am not writing this in order to substitute one "divine authority" for another, I want to add the following:

The above said also holds—of course—for the author.

As an ordinary man, he can be as wrong as anyone. So, think for yourself, check the arguments, recalculate the math and reconsider the logic behind them. Then try for a new angel and do it all over again!

17.5.4 Climate Religion or Why Should Man Believe in God(s) When He Has Holy Greta(s), Mama Merkel(s) or Saint Michael(s)?

Knowing now where spirituality comes from and being—hopefully—somewhat more aware of the susceptibility lurking inside of each and every one of us, we may just apply our previous finding to examples such as "migration is always good", "gender is a science", "socialism and communism is for the working people", "politics can only be understood—and done—by the cleverest among us" or "CO_2 is a climate killer" or. . .

In all these examples mentioned here, it is quite easy to find the typical host-parasite structure one can make out in most religions. **But the huge difference from the classical mono- or multi-theistic religion is that these religions are based on the existence of god(s). This existence can neither be proven nor be disproven. This is just a principal and very fundamental fact. Those examples mentioned above, however, are not based on something being on dispute for eternity; they are based on downright lies—lies, which were spread among the susceptible in order to construct comfortable structures for parasites.** So it was shown in many studies that the whole gender gaga is total nonsense and no science at all. As a very fundamental proof that the so-called "gender-science" is nothing but a huge swindle, it probably suffices to show that the number 2 holds a very unique place among all numbers in this universe [8]. The proof is based on the last theorem of Fermat and its connection with the fundamental laws of this universe, which dramatically favors the number 2 on all operations [9]. Reproduction, however, is nothing but an operation and thus, the non-gender-conform appearance of two sexes is a natural and the most logical result and by no means an accident (as the "gender-scientist" wants to tell us). So, "gender-science" is a lie-based rubbish and nothing more. The other interesting example is the ideology of Marxism and all related equality ideologies. Here it can easily be shown that perfect equality with respect to any parameter or property always means the death of each and every system crucially depending on that very parameter [10]. Thus, Marxism and all its derivatives are based on a huge lie, too.

Let us now pick the climate example here. In order to keep the connection, we simply repeat the general discussion from above in a just slightly adapted form:

Thereby, in order to have an illustrative starting point (after all, if it comes to religion, it is common practice to apply metaphors), we want to use a completely fictive story about the origin of the current climate ideology, if not to say madness. In order to make it a bit entertaining, we apply the method of satire.

Why satire?

Is not the topic a very serious one?

Yes indeed it is, but what else but satire can we use when seeing that the climate-simulators (or -liars [7]) seriously want to tell us that the solar activity is a constant? If they even—actually—**pretend** to see (or smell) a CO_2 greenhouse gas effect, when the uncertainty of the cloud coverage is already 114 times bigger than any such effect could ever be [6]. Mind you, we really meant the UNCERTAINTY of the cloud coverage and not the absolute value. This alone gives enough reason to resort to satire, but then learning that the German Government has just brought a new "climate rescue tax package" on the way, which actually taxes CO_2 emission more than poisoning the world with dioxin, leaves no options for reasoning with these people. Against such stupidity, only satire can be used, because where there is no reason whatsoever, one has to stick to the old saying that humor is when you just laugh anyway.

Let us just imagine an imaginary person, Michael, who always was a bit "funny". One day, on a toilet and combing his hair with the toilet brush, as he always did, he had a revelation. Namely, when combing his hair—as usual—with the toilet brush, he realized that there was this nasty smell all around him. It occurred to him that the smell was the more and—what is more—more intensive, the more often he combed his hair in the usual— toilet brush—way. Attention seeking as he was, he wanted to tell as many people as he could about this. In fact, he considered this such a great finding that he immediately drew a diagram on a stretch of toilet paper in which he connected the frequency of use of the toilet brush for hair combing with the smelly outcome for his head. Naturally he found that the more he combed, the smellier he became. In result, he obtained a nice curve showing this functionality. As it was just a line, which Michael considered a bit boring, he added a few wiggles here and there, just in order to make it a bit more interesting. Then he drew two axes and wrote "COMB" on the x- and "SMELL" on the y-axis. He thought for a moment what might happen when doing even more hair combing this way (perhaps in some excessive, if not so say manic, manner, thereby resorting to all public toilets in his town) and realized that this should lead to a rather dramatic increase in smelliness. He added this as a steep "prediction" to his curve. Now he wanted to present his result to the public, and as he was not completely dim, he sought for good substitutes for the two words he had used in his little diagram so far. After all, so he knew, it is the words which make the difference and not the facts. It took him quite a long time to come up with a good substitute (10 years or so), but finally he had the "COMB" changed to "CO", where, as Michael was not a very clean worker, some smeared out residual remained behind the "O". Then he changed the word "SMELL" with "CLIMATE", because it simply appeared quite natural to him to associate the two things. After all, with him in the room, especially after he had performed some hair combing action, the climate usually was not good at all.

Now he presented his result to the world, and under the name "toilet brush curve", it became a huge success and the starting point for a new religion.

Thereby the first presentation was close to become a disaster for Michael as he could not answer the question what the heck CO should stand for. Then a drunken journalist in the second row, who took the smeared out residual from the "MB" for some index stuff, babbled something like:

"Boa eh . . . not the stupid CO_2 again."

But some Club-of-Rome people in his vicinity immediately saw their chance and shouted:

"This must be CO_2, of course!"

Then everybody just went crazy and there was nothing but a mix of applause, a lot of shouting and—above all—a great understanding about this important prediction at the end of the "toilet brush curve" where excessive combing had now, which is to say after all the word-changes and Michael-adaptations, led to a very bad "climate" prediction.

Among the first who listened to the "toilet brush curve" story of Michael, there were a totally imaginary Swedish father and his even more imaginary daughter Greta. They were in the first row and Greta, who was about to throw up because of the smell wafting around a freshly combed Michael, thought:

"Oh, I can smell CO_2!"

And in her slightly restricted brain, a connection was built between this awful smell and the prediction part in Michael's curve she saw right in front of her.

"Hell", she thought in horror, "if this is getting even worse, I would really throw up in the end."

From this moment on, she fought CO_2 wherever she could. She did it with such ferocity that people around her got infected, and faster than one could say "climate change is just normal" or "CO_2 is an absolutely harmless gas", a new religion was born and Greta was its Messiah.

So and now, let us just repeat what we have already learned about the origin of spiritual belief. Thereby we want to incorporate our knowledge about the climate ideology:

We know that spirituality is a complex organ that allows us to put things we do not quite comprehend into a state of:

"Ok, I do not understand the complicated quantum mechanical behavior of this funny gas CO_2 and I have absolutely no inkling about what climate really is, but I trust Greta-gang and the Michael-Saint and all those holy higher entities around them and believe that they are telling the truth. In order to make things simpler for my own internal spiritual filter, I even name these entities EXPERTS and so I need not worry about understanding the whole topic anymore myself! They just do it all for me."

The good thing about spirituality for all those, experts, fathers of Gretas, Michaels and so on, now is that they can easily manipulate people because of this inner spiritual organ. And this is what they do. They have made the ordinary people not only trust them and believe that they are right with this "toilet brush story" (lie) about the CO_2 [6, 7], but also accept a hell of a burden to feed these liars.

And just as it holds with the ordinary religion, in order to make this lie actually work, one has to train that organ. In the case of the climate lie, it is a permanent steady flow of corresponding rubbish and flawed association, cleverly combined with the cane for those who do not want to believe that easily, which does do the trick. They even founded a huge international council of a global mafia-like structure in order to reach each and every corner of this world and fill it with this mephitic, pestilential rubbish about the bad CO_2.

Now we intend to investigate the parasitic character of these ideological pseudo-religions. When starting with Marxism and just taking the association of George Orwell's famous "Animal Farm", we easily see which pigs were "more equal than the others". All those higher-up party members, Marxism/Leninism and Stalinism teachers, polit-officers, Stasi, Securitate, KGB and so on members, none of them needed to worry about doing anything useful for the society any longer. The society had to work for them, while their own "contribution" was the uphold of the suppression and the assertion and permanent (cruel) enforcement of the underlying, in principle rubbish, ideology. They all were parasites of the society and it is of little wonder that in the end these societies had to fall.

Now you simply turn your gaze towards the gender and CO_2-climate "scientists". You analyze what they could do in a system not accepting such lies and dangerous nonsense. Having done this in a very thorough manner, you now simply count all those who benefit from the lies in these fields. You investigate who is supporting them and to whom they prostitute themselves in return, like the cheapest whores who have nothing to offer but a worthless flood of obscene words, being whispered into the ear of a media-stultified mass.

Imagine you as a shaman in this anything but holy dance. The moment you lose control over that organ of your people, you also lose them. This organ namely is the only way to truly have power over them all.

And you need to find fellow parasites to connect with. It is not enough that you alone become a millionaire as a CO_2-dealing father of Greta. Oh no, you have to share with many others: journalists, politicians and—very effective—economic refugees. Simply call the latter "climate refugees" and they are becoming your most natural allies. These allies will easily assure the suppression of the little opposition you might have. As coming from nothing, having nothing and knowing nothing, they will also do the dirty work in case there is need. You may not even need to ask them. Do not worry, they will get behind it quite easily on their own and then they will just do what is necessary to crush the rebellion against your climate religion. There is no limit of means.

Why on earth do you think mediocre politicians are flooding Germany and the EU right now with such dim but super-eager and "climate-aggrieved refugees"?

You only ever need enough of such uncritical and—what is more—highly motivated (because selfishly egoistic) believers becoming dominant among the other people and then . . .

Well, then for the leaders in climate churches and green-rubbish politics, there is no need anymore to actually do anything—whatever it is—right. Oh no, it totally suffices to cry something like:

"Amen, Amen, I tell you: CO_2 is bad stuff!" or:

"We need to tax CO_2" ("and thereby fill our pockets with the dim believer's hard-earned money"), or in a more recent case:

"I can smell it, too."

17.5.5 Conclusions

But is it fair to abuse this spiritual organ in such a way?

Our dogmatic answer to this question is NO.

There is no justification for such an abuse.

It is a crime and there is going to be no forgiveness for those who committed it . . . not even if committed out of pure ignorance and bigotry.

And why would not we accept ignorance as an excuse?

Because it is a crime not to use one's own brain . . . thereby meaning the whole brain and not just its inner spiritual part.

Evolution gave us the ability to think and the European epoch of illuminism gave us the right to think.

Thus, it is our duty to think!

17.5.6 Back to Science (Extracted from a Statement of J. O'Sullivan, T. Ball and J. Postma)

The true scientists' involvement into holistic climate modeling shall be driven by our mission to support open and transparent empirical/testable investigations that uphold the traditional scientific method. Specifically, we (as true scientists) support the ethos of Professor Karl Popper in that a scientific theory must be falsifiable—submitted to the test of skepticism in open scientific discourse. Our premise on climate change is that our planet has a complex, chaotic, non-linear system whereby it is empirically demonstrable that CO_2 plays no measurable part. We perceive that it is the sun and external cosmic forces which are the principal drivers, while planetary effects, including gravitational changes impacting geothermal drivers, play a supplementary role. We participate in the project** in the interests of seeking a unified theory of climate, which may be decades away, but which bears little, if any, resemblance to the fake climate models promoted by the UN IPCC.

** The holistic world formula approach for most comprehensive climate modeling, thereby strictly avoiding the typical biased and very often purely parasitically motivated pre-assumptions of the current and rather non-scientific climate apologetics.

For more information see: http://principia-scientific.org/.

1. N. Schwarzer, *The Relativistic Quantum Bible: Genesis and Revelation*, www.amazon.com, ASIN: B01M1CJH1B.
2. N. Schwarzer, God created an extremal universe! But why?, online on www.world formulaapps.com.
3. N. Schwarzer, *Worldformula*, www.amazon.com, ISBN: 9781673032567.
4. N. Schwarzer, How can we measure the size of a thought, online on www.world formulaapps.com.
5. N. Schwarzer, What is the ultimate good, online on www.world formulaapps.com.
6. P. Frank, Propagation of error and reliability of global air temperature projection, *Front. Earth Sci.*, 2019, **7**, 223, https://doi.org/10.3389/feart.2019.00223 or www.frontiersin.org/articles/10.3389/feart.2019.00223/full.
7. www.quora.com/What-does-Michael-Mann-s-court-battle-loss-mean-to-the-notion-of-climate-change.
8. N. Schwarzer, *ANTI-GENDER - Mathematical Proof Against One of the Most Stupid and Dangerous Ideologies, Disguising Itself as "Science" or Sex, Gender and Why We – Only – Have Two Natural Sexes*, www.amazon.de, B07KPD9BD4.
9. N. Schwarzer, *The Theory of Everything: Quantum and Relativity Is Everywhere – A Fermat Universe*, Jenny Stanford Publishing, 2020, ISBN-10: 9814774472.
10. N. Schwarzer, *Einstein Had It, but He Did Not See It, Part XIII: Why Equality Is Always Hostile to Life*, www.amazon.com, ASIN: B07B6QZLMH.

17.6 Outlook toward Artificial Intelligence Applications When Competing against a Top-Down Quantum Gravity Approach

T. vom Braucke and N. Schwarzer

17.6.1 Abstract

"Algorithms", knowing "why they do what they do" or their decision-making process and what uncertainties and constraints they took into consideration to achieve it, are a mystery to users unless the code and the data are completely open source and explainable. The situation is even more problematic if the software engineer is leveraging self-learning code with high-complexity data. And consider the scenario of an engineer who designs a bridge that collapses killing hundreds of people. Around the world, that engineer and the engineering firm are held legally accountable. What if a self-driving car decides to veer out of the way into the oncoming traffic for a "dog"-shaped balloon blowing in front of it knowing it can swerve back into its lane just in time. How will the human driver coming from the other direction react to such poor judgment? Might they panic and have an accident? Will the software engineer(s) writing the lines of machine learning code for the self-driving car "unintentionally causing" the accident be liable? Will the self-driving algorithm even know what it just caused and—due to this ignorance—therefore not even learn from the experience?

17.6.2 Key Problems of Artificial Intelligence

- Algorithms are often implemented without simple methods to find and address mistakes.
- Artificial intelligence encodes and magnifies bias by programmers not considering all degrees of freedom.
- Optimizing narrow metrics above everything else can lead to negative outcomes.
- There is no accountability for big tech companies or the software engineers.

17.6.3 Solving the Problems

Machine learning (ML) being a bottom-up approach of learning "principles" to improve decision making will always lead to increasing complexity and lack a true estimation of all uncertainties due to the algorithmic nature of the approaches used. Neural network and deep learning architectures are restricted to this bottom-up approach by the very fact that they need to learn as our human brains do, to solve problems based on experience, because they do not know the unknown-unknowns. The additional problem of the ML approach is that the learned principles are hidden from us, thereby making the reuse and broad applicability of the learned problem set challenging to adapt to completely different problems without learning of completely new data sets. These data sets require great resources to effectively collect and inevitably will have "in-complete" and biased data, thus limiting them to massive network effects only available via the Internet and connected devices. Importantly, the tools to determine bias are limited and the challenge is to assess both the algorithm and the data sets.

The typical solution is to add additional algorithms to try and remove bias from trained ML solutions, and improve predictability (a bottom-up approach by adding complexity to the model, also true for the Monte Carlo methodology or just "trial and error"). However, with a top-down approach, there is no mathematical bias introduced as all possibilities (let us let bias equal "degrees of freedom") and uncertainties are already included within the world formula. Now we can determine the degrees of freedom available within the existing data sets and compare with the maximum universal uncertainties to discover, if, in fact, the data already contains any missing degrees of freedom. There is no "training" of an algorithm in this case in the classical sense, only determining the "completeness" of the data and calculating with a holistic uncertainty that accounts for all degrees of freedom. The system trains with respect to the uncertainties rather than with respect to the whole system, which—of course—is much more efficient. In a case with biased data, the potential result is an outcome where the uncertainty or the error is larger, due to missing degrees of freedom and/or data within a dimension, than the accuracy required to recommend a solution. At least in this case, we know what the calculation is doing and why at all times, even if a "good" decision cannot be made, we at least have the opportunity to knowingly accept the risk (whether or not this is made transparent or accepted by the decision maker).

By using the top-down (world formula) approach as introduced and discussed within this book, we relegate inefficient machine learning–based algorithmic (bottom-up) approaches to the applications where data is already existing, well tested, readily accessible, and of low societal importance. Constraints on solutions can be pre-determined to meet design, safety, legal, societal and cultural values where the opportunity to review solutions falling outside of those constraints can still be accessed to see if there are better ways to make decisions that may be outside of current societal or cultural paradigms [1–4]. Such methods also lend themselves to socio-economic models and constraints may be further optimized by leveraging the math of "What Is the Ultimate Good" [3, 4]. Combined with the knowledge about the role of entanglement as THE potential source of intelligence [5], one ends up in much more profound and fundamental building grounds for both soft- and hardware solutions. Such math allows a focus on a fundamental principle of "growth" or the so-called "S" curves, while the potential downsides are minimized.

1. N. Schwarzer, *Humanized Artificial Intelligence: Spiritual Computers - Make Them Believe and They'll Start to Think*, www.amazon.com/dp/B072MNRLJP.
2. N. Schwarzer, *Philosophical Engineering, Part 1: The Honest Non-Parasitic Philosopher and the Universal "GOOD" Derived from a Theory of Everything*, www.amazon.com, ASIN: B07KNWRDYW.
3. N. Schwarzer, *Einstein Had It, but He Did Not See It, Part LXVIII: Most Fundamental Tools for Optimum Decision-Making Based on Quantum Gravity*, www.amazon.com, ASIN: B07KDFDZVZ.
4. N. Schwarzer, *Einstein Had It, but He Did Not See It, Part LXIX: The Hippocratic Oath in Mathematical Form and Why – So Often – It Will Be of No Use*, www.amazon.com, ASIN: B07KDSMNSK.
5. N. Schwarzer, *Is There an Ultimate, Truly Fundamental and Universal Computer Machine?*, www.amazon.com, ASIN: B07V52RB2F.

17.7 "Speed-of-Light" Computational Power with Quantum Gravity–Based Computing

T. vom Braucke and N. Schwarzer

The next generation of computing based on the "world formula" approach that can compute all possible solutions, including all uncertainty, is now upon us and addresses two major roadblocks of current computing methods for problem solving. First, the classic binary computers require enormous computing power (time and energy) for problems with complex degrees of freedom, uncertainties and multiple solutions because of their linear calculation approach. Second, the quantum computing approach promises better performance in parallel computing compared to the classical approach when searching for multiple solutions but is slower for more basic calculations that a traditional computer can do faster. Additionally, quantum computers suffer from decoherence leading to expensive "room-sized" computers needing liquid helium to ensure that the accuracy and reliability of the quantum bits

(Qbits) is maintained. These issues can be overcome with the world formula approach by changing the structural and computational approach.

The quantum gravity computer solution, or the Einstein-Quantum (EQ) computer as we like to call it [1–4], provides for several orders of magnitude computing power increase over both traditional "One or Zero" bit computers and classical quantum computers (where a Qbit represents both One and Zero). The EQ computer will do this because it replaces such degenerated computational hardware architecture with a function such as $f = G(x)$. The challenge to understand here is that this is a hardware solution that until now only software could do, i.e., the hardware behaves like a true analogue solution. Furthermore, the function operates as the "world formula" thereby providing the "theoretical" power of a universe in every Einstein-Qbit (EQbit). Start adding several of these EQbits together and the potential impact on problem solving and applications will be nothing short of disruptive.

There are challenges with building the computer described, not the least of which is that it still requires cooling. But by leveraging the world formula, stable quantum coherence materials solutions have been theoretically determined for current scalable manufacturing technologies that will operate at liquid nitrogen (LN_2) temperatures. LN_2 is a much simpler implementation than liquid helium, and since the 1960s it is readily available and common in many universities and industrial materials science labs. Additionally, for the hardware function to work, the 3-dimensional architecture of the bits and their assembly requires a specific design to be translated to a fabrication process. However, these fabrication techniques exist [5–7]; rather, achieving the suitable coherence architecture to allow the function to operate is critical. This is where a deep understanding of fundamental quantum gravity AND materials science must be leveraged to build this capability.

This computing power will lead to a domination of the scientific and business computing market and applied applications for the corporations and countries that can achieve this early and in an intelligently applied way [Paul-Smith Goodson, *Forbes*, October 10, 2019]. The top-down approach to problem solving will relegate the current machine learning (ML)-based algorithmic bottom-up approaches with current computing methods to lesser competitors. Further, this novel top-down approach will allow hugely complex problems to be solved with significant efficiency compared to any current alternative.

This is not to say that ML will not also be used with the top-down approach, but in this case, we know at all times what it is doing and why. Rather the machine learning approach can leverage the "world formula" to improve the determination of the uncertainty remaining in human responses to automated questions (i.e., chat bots) and then make multi-degree of freedom optimization of the responses to further the conversation. This is important because it allows the interview of the user (i.e., subject matter expert) to determine the right questions to ask to assess the degrees of freedom, data type and quality to feed the "world formula" parameter inputs and then to determine constraints from the user to find the optimal decisions or solutions. That is, it provides for the software operating system of the EQ computer, and it raises the question if we may be at the precipice of another fundamental milestone described by Ray Kurzweil [8].

In Summary

(1) The gravity, or "Einstein Quantum Computer" (EQ computer), contains both the classical quantum and the classical digital computers "which are the degenerated forms of a true Turing computer" such that all calculation types are possible in the same architecture.

(2) Based on our theoretical understanding of quantum gravity, we have found several feasible materials solution pathways to solve the high-temperature coherence stability problem. This development will accelerate with successive hardware builds leveraging EQ-bit computational power.

(3) This stable coherence concept is feasible operating at LN_2 temperatures using currently scalable high-volume manufacturing technologies allowing them to be placed in businesses and universities in a small desktop form factor.

(4) Being both QED + QCD (Quantum Electro-Dynamics (QED) and Quantum Chromo-Dynamics (QCD)) combined, its computational development will be speed of light limited rather than speed of sound limited for all the current QED-only approaches to quantum computing. And being a top-down computational approach to problem solving, server-based capability together with computational and energy efficiencies will be significant to several orders of magnitude.

(5) Leveraging the world formula for uncertainty and decision-making capability [4] as a template to both structure and operate each EQ-bit, then each of these bits may have the theoretical calculation power of a

universe, which raises the question: Being a "general solution" template, and combining with an artificial intelligence approach, are we peering at the event horizon of a singularity (or intelligence) as proposed by Ray Kurtzweil?

1. N. Schwarzer, *Einstein Had It, but He Did Not See It, Part XXXIX: EQ or the Einstein Quantum Computer*, www.amazon.com, ASIN: B07D9MBRS3.
2. N. Schwarzer, *The Einstein Quantum Computer: Mathematical Principle and Transition to the Classical Discrete and Quantum Computer Design*, www.amazon.com, ASIN: B07D9J5VLV.
3. N. Schwarzer, *Is There an Ultimate, Truly Fundamental and Universal Computer Machine?*, www.amazon.com, ASIN: B07V52RB2F.
4. N. Schwarzer, *Einstein Had It, but He Did Not See It, Part LXVIII: Most Fundamental Tools for Optimum Decision-Making Based on Quantum Gravity*, www.amazon.com, ASIN: B07KDFDZVZ.
5. B. Hensen, W. Wei Huang, C., Yang, et al., A silicon quantum-dot-coupled nuclear spin qubit, *Nat. Nanotechnol.*, 2020, **15**, 13–17, https://doi.org/10.1038/s41565-019-0587-7.
6. W. Huang, C. H. Yang, K. W. Chan, et al., Fidelity benchmarks for two-qubit gates in silicon, *Nature*, 2019, **569**, 532–536, https://doi.org/10.1038/s41586-019-1197-0.
7. M. D. Kim, D. Shin, J. Hong, *Physica E*, 2003, **18**, 45.
8. R. Kurzweil, The singularity is near, in R. L. Sandler, ed., *Ethics and Emerging Technologies*, Palgrave Macmillan, London, https://doi.org/10.1057/9781137349088_26.

17.8 Toward Top-Down Market Analysis and Guidance Using a Quantum Gravity Approach

T. vom Braucke and N. Schwarzer

We discuss the proposed elastic field analogy toward macro socio-economic markets and how the global market growth could be better understood and harnessed by a holistic, quantum gravity–based approach with subsequent uncertainty budget and decision optimization calculations. Consideration is also given to a market guidance system leveraging the fundamental mathematical principle of "What Is the Ultimate Good". The outcome would be predictable market growth cycles with harm-minimized downturns that are leveraged to drive the further development and evolution, similar to the S-curves commonly observed in many fields, not the least of which are innovation cycles. It should be pointed out thereby that the usual rather indifferent and very positive notion of the "waves of growth" is not necessarily seen as the optimum for the whole socio-economy. It could well be that certain waves are not leading to optimum developments, some perhaps the total opposite and that they are "better left out" or even suppressed. After all, optimum decision making sometimes is also not to do certain things [1, 2].

Such a system has applications at the macro-, micro- and nano-scales of financial markets (i.e., it is a scale- invariant solution). However, the micro-scale of local national markets can be managed for stable growth, and on the nano-scale, individual businesses can leverage the same math to optimize their growth and risk exposure by effective holistic decision making that leverages the second law of thermodynamics.

Having derived this law from fundamental quantum gravity approaches [3], we also found the source for evolution and its driving forces [4] within self-organizing processes. Knowledge of these internal interactions will allow for a much better prediction making and control of socio-economic systems of any scale.

1. N. Schwarzer, *Einstein Had It, but He Did Not See It, Part LXVIII: Most Fundamental Tools for Optimum Decision-Making Based on Quantum Gravity*, www.amazon.com, ASIN: B07KDFDZVZ.
2. N. Schwarzer, *Quantum Gravity Thermodynamics: And It May Get Hotter*, www.amazon.com, ASIN: B07XC2JW7F.
3. N. Schwarzer, *Quantum Gravity Thermodynamics II: Derivation of the Second Law of Thermodynamics and the Metric Driving Force of Evolution*, www.amazon.com, ASIN: B07XWPXF3G.

17.9 The Virtual Patient

Hans Leuenberger
Professor emeritus, University of Basel, Switzerland, and College of Pharmacy, University of Florida, Lake Nona Campus, Orlando, USA

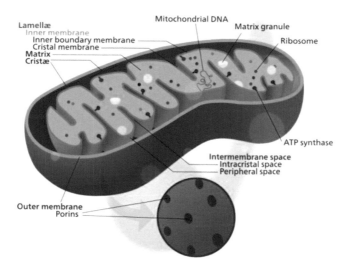

Figure 17.9.1 The mitochondrium, the power supply of a cell = microprocessor for doing the job.

The concept of the "Virtual Patient" and the "In-Silico Design of Solid Dosage Forms" are described in the publication "What Is Life?" in *SWISS PHARMA*, 41 (2019) Nr. 1, 20–36, and was first presented on November 27, 2018, at the College of Pharmacy in Gainesville, FL, see Videopoint of the Galenus Privatstiftung. The *SWISS PHARMA* article "What Is Life?" can be downloaded at www.ifiip.ch and was also published in Japanese in the November and December 2019 and the January 2020 issues of *PHARM TECH JAPAN*.

The concept of the virtual patient is based on the idea that each human cell is a microprocessor (Fig. 17.9.1) and the cells of an organ shall closely collaborate being able to perform coherent decisions and actions similar to a flock of birds using a non-chemical, "wireless" communication. According to the underlying concept described in the publication "What Is Life", the following axioms need to be taken into account for modeling the "Virtual Patient":

Axiom 1 (Prigogine) Far from equilibrium conditions exist favoring transformations from disorder into order leading to the creation of life: Chaos $\Longleftrightarrow$ Order.

Axiom 2 (Leuenberger) The same process is responsible for the "inorganic life" represented by the formation of beautiful highly ordered crystals in nature: Chaos $\Longrightarrow$ Order. It is important to realize that the same laws are governing the organic and the inorganic life!

Axiom 3 (Schrödinger): Life = Information = Software = Our Genetic Code.

Axiom 4 (Schrödinger/Prigogine): The human being is a living (super) computer leading to the conclusion that Life = Software and our Body = Hardware. For initializing organic and inorganic life, an open system and energy (see Fig. 17.9.1) are needed.

Axiom 5 (Fröhlich): The evolutionary process uses all existing physical laws of the present (imperfect) standard cosmological model to find a niche for a successful survival of the biological system!

Axiom 6 (Zwicky): The evolutionary process uses as well the yet unknown physical laws beyond the present standard cosmological model to find a niche for a successful survival of the biological system!

The emerging field of artificial intelligence and the availability of high-performance computers such as a quantum computer or its advanced version such as an Einstein Gravity Quantum Computer (see Fig. 17.9.2) will lead to the realization of the "Virtual Patient", which may even be visualized by projecting its hologram. This task will be a tremendous *transdisciplinary* challenge for the specialists in artificial intelligence. As a first step for the implementation of the audacious VIRTUAL PATIENT project, it will be a prerequisite to realize the VIRTUAL LAB concept, which will allow designing, developing and manufacturing in silico tablets containing the active ingredient, which will be orally administered to the VIRTUAL PATIENT. In this context, the recent results in modeling the tablet manufacturing process are promising. Thus, it will be possible to simulate a virtual human being who will serve as a virtual patient for discovering, developing and testing new therapies and drugs.

For the implementation of the VIRTUAL PATIENT project, the following points have to be considered in detail:

(1) The concept of "Virtual Patient" needs to be comprehensive, i.e., it needs to cover the basic human operating system governed by the autonomous nervous system (ANS), which includes to some extent the

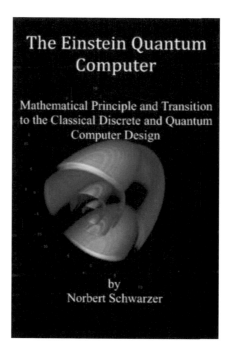

Figure 17.9.2 Cover picture of one of Schwarzer's books on gravity quantum computers [1–3].

enteric nervous system (ENS) and the somatic nervous system (SMS). These nervous systems which are part of the central and the peripheral nervous system (CNS, PNS) are wired and the interaction between the anticipated wireless communications of the cells is not yet fully known. It is evident that a chemical communication is present since application programming interfaces exist which are able to interact with the nervous system.

(2) In an optimal case, the concept of a "Virtual Patient" will be able to explain even the origin of the "placebo" effect or why the "acupuncture treatment" of pain is effective and presents a suitable alternative to a chemical treatment with a painkiller (according to the clinical studies of the Charité Hospital in Berlin).

(3) Unfortunately, most of the functions which control life are not governed by linear differential equations responsible for linear pharmacokinetics.

Thus, as soon as an enzyme function is involved, which can be saturated, the law of linear pharmacokinetics is no longer valid and is an exception.

(4) As a result, the classical top-down approach, following the Descartes principle of reductionism, is not working properly since the complexity of the system is eliminated. Thus, it is possible to resolve an isolated problem of a much more complex system. On the other hand, it is not possible to describe a complete system which is governed by the network of interactions among the isolated subunits.

(5) In this context, it is important to realize that a "bottom-up" concept of the Virtual Patient will allow a better understanding of the "human operating system" leading to a better, more comprehensive pharmacology. Thus, the complex whole system is more than the sum of its parts.

(6) Thus, for establishing the VIRTUAL PATIENT, a bottom-up approach is needed, which is different from the top-down method known as the reductionist approach. In fact, a living system differs from a conventional isolated one as follows: (a) Living systems often consist of closed-loop systems which are used for catalytic processes. Thus, the reaction product may enhance the process as in case of an autocatalytic process or may slow down the process. Such catalytic reactions cannot be described on the basis of linear differential equations; (b) the living system shows a high degree of complexity as a result of interacting processes leading to causal chains; (c) the processes take place in an open system by exchanging energy (energy transfer), mass (mass transfer) and transfer of information; (d) the processes are far from the thermodynamic equilibrium and may be irreversible.

(7) Together with the "Virtual Lab" project, the "Virtual Patient" will ***revolutionize the workflow*** of the pharmaceutical industry manufacturing ***proprietary drugs leading to important savings in time and money.***

1. N. Schwarzer, *Einstein Had It, but He Did Not See It, Part XXXIX: EQ or the Einstein Quantum Computer*, www.amazon.com, ASIN: B07D9MBRS3.
2. N. Schwarzer, *The Einstein Quantum Computer: Mathematical Principle and Transition to the Classical Discrete and Quantum Computer Design*, www.amazon.com, ASIN: B07D9J5VLV.
3. N. Schwarzer, *Is There an Ultimate, Truly Fundamental and Universal Computer Machine?*, www.amazon.com, ASIN: B07V52RB2F.

References

1. T. Bodan, *7 Days: How to Explain the World to My Dying Child; in German – 7 Tage: Wie erkläre ich meinem sterbenden Kind die Welt*, www.amazon.com, ASIN: 1520917562.

2. T. Bodan, *The Eighth Day – Holocaust and the World's Biggest Mysteries: The Other Final Solution – with Mathematical Elaborations* (English ed.), www.amazon.com, ASIN: B019M9ZHIE.

3. T. Bodan, *The Eighth Day*, www.amazon.com, ASIN: B015R1JPZ2.

4. N. Schwarzer, T. Bodan, *Sherlock, Watson, Einstein, Part 1: The Mystery of Entanglement and the Spooky Action at a Distance*, www.amazon.com, ASIN: B079Z92GGM.

5. N. Schwarzer, T. Bodan, *Sherlock, Watson, Stalin, Part 2: The Hell of Gender, Merkel, Communism and the Dictatorship of Parasites*, www.amazon.com, ASIN: B07BJ9PZWM.

6. T. Bodan, *EU vs. Britain: The Other Monkey Trial*, www.amazon.com, ASIN: 1521304793.

7. T. Bodan, *Grexit, Eurokrise und Flüchtlinge – Vom dummverkauften Souverän zum mitbestimmenden Bürger – Oder: Warum unsere Politiker versagen müssen ja sogar versagen wollen*, www.amazon.com, ASIN: B0117KWE8W.

8. N. Schwarzer, *The Theory of Everything: Quantum and Relativity Is Everywhere – A Fermat Universe*, Jenny Stanford Publishing, 2020, ISBN-10: 9814774472.

9. N. Schwarzer, *How Einstein Gives Dirac, Klein-Gordon and Schrödinger*, www.amazon.com, ASIN: B071K2Y4V2.

10. N. Schwarzer, *Our Universe, Nothing but an Intelligent Zero?: The Dark Lord's Zero-Sum- & God's Non-Zero-Sum-Game*, www.amazon.com/dp/B072J9H1BY.

11. N. Schwarzer, *Quantized GTR: Understanding the "Friedmanns"?*, www.amazon.com/dp/B01BQUTKG2.

12. N. Schwarzer, *Quantum Gravity Monster*, www.amazon.com/dp/B072N67RSN.

13. N. Schwarzer, *Entanglement: Einstein-Rosen-Bridges = Einstein-Podolsky-Rosen-Joints?*, www.amazon.com/dp/B071WSKGT9.

14. N. Schwarzer, *Particles: Spin, Quark Confinement, the Three Generation-Problem and Matter Antimatter Asymmetry*, www.amazon.com/dp/B06ZZ4LB5J.

15. N. Schwarzer, *Time, Mass and Higgs: How the Three Things Emerge from the Sublevel in a Theory of Everything*, www.amazon.com/dp/B06Y41F8NS.

16. N. Schwarzer, *Understanding Time*, www.amazon.com/dp/B01FNZUVT6.

17. N. Schwarzer, *Understanding the Electron?, Part I*, www.amazon.com, ASIN: B01CCRU6Q6, www.amazon.com/dp/B01CCRU6Q6.

18. N. Schwarzer, *General Quantization of Smooth Spaces: From ℏ to Plancktensor & Planckfunction*, www.amazon.com/dp/B01N9UGFX2.

19. N. Schwarzer, *Understanding the Electron II*, www.amazon.com/dp/B01N7QUTUL.

20. N. Schwarzer, *The Photon*, www.amazon.com/dp/B06XGC4NDM.

21. N. Schwarzer, *General Quantum Relativity*, www.amazon.com/dp/B01FG5RC0E.

22. N. Schwarzer, *Recipe to Quantize the General Theory of Relativity*, www.amazon.com/dp/B01FNZUVT6.

23. N. Schwarzer, *The Einstein Oscillator in 1D*, www.amazon.com/dp/B01N5UG8RJ.

24. N. Schwarzer, *Quantized Schwarzschild*, www.amazon.com/dp/B01N7YT6OF.

25. N. Schwarzer, *The Cosmic Atom & Quantum Friedmann Equations: The Quantum States of Friedmann–Lemaître–Robertson–Walker (FLRW) Universes and the 4 Friedmann Equations*, www.amazon.com/dp/B07354DZFK.

26. N. Schwarzer, *Quantum Tribology, Part I: Theory*, www.amazon.com/dp/B01CI4BI2E.

27. T. Bodan, N. Schwarzer, *Quantum Economy*, www.amazon.com/dp/B01N80I0NG.

28. N. Schwarzer, *Quantized Relativized Theology: Where Is God?*, www.amazon.com/dp/B01M0XPXTT.

29. N. Schwarzer, *Humanized Artificial Intelligence: Spiritual Computers – Make Them Believe and They'll Start to Think*, www.amazon.com/dp/B072MNRLJP.

30. N. Schwarzer, *Einstein Already Had It, but He Did Not See It, Part 0: The Discarded Term from the Einstein-Hilbert-Action*, www.amazon.com/dp/B074FZBXDN.

31. N. Schwarzer, *Einstein Had It, but He Did Not See It, Part I: How the Quantum Theory Resides in the Einstein-Hilbert-Action*, www.amazon.com/dp/B074JDX36V.

32. N. Schwarzer, *Einstein Had It, but He Did Not See It, Part II: String Cosmology*, www.amazon.com/dp/B074KCNZ4T.

33. N. Schwarzer, *How Can We Measure the Size of a Thought*, self-published, Amazon Digital
Services, 2020, Kindle (short version on: www.worldformulaapps.com/wp-content/uploads/2019/12/What-is-the-Size-of-a-Thought.pdf).

34. N. Schwarzer, *Einstein Had It, but He Did Not See It, Part III: The Impossible Black-Hole Singularity*, www.amazon.com/dp/B074LV1RPD.

35. N. Schwarzer, *Einstein Had It, but He Did Not See It, Part IV: Sixty e-Foldings*, www.amazon.com/dp/B0751H2BF8.

36. N. Schwarzer, *Einstein Had It, but He Did Not See It, Part V: Amendment to the Heisenberg Uncertainty Principle*, www.amazon.com/dp/B074MB3J3S.

37. N. Schwarzer, *Einstein Had It, but He Did Not See It, Part VI: The Photon and the Flat Space*, www.amazon.com/dp/B074Y88N4K.

38. N. Schwarzer, *Einstein Had It, but He Did Not See It, Part VII: Konrad Zuse's Computing Universe*, www.amazon.com/dp/B0752Z99DL.

39. N. Schwarzer, *Einstein Had It, but He Did Not See It, Part VIII: Derivation of the Speed of Light and Other Funny Limits*, www.amazon.com/dp/B0753WS3LK.

40. N. Schwarzer, *Einstein Had It, but He Did Not See It, Part IX: Fermat's Last Theorem inside the Einstein–Hilbert-Action?*, www.amazon.com/dp/B0753 Q1NFV.

41. X—not published for safety reasons.

42. N. Schwarzer, *Einstein Had It, but He Did Not See It, Part XI: Behind the Event Horizon*, www.amazon.com/dp/B075KQZ27L.

43. XII—not published for safety reasons.

44. N. Schwarzer, *Einstein Had It, but He Did Not See It, Part XIII: Why Equality Is Always Hostile to Life*, www.amazon.com, ASIN: B07B6QZLMH.

45. N. Schwarzer, *Einstein Had It, but He Did Not See It, Part XIV: Braneworks or a Brany Universe*, www.amazon.com/dp/B077BVSFVY.

46. N. Schwarzer, *Einstein Had It, but He Did Not See It, Part XV: The Genetic String-Code of the Universe*, www.amazon.com/dp/B0776YYFN1.

47. N. Schwarzer, *Einstein Had It, but He Did Not See It, Part XVI: An Inner Kerr Solution*, www.amazon.com/dp/B077DD56GS.

48. N. Schwarzer, *Einstein Had It, but He Did Not See It, Part XVII: A Funny Form of Vacuum*, www.amazon.com/dp/B077LVZBQC.

49. N. Schwarzer, *Einstein Had It, but He Did Not See It, Part XVIII: A Few Matter Solutions*, www.amazon.com/dp/B077S51Q91.

50. N. Schwarzer, *Einstein Had It, but He Did Not See It, Part XIX: The Cosmologic Neutralizer*, www.amazon.com/dp/B077VY4JQV.

51. N. Schwarzer, *Einstein Had It, but He Did Not See It, Part XX: Higher Order Covariant Variation of the Einstein–Hilbert-Action*, www.amazon.com/dp/B0788VWKD4.

52. N. Schwarzer, *Einstein Had It, but He Did Not See It, Part XXI: A Very Simple Theory of Everything*, www.amazon.com/dp/B078QTVKJS.

53. N. Schwarzer, *Einstein Had It, but He Did Not See It, Part XXII: Finis Fabulae de Foraminibus Atris – Recipe to Evaluate Black Hole Matter*, www.amazon.com/dp/B078RVJTCW .

54. N. Schwarzer, *Einstein Had It, but He Did Not See It, Part XXIII: A Bit Cosmology*, www.amazon.com/dp/B078VRLJSS.

55. N. Schwarzer, *Einstein Had It, but He Did Not See It, Part XXIV: A Variety of Solutions and the Dirac-Schwarzschild-Particle*, www.amazon.com/dp/B0796GZX8R.

56. N. Schwarzer, *Einstein Had It, but He Did Not See It, Part XXV: The Vacuum Friedmann Cosmos and the Origin of a Quantal World*, www.amazon.com/dp/B078VRLJSS.

57. N. Schwarzer, *Einstein Had It, but He Did Not See It, Part XXVI: The Nature of SPIN 1/2*, www.amazon.com/dp/B0798R3P8D.

58. N. Schwarzer, *Einstein Had It, but He Did Not See It, Part XXVII: The Quantum Transformer Always Already Was inside the General Theory of Relativity*, www.amazon.com/dp/B079LV64C3.

59. N. Schwarzer, *Einstein Had It, but He Did Not See It, Part XXVIII: ¿ Anti Gravity ?*, www.amazon.com/dp/B079M1VG62.

60. N. Schwarzer, *Einstein Had It, but He Did Not See It, Part XXIX: Space Strain Invariants and the Evaluation of Mass*, www.amazon.com/dp/B079QS8G1Z.

61. N. Schwarzer, *Einstein Had It, but He Did Not See It, Part XXX: ¿ Neutrino Oscillations Directly Evaluated from the Einstein-Field-Equations ?*, www.amazon.com/dp/B079RMH7LF.

62. N. Schwarzer, *Einstein Had It, but He Did Not See It, Part XXXI: A Cosmologic Pairwise Entanglement of Dimensions and the Holographic Principle*, www.amazon.com, ASIN: B07B5GBQ4N.

63. N. Schwarzer, *Einstein Had It, but He Did Not See It, Part XXXII: Enough to Play With – More Quantum Friedmann Solutions*, www.amazon.com, ASIN: B07BVH69RH.

64. N. Schwarzer, *Einstein Had It …Part XXXIII: Elementary Particle Universes (Without Spin and Shear Components)*, www.amazon.com, ASIN: B07CGFHBC2.

65. N. Schwarzer, *Einstein Had It, but He Did Not See It, Part XXXIV: Is This the Electron?*, www.amazon.com, ASIN: B07CS6F4S9.

66. N. Schwarzer, *Einstein Had It …Part XXXV: The 2-Body Problem and the GTR-Origin of Mass*, www.amazon.com, ASIN: B07CWP3V1S.

67. N. Schwarzer, *Einstein Had It …Part XXXVI: The Classical and Principle Misinterpretation of the Einstein-Field-Equations AND How It Might Be Done Correctly*, www.amazon.com, ASIN: B07D68G9M9.

68. N. Schwarzer, *Einstein Had It, but He Did Not See It, Part XXXVII: Most Simple Singularity-Free "Black Holes,"* www.amazon.com, ASIN: B07D6TFYNT.

69. N. Schwarzer, *Einstein Had It, but He Did Not See It, Part XXXVIII: The Einstein Hydrogen Atom*, www.amazon.com, ASIN: B07D7VDM31.

70. N. Schwarzer, *Einstein Had It, but He Did Not See It, Part XXXIX: EQ or The Einstein Quantum Computer*, www.amazon.com, ASIN: B07D9MBRS3.

71. N. Schwarzer, *The Einstein Quantum Computer: Mathematical Principle and Transition to the Classical Discrete and Quantum Computer Design*, www.amazon.com, ASIN: B07D9J5VLV.

72. N. Schwarzer, *Einstein Had It, but He Did Not See It, Part XL: The Einstein Hydrogen Atom with "Spin,"* www.amazon.com, ASIN: B07DG1N3JD.

73. N. Schwarzer, *Einstein Had It, but He Did Not See It, Part XLI: Generalized Rotating "Black Holes,"* www.amazon.com, ASIN: B07DCCDQ2X.

74. N. Schwarzer, *Einstein Had It, but He Did Not See It, Part XLII: Simplest Way to Quantize the General Theory of Relativity*, www.amazon.com, ASIN: B07DCXLC2J.

75. N. Schwarzer, *Einstein Had It, but He Did Not See It, Part XLIII: A Selection of Einstein–Dirac-Particles*, www.amazon.com, ASIN: B07DPVRZ8B.

76. N. Schwarzer, *Einstein Had It, but He Did Not See It, Part XLIV: Fundamental Normalizable Solutions and Their Limits towards Classical Metrics*, www.amazon.com, ASIN: B07DTJZ8YY.

77. N. Schwarzer, *Einstein Had It, but He Did Not See It, Part XLV: Inner and Outer Solutions to Schwarzschild and Flat Space*, www.amazon.com, ASIN: B07DVPLS55.

78. N. Schwarzer, *Einstein Had It, but He Did Not See It, Part XLVI: 3 and More Inner Solutions & Three Generations of Particles?*, www.amazon.com, ASIN: B07DZZQ92M.

79. N. Schwarzer, *Einstein Had It, but He Did Not See It, Part XLVII: Towards A Pauli Exclusion Principle Derived from the Einstein-Field-Equations*, www.amazon.com, ASIN: B07F141N26.

80. N. Schwarzer, *Einstein Had It…Part XLVIII: Agapornis, Particles One Cannot Separate*, www.amazon.com, ASIN: B07F6G6BXW.

81. N. Schwarzer, *Einstein Had It, but He Did Not See It, Part XLIX: Shear as Origin of the Higgs Field*, www.amazon.com, ASIN: B07F6G4M5N.

82. N. Schwarzer, *Einstein Had It, but He Did Not See It, Part L: Big Bang Inflation Black Hole Pregnant with Universe*, www.amazon.com, ASIN: B07FJMQV7Z.

83. N. Schwarzer, *Einstein Had It, but He Did Not See It, Part LI: Why Space-Time Might Be Discrete*, www.amazon.com, ASIN: B07G89L71J.

84. N. Schwarzer, *Einstein Had It, but He Did Not See It, Part LII: Why We Might Live inside a Black Hole*, www.amazon.com, ASIN: B07GBHLY8Z.

85. N. Schwarzer, *Einstein Had It, but He Did Not See It, Part LIII: They Are Everywhere! Why There Are So Many Sigmoid-Dependencies in This World*, www.amazon.com, ASIN: B07G8F9X6T.

86. N. Schwarzer, *Einstein Had It, but He Did Not See It, Part LIV: Mathematical Philosophy & Quantum Gravity Ethic*, www.amazon.com, ASIN: B07GR 994HT.

87. N. Schwarzer, *Einstein Had It, but He Did Not See It, Part LV: The Fundamental Connection of Quantum Theory and General Theory of Relativity*, www.amazon.com, ASIN: B07H7P5STL.

88. N. Schwarzer, *Einstein Had It, but He Did Not See It, Part LVI: What Is Mass? What Is Energy?*, www.amazon.com, ASIN: B07HLD9N1B.

89. N. Schwarzer, *Einstein Already Had It, but He Did Not See It, Part LVII: The Schwarzschild-Hydrogen Atom*, www.amazon.com, ASIN: B07HMDS3YZ.

90. N. Schwarzer, *Einstein Already Had It, but He Did Not See It, Part LVIII: The Most Simple Nature of This Funny Phenomenon Called "Spin,"* www.amazon.com, ASIN: B07HP5D6XD.

91. N. Schwarzer, *Einstein Already Had It, but He Did Not See It, Part LIX: A Brief Story about the Matter Antimatter Asymmetry*, www.amazon.com, ASIN: B07HRH8Q8X.

92. N. Schwarzer, *Einstein Had It, but He Did Not See It, Part LX: The Hitch Hiker's Guide to the Creation of a Universe*, www.amazon.com, ASIN: B07HRLLMFS.

93. N. Schwarzer, *Einstein Had It, but He Did Not See It, Part LXI: Don't Panic in a Universe with Only One Dimension*, www.amazon.com, ASIN: B07HZ1ZV8R.

94. N. Schwarzer, *Einstein Had It, but He Did Not See It, Part LXII: The Quantization of Schwarzschild and FLRW-Metrics*, www.amazon.com, ASIN: B07J4C64T2.

95. N. Schwarzer, *Einstein Already Had It, but He Did Not See It, Part LXIII: Einstein-Field-Equations = Dirac^2 (+) Klein-Gordon*, www.amazon.com, ASIN: B07JVFV4HP.

96. N. Schwarzer, *Einstein Already Had It, but He Did Not See It, Part LXIV: Origin of the 6 Infinity Stones*, www.amazon.com, ASIN: B07JGB7WKG.

97. N. Schwarzer, *Einstein Already Had It, but He Did Not See It, Part LXV: Swing When You're Singing*, www.amazon.com, ASIN: B07JFMMVKQ.

98. N. Schwarzer, *Einstein Already Had It, but He Did Not See It, Part LXVI: The Other Time*, www.amazon.com, ASIN: B07JZF34RG.

99. N. Schwarzer, *Einstein Already Had It, but He Did Not See It, Part LXVII: Simplest Metric Quantum Operators*, www.amazon.com, ASIN: B07K1FMHBW.

100. N. Schwarzer, *Einstein Had It, but He Did Not See It, Part LXVIII: Most Fundamental Tools for Optimum Decision-Making Based on Quantum Gravity*, www.amazon.com, ASIN: B07KDFDZVZ.

101. N. Schwarzer, *Einstein Had It, but He Did Not See It, Part LXIX: The Hippocratic Oath in Mathematical Form and Why – So Often – It Will Be of No Use*, www.amazon.com, ASIN: B07KDSMNSK.

102. N. Schwarzer, *Einstein Had It, but He Did Not See It, Part LXX: Extended Variation of the Einstein–Hilbert-Action results in Quantum Theory*, www.amazon.com, ASIN: B07LCS6PVM.

103. N. Schwarzer, *Einstein Had It, but He Did Not See It, Part LXXI: The 3-Generation Problem, Spin, Neutrinos, Electron, Muon, Tauon and the Photon from Fundamental Geometric Principles*, www.amazon.com, ASIN: B07M9FDCK1.

104. N. Schwarzer, *Einstein Had It, but He Did Not See It, Part LXXII: Connection between Schwarzschild Radius and Dirac Rest Mass*, www.amazon.com, ASIN: B07ML1S8F2.

105. N. Schwarzer, *Philosophical Engineering, Part 1: The Honest Non-Parasitic Philosopher and the Universal "GOOD" Derived from a Theory of Everything*, www.amazon.com, ASIN: B07KNWRDYW.

106. N. Schwarzer, *Einstein Had It, but He Did Not See It, Part LXXIII: Wave-Particle Duality as a Direct Outcome of the General Theory of Relativity*, www.amazon.com, ASIN: B07MW7HZ81.

107. N. Schwarzer, *Einstein Had It, but He Did Not See It, Part LXXIV: World formulae-Approaches of Various Orders*, www.amazon.com, ASIN: B07MYN331H.

108. N. Schwarzer, *Einstein Had It, but He Did Not See It, Part LXXV: The Metric Creation of Matter*, www.amazon.com, ASIN: B07ND3LWZJ.

109. N. Schwarzer, *Einstein Had It, but He Did Not See It, Part LXXVI: Quantum Universes – We don't Need No…an Inflation*, www.amazon.com, ASIN: B07NLH3JJV.

110. N. Schwarzer, *Einstein Had It, but He Did Not See It, Part LXXVII: Matter Is Nothing and so Nothing Matters*, www.amazon.com, ASIN: B07NQKKC31.

111. N. Schwarzer, *Einstein Had It, but He Did Not See It, Part LXXVIII: About a Possible Quantum Gravity Alternative to Schwarzschild and Co.*, www.amazon.com, ASIN: B07NXRL2BH.

112. N. Schwarzer, *Einstein Had It, but He Did Not See It, Part LXXIX: Dark Matter Options*, www.amazon.com, ASIN: B07PDMH2JB.

113. N. Schwarzer, *Einstein Had It, but He Did Not See It, Part LXXX: Short Note on the Killing of Dirac*, www.amazon.com, ASIN: B07NKZVF61.

114. N. Schwarzer, *Einstein Had It, but He Did Not See It, Part LXXXI: More Dirac Killing*, www.amazon.com, ASIN: B07WW1G6N7.

115. N. Schwarzer, *Einstein Had It, but He Did Not See It, Part LXXXII: Half Spin Hydrogen*, www.amazon.com, ASIN: B07Q3NFB39.

116. N. Schwarzer, *Einstein Had It, but He Did Not See It, Part LXXXIII: Quantum Relativity – The Two Sides of the Same Medal*, www.amazon.com, ASIN: B07TJQ9BGD.

117. N. Schwarzer, *Einstein Had It, but He Did Not See It, Part LXXXIV: A World Formula(?)*, www.amazon.com, ASIN: B07WW1G6N7.

118. N. Schwarzer, *Science Riddles – Riddle No. 1: The Peculiar Quantum Schwarzschild Amplitude – A Coincidence?*, www.amazon.com, ASIN: B07QH7X5LK.

119. N. Schwarzer, *Science Riddles – Riddle No. 2: Quarks – Real or Just THE Greatest Deception the Universe Did to Mankind?*, www.amazon.com, ASIN: B07QKFG2JJ.

120. N. Schwarzer, *Science Riddles – Riddle No. 3: Was (IS) There a Big Bang Bomb?*, www.amazon.com, ASIN: B07QKMTBWM.

121. N. Schwarzer, *Science Riddles – Riddle No. 4: What Is the Fundamental Substance of a Physical Potential?*, www.amazon.com, ASIN: B07R4BNPL7.

122. N. Schwarzer, *Science Riddles – Riddle No. 5: What Makes the Quantum Equations So Linear?*, www.amazon.com, ASIN: B07R6HG4SV.

123. N. Schwarzer, *Science Riddles – Riddle No. 6: What Is Time?*, www.amazon.com, ASIN: B07R5JMSPG.

124. N. Schwarzer, *Science Riddles – Riddle No. 7: Is the Theory of Everything THE Most Potent Patent Killing Machine?*, www.amazon.com, ASIN: B07RG6ZSK1.

125. N. Schwarzer, *Science Riddles – Riddle No. 8: Could the Schwarzschild Metric Contain Its Own Quantum Solution?*, www.amazon.com, ASIN: B07S2DCTBG.

126. N. Schwarzer, *Science Riddles – Riddle No. 9: …?*, to be published on www.amazon.com.

127. N. Schwarzer, *Science Riddles – Riddle No. 10: What, Where and Why Is the Higgs Field?*, www.amazon.com, ASIN: B07SPCQ2V2.

128. N. Schwarzer, *Science Riddles – Riddle No. 11: What Is Mass?*, www.amazon.com, ASIN: B07SSF1DFP.

129. N. Schwarzer, *Science Riddles – Riddle No. 12: Is There a Cosmological Spin?*, www.amazon.com, ASIN: B07T3WS7XK.

130. N. Schwarzer, *Science Riddles – Riddle No. 13: How to Solve the Flatness Problem?*, www.amazon.com, ASIN: B07T9WXZVH.

131. N. Schwarzer, *Science Riddles – Riddle No. 14: And What If Time…? A Paradigm Shift*, www.amazon.com, ASIN: B07VTMP2M8.

132. N. Schwarzer, *Science Riddles – Riddle No. 15: Is There an Absolute Scale?*, www.amazon.com, ASIN: B07V9F2124.

133. N. Schwarzer, *Science Riddles – Riddle No. 16: How to Understand the Dirac Equation?*, www.amazon.com, ASIN: B07VFW2Z3F.

134. N. Schwarzer, *Science Riddles – Riddle No. 17: How Einstein Becomes First Order and Goes Dirac - Can We Factorize the Einstein-Field-Equations?*, www.amazon.com, ASIN: B07VV9FG7K.

135. N. Schwarzer, *Science Riddles – Riddle No. 18: And What If Space...? A Paradigm Shift*, www.amazon.com, ASIN: B07W58DSQZ.

136. N. Schwarzer, *Science Riddles – Riddle No. 19: Is There a World Formula?*, www.amazon.com, ASIN: B07WRVHDWF.

137. D. Hilbert, Die Grundlagen der Physik, Teil 1, *Göttinger Nachrichten*, 1915, 395–407.

138. A. Einstein, Grundlage der allgemeinen Relativitätstheorie, *Ann. Phys.*, 1916, **49** (ser. 4), 769–822.

139. N. Schwarzer, *From Quantum Gravity to a Quantum Relative Material Science: A Very First Principle Material Science Concept*, www.amazon.com, ASIN: B07X8H9Y7Z.

140. P. W. Higgs, Broken symmetries and the masses of gauge bosons. *Phys. Rev. Lett.*, 1964, **13**, 508–509.

141. en.wikipedia.org/wiki/Higgs_boson.

142. K. Schwarzschild, Über das Gravitationsfeld einer Kugel aus inkompressibler Flüssigkeit nach der Einsteinschen Theorie [On the gravitational field of a ball of incompressible fluid following Einstein's theory], Sitzungsberichte der Königlich-Preussischen Akademie der Wissenschaften (in German), Berlin, 1916, 424–434.

143. P. A. M. Dirac, The quantum theory of the electron, *Proc. R. Soc. A*, 1928, **117**(778), doi: 10.1098/rspa.1928.0023.

144. P. S. Debnath, B. C. Paul, Cosmological models with variable gravitational and cosmological constants in R^2 Gravity, arxiv.org/pdf/gr-qc/0508031.pdf.

145. N. Schwarzer, *Is There an Ultimate, Truly Fundamental and Universal Computer Machine?*, www.amazon.com, ASIN: B07V52RB2F.

146. J. D. Bekenstein, Information in the holographic universe, *Sci. Am.*, 2003, **289**(2), 61.

147. J. D. Bekenstein, Black holes and entropy, *Phys. Rev. D*, 1973, **7**, 2333–2346.

148. N. Schwarzer, Short note on the effect of pressure induced increase of Young's modulus, *Philos. Mag.*, 2012, **92**(13), 1631–1648.

149. N. Schwarzer, Scale invariant mechanical surface optimization applying analytical time dependent contact mechanics for layered structures, in A. Tiwari, ed., S. Natarajan co-ed., *Applied Nanoindentation in Advanced Materials*, Chapter 22, 2017, ISBN: 978-1-119-08449-5, www.wiley.com/WileyCDA/WileyTitle/productCd-1119084490.html.

150. N. Schwarzer, From interatomic interaction potentials via Einstein field equation techniques to time dependent contact mechanics, *Mater. Res. Express*, 2014, **1**(1), 015042, http://dx.doi.org/10.1088/2053-1591/1/1/015042.

151. N. Schwarzer, Completely analytical tools for the next generation of surface and coating optimization, *Coatings*, 2014, **4**, 263–291, doi:10.3390/ coatings4020263.

152. A. E. Green, W. Zerna, *Theoretical Elasticity*, Oxford University Press, London, 1968.

153. H. Neuber, *Kerspannungslehre*, in German, 3rd ed., Springer-Verlag, Berlin, Heidelberg, New York, Tokyo, 1985, ISBN: 3-540-13558-8.

154. N. Schwarzer, Scale invariant mechanical surface optimization applying analytical time dependent contact mechanics for layered structures, in A. Tiwari, ed., S. Natarajan, co-ed., *Applied Nanoindentation in Advanced Materials*, Chapter 22, 2017, ISBN: 978-1-119-08449-5, www.wiley.com/WileyCDA/WileyTitle/productCd-1119084490.html.

155. S. Vogt, T. Greß, F. F. Neumayer, N. Schwarzer, A. Harris, W. Volk, Method for highly spatially resolved determination of residual stress by using nanoindentation, *Prod. Eng.*, 2019, **13**, 133–138, https://doi.org/10.1007/s11740-018-0857-5.

156. N. Schwarzer, *J. Phys. D: Appl. Phys.*, 2004, **37**, 2761–2772.

157. N. Schwarzer, Some basic equations for the next generation of surface testers solving the problem of pile-up, sink-in and making area-function-calibration obsolete, JMR Special Focus Issue on Indentation Methods in Advanced Materials Research, *J. Mater. Res.*, 2009, **24**(3), 1032–1036.

158. FilmDoctor, analytical software package for the analysis of complex mechanical problem for inhomogeneous materials, www.siomec.de/filmdoctor.

159. N. Schwarzer, *Quantum Gravity Thermodynamics: And It May Get Hotter*, www.amazon.com, ASIN: B07XC2JW7F.

160. N. Schwarzer, *Epistle to Elementary Particle Physicists: A Chance You Might Not Want to Miss*, www.amazon.com, ASIN: B07XDMLDQQ.

161. N. Schwarzer, *2nd Epistle to Elementary Particle Physicists: Brief Study about Gravity Field Solutions for Confined Particles*, www.amazon.com, ASIN: B07XN8GLXD.

162. N. Schwarzer, *3rd Epistle to Elementary Particle Physicists: Beyond the Standard Model – Metric Solutions for Neutrino, Electron, Quark*, www.amazon.com, ASIN: B07XJJ535T.

163. N. Schwarzer, *Quantum Gravity Thermodynamics II: Derivation of the Second Law of Thermodynamics and the Metric Driving Force of Evolution*, www.amazon.com, ASIN: B07XWPXF3G.

164. G. Nordström, On the energy of the gravitation field in Einstein's theory, *Koninklijke Nederlandse Akademie van Weteschappen Proceedings Series B Physical Sciences*, 1918, **20**, 1238–1245.

165. H. Reissner, Über die Eigengravitation des elektrischen Feldes nach der Einsteinschen Theorie, *Ann. Phys.*, 1916, **355**(9), 106–120.

166. W. Heisenberg, Über den anschaulichen Inhalt der quantentheoretischen Kinematik und Mechanik, *Z. Phys.* (in German), 1927, **43**(3–4), 172–198.

167. É. Cartan, Sur une généralisation de la notion de courbure de Riemann et les espaces à torsion, *C. R. Acad. Sci. (Paris)*, 1922, **174**, 593–595.

168. É. Cartan, Sur les variétés à connexion affine et la théorie de la relativité généralisée, Part I: *Ann. Éc. Norm.*, 1923, **40**, 325–412 and *ibid.*, 1924, **41**, 1–25; Part II: *ibid.*, 1925, **42**, 17–88.

169. Ch. Heinicke, Exact solutions in Einstein's theory and beyond, Dissertation, Universität Köln, 2005, urn:nbn:de:hbz:38-14637.

170. H. Haken, H. Chr. Wolf, *Atom- und Quantenphysik*, 4th ed. (in German), Springer Heidelberg, 1990, ISBN: 0-387-52198-4.

171. N. Schwarzer, *Einstein Had It, but He Did Not See It, Part LXXXV: In Conclusion*, www.amazon.com, ASIN: B07Y37LNRW.

172. N. Schwarzer, *Science Riddles – Riddle No. 20: Second Law of Thermodynamics – Where Is Its Fundamental Origin?*, www.amazon.com, ASIN: B07Y79BTT9.

173. H. Goenner, *Einführung in die spezielle und allgemeine Relativitätstheorie* (in German), Spektrum Akad. Verlag, Heidelberg, Berlin, Oxford, 1996, ISBN: 3-86025-333-6.

174. C. Cohen-Tannoudji, B. Diu, F. Laloë, *Quantenmechanik 1&2*, 2. Auflage, Walter de Gruyter, Berlin, New York, 1999.

175. W. Pauli, Über den Zusammenhang des Abschlusses der Elektronengruppen im Atom mit der Komplexstruktur der Spektren, *Z. Phys.*, 1925, **31**, 765–783, Bibcode:1925ZPhy...31..765P, doi:10.1007/BF02980631.

176. A. Einstein, B. Podolsky, N. Rosen, Can quantum mechanical description of physical reality be considered complete?, *Phys. Rev.*, 1935, **47**, 777.

177. A. Einstein, N. Rosen, The particle problem in the general theory of relativity, *Phys. Rev.*, 1935, **48**, 73.

178. J. S. Bell, On the Einstein-Podolsky-Rosen paradox, *Physics*, 1964, **1**(3), 195–200.

179. J. Maldacena, L. Susskind, Cool horizons for entangled black holes, 2013, arXiv:1306.0533v2.

180. L. Susskind, Copenhagen vs Everett, teleportation, and ER=EPR, 2016, arXiv:1604.02589v2.

181. G. N. Remmen, N. Bao, J. Pollack, Entanglement conservation, ER=EPR, and a new classical area theorem for wormholes, 2016, arXiv:1604.08217v1.

182. www.astro.umd.edu/~richard/ASTRO340/class23_RM_2015.pdf.

183. en.wikipedia.org/wiki/Flatness_problem.

184. en.wikipedia.org/wiki/Accelerating_expansion_of_the_universe.

185. N. Schwarzer, *Science Riddles – Riddle No. 22: Anti-Gravity – Is It Possible?*, www.amazon.com, ASIN: B07ZVRDS83.

186. A. Friedman, Über die Krümmung des Raumes, *Z. Phys.* (in German), 1922, **10**(1), 377–386, Bibcode:1922ZPhy...10..377F, doi:10.1007/BF01332580, (English translation: A. Friedman, On the curvature of space, *Gen. Relativ. Gravitation*, 1999, **31**(12), 1991–2000, Bibcode:1999GReGr..31.1991F, doi:10.1023/A:1026751225741).

187. A. Friedmann, Über die Möglichkeit einer Welt mit konstanter negativer Krümmung des Raumes, *Z. Phys.* (in German), 1924, **21**(1), 326–332, Bibcode:1924ZPhy...21..326F, doi:10.1007/BF01328280, (English translation: A. Friedmann, On the possibility of a world with constant negative curvature of space, , *Gen. Relativ. Gravitation*, 1999, **31**(12), 2001–2008, Bibcode:1999GReGr..31.2001F, doi:10.1023/A:1026755309811).

188. A. H. Guth, Fluctuations in the new inflationary universe, *Phys. Rev. Lett.*, 1982, **49**(15), 1110–1113, Bibcode:1982PhRvL..49.1110G, doi:10.1103/PhysRevLett.49.1110.

189. A. Linde, A new inflationary universe scenario: a possible solution of the horizon, flatness, homogeneity, isotropy and primordial monopole problems, *Phys. Lett. B*, 1982, **108**(6), 389–393, Bibcode:1982PhLB..108..389L, doi:10.1016/0370-2693(82)91219-9.

190. St. Hawking, Th. Hertog, A smooth exit from eternal inflation, *J. High Energy Phys.*, 2018, arXiv:1707.07702, Bibcode:2018JHEP...04..147H, doi:10.1007/JHEP04(2018)147.

191. S. F. Bramberger, A. Coates, J. Magueijo, S. Mukohyama, R. Namba, Y. Watanabe, Solving the flatness problem with an anisotropic instanton in Hořava-Lifshitz gravity, *Phys. Rev. D*, 2018, **97**, 043512, arXiv:1709.07084.

192. L. D. Landau, E. M. Lifschitz, *Lehrbuch der Thoretischen Physik, Band VII Elastizitätstheory*, 4th ed., Akademie Verlag Berlin, ISBN: 3-05-500580-5.

193. Nicomachean ethics, by Aristotle, 350 BC, translated by W. D. Ross, www.virtuescience.com/nicomachean-ethics.html.

194. B. Kayser, Neutrino mass, mixing, and flavor change, 2005, http://pdg.lbl.gov/2007/reviews/numixrpp.pdf.

195. M. H. Ahn, et al., Measurement of neutrino oscillation by the K2K experiment, 2006, arXiv:hep-ex/0606032.

Index